PRINCIPLES OF MICROELECTROMECHANICAL SYSTEMS

PRINCIPLES OF MICROELECTROMECHANICAL SYSTEMS

KI BANG LEE, Ph.D.
KB Lab

IEEE
IEEE PRESS

WILEY
A JOHN WILEY & SONS, INC., PUBLICATION

Published by John Wiley & Sons, Inc., Hoboken, New Jersey.
Published simultaneously in Canada.

Library of Congress Cataloging-in-Publication Data:

Principles of microelectromechanical systems / Ki Bang Lee.
p. cm.
ISBN 978-0-470-46634-6 (cloth)
1. Microelectromechanical systems. I. Lee, Ki Bang. II. Title: Principles of Microelectromechanical systems.
TK7875.P75 2010
621.381–dc22

2010007955

10 9 8 7 6 5 4 3 2 1

MIX
Paper from responsible sources
FSC® C013604

To

my parents, Won Eui Lee and Cha Won Cho

parents-in-law, Jong Hoon Suh and Soon Ja Chung

wife, Jeong Soo Seo

and children, Jong Eun Lee, Jong Min Lee, and Jong Hyeon Lee

CONTENTS

PREFACE

This book aims to describe the principles of microelectromechanical systems (MEMS) via *a unified approach and closed-form solutions* that I have newly developed. The book is designed for senior (or graduate) students in universities and engineers from industry who want to design MEMS without numerical simulation. Since readers will have diverse backgrounds (e.g., electrical engineering, mechanical engineering, physics, chemistry), the book is organized so that it will be easy to understand the theory and closed-form solutions presented, which are used to design micro- and even nanosystems.

Through eighteen years of professional experience on MEMS, I have come to realize that most systems are so complicated that their static and dynamic behavior is too difficult to be expressed in closed forms. MEMS are systems whose mechanical behavior is coupled with electrical or other behavior, such as optical behavior. As a result, the governing equations for many MEMS become complicated or nonlinear, and their solutions have generally been obtained using numerical methods. For example, parallel plates connected to a voltage source show highly nonlinear behavior, including the pull-in phenomenon (jumping at a critical voltage), and it was too difficult to obtain simple and exact solutions for the interplate gap, resonant frequency, and capacitance, and their sensitivity at arbitrary voltages. In order to develop theory on linear and nonlinear MEMS, I spent four years, and then another year, in writing this book, which is a compilation of my research notebooks. Some of these theories and solutions have been published in journals and others are published in this book for the first time. I believe that the theory and closed-form solutions will help readers to understand coupled or nonlinear behavior of micro- and nanosystems and to easily design systems using micro- and nanoelements.

The book consists of ten chapters and presents building blocks for MEMS design in each chapter. MEMS behavior can be understood by combining the building blocks so that the reader can design a MEMS device for a specific purpose. Chapter 1 is an introduction, in which MEMS are defined and the knowledge required to understand them is examined briefly. Also presented is dimensional analysis, which provides basic dimensionless parameters existing in large- and small-scale worlds which we can use to understand the forces and other parameters in the small-scale world. Chapter 2 covers basic microfabrication, which presents knowledge on common fabrication processes employed in designing realistic MEMS. Chapter 3 is focused on statics, which is concerned with forces and moments acting on mechanical structures in static equilibrium. The static response of such mechanical elements as a prismatic beam is described using the governing equations obtained from force and moment equilibria. Chapter 4 treats the static behavior of structures consisting of mechanical elements. Chapter 5 deals with dynamic responses of mechanical structures by solving linear as well as nonlinear governing equations. Chapter 6 presents fluid flow in MEMS and evaluates the damping force acting on moving structures. Chapter 7 covers basic equations of electromagnetics, which govern the electrical behavior of MEMS and by which movable elements are actuated or sensing elements work.

Subsequent chapters are devoted to combing MEMS building blocks to form actuators and sensors for specific purposes. In Chapter 8 I set up models for actuators using piezoelectric material and thermal expansion and find closed-form solutions from the models. Chapter 9 deals with electrostatic and electromagnetic actuators whose mechanical and electrical behavior is coupled. Chapter 10 provides sensors that employ mechanical and electrical elements and the actuators covered in Chapters 8 and 9. In each chapter, theoretical models are presented for the elements and systems, and solutions for the static and dynamic responses are obtained in closed form. For easy understanding of complex MEMS, I use analogies and illustrations.

I would like to acknowledge my parents, Won Eui Lee and Cha Won Cho, and parents-in-law, Jong Hoon Suh and Soon Ja Chung, who live in South Korea and have supported me through thick and thin. My lovely wife, Jeong Soo Seo, gave me the confidence to write. Finally, I would like to thank George J. Telecki, Lucy Hitz, Angioline Loredo, and many other staff members of John Wiley & Sons, Inc. for their support. The book would not exist had it not been for their continuous encouragement and patience. If I become a professor at a university, I will use this book as a textbook.

KI BANG LEE

kibanglee@hotmail.com

CHAPTER 1

INTRODUCTION

1.1 MICROELECTROMECHANICAL SYSTEMS

MEMS, *microelectromechanical systems*, are systems that consist of small-scale electrical and mechanical components for specific purposes. MEMS were translated into systems with electrical and mechanical components but have extended their boundaries to include optical, radio-frequency, and nano devices. As a result, depending on the components included and applications desired, MEMS have different names: for example, MOEMS (micro-optoelectromechanical systems) for optical applications, RF MEMS (radio-frequency MEMS) to refer to radio-frequency components and applications, and NEMS (nanoelectromechanical systems) if the systems include at least one component whose dimension is less than 1 μm. When MEMS use bio-related material (e.g., strands of DNA) to detect desired targets or to manipulate cells, the corresponding MEM system is currently called bioMEMS. Different names may refer to MEMS: *microsystems technology* (MST) in Europe and *micromachines* in Japan. Throughout this book, MEMS will be referred to as systems that include at least one set of electrical and mechanical components for a specific purpose. Depending on the specific purpose, more components, such as a reflective surface for a micromirror, can be added to a MEMS device. A typical dimension of a component of MEMS varies from 1 μm to a few hundred micrometers, and the overall size is approximately less than 1 mm. In this book we describe MEMS principles via a unified approach

Principles of Microelectromechanical Systems, By Ki Bang Lee

and newly developed closed-form solutions. Readers are assumed to be familiar with mathematical background at the third-year college and university level.

1.2 COUPLED SYSTEMS

MEMS are coupled systems since they consist of electrical and mechanical components; the mechanical behavior of MEMS are in general coupled with the electrical behavior. For example, let us consider the first electrostatic MEMS device (Fig. 1.1), presented by Nathanson et al. in the 1960s to filter or amplify electrical signals using the resonance of an electroplated cantilever. When an input signal (electrical signal) is applied across the end of the cantilever and the actuation electrode on a substrate, the electrical attractive force, given by Coulomb's law, actuates the cantilever, and a detection circuit formed under the cantilever detects the filtered or amplified electrical signal that is generated by the mechanical vibration of the cantilever.

Since the development of the first MEMS device, many other MEMS have been developed. For example, as one of the important components of MEMS, the parallel plate shown in Fig. 1.2 (similar to the cantilever of Fig. 1.1) is widely used in many microdevices that employ electrostatic forces for actuation of a microstructure or detection of a physical quantity. The typical parallel plate shown in Fig. 1.2 illustrates the basic knowledge that is required to understand MEMS behavior. The parallel plate consists of a movable plate suspended by flexures, a stationary plate, and a voltage source to supply voltage or electrical charge to the movable and stationary plates. The flexures are used to support the movable plate and act as a spring. The gap between plates can be adjusted when a force (e.g., electrostatic force or inertial force) acts on the plate.

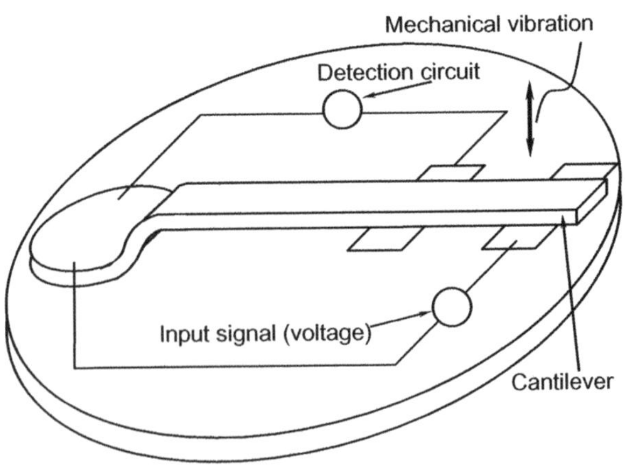

Figure 1.1 Resonant gate transistor.

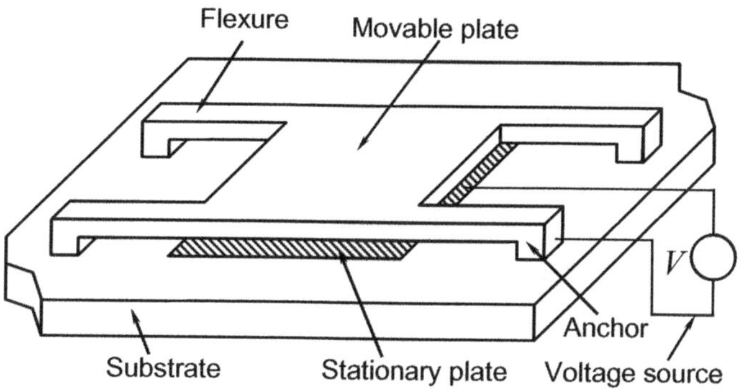

Figure 1.2 Parallel plate.

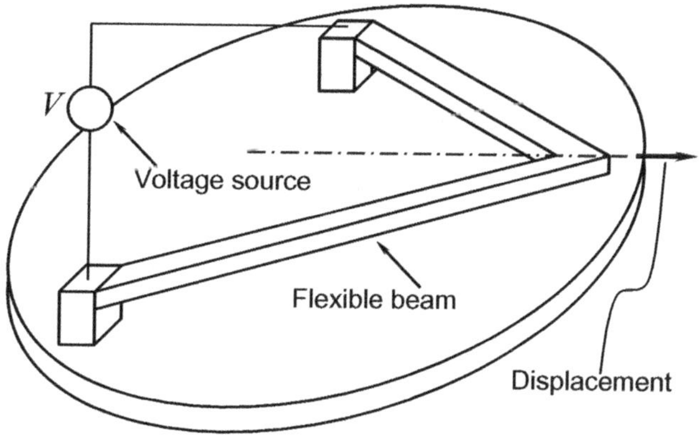

Figure 1.3 Electrothermal actuator.

Let us suppose that we apply a voltage across the movable and stationary plates. Upon applying the voltage, positive charges (or negative charges, depending on the electrical connection) are accumulated on the movable plate while opposite charges are accumulated on the stationary plate. As a result, the positive and negative charges on the plates generate an attractive force, the electrostatic force, which can push down the movable plate. The movable plate is displaced until the spring force (restoring force) due to the flexures balances the electrostatic force; that is, the displaced movable plate is in equilibrium while the voltage is applied. However, when the voltage is greater than a critical voltage called the *pull-in voltage*, the movable plate collapses into the lower plate.

A thermal actuator (Fig. 1.3) utilizes the thermal expansion due to Joule heating. As a voltage source supplies electrical current through the flexible beam that acts as a heater, heat is generated in the heater. The thermal

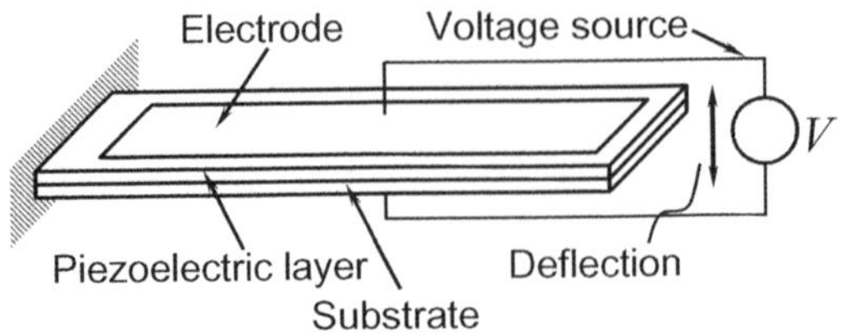

Figure 1.4 Piezoelectric actuator.

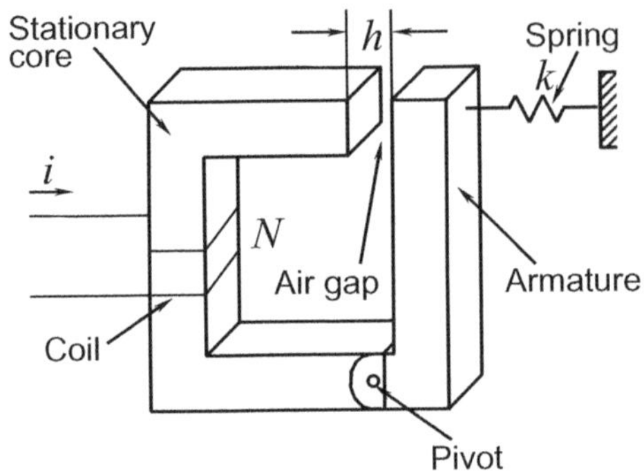

Figure 1.5 Electromagnetic relay.

expansion of the beam provides the displacement shown in the figure. The displacement depends on the voltage applied, the resistance of the beam, and the stiffness. Therefore, the mechanical behavior (e.g., displacement) of thermal actuators is coupled with the electrical and thermal behavior.

A piezoelectric actuator (Fig. 1.4) utilizes a piezoelectric material whose shape is deformed when exposed to an electric field. In Fig. 1.4 a piezoelectric layer is glued or deposited on a substrate. A thin conductive electrode is placed or deposited on the piezoelectric layer so that the layer is exposed to an electric field when a voltage source applies a voltage across the layer. In this situation, the layer expands or contracts, depending on the polarity of the voltage. For example, if the piezoelectric layer expands in the longitudinal direction, the right end of the actuator moves downward. The end of the actuator moves upward when the polarity of the voltage is reversed. The mechanical behavior of the piezoelectric actuator is then coupled with the piezoelectric constants that relate the voltage to the deformation of the piezoelectric layer, the mechanical properties (e.g., Young's modulus), and the layer geometry.

Electromagnetic force is also used to actuate microstructures. Figure 1.5 shows a model of an electromagnetic relay, one type of electromagnetic actuator. The relay consists of a movable bar (called an *armature*), a stationary core connected to the movable bar, a coil to generate magnetic field in the movable bar and stationary core, and a spring to provide the movable bar with a restoring force. When an electric current is applied to the coil, the relay is magnetized to generate an attractive force between the movable bar and the stationary core, and the movable bar is then attached to the stationary core. If the current is removed, the movable bar returns to its initial position under the restoring force of the spring. Thus, the mechanical behavior of electromagnetic actuators depends on the applied current, the magnetic and mechanical properties of the material used, the geometry of the actuator, and the stiffness of the spring.

As briefly discussed above, actuators use electricity to generate mechanical motion such as displacement, and the resulting mechanical behaviors are then coupled with electrical behavior, material properties, geometry, and so on. As a result of the coupling, the mechanical behavior is, in general, related nonlinearly to electric input (e.g., applied voltage) except in a few cases, or are expressed as complicated functions of electric input. To understand these nonlinear actuators and sensors, numerical analyses have been widely used. For example, to obtain the sensitivity to voltage of the capacitance of a parallel plate (Fig. 1.2), numerical analyses have been used to solve the equilibrium equation that governs the equilibrium position of the movable plate. Therefore, researchers, designers, and students have required commercial software to solve a problem or the skill to develop codes or programs that obtain the solution numerically. This book is designed to provide analytical closed-form solutions of both linear and nonlinear actuators in which mechanical behavior and electrical behavior are coupled. Since most MEMS-based sensors use actuators to measure physical quantities, this book can be used to design and analyze sensors.

1.3 KNOWLEDGE REQUIRED

As discussed in the foregoing section, MEMS are systems that consist of mechanical and electrical components and that may also involve other components, such as a reflective layer for a micromirror, depending on the purpose. Since the mechanical behavior of MEMS are coupled with other behavior, we should study interdisciplinary subjects in the fields of science and engineering to understand the coupled behaviors. Figure 1.6 shows an overview of the knowledge required for the research and development of MEMS and their derivatives. Because the most convenient and controllable energy is the electrical energy, electrical and electronic engineering covering electromagnetics, circuit theory, or signal processing is required to control phenomena associated with electric charge (i.e., electron, current). For example, from the point of view of electrical engineering, the parallel plate of Fig. 1.2 may be considered to be a capacitor consisting of movable and stationary plates, so knowledge of electrical engineering is necessary to calculate the capacitance of the

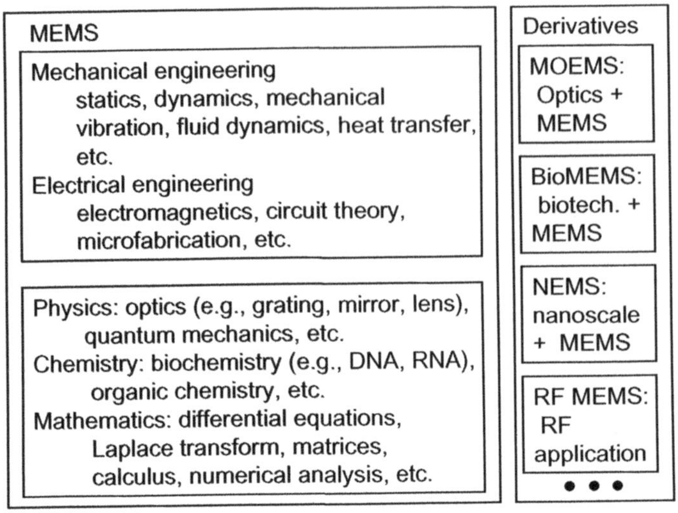

Figure 1.6 Knowledge required to understand MEMS.

parallel plate and to obtain the electrostatic force acting on the movable plate as a function of the interplate gap and the applied voltage. Similarly, since the magnetic relay of Fig. 1.5 is an electromagnet with a variable air gap, we use the magnetic energy that is stored in the electromagnet and calculate the magnetic force pulling the movable bar into the stationary core.

Physically, MEMS are mechanical structures that are designed for specific purposes. For desired functions, components of MEMS must be mechanically stable, vibrate if the mechanical resonance is utilized, and be deformed if deformation or displacement is needed. For the design and analysis of mechanical components, we need statics for the mechanical structure design, dynamics and vibration for resonance and mechanical vibration, heat transfer for thermal actuation, and fluid dynamics for the evaluation of damping due to the movement of microstructures. Let us consider the parallel plate in Fig. 1.2 as an example of a mechanical structure. Since the four flexures support the movable plate under the electrostatic force, we need statics to determine the flexure dimensions: the length, width, and thickness of the flexures. If the movable plate operates at resonance for a mechanical filter, we should use our knowledge of dynamics or mechanical vibration to design the resonant frequency desired. If we wish to set up a mechanical quality factor that affects the bandwidth of a mechanical filter, we should evaluate the damping force or damping coefficient that can be provided by fluid dynamics. If we wish to design an accelerometer or acceleration switch using the parallel plate shown in Fig. 1.2, we need to know the dynamics and mechanical vibration.

All the above-mentioned knowledge is coupled, so the design of a parallel plate for a specific application is very complicated even though the parallel plate in Fig. 1.2 looks simple. In addition to these complexities, we may need physics, chemistry, mathematics, and other subject areas, and MEMS may have different names, as described in Fig. 1.6: MOEMS if a MEMS device involves at least one optical component; RF MEMS if a MEMS device is designed for a radio-frequency application such as an RF filter; bioMEMS if a MEMS device is used for biological applications such as the detection of DNA strands; NEMS if at least one dimension of the mechanical structure is less than 1 μm; and perhaps other names in future applications if mechanical structures with electrical components are used for a specific purpose.

1.4 DIMENSIONAL ANALYSIS

Dimensional analysis and dimensionless numbers allow us to investigate complicated or coupled systems such as the MEMS described in Section 1.3. Using dimensional analysis and experimental results (or numerical simulation), we can find relationships between variables that are involved in a problem or system. If we apply dimensional analysis to a governing equation that describes a physical phenomenon and cannot be solved due to its nonlinearity, we can obtain useful dimensionless numbers that play crucial roles in describing the phenomenon. We begin with easy dimensionless numbers with which we are familiar.

Let us begin by considering the ratio of the circumference of a circle to its diameter, the well-known constant. Figure 1.7a, b, and c show a circular column, a rectangular column, and an arbitrarily shaped body, respectively. As the radius and height of the circular column (Fig. 1.7a) are represented by r and t, respectively, the perimeter l of the top view, the top-view area A, and the volume V are given by

$$l = 2\pi r = \pi d$$

$$A = \pi r^2 = \frac{\pi}{4} d^2$$

$$V = \pi r^2 t = \frac{\pi}{4} d^2 t$$

where π denotes the ratio of the circumference of a circle to its diameter and d represents the diameter of the column. It is worth noting that the perimeter, area, and volume are proportional to the diameter, the square of the diameter, and the product of the area and thickness, respectively. It is also noted that if the circular column become n times larger than its original dimensions d and t, the corresponding length, area, and volume will be nl, n^2A, and n^3V, respectively. Manipulating the equations above gives dimensionless forms as follows:

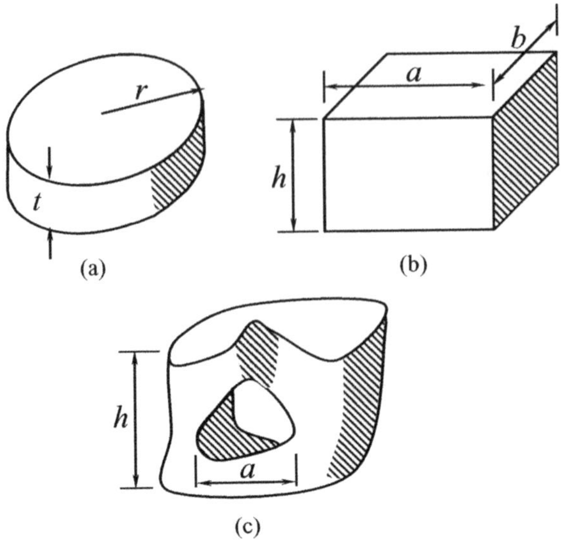

Figure 1.7 Various three-dimensional objects.

$$\frac{l}{d} = \pi$$

$$\frac{A}{r^2} = \pi \quad \text{or} \quad \frac{A}{d^2} = \frac{\pi}{4}$$

$$\frac{V}{d^2 t} = \frac{\pi}{4} \quad \text{or} \quad \frac{V}{d^3} = \frac{\pi}{4}\frac{t}{d}$$

In the preceding equations, the diameter d may be considered a characteristic length that represents the dimension of the circular column. The first two equations above give the constants (numbers) on their right-hand sides, and the third equation also yields a constant if the t/d remains unchanged. In this case, the dimensionless numbers l/d and A/d^2 remain unchanged even though the diameter becomes larger or smaller. However, the dimensionless number V/d^3 is proportional to the dimensionless number t/d. If the diameter and thickness become n times the original dimensions, the resulting length and area are, respectively, nl and n^2A, and the volume becomes n^3V since t/d does not change for a uniform transform (i.e., $nt/nd = t/d$). The preceding equations may be expressed in more general dimensionless forms as

$$f_1\left(\frac{l}{d}\right) = \frac{l}{d} - \pi = 0$$

$$f_2\left(\frac{A}{d^2}\right)=\frac{A}{d^2}-\frac{\pi}{4}=0$$

$$f_3=\left(\frac{V}{d^3},\frac{t}{d}\right)=\frac{V}{d^3}-\frac{\pi}{4}\frac{t}{d}=0$$

Similarly, for the rectangular column of Fig. 1.7b, the perimeter l of the top view, the top-view area A, and the volume V are given by

$$l=2(a+b)$$

$$A=ab$$

$$V=abh$$

We transform the equations above into dimensionless equations as follows:

$$f_4\left(\frac{l}{a},\frac{b}{a}\right)=\frac{l}{a}-2\left(1+\frac{b}{a}\right)=0$$

$$f_5\left(\frac{A}{a^2},\frac{b}{a}\right)=\frac{A}{a^2}-\frac{b}{a}=0$$

$$f_6\left(\frac{V}{a^3},\frac{b}{a},\frac{h}{a}\right)=\frac{V}{a^3}-\frac{b}{a}\frac{h}{a}=0$$

The dimensionless equations f_4, f_5, and f_6 represent functions for the perimeter and area of the top view of the rectangular column and the volume, respectively. It should be noted that the dimensionless length l/a, area A/a^2, and volume V/a^3 are expressed as functions of dimensionless variables b/a and h/a. This concept may be extended into more general cases.

Let us consider the complex three-dimensional structure shown in Fig. 1.7c. We wish to obtain the dimension a, the area, and the volume of the structure as functions of a characteristic length. The relations may be used to build a miniature or larger structure. Let l (not shown in Fig. 1.7c), h, A, and V represent a length, the height, the area, and the volume of a structure, respectively. The following equations can be written for a dimensional analysis:

$$f_7(a,l,h)=0$$

$$f_8(A,a,l,h)=0$$

$$f_9(V,a,l,h)=0$$

Let l be a characteristic length of a structure. Since a, l, and h have the dimensions of length and A and V have the dimensions of the square and cube of length, respectively, the dimensionless equations are given by

$$f_{10}\left(\frac{a}{l},\frac{h}{l}\right)=0$$

$$f_{11}\left(\frac{A}{l^2},\frac{a}{l},\frac{h}{l}\right)=0$$

$$f_{12}\left(\frac{V}{l^3},\frac{a}{l},\frac{h}{l}\right)=0$$

When the preceding equations are set up by dividing the arguments of the function by l, l_2, or l_3, the number of arguments is reduced by one in each equation. Rearranging the preceding equations yields the length a, the area A, and the volume V in dimensionless forms:

$$\frac{a}{l}=f_{13}\left(\frac{h}{l}\right)$$

$$\frac{A}{l^2}=f_{14}\left(\frac{a}{l},\frac{h}{l}\right)$$

$$\frac{V}{l^3}=f_{15}\left(\frac{a}{l},\frac{h}{l}\right)$$

For the enlargement or contraction of the structure, the ratio of linear dimensions, h/l, remains constant, and then the ratio a/l also becomes constant. The first equation above becomes $a/l = c_1$, where c_1 is a constant. Consequently, the first equation above gives a linear relation between a and l:

$$a=c_1 l \tag{1.1}$$

Similarly, the equations for the area A and the volume can be expressed as

$$A=c_2 l^2 \tag{1.2}$$

$$V=c_3 l^3 \tag{1.3}$$

where c_2 and c_3 denote constants. These three equations state that for any structures in three-dimensional space, if the shape of the structure remains unchanged for enlargement or contraction, the length from one point to another, the area of any portion of the structure, and the volume of the structure are proportional to a characteristic length and to the square and cube of the characteristic length, respectively. During derivation of equations (1.1) to (1.3), the characteristic length can be taken to be any dimension: for example, the width or the height. If the height h is selected as the characteristic length, the foregoing equations may be expressed as follows: $a = d_1h$, $A = d_2h^2$, *and* $V = d_3h^3$, where d_1, d_2, and d_3 represent constants.

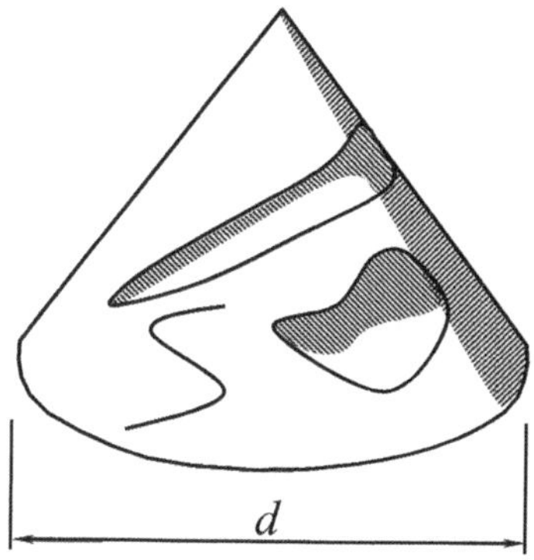

Figure 1.8 A piece of chocolate.

Example 1.1 A chocolate company decides to build an enlarged model of a piece of chocolate for an advertisement. The piece of chocolate shown in Fig. 1.8 will be enlarged n times. In other words, the dimension d will be nd in the model. In order to build the model and paint the outside, the company must calculate the length of the company logo S shown in Fig. 1.8 and the area and volume of the model. If the logo length is l_p, the outside area A_p, and the volume V_p, find the logo length of the model, the required volume of the piece of chocolate, and the outside area. If the company paints the outside to a thickness of t (wet paint), determine the volume of paint required.

As discussed above, the dimension, area, and volume of the model are proportional to a characteristic length and to the characteristic length squared and cubed, respectively. Let the logo length of model be l_m, the outside area A_m, and the volume V_m. For the length, area, and volume of the prototype (original) piece of chocolate, (1.1) to (1.3) give

$$l_p = c_1 d \quad \text{(a)}$$

$$A_p = c_2 d^2 \quad \text{(b)}$$

$$V_p = c_3 d^3 \quad \text{(c)}$$

where c_1, c_2, and c_3 denote constants for the relationships between the prototype and the model. The preceding equations also hold for the model as follows:

$$l_m = c_1 nd \quad \text{(d)}$$

$$A_m = c_2 (nd)^2 \quad \text{(e)}$$

$$V_m = c_3 (nd)^3 \quad \text{(f)}$$

From the preceding equations, we thus find relationships that will hold for both the prototype and the model:

$$l_m = nl_p \tag{g}$$

$$A_m = n^2 A_p \tag{h}$$

$$V_m = n^3 V_p \tag{i}$$

The volume of paint required for the model is given by $tA_m = n^2tA_p$. According to (g) to (i), if the dimensions of a model are n times those of a prototype (i.e., geometrically similar), the linear dimension, area, and volume of the model are increased to n, n^2, and n^3 times those of the prototype, respectively. For example, if a structure is magnified by a factor of 10, any length, area, and volume of the magnified structure become, respectively, 10, 10^2, and 10^3 times those of the original structure.

In the foregoing discussion and Example 1.1, the dimensional analysis has been described for similar structures. These concepts may be extended to involve more general cases that are related to force, stress, energy, or any other physical quantities. As a physical quantity, force such as the weight of a structure is measured in newtons (N) if we use SI units (an abbreviated form of the French term corresponding to "international system of units"). Weight may be measured in other units: for example, lb_f (pound-force). To avoid any confusion associated with force units such as N and lb_f, F is used to represent the force dimension in dimensional analysis. Similarly, L and T represent the dimensions of length and time, respectively. In many cases, force (F), length (L), and time (T) are used as the fundamental units if physical quantities involved in a problem are expressed using force, length, and time. If force, length, and time are used as the fundamental units, the system of units is called the *F–L–T system*. Other dimensions that can be derived from the fundamental units are known as ***derived units***. Derived units can be derived easily from basic equations. For example, the mass of a structure is a derived unit that is defined as $FL^{-1}T^2$, since the mass may be expressed as $m = F/a$, where F and a denote the force acting on the mass and the acceleration of the structure, respectively. If physical quantities under consideration cannot be derived from the preceding fundamental units (F–L–T units), the physical unit may be added to the list of fundamental units. For example, the temperature for thermal study and the electric charge for electric phenomena can be considered as additional units. As the fundamental units, the temperature and electric charge are represented by θ and Q, respectively. Fundamental and derived units that are widely used in scientific and engineering problems are shown in Table 1.1. In physics, researchers may use mass as a fundamental unit instead of force. In this case, the dimensions in Table 1.1 can be converted into an M–L–T system as MLT^{-2} is substituted for force F. For example, FL of the F–L–T system, representing energy, is converted to ML^2T^{-2} in the M–L–T system. The F–L–T and M–L–T systems yield the same results for the dimensional analysis of a physical problem.

TABLE 1.1 Dimensions of Physical Quantities

Physical Quantities	Dimensions	Physical Quantities	Dimensions
Force	F	Thermal conductivity	$FT^{-1}\theta^{-1}$
Length, displacement	L	Electric charge	Q
Time	T	Current	QT^{-1}
Mass	$FL^{-1}T^2$	Voltage	FLQ^{-1}
Density	$FL^{-4}T^2$	Resistance	$FLTQ^{-2}$
Velocity	LT^{-1}	Permittivity	$F^{-1}L^{-2}Q^2$
Acceleration	LT^{-2}	Capacitance	$F^{-1}L^{-1}Q^2$
Energy	FL	Magnetic field strength	$L^{-1}T^{-1}Q$
Stress, pressure	FL^{-2}	Magnetic flux density	$FL^{-1}T\,Q^{-1}$
Viscosity	$FL^{-2}T$	Permeability	FT^2Q^{-2}
Angle	dimensionless	Inductance	FLT^2Q^{-2}
Temperature	θ		

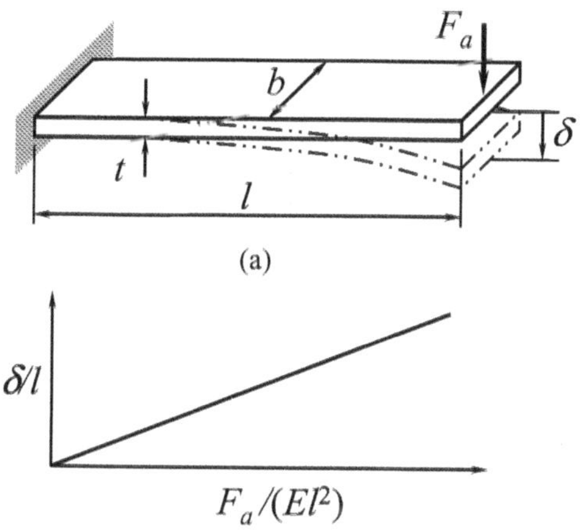

Figure 1.9 Cantilever subjected to a force.

Using the fundamental and derived units listed in Table 1.1, let us perform dimensional analysis that may be used to obtain a relation among many variables. Consider the cantilever beam of Young's modulus E (dimensions: N/m^2; a modulus relating stress to strain), length l, width b, and thickness t, as shown in Fig. 1.9a. For this simple problem we know from statics (that will be dealt with in Chapter 3) that a solution for the deflection δ at the end of a cantilever under an applied force F_a is given by

$$\delta = \frac{1}{3}\frac{F_a l^3}{EI} \tag{1.4}$$

where I denotes the moment of inertia of the cantilever, defined as $bt^3/12$. However, in order to know how to conduct the dimensional analysis of this problem, it is assumed that at this stage we don't know the solution. From Fig. 1.9a we know that F_a, E, δ, b, t, and l enter into the problem. The problem may be expressed as

$$f(F_a, E, \delta, b, t, l) = 0 \tag{1.5}$$

From Table 1.1 we know the dimensions of the variables as follows (E has the dimension of stress):

$$\begin{aligned} F_a &= [F] \\ E &= [F/L^2] \\ \delta &= [L], \quad b = [L], \quad t = [L], \quad l = [L] \end{aligned}$$

These variables involve the two fundamental dimensions F and L. If we divide the first equation by the second to eliminate the fundamental unit F, we have

$$\frac{F_a}{E} = \frac{[F]}{[F/L^2]} = [L^2]$$

which involves only the fundamental dimension L. We have δ, b, t, and l for the fundamental dimension of L, and any of them may be selected to represent L. Using l for a characteristic length representing the fundamental dimension L, the preceding equation can be converted into the following dimensionless number:

$$\frac{F_a}{El^2} = \frac{[L^2]}{[L^2]} = [0]$$

where [0] states that the number Fa/El^2 is dimensionless. Similarly, the other variables, δ, b, and t, can be converted into corresponding dimensionless numbers by dividing the variables by the characteristic length l, and the original problem can be expressed as

$$f\left(\frac{F_a}{El^2}, \frac{\delta}{l}, \frac{b}{l}, \frac{t}{l}\right) = 0$$

Since we wish to obtain an expression for the beam deflection δ, the foregoing equation may be expressed as follows:

$$\frac{\delta}{l} = f_1\left(\frac{F_a}{El^2}, \frac{b}{l}, \frac{t}{l}\right) \tag{1.6}$$

Equation (1.6) states that the deflection of the cantilever in Fig. 1.9a is expressed as the four dimensionless numbers Fa/El^2, δ/l, b/l, and t/l. Equation (1.6) is all that we can obtain from dimensional analysis. However, if we obtain more information from experiments or numerical simulations, the relationship (1.6) can be expressed in a more accurate expression. From an experiment or numerical analysis, we can find more relationships:

$$\frac{\delta}{l}=c_1\frac{F_a}{El^2} \qquad \frac{\delta}{l}=c_2\left(\frac{b}{l}\right)^{-1} \qquad \frac{\delta}{l}=c_3\left(\frac{t}{l}\right)^{-3} \tag{1.7}$$

These equations are substituted into equation (1.6) and we then have

$$\frac{\delta}{l}=f_1\left(\frac{F_a}{El^2},\frac{b}{l},\frac{t}{l}\right)=c\frac{F_a}{El^2}\left(\frac{b}{l}\right)^{-1}\left(\frac{t}{l}\right)^{-3}=c\frac{F_a l^2}{Ebt^3}$$

or

$$\delta=c\frac{F_a l^3}{Ebt^3}$$

where c is a constant that represents the product of c_1, c_2, and c_3 and can be obtained from the experiment or numerical analysis (e.g., Fig. 1.9b) and obtained easily from a graph of $F_a l^2/(Ebt^3)$ against δ/l. We know that the constant will be 4 when the equation above is compared with the analytical solution, (1.4).

In the procedure used to obtain the dimensionless equation above, note that after selecting variables as characteristic variables (E and l), the other variables were divided by the variables selected. Note also that the number of resulting dimensionless variables is reduced by the number of fundamental units. In the cantilever problem in Fig. 1.9a, the number of dimensional variables of (1.5) was 6, but in the dimensionless form, the number of dimensionless variables was reduced to 4 (6 – 2, where 2 is the number of fundamental units, F and L in this case). This procedure is generalized by *Buckingham's π-theorem* (Buckingham, 1914), which may be stated as follows:

If n variables (υ_1, υ_2, … , υ_n), which can be expressed by N fundamental units, are involved in a problem, the dimensional equation for the problem may be expressed as

$$f(\upsilon_1,\upsilon_2,\upsilon_3,\ldots,\upsilon_n)=0 \tag{1.8}$$

and the corresponding dimensionless equation can be a function of the $n - N$ dimensionless variables (π_1,π_2, π_3, … , π_{n-N}), as follows:

$$g(\pi_1,\pi_2,\pi_3,\ldots,\pi_{n-N})=0 \tag{1.9}$$

As discussed in the foregoing problem (Fig. 1.9a), equation (1.9) can give useful information and becomes more accurate if more information, such as data from an experiment or numerical analysis, is available.

The problem shown in Fig. 1.9a can also be solved using the π-theorem. In Fig. 1.9a we have six dimensional variables (F_a, E, δ, b, t, and l), and the problem of obtaining the deflection at the end of the cantilever can be expressed as

$$f(F_a, E, \delta, b, t, l) = 0$$

Since all the dimensional variables are expressed by the fundamental units of F and L, we expect that four dimensionless variables ($6 - 2 = 4$) will appear in a dimensionless equation:

$$g(\pi_1, \pi_2, \pi_3, \pi_4) = 0 \tag{1.10}$$

δ, b, t, and l have the dimension of L, and then we take the first three dimensionless variables as $\pi_1 = \delta/l$, $\pi_2 = b/l$, and $\pi_3 = t/l$. The fourth variable, π_4, may be expressed in the form

$$\pi_4 = F_a E^a l^b = [F][F/L^2]^a [L]^b = [F^{1+a} L^{-2a+b}]$$

for π_4 to be a dimensionless number (i.e., [0]), we have two equations,

$$1 + a = 0 \quad \text{and} \quad -2a + b = 0$$

From these equations, $a = -1$ and $b = 2a = -2$ are obtained. Substituting a and b into π_4 above gives

$$\pi_4 = F_a E^{-1} l^{-2} = \frac{F_a}{El^2}$$

The dimensionless variables (π_1, π_2, π_3 and π_4) are substituted into (1.10) to yield

$$g\left(\frac{\delta}{l}, \frac{b}{l}, \frac{t}{l}, \frac{F_a}{El^2}\right) = 0$$

The preceding equation is the same as (1.6). For more general applications of dimensional analysis, more examples are presented below.

Example 1.2 Parallel plates are used widely in MEMS to actuate microstructures and to sense physical quantities. As illustrated in Fig. 1.10, a parallel plate consists of an upper plate on a lower plate, and a voltage V may be applied across the plates, which are separated by a gap h. The length and width of the upper plate are l_1 and l_2 (Fig. 1.10) and the thickness and fringing field effect

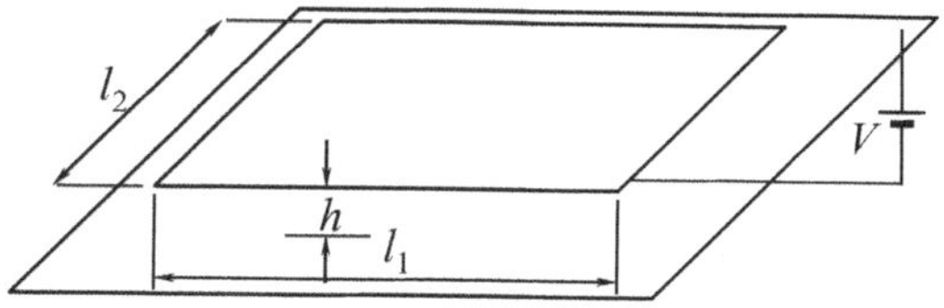

Figure 1.10 Dimensions of a parallel plate.

can be neglected if the thickness and the gap are much less than the length and width. The lower plate can be considered an infinite plate since its length and width are much larger than those of the upper plate. We wish to obtain a dimensionless equation that represents the electrostatic force F_e acting on the upper plate. Obtain the dimensionless equation in a simple form if from an experiment for $h << l_1$ and l_2, the electrostatic force F is proportional to the length and width of the upper plate, l_1 and l_2, and inversely proportional to the square of the gap h.

We derive the analytic solution of the electrostatic force for $h << l_1$ and l_2 in Chapter 7. At this stage it is assumed that we do not know the relation, in order to study the dimensional analysis for an electrostatic problem. Let ε represent the permittivity of the material (e.g., air) between the plates. The dimensional equation f for the problem may be expressed as

$$f(F_e,\varepsilon,V,l_1,l_2,h)=0 \tag{a}$$

and the dimension of the variables in equation (a) is written, with reference to Table 1.1, as follows:

$$\begin{aligned} F_e &= [F] \\ \varepsilon &= [Q^2/FL^2] \\ V &= [FL/Q] \\ l_1 = l_2 = h &= [L] \end{aligned} \tag{b}$$

We have three fundamental variables (F, Q, and L) and the six dimensional variables above. We thus expect three dimensionless variables and may write the dimensionless equation g as

$$g(\pi_1,\pi_2,\pi_3)=0 \tag{c}$$

where π_1, π_2 and π_3 are the dimensionless variables to be found. If l_1 is taken as the characteristic length that represents the length of the parallel plate, we have the following dimensionless variables:

$$\pi_1 = \frac{l_2}{l_1}$$

$$\pi_2 = \frac{h}{l_1} \qquad \text{(d)}$$

$$\pi_3 = F_e \varepsilon^c V^d l_1^e = [F]\left[\frac{Q^2}{FL^2}\right]^c \left[\frac{FL}{Q}\right]^d [L]^e = [F^{1-c+d} L^{-2c+d+e} Q^{2c-d}]$$

For π_3 to be dimensionless (i.e., [0]) we have a set of equations to construct the dimensionless variable π_3:

$$\begin{aligned} 1-c+d &= 0 \\ -2c+d+e &= 0 \\ 2c-d &= 0 \end{aligned} \qquad \text{(e)}$$

which gives $c = -1$, $d = -2$, and $e = 0$, and we find the third dimensionless variable as

$$\pi_3 = \frac{F_e}{\varepsilon V^2} \qquad \text{(f)}$$

Substituting equations (d) and (f) into (c), we have the dimensionless equation

$$g(\pi_1, \pi_2, \pi_3) = g\left(\frac{l_2}{l_1}, \frac{h}{l_1}, \frac{F_e}{\varepsilon V^2}\right) = 0$$

or

$$\frac{F_e}{\varepsilon V^2} = g_1\left(\frac{l_2}{l_1}, \frac{h}{l_1}\right) \qquad \text{(g)}$$

Equation (g) states that the ratio (i.e., π_3) of the electrostatic force acting on the upper plate to the product of the permittivity and the square of voltage is a function of the dimensionless ratios l_2/l_1 and h/l_1. From experiments of the parallel plate, it was found that the electrostatic force F is proportional to the length and width of the upper plate, l_1 and l_2, and inversely proportional to the square of the gap h. Using this information, equation (g) is simplified as follows:

$$\frac{F_e}{\varepsilon V^2} = c\frac{l_2}{l_1} \Big/ \left(\frac{h}{l_1}\right)^2 = c\frac{A}{h^2} \qquad \text{(h)}$$

where c represents a constant and A denotes the area of the upper plate, defined as $A = l_1 l_2$. The constant is also determined by the experimental data.

If we perform an experiment or numerical simulation and if we plot a graph of $F_e/\varepsilon V^2$ against A/h^2, the constant c is obtained from the slope of the graph. From the theoretical derivation described in Chapter 7, the solution for the electrostatic force acting on the parallel plate for $h \ll l_1$ and $h \ll l_2$ is given by

$$F_e = \frac{1}{2}\frac{\varepsilon A V^2}{h^2} \tag{i}$$

From equations (h) and (i) we know that the constant c must be 1/2.

The analysis to evaluate electrostatic force discussed in Example 1.2 may be extended to obtain the electrostatic force acting on more complicated structures across which a voltage is applied. Consider the structures in Fig. 1.11 under electrostatic force due to the applied voltage V. Let the dimensions of the structures be represented by $a, b, c, d, \ldots$ and the interstructure gap be denoted by h. When we repeat the dimensional analysis discussed in Example 1.2, the following equation is obtained:

$$\frac{F_e}{\varepsilon V^2} = g\left(\frac{h}{a}, \frac{b}{a}, \frac{c}{a}, \frac{d}{a}, \cdots\right) \tag{1.11}$$

In equation (1.11), for convenience, dimension a was taken as the characteristic dimension. Any physical dimensions, such as $b, c, \ldots$, can be selected as the characteristic dimension. If b is taken as the characteristic length, the argument of equation (1.11) may be $(h/b, a/b, c/b, d/b, \ldots)$. In many microstructures and nanostructures, only one dimension, such as h in Fig. 1.11, varies with time

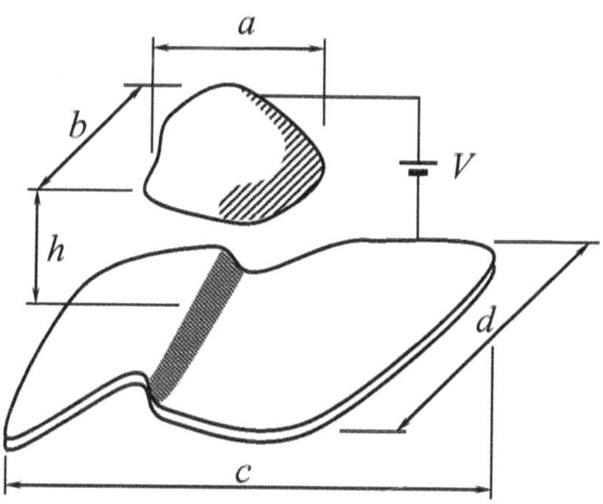

Figure 1.11 Arbitrarily shaped structures under electrostatic force.

or can be adjusted, while the other dimensions ($b, c, d, \ldots$ in Fig. 1.11) remain unchanged since the dimensions are formed during fabrication of the structures. Then the dimension ratios (b/a, c/a, ...) in equation (1.11) also remain unchanged, and we rewrite (1.11) in the simpler form

$$F_e = \varepsilon V^2 g_1\left(\frac{h}{a}\right) \tag{1.12}$$

Application of (1.11) is described in the following example.

Example 1.3 A conductive strip of width $a = 10\,\mu\text{m}$, thickness $b = 5\,\mu\text{m}$, and length c (much larger than a and b) is positioned over an infinite conductive plate as shown in Fig. 1.12a. The gap h of the strip can be adjusted from the plate and a voltage is applied across the strip and the plate. The length is much larger than the dimensions a, b, and h, so the problem in Fig. 1.12a can be considered to be a two-dimensional problem. Using commercial software for two-dimensional electrostatic analysis, we obtain the electrostatic forces (per unit meter) of 4.5685×10^{-3} N, 4.9595×10^{-5} N, 2.5008×10^{-6} N, and 7.4195×10^{-7} N at $h = 0.1 \times 10^{-6}$ m, 1×10^{-6} m, 5×10^{-6} m, and 10×10^{-6} m, respectively. It is desired to find a relation between the force F_e and the gap h in the form $F_e = ph^q$, where p and q are constants.

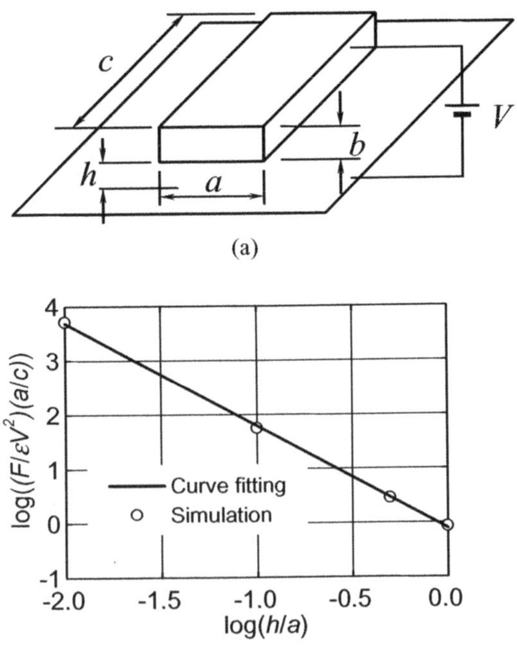

Figure 1.12 Electrostatic force acting on a strip configured over a plate.

In designing MEM devices, this type of problem will be encountered, which cannot be solved by an analytical method or whose analytical solution is in very a complicated form or expressed in an infinite series. In this example, since the thickness b and gap h are comparable to the width a, equation (i) of Example 1.2 for the parallel-plate force generates significant errors. Thus, we require an expression that can be used to evaluate the force. For this purpose, we can use (1.11), which was obtained from dimensional analysis.

Let us write (1.11) with the dimensions a, b, and c:

$$\frac{F_e}{\varepsilon V^2} = g\left(\frac{h}{a}, \frac{b}{a}, \frac{c}{a}\right) \tag{a}$$

Since c is much greater than a and b, the problem is considered a two-dimensional problem, and the electrostatic force F_e is then proportional to the length, c. Thus, equation (a) is written in the form

$$\frac{F_e}{\varepsilon V^2} = \frac{c}{a} g_1\left(\frac{h}{a}, \frac{b}{a}\right) \tag{b}$$

where g_1 is a new function of h/a and b/a. However, since b/a remains constant while h/a varies, equation (b) may be written as

$$\frac{F_e}{\varepsilon V^2} = \frac{c}{a} g_2\left(\frac{h}{a}\right)$$

or

$$\frac{F_e}{\varepsilon V^2}\frac{a}{c} = g_2\left(\frac{h}{a}\right) \tag{c}$$

where g_2 is a function of h/a when b/a remains unchanged. As shown in Fig. 1.12b, the dimensionless force defined on the left-hand side of equation (c) is plotted as circles against h/a on a log-log scale (base 10). The force–gap relation is linear in the log-log graph and can be expressed as a linear equation:

$$\log\left(\frac{F_e}{\varepsilon V^2}\frac{a}{c}\right) = c_1 \log\frac{h}{a} + c_2 \tag{d}$$

From a linear curve fitting (e.g., graphical method or least-squares method) using the force data given, we find that $c_1 = -1.8990$ and $c_2 = -0.1084$. Substituting the coefficients into equation (d) leads us to

$$\frac{F_e}{\varepsilon V^2}\frac{a}{c} = 10^{c_1 \log(h/a)+c_2} = 10^{c_2}\left(\frac{h}{a}\right)^{c_1} = 0.7791\left(\frac{h}{a}\right)^{-1.8990} \tag{e}$$

Rearranging equation (e) gives us the electrostatic force F_e in the form $F_e = ph^q$:

$$F_e = 0.7791\varepsilon V^2 \frac{c}{a}\left(\frac{a}{h}\right)^{1.8990} \tag{f}$$

It should be noted that equation (f) is valid for width $a = 10 \times 10^{-6}$ m, thickness $b = 5 \times 10^{-6}$ m, $a << c$, $b << c$, and $0.1 \times 10^{-6}\text{m} \le h \le 10 \times 10^{-6}$ m (the range of the input data). To validate equation (f) we can substitute $a = 10 \times 10^{-6}$ m, $b = 5 \times 10^{-6}$ m, $h = 5 \times 10^{-6}$ m, $c = 1$ m, $\varepsilon = 8.854 \times 10^{-12}$ F/m, and $V = 1$ V to give $F_e = 2.5727 \times 10^{-6}$ N, which is very close to 2.5008×10^{-6} N from numerical analysis. For $a = 10 \times 10^{-6}$ m, $b = 5 \times 10^{-6}$ m, $h = 5 \times 10^{-6}$ m, $c = 10a = 100 \times 10^{-6}$ m, $\varepsilon = 8.854 \times 10^{-12}$ F/m, and $V = 10$ V, equation (f) yields $F_e = 2.5727 \times 10^{-8}$ N.

It is worth noting that (1.11) holds for any structures that are geometrically similar; that is, (f) holds for any structures of $b/a = 0.5$, $c/a >> 1$, and $0.1 \le h/a \le 1$. For example, (f) can be used to evaluate the electrostatic force of the structure, which is suspended over a large plate and has the following dimensions: $a = 20 \times 10^{-6}$ m, $b = 10 \times 10^{-6}$ m, $h = 10 \times 10^{-6}$ m, and $c = 10a = 200 \times 10^{-6}$ m. The structure is the doubled structure of that above and generates the same magnitude of electrostatic force (2.5727×10^{-8} N) if the same voltage (10V) is applied.

If a microstructure immersed in a gas or liquid is moved, a drag force due to the viscosity is generated and acts as a damper that decays the energy stored in the microstructure. The drag force or damping force plays an important role since it is related to the damping coefficient or quality factor (this subject is dealt with in detail in Chapter 5). The drag force is also expressed in dimensionless form. The following example shows how to use dimensional analysis to obtain the damping force.

Example 1.4 A microplate of length l and width w (Fig. 1.13) moves at a velocity of u toward a lower plate that is much larger than the microplate. During the motion of the microplate, the upper plate, spaced by a gap h from the lower plate, squeezes air between the plates so that the air moves out to atmosphere and the pressure under the microplate is increased. If the microplate is moved up, the pressure is lowered and the air will move in from the atmosphere. The pressure acting on the microplate is a source of damping force. Experiments show that the damping force F_d depends on the velocity u, the viscosity μ, the length l, the width w, and the gap h. We wish to find a relation between the damping force and the variables. If the damping force F_d is inversely proportional to the cubic of the gap h, refine the relation.

This subject is covered in detail in Chapter 6. At this stage we use dimensional analysis to obtain the relation among the parameters. The dimensional equation of the problem may be expressed as

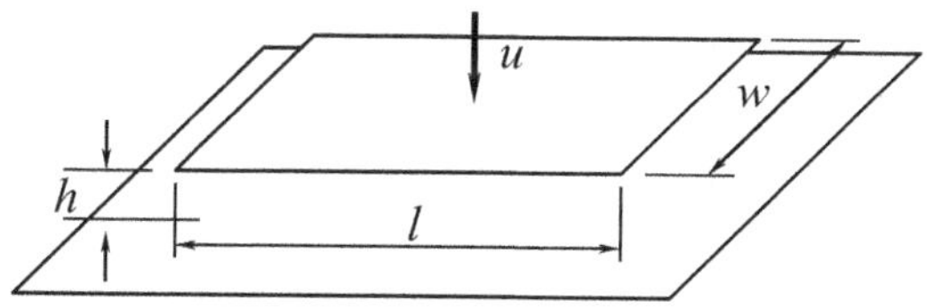

Figure 1.13 Parallel plate subjected to a damping force.

$$f(F_d, \mu, u, l, w, h) = 0 \tag{a}$$

and the dimension of the variables in equation (a) is written, with reference to Table 1.1, as follows:

$$\begin{aligned} F_d &= [F] \\ \mu &= [FT/L^2] \\ u &= [L/T] \\ l = w = h &= [L] \end{aligned} \tag{b}$$

We have three fundamental variables (F, L, and T) and the six dimensional variables above. Three dimensionless variables will then be involved in this problem and the corresponding dimensionless equation g may be written as

$$g(\pi_1, \pi_2, \pi_3) = 0 \tag{c}$$

where π_1, π_2, and π_3 are the dimensionless variables to be found. If l is taken as the characteristic length that represents the upper plate, we have the following dimensionless variables:

$$\begin{aligned} \pi_1 &= \frac{w}{l} \\ \pi_2 &= \frac{h}{l} \\ \pi_3 &= F_e \mu^c u^d l^e = [F]\left[\frac{FT}{L^2}\right]^c \left[\frac{L}{T}\right]^d [L]^e = [F^{1+c} L^{-2c+d+e} T^{c-d}] \end{aligned} \tag{d}$$

Setting $\pi_3 = [0]$ gives a set of equations to construct the dimensionless variable π_3:

$$\begin{aligned} 1+c&=0 \\ -2c+d+e&=0 \\ c-d&=0 \end{aligned} \tag{e}$$

which gives $c=-1$, $d=-1$, and $e=-1$, and we find the third dimensionless variable as

$$\pi_3 = \frac{F_d}{\mu u l} \tag{f}$$

Substituting the first two equations of (d) and (f) into (c), we have the dimensionless equation

$$g(\pi_1,\pi_2,\pi_3) = g\left(\frac{w}{l},\frac{h}{l},\frac{F_d}{\mu u l}\right) = 0$$

or

$$\frac{F_d}{\mu u l} = g_1\left(\frac{w}{l},\frac{h}{l}\right) \tag{g}$$

For refinement of equation (g), we use the inverse proportionality of the damping force F_d to the cubic of the gap h. Using the proportionality, equation (g) is rewritten in the form

$$\frac{F_d}{\mu u l} = \left(\frac{l}{h}\right)^3 g_2\left(\frac{w}{l}\right) \tag{h}$$

Even though equation (h) is valid, a slightly different form may be obtained for our convenience. Since the product of dimensionless numbers generates a dimensionless number, (g) can be written

$$\frac{F_d}{\mu u l} = g_3\left(\frac{w}{l},\frac{w}{l}\Big/\frac{h}{l}\right) = g_3\left(\frac{w}{l},\frac{w}{h}\right) \tag{i}$$

If proportionality is used, equation (i) is converted into

$$\frac{F_d}{\mu u l} = \left(\frac{w}{h}\right)^3 g_4\left(\frac{w}{l}\right) \tag{j}$$

After studying squeeze damping in Chapter 6, we obtain the following closed-form equation for the damping force:

$$F_d = \beta \frac{\mu u w^3 l}{h^3} \tag{k}$$

where β denotes a function of the ratio w/l. It is noted that if we define $\beta = g_4(w/l)$, (k) is the same as (j). It is noted that comparing (h) and (j) gives

$$g_4 = \left(\frac{l}{w}\right)^3 g_2\left(\frac{w}{l}\right)$$

The damping problem discussed in Example 1.4 may be extended to involve more dimensions that affect the damping force. If additional dimensions are represented by $a, b, c, \ldots$, we have the following dimensional equation:

$$f(F_d, \mu, u, l, w, h, a, b, c, \ldots) = 0 \tag{1.13}$$

When the procedure used in example 1.4 is repeated, the dimensionless equation for the general case is given by

$$\frac{F_d}{\mu u l} = \left(\frac{w}{h}\right)^3 g\left(\frac{w}{l}, \frac{a}{l}, \frac{b}{l}, \frac{c}{l}, \cdots\right)$$

or

$$F_d = \left(\frac{w}{h}\right)^3 \mu u l\, g\left(\frac{w}{l}, \frac{a}{l}, \frac{b}{l}, \frac{c}{l}, \cdots\right) \tag{1.14}$$

Equation (1.14) can be interpreted as follows: The damping force due to squeezed gas is proportional to the viscosity of the surrounding gas or fluid and to the velocity perpendicular to the plate. Furthermore, the damping force is inversely proportional to the cubic of the interplate gap. Other dimensions, such as the size of the perforation and the plate shape (e.g., rectangle or circle), also affect the damping force, and their effect on the force may be involved in a correction factor such as the function g of equation (1.14).

So far, to conduct dimensional analysis for problems, we have assumed that we did not know exact principles or governing equations but knew parameters involved in the problems. If we know governing equations of problems that cannot be solved due to their nonlinearity or for which it is difficult to obtain analytical solutions, we can also employ dimensional analysis to anticipate which parameters affect the solutions, to design experiments or numerical analysis, or to obtain closed-form expressions from experiments or numerical analysis. We begin by considering the cantilever deflection problem, which has already been studied in association with Fig. 1.9.

The governing equation for deflection y of the cantilever at distance x is given by (its derivation is dealt with in Chapter 3)

$$EI\frac{d^2y}{dx^2} = M \tag{1.15}$$

where E, I, and M denote Young's modulus, the moment of inertia of the beam defined as $bt^3/12$, and the moment acting on the beam cross section, respectively. The deflection y is easily obtained since the equation is linear and the moment M is defined as $F_a(l - x)$. To obtain a dimensionless equation from equation (1.15), we introduce the following dimensionless parameters:

$$X = \frac{x}{l}$$

$$Y = \frac{y}{\delta}$$

where the dimensionless parameters X and Y represent x and y, respectively, and l and δ are taken as the characteristic dimensions for the dimensionless parameters. It is noted that X and Y are normalized by the maximum values (i.e., l, δ) of the corresponding dimensional variables x and y, and then the maximum X and Y are unity. Recalling the chain rule of differentiation, the first and second derivatives of y with respect to x are obtained as

$$\frac{dy}{dx} = \frac{dX}{dx}\frac{dy}{dX} = \frac{1}{l}\frac{d\delta Y}{dX} = \frac{\delta}{l}\frac{dY}{dX}$$

$$\frac{d^2y}{dx^2} = \frac{d}{dx}\frac{dy}{dx} = \frac{1}{l}\frac{d}{dX}\left(\frac{\delta}{l}\frac{dY}{dX}\right) = \frac{\delta}{l^2}\frac{d^2Y}{dX^2}$$

Substituting the preceding equations into the governing equation with $M = F_a(l - x) = F_a l(1 - X)$, we have

$$EI\frac{\delta}{l^2}\frac{d^2Y}{dX^2} = F_a l(1-X)$$

or in dimensionless form,

$$\frac{d^2Y}{dX^2} = G(1-X) \tag{1.16}$$

where G is a dimensionless force defined as

$$G = \frac{F_a l^3}{EI\delta}$$

The foregoing governing equation in dimensionless form states that the dimensionless deflection Y is a function of the dimensionless force G and the dimensionless position X. This statement can be expressed as follows:

$$f(Y,G,X)=0$$

or

$$Y=g(G,X)$$

For a small deflection (i.e., $\delta/l << 1$) we can find from an experiment or numerical simulation that there is a linear relation between Y and G and that the following equation holds:

$$Y=Gg_1(X)$$

Substituting Y, G, and X defined earlier into the foregoing equation leads us to

$$\frac{y}{\delta}=\frac{F_a l^3}{EI\delta}g_1\left(\frac{x}{l}\right)$$

or

$$y=\frac{F_a l^3}{EI}g_1\left(\frac{x}{l}\right) \tag{1.17}$$

The exact solution of this problem, which is dealt with in Chapter 3, is given by

$$y=\frac{F_a l^3}{6EI}\left[-\left(\frac{x}{l}\right)^3+3\left(\frac{x}{l}\right)^2\right]$$

Comparing the preceding equation with (1.17), we find that

$$g_1=\frac{1}{6}\left[-\left(\frac{x}{l}\right)^3+3\left(\frac{x}{l}\right)^2\right]$$

If g_1 of (1.17) is not known, we may obtain the coefficient from experiment.

As shown, if we make a governing equation dimensionless, we can find the dimensionless parameters that are involved in the governing equation. Another example, involving manipulating governing equations to obtain dimensionless equations and variables, is presented below.

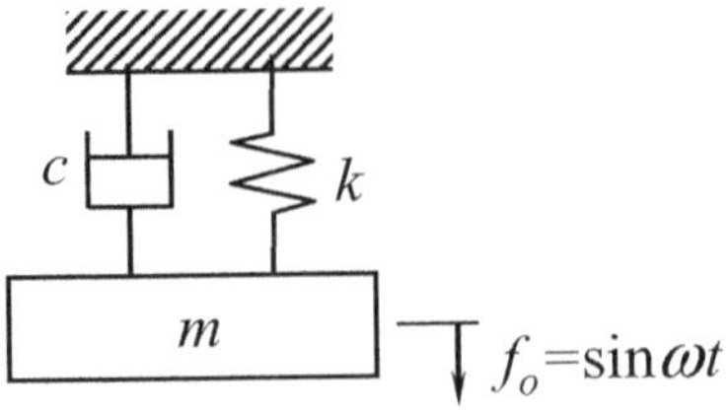

Figure 1.14 Vibrating system.

Example 1.5 In physics many vibrating systems consist of a mass m, a spring of stiffness k to generate a restoring force, and a damper with a damping coefficient c. The model for the systems is illustrated in Fig. 1.14, in which the mass is actuated by a time-varying force $f_o \sin \omega t$, where f_o, ω, and t denote the force amplitude, angular frequency, and time, respectively. The governing equation for the response of the mass to the force is given by

$$m\frac{d^2y}{dt^2} + c\frac{dy}{dt} + ky = f_0 \sin \omega t \tag{a}$$

where y is the response of the mass. We wish to obtain the dimensionless form of equation (a) and find dimensionless parameters that are involved in the problem.

To obtain the dimensionless equation, we first define dimensionless parameters for y and t as follows:

$$Y = \frac{y}{y_c} \qquad T = \frac{t}{\tau} \tag{b}$$

where Y and T represent the dimensionless displacement and time, and y_c and t denote the characteristic displacement and time, respectively, which will be defined later. Using the foregoing dimensionless parameters and recalling the chain rule of differentiation yields the time derivatives

$$\frac{d}{dt} = \frac{dT}{dt}\frac{d}{dT} = \frac{1}{\tau}\frac{d}{dT}$$

Similarly,

$$\frac{d^2}{dt^2} = \frac{d}{dt}\left(\frac{d}{dt}\right) = \frac{1}{\tau^2}\frac{d^2}{dT^2}$$

Substituting the preceding four equations into the original governing equation, we have

$$m\frac{1}{\tau^2}\frac{d^2(y_cY)}{dT^2}+c\frac{1}{\tau}\frac{d(y_cY)}{dT}+ky_cY=f_0\sin\omega\tau T$$

Dividing the foregoing equation by ky_c leads to

$$\frac{m}{k\tau^2}\frac{d^2Y}{dT^2}+\frac{c}{k\tau}\frac{dY}{dT}+Y=\frac{f_0}{ky_c}\sin\omega\tau T \tag{c}$$

In equation (c) the characteristic displacement y_c and time τ are arbitrary and can be taken to reduce equation (c) to simplest form. For this purpose we may set both the first coefficient on the left-hand side of equation (c) and the coefficient on the right-hand side equal to unity as follows:

$$\begin{aligned}\frac{m}{k\tau^2}&=1\\ \frac{f_0}{ky_c}&=1\end{aligned} \tag{d}$$

The two equations (d) give the following characteristic time and displacement to provide the simplest dimensionless form:

$$\begin{aligned}\tau&=\sqrt{\frac{m}{k}}=\frac{1}{\omega_n}\\ y_c&=\frac{f_0}{k}\end{aligned} \tag{e}$$

where ω_n denotes the natural frequency of the vibrating system, defined as $\sqrt{k/m}$, and y_c represents the static displacement. Using the definition of τ and y_c above and introducing the quality factor $Q=\sqrt{mk}/c$, we obtain the dimensionless governing equation as follows:

$$\frac{d^2Y}{dT^2}+\frac{1}{Q}\frac{dY}{dT}+Y=\sin\Omega T \tag{f}$$

where Ω is the dimensionless frequency, defined as $\Omega = \omega/\omega_n$. Therefore, the dimensionless response may be expressed as

$$g(Y,Q,\Omega,T)=0$$

or

$$Y = g_1(Q, \Omega, T) \qquad \text{(g)}$$

This equation shows clearly that the response Y is a function of the quality factor Q, which is defined by m, c, and k, the dimensionless angular frequency Ω, and the dimensionless time T. The corresponding analytic solution is dealt with in Chapter 5.

PROBLEMS

1.1 The size of the piece of chocolate in Fig. 1.8 is increased n times and its weight is proportional to the mass. If the density of the model is half that of the original piece, determine the weight of the model.

1.2 Suppose that an astronaut visits a planet on which the gravitational acceleration is one-tenth that on Earth. He finds a giant ant that consists of the same material as, but whose size is 100 times greater than, that of the ant on Earth. He returns to Earth with the giant and puts it in a cage. By dimensional analysis, describe what happens to the giant ant.

1.3 A fixed–fixed beam is subjected to a load q (N/m) that is distributed uniformly along the beam. The governing equation for deflection y at x from one end of the beam is given by

$$EI\frac{d^4y}{dx^4} - N_0\frac{d^2y}{dx^2} = q$$

where E, I, and N_o denote Young's modulus (N/m^2), the moment of inertia of the beam (m^4), and the tension (N) acting along the beam, respectively. Find the dimensionless equation and suggest an expression for the solution in dimensionless form.

1.4 A movable plate of length l and width w in a gas moves into an infinite stationary plate at a velocity of u (Fig. P1.4). The movable plate is perforated to reduce the damping force. The plate gap, the pitch, and the size of the perforation are denoted by h, p, and f, respectively. Find an expression for the damping force F_d in dimensionless form. If experiments show that the damping force is inversely proportional to the gap cubed, refine the expression.

1.5 The sound of a bell depends on the natural frequency of the bell structure. The bell is made of a material of Young's modulus E (N/m^2) and density ρ (kg/m^3). If the natural frequency of the original bell is f_p, find the natural frequency of an n-fold increased model that is geometrically similar to the original, but whose Young's modulus and density are changed from the original values.

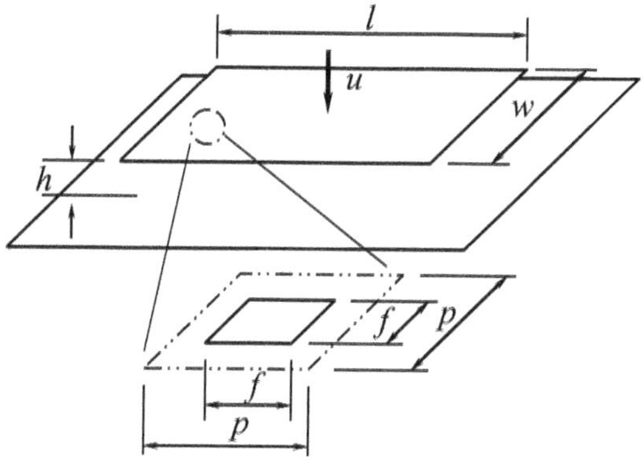

Figure P1.4 Perforated plate.

CHAPTER 2

MICROFABRICATION

The microfabrication technology for MEMS was adopted from microelectronic fabrication technology and has been extended to include other technologies. In this chapter we describe essentials of the microfabrication that are needed to design and fabricate MEMS. If readers need in-depth knowledge of microfabrication technology, they may consult a professional in the field of microfabrication or refer to microfabrication-related articles or books (e.g., Petersen, 1982; Jaeger, 1988; Kovacs, 1998).

2.1 BULK AND SURFACE MICROMACHINING

The terms *bulk micromachining* and *surface micromachining* may be used to refer to how to construct both MEMS structures and electronic devices. Readers not familiar with the microfabrication of electronic devices may have difficulty understanding micromachining. For easy understanding of bulk and surface micromachining, we shall use analogies or similar cases from macroscale systems. We begin by considering architectures from ancient cultures. Shown in Fig. 2.1 is a channel built by our ancestors to connect a reservoir to a river and an arched bridge over the channel. The channel shown was made simply by digging the earth. However, building the bridge was is much more difficult. To build the bridge, our ancestors developed the technology shown in Fig. 2.2. Sand and stones (Fig. 2.2b) that would serve as temporary or

Principles of Microelectromechanical Systems, By Ki Bang Lee

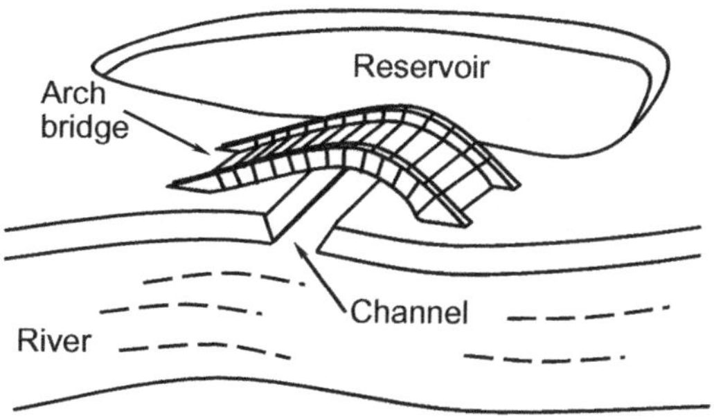

Figure 2.1 Construction of a channel and an arched bridge.

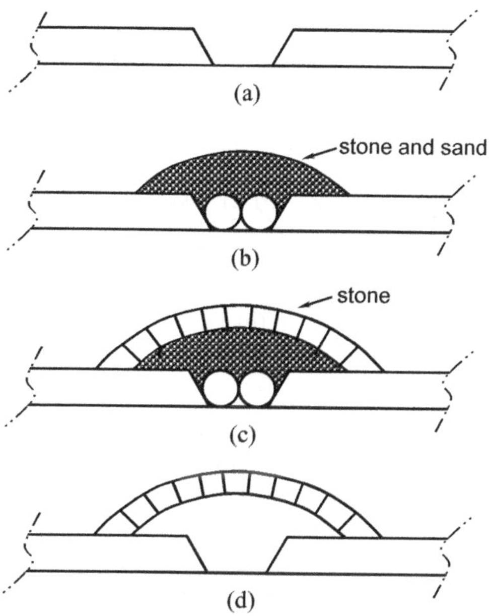

Figure 2.2 Construction method for an arched bridge.

sacrificial layers were first deposited over the channel, and stones was arranged to make the arch (Fig. 2.2c) that would be comprise the structure of the bridge. Afterward, the sacrificial layer was removed to form the bridge shown in Fig. 2.2d.

Let us turn our attention to current machining technology. If we wish to create the steel cantilever whose dimensions are shown in Fig. 2.3a, we may give the design to a machine shop and have a technician make the cantilever on a plate by milling bulk steel into the desired shape (Fig. 2.3a). If we use clay, a different process to form the cantilever may be developed (Fig. 2.3b).

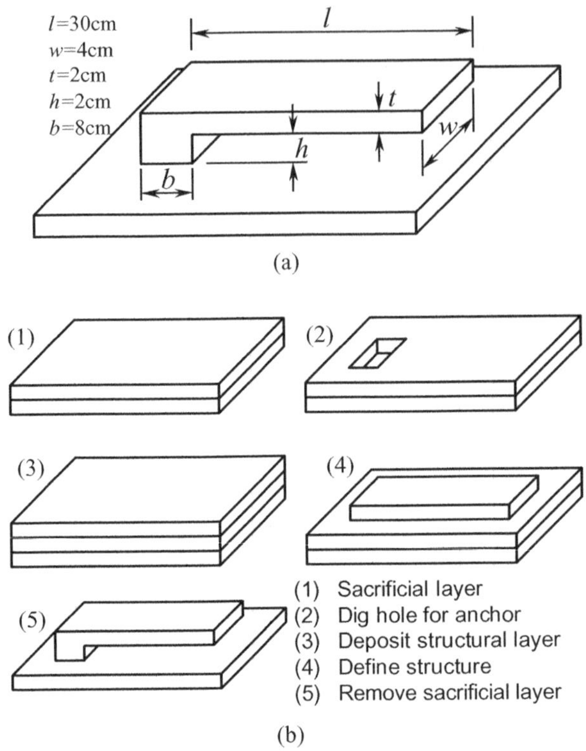

Figure 2.3 Construction of a cantilever on a substrate.

A temporary or sacrificial brick is deposited on a substrate, followed by digging a hole for an anchor. A clay blanket is then deposited on the temporary structure, and portions of the blanket not used to form the cantilever are cut by a knife and removed. The cantilever with the temporary brick is dried for a sufficient time and the brick is then removed to form a cantilever attached to a substrate. The entire structure may be fired at a high temperature if a hardened structure is needed.

So far, we have reviewed the historical and current technology used to build relatively large structures. These technologies may be classified into two groups: (1) direct machining of bulk material (e.g., earth or substrate), and (2) formation of a structural layer on a sacrificial structure and removal of the sacrificial layer to form a structure. In the area of MEMS or microelectronic devices, the first technology is called *bulk micromachining* and the second is referred to as *surface micromachining*. However, since MEMS or microelectronic devices are made of tiny movable or nonmovable parts less than 1 mm in size, the human work or machining to used form relatively large-scale structures must be replaced by other technologies. Currently, in order to fabricate MEMS and their derivatives (e.g., RF MEMS, NEMS),

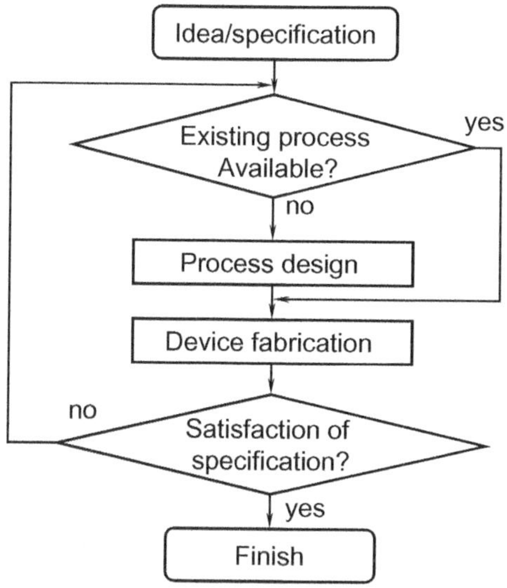

Figure 2.4 Flowchart for the design and fabrication of MEMS.

bulk and surface micromachining adopted from microelectronic fabrication technology are the dominating technologies and may be referred to as MEMS fabrication technology. For convenience, MEMS fabrication may be classified into lithography deposition and etching, and is described in subsequent sections of the chapter.

Before moving into the next sections, a flowchart used in the design and fabrication of MEMS is reviewed in Fig. 2.4. Like other designs and research, the concept behind a MEMS device is most important, with fabrication technology of almost equal importance since it brings the concept into reality. Following an idea for a MEMS device, specification, structure dimensions, and expected performance will be determined using the theory and closed-form solutions described in subsequent chapters or by using numerical analysis. An initial fabrication process can be carried out based on the design of the structure. At this stage, we may first check whether or not the present process is available. If one cannot find a fabrication process for a device or does not want to use the prevailing process, he or she may develop his or her own process. If foundry service is available, it may be used to fabricate MEMS devices.

Many foundry services for surface and bulk micromachining are currently available at an acceptable cost. A foundry service may be used to conduct a feasibility test of a device or to mass produce a device. We may use a search engine for Web sites or refer to a directory to find a suitable foundry service.

In the remaining sections of this chapter we describe microelectronic fabrication that is widely used to fabricate both microelectronic and MEMS devices and then add some technologies that are used for MEMS fabrication.

2.2 LITHOGRAPHY

For fabrication of MEMS devices, we prefer planar substrates that are suitable for a lithographic process. MEMS devices have been fabricated on a silicon wafer. In this section, silicon lattice and silicon wafer are described and then the lithographic process.

Silicon Wafer

Single-crystal silicon is widely used as a substrate for both microelectronic devices and MEMS fabrications, and its structure can be represented by a cubic lattice (Fig. 2.5a). Using Miller indices, we can define the planes and directions of the silicon crystal as shown in Fig. 2.5a–c. In Fig. 2.5a, the plane cutting the x axis can be defined as (100), where the parentheses represents the plane indicated by the argument 100. Similarly, the positive planes cutting the y and z axes can be denoted by (010) and (001). The planes facing the positive planes (100), (010), and (001) are defined as $(\bar{1}00)$, $(0\bar{1}0)$, and $(00\bar{1})$, respectively, where $\bar{1}$ denotes −1. Since the cubic crystal is symmetric about its center, all six planes provide the same information, and then a family of planes can be denoted by {100}, where the braces represents the family. Directions perpendicular to the planes above can be defined as [100] for (100), [010] for (010), and so on. As shown in Fig. 2.5a, the directions are perpendicular to the corresponding planes and act as a vector normal to the plane in the

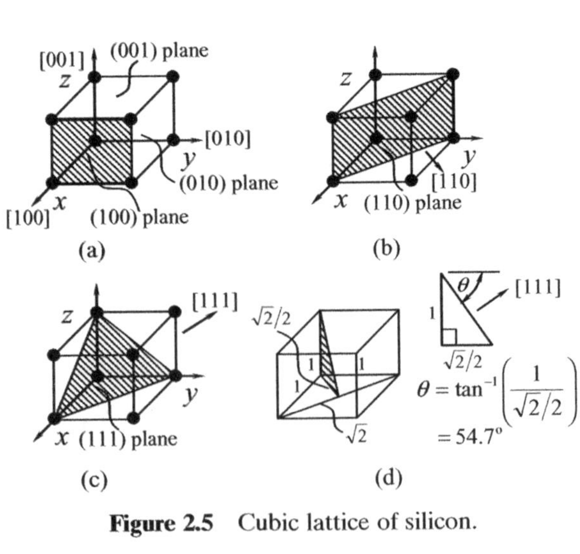

Figure 2.5 Cubic lattice of silicon.

rectangular coordinate system. Since there are six directions perpendicular to the planes, the notation <100> denotes a family of six directions. The (110) plane and the corresponding <110> direction are shown in Fig. 2.5b. Figure 2.5c defines the (111) plane and the corresponding <111> direction used in the bulk micromachining described in the remaining sections of the chapter. In bulk micromachining, the wet etching (using a wet etchant) proceeds along the (111) plane that makes an angle of θ with the upper {100} or (001) in Fig. 2.5c. From Fig. 2.5d, a simple geometric relation is used to give the angle θ:

$$\theta = \tan^{-1}\frac{1}{\sqrt{2}/2} = \tan^{-1}\sqrt{2} = 54.7^\circ$$

The silicon planes defined in Fig. 2.5 can be used in micromachining processes, and different types and concentrations of impurities to modify electrical properties of silicon are required to fabricate devices desired. As a result, an identification method of commercially available silicon wafer was set up that uses flats (Fig. 2.6). The flats provide information on wafer type and orientation of crystal and can be used with automatic equipment to recognize wafer orientation.

In macroscale machining, a tool such as a cutter is used to carve or cut materials into the shape desired. For example, in the fourth step of Fig. 2.3b, a cutter is used to define the cantilever from the clay blanket and to remove the remaining clay. However, in a different case in which the minimum feature (e.g., the beam width) is less than 10 μm, it is very difficult to carve or macromachine to form the cantilever shown in Fig. 2.3b. To form fine patterns whose minimum feature is less than 10 μm, scientists and engineers have developed, for printing and photography, lithography that uses engraved stones or plates, and have extended the technology to develop micro- and nanofabrication

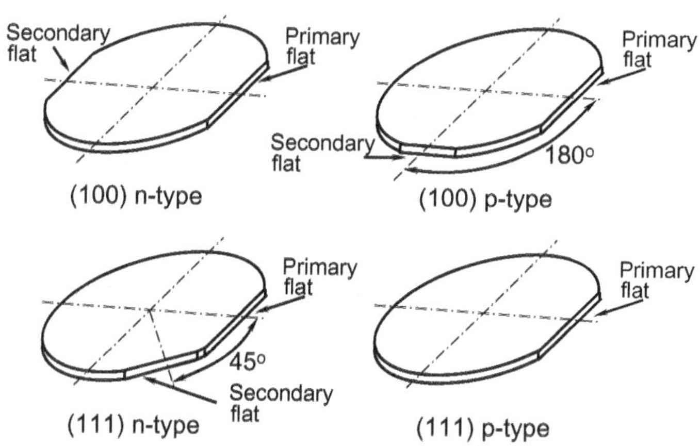

Figure 2.6 Flats of silicon wafers.

technology for integrated circuits (ICs), MEMS and NEMS. One of the most important steps in micro- and nanofabrication is to make a desired pattern on a material. For MEMS as well as microelectronic devices, the lithography is an essential technology for generating fine patterns on both structures and substrates.

The currently prevailing lithography, called *photolithography*, is based on photography using a light-sensitive material. As illustrated in Fig. 2.3b, a portion of layers must remain on a lower layer or the substrate, while portions that are not used should be removed. For defining tiny patterns to fabricate MEMS and microelectronic devices, we first construct a mask with patterns that are transparent or opaque to light, depending on our design, and then the patterns are transferred to the structural or sacrificial layers on a substrate by using photography technology. The perspective view shown in Fig. 2.7 can be used is to understand the lithography. The purpose of Fig. 2.7 is to show how to use photographic techniques to form a hole or window on a layer (Fig. 2.7b) and to leave a mechanical structure on a layer (Fig. 2.7c). Using these techniques we can generate most MEMS and NEMS: for example, the microversion of the cantilever shown in Fig. 2.3a. For convenience, we begin with a substrate that has a silicon dioxide (SiO_2) layer on the upper surface (Fig. 2.8a); the process to form the silicon dioxide layer is described in subsequent sections of the chapter. It is also assumed that we have two glass masks: one with an opaque metal block on transparent glass; another with an optical window on glass covered by a metal layer. A positive photosensitive material called *photoresist* is coated on the SiO_2 layer to generate the pattern. Afterward the masks are placed on top of the substrates and exposed to light to form latent images corresponding to the pattern on the masks. The latent images of Figs. 2.7b and 2.8c1 are invisible since they are formed by the chemical reaction of the resist. When exposed photoresist is removed by a chemical reaction, we can see the photoresist patterns transferred from the masks in Fig. 2.7.c or 2.8c2. Using photoresist as masks, the SiO_2 layer is etched to form SiO_2 patterns, and the photoresist is stripped to obtain the final window or structure

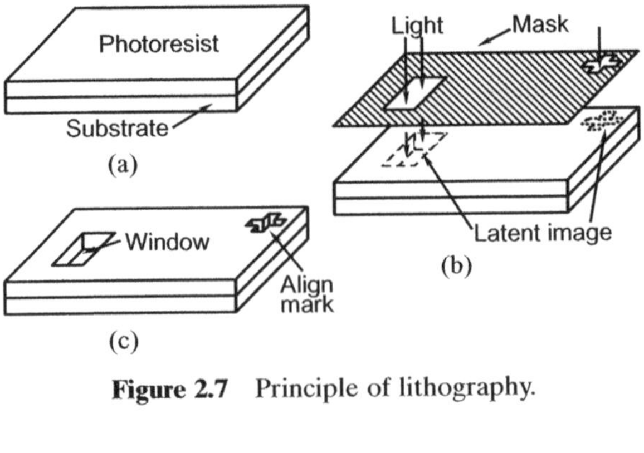

Figure 2.7 Principle of lithography.

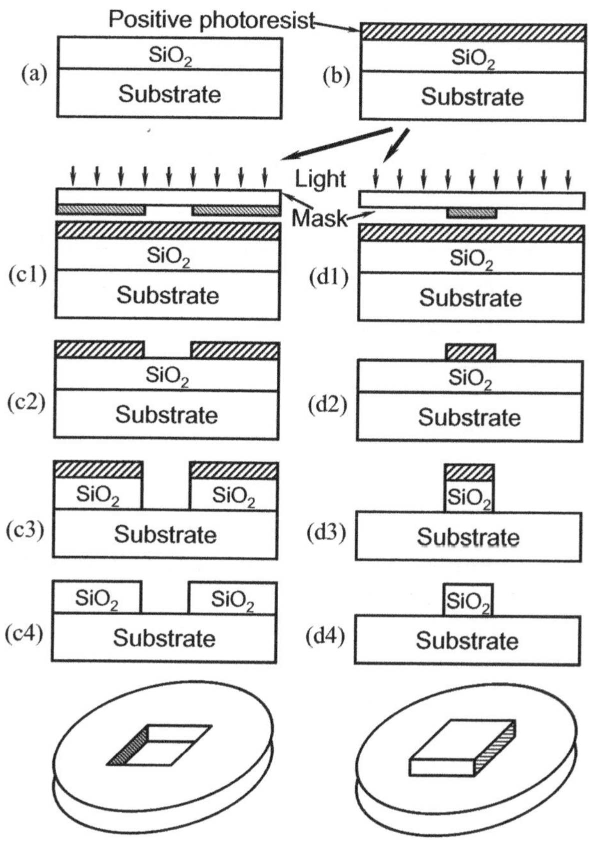

Figure 2.8 Lithography steps.

(Fig. 2.8c4) made of SiO_2. As shown in Figs. 2.7 and 2.8, during the lithography process, the patterns formed on the lower side of the glass mask are first transferred to the resist on SiO_2 layer in the form of latent images, and the SiO_2 pattern is obtained when the SiO_2 layer is etched and the remaining photoresist is stripped. In other words, the mask pattern is transferred to the structural or sacrificial layer (in this case, SiO_2) during the lithographic process. As shown, the lithographic process is used for pattern transfer from a mask to a desired material. More detailed steps of the photography in Fig. 2.8 are illustrated in Fig. 2.9. The steps are described in detail in subsequent paragraphs.

Wafer Cleaning

Wafers might be contaminated with unwanted chemical, such as an organic material that is potentially harmful to MEMS devices. The wafers can be

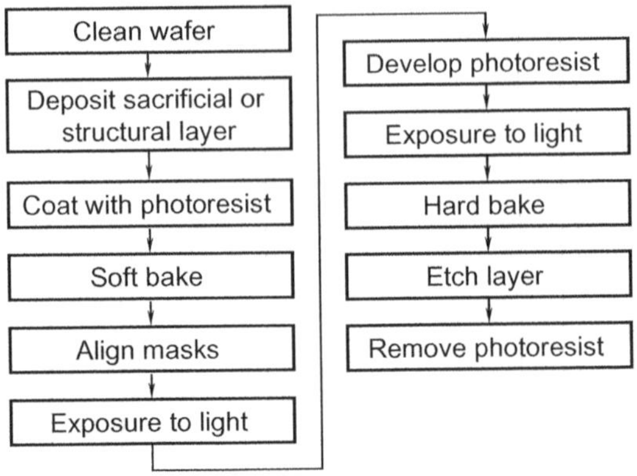

Figure 2.9 Flowchart for lithography.

cleaned using a cleaning procedure. The standard cleaning procedure (known as *RCA clean*; Kern and Poutinen, 1970) is widely used and consists of removal of residual organic/ionic contamination, oxide removal, and heavy metal cleaning. Depending on the layers already existing on the wafer, the cleaning procedure may be modified. For example, if silicon dioxide structures or layers already exist on the wafer, the oxide removal step can be omitted to protect the SiO_2 structures or layers. If different wafers, such as glass wafers are used, a different procedure is required to clean the wafers.

Photoresist

Photoresist is a light-sensitive material that can be applied to the top of a wafer while the wafer on the chuck of a spinner spins at 1000 to 5000 rpm. Photoresist-related works in Fig. 2.9, such as photoresist coating, soft baking to remove solvent and improve adhesion to the wafer, and hard baking, are recommended by the manufacturer's data sheet. Typical photoresist thickness ranges from 2.5 to 5 μm, depending on the spinning speed. After exposure to light, hard baking is conducted to harden photoresist that will serve as a mask in the etching process. After finishing etching a layer under the photoresist, the remaining photoresist is stripped to obtain the etched structure.

Mask Work

After design of a MEMS device, we generate a mask pattern using a computer-aided design (CAD) tool. A set of masks is made by a mask pattern generator. We can send the mask data in the form required to a mask fabrication company or department and the masks will be delivered to us. If we use the first mask to generate a pattern, we do not need the alignment. However, for other masks, we must align a mask to a previous pattern that may be formed on the structural or sacrificial layer. For the purpose of the alignment, we design an

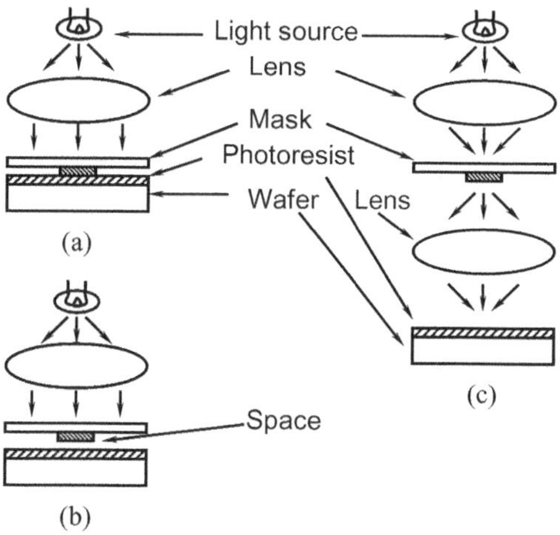

Figure 2.10 Printing techniques.

alignment mark on each mask that was shown in Fig. 2.7. The wafer with an aligned mask is placed under a light source and exposed to form latent images as shown in Figs. 2.7b and 2.10. Depending on the optical systems used for exposure, printing techniques (Fig. 2.10) to transfer the mask pattern to the wafer are classified into three groups: contact printing, proximity printing, and projection printing. Contact printing (Fig. 2.10a) provides high-resolution printing and is used in MEMS fabrication for research in many institutions. In contact printing (Fig. 2.10a), the mask is in contact with photoresist on a wafer while the light source is working. Thus, the mask is contaminated with photoresist, which may generate an unwanted pattern. To avoid contamination, proximity printing (Fig. 2.10b) is used, in which the mask is spaced by a gap from the photoresist, determined by the wavelength of the light source used to minimize the diffraction effect due to the pattern on the mask. For mass production of microelectronic devices and some MEMS, projection printing (Fig. 2.10c) is used, in which an optical system projects the mask image into a small image on the photoresist. For example, if the optical system has a reduction of 10:1, the image of the mask is tenfold greater than that formed on a wafer.

2.3 LAYER DEPOSITION

An upper layer may be deposited on a lower layer to form structural or sacrificial layers, and can be etched for a desired structure. The layers deposited may be formed by a physical or chemical method that results in mass transfer. Before describing the methods in detail, let us use analogies to explain the methods. Figure 2.11 shows four cases of analogies (a–d) and the pattern resulting (e) from those cases. If we wish to obtain a letter L on a board, we

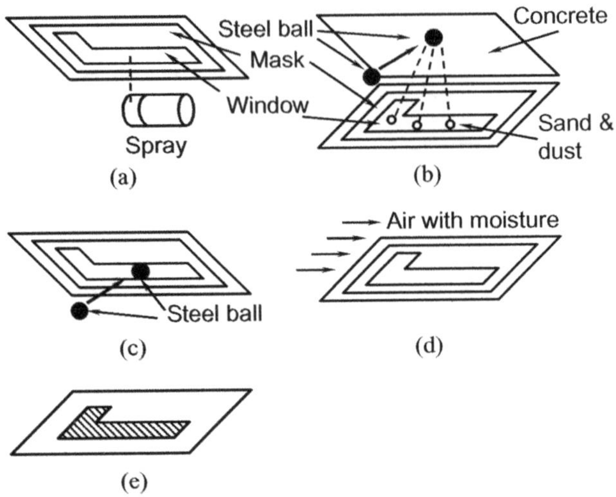

Figure 2.11 Macroscale deposition and removal of layers.

can consider the four cases shown. In Fig. 2.11a we spray paint on a masked board and obtain the letter L on the board after removing the mask. In Fig. 2.11b, the upper board, made of concrete, is exposed to a steel ball or sand blasting to generate a random stream of particles from the upper board, in turn forming the letter L on the masked lower board. In the third case (Fig. 2.11c) we throw steel balls or blast sand into the masked board, which was initially painted with a color, to obtain the letter L on the board. In Fig. 2.11d, the board, covered with a moisture-sensitive material that will change its color when exposed to moisture, is masked with a stencil. For a letter L on the board, the board with the stencil is exposed to moisture. All four processes involve a mass transfer to form a desired letter and are analogous to physical or chemical deposition. For example, the paint spray (Fig. 2.11a) is analogous to the thermal evaporation of a material in vacuum, throwing steel balls at materials (Fig. 2.11b and c) is analogous to the sputtering and ion milling (or ashing) of materials using argon atoms, and a chemical reaction (Fig. 2.11d) of material due to moisture is analogous to chemical vapor deposition. All of these methods are described in detail in subsequent paragraphs.

Thermal Evaporation and Electron Beam Evaporation

One of the simplest methods of depositing a layer is thermal evaporation of a material in vacuum (Fig. 2.12). This method is analogous to the paint spray method shown in Fig. 2.11a. When a boat filled with metal is heated by Joule heating using a current through the boat, the metal evaporates into vacuum (low pressure) and is directed toward the wafer. Since the metal beam can travel a long distance between the boat and the wafer in vacuum, the metal is

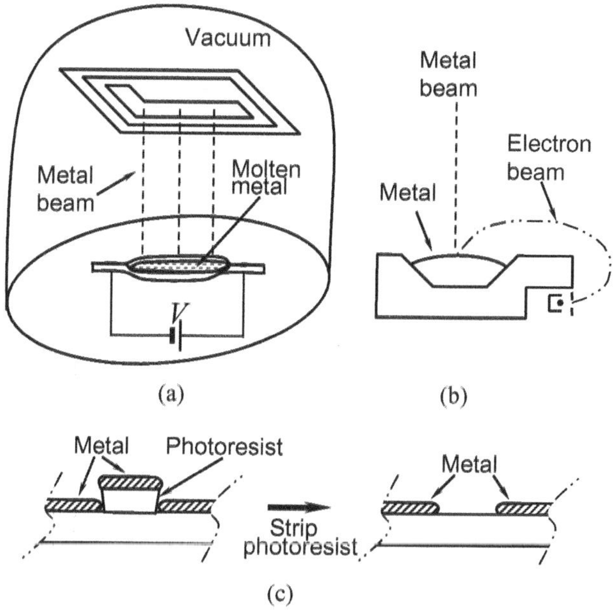

Figure 2.12 Metal evaporation.

deposited on the wafer. Instead of the boat, various heating elements, such as a filament, can be used as the Joule heating element. Thermal evaporation is generally used to deposit metals with a low temperature of evaporation (e.g., aluminum, copper, and gold). For a metal such as tungsten, which has a high temperature of evaporation, electron beam evaporation, which uses electrons of high velocity, may be employed. An electron beam of high velocity (Fig. 2.12b) is generated by an electron beam source and guided to the metal in a crucible. When the electrons hit the metal, kinetic energy due to the electron velocity is converted into heat energy, which increases the temperature of the metal to evaporate the metal into the wafer (not shown in Fig. 2.12b). The evaporated metal beam in vacuum tends to be directional, so the metal beam results in poor step coverage, as shown on the left in Fig. 2.12c. The process (Fig. 2.12c), known as *lift-off*, uses the poor coverage of the metal evaporation in vacuum and to form a metal pattern on a layer. In the lift-off process, a sacrificial pattern (e.g., photoresist) is formed on the layer, and a metal layer is then deposited by thermal evaporation. Upon removing the sacrificial layer, the upper metal on the sacrificial pattern is lifted off and only the lower metal pattern remains.

Sputtering

For easy clarification, we refer to the analogy of shooting balls at a concrete board shown in Fig. 2.11b. This sputtering is also observed in fluorescent lamps.

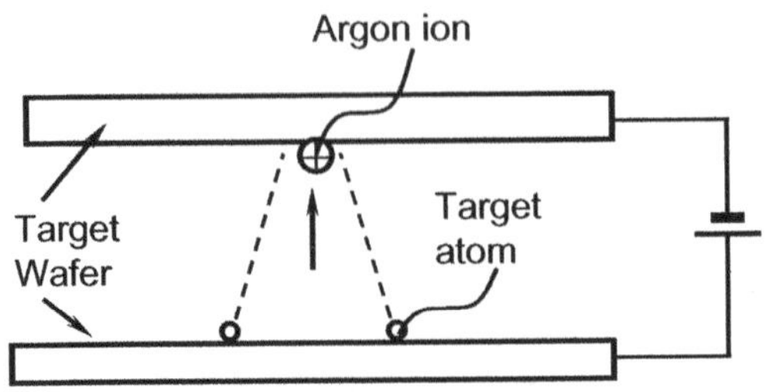

Figure 2.13 Sputtering.

Figure 2.13 shows a schematic of dc (direct current) sputtering in which an upper metal target used to generate metal particles serves as a cathode (which can supply electrons) and the substrate (wafer) placed under the target acts as an anode (which can accept electrons). A dc voltage is applied across the target and substrate when the dc sputtering systems are surrounded by argon (or other noble gas) at low pressure, ranging from 0.1 to 10 torr. Electrons from the target (cathode) flow toward the substrate (anode) and hit argon atoms, so that the argon is excited and ionized: $Ar \rightarrow Ar^+ + e^-$. Most ionized argons absorb electrons again, and then the energy difference between ionized and stable argon is converted into light energy. As a result of this process, we can observe glow from the dc sputtering system. Due to dc bias voltage, some portion of the ionized argon (Ar^+) is accelerated into the target and collides with the target, so that the kinetic energy of the ionized argon is transferred to the target. Bombardment of the ionized argon results in random ejection (scattering) of the target metal, and the scattered metal atoms then move into the substrate and are deposited on the substrate. The ejection of target is known as *sputtering*. Since the target (metal source) is large and scattering of the target is random, the sputtering has good step coverage, unlike the metal evaporation discussed earlier.

Dc sputtering is used only for a metal target since metal acts as a conductor and can transfer electrons to the ionized argon (Ar^+) reaching the target. If we wish to deposit an insulation layer, we can use RF (radio-frequency) sputtering, where ac (alternating current) voltage with dc bias voltage is applied across the target and substrate. The insulation target acts as a capacitor and the electrons can be transferred to the ionized argon. As a result of this process, RF sputtering works and deposits insulation layers as well as conductive layers.

Chemical Vapor Deposition

As discussed earlier, metal evaporation (thermal evaporation and electron beam evaporation) and sputtering employ physical methods using particles that travel at a velocity from particle source to substrate. For example, for thermal evaporation, the evaporated metal particle in vacuum travels from the metal source to the surface of a substrate. However, chemical vapor deposition

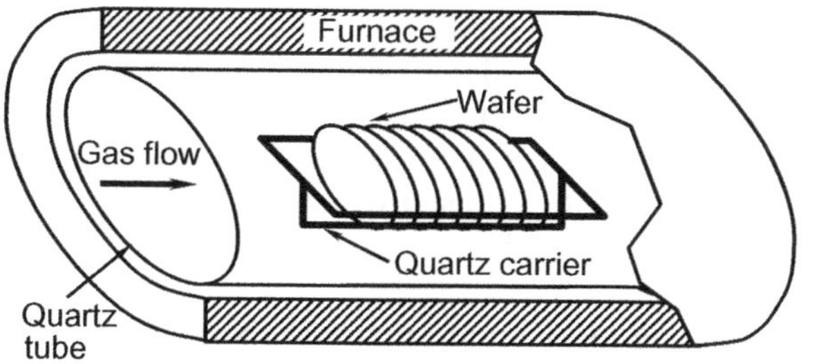

Figure 2.14 Chemical vapor deposition.

(CVD), uses thermal decomposition and/or reaction of chemicals to deposit a layer on a substrate. Figure 2.14 shows a schematic of LPCVD (low-pressure CVD) which is used widely for deposition of polycrystalline silicon (also known as polysilicon or polySi), silicon dioxide, and silicon nitride. In Fig. 2.14, reactant gases are introduced from the left side, flow over wafers, and are pumped out the other side. While the gases flow over the wafers at a temperature of 300 to 1150°C, the gases introduced are decomposed into a desired material that is deposited on the wafers. Using LPCVD, we can obtain uniform layers and can process several hundred wafers in a single run. The following are typical thin films used for the fabrication of MEMS and microelectronic devices.

Polycrystalline Silicon *(Polysilicon)* Silane (SiH_4) and nitrogen are used as reactant and dilution gases, respectively. Under temperature conditions ranging from 600 to 650°C, the following reaction occurs to form a polysilicon layer:

$$SiH_4 \xrightarrow{600\text{–}650^\circ C} Si + 2H_2$$

If we use this reaction, undoped polysilicon will be deposited. If we need doped silicon that acts as a conductor in a MEMS device, dopant gases such as phosphine (PH_3), arsine (AsH_3), and diborane (B_2H_6) can be added to silane so that polysilicon is doped with dopant atoms during the deposition. This process is known as *in situ doping*. Alternatively, the undoped polysilicon can be doped by diffusion or ion implantation (discussed later). Since the residual stress in the film depends on the process temperature, the temperature should be controlled.

Silicon Dioxide Silane (SiH_4) is also used to form a silicon dioxide (SiO_2) layer, but oxygen (O_2) is added to the gas flow. The following reaction occurs at 300 to 500°C for the deposition of silicon dioxide on a substrate:

$$SiH_4 + O_2 \xrightarrow{300\text{–}500^\circ C} SiO_2 + 2H_2$$

This process can be used to obtain a silicon dioxide layer over aluminum since the temperature is below the silicon–aluminum eutectic point of 577°C at which silicon and aluminum form a eutectic mixture. Like polysilicon, silicon dioxide is doped with dopants. Silicon dioxide doped with phosphorus, known as phosphosilicate glass (PSG), is used widely as a sacrificial layer for surface micromachining (see Section 2.5).

Silicon Nitride Silane (SiH_4) and ammonia (NH_3) react to produce silicon nitride (Si_3N_4) at a temperature between 700 and 900°C:

$$3SiH_4 + 4NH_3 \xrightarrow{700-900^\circ C} Si_3N_4 + 12H_2$$

A silicon nitride layer can be used as a mask for silicon oxide etching or a passivation layer to protect a silicon oxide layer during etching of unwanted silicon oxide.

Thermal Oxidation of Silicon

If a silicon wafer is exposed to oxygen at room temperature, the wafer oxidizes to form thin silicon dioxide, called *native oxide*, that is approximately 1 nm. For thicker silicon dioxide, LPCVD silicon oxide is formed. Alternatively, thermal oxidation is used to form silicon dioxide on silicon. Since the thermally grown silicon dioxide layer is a high-quality electrical insulator, the silicon dioxide is used to form the gate oxide that is modulating the conductance or resistance of the channel (the gate oxide or gate is analogous to the mechanical valve of a piping system). Silicon oxide is also used for several purposes (e.g., the mask layer) in the MEMS process.

Thermal oxidation (Fig. 2.15) can be achieved by heating a wafer to a high temperature between 900 and 1200°C in either pure oxygen or water vapor. The chemical reaction for the thermal oxidation in dry oxygen is written as

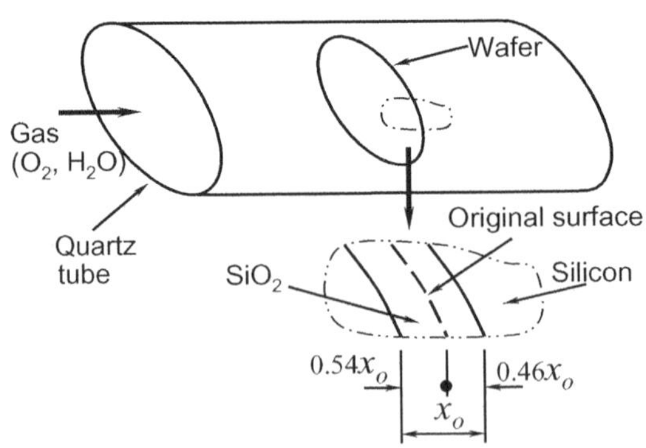

Figure 2.15 Thermal oxidation.

$$\mathrm{Si} + \mathrm{O}_2 \longrightarrow \mathrm{SiO}_2$$

and in water vapor as

$$\mathrm{Si} + 2\mathrm{H}_2\mathrm{O} \longrightarrow \mathrm{SiO}_2 + 2\mathrm{H}_2$$

The preceding oxidation processes consume silicon, and then the oxide expands as shown in the lower part of Fig. 2.15. The mechanism of the thermal oxidation above is given by the Deal–Grove model (Deal and Grove, 1965), as described below.

As shown in Fig. 2.16a, silicon is oxidized to form silicon dioxide on the silicon–silicon dioxide interface while the oxidant species (e.g., oxygen or water vapor) is supplied through the oxide layer already formed. The oxide thickness increases as time elapses. For example, as shown in Fig. 2.16a, the oxide grows by Δx_0 during an elapsed time Δt, so the oxide volume increases by $\Delta V = A\,\Delta x$, where A denotes the area of the silicon wafer. To obtain the oxide thickness x as a function of time, let us consider the oxygen concentration profile of Fig. 2.16b, where C_g, C_s, C_o, and C_i denote the concentration of the oxidant species in the number of the oxidant species per unit volume (e.g., atoms/cm^3) far from the wafer, in the vicinity of the oxide surface, at the oxide surface, and at the silicon–silicon dioxide interface, respectively.

Since the oxide thickness increases, the overall oxidizing process is a time-dependent process. However, if we assume that the transfer process of the oxidizing species is a quasi-steady-state process, the modeling is greatly

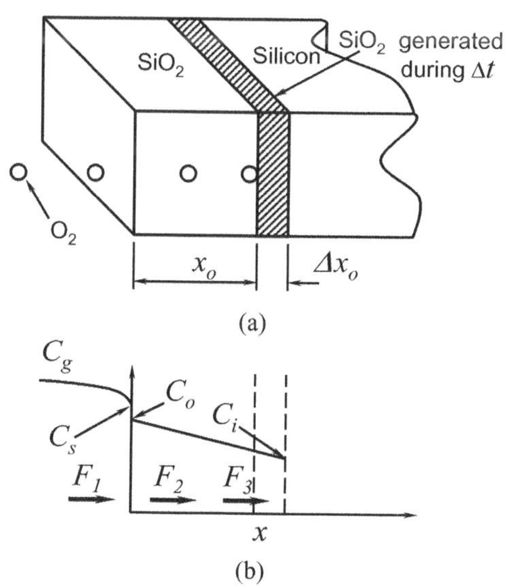

Figure 2.16 Oxidation mechanism of silicon.

simplified. Since all oxidant species should be supplied from the gas outside the oxide, we can consider the following three fluxes of the oxidant species in atoms/cm^2·s:

$$F_1 = h(C^* - C_o)$$
$$F_2 = -D\frac{dC}{dx} = -D\frac{C_i - C_o}{x} = D\frac{C_o - C_i}{x} \qquad (2.1)$$
$$F_3 = k_s C_i$$

where h, C^*, D, k_s, and x represent the mass transfer coefficient (cm/s), the equilibrium concentration of the oxidant species, the diffusion constant (or *diffusivity*) (cm^2/s), the rate constant for the chemical reaction (cm/s), and the thickness of the oxide (cm), respectively. C^* is proportional to the partial pressure of the gas of the concentration C_g. The second of equations (2.1) is obtained from *Fick's first law of diffusion*, stating that the particle flow per unit area is proportional to the concentration gradient of the particle and the particle flows from a region of high concentration to a region of low concentration. In addition to the preceding equations, a balance equation of the oxidant species is given by

$$F_3 A\,\Delta t = N_{\text{ox}} A\,\Delta x \qquad (2.2)$$

where N_{ox} denotes the density of the oxide (atoms/cm^3). The left- and right-hand sides of equation (2.2) represent the number of atoms of the oxidant that is supplied to the interface per unit area during time Δt and the number of atoms of the newly formed oxide in a volume of $A\,\Delta x$.

Setting $F = F_1 = F_2 = F_3$ under the assumption that the oxidant supplied is consumed at the interface, we have, from (2.1),

$$F = \frac{C^*}{1/h + x/D + 1/k_s}$$

Substituting the preceding equation into (2.2) gives a differential equation for the oxide thickness x:

$$\frac{dx}{dt} = \lim_{\Delta t \to 0}\frac{\Delta x}{\Delta t} = \frac{C^*}{N_{\text{ox}}}\frac{1}{1/h + x/D + 1/k_s} = \frac{C^* D}{N_{\text{ox}}}\frac{1}{x + D(1/h + 1/k_x)}$$

Integrating the foregoing equation and using the conditions $x = x_i$ at $t = 0$ and $x = x_0$ at t gives the quadratic equation

$$x_0^2 + Ax_o - B(t + \tau) = 0 \qquad (2.3)$$

where

$$A = 2D\left(\frac{1}{h} + \frac{1}{k_s}\right)$$

$$B = 2\frac{C^*D}{N_{ox}}$$

$$\tau = \frac{x_i^2}{B} + \frac{A}{B}x_i$$

The constant τ is the time corresponding to the initial oxide thickness x_i.

The quadratic equation yields the oxide thickness x_0 as follows:

$$x_0 = \frac{A}{2}\left[\sqrt{1 + \frac{4B}{A^2}(t+\tau)} - 1\right] \tag{2.4}$$

For a short oxidation time (i.e., $t + \tau << A^2/4B$), equation (2.4) reduces to

$$x_0 = \frac{B}{A}(t+\tau) \tag{2.5}$$

which states that the oxide thickness is proportional to time since the oxide thickness is thin and its resistance to the oxidant flow is negligible. The coefficient B/A is defined as the linear rate coefficient. For long oxidation times (i.e., $t + \tau >> A^2/4B$), the solution can be written in simpler form as

$$x_0 = \sqrt{B(t+\tau)} \approx \sqrt{Bt} \tag{2.6}$$

The coefficient B is called the *parabolic rate coefficient*. The right-hand side of equation (2.6) holds since τ, corresponding to the initial thin oxide (e.g., the native oxide of approximately 1 nm), is much less than time t.

Equation (2.6) states that the growth of thick oxide depends on diffusion coefficient D of the oxide and is independent of the orientation of the silicon crystal. The coefficients B/A and B, obtained from experiments, can be expressed by

$$\frac{B}{A} = \alpha \exp\left(-\frac{E_\alpha}{kT}\right) \tag{2.7}$$

$$B = \beta \exp\left(-\frac{E_\beta}{kT}\right) \tag{2.8}$$

where α, E_α, β, and E_β denote coefficients, and k and T represent Boltzmann's constant, defined as 8.6174×10^{-5} eV/K, and temperature in kelvin, respectively.

Table 2.1 shows the coefficients of *B/A* and *A* of a (100) silicon wafer and the initial oxide thickness for the calculations above. For easy calculation of the silicon dioxide thickness, the graph of silicon oxide growth shown in Fig. 2.17 is widely used.

Example 2.1 Determine the oxidation time for 0.5-μm-thick silicon dioxide on (100) silicon if we use wet oxidation at 1000°C.

From (2.7) and (2.8) and Table 2.1, $B/A = 9.70 \times 10^7 \exp(-2.05/[8.6174 \times 10^{-5} \times (1000 + 273)]) = 0.743\,\mu\text{m/h}$, $B = 386 \exp(-0.78/[8.6174 \times 10^{-5} \times (1000 + 273)]) = 0.315\,\mu\text{m}^2/\text{h}$. Substituting these values into (2.3) gives the oxidation time:

$$t = \frac{x_0^2 + Ax_0}{B} = \frac{x_0^2 + B/(B/A)x_0}{B} = \frac{0.5^2 + 0.315/0.743(0.5)}{0.315} = 1.47\text{ h}$$

Alternatively, when we enter the graph of Fig. 2.17 with $x_0 = 0.5\,\mu$m and temperature = 1000°C, we obtain approximately 1.5 h.

TABLE 2.1 Oxidation Coefficient Values of (100) Silicon

	B/A		*B*		
Oxidation	$\alpha(\mu\text{m/h})$	$E_\alpha(\text{eV})$	$\beta(\mu\text{m}^2/\text{h})$	$E_\beta(\text{eV})$	$x_i(\text{nm})$
Wet (H_2O vapor)	9.70×10^7	2.05	386	0.78	0
Dry (O_2)	3.71×10^6	2.00	772	1.23	25

Source: Data adapted from Deal (1978) and Jaeger (1988).

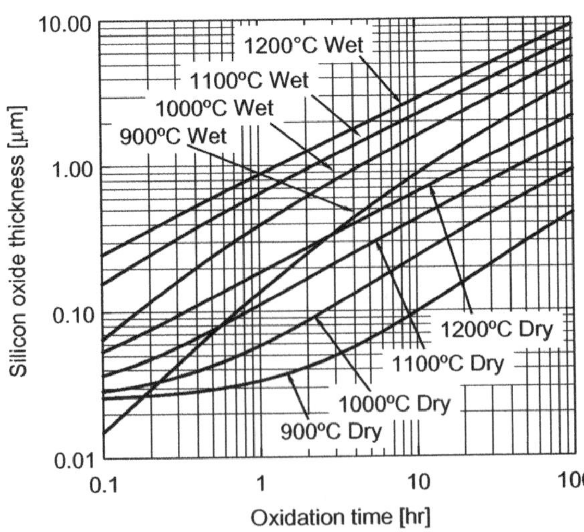

Figure 2.17 Silicon dioxide thickness of (100) silicon (drawn using the data in Table 2.1).

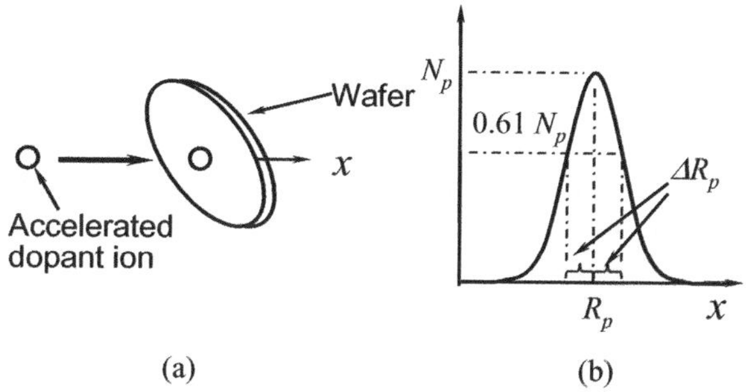

Figure 2.18 Ion implantation.

Ion Implantation

When we form a resistor (called a piezoresistor, discussed in Chapter 7) in a selective area on a silicon wafer, ion implantation and diffusion of dopant may be used. Silicon doped with a dopant such as boron or phosphorus acts as a resistor that can be used as a sensing element based on the piezoresistive effect. In an ion implanter, ionized dopant atoms are accelerated by an electric field and guided to a wafer that is masked selectively to form the resistor. Atoms hit the surface of an open area of the wafer at high velocity, as shown in Fig. 2.18a, penetrate the wafer, and are placed below the surface. The penetration depth depends on the energy of atoms and is typically less than 1 μm. Figure 2.18b shows the distribution N of implanted ions that may be approximated as a Gaussian distribution given by

$$N = N_p \exp\left[-\frac{(x-R_p)^2}{2(\Delta R_p)^2}\right] \tag{2.9}$$

where N_p, R_p, and ΔR_p are the peak dopant concentration (atoms/cm^3), the projected range, and the standard deviation of R_p, respectively. An ion implanter can count the implanted dose Q per unit area by measuring the electron current. The dose Q is implanted into the wafer and is in the Gaussian distribution defined by equation (2.9). From the fact that the implanted dose Q is equal to the integration of the distributed dose, that is,

$$Q = \int_0^\infty N(x)dx = \sqrt{2\pi} N_p \Delta R_p$$

where the right-hand side is obtained for $R_p/\Delta R_p > 2$, the peak dopant concentration N_p is given by

$$N_p = \frac{Q}{\sqrt{2\pi}\Delta R_p} \tag{2.10}$$

As seen in equation (2.10), if the known dose Q and the standard deviation ΔR_p are given, we can use the equation to calculate the peak dopant concentration N_p. For various dopant and wafer material, R_p and ΔR_p are functions of ion energy and are given in graphs (e.g., Muller and Kamins, 1986). As discussed already, the dopant atoms are accelerated by an accelerator, implanted mechanically into the wafer, and is placed randomly below the surface of the wafer. Annealing or a mild heat treatment is needed not only to activate the dopant but also to remove the potential implantation damage. For this purpose, the wafer may be heated at a temperature of 800 to 1000°C for 30 min.

Diffusion

The dopant diffusion in silicon is governed by Fick's second law of diffusion, which describes the conservative transport of mass (or dopant particles) in a material. Let us consider an infinitesimally thin element of area A as shown in Fig. 2.19. Dopant particles flow into and out of the element and then the conservation of the particles can be written

$$\Delta N A\,\Delta x = \left[-D\frac{\partial N}{\partial x}\right] A\Delta t - \left[-D\frac{\partial N}{\partial x} + \frac{\partial}{\partial x}\left(-D\frac{\partial N}{\partial x}\right)\Delta x\right] A\,\Delta t$$

where N and D denote the concentration of the dopant and the diffusion constant, respectively. The left-hand side of the preceding equation represents the increase in the number of particles in the volume $A\,\Delta x$ during time Δt, and the right-hand side is the net flow of the particles during the same time. In the limit as $\Delta t \to 0$, the equation above for constant D becomes *Fick's second law of diffusion*:

$$\frac{\partial N}{\partial t} = D\frac{\partial^2 N}{\partial x^2} \tag{2.11}$$

This linear equation can be solved for N if the initial and boundary conditions are given.

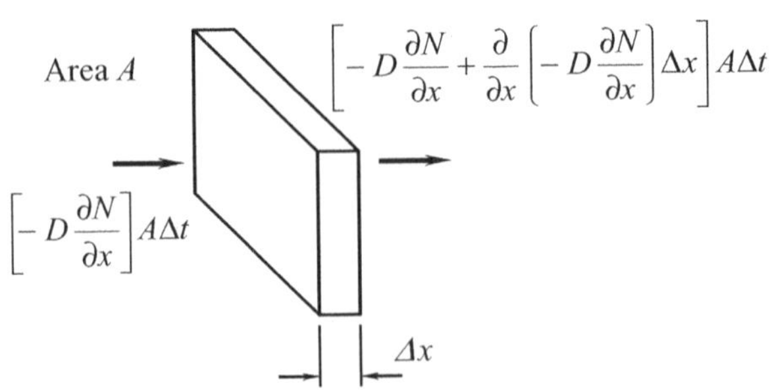

Figure 2.19 Diffusion of dopant into silicon.

We first consider constant-source diffusion. If the silicon wafer is in contact with a dopant source such as a dopant gas and a highly doped glass, whose concentration is higher than the solubility of the dopant in silicon, the concentration on the silicon surface is limited to the solubility and acts as a constant concentration for diffusion. Under this boundary condition, the solution of (2.11) is given by

$$N(x,t) = N_0 \operatorname{erfc}\left(\frac{x}{2\sqrt{Dt}}\right) \tag{2.12}$$

where erfc is the complementary error function, defined as

$$\operatorname{erfc}(u) = 1 - \operatorname{erf}(u) = 1 - \frac{2}{\sqrt{\pi}} \int_0^u \exp(-v^2)\,dv$$

which can be obtained from a mathematical table or a computer program.

Next, we place a limited dopant source on the surface of silicon and then redistribute the dopant in silicon by a thermal treatment. For example, for the redistribution, the diffusion process for the dopant redistribution can be conducted at a temperature of 900 to 1200°C. The dopant (e.g., implanted dopant) within a thin layer can be considered as the impulse function (known as *Dirac's delta function* in physics), so we obtain a solution to (2.11) to give the dopant distribution at time t as follows:

$$N(x,t) = \frac{Q}{\sqrt{\pi Dt}} \exp\left[-\left(\frac{x}{2\sqrt{Dt}}\right)^2\right] \tag{2.13}$$

Equation (2.13) shows that the concentration of the dopant is in Gaussian distribution. It should be noted that (2.13) gives infinite concentration as time t approaches zero. The infinite concentration is not acceptable and is due to the fact that the limit source is considered mathematically as the impulse function which yields infinity at $t = 0$. This infinity can be avoided if we consider a finite width of the dopant at time $t = t_1$ to be redistributed in successive time (this is discussed below).

Figure 2.20 shows the normalized complementary error function and Gaussian function for the preceding diffusion processes. The small inset figures in Fig 2.20 show the redistribution process of the constant limited source diffusions. The dopant of the constant source diffusion spreads as time increases. However, the dopant of the limited dopant source also spreads with time, but the maximum concentration decreases with time due to the limited amount of dopant.

The diffusion coefficient D that plays a crucial role in diffusion process is given by

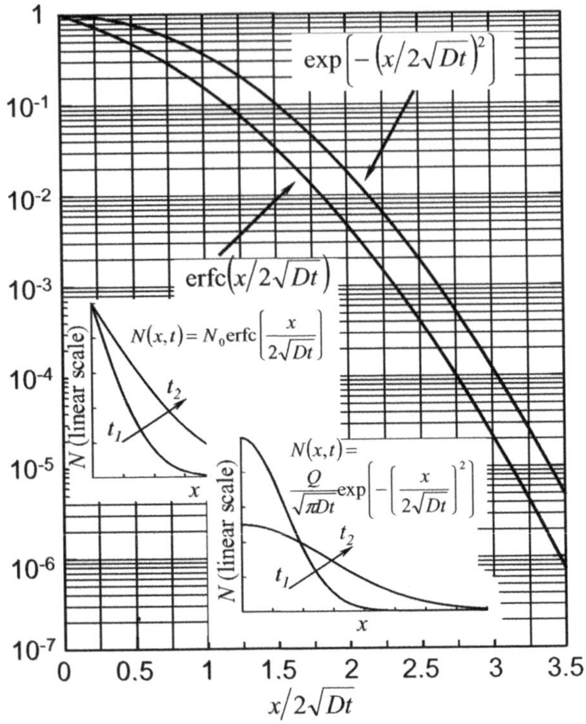

Figure 2.20 Diffusion profile of dopant.

$$D = D_0 \exp\left(-\frac{E_a}{kT}\right) \tag{2.14}$$

where D_0, E_a, k, and T denote a constant, the activation energy in electron volts, Boltzmann's constant defined as 8.6174×10^{-5} eV/K, and temperature in kelvin, respectively. Table 2.2 lists typical diffusion coefficient values for diffusion of some dopant in silicon. If dopant in silicon is exposed several times to high-temperature processes, the total product, $(Dt)_t$, can be obtained by summing the Dt products for all the high-temperature processes:

$$(Dt)_t = \sum_j (Dt)_j \tag{2.15}$$

where $(Dt)_j$ denotes the product at the jth processing step. Using the total product $(Dt)_t$ and the dopant distribution equation (2.12) or (2.13), we can obtain the final dopant distribution when the process is finished.

TABLE 2.2 Diffusion Coefficient Values

Element	D_0 (cm^2/s)	E_a (eV)
Boron	10.5	3.69
Phosphorus	10.5	3.69

Source: Data from Jaeger (1988).

Electroplating

Some MEMS processes use electroplating (also known as *electrodeposition*) to construct a structural layer on a seed layer that is a thin conducting layer. If a seed layer, masked by an insulator, and an inert metal (e.g., Pt) are in an ionic solution (electrolyte), and if a current flows from the inert metal (anode) and to the seed layer (cathode), metal ions in the solution are deposited on the seed layer since cathode can supply electrons to metal ions. After electroplating, the mold is removed to obtain desired structures. Typically, gold, silver, copper, nickel, and nickel ions are deposited. For example, if one wishes to deposit a copper layer, a mixture of copper sulfate ($CuSO_4$) of 120 g/L and sulfuric acid of 100g/L can be used for the electroplating solution. If magnetic or electromagnetic MEMS devices are required, one may consider a nickel–iron solution as an electrolyte. Formulas for electroplating solutions are available in many references (e.g., Kovacs, 1998). Using the standard lithography, a mold for electroplating can be formed. However, if high-aspect-ratio structures are needed, we can use x-rays as a high-energy source. The electroplating process using x-rays from a highly collimated synchrontron is known as LIGA [a German acronym for "lithographie, galvanoformung, abformung," that is, "(x-ray) lithography, electroplating, and molding"]. In this process, poly(methyl methacrylate) (PMMA), patterned using x-ray lithography, serves as the mold, and nickel is electroplated for microstructures.

2.4 LAYER ETCHING

The layer deposited as described in Section 2.3 may be etched by an etchant in order to form desired structures. Etching technologies are classified into two groups: wet etching and dry etching.

Wet Etching

Wet etching uses a wet etchant to form structures from a deposited layer or substrate or by removing undesired portions of a layer under a structure. Depending on the etchant and layers to be etched, wet etching is divided into isotropic and anisotropic wet etching.

Isotropic wet etching is achieved by using isotropic wet etchants that etch layers or substrate in all directions at almost the same rate. Isotropic etching

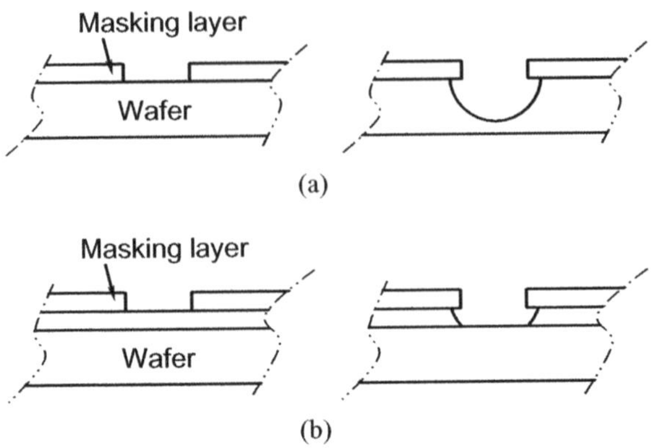

Figure 2.21 Isotropic etching.

of bulk silicon and a layer deposited on a substrate is characterized in Fig. 2.21. Bulk silicon in Fig. 2.21a is etched by an isotropic etchant to provide a channel. As a result, an undercut is formed under the masking layer. The channel shape may depend on the etching conditions, such as etchant type, temperature, and agitation. As shown in Fig. 2.21b, a sacrificial layer under a structural layer or masking layer is etched away by an isotropic etchant. Etchant can be selected based on information on the layer to be etched, the masking layer, and other materials exposed to etchants. Table 2.3 lists typical etchants that may be used to etch silicon and silicon-related material. For example, if we use buffered hydrofluoric acid to etch out a LPCVD silicon dioxide for surface micromachining, an etch rate of more than 440 nm/min is expected. More information on etchants may be found in an article by Williams et al. (2003).

Anisotropic etchants etch materials much faster in one direction than in others. For example, ethylenediamene pyrocatecol (EDP) and potassium hydroxide (KOH) (see Table 2.3) can be used to etch a (100) bulk silicon wafer since its etch rate in the direction <111> is almost zero compared to etch rates in other directions. Figure 2.22 illustrates how anisotropic etchants etch (100) bulk silicon. A mask (e.g., SiO_2 layer) formed on (100) silicon has a square and a rectangle to form pyramidal and rectangular pits. When we immerse the wafer into EDP solution, the etchant attacks silicon in the <100> direction but etches silicon in <111> very slowly (almost at a zero rate). As a result of the etching, we obtain the pyramidal and rectangular pits on the silicon surface shown in Fig. 2.22. Note that the (111) plane makes an angle of 54.7°, shown analytically in Fig. 2.5d, and that the etching stops when two (111) planes meet. If the sides of the square opening are large, the resulting shape will be a hole through the wafer. It is worth noting that the opening of the mask in Fig. 2.22 is aligned with the <110> direction. Experiments show that the convex surface of a (100) silicon wafer is attacked by anisotropic etchants such as EDP and KOH. If the opening is not aligned with the <110> direction or is convex, the

TABLE 2.3 Wet Etchants

Target Material	Etchant	Etch Rate (nm/min)
Silicon dioxide	Buffered hydrofluoric acid (vol. ratio, 40% NH_4F/49% HF = 5:1, ~20°C)	100 (thermally grown) >440 (LPCVD)
Polysilicon (LPCVD, in-situ, n^+)	KOH (30% by weight, 80°C)	>1000
Bulk and polysilicon	Isotropic silicon etchant "Trilogy Etch" (HNO_3/H_2O/ NH4F = 126:60:5, ~20°C)	100 for bulk silicon (100) >100 for polysilicon
Silicon nitride	Phosphoric acid (85% by weight, 160°C)	4.5
Silicon bulk (100)	Ethylenediamene pyrocatecol (EDP) (110°C)	1300
	KOH (30% by weight, 80°C)	1100

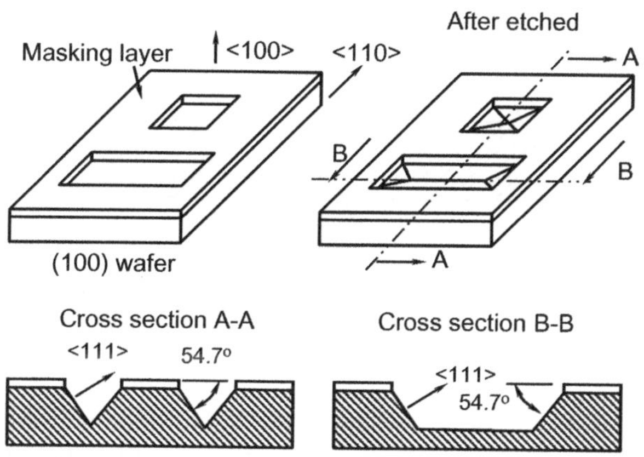

Figure 2.22 Anisotropic etching.

shape will differ from the desired pit or hole. For example, when a wafer with the opening shown in Fig. 2.23a is exposed to EDP solution, V grooves (Fig. 2.23b) are formed along the opening, the convex corners are attacked by the etchant, and the silicon under the protruding portion of the mask is etched away fully, to leaving an overhanging cantilever (Fig. 2.23c). This principle is extended to form a pit or hole by using general mask openings on a (100) silicon wafer, as shown in Fig. 2.23e. The rectangular pit or hole is defined by the family of <110> directions in Fig. 2.23e. If we are interested in forming a uniform thin layer or membrane using wet etching, a highly boron-doped (usually denoted by p^{++}) layer that is typically prepared using gaseous or solid boron diffusion is formed at the bottom of a (100) silicon wafer. The structure

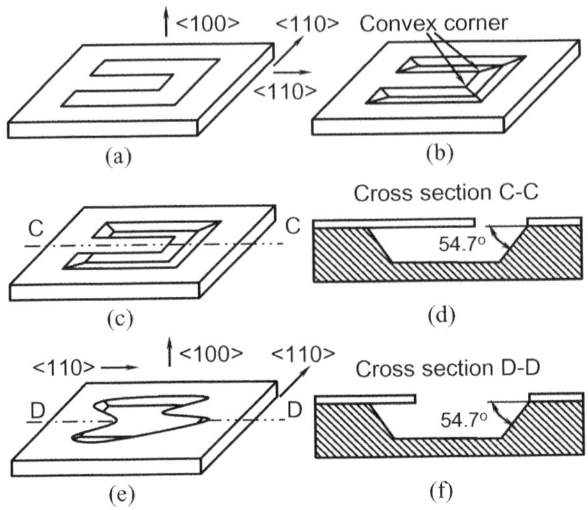

Figure 2.23 Wet etching using an arbitrarily shaped mask.

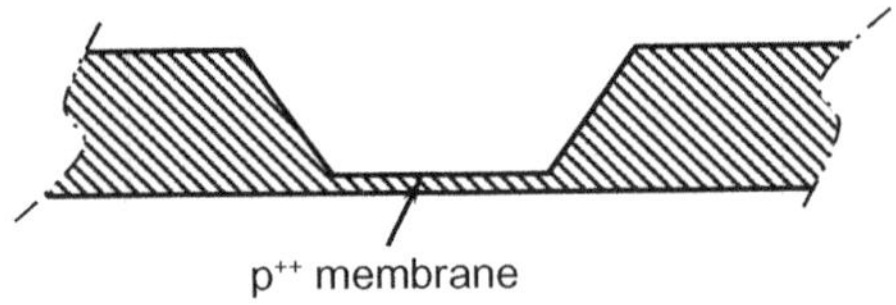

Figure 2.24 Fabrication of a thin silicon membrane using wet etching.

obtained using wet etching is shown in Fig. 2.24. Since wet etchants such as KOH and EDP do not attack the p^{++} layer, the resulting structure becomes a membrane of uniform thickness.

Dry Etching

Compared to wet etching, dry etching uses gas vapor or plasma to etch layers or structures. Dry etching can be classified into nonplasma and plasma etchings. Xenon difluoride etching is a nonplasma, isotropic dry etching process that uses xenon difluoride gas (XeF_2) as a dry etchant to selectively etch silicon or polysilicon without attack by silicon oxide, silicon nitride, photoresist, and metals such as aluminum. Although the etched surface tends to be rough, xenon difluoride etching is useful to etch the underlying silicon of MOS (metal–oxide semiconductor) devices. In the MEMS process, xenon difluoride etching may be used to etch away a polysilicon layer if the polysilicon layer is used as a sacrificial layer.

We may refer to the analogy of Fig. 2.11c, where the steel balls of a velocity hit the open area of a layer so that some portions of the layer is removed by the kinetic energy due to the velocity of the steel ball. If we substitute an inert ion in a plasma (e.g., Ar^+, discussed in Section 2.3) for the steel ball, the inert ion hits a layer, so the target layer is removed as a result of the hitting. This principle, called the *sputter effect*, also referred to as *ion milling*, removes the undesired portions of the layer in a manner similar to mechanical milling. If we use oxygen instead of an inert gas, the oxygen ions (O^{2+}) react with some species, such as photoresist, to produce water and carbon dioxide, which are volatile in vacuum. This process is called *ashing*.

Let us suppose that a reactive ion in plasma is substituted for a steel ball (Fig. 2.11c) so that the ion contacts a layer and reacts with the atom of the layer to form a volatile species. This process is known as *reactive ion etching* (RIE). For example, if sulfur hexafluoride (SF_6) is used as a reactive gas for silicon, the ionized sulfur hexafluoride etches away silicon. More gases are available for dry etching: for example, CF_4 and NF_3 for silicon; CHF_3 and $CF_4 + H_2$ for SiO2; and CF_4 and $CF_4 + H_2$ for silicon nitride.

Deep reactive ion etching (DRIE) is an important processes that allows us to fabricate high-aspect-ratio silicon structures. The DRIE process is a derivative of RIE and consists of successive silicon etching and protective polymer deposition. The DRIE process (Fig. 2.25) is a cross section of a structure where photoresist mask defines the etching area. Sulfur hexafluoride (SF_6) ions etch the silicon area and then octafluorocyclobutane (C_4F_8) plasma forms a polymer layer on the silicon. In the subsequent steps, SF_6 plasma removes the bottom layer of the deposited polymer to etch silicon, and afterward, C_4F_8 plasma

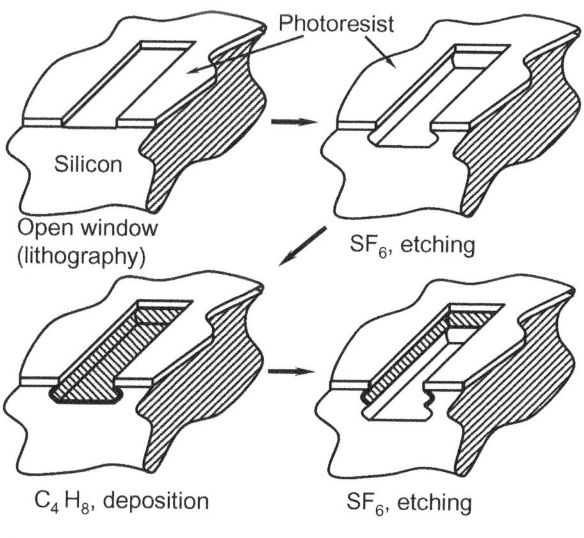

Figure 2.25 Mechanism of deep reactive ion etching.

deposits a polymer layer on the silicon. This sequence continues until a desired depth is obtained. As a result, DRIE provides aspect ratios of up to 30:1 and sidewall angles of $90 \pm 2°$.

2.5 FABRICATION PROCESS DESIGN

Fabrication is a combination of the processes discussed earlier used to fabricate a functional MEMS device. This is affected by many factors, such as temperature and chemicals, since later processes interfere with earlier processes. In Figs. 2.1 and 2.2, we examined a macroscale channel and arched bridge constructed by our ancestors. With reference to Fig. 2.3, we considered macroscale surface micromachining using clay for the construction of a cantilever on plate. These macroscale machining processes (e.g., digging a channel or cutting clay) cannot be used for microscale machining (micromachining) since it is very difficult to machine a tiny part, on the order of 1 μm. Furthermore, macroscale machining is not suitable for mass production of MEMS devices. In order to fabricate microscale parts and devices, a microelectronic fabrication process using chemical reactions and microscale physical phenomena was adopted from integrated-circuit (IC) fabrication. Before beginning our discussion of the micromachining process used for MEMS, we note that there are analogies between our ancestors' constructions and the micromachining process: Digging a channel is analogous to bulk micromachining, and construction of the bridge is analogous to surface micromachining. With this resemblance in mind, it is easy to understand bulk and surface micromachining.

Bulk Micromachining

The bulk micromachining process is relatively simpler than the surface micromachining process. The perspective view in Fig. 2.26 illustrates a groove or channel that is fabricated by bulk-micromachining using an anisotropic etchant and one mask. In the other portions of the figure, the bulk micromachining process is shown. We start with a (100) wafer since a V-shaped groove is needed on a substrate. A silicon dioxide layer is selected as a masking layer against anisotropic etchants. The silicon dioxide layer is grown thermally on the wafer using thermal oxidation (Section 2.3) or deposited by LPCVD silicon dioxide. The thickness of the silicon dioxide layer depends on the etchant and the groove depth. If a 200-μm-deep groove is needed and if the etch rate of (100) silicon and the oxide in KOH are 1100 and 7.7 nm/mm, respectively, the silicon dioxide should be thicker than 140 nm (0.14 μm) from a simple calculation [$200/(1.1)(7.7) = 140$ nm]. A 0.4-μm-thick silicon oxide is sufficient to protect the masked silicon from the etchant. The lithography discussed in Section 2.2 is used to open a window for wet etching. Lithography with a mask first transfers the pattern of the mask to photoresist, and buffered hydrofluoric acid (BHF) etching may be used to transfer the photoresist

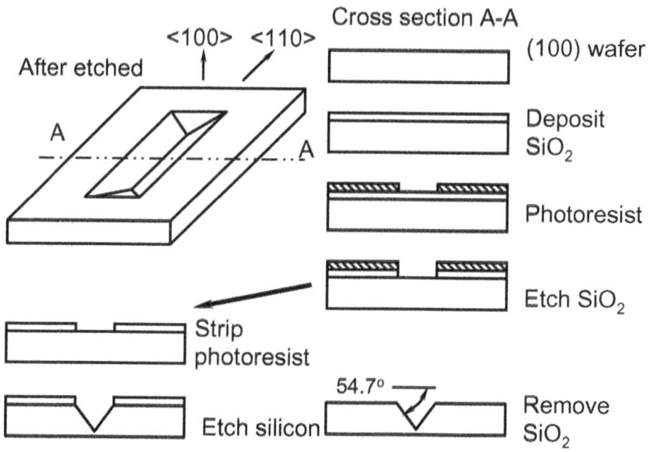

Figure 2.26 Bulk micromachining used to form a groove on a silicon wafer.

pattern to the oxide mask. Afterward, the photoresist is stripped and the silicon is etched in KOH solution. We obtain the final shape of the groove in Fig. 2.26 when the oxide is etched away. The bulk micromachining process of Fig. 2.26 is a one-mask process. Various etching processes may be used for bulk micromachining (e.g., a DRIE process for high-aspect-ratio structures).

Surface Micromachining

The process of surface micromachining is, in general, more complicated than that of bulking micromachining. The number of masks used for a process is a measure of the complexity of the process. The number of masks used in surface micromachining is, in general, greater than that used in bulk micromachining. Let us consider the simple surface micromachining process shown in Fig. 2.27, which is analogous to the clay process used to construct a clay cantilever. The macroscale cantilever of Fig. 2.3 is reduced to the microscale cantilever in Fig. 2.27. For convenience, the major steps are shown in Fig. 2.27, but common steps such as lithography described in Section 2.2 are not shown. A cantilever (Fig. 2.27) is desired for a MEMS device, so we design a fabrication process for the cantilever. We can draw major steps, shown in Fig. 2.27(1)–(5), which are very similar to the steps in the clay process shown in Fig. 2.3. In Fig. 2.27, polysilicon (typical thickness 2 μm) and silicon dioxide (typical thickness 2 μm) are used as the structural and sacrificial layers, respectively (Tang et al., 1989). We start with a silicon wafer with a first sacrificial layer. Using a first mask [Fig. 2.27(2)], a window is defined as an the anchor for the cantilever, followed by deposition of a 2-μm-thick LPCVD polysilicon (undoped or doped). A second mask [Fig. 2.27(4)] and reactive ion etching (RIE) are used to define

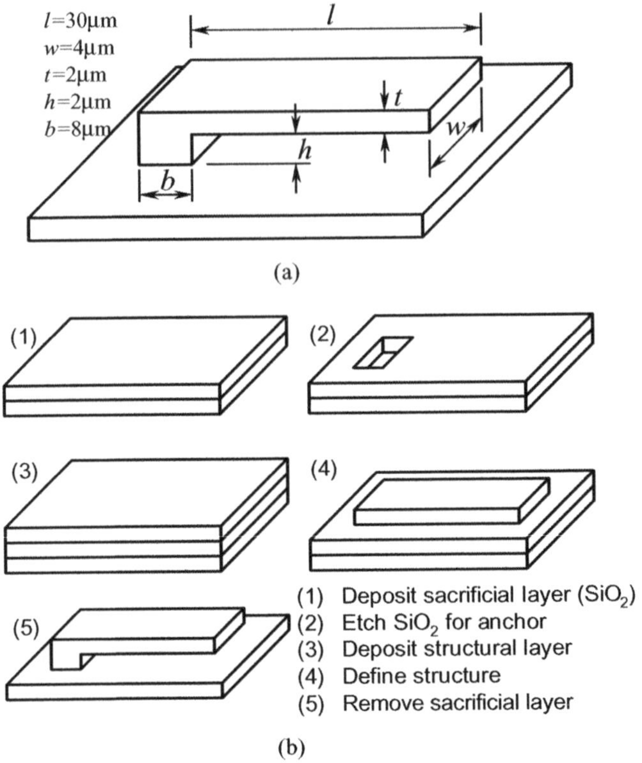

Figure 2.27 Surface micromachining to form a microscale cantilever.

the cantilever structure, a second sacrificial silicon dioxide layer [not shown in Fig. 2.27(4)] is deposited on the polysilicon, and the first and second sacrificial layers are etched away as shown in Fig. 2.27(5). The major steps in Fig. 2.27b are converted into the detailed process shown in Fig. 2.28. The detailed process includes masks, deposition of the first and second PSGs, which serve as a dopant source and also as a sacrificial layer, and annealing and drive-in (diffusion) steps, since it is assumed that undoped polysilicon layers are deposited. The undoped 2-μm-thick polysilicon is doped during the annealing and then converted as a conductor that can carry electric current. For the polysilicon structure release [Fig. 2.27(5)], we may use wet etching (e.g., a hydrofluoric acid solution). If the cantilever is very long compared to its thickness and we are concerned about the stiction of the released cantilever to the substrate due to the surface tension of liquid during release, we may use a sublimation material that changes from solid to gas during the release step. Since the liquid phase does not exist during sublimation, microstructures do not experience the stiction problem. The surface micromachining of Figs. 2.27 and 2.28 is a process that forms only the mechanical structure on a wafer. If additional

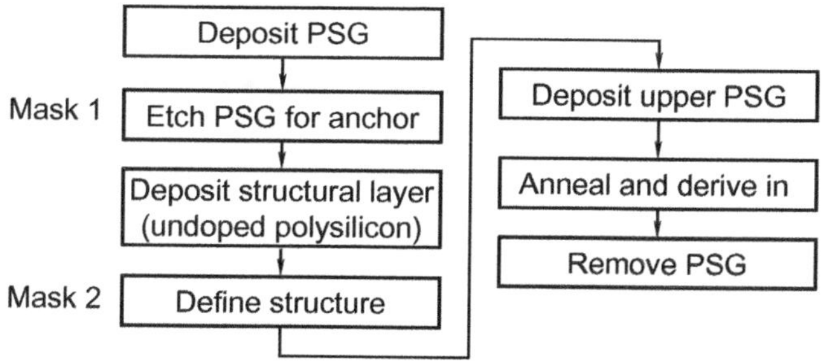

Figure 2.28 Flowchart for surface micromachining.

functions are needed such as electrical connection between structures, more layers may be used for the functions (e.g., refer to Problem 2.4). So far, typical bulk and surface micromachining processes have been described. As discussed, the micromachining process depends on the design of MEMS devices, and vice versa.

PROBLEMS

2.1 We need 1.0-μm-thick silicon dioxide on a (100) silicon wafer. If wet oxidation at a temperature of 1000°C is used, find the oxidation time.

2.2 We wish to form a square pyramid on a (100) silicon wafer using wet etching. On a silicon dioxide mask, a 20-μm-wide square window is formed whose sides are aligned with the <110> directions. If the undercut is negligible, find the length of the side (corresponding to the mask window) of the pyramid after wet etching. If the sides of the mask window make an angle of 45° with the <110> directions, determine the length of the side of the pyramid after finishing the etching.

2.3 A 20-μm-thick p^{++} membrane (Fig. 2.24) is desired for a microdevice. Using a 500-mm-thick (100) silicon wafer and KOH as an etchant, determine the etching time to finish the wet etching.

2.4 A surface-micromachined cantilever (Fig. 2.27a) does not have electrical connections between the cantilever and other structures (not shown in Fig. 2.27). For the electrical connection and insulation desired, more layers may be deposited and followed by etching. Draw a flowchart of the surface micromachining.

CHAPTER 3

STATICS

As a branch of mechanics, statics is concerned with force and moment on mechanical structures in static equilibrium. Using statics, we can analyze forces and moments acting on mechanical structures and evaluate stress, defined as force per unit area, displacement, or deformation of a structural element. Structural elements are essential for MEMS since every component of a MEMS device consists of mechanical structures such as plates and flexures. For a desired function, mechanical structures must be stable and move properly.

3.1 STATIC EQUILIBRIUM

Let us consider a system that consists of an object subjected to forces and moments. If the system produces no motion, the system is in static equilibrium. For example, consider a rock that is placed on the ground as shown in Fig. 3.1a. Since the gravitational force is an attractive force, the rock in Fig. 3.1a weighs mg, where m and g denote the mass of the rock and the gravitational acceleration, defined as approximately 9.8 m/s^2 on the surface of the Earth. As shown in Fig. 3.1b, the weight mg, also known as the *gravitational force*, acts on the center of the rock and the ground generates a counterforce to support the rock. The counterforce N must be mg to maintain equilibrium. If a diagram like that of Fig. 3.1b shows all external forces acting on an isolated body, the diagram is called a *free-body diagram*. Since the force is a vector quantity that

Principles of Microelectromechanical Systems, By Ki Bang Lee

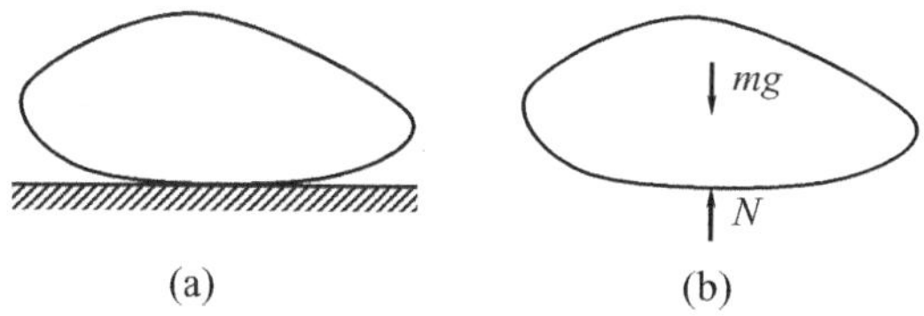

Figure 3.1 Rock in equilibrium.

is represented by its magnitude and direction, we can use the vector calculation to determine unknown forces. If an object is in equilibrium, the following force equilibrium condition must be satisfied:

$$\sum \mathbf{F} = \mathbf{F}_1 + \mathbf{F}_2 + \cdots + \mathbf{F}_n = 0 \tag{3.1}$$

where $\mathbf{F}_1, \mathbf{F}_2, \ldots, \mathbf{F}_n$ are the forces acting on the object.

We can apply the force equilibrium condition to obtain the counterforce N in Fig. 3.1b. The force equilibrium equation is given by

$$\sum \mathbf{F} = N\mathbf{j} - mg\mathbf{j} = 0$$

where $\mathbf{j}$ is a unit vector parallel to the force mg. The preceding equation is rewritten in the form

$$(N - mg)\mathbf{j} = 0$$

Since $\mathbf{j}$ is the unit vector and is not zero, we obtain the counterforce as

$$N = mg$$

Even though forces acting on an object are balanced, the object may tend to rotate about a pivot. To avoid the rotation and to maintain equilibrium, the *moment*, which is a measure of the tendency of a force to cause rotation of an object about an axis, must also be balanced. If a force vector $\mathbf{F}$ (Fig. 3.2a) acts at a point P in space and a displacement vector $\mathbf{r}$ from an origin O to point P is given, the moment vector due to the force $\mathbf{F}$ is defined as

$$\mathbf{M} = \mathbf{r} \times \mathbf{F}$$

where the symbol $\times$ represents the cross product of vectors. The direction of the moment vector $\mathbf{M}$ is determined by the right-hand rule. As a vector quantity, the moment $\mathbf{M}$ is perpendicular to a plane in which $\mathbf{r}$ and $\mathbf{F}$ lie. The moment equilibrium condition of an object can then be written

$$\sum \mathbf{M} = \mathbf{M}_1 + \mathbf{M}_2 + \cdots + \mathbf{M}_n = 0 \tag{3.2}$$

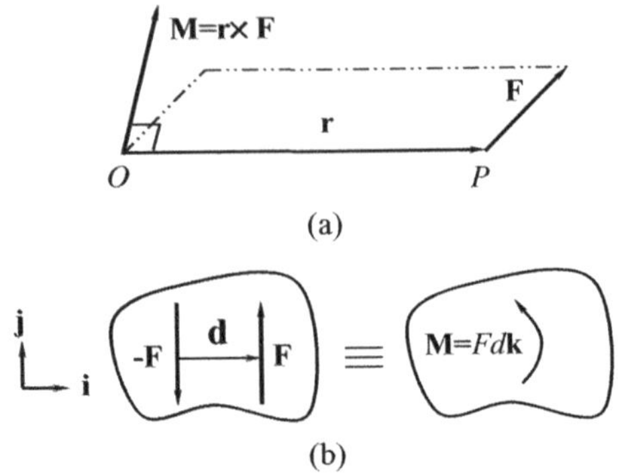

Figure 3.2 Moment due to a force.

where $\mathbf{M}_1, \mathbf{M}_2, \ldots, \mathbf{M}_n$ are the moments due to forces acting on the object.

We consider an interesting configuration of moments, known as a *couple*, that is shown in Fig. 3.2b. One force acting in the direction of the unit vector **j** and another force of the same magnitude but opposite direction are applied to a body. They are spaced a distance *d*. The moment due to the forces is written

$$\mathbf{M} = \mathbf{r} \times (-F\mathbf{j}) + (\mathbf{r} + d\mathbf{i}) \times F\mathbf{j} = d\mathbf{i} \times F\mathbf{j} = Fd\mathbf{k}$$

where **r** is an arbitrary position vector, starting at a point and ending at another point on a line that coincides with the line of action of the left force in Fig. 3.2b, and **k** is the unit vector perpendicular to the unit vectors **i** and **j** (i.e., **k** is perpendicular to the page and directed toward the reader). We note that the couple in Fig. 3.2b generates a moment **M** and that the forces cancell out since the force configuration satisfies the force equilibrium requirement (3.1). As a result, the couple may be thought of as a pure moment. To prevent rotation of the body in Fig. 3.2b, according to (3.2), a countermoment $\mathbf{M}_c$ (not shown in the figure) must be applied to the body as follows:

$$\sum \mathbf{M} = \mathbf{M}_c + \mathbf{M} = 0$$

or

$$\mathbf{M}_c = -\mathbf{M} = -Fd\mathbf{k}$$

Let us suppose that the cantilever beam in Fig. 3.3 (upper figure) is subjected to a force in the *y* direction. The *x* axis coincides with the longitudinal axis of the beam. For convenience, the force *F* directed in the *y* direction is defined to be positive. We wish to evaluate the forces and moment at *x* from the clamped end of the cantilever beam. The lower figure in Fig. 3.3 illustrates a

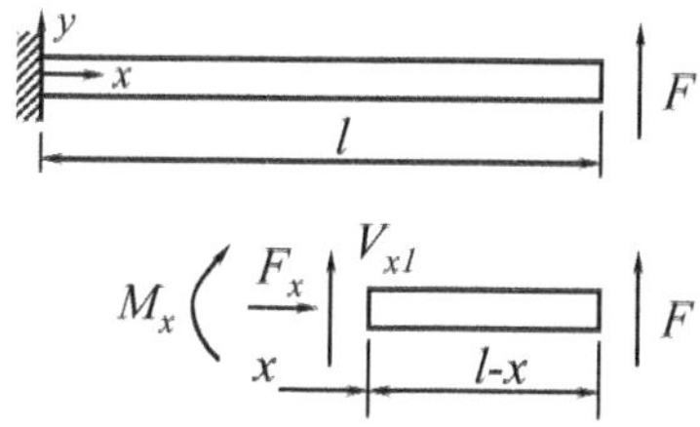

Figure 3.3 Cantilever beam under a force.

free-body diagram of a segment that is cut at x and isolated. To maintain the equilibrium of the segment, forces in the x and y directions and a moment are drawn in the diagram. Since the magnitude and direction of the forces and moment are not yet determined, arbitrary directions for the forces and moment (a couple) are assigned as shown in Fig. 3.3. Two forces, F_x and V_{x1}, and a moment, M_x, appear as unknowns to be determined. The force equilibrium equation (3.1) gives

$$\sum \mathbf{F} = F_x\mathbf{i} + F\mathbf{j} + V_{x1}\mathbf{j} = F_x\mathbf{i} + (F + V_{x1})\mathbf{j} = 0$$

where **i** and **j** represent unit vectors in the x and y directions. From the preceding equation, the coefficients of the unit vectors must vanish to give

$$F_x = 0$$
$$V_{x1} = -F$$

Even though the forces are balanced, the segment tends to rotate at a point since a moment due to the force F is generated. To obtain the moment M_x that prevents rotation of the segment, we write the moment equilibrium equation (3.2):

$$\sum \mathbf{M} = -M_x\mathbf{k} + (l - x)\mathbf{i} \times F\mathbf{j} = [-M_x + (l - x)F]\mathbf{k} = 0$$

which gives the bending moment at the cut surface x:

$$M_x = (l - x)F$$

The bending moment (or resisting moment) M_x may be considered as a couple to balance the cut segment. To maintain equilibrium of the segment subjected to the positive shear force F in Fig. 3.3, the bending moment M_x applied to the left face of the segment is positive in the sense shown in Fig. 3.3, but the shear force V_{x1} has a negative value of $-F$. It is very convenient that the shear force on the left face of the segment appears positive when a positive shear force is applied at the end of the cantilever beam.

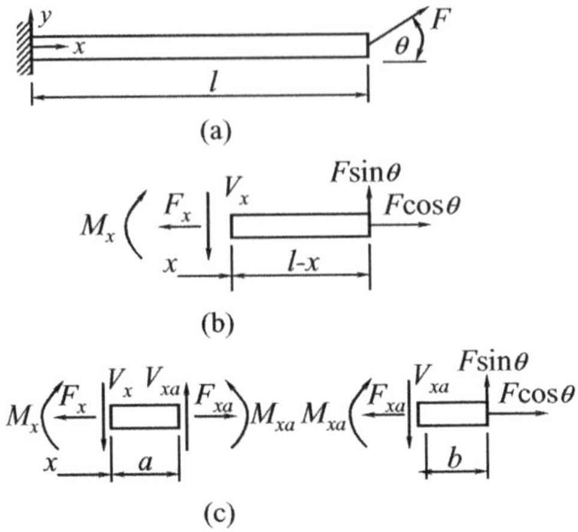

Figure 3.4 Cantilever beam subjected to an arbitrary force at the end.

A more general case of the cantilever beam problem is shown in Fig. 3.4a, where a force making an angle θ with the x axis is applied at the end of the cantilever. For a positive shear force at the cut surface, the shear force V_x is downward, and for the same reason the force acting along the cantilever beam is leftward in Fig. 3.4b. When we repeat the foregoing procedure for the free-body diagram (Fig. 3.4b), we have

$$F_x = F\cos\theta$$

$$V_x = F\sin\theta$$

$$M_x = (l-x)F\sin\theta$$

We noted that all force and moment acting on the left face appear positive when the x and y components of the force F is positive in the x and y directions, respectively. If we cut the segment of Fig. 3.4b into two small segments as shown in Fig. 3.4c and repeat the procedure, the forces and moment on the new cut face are given by

$$F_{xa} = F\cos\theta$$

$$V_{xa} = F\sin\theta$$

$$M_{xa} = bF\sin\theta$$

As shown in the foregoing equations, all force and moment at the cut faces in Fig. 3.4c appear positive, as positive forces are applied in the x and y directions at the end of the cantilever beam. Referring to Fig. 3.4c, we can therefore define the following rule, associated with force and moment in a beam analysis, which is used throughout the book:

1. A segment cut from a beam has two faces perpendicular to the longitudinal axis (x axis) of the beam: a positive face if the face is located far from the origin of the axis, a negative face if the face is located close to the origin.
2. If a tensile force is exerted along the length of the beam, the tensile force is said to be a positive normal force. A compressive force is defined as a negative normal force.
3. A shear force acting on the positive face is said to be positive when the shear force is directed to the positive y direction perpendicular to the x axis. If a shear force on the negative face is in the negative y direction, the shear force is also positive.
4. If a moment (couple) acting on the positive face is directed in the positive z direction determined by the right-hand rule, the moment is said to be positive. When a moment directed in the negative z direction is applied to the negative face, the moment is also positive.

The positive normal and shear force and moment according to the definition above are shown on the left in Fig. 3.4c. It should be noted that other researchers or books may use a different rule to determine the sign of the force and moment. Using the definition described above, we obtain the forces and moment (Fig. 3.5) at x when a pure moment (couple) is applied at the end of a cantilever beam. Applying (3.1) and (3.2) to the free-body diagram in Fig. 3.5 gives the normal force, shear force, and moment at x:

$$F_x = V_x = 0$$

$$M_x = M$$

A pure moment (couple) exerted on the end of a cantilever beam generates only a pure moment in the cantilever beam; normal and shear forces do not appear.

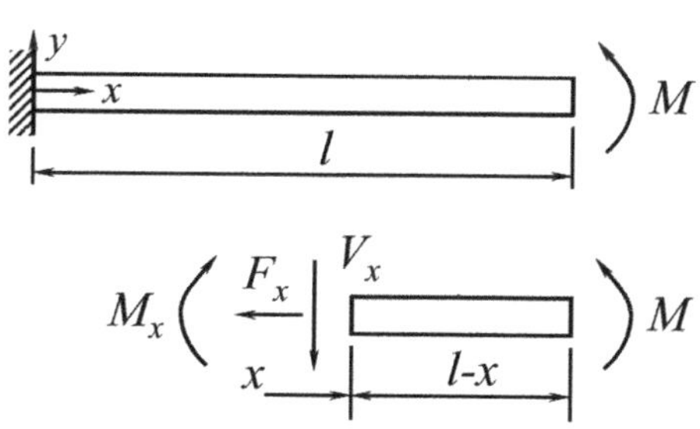

Figure 3.5 Cantilever beam subjected to a moment.

3.2 STRESS–STRAIN RELATIONSHIP

In Section 3.1 we described the normal and shear forces exerted on a body and defined the positive direction of the forces and moment acting on a beam. Let us turn our attention to the state of stress and strain in a body. A slender (narrow) beam under a tensile force (positive normal force) is illustrated in Fig. 3.6a, where the dashed and solid beams are those before and after stretching. From our experience we know that since the beam is elastic, the beam in Fig. 3.6a elongates under the tension and the cross section is reduced. According to Hooke's law, the elongation of the beam is linearly proportional to the force. Furthermore, the elongation is proportional to the initial length of the beam and inversely proportional to the initial cross-sectional area. As a result, the relationship between the force and elongation may be given by

$$\frac{F}{A} \propto \frac{\delta}{l}$$

where A and l denote the cross-sectional area and length of the beam before the beam is stressed. For convenience, F/A is defined as the stress σ_x (force

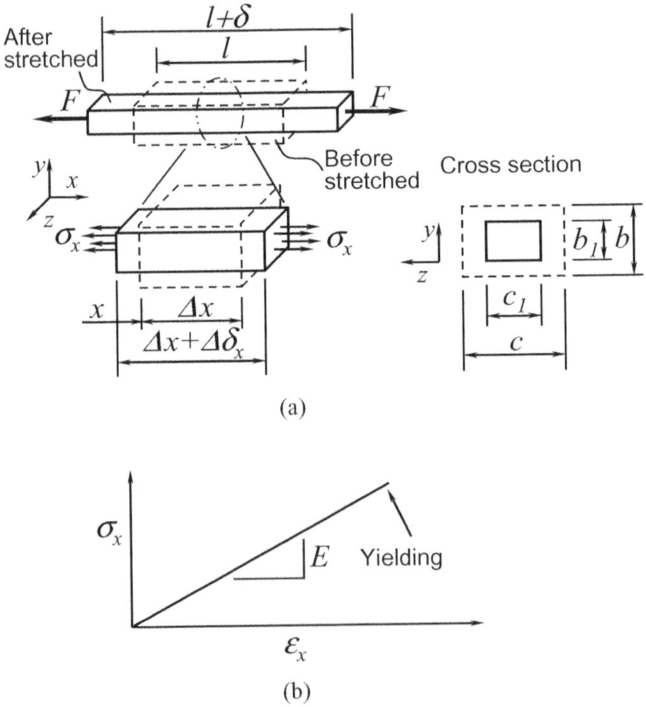

Figure 3.6 Slender beam subjected to a tensile force.

per unit area) and δ/l is defined to be the strain ε_x (elongation per unit length). Using these definitions, the foregoing equation is rewritten as

$$\sigma_x = E\varepsilon_x \tag{3.3}$$

where E is a proportional constant known as *Young's modulus* or the *modulus of elasticity*, and the subscript x denotes the component in the x direction in Fig. 3.6a. Young's modulus is a material property that depends on material. The stress–strain relationship of a brittle material is shown in Fig. 3.6b. Brittle materials (e.g., bulk silicon, glass) are broken when the stress in a beam reaches the yielding stress (labeled "Yielding" in Fig. 3.6b). Ductile materials (e.g., nickel) show a straight stress–strain relationship at relatively low stress but a nonlinear stress–strain relationship around the yielding stress. However, since most microstructures in MEMS and NEMS undergo a stress much less than the yielding stress, it is assumed that the linear stress–strain relationship (3.3) holds for all materials that are used in MEMS and NEMS. Since experiments show almost the same Young's modulus in a compressive stress, we assume that the linear stress–strain relationship (3.3) holds for all materials and for tensile and compressive stresses.

We noted above that a beam under a tensile force elongates and that the elongation or strain can be evaluated using the stress–stress relationship (3.3). We must also note that the width and thickness of the cross-sectional area shrinks from its initial values as shown on the lower right in Fig. 3.6a. Experiments show that the strain in the direction perpendicular to the beam length is proportional to the strain along the beam. The relationship may be expressed as

$$\begin{aligned} \varepsilon_y &= -\nu\varepsilon_x \\ \varepsilon_z &= -\nu\varepsilon_x \end{aligned} \tag{3.4}$$

where

$$\begin{aligned} \varepsilon_x &= \frac{\delta}{l} = \frac{\Delta\delta_x}{\Delta x} \\ \varepsilon_y &= \frac{b_1 - b}{b} \\ \varepsilon_z &= \frac{c_1 - c}{c} \end{aligned}$$

denotes the strains in the x, y, and z directions and ν is a proportional constant, known as *Poisson's ratio*. The negative sign in the preceding equations means that the dimensions of the cross-sectional area are reduced as the beam is stretched. Like Young's modulus, Poisson's ratio υ is a material property and is measured from experiment. For an isotropic material, Poisson's ratio remains constant in all directions.

Effective Young's Modulus

We have described Young's modulus, Poisson's ratio of a material, and normal stress and strain in a slender beam. Young's modulus remains constant in all directions at a point in a body of isotropic material. However, experiments show that elongation is also related to geometry. Let us consider the wide beam shown in Fig. 3.7. A beam is defined as a wide beam if the length l (Fig. 3.7) is much greater than the thickness t and the width w is equal to or greater than five times the thickness. Experiments show that for a wide beam satisfying the foregoing condition, contraction in the width of the beam is negligible, as shown in the second figure in Fig. 3.7. In other words, the strain in the width direction can be considered to be zero. As a result, the strain of a wide beam is smaller than that of a narrow beam when axial stress is applied to the beam. This phenomenon results in an apparent Young's modulus known as the *effective Young's modulus*, which is higher than Young's modulus E.

Let us consider the cross sections of Fig. 3.7a and b to obtain the effective Young's modulus. Since the material of Young's modulus E and Poisson's ratio υ is linear, we use the superposition principle that holds for linear systems. The

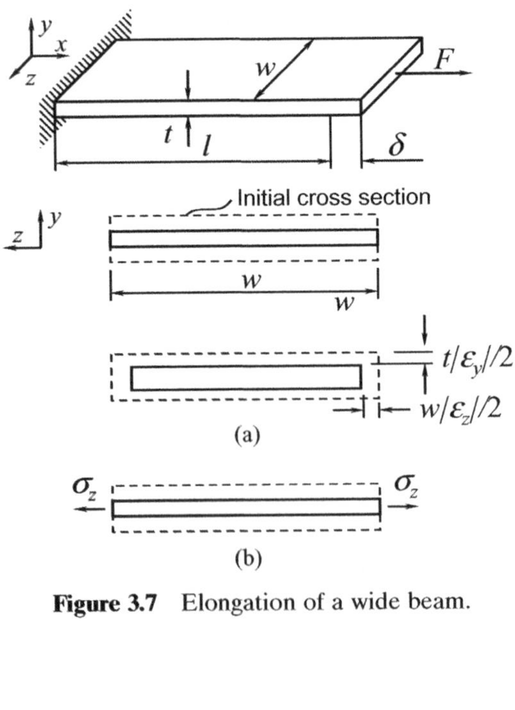

Figure 3.7 Elongation of a wide beam.

wide beam is stretched in the axial direction without the geometric effect in Fig. 3.7a (like the narrow beam in Fig. 3.6), and then the beam is exposed to a stress to obtain zero strain in the width direction. When we apply a force F in the x direction in Fig. 3.7a, the force–elongation relationship is given by (3.3)

$$\frac{F}{A} = E\frac{\delta_{xe}}{l}$$

where A, δ_{xe}, and l denote the cross-sectional area of the beam (i.e., wt), the elongation, and the length of the beam. The equation above gives the elongation

$$\delta_{xe} = \frac{Fl}{EA}$$

$$\varepsilon_{xe} = \frac{\delta_{xe}}{l} = \frac{F}{EA}$$

in the x direction. Due to the axial force F, the width decreases and we can evaluate the strain ε_z from (3.4) as follows:

$$\varepsilon_z = -\nu\varepsilon_{xe} = -\nu\frac{F}{EA}$$

where the negative sign denotes the contraction. To make the overall strain zero in the z direction, the beam is subjected to a strain given by

$$\varepsilon_{zr} = |\varepsilon_z| = -\varepsilon_z = \nu\varepsilon_{xe} = \nu\frac{F}{EA}$$

and the corresponding stress in the z direction is obtained as

$$\sigma_r = E\varepsilon_{zr} = \nu\frac{F}{A}$$

The stress σ_r in the width direction also generates the strain in the longitudinal direction of the length of the beam, which is given by (3.4)

$$\varepsilon_{xc} = -\nu\varepsilon_{zr} = -\nu(\nu\varepsilon_x) = -\nu^2\frac{F}{EA}$$

where the negative sign represents the contraction of the length. The equation above gives the contracted length due to the zero strain of the wide beam in the width direction:

$$\delta_{xc} = \varepsilon_{xc} l = -\nu^2 \frac{Fl}{EA}$$

The overall elongation due to the force F exerted on a wide beam is given by the sum of the elongations:

$$\delta = \delta_{xe} + \delta_{xc} = \frac{Fl}{EA} - \nu^2 \frac{Fl}{EA} = \frac{Fl}{EA}(1 - \nu^2)$$

Note that δ_{xe} is due to the axial force F, while δ_{xc} is caused by the geometric effect of the wide beam. If the stress–strain relation (3.3) and the equation above are used, the effective Young's modulus $\tilde{E}$ of the wide beam is obtained as

$$\tilde{E} = \frac{\sigma_x}{\varepsilon_x} = \frac{F/A}{\delta/l} = \frac{F/A}{(Fl/EA)(1-\nu^2)/l} = \frac{E}{1-\nu^2}$$

The effective Young's modulus $\tilde{E}$ holds for wide beams satisfying the condition $w \geq 5t$ or $t \geq 5w$. Even though the effective Young's modulus is a continuous function of w/t, for convenience, the effective Young's modulus $\tilde{E}$ may be defined as follows (Osterberg and Senturia, 1997): For a narrow beam, $w < 5t$ and $t < 5w$,

$$\tilde{E} = E \tag{3.5}$$

For a wide beam, $w \geq 5t$ or $t \geq 5w$,

$$\tilde{E} = \frac{E}{1-\nu^2} \tag{3.6}$$

The effective Young's modulus defined above is used in the analysis of a beam subjected to normal stress along the beam length (i.e., the axial force and bending moment are exerted on the beam). In the analysis of plates, diaphragms, and membranes, the effective Young's modulus (3.6) appears in the governing equations since the width is much greater than the thickness.

Shear Stress

Shear stress is defined as a shear force per unit area. Consider the elastic rectangular parallelepiped in Fig. 3.8, which is sandwiched between relatively rigid plates. When a shear force F is applied in the x direction, the upper plate is displaced from its original position by u. Experiments show that the displacement u is proportional to the force and the thickness b of the elastic parallelepiped and inversely proportional to the area of the plate. This relationship may be expressed as

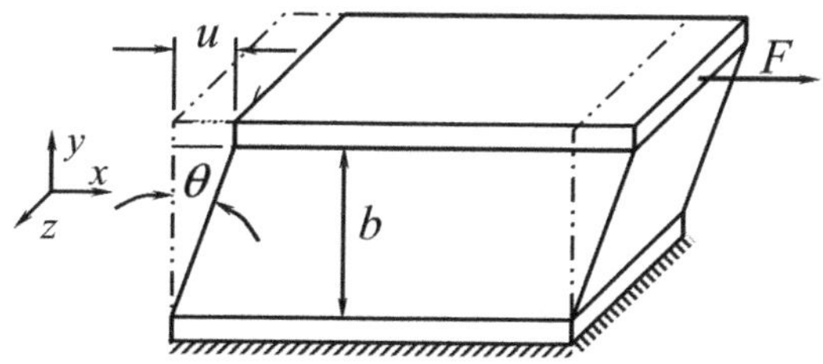

Figure 3.8 Elastic rectangular parallelepiped under shear force.

$$u \propto \frac{F}{A} b$$

The equation above can be rewritten in the convenient form

$$\tau = G\gamma \tag{3.7}$$

where $\tau = F/A$, $\gamma = u/b$, and G are defined as the shear stress, the shear strain, and the shear modulus, respectively. For $u << b$, the angular displacement (shear angle) θ in radians is written

$$\theta \cong \gamma = \frac{u}{b}$$

If the material is a linear isotropic material, we can show that the shear modulus G is expressed in terms of Young's modulus and Poisson's ratio. The relationship between G and E will be derived later in this section.

Plane Stress

Even though stress distribution is generally expressed with respect to a three-dimensional coordinate system, stresses in many engineering and scientific problems may be considered as two-dimensional stresses. Let us consider a thin elastic body that is subjected to forces in the x–y plane, as shown in the upper figure in Fig. 3.9. All forces are acting in a plane parallel to the x–y plane, so the normal and shear stress are assumed to act in the x–y plane. Change in the stress in the z direction is negligible. Under this assumption, the normal and shear stresses act in the x–y plane; it is illustrated in the lower right figure in Fig. 3.9. A small rectangular element is taken to show the stress distribution in the figure. We first define the faces of the element: The right face is the positive x face because the face crosses the x axis and is far from the origin of x axis; the left face is the negative x face since it is close to the origin of the x axis; similarly, the upper face is the positive y face and the lower face is the negative y face. Similar to the definition of the positive normal and shear force

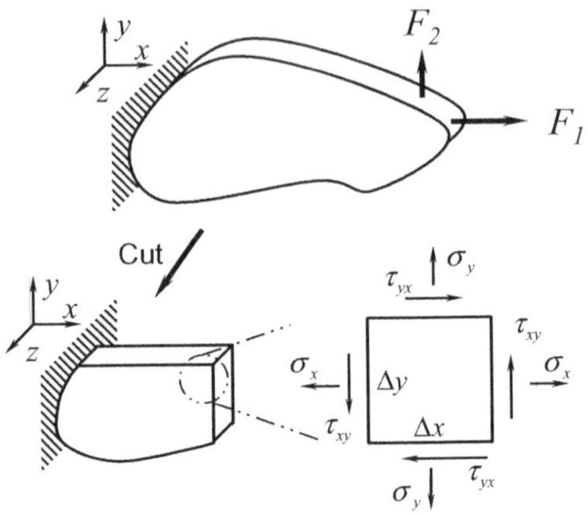

Figure 3.9 Plane stress.

in Section 3.1, all stresses in the right figure in Fig. 3.9 are defined as positive. For example, the normal stress on the positive x face (i.e., σ_x) is positive since the force is applied on the positive x face and directed in the positive x direction. The shear stress (τ_{xy}) on the positive x face is positive since the stress acts on the positive x face and is directed in the positive y direction. For the element to remain in equilibrium, the moment equilibrium requirement (3.2) at its center must be satisfied as follows:

$$\sum M = \tau_{xy}\Delta y\,\Delta z\frac{\Delta x}{2} + \tau_{xy}\Delta y\,\Delta z\frac{\Delta x}{2} - \tau_{yx}\Delta x\,\Delta z\frac{\Delta y}{2} - \tau_{yx}\Delta x\,\Delta z\frac{\Delta y}{2} = 0$$

where Δx, Δy, and Δz denote the increments in x, y, and z, respectively. The preceding equation gives

$$\tau_{yx} = \tau_{xy} \tag{3.8}$$

Note from equation (3.8) that in the plane stress, the shear stress τ_{yx} on the y faces is always equal to the other shear stress, τ_{xy}, on the x faces.

Stress with Respect to a Set of Rotated Axes

In many engineering and scientific problems, it is necessary to evaluate stress components with respect to a set of new axes that are rotated from the original coordinate axes. We first examine a simple case and then deal with more complicated cases. Consider the cantilever beam in Fig. 3.10a, which is subjected to a tensile force F in the longitudinal direction. From the force equilibrium of the cut segment in the x' and y' directions in Fig. 3.10b,

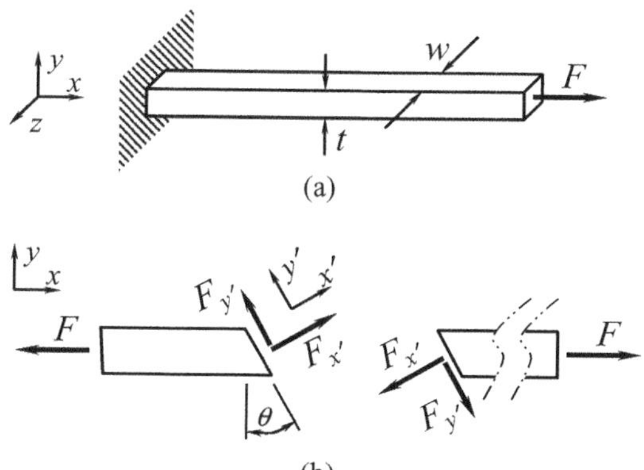

Figure 3.10 Normal and shear forces in an inclined cross section of a cantilever.

$$\sum F_{x'} = F\cos\theta - F_{x'} = 0$$
$$\sum F_{y'} = -F\sin\theta - F_{y'} = 0$$

we have

$$F_{x'} = F\cos\theta$$
$$F_{y'} = -F\sin\theta$$

The cross-sectional area of the inclined surface is given by

$$A' = w\frac{t}{\cos\theta} = \frac{A}{\cos\theta}$$

where A represents the cross-sectional area at $\theta = 0$. The normal and shear stresses are given, respectively, by

$$\sigma_{x'} = \frac{F_{x'}}{A'} = \frac{F\cos\theta}{A/\cos\theta} = \frac{F}{A}\cos^2\theta = \sigma_x\cos^2\theta$$
$$\tau_{x'y'} = \frac{F_{y'}}{A'} = \frac{-F}{A/\cos\theta} = -\frac{F}{A}\sin\theta\cos\theta = -\sigma_x\sin\theta\cos\theta$$

It is worth noting that even though the cantilever beam is subjected to a positive axial force, the normal and shear forces on the inclined cross-sectional area are generated to maintain equilibrium of the cantilever beam, which, in turn, provides the normal and shear stresses on the inclined surface.

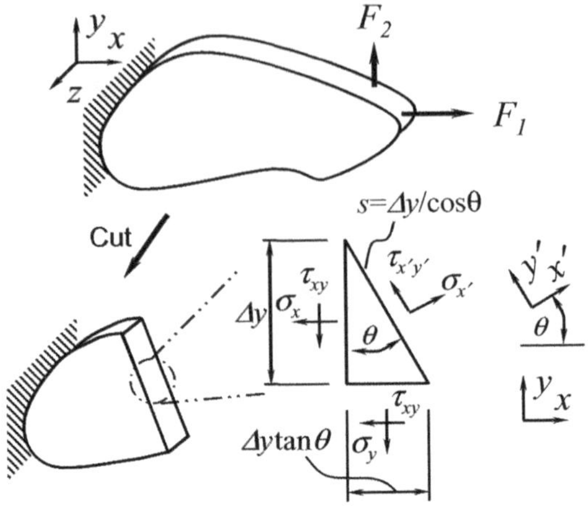

Figure 3.11 Stress components in a set of axes rotated with respect to the original axes.

This idea can be extended to give the stress in a set of axes rotated with respect to the original axes. In Fig. 3.11, under the assumption that we know the stress components σ_x and τ_{xy} in the x–y coordinates, we find the stress components $\sigma_{x'}$ and $\tau_{x'y'}$ in the x' and y' axes shown in Fig. 3.11. The rotated face makes an angle of θ with the y direction, and the wedge cut from the elastic body is in equilibrium. The force equilibrium in the x' direction gives

$$\begin{aligned}\sum F_{x'} &= \sigma_{x'}\frac{\Delta y}{\cos\theta} - \tau_{xy}\Delta y\sin\theta - \sigma_x\Delta y\cos\theta \\ &\quad + \tau_{xy}\Delta y\tan\theta\cos\theta - \sigma_y\Delta y\tan\theta\sin\theta = 0\end{aligned}$$

Rearrangement of the terms above yields the normal stress in the x' direction as follows:

$$\sigma_{x'} = \sigma_x\cos^2\theta + \sigma_y\sin^2\theta + 2\tau_{xy}\sin\theta\cos\theta \tag{3.9}$$

Similarly, the force equilibrium equation in the y' direction is written as

$$\begin{aligned}\sum F_{y'} &= \tau_{x'y'}\frac{\Delta y}{\cos\theta} - \tau_{xy}\Delta y\cos\theta + \sigma_x\Delta y\sin\theta \\ &\quad + \tau_{xy}\Delta y\tan\theta\sin\theta - \sigma_y\Delta y\tan\theta\cos\theta = 0\end{aligned}$$

Rearranging the equation above gives

$$\tau_{x'y'} = -(\sigma_x - \sigma_y)\sin\theta\cos\theta + \tau_{xy}\left(\cos^2\theta - \sin^2\theta\right) \tag{3.10}$$

The normal force in the y' direction can result from substituting $\sigma_{y'}$ for σ_x and $\theta + \pi/2$ for θ into (3.9):

$$\sigma_{y'} = \sigma_x \sin^2\theta + \sigma_y \cos^2\theta - 2\tau_{xy} \sin\theta\cos\theta \tag{3.11}$$

Let us apply the preceding equations to obtain the normal and shear stresses at the center of the inclined cross section in Fig. 3.10b. If we substitute $\sigma_y = 0$ and $\tau_{xy} = 0$ into (3.9) and (3.10), we find that

$$\sigma_{x'} = \sigma_x \cos^2\theta$$

$$\tau_{x'y'} = -\sigma_x \sin\theta\cos\theta$$

which are the same as those obtained previously from the force equilibrium of Fig. 3.10b.

Strains

When an elastic material is exposed to stresses, the material is deformed and points in the material are displaced as shown in Fig. 3.12. Consider a rectangular element under stresses in the x–y plane in Fig. 3.12a. The element of

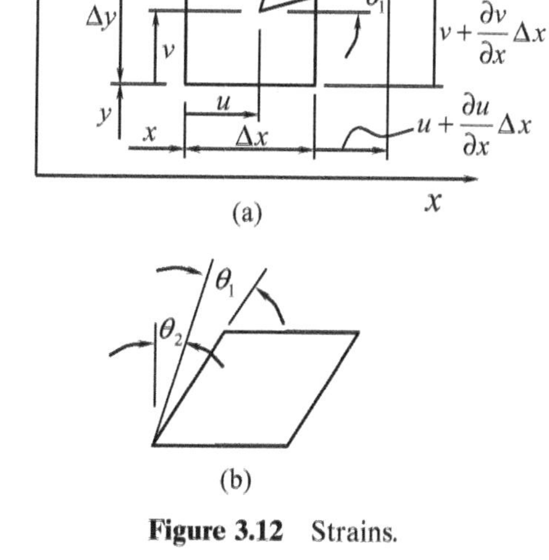

Figure 3.12 Strains.

length Δx and height Δy is placed at the initial position and is displaced to the new position (diamond shape) when exposed to the stresses. Let u and v represent the displacements of the particles in the x and y directions. The normal strains in the x and y directions are defined as follows:

$$\varepsilon_x = \lim_{\Delta x \to 0} \frac{1}{\Delta x}\left(\Delta x + \frac{\partial u}{\partial x}\Delta x - \Delta x\right) = \frac{\partial u}{\partial x} \tag{3.12}$$

$$\varepsilon_y = \lim_{\Delta y \to 0} \frac{1}{\Delta y}\left(\Delta y + \frac{\partial v}{\partial y}\Delta y - \Delta y\right) = \frac{\partial v}{\partial y} \tag{3.13}$$

The shear strain can be defined as the total amount of shear angle as described in (3.7) and Fig. 3.8. The shear angles appear as θ_1 and θ_2 in Fig. 3.12a, and the total angle (Fig. 3.12b) is given by

$$\begin{aligned}\gamma_{xy} &= \theta_1 + \theta_2 \\ &= \lim_{\Delta x \to 0} \frac{1}{\Delta x}\left(v + \frac{\partial v}{\partial x}\Delta x - v\right) + \lim_{\Delta y \to 0} \frac{1}{\Delta y}\left(u + \frac{\partial u}{\partial y}\Delta y - u\right) \\ &= \frac{\partial v}{\partial x} + \frac{\partial u}{\partial y}\end{aligned} \tag{3.14}$$

The shear stress τ_{xy} according to (3.7) is given by

$$\tau_{xy} = G\gamma_{xy}$$

Relationship Between Young's Modulus and Shear Modulus

When an isotropic elastic body is exposed to a shear stress as shown in Fig. 3.8, the shear strain can be given by the shear stress and strain relationship (3.7). The shear modulus is related to Young's modulus and Poisson's ratio, and the relation between the moduli can be derived theoretically. We derive the shear modulus–Young's modulus relationship from the stress–strain relationship and a geometric relationship. We first examine a particular situation, shown in the left figure in Fig. 3.13, in which a small square is subjected to a tensile stress σ in the x direction and $-\sigma$ in the y direction. We are interested in the stress state on the plane that makes an angle of $-45°$ with the y direction. Equations (3.9) and (3.10) yield

$$\sigma_{x'} = 0$$

$$\tau_{x'y'} = \sigma$$

The stress state above states that there is no normal stress in the x' direction but that a shear stress $\tau_{x'y'} = \sigma$ is generated in the y' direction. As shown in the lower figure in Fig. 3.13, the element in the x'–y' plane undergoes a pure

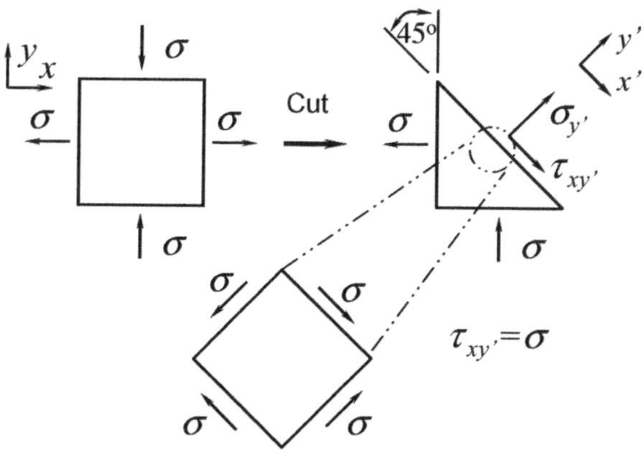

Figure 3.13 Stress state under the tensile and compressive stress in the x and y directions.

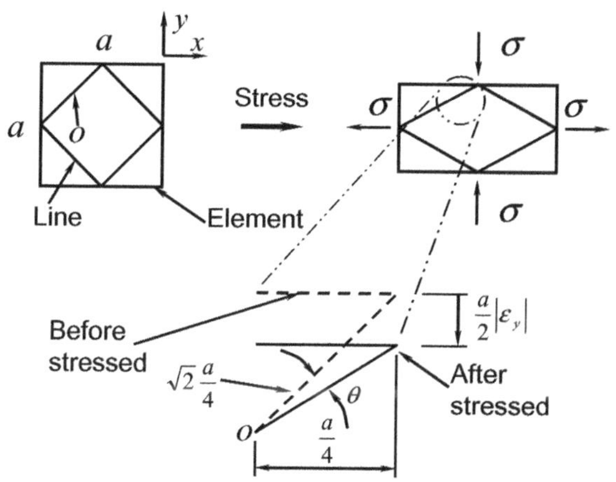

Figure 3.14 Displacement of a rectangle under tensile and compressive stresses.

shear stress like the shear stress in Fig. 3.8. As a result, the shear strain is obtained as

$$\gamma_{x'y'} = \frac{\tau_{x'y'}}{G}$$

Let us trace displacements of the square element of side length a, shown in the left figure in Fig. 3.14. To trace the displacement of element particles, we draw lines to form a diamond shape on the element as shown in the left figure in Fig. 3.14. Afterward, the element is exposed to tensile and compressive stresses in the x and y directions, which are shown in the right figure in Fig. 3.14. The element is deformed as shown in the figure and the lines on the diamond are displaced. However, the center of the element is not displaced because of the symmetry of the element and the stresses about the center. The strains in the x and y directions are obtained from (3.3) and (3.4) as follows:

$$\varepsilon_x = \frac{\sigma}{E} - v\left(\frac{-\sigma}{E}\right) = \frac{\sigma}{E}(1+v)$$

$$\varepsilon_y = \frac{-\sigma}{E} - v\left(\frac{\sigma}{E}\right) = -\frac{\sigma}{E}(1+v) = -\varepsilon_x$$

The deformed element is illustrated in the right figure in Fig. 3.14. It is of interest to note that in this situation, the point labeled o on the left figure in Fig. 3.14 remains undisplaced. The situation is illustrated in the lower figure in Fig. 3.14, where the dashed line is the initial line and the solid line represents the line after the square is stressed. From the geometry in the lower figure in Fig. 3.14, the shear strain $\gamma_{x'y'}$ on the diamond is obtained as the sum of the shear angles (it is similar to the situation in Fig. 3.12b):

$$\gamma_{x'y'} = \theta + \theta = 2\theta$$

The shear angle θ is defined from the geometrical relationship as follows:

$$\theta \cong \frac{(a|\varepsilon_y|/2)/\sqrt{2}}{\sqrt{2}\,a/4} = |\varepsilon_y|$$

Combining the two equations above yields the relationship between the normal and shear strains,

$$\gamma_{x'y'} = 2|\varepsilon_y|$$

Substituting definition (3.7) of the shear strain and the strain $\varepsilon_y = -\sigma(1 + v)/E$ described above, into the preceding equation leads to

$$\frac{\tau_{x'y'}}{G} = 2\frac{\sigma}{E}(1+v)$$

Recalling that $\tau_{x'y'} = \sigma$ from Fig. 3.13, we obtain the shear modulus G in terms of Young's modulus E and Poisson's ratio v as follows:

$$G = \frac{E}{2(1+v)} \tag{3.15}$$

This relationship will be used to obtain the torsional stiffness of a microstructure subjected to the torsional moment.

Stress–Strain Relationships

With reference to Figs. 3.6, 3.8, and 3.9, we have described the plane stress that appears often in engineering. Generally, stress distribution is three-dimensional. For a linear isotropic material under a constant temperature, their normal strains are expressed as the algebraic sum of strain contributions due to normal strains, and a shear strain depends on the corresponding shear stress. The stress–strain relationships with respect to a Cartesian coordinate system (also known as a rectangular coordinate system) are given by

$$\begin{aligned}
\varepsilon_x &= \frac{1}{E}[\sigma_x - \nu(\sigma_y + \sigma_z)] \\
\varepsilon_y &= \frac{1}{E}[\sigma_y - \nu(\sigma_z + \sigma_x)] \\
\varepsilon_z &= \frac{1}{E}[\sigma_z - \nu(\sigma_x + \sigma_y)] \\
\gamma_{xy} &= \frac{\tau_{xy}}{G} \\
\gamma_{yz} &= \frac{\tau_{yz}}{G} \\
\gamma_{zx} &= \frac{\tau_{zx}}{G}
\end{aligned} \tag{3.16}$$

The first equation represents the strain ε_x in the x direction as the stresses under σ_x, σ_y, and σ_z. The strain ε_x is an algebraic sum of σ_x/E, (3.3), and the strain contributions due to σ_y and σ_z. Other normal strains are expressed in similar forms. The shear strains are affected by the shear stress in the same faces and directions. A simple case of the shear stress was studied in relation to Fig. 3.8.

However, if the temperature of the material is not constant but varies from one point to another, the thermal expansion enters into the preceding equation. When a material undergoes a change in temperature during a time period, the material also experiences a change in stress or strain. To evaluate the effect of thermal expansion on the stress and strain, we consider a simple case that is illustrated in Fig. 3.15, where a rectangular parallelepiped undergoes a temperature change from T_0 to T. During the temperature change, the dimensions in the x, y, and z directions are changed from the original dimensions (x, y, and z) to the expanded dimensions ($x + \Delta x$, $y + \Delta y$, and $z + \Delta z$). Using the definition of the coefficient of thermal expansion, denoted by α, we obtain the expanded dimension Δx as follows:

$$\Delta x = \alpha x (T - T_0)$$

or

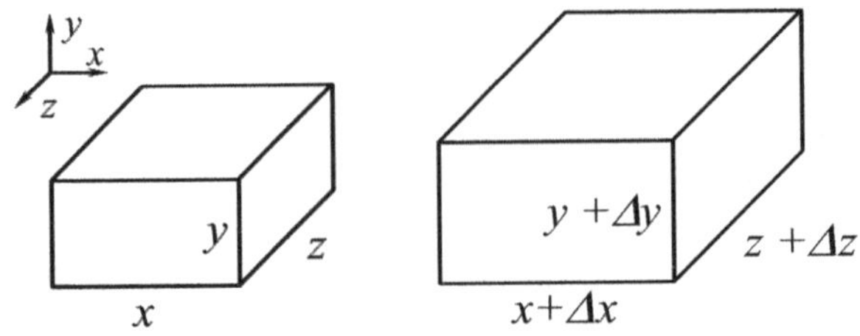

Figure 3.15 Dimension change of a parallelepiped exposed to a temperature change.

$$\varepsilon_x = \frac{\Delta x}{x} = \alpha(T - T_0)$$

Similarly, the strains in the y and z directions are given by

$$\varepsilon_y = \varepsilon_z = \alpha(T - T_0)$$

The equations above state that the dimensions of a material experience the same amount of strain in all directions when the material is exposed to a temperature change. It may be interpreted that every small element in the material is expanded by a displacement corresponding to the strain $\alpha(T - T_0)$ in all directions. In other words, the material undergoes normal expansion except shear strain. Then the material does not experience shear stress due to the thermal expansion. Therefore, we add the thermal effect to (3.16) as follows:

$$\begin{aligned}
\varepsilon_x &= \frac{1}{E}[\sigma_x - \nu(\sigma_y + \sigma_z)] + \alpha(T - T_0) \\
\varepsilon_y &= \frac{1}{E}[\sigma_y - \nu(\sigma_z + \sigma_x)] + \alpha(T - T_0) \\
\varepsilon_z &= \frac{1}{E}[\sigma_z - \nu(\sigma_x + \sigma_y)] + \alpha(T - T_0) \\
\gamma_{xy} &= \frac{\tau_{xy}}{G} \\
\gamma_{yz} &= \frac{\tau_{yz}}{G} \\
\gamma_{zx} &= \frac{\tau_{zx}}{G}
\end{aligned} \tag{3.17}$$

When a body (Fig. 3.16) without physical constraint and initial stresses is exposed to a temperature change of $T - T_0$, the body expands freely in all

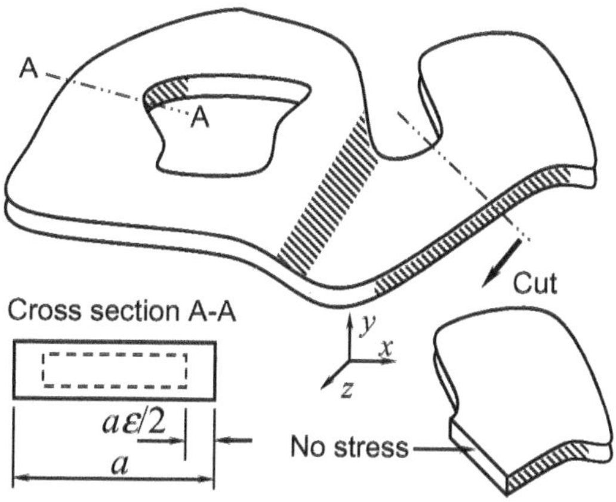

Figure 3.16 Body under a temperature change.

directions according to the preceding equations. However, if we impose any constraint on the body of Fig. 3.16, the strains of equation (3.17) are changed. The constraint then generates stresses given by the preceding equations. The stress due to a temperature change and a constraint on a body is called *thermal stress*.

3.3 THERMAL STRESS

Some MEMS devices operate based on the thermal strain or thermal stress that results from the thermal expansion of an element. We consider a simple case (the first figure in Fig. 3.17), where both ends of a slender beam of a coefficient of thermal expansion α, Young's modulus E, cross-sectional area A, and length l is fixed between rigid walls that are not expanded during a heating process. When the temperature of the beam is changed from T_0 to T, thermal stress is generated on both ends to push the rigid wall. This stress is evaluated using the cantilever beam shown in the second to fourth figures of Fig. 3.17. The cantilever beam is under free stress at a temperature T_0 in the second figure, and afterward the temperature is elevated to T. As shown in the third figure, the beam is expanded by $\Delta l = \alpha l(T - T_0)$ due to the thermal expansion. If we push the right end of the beam so that the end returns to its original position, (3.3) can be used to evaluate the stress and force acting on the beam:

$$\sigma_x = E\varepsilon_x = E\frac{\alpha l(T-T_0)}{l} = E\alpha(T-T_0) \qquad \text{compressive stress}$$

$$F = \sigma_x A = EA\alpha(T-T_0) \qquad \text{compressive force}$$

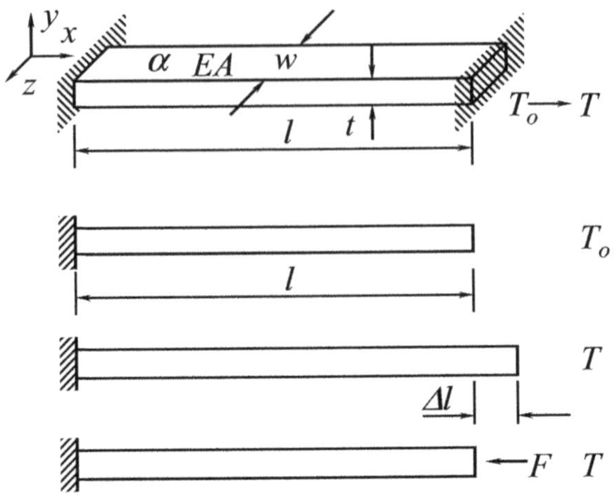

Figure 3.17 Slender beam under a change in temperature.

This result is obtained for a narrow beam. The same results can be obtained using the stress–strain relationships (3.17). Let x represent the axis along the beam length and the y and z axes perpendicular to the x axis. During the heating process, the beam is constrained by the walls, so that the strain ε_x is zero but stresses may be changed. To obtain the thermal stress σ_x, the first equation of (3.17) is written as follows:

$$\varepsilon_x = \frac{1}{E}[\sigma_x - \nu(\sigma_y + \sigma_z)] + \alpha(T - T_0) = 0$$

Since the stresses σ_y and σ_z are zero, the preceding equation yields

$$\sigma_x = -E\alpha(T - T_0)$$

and

$$F = \sigma_x A = -EA\alpha(T - T_0)$$

where the negative sign indicates the compressive stress and force acting on the beam. Consequently, we obtain the same stress and force as those obtained by superposition. If a wide beam is used in Fig. 3.17, $E/(1 - \nu^2)$ can be substituted for E in the equation above. As a wide beam is used, the substitution, $E/(1 - \nu^2)$ for E, is also validated by using (3.17). For example, we can consider the thermal expansion case of Fig. 3.17. With the constraints $\varepsilon_x = 0$, $\varepsilon_z = l(T - T_0)$, and $\sigma_y = 0$, the first and third equations of (3.17) for a wide beam of width w are written as

$$\varepsilon_x = \frac{1}{E}[\sigma_x - \nu(0+\sigma_z)] + \alpha(T-T_0) = 0$$

$$\varepsilon_z = \frac{1}{E}[\sigma_z - \nu(\sigma_x+0)] + \alpha(T-T_0) = \alpha(T-T_0)$$

The equations above yield

$$\sigma_x = -\frac{E}{1-\nu^2}\alpha(T-T_0)$$

$$\sigma_z = -\upsilon\frac{E}{1-\nu^2}\alpha(T-T_0)$$

The strain in the y direction is evaluated as

$$\begin{aligned}\varepsilon_y &= \frac{1}{E}[\sigma_y - \nu(\sigma_z+\sigma_x)] + \alpha(T-T_0)\\ &= \frac{1}{E}\left\{0-\nu\left[-\nu\frac{E}{1-\nu^2}\alpha(T-T_0) - \frac{E}{1-\nu^2}\alpha(T-T_0)\right]\right\} + \alpha(T-T_0)\\ &= \left(\frac{\nu^2}{1-\nu^2} + \frac{\nu}{1-\nu^2} + 1\right)\alpha(T-T_0) = \frac{\alpha(T-T_0)}{1-\nu}\end{aligned}$$

The stress σ_x along the beam length is the same as that obtained from the superposition. It is noted from Fig. 3.17 that if the wall is removed, the thermal stress vanishes in the beam and the beam expands by Δl, as shown in the third figure in Fig. 3.17. As a result, a cantilever beam does not incur thermal stress as well as residual stress in the direction of the length of the cantilever beam. If a strain gradient exists in the direction perpendicular to the longitudinal direction, the cantilever beam will be curved upward or downward, depending on the strain gradient. This issue is described in Chapter 8.

We have evaluated the thermal stress in Fig. 3.17, where the temperature remains constant along the beam direction before and after temperature change. However, in general, as shown in Fig. 3.18, the temperature of the beam may be expressed as a function of the distance from the left fixed end. We define a temperature distribution of the beam as follows:

$$T - T_0 = f_T(x)$$

To calculate the thermal stress, we remove the right rigid wall and the beam will become a cantilever beam (Fig. 3.18c) with the temperature distribution noted above. Consider a small element of the cantilever beam, as shown in the left figure in Fig. 3.18d, whose initial temperature is T_0. The temperature is elevated to be T, while both ends of the cantilever remain at the initial temperature T_0. During the temperature change, the left element in Fig. 3.18d at x is expanded and moved into a new position, $x + \delta$, and the thermal expansion is given by

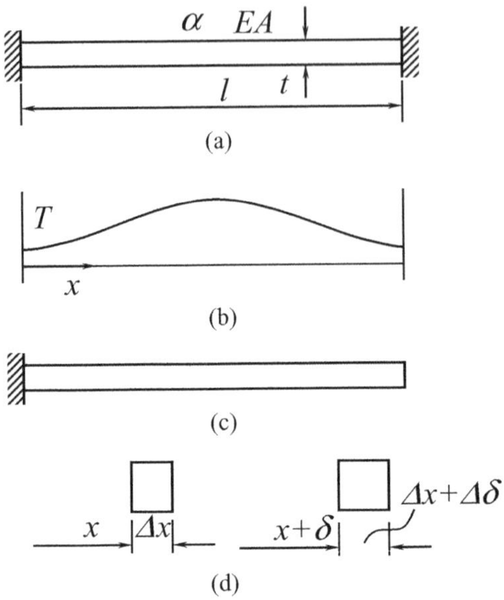

Figure 3.18 Fixed–fixed beam subjected to a thermal stress.

$$\Delta\delta = \alpha\Delta x(T - T_0)$$

where Δx and $\Delta\delta$ denote the length of the small element and its thermal expansion. The total thermal expansion Δl of a beam of length l is expressed as an integration of the thermal expansion of the elements:

$$\Delta l = \delta = \int_0^{\delta} d\delta = \int_0^{l} \alpha(T - T_0)dx \tag{3.18}$$

As discussed earlier, the cantilever beam does not exert an axial force and then there is no thermal stress in the cantilever beam. At this stage we apply a normal stress σ_x in the longitudinal direction to cancel the thermal expansion Δl:

$$\sigma_x = \tilde{E}\varepsilon_x = \tilde{E}\frac{-\Delta l}{l} = -\frac{\tilde{E}}{l}\int_0^{l} \alpha(T - T_0)dx \tag{3.19}$$

where the negative sign means that a thermal expansion generates a compressive stress. Consequently, the axial force exerted by the beam on the right wall (Fig. 3.18a) is given by

$$F = \sigma_x A = -\frac{\tilde{E}A}{l}\int_0^{l} \alpha(T - T_0)dx$$

Example 3.1 In many MEMS devices, the thermal stress is generated in a beam in Fig. 3.18 when the beam is heated up. The temperature distribution depends on the heating method and the heat transfer coefficient of the beam material and surrounding gas. If the beam element is heated up by applying an electric current through the beam, and if the heat transfer to the surrounding gas is negligible, we can use the following temperature distribution $T - T_0$, which is derived in Chapter 8:

$$T - T_0 = 4\frac{(T - T_0)_{\max}}{l^2}\left[\left(\frac{l}{2}\right)^2 - \left(x - \frac{l}{2}\right)^2\right]$$

where $(T - T_0)_{\max}$ is the maximum temperature at the midspan of the beam and l is the length of the beam. We wish to determine the thermal stress in the beam.

Substituting the temperature distribution above into (3.19) yields the thermal stress as follows:

$$\begin{aligned}\sigma_x &= -\frac{\tilde{E}}{l}\int_0^l \alpha(T - T_0)\,dx \\ &= -\frac{\tilde{E}}{l}\int_0^l 4\alpha\frac{(T - T_0)_{\max}}{l^2}\left[\left(\frac{l}{2}\right)^2 - \left(x - \frac{l}{2}\right)^2\right]dx \\ &= -\frac{2}{3}\tilde{E}\alpha(T - T_0)_{\max}\end{aligned}$$

In the calculation above, we assume that the coefficient of thermal expansion α is a constant. If the heat transfer to the surrounding gas plays a role, the temperature distribution and resulting thermal stress are more complicated than those given above.

As described in Example 3.1, the temperature distribution is necessary to calculate the thermal expansion or stress. We seek a governing equation for the temperature distribution of the beam illustrated in Fig. 3.19. The beam is fixed at anchors, attached on a substrate, and it is spaced a gap h from the substrate. When an electric current flows through the beam that serves as electrical resistance, the beam generates $\dot{Q}_t$ (heat per unit length per unit time), and in turn, the heat transfers from the beam to the anchors and to the surrounding gas. The heat transfer due to conduction is governed by Fourier's law of heat conduction, in which the heat transfer rate q along the beam of cross-sectional area A is proportional to the temperature gradient $\partial T/\partial x$, given by

$$q = -k_s A\frac{\partial T}{\partial x} \tag{3.20}$$

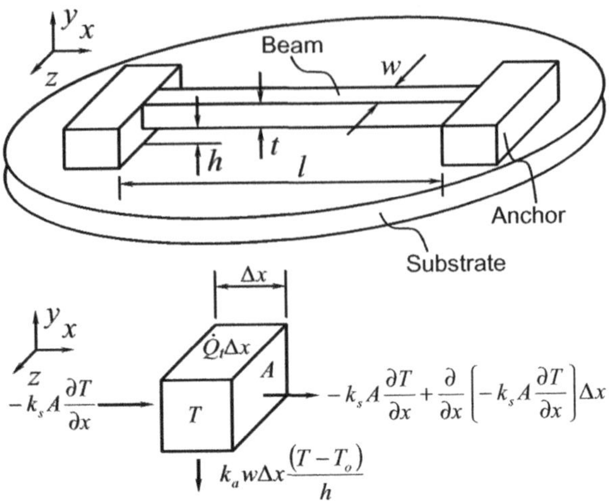

Figure 3.19 Fixed–fixed beam generating heat.

where k_s is the *heat transfer coefficient* (also known as *thermal conductivity*). The negative sign in equation (3.20) states that the heat flows from a higher temperature to a lower temperature. Using heat transfer equation (3.20), we can write the energy balance equation for the element in the lower figure in Fig. 3.19 as follows:

$$\text{energy in} = \text{change in internal energy} + \text{energy out}$$

$$-k_s A\frac{\partial T}{\partial x} + \dot{Q}_t \Delta x = \rho c A \Delta x \frac{\partial T}{\partial t_1} - k_s A\frac{\partial T}{\partial x} + \frac{\partial}{\partial x}\left(-k_s A\frac{\partial T}{\partial x}\right)\Delta x + k_a w \Delta x \frac{T - T_0}{h}$$

where ρ, c, t_1, k_a, w, T_0, and h denote the density and the specific heat of the beam material, time, the conduction heat transfer coefficient of the surrounding gas, the width of the beam, the temperature of the substrate, and the gap, respectively. Note that the radiation term is neglected in the preceding equation since a moderate temperature (e.g., lower than 1000 K) is being considered. The convective term is also negligible because the buoyant force of the surrounding gas due to the temperature change in the vicinity of the beam is much smaller than the gravitational force. The equation above is simplified to give a one-dimensional thermal conduction equation:

$$\frac{\partial}{\partial x}\left(k_s \frac{\partial T}{\partial x}\right) - \frac{k_a}{th}(T - T_0) + \frac{\dot{Q}_t}{A} = \rho c \frac{\partial T}{\partial t} \tag{3.21}$$

To nondimensionalize the equation (3.21), we define the following dimensionless numbers:

$$X = \frac{x}{l}$$

$$T_1 = \frac{t}{\tau} \tag{3.22}$$

$$\Theta = \frac{T}{T_c}$$

where τ and T_c denote the characteristic time and temperature, respectively. We then obtain the corresponding dimensionless form as follows:

$$\frac{\partial^2\Theta}{\partial X^2} - \frac{k_a}{k_s}\frac{l^2}{th}(\Theta - \Theta_0) + \frac{l^2\dot{Q}_t}{k_s T_c A} = \frac{\rho c l^2}{k_s \tau}\frac{\partial\Theta}{\partial T_1} \tag{3.23}$$

The dimensionless coefficient on the right-hand side of equation (3.23) represents the dimensionless heat transfer time, in which the heat in the beam transfers from one point to another. The constant may be written in a different form:

$$\frac{\rho c l^2}{k_s \tau} = \frac{l^2}{\alpha_1 \tau}$$

where $\alpha_1 = k_s/\rho c$ is known as the *thermal diffusivity* of the material. We can then define a characteristic time as

$$\tau = \frac{l^2}{\alpha_1}$$

For silicon (one of common materials in MEMS), the thermal diffusivity and the typical length of the beam are $\alpha_1 = 9 \times 10^{-5}\,\text{m}^2/\text{s}$ and $l = 100\,\mu\text{m}$. Substituting the thermal diffusivity and the typical length into the preceding equation, we have the characteristic time $\tau = 1.1 \times 10^{-4}/\text{s}$. If we are interested in a step response (e.g., a constant current flows through the heating element) or if the heating frequency is much less than the characteristic frequency ($f_c = 1/\tau$), the dimensionless coefficient on the right-hand side of the preceding thermal conduction equation is negligible and the dimensionless governing equation of the heat transfer in the beam is given by

$$\frac{d^2\Theta}{dX^2} - K_h(\Theta - \Theta_0) + L = 0 \tag{3.24}$$

where

$$K_h = \frac{k_a}{k_s}\frac{l^2}{th} \tag{3.25}$$

$$L = \frac{l^2 \dot{Q}_t}{k_s T_c A} \tag{3.26}$$

represent the ratio of the conduction heat transfer from the beam to the air to the conduction heat transfer along the beam, and a heat source term, respectively. It is worth noting that if the dimensional parameter K_h is much less than unity, the second term in (3.24) can be negligible since heat transfer via the surrounding gas is much less than heat transfer along the beam. For example, if the heat transfer occurs in vacuum or in rarefied gas, the second term can be neglected. The dimensional equation corresponding to (3.24) is given by

$$k_s \frac{d^2T}{dx^2} - \frac{k_a}{th}(T - T_0) + \frac{\dot{Q}_t}{A} = 0 \tag{3.27}$$

In Chapter 8 we solve heat transfer equation (3.27) to obtain the temperature distribution.

3.4 BEAM BEHAVIOR SUBJECTED TO A TORSIONAL MOMENT

The rectangular beams shown in Fig. 3.20a are widely used in MEMS as torsional springs, while the circular beams in Fig. 3.20b are used in macro scale machines. When a torsional moment M_t is applied to the free end of the

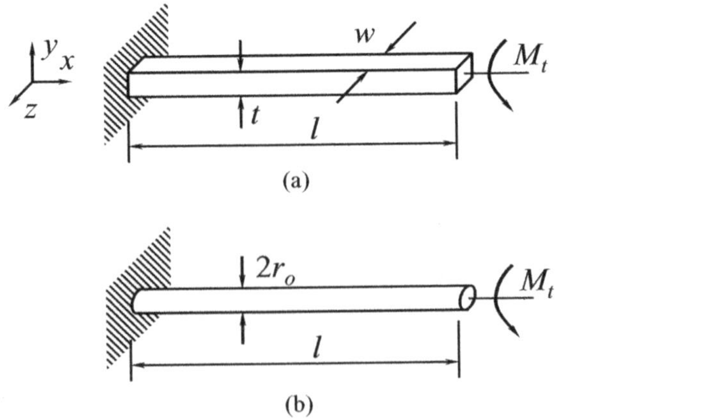

Figure 3.20 Rectangular and circular beams subjected to a torsional moment.

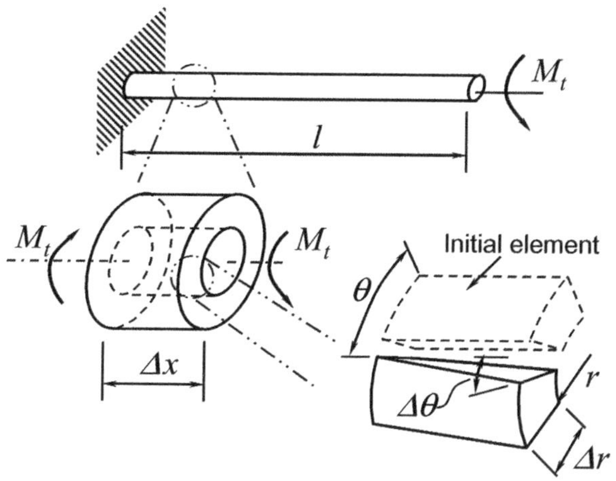

Figure 3.21 Circular torsional beam.

cantilever beams shown in Fig. 3.20a and b, the beams are twisted by an angle that is proportional to the applied moment. The stiffness of these springs, known as *torsional springs*, depend on the material properties and geometry. Even though rectangular torsional springs are used in MEMS, circular torsional springs are described first, due to the simplicity of the theory. Afterward, the rectangular torsional spring is discussed.

Let us consider a circular torsional spring in Fig. 3.21 that is subjected to a torsional moment M_t. According to the moment equilibrium condition (3.2), the beam element, shown in the lower left figure in Fig. 3.21, is subjected to the applied moment M_t and it is twisted. For convenience, we use the cylindrical coordinate to describe the stress and strain in the element. Since the initial element is deformed into the element drawn indicated in solid line in the lower left figure in Fig. 3.21, we can define the shear strain and stress on the positive x face as follows:

$$\gamma_{x\theta} = \lim_{\Delta x \to 0} \frac{r\Delta\theta}{\Delta x} = \frac{rd\theta}{dx} \tag{3.28}$$

$$\tau_{x\theta} = G\gamma_{x\theta} = G\frac{rd\theta}{dx} \tag{3.29}$$

where r, θ, and G_s denote the radius, the angle (angular displacement) due to the applied moment, and the shear modulus defined in (3.15), respectively. The moment due to the shear stress on the element (Fig. 3.22) is expressed as

$$dM = dFr = \tau_{x\theta} rdAr = \tau_{x\theta} r^2 dA$$

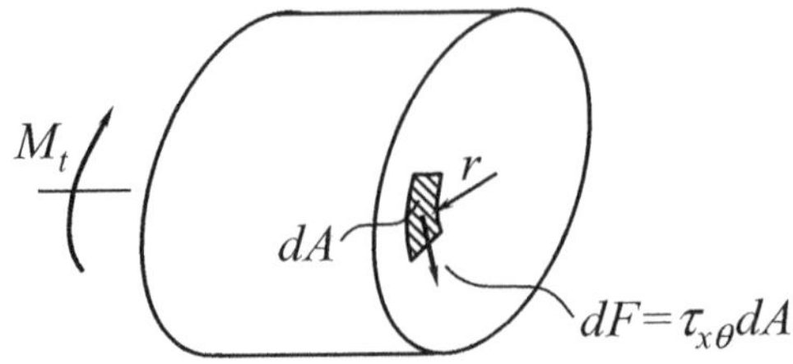

Figure 3.22 Torsional moment due to shear stress on the cross section.

and the moment acting on the cross section perpendicular to the beam length is given by the integration of the moment on the element:

$$M_t = \int dM = \int_A \tau_{x\theta} r^2 \, dA = G\frac{d\theta}{dx}\int_A r^2 \, dA = G\frac{d\theta}{dx}J \tag{3.30}$$

where

$$J = \int_A r^2 \, dA$$

is the polar moment of inertia of the circular beam. Referring to the infinitesimal area of $dA = r\,dr\,d\theta$, we obtain the polar moment,

$$J = \int_0^{2\pi}\int_0^{r_0} r^3 \, dr\, d\theta = \frac{\pi r_0^4}{2}$$

where r_0 denotes the radius of the circular beam. The moment–angular displacement relation (3.30) gives

$$\frac{d\theta}{dx} = \frac{M_t}{GJ}$$

Integration of the preceding equation yields the angular displacement of the end of the beam as follows:

$$\theta = \int_0^l \frac{M_t}{GJ} dx = \frac{M_t l}{GJ} \tag{3.31}$$

Therefore, the stiffness of the circular beam is defined as

$$k_t = \frac{M_t}{\theta} = \frac{GJ}{l} \tag{3.32}$$

This equation states that the torsional stiffness of the circular beam is proportional to the product of the shear modulus (3.15) and the polar moment of inertia of the circular cross section and inversely proportional to the beam length. For slender beams with other cross sections, such as the rectangular cross section shown in Fig. 3.20a, the preceding two equations still hold, but the polar moment of inertia J should be changed. The polar moment of inertia for the rectangular cross section is given in an infinite series as follows (Timoshenko and Goodier, 1970):

$$J = \frac{wt^3}{3}\left(1 - \frac{192}{\pi^5}\frac{t}{w}\sum_{n=1}^{\infty}\frac{1}{n^5}\tanh\frac{n\pi w}{2t}\right)$$

which holds for $w \geq t$.

For convenience, using a curve-fitting technique, we can express the polar moment of inertia above as a closed-form equation. For the rectangular beam of $w \geq t$, the polar moment of inertia is expressed as

$$J = \frac{wt^3}{3}\left[1 - 0.616\left(\frac{t}{w}\right) - 0.087\left(\frac{t}{w}\right)^2 + 0.125\left(\frac{t}{w}\right)^3\right] \tag{3.33}$$

where t and w denote the thickness and width of the rectangular beam shown in Fig. 3.20a, respectively. For a rectangular beam of $w \leq t$, w and t in the preceding equation switch places, so we obtain the following equation

$$J = \frac{tw^3}{3}\left[1 - 0.616\left(\frac{w}{t}\right) - 0.087\left(\frac{w}{t}\right)^2 + 0.125\left(\frac{w}{t}\right)^3\right] \tag{3.34}$$

These equations are in excellent agreement with the preceding J in infinite series with an error of 0.08%.

3.5 MOMENT–CURVATURE RELATIONSHIP

A slender beam can carry a bending moment and transverse shear force. For example, a slender beam in Fig. 3.23 is subjected to pure bending (i.e., the couple discussed in Section 3.1). The beam equation, known as the Euler–Bernoulli equation, should be solved for the deflection of the beam. We shall derive the beam equation and solve the beam equation with boundary conditions that will be used to understand the nonlinear behavior of MEMS.

To obtain the beam equation, let us begin by considering the a slender beam shown in Fig. 3.23, whose cross section is symmetric about the y axis. The x axis is aligned parallel to the longitudinal axis of the beam, and the z axis is perpendicular to the x and y axes. The location of the x axis will be determined

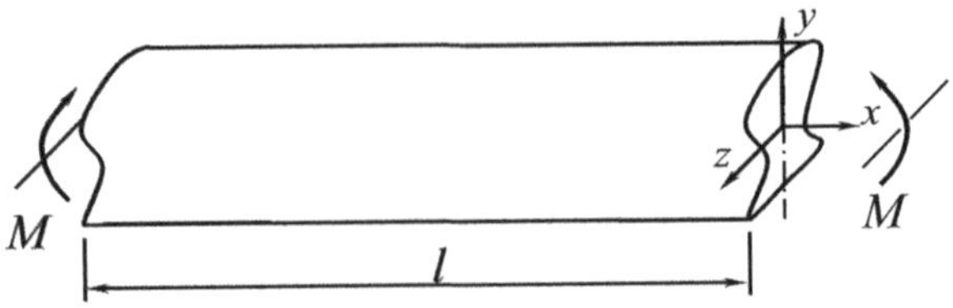

Figure 3.23 Slender beam subjected to a pure bending moment.

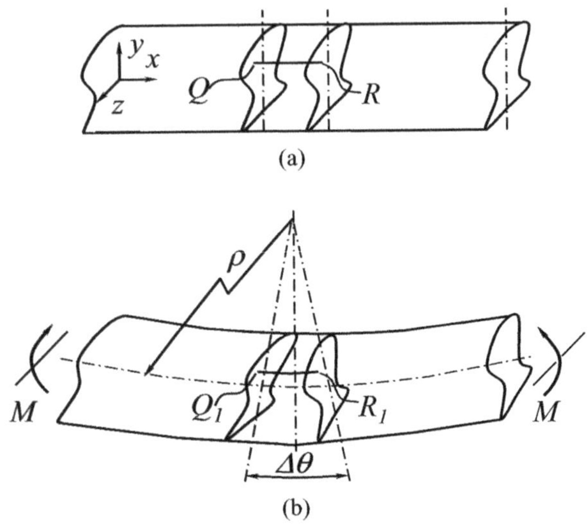

Figure 3.24 Deflection of a beam subjected to pure bending.

later, but at this stage we assume that the x axis lies in a neutral surface (x–z plane in Fig. 3.23) in which there is no normal stress along the x axis. We shall examine the strain and stress in the beam and then obtain the moment–stress relationship and the beam equation. To examine the strain in the beam, consider a straight line QR parallel to the x axis of the beam without the bending moment, as shown in Fig. 3.24a. When a bending moment M, defined as positive, is applied to the ends of the beam, the beam is deflected as shown in Fig. 3.24b, and the radius of curvature ρ is positive in the sense shown in Fig. 3.24. As the moment is applied, the original line QR is changed to the new curve Q_1R_1 shown in Fig. 3.24b. The situation is detailed in Fig. 3.25. After applying the moment M, the length of the line, which coincides with the x axis in Fig. 3.24a, remains unchanged, as shown in the dashed-dotted curve in Fig. 3.24b. The strain along the curve is zero and the x axis is thus referred to as the neutral axis. However, the line QR (length Δx) at y from the neutral axis is changed to the new length $(\rho - y)\,\Delta\theta$ after a bending moment is applied. We then find the strain in the beam,

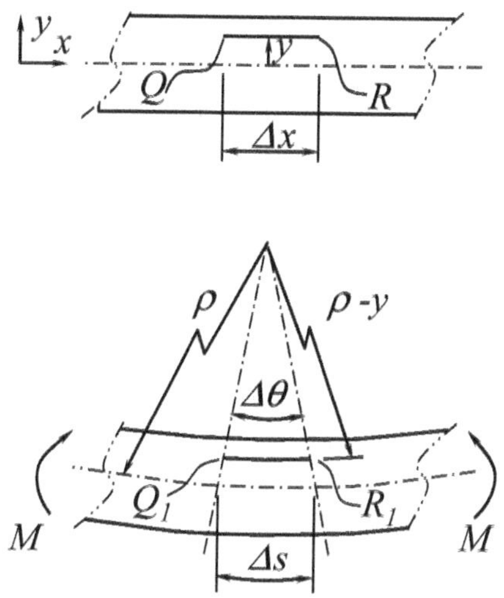

Figure 3.25 Deformation of a line in a beam subjected to a bending moment.

$$\varepsilon_x = \frac{(\rho - y)\Delta\theta - \Delta x}{\rho\Delta\theta} = -\frac{y}{\rho} \tag{3.35}$$

where $\Delta x \cong \Delta s = \rho\Delta\theta$ from the geometric relationship. Note from equation (3.35) that the upper portion of the beam (i.e., positive y) is compressed and the lower portion (i.e., negative y) is stretched. Also note that the moment in any cross section along the length of the beam remains constant and that the strain in the beam is only a function of the coordinate y from the neutral axis and the radius of curvature ρ. In other words, a beam subjected to pure bending deforms into a curved beam with a radius of curvature ρ. Since the length of the arc is obtained as $\Delta s = \rho\Delta\theta$, the curvature is defined as

$$\frac{1}{\rho} = \lim_{\Delta s \to 0} \frac{\Delta\theta}{\Delta s} = \frac{d\theta}{ds} \tag{3.36}$$

Substituting equation (3.36) into (3.35), we have

$$\varepsilon_x = -\frac{d\theta}{ds} y \tag{3.37}$$

As shown in Fig. 3.26, the shear stress on the y–z plane vanishes because the cross section does not carry transverse shear force. The normal stress due to the normal strain (3.37) generates a moment that maintains the equilibrium of the beam segment in Fig. 3.26. Applying the moment equilibrium (3.2) to

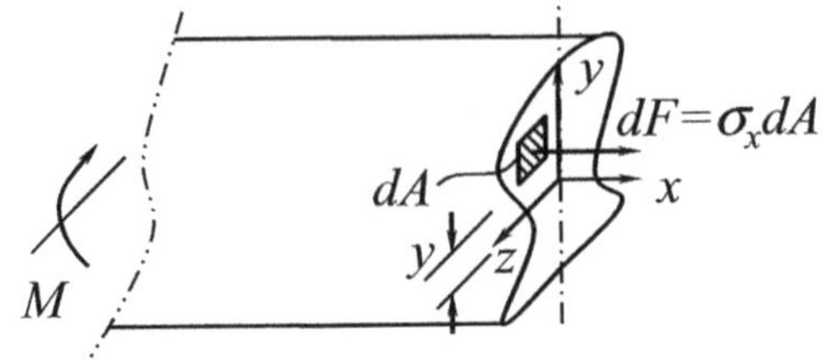

Figure 3.26 Moment due to normal stress.

the beam segment leads to the relation between the applied bending moment and the curvature:

$$M = -\int_A y\sigma_x\, dA = -\int_A y\tilde{E}\varepsilon_x\, dA = -\int_A y\tilde{E}\left(-\frac{d\theta}{ds}y\right)dA$$
$$= \tilde{E}\frac{d\theta}{ds}\int_A y^2\, dA = \frac{d\theta}{ds}\tilde{E}I \tag{3.38}$$

where

$$I = \int_A y^2\, dA \tag{3.39}$$

is the moment of inertia of the cross-sectional area of the beam and its integration is over the cross-sectional area. The effective Young's modulus $\tilde{E}$, defined by (3.5) and (3.6), is used to generalize the moment–curvature relationship (3.38), which can be used for the narrow as well as the wide beams. From geometry, we can obtain the angle θ at x as $\theta = dy/dx$, which is substituted into (3.38) to yield

$$M = \tilde{E}I\frac{d\theta}{ds} \cong \tilde{E}I\frac{d\theta}{dx} = \tilde{E}I\frac{d^2y}{dx^2}$$

or

$$\tilde{E}I\frac{d^2y}{dx^2} = M \tag{3.40}$$

where $\tilde{E}I$ is called the *flexural rigidity* or *bending modulus*. This moment–curvature relationship plays a crucial role in developing the beam theory. Combining $\sigma_x = \tilde{E}\varepsilon_x$, (3.37) and (3.38), we find the normal strain and the corresponding stress acting on the cross section as follows:

$$\varepsilon_x = -\frac{M}{\tilde{E}I}y \qquad \sigma_x = -\tilde{E}\frac{M}{\tilde{E}I}y = -\frac{M}{I}y \tag{3.41}$$

Since the moment of inertia I is calculated based on the coordinate y from the neutral surface, we need to find the location of the surface. Applying the force equilibrium condition (3.1) to the beam segment of Fig. 3.26, we have

$$\sum F_x = \int_A \sigma_x \, dA = \int_A \tilde{E}\varepsilon_x \, dA = \int_A \tilde{E}\left(-\frac{d\theta}{ds}y\right) dA$$

$$= -\tilde{E}\frac{d\theta}{ds}\int_A y \, dA = 0$$

or

$$\int_A y \, dA = 0 \tag{3.42}$$

where the integration above is taken over the crosssection. The coordinate y from the neutral axis may be expressed as

$$y = y_1 - y_c$$

where y_1 is the distance from the bottom of the beam and y_c denotes the coordinate of the neutral surface, measured from the bottom of the beam. With the foregoing definition of the coordinate y, performing the integration of (3.42) gives

$$y_c = \frac{\int_A y_1 \, dA}{\int_A dA} = \frac{\int_A y_1 \, dA}{A} \tag{3.43}$$

It is worth noting that the definition of the location of the neutral surface above is equal to the centroid of the cross-sectional area of the beam. In other words, the neutral surface, defined as a plane in which there is no strain, passes through the centroid of the cross-sectional area of a beam. For example, the neutral surface of a rectangular beam is located at a half thickness from the bottom of the beam. Similarly, for a circular beam, the neutral surface passes through the center of the diameter of the beam.

For a rectangular beam of width w_b and thickness t, the moment of inertia I is obtained from (3.39) as

$$I = \int_A y^2 \, dA = \int_{-t/2}^{t/2} y^2 w_b \, dy = \int_{-t/2}^{t/2} y^2 w_b \, dy = \frac{w_b t^3}{12} \tag{3.44}$$

Similarly, for a circular beam of radius r_0, the moment of inertia I is obtained from (3.39) as

$$I = \int_A y^2\, dA = \int_0^{2\pi} \int_0^{r_0} (r \sin\theta)^2\, r\, dr\, d\theta = \frac{\pi r_0^4}{4} \tag{3.45}$$

For an arbitrary cross-sectional area of a beam, the location of the neutral surface and the moment of inertia of the cross-sectional area about the z axis lying in the neutral surface are given by equations (3.43) and (3.39). A calculation for a general shape is given in the following example.

Example 3.2 In MEMS, a bulk-micromachined silicon beam is widely used to support a microstructure or to act as a spring. To calculate the stiffness of the beam (that is described in the later portion of this chapter), we need the location of the neutral surface and the moment of inertia of the cross-sectional area shown in Fig. 3.27.

Since the anisotropic etchant etches the bulk silicon along the (111) plane as discussed in Chapter.2, the angle in Fig. 3.27 is given by $\theta = \tan^{-1}\sqrt{2}$. We shall first obtain the location of the neutral axis of the general trapezoidal shape and the moment of inertia about the neutral axis. We then obtain the location and the moment of inertia by substituting $\tan^{-1}\sqrt{2}$ for θ for the bulk-micromachined beam. From the geometry, we have the width at y measured from the coordinate y_c that will be obtained:

$$\frac{t}{w_2 - w_1} = \frac{y_1}{w_2 - w}$$

or

$$w = w_2 - \frac{w_2 - w_1}{t} y_1 \tag{a}$$

Furthermore, the following relationship holds for the shape in Fig. 3.27:

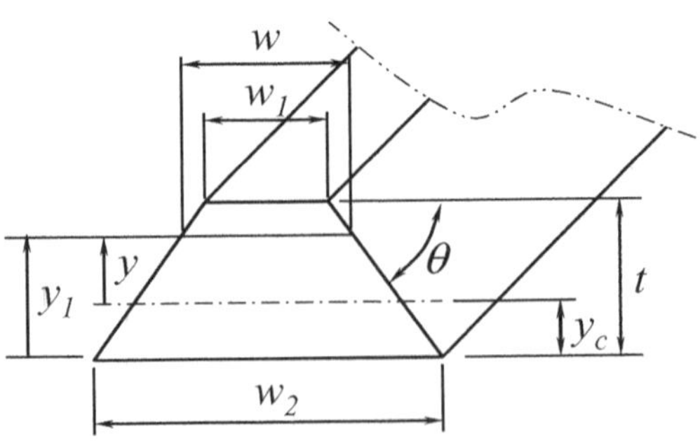

Figure 3.27 Cross section of bulk-micromachined silicon beam.

$$w_2 = w_1 + 2t\frac{1}{\tan\theta} \tag{b}$$

The area is given by

$$A = \frac{1}{2}t(w_1 + w_2) \tag{c}$$

Substitution of (a) into (3.43) yields the location of the neutral axis:

$$\begin{aligned} y_c &= \frac{1}{A}\int_A y_1\, dA = \frac{1}{A}\int_A y_1 w\, dy_1 \\ &= \frac{1}{A}\int_A y_1\left(w_2 - \frac{w_2 - w_1}{t}y_1\right)dy_1 = \frac{1}{A}\frac{t^2}{6}(w_2 + 2w_1) \\ &= \frac{t}{3}\frac{w_2 + 2w_1}{w_2 + w_1} \end{aligned} \tag{d}$$

If we use (b), equation (d) is expressed as a function of t, w_1/t, and θ:

$$y_c = \frac{t}{3}\frac{1+(3/2)(w_1/t)\tan\theta}{1+(w_1/t)\tan\theta} \tag{e}$$

Substituting $\sqrt{2}$ for $\tan\theta$ gives the location of the neutral axis of the bulk-micromachined beam as follows:

$$y_c = \frac{t}{3}\frac{1+\left(3\sqrt{2}/2\right)(w_1/t)}{1+\sqrt{2}\,(w_1/t)} \tag{f}$$

The moment of inertia (3.39) can be expressed in terms of the coordinate y_1 from the bottom of the beam and the location of the neutral axis y_c:

$$\begin{aligned} I_{y_c} &= \int_A y^2\, dA = \int_A (y_1 - y_c)^2\, dA = \int_A \left(y_1^2 - 2y_1y_c + y_c^2\right)dA \\ &= \int_A y_1^2\, dA - 2y_c\int_A y_1\, dA + y_c^2\int_A dA \end{aligned}$$

The first term on the right-hand side of the foregoing equation represents the moment of inertia about the bottom of the beam, which is denoted by

$$I_0 = \int_A y_1^2\, dA \tag{g}$$

Recalling the definition of the location of the neutral axis,

$$y_c = \frac{1}{A}\int_A y_1\, dA$$

the moment of inertia about the location of the neutral axis may be written as

$$I_{y_c} = I_0 - 2y_c^2 A + y_c^2 A = I_0 - y_c^2 A \tag{h}$$

Equation (h), is known as the *parallel axis theorem*, is used to determine the moment of inertia about the centroidal axis (i.e., the neutral axis in our case) if the moment of inertia is given about a parallel axis from the centroidal axis, and vice versa. The moment of inertia I_0 is obtained as

$$\begin{aligned} I_0 &= \int_A y_1^2\, dA = \int_0^t y_1^2 w\, dy_1 \\ &= \int_0^t y_1^2 \left(w_2 - \frac{w_2 - w_1}{t} y_1 \right) dy_1 \\ &= \frac{t^3}{12}(3w_1 + w_2) \end{aligned} \tag{i}$$

Substituting I_0 into (h), we find the moment of inertia I_{yc} about the neutral axis as follows:

$$\begin{aligned} I_{y_c} &= I_0 - y_c^2 A = \frac{t^3}{12}(3w_1 + w_2) - \left(\frac{t}{3} \frac{w_2 + 2w_1}{w_2 + w_1} \right)^2 \frac{1}{2} t(w_1 + w_2) \\ &= \frac{t^3}{36} \frac{w_1^2 + 4w_1 w_2 + w_2^2}{w_1 + w_2} \end{aligned} \tag{j}$$

Substituting w_2 of (b) into (j) leads to

$$\begin{aligned} I_{y_c} &= \frac{t^3}{36} \frac{w_1^2 + 4w_1[w_1 + 2t(1/\tan\theta)] + [w_1 + 2t(1/\tan\theta)]^2}{w_1 + [w_1 + 2t(1/\tan\theta)]} \\ &= \frac{t^4}{18} \frac{1}{\tan\theta} \frac{w_1^2 + 4w_1[w_1 + 2t(1/\tan\theta)] + [w_1 + 2t(1/\tan\theta)]^2}{w_1 + [w_1 + 2t(1/\tan\theta)]} \end{aligned} \tag{k}$$

which is a function of the upper width w_1 and the angle θ. For the bulk-micromachined silicon, we substitute $\sqrt{2}$ for $\tan\theta$ to give

$$I_{y_c} = \frac{\sqrt{2}}{36} \frac{1 + 3\sqrt{2}(w_1/t) + 3(w_1/t)^2}{1 + \sqrt{2}(w_1/t)} t^4 \tag{l}$$

If $w_1/t >> 1$, I_{yc} approaches the moment of inertia of the rectangular beam (i.e., $I_{yc} = w_1 t^3/12$).

3.6 BEAM EQUATION

We consider the slender beam shown in Fig. 3.28, whose ends are fixed at rigid walls and which is loaded by a distributed load q (load per unit length). An element cut from the beam is illustrated in the lower figure in Fig. 3.28 to show the forces and moments, which are defined as positive. To obtain the beam equation associated with a MEMS device, we combine the moment–curvature equation (3.40) and the moment–load relationships described below. The infinitesimal beam element of Fig. 3.28 is subjected to the bending moment M, the axial force N that remains constant along the length of the beam, and the transverse shear force that is induced by the applied load q. In the figure, θ and Δ denote the angle that the beam makes with the x axis, and the increment of the variables, respectively. The force and moment equilibrium conditions (3.1) and (3.2) are applied to the infinitesimal beam element shown in Fig. 3.28. The force equilibrium equation (3.1) in the y direction is written as

$$-N\sin\theta + N\sin(\theta+\Delta\theta) + V + \Delta V - V + q\Delta x = 0$$

and can be simplified for small $\theta + \Delta\theta$ as follows:

$$\frac{dV}{dx} = -q - N\frac{d\theta}{dx} = -q - N\frac{d^2y}{dx^2} \qquad (3.46)$$

To obtain the right-hand side of equation (3.46), $\theta = dy/dx$ is used. The moment equilibrium condition (3.2) gives the relationship between the bending moment and the transverse shear force,

$$(V+\Delta V)\frac{\Delta x}{2} + V\frac{\Delta x}{2} - M + M + \Delta M = 0$$

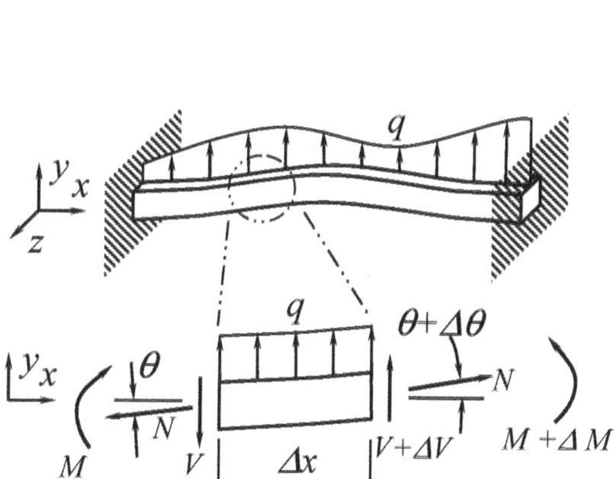

Figure 3.28 Fixed–fixed beam loaded by a distributed load.

or

$$\frac{dM}{dx} = -V \tag{3.47}$$

Substituting (3.46) into equation (3.47) yields the relationship among the bending moment M, the distributed load q, and the axial force N:

$$\frac{d^2M}{dx^2} = -\frac{dV}{dx} = -\left(-q - N\frac{d^2y}{dx^2}\right) = q + N\frac{d^2y}{dx^2} \tag{3.48}$$

Substituting equation (3.48) into the moment–curvature relationship (3.40), we have the following beam equation:

$$\frac{d^2}{dx^2}\left(\tilde{E}I\frac{d^2y}{dx^2}\right) - N\frac{d^2y}{dx^2} = q \tag{3.49}$$

If $\tilde{E}I$ is constant along the length of the beam, equation (3.49) is expressed as

$$\tilde{E}I\frac{d^4y}{dx^4} - N\frac{d^2y}{dx^2} = q \tag{3.50}$$

$\tilde{E}I\, d^4y/dx^4 = q$ is known as the *Euler–Bernoulli beam equation*, the beam equation that will serve as a governing equation to solve beam problems in MEMS.

Note that the axial force N may be classified into two forces: (1) an axial force due to the residual stress that may be generated in a beam when the beam is fabricated, and (2) a nonlinear force due to the geometric nonlinearity that is caused by the large deflection of beam. We first focus on the geometric nonlinearity in beam and then the obtain the overall axial force, which includes the residual stress and the geometric nonlinearity. The initially straight beam element is deformed into the element shown in the lower figure in Fig. 3.28. In this case, the length of the element after deformation may be obtained as

$$\Delta s = \sqrt{(\Delta y)^2 + (\Delta x)^2} = \sqrt{1 + \left(\frac{\Delta y}{\Delta x}\right)^2}\,\Delta x \cong \left[1 + \frac{1}{2}\left(\frac{\Delta y}{\Delta x}\right)^2\right]\Delta x$$

where the last expression is obtained for $\Delta y/\Delta x << 1$. The elongated length of the beam element is written as

$$\Delta\delta = \Delta s - \Delta x = \frac{1}{2}\left(\frac{\Delta y}{\Delta x}\right)^2 \Delta x$$

The elongation of the beam is expressed as an integration of the elongated length of the element:

$$\delta = \frac{1}{2}\int_0^l \left(\frac{dy}{dx}\right)^2 dx \tag{3.51}$$

From the stress–strain relationship, $N_n/A = E\delta/l$, where N_n is the nonlinear axial force due to the beam elongation, we find that

$$N_n = \frac{\tilde{E}A}{l}\delta = \frac{1}{2}\frac{\tilde{E}A}{l}\int_0^l \left(\frac{dy}{dx}\right)^2 dx \tag{3.52}$$

Since the axial force is the sum of the initial axial force N_0 due to the residual stress and the axial force N_n caused by the geometric nonlinearity, we find that

$$N = N_0 + N_n = N_0 + \frac{1}{2}\frac{\tilde{E}A}{l}\int_0^l \left(\frac{dy}{dx}\right)^2 dx \tag{3.53}$$

If we substitute equation (3.53) into (3.49), we obtain the following beam equation, which will be used to solve beam problems in MEMS:

$$\frac{d^2}{dx^2}\left(\tilde{E}I\frac{d^2y}{dx^2}\right) - \left[N_0 + \frac{1}{2}\frac{\tilde{E}A}{l}\int_0^l \left(\frac{dy}{dx}\right)^2 dx\right]\frac{d^2y}{dx^2} = q$$

or

$$\tilde{E}I\frac{d^4y}{dx^4} - \left[N_0 + \frac{1}{2}\frac{\tilde{E}A}{l}\int_0^l \left(\frac{dy}{dx}\right)^2 dx\right]\frac{d^2y}{dx^2} = q \tag{3.54}$$

where $\tilde{E}$ is the effective Young's modulus, defined in (3.5) and (3.6), to give a general expression of the beam equation. Equation (3.54) holds for $\tilde{E}I$ = constant. The axial force may be written as $N_0 = \sigma_0 A$, where σ_0 is the residual stress in the beam, defined as $\sigma_0 = \tilde{E}\varepsilon_0$, where ε_0 is a strain that caused the residual stress and A is the cross-sectional area. The distributed load q, defined as force per unit length, may be the electrostatic force, which we deal with in Chapter 9. If the beam is a cantilever beam, the terms for the axial force [i.e., the terms in brackets in equation (3.54)] vanish. If we use a beam carrying an axial force, the terms in brackets play a role. Further analysis shows that the nonlinear term (the second term in brackets) is negligible if the ratio of the beam deflection of the midspan of a fixed–fixed beam to the thickness (i.e., the beam dimension in the direction of the beam deflection) is much less than 0.85 (this value is for $N_0 = 0$; the full condition is described in Section 3.7). In the case where the nonlinear term is neglected, the preceding beam equation is reduced to

$$\tilde{E}I\frac{d^4y}{dx^4} - N_0\frac{d^2y}{dx^2} = q \tag{3.55}$$

Solutions of Beam Equation

We shall show how to solve the linear beam equation (3.55) for the deflection y. Let us begin by considering a cantilever beam of flexural rigidity $\tilde{E}I$ subjected to a transverse force F, shown in Fig. 3.29a. We can use the moment–curvature relationship (3.40) or the beam equation (3.54) to obtain the deflection y of the beam. When we draw the free-body diagram of the beam cut at x from the fixed end in Fig. 3.29b, we have the transverse shear force and the moment from (3.1) and (3.2) to maintain the equilibrium of the cut beam as follows:

$$M = F(l-x)$$

$$V = F$$

For a slender beam such as the cantilever in Fig. 3.29, the deflection of the beam is affected primarily by the moment in the cross section, and the shear force contribution to the deflection is negligible (the shear force effect on the slender beam is discussed in Section 3.8). As a result, the beam deflection is governed by the moment–curvature relationship (3.40) and the beam equation (3.54). For easy understanding, we first use the moment–curvature relationship (3.40) to give

$$\tilde{E}I\frac{d^2y}{dx^2} = M = F(l-x)$$

which must be solved with the boundary conditions: $y = 0$ and $dy/dx = 0$ at $x = 0$. Integrating the foregoing equation yields

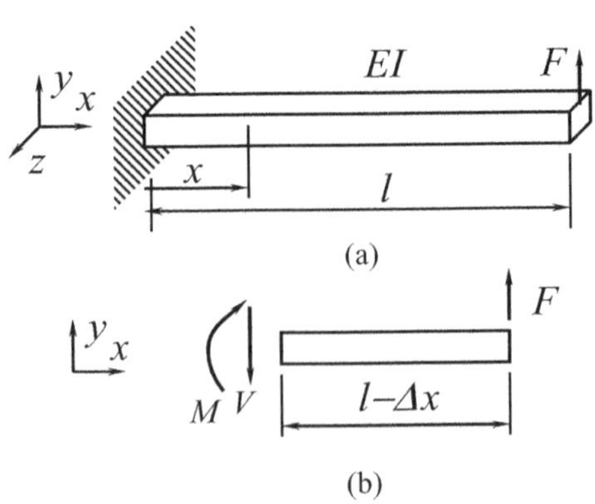

Figure 3.29 Cantilever beam subjected to a transverse shear force.

$$\tilde{E}I\frac{dy}{dx} = F\left(lx - \frac{x^2}{2} + c_1\right)$$

$$\tilde{E}Iy = F\left(l\frac{x^2}{2} - \frac{x^3}{6} + c_1x + c_2\right)$$

Applying the boundary conditions to the preceding two equations, we find that $c_1 = 0$ and $c_2 = 0$ and the deflection y of the cantilever beam subjected to the transverse force is given by

$$y = \frac{F}{6\tilde{E}I}(3lx^2 - x^3) \tag{3.56}$$

The maximum deflection δ at $x = l$ is given by

$$\delta = y|_{x=l} = \frac{Fl^3}{3\tilde{E}I} \tag{3.57}$$

The stiffness of the cantilever beam subjected to a concentrated load is defined as

$$k = \frac{F}{\delta} = \frac{3\tilde{E}I}{l^3} \tag{3.58}$$

To become familiar with how to use the beam equation (3.54), we attack the same problem. Since the cantilever beam is not subjected to any axial force, the terms in brackets in (3.54) vanish to provide the following equation:

$$\tilde{E}I\frac{d^4y}{dx^4} = q = 0 \tag{3.59}$$

It is noted from the foregoing equation that we consider the same beam in Fig. 3.29a, where no external load applies to the beam (i.e., $q = 0$). The following boundary conditions are imposed on the problem: At $x = 0$, $y = 0$ and $dy/dx = 0$, and at $x = l$, $M = 0$ and $V = F$. Using the definition of the positive force and moment (Section.3.1), the moment–curvature relationship (3.40), and the moment–force relationship (3.47), we obtain the differential equation forms of the boundary conditions: At $x = l$,

$$M = \tilde{E}I\frac{d^2y}{dx^2} = 0$$

and since

$$V = -\frac{dM}{dx} = -\frac{d}{dx}\left(\tilde{E}I\frac{d^2y}{dx^2}\right) = -\tilde{E}I\frac{d^3y}{dx^3} = F$$

at $x = l$,

$$\tilde{E}I\frac{d^3y}{dx^3} = -F$$

A successive integration of beam equation (3.59) gives a set of linear equations with four unknowns as follows:

$$\tilde{E}I\frac{d^3y}{dx^3} = c_1$$

$$\tilde{E}I\frac{d^2y}{dx^2} = c_1x + c_2$$

$$\tilde{E}I\frac{dy}{dx} = c_1\frac{x^2}{2} + c_2x + c_3$$

$$\tilde{E}Iy = c_1\frac{x^3}{6} + c_2\frac{x^2}{2} + c_3x + c_4$$

Applying the boundary conditions at $x = 0$ gives $c_3 = 0$ and $c_4 = 0$. We use the boundary conditions at $x = l$ to give

$$\tilde{E}I\frac{d^3y}{dx^3} = c_1 = -F$$

$$\tilde{E}I\frac{d^2y}{dx^2} = c_1l + c_2 = 0$$

To satisfy the two equations above, the coefficients must be $c_1 = -F$ and $c_2 = F_l$. Substituting the coefficients into the last of the set of equations, we find the deflection of the cantilever beam as

$$y = \frac{F}{6\tilde{E}I}(3lx^2 - x^3)$$

which is the same as (3.56). Therefore, we can use the moment–curvature equation (3.40) or the beam equation (3.54) to obtain the deflection of a beam subjected to a distributed or concentrated load.

Many MEMS devices use a movable structure which is suspended by straight flexures over a substrate as shown in Fig. 3.30. The flexures are anchored on the substrate and can be modeled as a straight beam of flexural

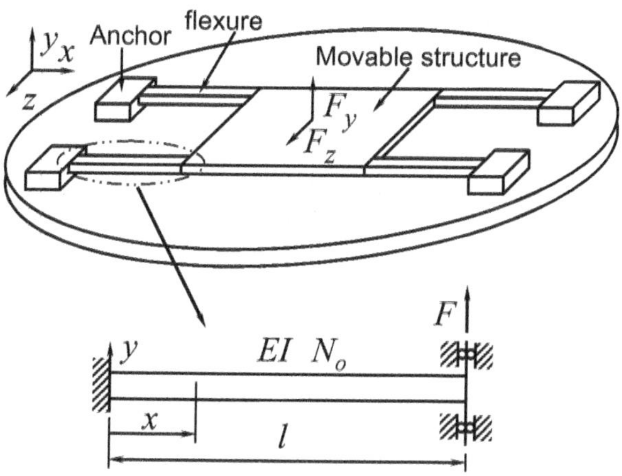

Figure 3.30 Flexures to suspend a movable structure.

rigidity $\tilde{E}I$, initial tension N_0, and length l, as shown in the lower figure in Fig. 3.30. When a transverse force F is applied to the right end, the beam is deflected. We first solve the beam equation without the initial tension and then obtain the solution with the initial tension. If we assume that the initial tension is zero, we can use the beam equation (3.55) with the effective Young's modulus,

$$\tilde{E}I\frac{d^4y}{dx^4} = q = 0$$

which must be solved with the following boundary conditions: At $x = 0$, $y = 0$ and $dy/dx = 0$, and at $x = l$, $dy/dx = 0$ and $V = F$. Using (3.40) and (3.47), the last boundary condition can be written in differential equation form as follows:

$$V = -\frac{dM}{dx} = -\frac{d}{dx}\left(\tilde{E}I\frac{d^2y}{dx^2}\right) = -\tilde{E}I\frac{d^3y}{dx^3} = F$$

or

$$\tilde{E}I\frac{d^3y}{dx^3} = -F$$

Integrating the beam equation and applying the four boundary conditions leads to the beam deflection y as follows:

$$y = \frac{F}{12\tilde{E}I}(3lx^2 - 2x^3) \tag{3.60}$$

The maximum deflection at $x = l$ is given by

$$\delta = y|_{x=l} = \frac{Fl^3}{12\tilde{E}I} \tag{3.61}$$

The stiffness of the beam shown in the lower figure in Fig. 3.30 is obtained as

$$k = \frac{F}{\delta} = \frac{12\tilde{E}I}{l^3} \tag{3.62}$$

We can solve the same problem with the initial axial force N_0. The beam equation with axial force and zero load along the beam is written as

$$\tilde{E}I\frac{d^4y}{dx^4} - N_0\frac{d^2y}{dx^2} = 0 \tag{3.63}$$

The boundary conditions are the same as those of the previous case with the zero axial force: At $x = 0$, $y = 0$ and $dy/dx = 0$, and at $x = l$, $dy/dx = 0$ and $EI(d^3y/dx^3) = -F$. Setting $y = e^{\lambda_1 x}$ and substituting y into the beam equation above yields the following equation:

$$\lambda_1^4 - \frac{N_0}{\tilde{E}I}\lambda_1^2 = 0$$

which leads to four roots:

$$\lambda_1 = 0\,(\text{double root})$$
$$\lambda_1 = \pm\beta$$

where

$$\beta = \sqrt{\frac{N_0}{\tilde{E}I}}$$

We then obtain a general solution of the beam equation as follows:

$$y = A\cosh\beta x + B\sinh\beta x + C + Dx \tag{3.64}$$

where A, B, C, and D are constants that can be determined by the boundary conditions. To apply the boundary conditions, equation (3.64) is successively differentiated three times:

$$\frac{dy}{dx} = A\beta\sinh\beta x + B\beta\cosh\beta x + D$$

$$\frac{d^2y}{dx^2} = A\beta^2 \cosh \beta x + B\beta^2 \sinh \beta x$$

$$\frac{d^3y}{dx^3} = A\beta^3 \sinh \beta x + B\beta^3 \cosh \beta x$$

Using the boundary conditions at $x = 0$, we have

$$y|_{x=0} = A + C = 0$$

$$\left.\frac{dy}{dx}\right|_{x=0} = B\beta + D = 0$$

These give

$$C = -A$$

$$D = -B\beta$$

Applying the boundary conditions $dy/dx(l) = 0$ and $\tilde{E}I\, d^3y/dx^3 = -F$ at $x = l$, we find that

$$\left.\frac{dy}{dx}\right|_{x=l} = A\beta \sinh \beta l + B\beta \cosh \beta l + D = 0$$

$$\left.\frac{d^3y}{dx^3}\right|_{x=l} = A\beta^3 \sinh \beta l + B\beta^3 \cosh \beta l = -\frac{F}{\tilde{E}I}$$

Substituting $D = -B\beta$ into the equations above leads to

$$A \sinh \beta l + B(\cosh \beta l - 1) = 0$$

$$A \sinh \beta l + B \cosh \beta l = -\frac{F}{\beta^3 \tilde{E}I}$$

The two equations are solved for coefficients A and B,

$$A = -\frac{1 - \cosh \beta l}{\sinh \beta l} \frac{F}{\beta^3 \tilde{E}I}$$

$$B = -\frac{F}{\beta^3 \tilde{E}I}$$

Upon substituting the coefficients A, B, C, and D into y, we have

$$y = A \cosh \beta x + B \sinh \beta x + C + Dx$$

$$= A(\cosh \beta x - 1) + B(\sinh \beta x - \beta x)$$

or

$$\begin{aligned} y &= -\frac{1-\cosh\beta}{\sinh\beta l}\frac{F}{\beta^3\tilde{E}I}(\cosh\beta x - 1) - \frac{F}{\beta^3 EI}(\sinh\beta x - \beta x) \\ &= \frac{F}{\beta^3\tilde{E}I}\frac{1}{\sinh\beta l}[(1-\cosh\beta l)(1-\cosh\beta x) + \sinh\beta l(\beta x - \sinh\beta x)] \end{aligned} \tag{3.65}$$

$$\begin{aligned} \delta = y|_{x=l} &= \frac{F}{\beta^3\tilde{E}I}\frac{1}{\sinh\beta l}\left[(1-\cosh\beta l)^2 + \sinh\beta l(\beta l - \sinh\beta l)\right] \\ &= \frac{F}{\beta^3\tilde{E}I}\frac{\beta l\sinh\beta l + 2(1-\cosh\beta l)}{\sinh\beta l} \end{aligned} \tag{3.66}$$

The stiffness k_1 of the beam with the axial force, shown in the lower figure in Fig. 3.30, is defined as

$$k_1 = \frac{F}{\delta} = \beta^3\tilde{E}I\frac{\sinh\beta l}{\beta l\sinh\beta l + 2(1-\cosh\beta l)} \tag{3.67}$$

This stiffness can be written in a convenient form:

$$k_1 = \frac{\tilde{E}I}{l^3}\frac{\lambda^3\sinh\lambda}{\lambda\sinh\lambda + 2(1-\cosh\lambda)} \tag{3.68}$$

where

$$\lambda = \beta l = l\sqrt{\frac{N_0}{\tilde{E}I}} \tag{3.69}$$

is a dimensionless number that represents the axial force N_0. It is noted that the stiffness is expressed as a function of the flexural rigidity $\tilde{E}I$, the beam length l, and the dimensionless axial force λ. It is worth noting that the stiffness above is nonlinear with respect to the dimensionless axial force λ. The stiffness k_1 can also be expressed as a function of the axial force N_0. From equation (3.69), we have $\tilde{E}I = N_0 l^2 \lambda^2$ and express the stiffness as follows:

$$k_1 = \frac{N_0}{l}\frac{\lambda\sinh\lambda}{\lambda\sinh\lambda + 2(1-\cosh\lambda)} \tag{3.70}$$

It should be noted from (3.68) and (3.70) that the beam under a tensile force (i.e., $N_0 > 0$) could act as a beam of flexural rigidity $\tilde{E}I$ or as a string of tension

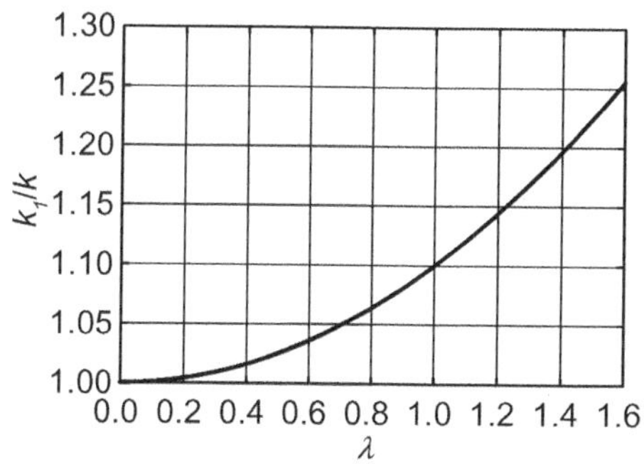

(a)

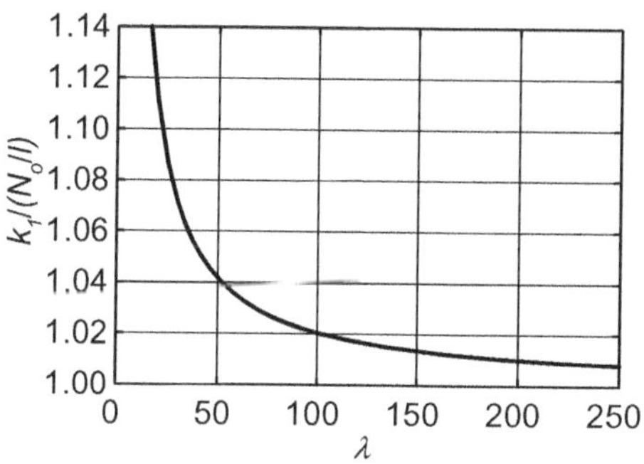

(b)

Figure 3.31 Stiffness as a function of a dimensionless axial force.

N_0. The graphs in Fig. 3.31 show both the cases: bending moment and tension dominant cases. In Fig. 3.31a, the stiffness of the beam is very close to unity for $\lambda \geq 0$ and $\lambda << 1$, and the beam in the lower figure in Fig.3.30 may be considered as a beam of stiffness k (3.62). For example, for $\lambda = 0.5$, the error between the stiffness (3.68) and (3.62) is estimated as 2.5%. In the tension-dominant case shown in Fig. 3.31b, the stiffness becomes N_0/l as λ approaches infinity:

$$k_1 = \lim_{\lambda \to \infty} \left[\frac{N_0}{l} \frac{\lambda \sinh \lambda}{\lambda \sinh \lambda + 2(1 - \cosh \lambda)} \right] = \frac{N_0}{l} \tag{3.71}$$

For $\lambda \geq 50$, the stiffness k_1 (3.70) is in good agreement with the stiffness N_0/l within an error of 4%. In this case, a narrow beam can be thought of as a string and a wide beam can be considered as a membrane that cannot resist a bending moment. In most MEMS devices, the axial tension N_0/l is much smaller than $\tilde{E}I/l^3$, so the stiffness (3.68) expressed in terms of $\tilde{E}I$ is convenient for designing flexures subjected to an axial force (tension or compression).

To obtain an approximate expression, let us examine the stiffness (3.68). The stiffness is expressed in dimensionless form,

$$\frac{k_1}{k} = \frac{1}{12} \frac{\lambda^3 \sinh \lambda}{\lambda \sinh \lambda + 2(1 - \cosh \lambda)} \tag{3.72}$$

where

$$k = \frac{12\tilde{E}I}{l^3} \tag{3.73}$$

is given by (3.62), the stiffness of the beam with zero axial force. For $\lambda << 1$, substituting $\sinh\lambda = \lambda + \lambda^3/3! + \lambda^5/5! + \cdots$ and $\cosh\lambda = 1 + \lambda^2/2! + \lambda^4/4! + \cdots$ into (3.72) and using the definition of the dimensionless axial force, (3.69), we have the following simple approximate expression:

$$\begin{aligned}\frac{k_1}{k} &= 1 + \frac{1}{10}\lambda^2 - \frac{1}{8400}\lambda^4 + \text{high-order terms} \\ &\cong 1 + \frac{1}{10}\lambda^2 = 1 + \frac{1}{10}\frac{l^2 N_0}{EI}\end{aligned} \tag{3.74}$$

The approximate stiffness above is sufficiently accurate to be used for a straight beam under compressive and tensile force, as shown in Fig. 3.30. For example, the preceding stiffness gives an error of 0% (i.e., exact solution), 1.7%, and 5.6% at $\lambda = 0, 5$, and 10 (all tensile forces), respectively. Substituting the stiffness k, (3.73), into the preceding equation, we obtain the approximate linear stiffness as a function of the flexural rigidity $\tilde{E}I$, the length l, and the axial force N_0:

$$k_1 = \frac{12\tilde{E}I}{l^3} + \frac{6}{5}\frac{N_0}{l} \tag{3.75}$$

It is worth noting that in the bending moment–dominant case, the contribution of the axial force to the stiffness k_1 is 6/5 times N_0/l.

The stiffness equation (3.72) for the beam shown in Fig. 3.30 can be used to obtain the buckling axial force. In macroscale structures such as a column to support a bridge, the buckling of the column is defined as a sudden failure mode when the structure is subjected to a compressive force that reaches a certain level known as the *critical load*. Buckling occurs in structures if the bending moment in the structures cannot balance the bending moment due to the compressive force; that is, a structure becomes unstable at the critical load. In other words, the stiffness (3.72) becomes zero at the critical load. Let us rotate the beam shown in the lower figure in Fig. 3.30 by 90° counterclockwise to obtain the vertical column of Fig. 3.32a. Upon substituting $N_0 = -P$ into the beam equation (3.63), we obtain the beam equation for the buckling analysis:

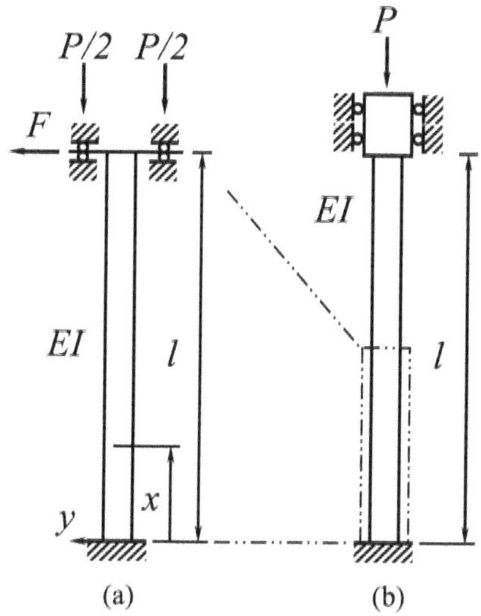

Figure 3.32 Vertical columns subjected to a compressive force.

$$\tilde{E}I\frac{d^4y}{dx^4}+P\frac{d^2y}{dx^2}=0 \tag{3.76}$$

where P is positive for compressive force. If we repeat the procedure that was used to give the stiffness (3.68), we obtain the stiffness of the beam (Fig. 3.32a) while the compressive force P is exerted on the beam along its length direction:

$$\frac{k_1}{k}=\frac{P}{\delta}=\frac{1}{12}\frac{\lambda_n^3\sin\lambda_n}{2(1-\cos\lambda_n)-\lambda_n\sin\lambda_n} \tag{3.77}$$

where P, δ, and λ_n denote the compressive force and displacement of the upper end of the beam and a dimensionless force defined as

$$\lambda_n=l\sqrt{\frac{P}{\tilde{E}I}} \tag{3.78}$$

The stiffness (3.77) may be obtained directly from the stiffness (3.72) that was obtained for tensile force. Substituting $N_0 = -P$ (i.e., $\lambda = j\lambda_n$) and recalling that

$$\cosh\lambda=\cosh(j\lambda_n)=\cos\lambda_n$$

$$\sinh\lambda=\sinh(j\lambda_n)=j\sin\lambda_n$$

where $j=\sqrt{-1}$, we obtain the same stiffness (3.77). Since setting $k_1 = 0$ in (3.77) gives $\sin\lambda_n = 0$ or $\lambda_n = \pi$, we then obtain the first critical load for the column of Fig. 3.32a from (3.78) as follows:

$$P_{\mathrm{cr}} = \frac{\pi^2 \tilde{E}I}{l^2} \tag{3.79}$$

which is equal to the critical force (load) first derived by Leonhard Euler (Crandall et al., 1978). Since we can consider the column of Fig. 3.32a as half the column in Fig. 3.32b, we obtain the critical load for the column of Fig. 3.31b.

$$P_{\mathrm{cr}} = \frac{\pi^2 \tilde{E}I}{(l/2)^2} = \frac{4\pi^2 \tilde{E}I}{l^2} \tag{3.80}$$

Note that approximate equations (3.74) and (3.75) also hold for the compressive load. For example, we can obtain an approximate critical load corresponding to (3.79) when we set the stiffness (3.75) equal to zero:

$$N_{0,\mathrm{cr}} = -\frac{10\tilde{E}I}{l^2} \tag{3.81}$$

where the negative sign indicates a compressive force. The preceding load is only 1.3% greater than the exact critical load (3.79).

3.7 GALERKIN'S METHOD

We have obtained solutions for the deflections and stiffness of beams when the geometric nonlinearity of the beam equation (3.54) is negligible. Since the beam equation (3.54) is nonlinear, we cannot derive its exact solution but can obtain the approximate solutions with sufficient accuracy. For an approximate solution Galerkin's method (Fung, 1965) is introduced, which has been used in scientific and engineering analyses, including finite element analysis (Bathe, 1982). In this section, we shall introduce Galerkin method and its application to a nonlinear structure, and in the next section an energy method corresponding to the Galerkin method used to obtain approximate solutions for the nonlinear behavior of MEMS structures is described. The solutions developed will be used to understand the nonlinear behavior of MEMS devices described in subsequent chapters.

We first rewrite the beam equation (3.54) with the nonlinear term as follows:

$$\tilde{E}I\frac{d^4y}{dx^4} - \left[N_0 + \frac{1}{2}\frac{\tilde{E}A}{l}\int_0^l \left(\frac{dy}{dx}\right)^2 dx\right]\frac{d^2y}{dx^2} = q \tag{3.82}$$

where the force term q may be linear or nonlinear. For an approximate solution to the deflection y, Galerkin's method uses minimization of errors by orthogonalizing with respect to a set of basis functions (Fung, 1965):

$$y_N = \sum_{k=1}^{N} a_k \psi_k \tag{3.83}$$

where y_N, a_k, ψ_k, and N denote the approximate solution, coefficients, and the basis function at k, and the number of the functions, respectively. To obtain a closed-form solution from the nonlinear equation (3.82), we use a function y_1:

$$y_1 = y_m \psi_1 \tag{3.84}$$

where y_m is a coefficient to be determined and ψ_1 is a basis function (mode shape) that represents the deflection of a beam. The foregoing equation may be written in a more convenient form that is suitable for integration:

$$y_1 = y_m \psi_1 = y_m \eta \psi \tag{3.85}$$

where η is a constant and ψ is a mode shape, which is defined as a function of the dimensionless coordinate X:

$$\psi = f(X) \tag{3.86}$$

$$X = \frac{x}{l} \tag{3.87}$$

For example, the deflection of the fixed–fixed beam under a uniformly distributed load can be expressed as (described later)

$$y_1 = y_m 16\left(\frac{x}{l}\right)^2\left(1-\frac{x}{l}\right)^2 = y_m \psi_1$$

where y_m is the midspan deflection and ψ_1 is the mode shape of the beam, given by

$$\psi_1 = 16\left(\frac{x}{l}\right)^2\left(1-\frac{x}{l}\right)^2 = \eta\psi(X) \tag{3.88}$$

$$\psi(X) = \left(\frac{x}{l}\right)^2\left(1-\frac{x}{l}\right)^2 = X^2(X-1)^2$$

η is obtained as 16 for the mode shape. It is noted that the deflection of the fixed–fixed beam is normalized to unity by dividing the deflection by its maximum deflection at the midspan.

Substituting (3.85) into beam equation (3.82), multiplying by the mode shape ψ_1, and integrating the resulting equation along the beam gives the following force equilibrium equation that is a function of the unknown y_m:

$$k_1 y_m + k_3 y_m^3 = f_q \tag{3.89}$$

where k_1, k_3, and f_q represent a linear stiffness, a cubic stiffness, and a force, and they are given by the following equations:

$$k_1 = \tilde{E}I\int_0^l \frac{d^4\psi_1}{dx^4}\psi_1\,dx - N_0\int_0^l \frac{d^2\psi_1}{dx^2}\psi_1\,dx \tag{3.90}$$

$$k_3 = -\frac{1}{2}\frac{\tilde{E}A}{l}\int_0^l\left(\frac{d\psi_1}{dx}\right)^2 dx\int_0^l \frac{d^2\psi_1}{dx^2}\psi_1\,dx \tag{3.91}$$

$$f_q = \int_0^l q\psi_1\,dx \tag{3.92}$$

where $\tilde{E}I$, N_0, and $\tilde{E}A$ are assumed as constants. Equation (3.89) may be interpreted as a one-degree-of-freedom force equilibrium equation. If the equation is solved for y_m, we can obtain deflection of the beam from (3.85). Using (3.84), we can convert the preceding dimensional equations into dimensionless forms that are suitable for integration, as described below. It is assumed that $\tilde{E}I$ and N_0 are constants:

$$Y_m + k_r Y_m^3 = F_q \tag{3.93}$$

where

$$Y_m = \frac{y_m}{y_c} \tag{3.94}$$

$$k_r = \frac{k_3 y_c^2}{k_1} \tag{3.95}$$

$$F_q = \frac{f_q}{k_1 y_c} \tag{3.96}$$

where the dimensionless Y_m, k_r, and F_q represent the deflection of interest, the geometric nonlinearity, and the load, respectively. y_c is a characteristic dimension: for example, the thickness of the beam. It is noted that k_r can be used as a measure of the geometric nonlinearity of the beam. We thus define k_r as a nonlinear factor for further reference. The stiffness and force (3.90)–(3.92) are expressed as functions of the mode shape ψ as follows:

$$k_1 = \eta^2 \frac{\tilde{E}I}{l^3} \int_0^1 \frac{d^4\psi}{dX^4} \psi \, dX - \eta^2 \frac{N_0}{l} \int_0^1 \frac{d^2\psi}{dX^2} \psi \, dX \tag{3.97}$$

$$k_3 = -\frac{\eta^4}{2} \frac{\tilde{E}A}{l^3} \int_0^1 \left(\frac{d\psi}{dX} \right)^2 dX \int_0^1 \frac{d^2\psi}{dX^2} \psi \, dX \tag{3.98}$$

$$f_q = \eta q_m l \int_0^1 \psi_q \psi \, dX \tag{3.99}$$

where q_m is the load at the coordinate x corresponding to y_m and ψ_q is a dimensionless load distribution, defined as

$$\psi_q = \frac{q}{q_m} \tag{3.100}$$

The first integral in (3.97) represents stiffness due to the bending effect of the beam. Using integration by parts and the boundary conditions gives a simple form for the first integral of (3.97):

$$\begin{aligned} \int_0^1 \frac{d^4\psi}{dX^4} \psi \, dX &= \left. \frac{d^3\psi}{dX^3} \psi \right|_0^1 - \int_0^1 \frac{d^3\psi}{dX^3} \frac{d\psi}{dX} dX \\ &= \left. \frac{d^3\psi}{dX^3} \psi \right|_0^1 - \left. \frac{d^2\psi}{dX^2} \frac{d\psi}{dX} \right|_0^1 + \int_0^1 \frac{d^2\psi}{dX^2} \frac{d^2\psi}{dX^2} dX \\ &= \int_0^1 \left(\frac{d^2\psi}{dX^2} \right)^2 dX \end{aligned} \tag{3.101}$$

To obtain the last integral, we assumed that all other terms except the final integral vanish at the boundaries. For example, the fixed–fixed beam gives the boundary conditions $y = 0$ and $dy/dx = 0$, so the terms associated with the boundary in equation (3.101) vanish to give the final integral. Similarly, a cantilever beam under a distributed load gives the final integral since $y = 0$ and $dy/dx = 0$ at $x = 0$ and $d^2y/dx^2 = 0$ [i.e., zero moment; see (3.40)] and $d^3y/dx^3 = 0$ [i.e., zero transverse shear force; see (3.47)]. The second integral in (3.97) describes stiffness caused by the axial force N_0 and the same procedure is used to yield

$$\int_0^1 \frac{d^2\psi}{dX^2} \psi \, dX = \left. \frac{d\psi}{dX} \psi \right|_0^1 - \int_0^1 \frac{d\psi}{dX} \frac{d\psi}{dX} dX = -\int_0^1 \left(\frac{d\psi}{dX} \right)^2 dX \tag{3.102}$$

Substituting equations (3.101) and (3.102) into (3.97) and (3.98), we obtain a convenient form of the stiffness:

$$k_1 = \eta^2 \frac{\tilde{E}I}{l^3} \int_0^1 \left(\frac{d^2\psi}{dX^2} \right)^2 dX + \eta^2 \frac{N_0}{l} \int_0^1 \left(\frac{d\psi}{dX} \right)^2 dX \tag{3.103}$$

$$k_3 = \frac{1}{2}\eta^4 \frac{\tilde{E}A}{l^3} \left[\int_0^1 \left(\frac{d\psi}{dX} \right)^2 dX \right]^2 \tag{3.104}$$

Example 3.3 To become familiar with the above-mentioned procedure for obtaining the force equilibrium equation (3.89), let us consider the flexure in Fig. 3.30, whose linear stiffness k_1 was obtained in (3.75). It is desired to obtain the linear and nonlinear stiffness of the lower figure in Fig. 3.30 and to evaluate the nonlinear factor.

We first define the boundary conditions for the beam of the Fig. 3.30 as follows: At $x = 0$,

$$y = 0 \quad \text{and} \quad \frac{dy}{dx} = 0 \tag{a}$$

and at $x = l$,

$$\frac{dy}{dx} = 0 \quad \text{and} \quad V = F = -EI\frac{d^3y}{dx^3} = -EIy_m \frac{d^3\psi_1}{dx^3} \tag{b}$$

Since the terms related to the boundary condition above do not vanish in (3.101), we cannot use equations (3.97) and (3.98). We must then obtain the solution from the beam equation (3.54). If we consider the beam from $x = 0$ to $x = l - \Delta l$, where Δl approaches zero, we obtain the beam equation (3.54) with $q = 0$ as follows:

$$\tilde{E}I\frac{d^4y}{dx^4} - \left[N_0 + \frac{1}{2}\frac{\tilde{E}A}{l}\int_0^l \left(\frac{dy}{dx} \right)^2 dx \right] \frac{d^2y}{dx^2} = 0 \tag{c}$$

Under the assumption that the beam is bending moment–dominant (i.e., the dimensionless axial force λ is relatively small), we take the deflection (3.60) of the flexure as the mode shape:

$$y_1 = \frac{F}{12EI}(3lx^2 - 2x^3) \tag{d}$$

Dividing equation (d) by the maximum deflection y_m at $x = l$, we find the approximate solution, given by

$$y_1 = y_m\left[3\left(\frac{x}{l}\right)^2 - 2\left(\frac{x}{l}\right)^3\right] = y_m\psi_1 = y_m\eta\psi \tag{e}$$

where y_m is the end deflection to be determined and ψ_1 is the mode shape of the beam, given by

$$\psi_1 = 3\left(\frac{x}{l}\right)^2 - 2\left(\frac{x}{l}\right)^3 = \eta\psi(X) \tag{f}$$

$$\psi = 3X^2 - 2X^3 \tag{g}$$

and η is 1. Substituting the approximate solution y_1 for y into (c), multiplying by ψ_1, and integrating the resulting equation leads to

$$y_m\tilde{E}I\int_0^l \frac{d^4\psi_1}{dx^4}\psi_1\,dx - y_mN_0\int_0^l \frac{d^2\psi_1}{dx^2}\psi_1\,dx - \frac{y_m^3\tilde{E}A}{2l}\int_0^l\left(\frac{d\psi_1}{dx}\right)^2 dx\int_0^l \frac{d^2\psi_1}{dx^2}\psi_1\,dx = 0 \tag{h}$$

The first integral in (h) can be written

$$\begin{aligned}\int_0^l \frac{d^4\psi_1}{dx^4}\psi_1\,dx &= \frac{d^3\psi_1}{dx^3}\psi_1\bigg|_0^l - \frac{d^2\psi_1}{dx^2}\frac{d\psi_1}{dx}\bigg|_0^l + \int_0^l \frac{d^2\psi_1}{dx^2}\frac{d^2\psi_1}{dx^2}dx \\ &= \left(-F\frac{1}{EIy_m}\psi_1\right)_{x=l} + \int_0^l\left(\frac{d^2\psi_1}{dx^2}\right)^2 dx \\ &= -F\frac{1}{EIy_m} + \int_0^l\left(\frac{d^2\psi_1}{dx^2}\right)^2 dx\end{aligned} \tag{i}$$

Similarly,

$$\begin{aligned}\int_0^l \frac{d^2\psi_1}{dx^2}\psi_1\,dx &= \frac{d\psi_1}{dx}\psi_1\bigg|_0^l - \int_0^l \frac{d\psi_1}{dx}\frac{d\psi_1}{dx}dx \\ &= -\int_0^l\left(\frac{d\psi_1}{dx}\right)^2 dx\end{aligned} \tag{j}$$

Substituting (i) and (j) into (h), using the definition of ψ_1, (f), and rearranging terms, we find that

$$k_1y_m + k_3y_m^3 = F \tag{k}$$

where

$$k_1 = \eta^2 \frac{\tilde{E}I}{l^3}\int_0^1 \left(\frac{d^2\psi}{dX^2}\right)^2 dX + \eta^2 \frac{N_0}{l}\int_0^1 \left(\frac{d\psi}{dX}\right)^2 dX \tag{l}$$

$$k_3 = \frac{1}{2}\eta^4 \frac{\tilde{E}A}{l^3}\left[\int_0^1 \left(\frac{d\psi}{dX}\right)^2 dX\right]^2 \tag{m}$$

are the linear stiffness due to the bending moment and axial force and the cubic stiffness due to the geometric nonlinearity, respectively. It is noted that equation (k) is similar to the force equilibrium equation (3.89) for a beam subjected to a distributed load and that the linear and cubic stiffness above are expressed in the same form as the stiffness (3.103) and (3.104). Let us determine the linear and cubic stiffness. Substitution of the mode shape (g) and the constant $\eta = 1$ into the linear and cubic stiffness (l) and (m) gives

$$\begin{aligned} k_1 &= \frac{\tilde{E}I}{l^3}\int_0^1 [6(1-2X)]^2 dX + \frac{N_0}{l}\int_0^1 [6(X-X^2)]^2 dX \\ &= 12\frac{\tilde{E}I}{l^3} + \frac{6}{5}\frac{N_0}{l} \end{aligned} \tag{n}$$

$$\begin{aligned} k_3 &= \frac{1}{2}\frac{\tilde{E}A}{l^4}\left[\int_0^1 [6(X-X^2)]^2 dX\right]^2 \\ &= \frac{18}{25}\frac{\tilde{E}A}{l^3} \end{aligned} \tag{o}$$

It should be stressed that the linear stiffness (n) is equal to the approximate stiffness (3.75) that is obtained from the Taylor series expansion of the exact solution (3.72) of the beam equation without geometric nonlinearity. The cubic stiffness (o) as well as the linear stiffness (n) is validated by comparing the nonlinear force (k) with the force obtained from Abaqus, a nonlinear simulation program based on the finite element method. Figure 3.33 compares the nonlinear theoretical and simulated forces of a beam of length $l = 150\,\mu\text{m}$, width $w = 2\,\mu\text{m}$, thickness $t = 2\,\mu\text{m}$, Young's modulus $E = 110\,\text{GPa}$, and three residual stresses: (initial stress) $\sigma_0 = 0$, 64 MN (tensile stress), and –64 MN (compressive stress); the corresponding axial force is obtained as $N_0 = \sigma_0 wt$. In Fig. 3.33, the nonlinear force (k) is in good agreement with that from numerical analysis. It is noted from Fig. 3.33 that the nonlinear force (k) becomes more accurate as the compressive stress is applied to the beam. It is also noted that the force (k) is applicable to the buckled beam, as shown in the lower force curve in Fig. 3.33, which is subjected to the residual compressive stress that is beyond the buckling stress σ_{cr}, calculated as –16 MPa for the beam; note that $\sigma_{cr} = -P_{cr}/A$, where P_{cr} is the buckling load (3.79) and A is the beam cross-sectional area. The nonlinear restoring force (k) is written

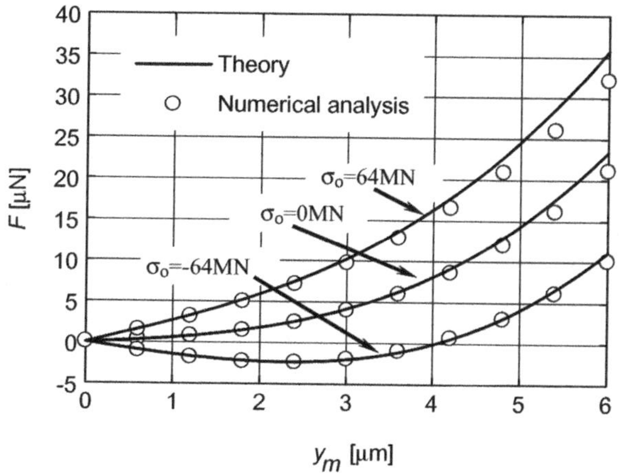

Figure 3.33 Validation of linear and cubic stiffness.

$$k_1 y_m + k_3 y_m^3 = k_1 y_m \left(1 + \frac{k_3}{k_1} y_m^2\right) = k_1 y_m (1 + k_r) = f_q \tag{p}$$

The second term in parentheses can be defined as the dimensionless beam nonlinear factor, k_r, which may be used as a measure of the geometric nonlinearity of the beam:

$$k_r = \frac{k_3}{k_1} y_m^2 \tag{q}$$

Substituting the linear and cubic stiffness k_1 and k_3 into equation (q) and noting the area $A = wt$ and the moment of inertia $I = wt^3/12$, we have the nonlinear factor of the beam shown in Fig. 3.30:

$$\begin{aligned} k_r &= \left[\frac{3}{50}\frac{A y_m^2}{I}\right] \Big/ \left[1 + \frac{1}{10}\frac{N_0 l^2}{\tilde{E} I}\right] \\ &= \left[\frac{18}{25}\left(\frac{y_m}{t}\right)^2\right] \Big/ \left[1 + \frac{6}{5}\frac{\sigma_0}{\tilde{E}}\left(\frac{l}{t}\right)^2\right] \end{aligned} \tag{r}$$

If the nonlinearity k_r is much smaller than unity, the nonlinearity term in (k) is negligible. Therefore, we find a condition for neglecting the nonlinearity as follows:

$$\frac{y_m}{t} \ll \sqrt{\frac{25}{18}\left[1 + \frac{6}{5}\frac{\sigma_0}{\tilde{E}}\left(\frac{l}{t}\right)^2\right]} \cong 1.1785\sqrt{1 + \frac{6}{5}\frac{\sigma_0}{\tilde{E}}\left(\frac{l}{t}\right)^2} \tag{s}$$

which gives $k_r << 1$. If the preceding condition is satisfied, the geometric nonlinearity can be neglected and we can write the force equilibrium equation (p) as

$$k_1 y_m \cong f_q \qquad \text{(t)}$$

which gives the maximum deflection:

$$y_m = \frac{f_q}{k_1} \qquad \text{(u)}$$

If condition (s) is not satisfied, we should solve nonlinear force equation (k) for y_m. In subsequent chapters, (k) will be used to develop further theories for MEMS structures.

Fixed–Fixed Beam Under a Uniformly Distributed Load

Many micro- and nanostructures employ the fixed–fixed beam shown in Fig. 3.34. For example, an electrostatic fixed–fixed beam actuator (described in Chapter 9) is subjected to the electrostatic force, which is nonlinear. For the nonlinear fixed–fixed analysis in subsequent chapters, we obtain the mode shape of the beam and the corresponding linear and cubic stiffness. To determine the mode shape and stiffness, we follow the procedure described in this chapter.

Since the boundary conditions ($y = 0$ and $dy/dx = 0$ at $x = 0$ and $x = l$) hold, equations (3.89), (3.99), (3.103), and (3.104) can be used to obtain the stiffness and the force equilibrium equation. First we obtain the mode shape of the fixed–fixed beam, which will be used for further derivation. Neglecting the axial force and the geometric nonlinearity, the beam equation (3.82) may be written as

$$\tilde{E}I \frac{d^4 y}{dx^4} = q \qquad (3.105)$$

Integrating the preceding equation four times gives the following set of equations:

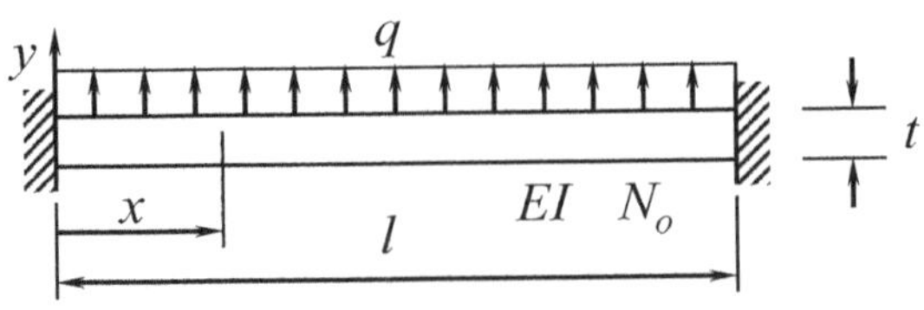

Figure 3.34 Fixed–fixed beam under a uniformly distributed load.

$$\tilde{E}I\frac{d^3y}{dx^3} = qx + c_1$$

$$\tilde{E}I\frac{d^2y}{dx^2} = q\frac{x^2}{2} + c_1x + c_2$$

$$\tilde{E}I\frac{dy}{dx} = q\frac{x^3}{6} + c_1\frac{x^2}{2} + c_2x + c_3$$

$$\tilde{E}Iy = q\frac{x^4}{24} + c_1\frac{x^3}{6} + c_2\frac{x^2}{2} + c_3x + c_4$$

Applying the boundary conditions ($y = 0$ and $dy/dx = 0$ at $x = 0$ and $x = l$) to the equations above leads to the following beam deflection:

$$y = \frac{q}{24\tilde{E}I}\left(x^4 - 2lx^3 + l^2x^2\right)$$

Since the deflection has a maximum at midspan, we write the preceding equation in the form

$$y = 16y_m\frac{x^4 - 2lx^3 + l^2x^2}{l^4}$$

where y_m is the maximum deflection and will serve as a characteristic deflection for the mode shape described below. The preceding deflection is expressed in more convenient form as follows:

$$y_1 = y = y_m\psi_1 = y_m\eta\psi \tag{3.106}$$

where η is a constant, 16, and ψ is a mode shape, which is defined as a function of the dimensionless coordinate $X = x/l$:

$$\psi(X) = \left(\frac{x}{l}\right)^2\left(1 - \frac{x}{l}\right)^2 = X^2(X-1)^2 \tag{3.107}$$

Substituting the preceding mode shape ψ and $\eta = 16$ into (3.103) and (3.104), we find the linear and cubic stiffness for a fixed–fixed beam in which the bending moment is dominant as follows:

$$k_1 = \eta^2\frac{\tilde{E}I}{l^3}\frac{4}{5} + \eta^2\frac{N_0}{l}\frac{2}{105} = \frac{1024}{5}\frac{\tilde{E}I}{l^3} + \frac{512}{105}\frac{N_0}{l} \tag{3.108}$$

$$k_3 = \eta^4\frac{\tilde{E}A}{l^3}\left(\frac{2}{105}\right)^2 = \frac{262{,}144}{11{,}025}\frac{\tilde{E}A}{l^3} \tag{3.109}$$

With $q_m = q$ and $\psi_q = 1$, the force f_q (3.99) is given by

$$f_q = \eta q_m l \int_0^l \psi_q \psi \, dx = \eta q l \frac{1}{30} = \frac{8}{15} q l \tag{3.110}$$

Therefore, force equilibrium equation (3.89) can be written as

$$k_1 y_m + k_3 y_m^3 = k_1 y_m (1 + k_r) = f_q \tag{3.111}$$

where the nonlinear factor representing the nonlinearity of the beam is given by

$$k_r = \frac{k_3}{k_1} y_m^2 \tag{3.112}$$

Upon substituting the linear and cubic stiffness k_1 and k_3 into equation (3.112) we have

$$\begin{aligned} k_r &= \left(\frac{256}{2205} \frac{A y_m^2}{I} \right) \bigg/ \left(1 + \frac{1}{42} \frac{N_0 l^2}{\tilde{E} I} \right) \\ &= \left(\frac{1024}{735} \left(\frac{y_m}{t} \right)^2 \right) \bigg/ \left(1 + \frac{2}{7} \frac{\sigma_0}{\tilde{E}} \left(\frac{l}{t} \right)^2 \right) \end{aligned} \tag{3.113}$$

To obtain the last expression in equation (3.113), the area $A = wt$ and the moment of inertia $I = wt^3/12$ are used, where w is the width of the beam. If $k_r << 1$, the nonlinearity term in (3.111) is negligible. We set up a condition for neglecting the nonlinearity of the fixed–fixed beam under a uniformly distributed load q, as follows:

$$\frac{y_m}{t} << \sqrt{\frac{735}{1024} \left[1 + \frac{2}{7} \frac{\sigma_0}{\tilde{E}} \left(\frac{l}{t} \right)^2 \right]} \cong 0.8472 \sqrt{1 + \frac{2}{7} \frac{\sigma_0}{\tilde{E}} \left(\frac{l}{t} \right)^2} \tag{3.114}$$

Equation (3.114) states that if the ratio of the midspan deflection of the fixed–fixed beam to the thickness (the dimension measured in the direction of the load q) satisfies the preceding condition, the geometric nonlinearity can be neglected and we may write force equilibrium equation (3.111) as

$$k_1 y_m \cong f_q \tag{3.115}$$

Under condition (3.114), the maximum deflection at the midspan of the beam is then given by

$$y_m \cong \frac{f_q}{k_1} \tag{3.116}$$

With the maximum deflection above, we can use beam deflection equation (3.106) to determine the deflection the beam at x. We note that satisfaction of condition (3.114) in the fixed–fixed beam problem greatly simplifies the procedure for determining the beam deflection at x. If condition (3.114) is not satisfied, we should solve the nonlinear equation (3.111) for y_m.

3.8 ENERGY METHOD

In the preceding sections we solved the beam equations to obtain the deflection of beams (e.g., the fixed–fixed beam) even though we were only interested in the deflection of a point in the beams. Solving the differential equations to obtain exact solutions is time-consuming work in many cases—even impossible if the equations involve nonlinear terms. In this section, energy methods are described that can be used to easily obtain the displacement of a point in an elastic body when a force or moment is applied to the point. Let us consider a coil spring (Fig. 3.35a) such as that found in many machines (e.g., a car). The spring consists of a coiled steel and has stiffness (also known as a *spring constant*), k, which is measured in force per unit length (e.g., N/m). A coiled spring or any elastic members are usually represented as shown in Fig. 3.35b. If the spring is suspended from a support as shown in Fig. 3.35b and is subjected to a force F in the direction of the spring length, the spring will be stretched by an elongation δ. The restoring force generated in the spring balances the external force. The stiffness is then defined as

$$k = \frac{F}{\delta} \tag{3.117}$$

If the force is nonlinear, it is expressed as a function of the elongation δ as follows:

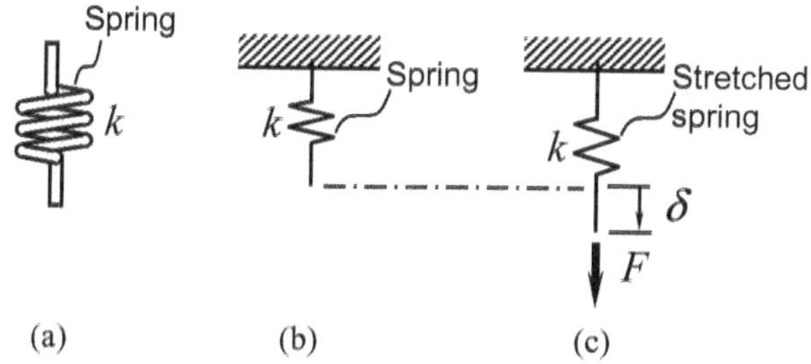

Figure 3.35 Spring subjected to an axial force.

$$F = f(\delta)$$

Referring to Fig. 3.36a, we can write the following increment in the elastic energy due to an infinitesimal elongation,

$$dU = \mathbf{F} \cdot d\boldsymbol{\delta} = F\,d\delta$$

where the dot denotes the scalar product of the force vector and the displacement vector. During the elongation, the total elastic energy stored in the spring is given by

$$U = \int_0^\delta F\,d\delta \tag{3.118}$$

which corresponds to the shaded area under the force curve in Fig. 3.36b. For convenience, we also define the *complementary energy* U^*, also called the *co-energy*, as shown in Fig. 3.36a. For infinitesimal increment in force, the complementary energy is written as

$$dU^* = \boldsymbol{\delta} \cdot d\mathbf{F} = \delta\,dF \tag{3.119}$$

As the force increases from zero to F, the complementary energy is given by

$$U^* = \int_0^F \delta\,dF \tag{3.120}$$

If the elongation is given as a function of the applied force,

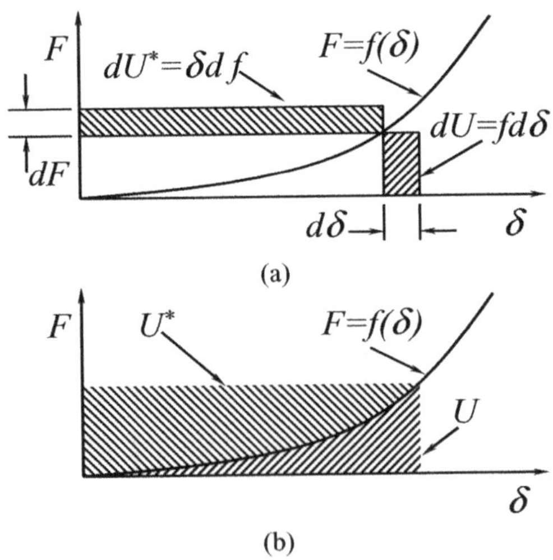

Figure 3.36 Elastic energy and complementary energy.

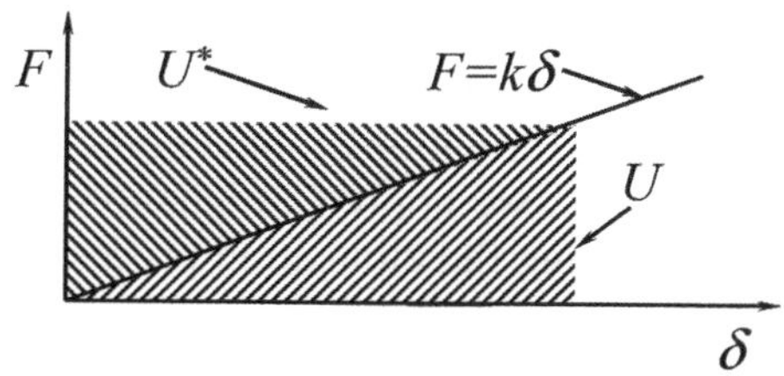

Figure 3.37 Elastic energy and complementary energy of a linear material.

$$U^* = f(F) \tag{3.121}$$

we can perform the integration above for evaluation of the complementary energy. Note that if the spring is linear (i.e., the force curve becomes a straight line, as shown in Fig. 3.37), the stored energy due to the elongation is equal to the corresponding complementary energy:

$$U^* = U$$

If we have the complementary energy U^* of (3.118), the elongation δ due to the force can be obtained using energy equation (3.119):

$$\delta = \frac{\partial U^*}{\partial F} \tag{3.122}$$

If the spring is linear, the elastic energy U can be used as follows:

$$\delta = \frac{\partial U}{\partial F} \tag{3.123}$$

In many problems, the elastic behavior of structures can be assumed to be linear. Under this assumption, equation (3.123) will be used in subsequent chapters to obtain deformation of elastic structures. Equation (3.123), known as *Castigliano's theorem*, is widely used in evaluating the deformation of structures. To become familiar with Castigliano's theorem, let's apply it to the spring problem in Fig. 3.35. From the force equilibrium, we find the force in the spring, given by

$$F = k\delta$$

From the preceding equation, we obtain straightforwardly the following elongation of the spring:

$$\delta = \frac{F}{k}$$

It is a special case for which we do not need a complicated calculation. However, for a complicated structure (e.g., a bridge), it is difficult using a simple calculation to obtain deformation at some point in a structure. To reach the same expression as above, we apply Castigliano's theorem to the same spring. The elastic energy during the elongation process is obtained as

$$U = \int_0^{\delta} F\,d\delta = \int_0^{\delta} k\delta\,d\delta = \frac{k\delta^2}{2}$$

which is expressed as a function of the elongation δ. To use Castigliano's theorem, the preceding energy should be expressed in terms of the applied force F. The elongation $\delta = F/k$ is substituted into the foregoing equation to give the energy as a function of the applied force:

$$U^* = U = \frac{k\delta^2}{2} = \frac{k}{2}\left(\frac{F}{k}\right)^2 = \frac{F^2}{2k}$$

Applying Castigliano's theorem, we obtain the elongation of the spring:

$$\delta = \frac{\partial U^*}{\partial F} = \frac{\partial U}{\partial F} = \frac{\partial}{\partial F}\left(\frac{F^2}{2k}\right) = \frac{F}{k}$$

The elongation above is the same as that obtained from the force equilibrium. Castigliano's theorem is extended to evaluate deformations at points in structures where forces are applied to the points, or to calculate angular displacement of points at which moments are applied. Consider an elastic body in Fig. 3.38, where forces $F_1, F_2, \ldots, F_n$ are applied to the body. Since the body is elastic, the points at which the forces are applied are displaced from their original points to new points, where the body is in equilibrium. Let δ_i represent the inline displacement due to the ith force F_i. For the elastic body, the complementary energy may be written as

$$U^* = f(F_1, F_2, \ldots, F_i, \ldots, F_n) = \sum_i \int \delta_i \, dF_i \tag{3.124}$$

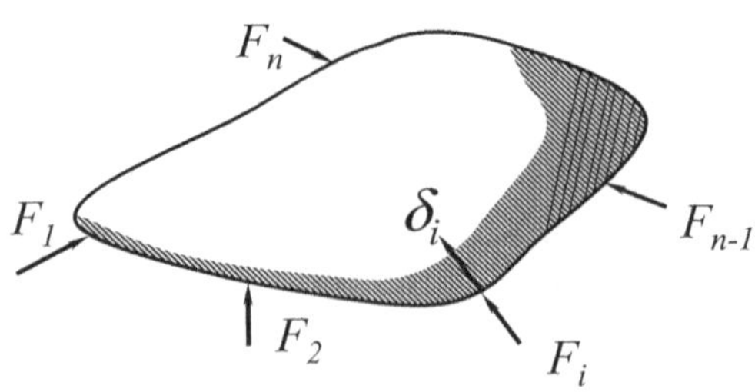

Figure 3.38 Elastic body subjected to multiple forces.

The energy increment due to F_i is given by

$$dU^* = \delta_i \, dF_i$$

which leads to the extended Castigliano's theorem for the multiple forces that are applied to the elastic body:

$$\delta_i = \frac{\partial U^*}{\partial F_i} \tag{3.125}$$

Similarly, if moments M_1, M_2, ... M_i, ... , M_n are applied to an elastic body, the angular displacement ϕ_i at the point of application of the ith moment is given by

$$\phi_i = \frac{\partial U^*}{\partial M_i} \tag{3.126}$$

where

$$U^* = f(M_1, M_2, \ldots, M_i, \ldots, M_m)$$

is the complementary energy due to the moments. As mentioned previously, the complementary energy can be replaced by the elastic energy if the elastic body is linear.

We can use Castigliano's theorem to obtain the displacement at the point of application of the external force in Fig. 3.39a. Before using the theorem, we obtain the force in the first and second springs. Upon applying the force equilibrium condition (3.1) to Fig. 3.39b, we have $F = F_1$, and the elongations in the springs are given by $\delta_1 = F/k_1$ and $\delta - \delta_1 = F/k_2$. These relationships give the displacement of the end of the spring:

$$\delta = \delta_1 + \frac{F}{k_2} = F\left(\frac{1}{k_1} + \frac{1}{k_2}\right)$$

It is noted that the displacement above is obtained from force equilibrium condition (3.1) and the geometric relationship. Using Castigliano's theorem, we obtain the same relationship. We first obtain the elastic energy as follows:

$$U = U_1 + U_2 = \frac{F^2}{2k_1} + \frac{F^2}{2k_2} = \frac{F^2}{2}\left(\frac{1}{k_1} + \frac{1}{k_2}\right)$$

Applying Castigliano's theorem, we find the displacement of the application of a force F as follows:

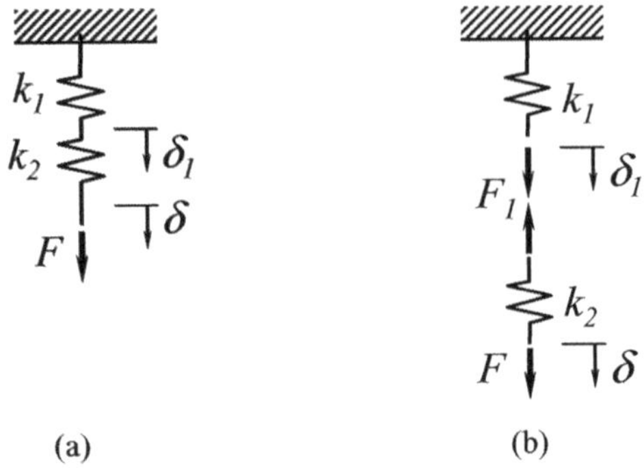

Figure 3.39 Two springs in series.

$$\delta = \frac{\partial U^*}{\partial F} = \frac{\partial U}{\partial F} = F\left(\frac{1}{k_1}+\frac{1}{k_2}\right)$$

We may define an effective stiffness due to two springs in series,

$$k = \frac{F}{\delta} = \left(\frac{1}{k_1}+\frac{1}{k_2}\right) \tag{3.127}$$

Similarly, the effective stiffness due to two springs in parallel is given by

$$k = k_1 + k_2 \tag{3.128}$$

where k_1 and k_2 denote the stiffness of the first and second springs, respectively.

In MEMS devices, many slender beams are used for mechanical elements, whose general configuration is shown in Fig. 3.40. Due to the microfabrication process discussed in Chapter 2, most beams are rectangular, as shown in Fig. 3.40a. However, for convenience of explanation, we first describe the energy method for the circular beam in Fig. 3.40b and then discuss the energy method for the rectangular beam. Let us suppose that the circular beam shown in Fig. 3.41 has area A, length l, and Young's modulus E. When an axial force is applied to the free end of the beam, the beam is elongated as illustrated in Fig. 3.41. The elongation of the end of the beam is obtained from the stress–strain relationship (3.3), given by $d\delta = F/EA\, dx$. Integrating the equation along the beam gives

$$\delta = \int_0^l \frac{F}{EA}\,dx = \frac{Fl}{EA}$$

Using the force–elongation equation above, we obtain the elastic energy due to the axial force F as follows:

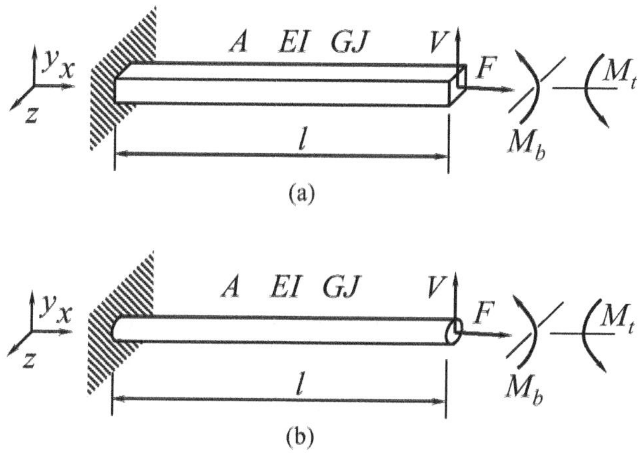

Figure 3.40 Slender beams under moments and forces.

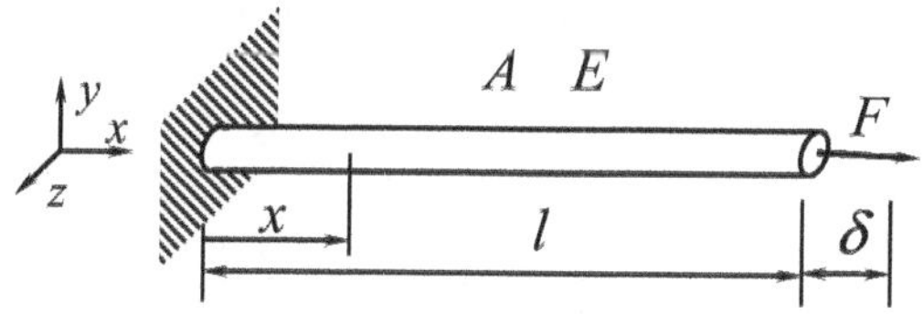

Figure 3.41 Slender beam subjected to an axial force.

$$U = \int_0^l F\,d\delta = \int_0^l \frac{EA}{l}\delta\,d\delta = \frac{EA}{2l}\delta^2 = \frac{F\delta}{2} \tag{3.129}$$

The preceding elastic energy can be rewritten as follows:

$$U = \frac{EA}{2l}\delta^2 = \frac{1}{2}\left(E\frac{\delta}{l}\right)\frac{\delta}{l}Al = \frac{1}{2}\sigma\varepsilon V$$

where σ, ε, and V denote the stress and strain in the beam and the volume of the beam, respectively. From the equation above, we define $\sigma\varepsilon/2$ as the elastic energy density (energy/unit volume) due to the normal strain. Using the preceding definition, we express the elastic energy (3.129) in a different form,

$$U = \int_V \frac{1}{2}\sigma\varepsilon\,dV \tag{3.130}$$

If the stress and strain in a elastic body are known, equation (3.130) can be used to obtain the elastic energy U.

When the circular beam in the upper figure of Fig. 3.42 carries a bending moment M, the bending moment in the beam is generated to maintain the

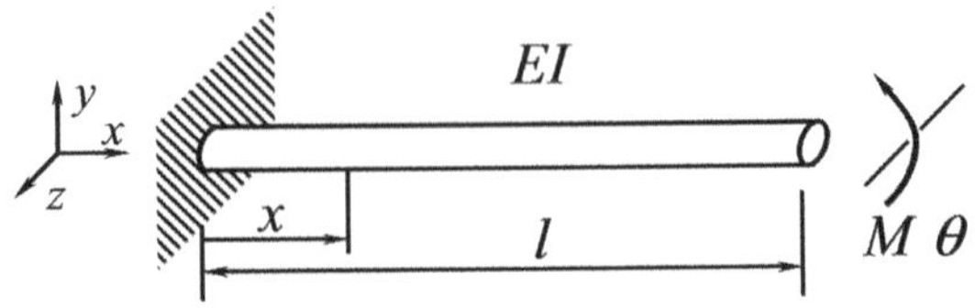

Figure 3.42 Slender beam under a pure bending moment.

beam equilibrium. The angle θ due to the moment varies along the beam length. To obtain the elastic energy due to the pure bending moment at x, we use the moment–curvature relationship (3.40):

$$EI\frac{d^2y}{dx^2} = EI\frac{d\theta}{dx} = M$$

where $\theta = dy/dx$ is the angle at x. Integration of the preceding equation gives the angular displacement $\theta = Ml/EI$ at the point of application of the moment M. Since the bending moment at x in Fig. 3.42 is the same as that applied at the end of the beam, the rate of angle change $d\theta/dx$ is a constant along the beam. If a moment about a point rotates by an angle, the energy done by the moment is defined as the product of the moment and the angle (in radians). We then write the energy done by the moment acting on the element at x as follows:

$$dU = Md\theta = M\left(\frac{M}{EI}\right)dx = \frac{M^2}{EI}dx$$

Integrating the preceding equation along the beam yields the elastic energy due to the bending moment:

$$U = \int dU = \int_0^\theta \frac{EI}{l}\theta\, d\theta = \frac{EI}{2l}\theta^2 = \frac{EI}{2l}\left(\frac{Ml}{EI}\right)^2 = \frac{M^2 l}{2EI}$$

Since the pure bending moment is constant along the beam, the elastic energy may be written

$$U = \int_0^l \frac{M^2}{2EI}dx \tag{3.131}$$

Equation (3.131) also holds when the moment varies along the beam. As discussed earlier, the bending moment in the beam is due to normal stress acting on the beam cross section. We then express the elastic energy (3.131) in terms of normal stress and strain. The elastic energy due to the normal stress is given by (3.130). Substituting the stress–strain relationship (3.3) into the elastic energy equation (3.130) and using the stress–moment equation (3.41), we again obtain the elastic energy (3.131) due to the bending moment:

$$\begin{aligned} U &= \int_V \frac{1}{2}\sigma_x \varepsilon_x \, dV = \int_V \frac{1}{2}\sigma_x \frac{\sigma_x}{E} dV = \int_V \frac{1}{2}\frac{\sigma_x^2}{E} dV \\ &= \int_V \frac{1}{2}\frac{1}{E}\left(-\frac{My}{I}\right)^2 dV = \int_x \frac{1}{2}\frac{1}{E}\frac{M^2}{I^2}\left(\int_A y^2 \, dA\right) dx \\ &= \int_x \frac{1}{2}\frac{1}{E}\frac{M^2}{I^2} I \, dx = \int_x \frac{M^2}{2EI} dx \end{aligned}$$

To obtain the preceding equation, the definition of the moment of inertia (3.39) is used.

The beam can carry a transverse shear force as shown in Fig. 3.40. If the shear force exists on the beam cross section, the bending moment varies along the length of the beam to maintain beam equilibrium. Even though the bending moment varies, the elastic energy (3.131) due to the bending moment still holds. However, the transverse force generates a shear stress that may vary along the beam length and with location on the cross section. Let us evaluate the elastic energy stored in the beam due to stress acting on the beam cross section. An element in Fig. 3.43 is subjected to a shear force F in the x direction. The elastic element is deformed as shown in the figure, according to the shear stress–strain relationship (3.7). Since the increment in the elastic energy due to the shear force in Fig. 3.43 is expressed as $dU = F dx$, we obtain the elastic energy due to the shear force as follows:

$$\begin{aligned} U &= \int dU = \int_x F dx = \int_0^\delta GA \frac{x}{l} dx = GA \frac{\delta^2}{2l} \\ &= \frac{1}{2} GA \frac{\delta}{l}\delta = \frac{1}{2}\tau A \delta = \frac{1}{2} F\delta \end{aligned}$$

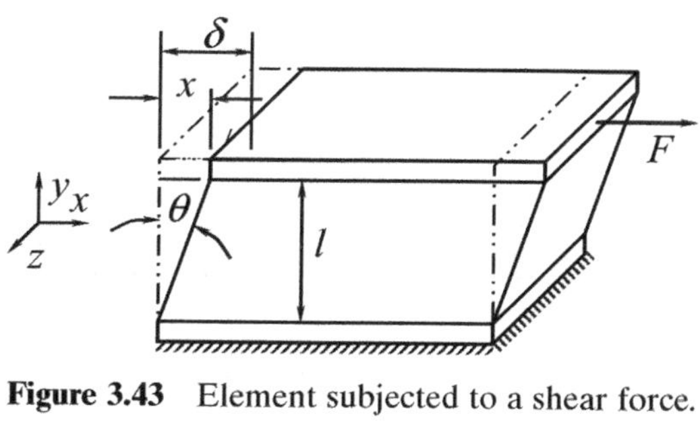

Figure 3.43 Element subjected to a shear force.

where G and A denote the shear modulus and the area of the top in Fig. 3.43. Using the shear stress–strain relationship (3.7),

$$\delta = \frac{F}{GA} l$$

the energy equation stored in the small element may be expressed in terms of the shear and strain as follows:

$$U = \frac{1}{2} F\delta = \frac{1}{2} F \frac{F}{GA} l = \frac{1}{2} \frac{F^2}{GA} l$$

If the stress varies in the beam, the preceding equation may be integrated and can be expressed in terms of the shear stress and strain as follows:

$$\begin{aligned} U &= \int_0^l \frac{1}{2} \frac{F^2}{GA} dx = \int_0^l \frac{1}{2} \frac{F}{A} \frac{F}{GA} A\, dx \\ &= \int_0^l \frac{1}{2} \frac{F}{A} \frac{F}{GA} A\, dx = \int_V \frac{1}{2} \tau\gamma\, dV \end{aligned} \tag{3.132}$$

A torsional moment M_t may be subjected to the slender beam in Fig. 3.44. Using the torsional moment–angular displacement relationship (3.30), we obtain the energy due to the torsional moment as follows:

$$dU = M_t\, d\theta = \frac{GJ}{l} \theta\, d\theta$$

Integrating the equation above along the beam, we find the elastic energy in the slender beam due to the torsional moment,

$$U = \int_0^\theta \frac{GJ}{l} \theta\, d\theta = \frac{GJ}{2l} \theta^2 = \frac{GJ}{2l} \left(\frac{M_t}{GJ} l \right)^2 = \frac{M_t^2}{2GJ} l$$

where $\theta = M_t l / GJ$ from (3.31) is used to obtain this equation. If the torsional moment M_t varies along the beam, the integration form of the foregoing equation may be used:

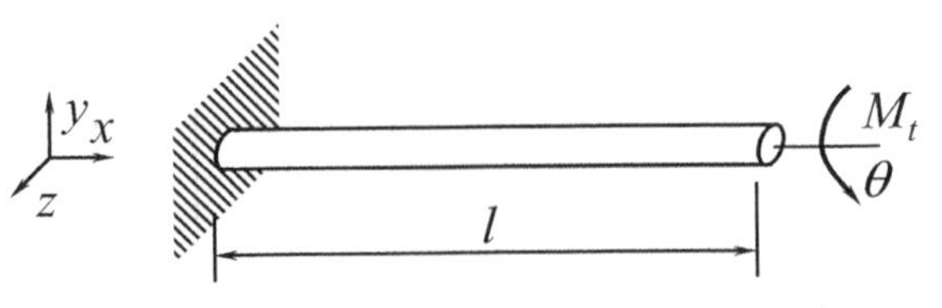

Figure 3.44 Slender beam subjected to a torsional moment.

$$U = \int_0^l \frac{M_t^2}{2GJ} dx \tag{3.133}$$

Note that since the torsional moment generates the shear stress and strain on the beam cross section, equation (3.133) may be obtained from (3.132) as follows:

$$U = \int_V \frac{1}{2} \tau_{x\theta} \gamma_{x\theta} \, dV = \int_V \frac{1}{2} \frac{(\tau_{x\theta})^2}{G} dV = \int_V \frac{1}{2} \frac{1}{G} \left(Gr \frac{d\theta}{dx} \right)^2 dV$$
$$= \int_0^l \frac{1}{2} G \left(\frac{d\theta}{dx} \right)^2 \left[\int_A r^2 \, dA \right] dx = \int_0^l \frac{1}{2} GJ \left(\frac{d\theta}{dx} \right)^2 dx = \int_0^l \frac{M_t^2}{2GJ} dx$$

where $\tau_{x\theta}$ and $\gamma_{x\theta}$ denote the shear stress and the shear strain acting on the beam cross section, defined by (3.28) and (3.29), respectively. In the preceding equation, the quantity in the brackets is defined as the polar moment of inertia, J. To obtain the last integral, the torsional moment–angular displacement relationship (3.30) is used. Energy equation (3.133) holds for a beam of an arbitrary cross section, including the rectangular cross section as well as the circular cross section.

Now we obtain the following elastic energy stored in a slender beam when the beam is subjected to the bending moment M, the torsional moment M_t, the axial force F, and the transverse shear force V:

$$U = \int_0^l \frac{M^2}{2EI} dx + \int_0^l \frac{M_t^2}{2GJ} dx + \int_0^l \frac{F^2}{2EA} dx + c \int_0^l \frac{V^2}{2GA} dx \tag{3.134}$$

where the last integral is the elastic energy due to the shear force under an assumption that the shear stress on the cross section perpendicular to the length of the beam is a constant, given by V/A, where A is the cross-sectional area of the beam. In the last term, c denotes a correction factor, whose order of magnitude is 1, used to give the elastic energy due to the shear stress. For example, the factor c for the rectangular beam in Fig. 3.40a is 3/5 since the shear stress is given by a parabolic distribution along the y direction in Fig. 3.40a; the shear stress distribution is not covered in this book. It should be noted that the bending moment M may be the pure bending moment or a moment due to the transverse shear force V as shown in Fig. 3.40. Since the shear force V generates the bending moment, we have two integrals affected by the shear force in equation (3.134). It is worth noting the orders of magnitude of the first integral (energy due to the bending moment) and the last integral. To compare the others to the magnitude of the first and last integrals, we define a dimensionless number as follows:

$$\Psi = \frac{\left[\int_0^l (V^2/2GA) dx \right]}{\left[\int_0^l (M^2/2EI) dx \right]} = \frac{(V^2/GA)l}{(M^2/EI)l} = \frac{EI}{GA} \frac{V^2}{M^2} = \frac{EI}{GA} \frac{V^2}{(Vl)^2} = \frac{EI}{GAl^2}$$

where the brackets represent the order of magnitude of a physical quantity in the brackets, and the length l is used as the characteristic length. To obtain this equation, the maximum moment Vl is used as the moment M. If we use the shear modulus–Young's modulus relationship $G = E/[2(1 + \nu)]$ for the isotropic beam material, and if we note that the moment of inertia can be expressed as $I = c_1At^2$, where c_1 is a constant (e.g., for the rectangular cross section $I = wt^3/12 = At^2/12$), the preceding dimensionless number can be expressed as

$$\Psi = 2(1+\nu)\frac{c_1At^2}{Al^2} = 2(1+\nu)c_1\left(\frac{t}{l}\right)^2 \tag{3.135}$$

The dimensionless number above represents the ratio of the elastic energy due to the shear stress, caused by a transverse shear force acting on the beam, to the elastic energy due to the bending moment caused by the shear force. Since the dimensionless number Ψ is proportional to the square of the ratio of the thickness t to the beam length l, the number Ψ for a slender beam (i.e., $t/l << 0$) is much smaller than unity and we can neglect the contribution of the shear stress to the elastic energy (3.134). Consequently, for a slender beam, the elastic energy (3.134) can be expressed as

$$U = \int_0^l \frac{M^2}{2\tilde{E}I}dx + \int_0^l \frac{M_t^2}{2GJ}dx + \int_0^l \frac{F^2}{2\tilde{E}A}dx \tag{3.136}$$

where Young's modulus E is replaced by the effective Young's modulus $\tilde{E}$ to include the wide beam. Furthermore, if a force generates the bending moment and the axial force (i.e., a force making an angle with the beam), the order of magnitude of the last integral in equation (3.136) is, in general, much less than that of the first integral. For a slender beam that is subjected to bending and torsional moments and in which the contribution of the axial force to the elastic energy is negligible, the elastic energy is given by

$$U = \int_0^l \frac{M^2}{2\tilde{E}I}dx + \int_0^l \frac{M_t^2}{2GJ}dx \tag{3.137}$$

The elastic energy in terms of stresses and strains (3.130) and (3.132) may be expressed as an integral with respect to three-dimensional stresses and strains. If an elastic body is under a general distribution of stress, the elastic energy due to the stress is, in general, expressed as

$$U = \frac{1}{2}\int_V (\sigma_x\varepsilon_x + \sigma_y\varepsilon_y + \sigma_z\varepsilon_z + \tau_{xy}\gamma_{xy} + \tau_{yz}\gamma_{yz} + \tau_{zx}\gamma_{zx})dV \tag{3.138}$$

where σ, ε, τ, and γ denote the normal stress and strain, and the shear stress and strain, respectively, and x, y, and z represent the coordinates in a rectangular coordinate system. The integral above is over the volume of the elastic body. Use of the energy method is described in detail in subsequent chapters.

3.9 ENERGY METHOD FOR BEAM PROBLEMS

In Section 3.7, the Galerkin method was used to determine the deflection of beams subjected to loads. In this section we derive an energy equation by applying Galerkin's method to the beam equation. In some cases, the energy method can be used to obtain beam deflections since it is simpler than the Galerkin method. Multiplying the beam equation (3.54) by the mode shape ψ_1 and integrating the result along the beam leads to

$$y_m \tilde{E}I \int_0^l \frac{d^4\psi_1}{dx^4}\psi_1\,dx - y_m N_0 \int_0^l \frac{d^2\psi_1}{dx^2}\psi_1\,dx - y_m^3 \frac{\tilde{E}A}{2l}\int_0^l \left(\frac{d\psi_1}{dx}\right)^2 dx \int_0^l \frac{d^2\psi_1}{dx^2}\psi_1\,dx = \int_0^l q\psi_1\,dx \tag{3.139}$$

where y_m is the characteristic deflection, defined by the approximate solution. The first integral of equation (3.139) may be expressed as

$$\int_0^l \frac{d^4\psi_1}{dx^4}\psi_1\,dx = \frac{d^3\psi_1}{dx^3}\psi_1\bigg|_0^l - \frac{d^2\psi_1}{dx^2}\frac{d\psi_1}{dx}\bigg|_0^l + \int_0^l \frac{d^2\psi_1}{dx^2}\frac{d^2\psi_1}{dx^2}dx$$

and second integral is given by

$$\int_0^l \frac{d^2\psi_1}{dx^2}\psi_1\,dx = \frac{d\psi_1}{dx}\psi_1\bigg|_0^l - \int_0^l \frac{d\psi_1}{dx}\frac{d\psi_1}{dx}dx$$

Substituting the preceding two equations into (3.139) and recalling the relationships

$$\tilde{E}I\frac{d^2y}{dx^2} = M$$

$$\frac{dM}{dx} = \tilde{E}I\frac{d^3y}{dx^3} = -V$$

$$y_1 = y_m\psi_1$$

we find a force equation that represents the force equilibrium of the beam at the mode shape y_1 defined by the equation above; for example, for the

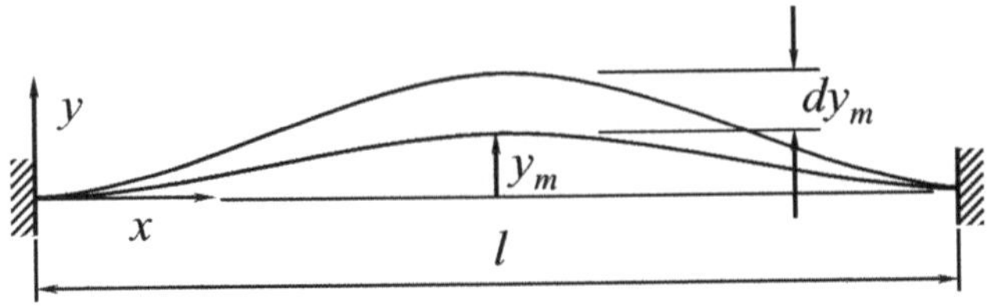

Figure 3.45 Beam deflection.

fixed–fixed beam in equilibrium, the beam deflection is shown in the lower curve in Fig. 3.45.

$$
\begin{aligned}
&-V_l\psi_{1,l}+V_0\psi_{1,0}-M_l\left(\frac{d\psi_1}{dx}\right)_l+M_0\left(\frac{d\psi_1}{dx}\right)_0+\tilde{E}Iy_m\int_0^l\left(\frac{d^2\psi_1}{dx^2}\right)^2dx\\
&-N_0\left[\left(\frac{dy_1}{dx}\right)_l\psi_{1,l}-\left(\frac{dy_1}{dx}\right)_0\psi_{1,0}\right]+N_0y_m\int_0^l\left(\frac{d\psi_1}{dx}\right)^2dx\\
&-\frac{1}{2}\frac{EA}{l}\int_0^l\left(\frac{dy_1}{dx}\right)^2dx\left[\left(\frac{dy_1}{dx}\right)_l\psi_{1,l}-\left(\frac{dy_1}{dx}\right)_0\psi_{1,0}\right]\\
&-\frac{1}{2}\frac{EA}{l}y_m^3\left[\int_0^l\left(\frac{d\psi_1}{dx}\right)^2dx\right]^2=\int_0^l q\psi_1\,dx
\end{aligned}
$$

where the subscripts 0 and l denote $x = 0$ and $x = l$. If we multiply both sides of the preceding equation by dy_m, as shown in Fig. 3.45, and integrate the resulting equation from 0 to y_m, we obtain the following energy equation, which represents the elastic energy stored in the beam and the energy supplied to the beam during the deflection:

$$
\frac{1}{2}\tilde{E}I\int_0^l\left(\frac{d^2y_1}{dx^2}\right)^2dx+\frac{1}{2}N_0\int_0^l\left(\frac{dy_1}{dx}\right)^2dx+\frac{1}{8}\frac{EA}{l}\left[\int_0^l\left(\frac{dy_1}{dx}\right)^2dx\right]^2=\int_0^l qy_1\,dx+R \tag{3.140}
$$

where the first, second, and third terms on the left-hand side denote the elastic energies due to the bending moment, the axial force, and the geometric nonlinearity. The first term on the right-hand side represents the energy supplied by the distributed load, and the second term, R, is the energy supplied by the forces and moments at the boundaries, given by

$$
\begin{aligned}
R=&V_ly_{1,l}-V_0y_{1,0}+M_l\theta_l-M_0\theta_0\\
&+\left[N_0+\frac{1}{2}\frac{EA}{l}\int_0^l\left(\frac{dy_1}{dx}\right)^2dx\right](\theta_ly_{1,l}-\theta_0y_{1,0})
\end{aligned} \tag{3.141}
$$

Using the definition of the characteristic deflection y_m and the mode shape $y_1 = y_m\eta\psi$, the energy equation for beams may be expressed as

$$k_1\frac{y_m^2}{2}+k_3\frac{y_m^4}{4}=f_q y_m+R \tag{3.142}$$

where

$$k_1=\eta^2\frac{\tilde{E}I}{l^3}\int_0^1\left(\frac{d^2\psi}{dX^2}\right)^2 dX+\eta^2\frac{N_0}{l}\int_0^1\left(\frac{d\psi}{dX}\right)^2 dX \tag{3.143}$$

$$k_3=\frac{1}{2}\eta^4\frac{\tilde{E}A}{l^3}\left[\int_0^1\left(\frac{d\psi}{dX}\right)^2 dX\right]^2 \tag{3.144}$$

$$f_q=\eta q_m l\int_0^l \psi_q\psi\, dx \tag{3.145}$$

$$\begin{aligned} R=&V_l y_{1,l}-V_0 y_{1,0}+M_l\theta_l-M_0\theta_0\\ &+\left[N_0+\frac{1}{2}\eta^2\frac{EA}{l}y_m^2\int_0^1\left(\frac{d\psi}{dX}\right)^2 dX\right](\theta_l y_{1,l}-\theta_0 y_{1,0}) \end{aligned} \tag{3.146}$$

where q_m is the characteristic load that is defined by $q = q_m\psi_q$, where ψ_q is the dimensionless distribution function of the distributed load along the beam. Note that the first and second terms on the left-hand side of (3.142) represent the stored elastic energies due to the linear and cubic stiffness of the beam. Using the energy equation (3.142), we can obtain the approximate solution to the nonlinear beam problems discussed earlier in the chapter.

For example, for the beam problem (Example 3.3 and Fig. 3.30), we use the mode shape $\psi = 3X^2 - 2X^3$, the constant $\eta = 1$, and $V_l = F$, $y_{1,0} = 0$, $\theta_l = \theta_0 = 0$, $q_m = 0$. With $y_m = y_{1,l}$, we find that

$$f_q=\eta q_m l\int_0^l \psi_q\psi\, dx=0$$

$$R=V_l y_{1,l}=F y_m$$

The energy equation is given by

$$k_1\frac{y_m^2}{2}+k_3\frac{y_m^4}{4}=F y_m$$

Differentiating this equation with respect to y_m, we have the force equilibrium for the beam:

$$k_1 y_m+k_3 y_m^3=F \tag{3.147}$$

Equation (3.147) is the same as (k) in Example 3.3.

For a fixed–fixed beam (Fig. 3.34) subjected to a uniformly distributed load q, the term R of (3.141) becomes zero. Differentiating the energy equation (3.142) with respect to y_m then gives the following force equilibrium, which is the same as that obtained by Galerkin's method (3.89):

$$k_1 y_m + k_3 y_m^3 = f_q \tag{3.148}$$

Equations (3.143)–(3.145) are the same as (3.97)–(3.99). Since a cantilever beam under a uniformly distributed load q gives $R = 0$, equation (3.148) as well as (3.143)–(3.145) hold. However, since the axial force of the cantilever beam vanishes, the second term in the linear stiffness (3.143) also vanishes, and furthermore, the cubic stiffness becomes zero. Consequently, the energy method derived from Galerkin's method can be used to obtain analytic solutions to linear and nonlinear beam problems.

PROBLEMS

3.1 A cantilever beam of flexural rigidity EI and length l is under a uniformly distributed load q (force per unit length). Find the beam deflection as a function of the coordinate x from the fixed end.

3.2 A uniformly distributed load q (force per unit length) is exerted on a simply supported beam whose left end is placed on an edge and whose right end is on another edge riding a roller (Fig. P3.2). As a result, the beam does not generate moments at both ends or any axial force. Find the beam deflection at x from the left end.

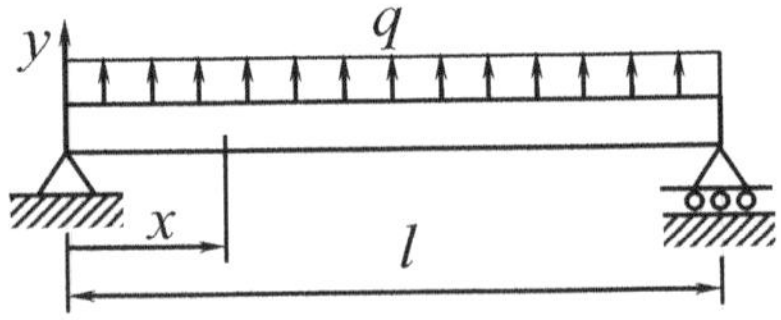

Figure P3.2

3.3 If the temperature in a fixed–fixed beam (Fig. 3.18a) of Young's modulus E, cross-sectional area A, length l, and the coefficient of thermal expansion α is given by $T - T_0 = T_{\max} \sin(\pi x/l)$, where $T_{\max}$ is a constant, x is the coordinate from the left end of the beam, and $T - T_0$ is the temperature difference. Determine the thermal stress in the beam.

3.4 Two springs of stiffness k_1 and k_2 are hung from a ceiling (Fig. P3.4). When a horizontal bar connecting the springs moves perpendicular to the ceiling, find the effective stiffness as a function of the spring's stiffness.

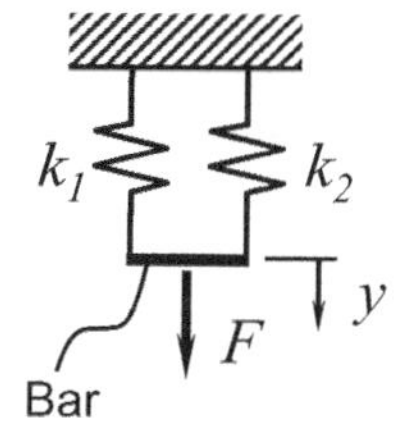

Figure P3.4

3.5 A beam of flexural modulus EI, initial force N_0, and length l is subjected to a force F (Fig. P3.5). If the cross section and dimensions are as defined in the figure, find (a) the location of the neutral surface from the bottom of the cross section, (b) the moment of inertia about the neutral surface, and (c) the stiffness of the beam. If $\tan\theta = \sqrt{2}$ is given, simplify the moment of inertia that may be used to design the bulk-micromachined beam.

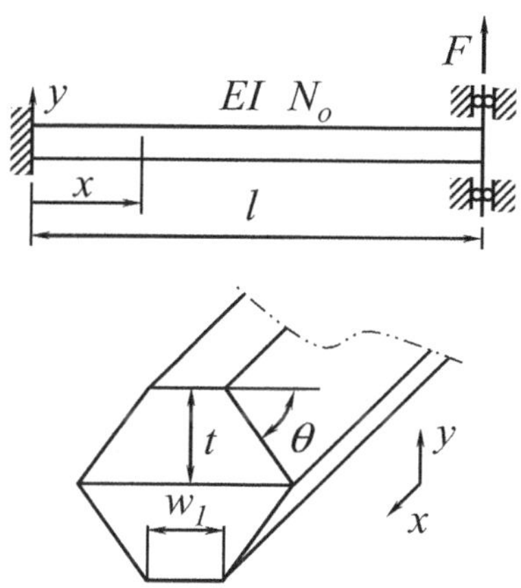

Figure P3.5

CHAPTER 4

STATIC BEHAVIOR OF MICROSTRUCTURES

In Chapter 3 we considered the basic equations for static equilibrium and described approaches to solutions for static behavior. Based on the equations and solutions that were derived, static behaviors of microstructures are described in detail in this chapter. The static behavior of elements of microstructures is dealt with first, then that of microstructures consisting of the elements we described. Afterward, nonlinear behavior of microstructures is covered.

4.1 ELEMENTS OF MICROSTRUCTURES

In the early stage of the development of MEMS, some researchers were interested in rotating machines that were similar to large-scale engines or electric motors. However, they could not develop an efficient rotating machine since the friction between the microstructures plays a critical role. In other words, the input energy is dissipated into heat due to the friction. This phenomenon may be verified by using dimensional analysis, which was covered in Chapter 1; the frictional force is proportional to the contact area between microstructures, and the frictional force in microscale machines becomes dominant. A similar argument holds for the viscous (damping) force due to the viscosity of gas surrounding the microstructures, since the viscous force is also proportional to the area exposed to the surrounding gas. For this reason, most MEMS

Principles of Microelectromechanical Systems, By Ki Bang Lee

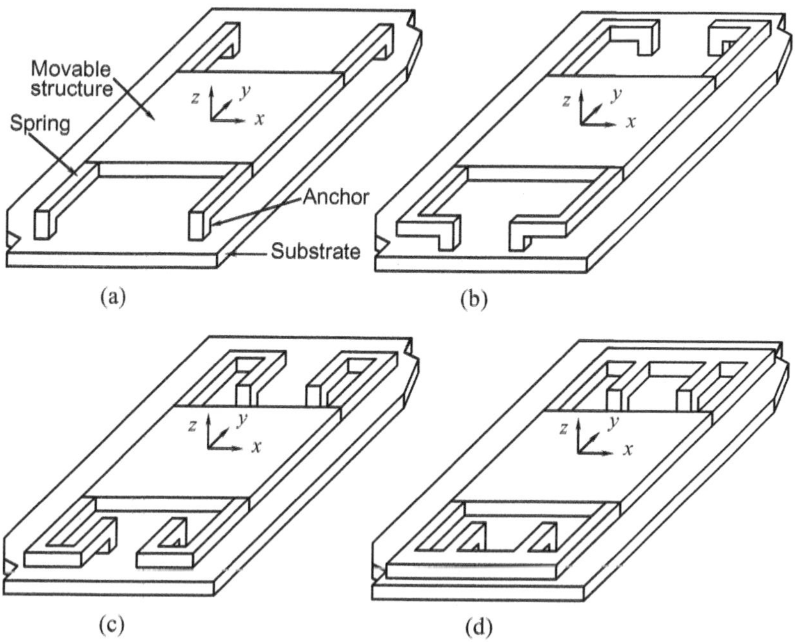

Figure 4.1 Typical microstructures.

consist of movable structures whose ends are anchored on a substrate and whose portions are flexible for springlike elements or relatively rigid for mass-like elements, depending on purpose. If needed, stationary structures may be placed on or near the substrate to interact with the movable structures. With this configuration, more efficient and reliable MEMS have been achieved.

Typical configurations of microstructures widely used in MEMS are illustrated in Fig. 4.1. Movable structures are suspended over a substrate by springs (flexures) one end of which is anchored on a substrate. The springs may be designed to allow motion of the movable structure in specific directions or to impose constraints on the movable structures in Fig. 4.1. It is noted in the figure that all suspended microstructures are symmetric about the center of the structures (i.e., about x and y axes). These symmetric configurations greatly simplify the design of the microstructures that will be detailed in subsequent sections of this chapter. However, if an asymmetric configuration is required for a specific purpose, a asymmetric movable plate or asymmetric configurations of springs may be used.

Spring Connections

To achieve a specific purpose, several springs can be used, as shown in Fig. 4.1, and the stiffness of the springs may be expressed in simple forms if the

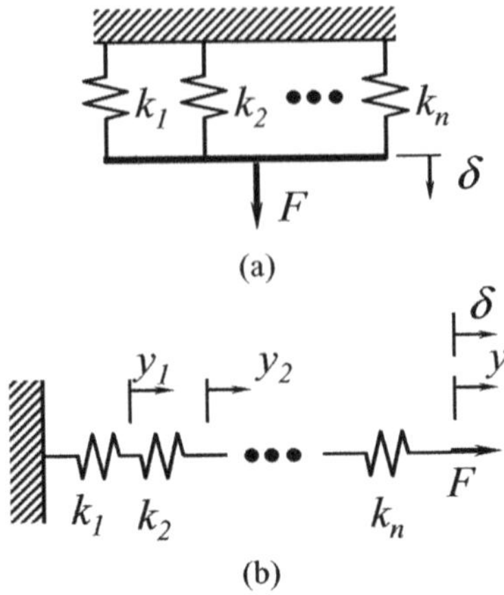

Figure 4.2 Spring connections.

movable structures move without angular displacement. To obtain the effective stiffness of connected springs, we use the force equilibrium in springs, the displacement of springs, and relations between the spring force and the displacement. Let us consider springs in parallel, as shown in Fig. 4.2a. The stiffness of the springs are denoted by $k_1, k_2, \ldots, k_n$. If all the springs are connected to a bar, and if the bar moves downward without angular displacement, the elongations of the springs are equal and the geometric relation may be expressed as follows:

$$\delta = y_1 = y_2 = \cdots = y_n$$

where $y_1, y_2, \ldots, y_n$ denote the displacement of the springs. From a free-body diagram of the bar, the force equilibrium equation is given by

$$F = F_1 + F_2 + \cdots + F_n$$

where $F_1, F_2, \ldots, F_n$ represent the spring forces. Hooke's law gives the relation between the force acting on the springs and the displacement:

$$F_1 = k_1 y_1, \quad F_2 = k_2 y_2, \quad \ldots, \quad F_n = k_n y_n$$

Substituting the preceding equation into the force equilibrium equation and recalling the geometric relation among the spring displacements gives the spring force due to the parallel springs:

$$F = k_{\text{eff}} \delta$$

where k_{eff} is known as the *effective stiffness* or the *equivalent spring constant.* The effective stiffness of the springs in parallel is then expressed as

$$k_{\text{eff}} = k_1 + k_2 + \cdots + k_n \tag{4.1}$$

Note from equation (4.1) that the effective stiffness of parallel springs is expressed as the sum of the stiffness of the springs. As a result, the effective stiffness of parallel springs increases as more springs are connected in parallel.

Springs may be connected in series as shown in Fig. 4.2b, where the displacement δ of the last spring is expressed as the sum of the relative displacements $(y_1, y_2, \ldots, y_n)$ of the springs. The geometric relation among the springs in series is then given by

$$\delta = y_1 + y_2 + \cdots + y_n$$

and Hooke's law gives the spring force due to the relative displacement as

$$F_1 = k_1 y_1, \quad F_2 = k_2 y_2, \quad \ldots, \quad F_n = k_n y_n$$

When we draw free-body diagrams of springs, we find that all springs are subjected to the constant force F. The force equilibrium equations are then given by

$$F = F_1 = F_2 = \cdots = F_n$$

Substituting the force–displacement relationship into the geometric relation from the springs and using the preceding equation, we obtain the force–displacement relation of the springs in series as follows:

$$F = k_{\text{eff}} \delta$$

Thus, the effective stiffness of the springs in series is given by

$$\frac{1}{k_{\text{eff}}} = \frac{1}{k_1} + \frac{1}{k_2} + \cdots + \frac{1}{k_n} \tag{4.2}$$

As expressed in equation (4.2), the effective stiffness of springs in series is less than the lowest stiffness of the springs. The effective stiffness of springs in parallel or in series is valid for simple structures such as the symmetric structures in Fig. 4.1, where the movable structures move in the x–y plane or along the z direction. However, in general cases where the movable structures move linearly and angularly, the preceding equations for the effective stiffness in the directions are not valid for stiffness estimation, and we need to develop more sophisticated methods by which the stiffness of structures supported by arbitrarily shaped springs can be evaluated.

4.2 STIFFNESS OF COMMONLY USED BEAMS

The typical microstructures shown in Fig. 4.1 are planar structures that are usually fabricated by the fabrication process described in Chapter 2. The widely used springs based on the planar structures are classified into four groups: the straight beam (Fig. 4.1a), the L-shaped beam (Fig. 4.1b), the folded beam (Fig. 4.1c), and the connected folded beam (Fig. 4.1d), which consists of two folded beams connected to each other.

Stiffness of a Straight Beam

As a spring, the straight beam in Fig. 4.1a gives low stiffness in the direction perpendicular to the longitudinal axis of the beam (i.e., the x and z directions in the figure), but it provides relatively high stiffness in its longitudinal direction (i.e., the y direction). This difference in the stiffness in the orthogonal directions is due to the different causes of the stiffness (i.e., bending moment for stiffness in the x direction, axial force for stiffness in the y direction). To obtain the stiffness of a straight beam, let us consider the cantilever beam in Fig. 4.3a, one end of which is fixed at a support, with the other end subjected to three forces in three orthogonal directions and three moments in three angular directions. It is noted that since the free end of the beam has six degrees of freedom, three orthogonal directions and three angular directions are required to describe the state of the displacements of the free end. For consistency with the typical configuration of microstructures shown in Fig. 4.1, the y axis coincides with the longitudinal axis of the straight beam and two other axes are taken as shown in Fig. 4.3a: the horizontal axis as the x axis and the vertical axis as the z axis. Let the length, width, and thickness of the beam be denoted by l, w, and t, respectively. The displacements in the x, y, and z directions are represented by δ_x, δ_y, and δ_z, respectively; the forces corresponding to the directions are F_x, F_y, and F_z, respectively. Similarly, the angles (angular displacements) about the x, y, and z axes is defined as ϕ, η, and θ, respectively, shown in Fig. 4.3a, and the corresponding moments are given as M_ϕ, T, and M_θ, respectively. These definitions of the displacements, angles, forces, and moments for the stiffness evaluation will be used in the remaining sections of this chapter.

A free-body diagram of the top view of a beam segment is drawn in Fig. 4.3b, and another free-body diagram of the side view of the beam segment is shown in Fig. 4.3c. In the free-body diagram, ζ, F_ζ, V_ζ, M_ζ, and T represent the dummy variables that denote the segment length, the axial and shear forces, and the bending and torsional moments at ζ, respectively. From the force and moment equilibrium in Fig. 4.3b, the shear and axial forces at the cut face are given by

$$V_\zeta = F_x \qquad F_\zeta = F_y$$

and the bending and torsional moments are obtained as

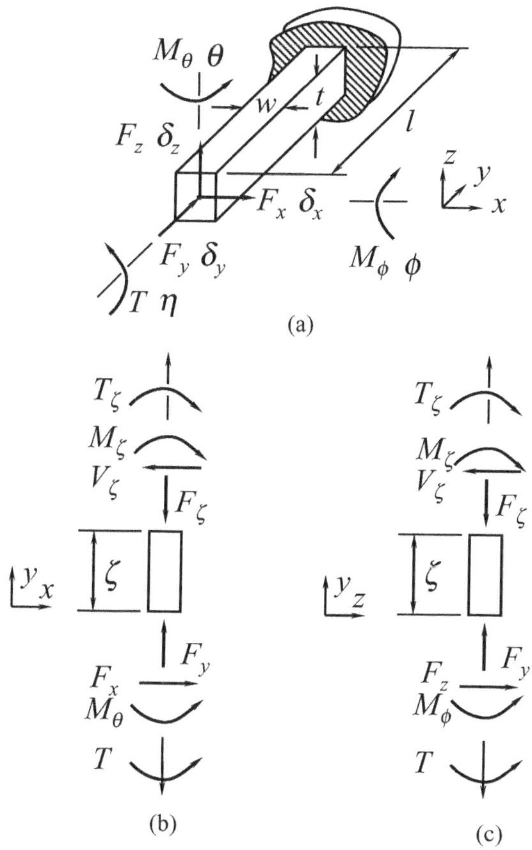

Figure 4.3 Straight beam.

$$M_\xi = F_x\zeta + M_\theta \qquad T_\zeta = T$$

Substituting the preceding forces and moments into the elastic energy equation for slender beams, (3.136), we have the energy stored in the straight beam as follows:

$$\begin{aligned} U &= \int_0^l \frac{M_\zeta^2}{2\tilde{E}I_z} d\zeta + \int_0^l \frac{T_\zeta^2}{2GJ} d\zeta + \int_0^l \frac{F_\zeta^2}{2\tilde{E}A} d\zeta \\ &= \int_0^l \frac{(F_x\zeta + M_\theta)^2}{2\tilde{E}I_z} d\zeta + \int_0^l \frac{T^2}{2GJ} d\zeta + \int_0^l \frac{F_y^2}{2\tilde{E}A} d\zeta \end{aligned}$$

where $\tilde{E}$ is the effective Young's modulus defined in (3.5) and (3.6), and G is the shear modulus, (3.15). For the rectangular cross section in Fig. 4.3a, I_z, J, and A are the moment of inertia, defined as $tw^3/12$; the polar moment of inertia, given by (3.33) and (3.34); and the cross-sectional area of the beam, wt, respectively. Applying Castigliano's theorem, (3.125) and (3.126), to the energy equation above, we obtain the displacements δ_x and δ_y and the angles θ and η at a distance ζ from the free end of the beam:

$$\begin{aligned}
\delta_x &= \frac{\partial U}{\partial F_x} = \int_0^l \frac{F_x\zeta^2 + M_\theta\zeta}{\tilde{E}I_z} d\zeta = \frac{F_x l^3}{3\tilde{E}I_z} + \frac{M_\theta l^2}{2\tilde{E}I_z} \\
\delta_y &= \frac{\partial U}{\partial F_y} = \int_0^l \frac{F_y}{\tilde{E}A} d\zeta = \frac{F_y l}{\tilde{E}A} \\
\theta &= \frac{\partial U}{\partial M_\theta} = \int_0^l \frac{F_x\zeta + M_\theta}{\tilde{E}I_z} d\zeta = \frac{F_x l^2}{2\tilde{E}I_z} + \frac{M_\theta l}{\tilde{E}I_z} \\
\eta &= \frac{\partial U}{\partial T} = \int_0^l \frac{T}{GJ} d\zeta = \frac{Tl}{GJ}
\end{aligned} \tag{4.3}$$

Referring to the free body of Fig. 4.3c, we find the forces and moments in the y–z plane as follows:

$$V_\zeta = F_z \qquad F_\zeta = F_y \qquad M_\xi = F_z\zeta + M_\phi \qquad T_\zeta = T$$

The energy stored in the slender beam due to the forces and moments above is expressed as

$$U = \int_0^l \frac{(F_z\zeta + M_\phi)^2}{2\tilde{E}I_x} d\zeta + \int_0^l \frac{T^2}{2GJ} d\zeta + \int_0^l \frac{F_y^2}{2\tilde{E}A} d\zeta$$

where I_x is the moment of inertia, given by $wt^3/12$. Following the procedure cited above, we find that

$$\begin{aligned}
\delta_z &= \frac{\partial U}{\partial F_z} = \int_0^l \frac{F_z\zeta^2 + M_\phi\zeta}{\tilde{E}I_x} d\zeta = \frac{F_z l^3}{3\tilde{E}I_x} + \frac{M_\phi l^2}{2\tilde{E}I_x} \\
\delta_y &= \frac{\partial U}{\partial F_y} = \int_0^l \frac{F_y}{\tilde{E}A} d\zeta = \frac{F_y l}{\tilde{E}A} \\
\phi &= \frac{\partial U}{\partial M_\phi} = \int_0^l \frac{F_z\zeta + M_\phi}{\tilde{E}I_x} d\zeta = \frac{F_z l^2}{2\tilde{E}I_x} + \frac{M_\phi l}{\tilde{E}I_x} \\
\eta &= \frac{\partial U}{\partial T} = \int_0^l \frac{T}{GJ} d\zeta = \frac{Tl}{GJ}
\end{aligned} \tag{4.4}$$

Note that (4.3) and (4.4) give eight equations, which include two identical equations for δ_y and η. As a result, we can take the six independent equations from (4.3) and (4.4). To construct a compliance matrix which represents the linear and angular displacements as the forces and moments are applied to the beam, the first three equations of (4.3) are selected and expressed in matrix form as follows:

$$\begin{Bmatrix} \delta_x \\ \delta_y \\ \theta \end{Bmatrix} = \begin{bmatrix} \dfrac{l^3}{3\tilde{E}I_z} & 0 & \dfrac{l^2}{2\tilde{E}I_z} \\ 0 & \dfrac{l}{\tilde{E}A} & 0 \\ \dfrac{l^2}{2\tilde{E}I_z} & 0 & \dfrac{l}{\tilde{E}I_z} \end{bmatrix} \begin{Bmatrix} F_x \\ F_y \\ M_\theta \end{Bmatrix} \tag{4.5}$$

or in compact form,

$$\boldsymbol{\delta}_1 = \mathbf{C}_1 \mathbf{F}_1 \tag{4.6}$$

where

$$\boldsymbol{\delta}_1 = \{\delta_x \quad \delta_y \quad \theta\}^{\mathrm{T}} \tag{4.7}$$

$$\mathbf{F}_1 = \{F_x \quad F_y \quad M_\theta\}^{\mathrm{T}} \tag{4.8}$$

$$\mathbf{C}_1 = \begin{bmatrix} \dfrac{l^3}{3\tilde{E}I_z} & 0 & \dfrac{l^2}{2\tilde{E}I_z} \\ 0 & \dfrac{l}{\tilde{E}A} & 0 \\ \dfrac{l^2}{2\tilde{E}I_z} & 0 & \dfrac{l}{\tilde{E}I_z} \end{bmatrix} \tag{4.9}$$

denote a generalized displacement vector containing the linear and angular displacements, a generalized force vector that includes the force and moment applied at the free end of the beam, and a compliance matrix that represents the static response to $\mathbf{F}_1$, respectively. The superscript T in (4.7) and (4.8) denotes the transpose of a vector. In general cases, the compliance matrix of planar structures may be expressed as

$$\mathbf{C}_1 = \begin{bmatrix} c_{xx} & c_{xy} & c_{x\theta} \\ c_{yx} & c_{yy} & c_{y\theta} \\ c_{\theta x} & c_{\theta y} & c_{\theta\theta} \end{bmatrix} \tag{4.10}$$

where $c_{xx}, c_{xy}, \ldots, c_{\theta\theta}$ are the components of the compliance matrix and represent displacement corresponding to the forces and moment. For example, c_{xx} denotes displacement in the x direction due to the force F_x and c_{xy} represents the displacement increase due to the force F_y. We can obtain the force vector

$\mathbf{F}_1$ if a displacement vector is given. Multiplying (4.5) by the inverse matrix C^{-1}, we obtain the following force $\mathbf{F}_1$ to maintain the displacement:

$$\begin{Bmatrix} F_x \\ F_y \\ M_\theta \end{Bmatrix} = \begin{bmatrix} \dfrac{l^3}{3\tilde{E}I_z} & 0 & \dfrac{l^2}{2\tilde{E}I_z} \\ 0 & \dfrac{l}{\tilde{E}A} & 0 \\ \dfrac{l^2}{2\tilde{E}I_z} & 0 & \dfrac{l}{\tilde{E}I_z} \end{bmatrix}^{-1} \begin{Bmatrix} \delta_x \\ \delta_y \\ \theta \end{Bmatrix} \tag{4.11}$$

or in compact form,

$$\mathbf{F}_1 = \mathbf{C}_1^{-1}\boldsymbol{\delta}_1 = \mathbf{K}_1\boldsymbol{\delta}_1 \tag{4.12}$$

where

$$\mathbf{K}_1 = \mathbf{C}_1^{-1} \tag{4.13}$$

is defined as the stiffness matrix. Referring to the Appendix for the inverse matrix calculation, we find that

$$\mathbf{K}_1 = \begin{bmatrix} \dfrac{12\tilde{E}I_z}{l^3} & 0 & -\dfrac{6\tilde{E}I_z}{l^2} \\ 0 & \dfrac{\tilde{E}A}{l} & 0 \\ -\dfrac{6\tilde{E}I_z}{l^2} & 0 & \dfrac{4\tilde{E}I_z}{l} \end{bmatrix} \tag{4.14}$$

In general cases, the stiffness matrix of the planar springs may be expressed as follows:

$$\mathbf{K}_1 = \begin{bmatrix} k_{xx} & k_{xy} & k_{x\theta} \\ k_{yx} & k_{yy} & k_{y\theta} \\ k_{\theta x} & k_{\theta y} & k_{\theta\theta} \end{bmatrix} \tag{4.15}$$

It should be noted from (4.9) that the compliance matrix $\mathbf{C}_1$ is symmetric and that the stiffness matrix $\mathbf{K}_1$, the inverse of the symmetric compliance matrix, is also symmetric. This symmetric relation of compliance matrix, known as the *reciprocal theorem*, holds for linear elastic materials. If a geometric constraint is imposed on the free end of the beam in Fig. 4.3a, the corresponding geometric nonlinearity is involved in the stiffness. For example, if the free end of the straight beam in Fig. 4.3a is attached to the movable structure as shown in Fig. 4.1a and a force is applied in the x direction, the restoring force generated in the beam in the x direction is evaluated as $F_x = k_{xx}\delta_x$ from (4.12) and (4.15), where δ_x is the displacement of the movable structure in the x direction. Under this geometric constraint, the restoring force F_x is valid for the straight beam of zero residual stress, satisfying the condition $\delta_x << 1.1785w$, which is given by (s) in Example 3.3 in Section 3.7. If the displacement δ_x is comparable to or greater than the width w, the geometric nonlinearity enters into the stiffness that becomes nonlinear. This type of nonlinear stiffness was described in Example 3.3 and will be dealt with further in Section 4.7. Thus, the stiffness (4.14) of the straight beam is valid if the geometric nonlinearity can be neglected. Similarly, the three independent equations of (4.3) may be written in matrix form as follows:

$$\begin{Bmatrix} \delta_z \\ \phi \\ \eta \end{Bmatrix} = \begin{bmatrix} \dfrac{l^3}{3\tilde{E}I_x} & \dfrac{l^2}{2\tilde{E}I_x} & 0 \\ \dfrac{l^2}{2\tilde{E}I_x} & \dfrac{l}{\tilde{E}I_x} & 0 \\ 0 & 0 & \dfrac{l}{GJ} \end{bmatrix} \begin{Bmatrix} F_z \\ M_\phi \\ T \end{Bmatrix} \tag{4.16}$$

or in compact form,

$$\boldsymbol{\delta}_2 = \mathbf{C}_2\mathbf{F}_2 \tag{4.17}$$

where

$$\boldsymbol{\delta}_2 = \{\delta_z \quad \phi \quad \eta\}^{\mathrm{T}} \tag{4.18}$$

$$\mathbf{F}_2 = \{F_z \quad M_\phi \quad T\}^{\mathrm{T}} \tag{4.19}$$

denote a generalized displacement vector and a generalized force vector. The compliance matrix is then defined as

$$\mathbf{C}_2 = \begin{bmatrix} \dfrac{l^3}{3\tilde{E}I_x} & \dfrac{l^2}{2\tilde{E}I_x} & 0 \\ \dfrac{l^2}{2\tilde{E}I_x} & \dfrac{l}{\tilde{E}I_x} & 0 \\ 0 & 0 & \dfrac{l}{GJ} \end{bmatrix} \tag{4.20}$$

The force vector $\mathbf{F}_2$ is obtained from (4.17) as follows:

$$\mathbf{F}_2 = \mathbf{C}_2^{-1}\boldsymbol{\delta}_2 = \mathbf{K}_2\boldsymbol{\delta}_2 \tag{4.21}$$

or is expressed in a different form,

$$\begin{Bmatrix} F_z \\ M_\phi \\ T \end{Bmatrix} = \mathbf{K}_2 \begin{Bmatrix} \delta_z \\ \phi \\ \eta \end{Bmatrix} \tag{4.22}$$

where $\mathbf{K}_2 = \mathbf{C}_2^{-1}$ is the stiffness, which is defined as

$$\mathbf{K}_2 = \begin{bmatrix} \dfrac{12\tilde{E}I_x}{l^3} & -\dfrac{6\tilde{E}I_x}{l^2} & 0 \\ -\dfrac{6\tilde{E}I_x}{l^2} & \dfrac{4\tilde{E}I_x}{l} & 0 \\ 0 & 0 & \dfrac{GJ}{l} \end{bmatrix} \tag{4.23}$$

In general, cases such as the microstructures in Fig. 4.1, the compliance matrix $\mathbf{C}_2$, and the stiffness matrix $\mathbf{K}_2$ can be expressed as

$$\mathbf{C}_2 = \begin{bmatrix} c_{zz} & c_{z\phi} & c_{z\eta} \\ c_{\phi z} & c_{\phi\phi} & c_{\phi\eta} \\ c_{\eta z} & c_{\eta\phi} & c_{\eta\eta} \end{bmatrix} \tag{4.24}$$

$$\mathbf{K}_2 = \begin{bmatrix} k_{zz} & k_{z\phi} & k_{z\eta} \\ k_{\phi z} & k_{\phi\phi} & k_{\phi\eta} \\ k_{\eta z} & k_{\eta\phi} & k_{\eta\eta} \end{bmatrix} \tag{4.25}$$

As shown in the compliance and stiffness matrices (4.20) and (4.23) for the straight beam, the compliance and stiffness matrices, (4.24) and (4.25), for general planar springs are also symmetric. In the preceding equations, the compliance and stiffness matrices are expressed as a function of the effective Young's modulus $\tilde{E}$, defined in (3.5) for narrow beams and in (3.6) for wide beams. In the remaining portion of this chapter, we use E instead of $\tilde{E}$ for notation convenience, so E is automatically understood to be $\tilde{E}$ unless the contrary is stated explicitly.

Stiffness of Folded Beams

The straight beam described above can be folded or bent in order to reduce the geometric nonlinearity and achieve lower stiffness. Typical springs using

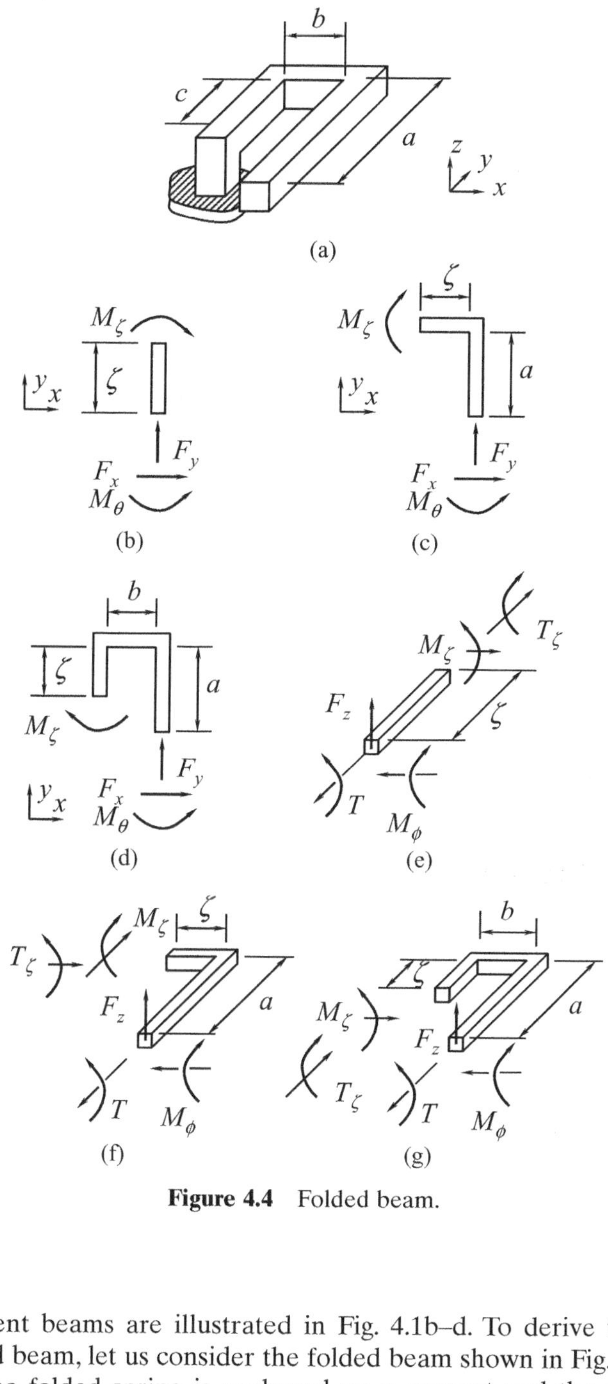

Figure 4.4 Folded beam.

folded or bent beams are illustrated in Fig. 4.1b–d. To derive the stiffness of the folded beam, let us consider the folded beam shown in Fig. 4.4a, where an end of the folded spring is anchored on a support and the other end is subjected to the forces and moments. Beam segments of length a, b, and c

will be referred to as beams a, b, and c, respectively. For consistency and convenience, the coordinate system, forces, moments, displacements, and angles illustrated in Fig. 4.3a will be used for further derivation of the stiffness of folded beams. Figure 4.4b–d are free-body diagrams of the beam segment used to calculate the bending moment that balances forces and moment-causing motions in the x–y plane. Free-body diagrams of the beam segments are shown in Fig. 4.4e–g to obtain bending and torsional moments that result in displacements out of the x–y plane. In the free-body diagrams of Fig. 4.4, the axial and shear forces acting on the cut face of the segment are needed to balance the segments but are not drawn in the free-body diagrams since they are not needed to obtain the bending and twisting moments at the cut face (the contribution of the axial and shear forces to the elastic energy in the folded beam can be negligible, as discussed in Chapter 3). From the moment equilibrium about the cut face in Fig. 4.4b–d, the following bending moments in the segments are obtained: For the segment of beam a,

$$M_\xi = F_x\zeta + M_\theta$$

For the segment of beam b,

$$M_\xi = F_x a + F_y\zeta + M_\theta$$

For the segment of beam c,

$$M_\xi = F_x(a-\zeta) + F_y b + M_\theta$$

The elastic energy stored in the folded beam of Fig. 4.4a is then expressed as

$$U = \int_0^a \frac{(F_x\zeta + M_\theta)^2}{2EI_{za}}d\zeta + \int_0^b \frac{(F_x a + F_y\zeta + M_\theta)^2}{2EI_{zb}}d\zeta + \int_0^c \frac{(F_x(a-\zeta) + F_y b + M_\theta)^2}{2EI_{zc}}d\zeta \qquad (4.26)$$

where E is the effective Young's modulus defined in (3.5) and (3.6); G is the shear modulus, (3.15); and I_{za}, I_{zb}, and I_{zc} are the moments of inertia of beams a, b, and c, which are defined as $tw_a^3/12$, $tw_b^3/12$, and $tw_c^3/12$, where t, w_a, w_b, and w_c denote the thickness of the folded beam and the widths of beams a, b, and c, respectively. Applying Castigliano's theorem to the energy equation above, we find the displacements δ_x and δ_y and the angles θ at the lower end of the folded beam:

$$\begin{aligned}
\delta_x &= \frac{\partial U}{\partial F_x} = \left[\frac{a^3}{3EI_{za}} + \frac{a^2 b}{EI_{zb}} + \frac{1}{3EI_{zc}}\left(a^3 - (a-c)^3\right)\right]F_x \\
&\quad + \left[\frac{ab^2}{2EI_{zb}} + \frac{b}{2EI_{zc}}\left(a^2 - (a-c)^2\right)\right]F_y \\
&\quad + \left[\frac{a^2}{2EI_{za}} + \frac{ab}{EI_{zb}} + \frac{1}{2EI_{zc}}\left(a^2 - (a-c)^2\right)\right]M_\theta \\
\delta_y &= \frac{\partial U}{\partial F_y} = \left[\frac{ab^2}{2EI_{zb}} + \frac{b}{2EI_{zc}}\left(a^2 - (a-c)^2\right)\right]F_x \\
&\quad + \left[\frac{b^3}{3EI_{zb}} + \frac{b^2 c}{EI_{zc}}\right]F_y + \left[\frac{b^2}{2EI_{zb}} + \frac{bc}{EI_{zc}}\right]M_\theta \\
\theta &= \frac{\partial U}{\partial M_\theta} = \left[\frac{a^2}{2EI_{za}} + \frac{ab}{EI_{zb}} + \frac{1}{2EI_{zc}}\left(a^2 - (a-c)^2\right)\right]F_x \\
&\quad + \left[\frac{b^2}{2EI_{zb}} + \frac{bc}{EI_{zc}}\right]F_y + \left[\frac{a}{EI_{za}} + \frac{b}{EI_{zb}} + \frac{c}{EI_{zc}}\right]M_\theta
\end{aligned} \tag{4.27}$$

The preceding equation may be expressed in the simple form

$$\boldsymbol{\delta}_1 = \mathbf{C}_1 \mathbf{F}_1 \tag{4.28}$$

where

$$\boldsymbol{\delta}_1 = \{\delta_x \quad \delta_y \quad \theta\}^{\mathrm{T}} \tag{4.29}$$

$$\mathbf{C}_1 = \begin{bmatrix} c_{xx} & c_{xy} & c_{x\theta} \\ c_{yx} & c_{yy} & c_{y\theta} \\ c_{\theta x} & c_{\theta y} & c_{\theta\theta} \end{bmatrix} \tag{4.30}$$

$$\mathbf{F}_1 = \{F_x \quad F_y \quad M_\theta\}^{\mathrm{T}} \tag{4.31}$$

As described in the preceding straight beam problem, it is desired to express the forces F_x and F_y and the bending moment M_θ as functions of displacements δ_x and δ_y and the angle θ. For convenience, we introduce the following parameters:

$$\begin{aligned}
&D_x = \frac{6EI_{za}}{a^3}\delta_x \qquad D_y = \frac{6EI_{za}}{a^3}\delta_y \qquad D_\theta = \frac{6EI_{za}}{a^2}\theta \\
&\beta = \frac{b}{a} \qquad \gamma = \frac{c}{a} \\
&p = \frac{EI_{za}}{EI_{zb}} \qquad q = \frac{EI_{za}}{EI_{zc}}
\end{aligned} \tag{4.32}$$

From the definition above, D_x, D_y, and D_θ have the dimensions of force and represent the displacements δ_x, δ_y, and θ. The dimensionless parameters β, γ, p, and q denote the ratios of beam lengths and flexural rigidities, respectively. Using the definitions above, we have a different form of (4.27):

$$\begin{Bmatrix} D_x \\ D_y \\ D_\theta \end{Bmatrix} = \mathbf{C}_{1n} \begin{Bmatrix} F_x \\ F_y \\ M_\theta / a \end{Bmatrix} \tag{4.33}$$

where

$$\mathbf{C}_{1n} = \begin{bmatrix} 2\{1+3\beta p+q[1-(1-\gamma)^3]\} & 3\beta\{\beta p+q[1-(1-\gamma)^2]\} & 3\{1+2\beta p+q[1-(1-\gamma)^2]\} \\ 3\beta\{\beta p+q[1-(1-\gamma)^2]\} & 2\beta^2(\beta p+3\gamma q) & 3\beta(\beta p+2\gamma q) \\ 3\{1+2\beta p+q[1-(1-\gamma)^2]\} & 3\beta(\beta p+2\gamma q) & 6(1+\beta p+\gamma q) \end{bmatrix} \tag{4.34}$$

is a symmetric compliance matrix in dimensionless form, which is more convenient to use to obtain the corresponding stiffness matrix in closed form. From the displacement–force relation (4.33), we obtain the force–displacement relation as follows:

$$\begin{Bmatrix} F_x \\ F_y \\ M_\theta / a \end{Bmatrix} = \mathbf{K}_{1n} \begin{Bmatrix} D_x \\ D_y \\ D_\theta \end{Bmatrix} \tag{4.35}$$

where

$$\mathbf{K}_{1n} = \mathbf{C}_{1n}^{-1} \tag{4.36}$$

is a dimensionless stiffness that relates the force vector to the displacement vector. Substitution of (4.32) into (4.35) and rearranging the resulting terms leads to a different form of (4.35):

$$\mathbf{F}_1 = \mathbf{K}_1 \boldsymbol{\delta}_1 \tag{4.37}$$

where

$$\boldsymbol{\delta}_1 = \{\delta_x \quad \delta_y \quad \theta\}^{\mathrm{T}} \tag{4.38}$$

$$\mathbf{F}_1 = \{F_x \quad F_y \quad M_\theta\}^{\mathrm{T}} \tag{4.39}$$

are the displacement and force vectors, respectively. We note that the stiffness matrix $\mathbf{K}_{1n}$ of (4.36) is dimensionless and symmetric and that the matrix can be converted into the following dimensional stiffness:

$$\mathbf{K}_1 = \begin{bmatrix} \frac{6EI_{za}}{a^3}K_{1n,xx} & \frac{6EI_{za}}{a^3}K_{1n,xy} & \frac{6EI_{za}}{a^2}K_{1n,x\theta} \\ & \frac{6EI_{za}}{a^3}K_{1n,yy} & \frac{6EI_{za}}{a^2}K_{1n,y\theta} \\ \text{sym.} & & \frac{6EI_{za}}{a}K_{1n,\theta\theta} \end{bmatrix} \tag{4.40}$$

where $K_{1n,xx}$, $K_{1n,xy}$, … , $K_{1n,\theta\theta}$ denote the components of stiffness matrix (4.36) and "sym." indicates a symmetric matrix.

We shall obtain the compliance and stiffness, which are associated with the out-of-plane motion. We first obtain the bending and torsional moments associated with F_z, M_ϕ, and T in Fig. 4.4e–g. Considering moment equilibrium at the cut face of the segments of Fig. 4.4e–g, we have, for the segment of beam a,

$$M_\xi = F_z\zeta + M_\phi \qquad T_\xi = T$$

for the segment of beam b,

$$M_\xi = F_z\zeta + T \qquad T_\xi = F_z a + M_\phi$$

and for the segment of beam c,

$$M_\xi = F_z(a-\zeta) + M_\phi \qquad T_\xi = F_z b + T$$

We write the elastic energy stored in the folded beam as

$$U = \int_0^a \frac{(F_z\zeta + M_\phi)^2}{2EI_{xa}}d\zeta + \int_0^a \frac{T^2}{2GJ_a}d\zeta + \int_0^b \frac{(F_z\zeta + T)^2}{2EI_{yb}}d\zeta + \int_0^b \frac{(F_z a + M_\phi)^2}{2GJ_b}d\zeta$$
$$+ \int_0^c \frac{(F_z(a-\zeta) + M_\phi)^2}{2EI_{xc}}d\zeta + \int_0^c \frac{(F_z b + T)^2}{2GJ_c}d\zeta \tag{4.41}$$

where I_{xa}, I_{yb}, and I_{xc} denote the moments of inertia that are defined to be $w_a t^3/12$, $w_b t^3/12$, and $w_c t^3/12$, respectively, and w_a, w_b, and w_c are the widths of beam segments a, b, and c. The parameters J_a, J_b, and J_c represent the polar moments of inertia of beam segments a, b, and c, which are defined in (3.33) and (3.34), respectively. Applying Castigliano's theorem to the foregoing equation yields the displacement and angles due to the applied force and moments as follows:

$$\begin{aligned}
\delta_z &= \frac{\partial U}{\partial F_z} = \left[\frac{a^3}{3EI_{xa}} + \frac{b^3}{3EI_{yb}} + \frac{a^2 b}{GJ_b} + \frac{a^3-(a-c)^3}{3EI_{xc}} + \frac{b^2 c}{GJ_c}\right]F_z \\
&\quad + \left[\frac{a^2}{2EI_{xa}} + \frac{ab}{GJ_b} + \frac{a^2-(a-c)^2}{2EI_{xc}}\right]M_\phi + \left[\frac{b^2}{2EI_{yb}} + \frac{bc}{GJ_c}\right]T \\
\phi &= \frac{\partial U}{\partial M_\phi} = \left[\frac{a^2}{2EI_{xa}} + \frac{ab}{GJ_b} + \frac{a^2-(a-c)^2}{2EI_{xc}}\right]F_z \\
&\quad + \left[\frac{a}{EI_{xa}} + \frac{b}{GJ_b} + \frac{c}{EI_{xc}}\right]M_\phi \\
\eta &= \frac{\partial U}{\partial T} = \left[\frac{b^2}{2EI_{yb}} + \frac{bc}{GJ_c}\right]F_z + \left[\frac{a}{GJ_a} + \frac{b}{EI_{yb}} + \frac{c}{GJ_c}\right]T
\end{aligned} \tag{4.42}$$

The equations above can be expressed in matrix form as follows:

$$\boldsymbol{\delta}_2 = \mathbf{C}_2\mathbf{F}_2 \tag{4.43}$$

where

$$\boldsymbol{\delta}_2 = \{\delta_z \quad \phi \quad \eta\}^{\mathrm{T}} \tag{4.44}$$

$$\mathbf{C}_2 = \begin{bmatrix} c_{zz} & c_{z\phi} & c_{z\eta} \\ c_{\phi x} & c_{\phi\phi} & c_{\phi\eta} \\ c_{\eta z} & c_{\eta\phi} & c_{\eta\eta} \end{bmatrix} \tag{4.45}$$

$$\mathbf{F}_2 = \{F_z \quad M_\phi \quad T\}^{\mathrm{T}} \tag{4.46}$$

In (4.45), the components of the matrix are the coefficients of equation (3.42). For example, $c_{z\phi}$ denotes the coefficient of M_ϕ in the first equation of (3.42).

We introduce the following parameters to obtain the compliance and stiffness:

$$\begin{aligned}
&D_z = \frac{6EI_{xa}}{a^3}\delta_z \qquad D_\phi = \frac{6EI_{xa}}{a^2}\phi \qquad D_\eta = \frac{6EI_{xa}}{a^2}\eta \\
&\beta = \frac{b}{a} \qquad \gamma = \frac{c}{a} \\
&p_1 = \frac{EI_{xa}}{EI_{xb}} \qquad q_1 = \frac{EI_{xa}}{EI_{xc}} \\
&s_a = \frac{EI_{xa}}{GJ_a} \qquad s_b = \frac{EI_{xa}}{GJ_b} \qquad s_c = \frac{EI_{xa}}{GJ_c}
\end{aligned} \tag{4.47}$$

Referring to the definition of parameters above, we have equilibrium equation (4.36) in matrix form:

$$\begin{Bmatrix} D_z \\ D_\phi \\ D_\eta \end{Bmatrix} = \mathbf{C}_{2n} \begin{Bmatrix} F_z \\ M_\phi/a \\ T/a \end{Bmatrix} \tag{4.48}$$

where

$$\mathbf{C}_{2n} = \begin{bmatrix} 2\{1+\beta^3 p_1 + 3\beta s_b + [1-(1-\gamma)^3]q_1 + 3\beta^2\gamma s_c\} & 3\{1+2\beta s_b + [1-(1-\gamma)^2]q_1\} & 3\beta(\beta p_1 + 2\gamma s_c) \\ 3\{1+2\beta s_b + [1-(1-\gamma)^2]q_1\} & 6(1+\beta s_b + \gamma q_1) & 0 \\ 3\beta(\beta p_1 + 2\gamma s_c) & 0 & 6(s_a + \beta p_1 + \gamma s_c) \end{bmatrix} \tag{4.49}$$

is the dimensionless compliance matrix. Multiplying (4.48) by the inverse of $\mathbf{C}_{2n}$, we have the following force–displacement relation:

$$\begin{Bmatrix} F_z \\ M_\phi/a \\ T/a \end{Bmatrix} = \mathbf{K}_{2n} \begin{Bmatrix} D_z \\ D_\phi \\ D_\eta \end{Bmatrix} \tag{4.50}$$

where

$$\mathbf{K}_{2n} = \mathbf{C}_{2n}^{-1} \tag{4.51}$$

is the dimensionless stiffness. Using the displacement definition of (4.47), the preceding force–displacement relation can be written in dimensional form:

$$\mathbf{F}_2 = \mathbf{K}_2 \boldsymbol{\delta}_2 \tag{4.52}$$

where

$$\mathbf{K}_2 = \begin{bmatrix} \frac{6EI_{xa}}{a^3} K_{2n,zz} & \frac{6EI_{xa}}{a^2} K_{2n,z\phi} & \frac{6EI_{xa}}{a^2} K_{2n,z\eta} \\ & \frac{6EI_{xa}}{a} K_{2n,\phi\phi} & \frac{6EI_{xa}}{a} K_{2n,\phi\eta} \\ \text{sym.} & & \frac{6EI_{xa}}{a} K_{2n,\eta\eta} \end{bmatrix} \tag{4.53}$$

represents the dimensional stiffness matrix.

Stiffness of L-Shaped Beam

The L-shaped beam shown in Figs. 4.1b and 4.5 can be considered as a particular case of the folded beam shown in Figs. 4.1c and 4.4a, where the length of beam c approaches zero. Setting $c = 0$ (i.e., $\gamma = 0$) in (4.34), the compliance matrix, (4.34), can be written as follows:

$$\mathbf{C}_{1n} = \begin{bmatrix} 2(1+3\beta p) & 3\beta^2 p & 3(1+2\beta p) \\ 3\beta^2 p & 2\beta^3 p & 3\beta^2 p \\ 3(1+2\beta p) & 3\beta^2 p & 6(1+\beta p) \end{bmatrix} \tag{4.54}$$

The corresponding stiffness matrix is then given by

$$\mathbf{K}_{1n} = \begin{bmatrix} 2(1+3\beta p) & 3\beta^2 p & 3(1+2\beta p) \\ 3\beta^2 p & 2\beta^3 p & 3\beta^2 p \\ 3(1+2\beta p) & 3\beta^2 p & 6(1+\beta p) \end{bmatrix}^{-1} \tag{4.55}$$

which can be evaluated either numerically or analytically. If we are interested only in displacement in the x direction in Fig. 4.1b, we need k_{xx} for design of the microstructures. In this case, closed-form stiffness is more convenient for the spring design. Let us obtain closed-form stiffness, $\mathbf{K}_1$, from the preceding compliance matrix. Referring to the inverse of a 3×3 matrix (see the Appendix), we write the following equation for K_{xx} and simplify it to obtain the final expression below:

$$\begin{aligned} \frac{1}{K_{1n,xx}} &= 2(1+3\beta p) - 3\beta^2 p \frac{\begin{vmatrix} 3\beta^2 p & 3\beta^2 p \\ 3(1+2\beta p) & 6(1+\beta p) \end{vmatrix}}{\begin{vmatrix} 2\beta^3 p & 3\beta^2 p \\ 3\beta^2 p & 6(1+\beta p) \end{vmatrix}} \\ &\quad + 3(1+2\beta p) \frac{\begin{vmatrix} 3\beta^2 p & 2\beta^3 p \\ 3(1+2\beta p) & 3\beta^2 p \end{vmatrix}}{\begin{vmatrix} 2\beta^3 p & 3\beta^2 p \\ 3\beta^2 p & 6(1+\beta p) \end{vmatrix}} \\ &= 2(1+3\beta p) - 3\beta^2 p \frac{9\beta^2 p}{3\beta^3 p} \frac{2(1+\beta p)-(1+2\beta p)}{4(1+\beta p) - 3\beta p} \\ &\quad + 3(1+2\beta p) \frac{3\beta^3 p}{3\beta^3 p} \frac{3\beta p - 2(1+2\beta p)}{4(1+\beta p) - 3\beta p} \\ &= \frac{2(1+\beta p)}{4+\beta p} \end{aligned}$$

Similarly, we obtain the others and can construct the dimensionless stiffness matrix $\mathbf{K}_{1n}$ in closed form as follows:

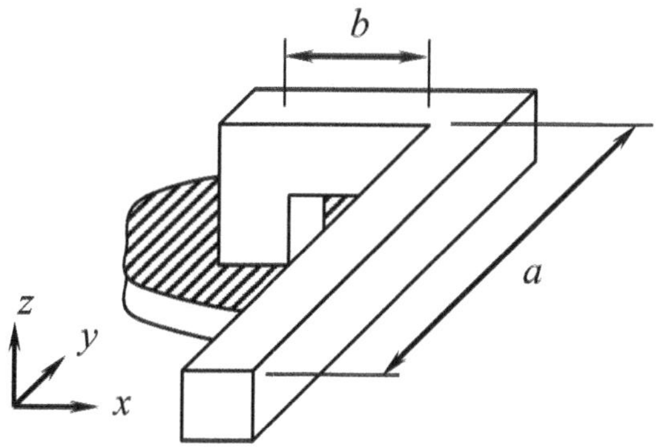

Figure 4.5 L-shaped beam.

$$\mathbf{K}_{1n} = \begin{bmatrix} \frac{4+\beta p}{2(1+\beta p)} & \frac{-3}{2\beta(1+\beta p)} & -\frac{2+\beta p}{2(1+\beta p)} \\ \frac{-3}{2\beta(1+\beta p)} & \frac{1}{2\beta^3 p}\frac{1+4\beta p}{1+\beta p} & \frac{1}{2\beta(1+\beta p)} \\ -\frac{2+\beta p}{2(1+\beta p)} & \frac{1}{2\beta(1+\beta p)} & \frac{4+3\beta p}{6(1+\beta p)} \end{bmatrix} \tag{4.56}$$

With reference to (4.40), we obtain the dimensional form of the preceding stiffness:

$$\mathbf{K}_1 = \begin{bmatrix} \frac{3EI_{za}}{a^3}\frac{4+\beta p}{1+\beta p} & -\frac{9EI_{za}}{a^3}\frac{1}{\beta(1+\beta p)} & -\frac{3EI_{za}}{a^2}\frac{2+\beta p}{1+\beta p} \\ & \frac{3EI_{za}}{a^3}\frac{1+4\beta p}{\beta^3 p(1+\beta p)} & 3\frac{EI_{za}}{a^2}\frac{1}{\beta(1+\beta p)} \\ \text{sym.} & & \frac{EI_{za}}{a}\frac{4+3\beta p}{1+\beta p} \end{bmatrix} \tag{4.57}$$

When the displacement vector of the lower end of the L-shaped spring (Fig. 4.5) is given, the force vector to maintain the displacements is obtained from the force–displacement (4.37).

The stiffness $\mathbf{K}_2$ for an L-shaped spring will be obtained from a similar procedure. Referring to (4.49), we write the dimensionless compliance $\mathbf{C}_{2n}$ with $\gamma = 0$ as follows:

$$\mathbf{C}_{2n} = \begin{bmatrix} 2(1+\beta^3 p_1 + 3\beta s_b) & 3(1+2\beta s_b) & 3\beta^2 p_1 \\ 3(1+2\beta s_b) & 6(1+\beta s_b) & 0 \\ 3\beta^2 p_1 & 0 & 6(s_a + \beta p_1) \end{bmatrix} \tag{4.58}$$

Using the procedure to construct an inverse matrix (see the Appendix), we have the dimensionless stiffness:

$$\mathbf{K}_{2n} = \mathbf{C}_{2n}^{-1}$$

which can be expressed in dimensional form as follows:

$$\mathbf{K}_2 = \begin{bmatrix} k_{2,zz} & k_{2,z\phi} & k_{2,z\eta} \\ & k_{2,\phi\phi} & k_{2,\phi\eta} \\ \text{sym.} & & k_{2,\eta\eta} \end{bmatrix} = \begin{bmatrix} \frac{6EI_{xa}}{a^3}K_{2n,zz} & \frac{6EI_{xa}}{a^2}K_{2n,z\phi} & \frac{6EI_{xa}}{a^2}K_{2n,z\eta} \\ & \frac{6EI_{xa}}{a^2}K_{2n,\phi\phi} & \frac{6EI_{xa}}{a^2}K_{2n,\phi\eta} \\ \text{sym.} & & \frac{6EI_{xa}}{a}K_{2n,\eta\eta} \end{bmatrix} \tag{4.59}$$

where

$$\frac{1}{K_{2n,zz}} = 2(1+\beta^3 p_1 + 3\beta s_b) - \frac{3}{2}\frac{(1+2\beta s_b)^2}{1+\beta s_b} - \frac{3}{2}\beta^4 p_1^2 \frac{1}{s_a+\beta p_1}$$

$$\frac{1}{K_{2n,z\phi}} = -4(1+\beta^3 p_1 + 3\beta s_b)\frac{1+\beta s_b}{1+2\beta s_b} + 3(1+2\beta s_b) + 3\beta^4 p_1^2 \frac{1+\beta s_b}{(1+2\beta s_b)(s_a+\beta p_1)}$$

$$\frac{1}{K_{2n,z\eta}} = -\frac{4}{\beta^2 p_1}(1+\beta^3 p_1 + 3\beta s_b)(s_a+\beta p_1) + \frac{3}{\beta^2 p_1}\frac{(1+2\beta s_b)^2 (s_a+\beta p_1)}{(1+\beta s_b)} + 3\beta^2 p_1$$

$$\frac{1}{K_{2n,\phi\phi}} = -18\frac{(1+2\beta s_b)^2 (s_a+\beta p_1)}{4(1+\beta^3 p_1 + 3\beta s_b)(s_a+\beta p_1) - 3\beta^4 p_1^2} + 6(1+\beta s_b)$$

$$\frac{1}{K_{2n,\phi\eta}} = -6\beta^2 p_1 \frac{1+\beta s_b}{1+2\beta s_b}$$
$$+ \frac{2}{\beta^2 p_1}\frac{s_a+\beta p_1}{1+2\beta s_b}\left[4(1+\beta^3 p_1 + 2\beta s_b)(1+\beta s_b) - 3(1+2\beta s_b)^2\right]$$

$$\frac{1}{K_{2n,\eta\eta}} = -18\beta^4 p_1^2 \frac{1+\beta s_b}{4(1+\beta^3 p_1 + 3\beta s_b)(1+\beta s_b) - 3(1+2\beta s_b)^2} + 6(s_a+\beta p_1)$$

Stiffness of Folded Beams Consisting of Two Identical Beams

For design convenience, we can design a folded beam in which beam lengths a and c of the folded beam in Fig. 4.4a are identical; furthermore, the cross-sectional areas of the beams are also identical. In this case, the compliance and stiffness matrices are expressed in simpler forms. Substituting $\gamma = 1$ and $q = 1$ into (4.34), we have the dimensionless compliance matrix for in-plane motion:

$$\mathbf{C}_{1n} = \begin{bmatrix} 2(2+3\beta p) & 3\beta(1+\beta p) & 6(1+\beta p) \\ 3\beta(1+\beta p) & 2\beta^2(3+\beta p) & 3\beta(2+\beta p) \\ 6(1+\beta p) & 3\beta(2+\beta p) & 6(2+\beta p) \end{bmatrix} \tag{4.60}$$

Using the same procedure as that described to obtain the inverse matrix, we find the dimensionless stiffness:

$$\mathbf{K}_{1n} = \mathbf{C}_{1n}^{-1} = \begin{bmatrix} \dfrac{2+\beta p}{2(1+2\beta p)} & 0 & -\dfrac{1+\beta p}{2(1+2\beta p)} \\ 0 & \dfrac{2}{\beta^2(6+\beta p)} & \dfrac{-1}{\beta(6+\beta p)} \\ -\dfrac{1+\beta p}{2(1+2\beta p)} & \dfrac{-1}{\beta(6+\beta p)} & \dfrac{15+26\beta p+3(\beta p)^2}{6\left[6+13\beta p+2(\beta p)^2\right]} \end{bmatrix} \tag{4.61}$$

According to (4.40), the foregoing equation is converted to the dimensional form

$$\mathbf{K}_1 = \begin{bmatrix} \dfrac{3EI_{za}}{a^3}\dfrac{2+\beta p}{1+2\beta p} & 0 & -\dfrac{3EI_{za}}{a^2}\dfrac{1+\beta p}{1+2\beta p} \\ 0 & \dfrac{12EI_{za}}{a^3}\dfrac{2}{\beta^2(6+\beta p)} & -\dfrac{3EI_{za}}{a^2}\dfrac{1+\beta p}{1+2\beta p} \\ -\dfrac{3EI_{za}}{a^2}\dfrac{1+\beta p}{1+2\beta p} & -\dfrac{3EI_{za}}{a^2}\dfrac{1+\beta p}{1+2\beta p} & \dfrac{EI_{za}}{a}\dfrac{15+26\beta p+3(\beta p)^2}{6+13\beta p+2(\beta p)^2} \end{bmatrix} \tag{4.62}$$

Upon substituting $\gamma = 1$, $q_1 = 1$, and $s_c = s_a$ into (4.49), we have the compliance matrix for out-of-plane motion as follows:

$$\mathbf{C}_{2n} = \begin{bmatrix} 2(2+\beta^3 p_1+3\beta s_b+3\beta^2 s_a) & 6(1+\beta s_b) & 3\beta(\beta p_1+2s_a) \\ 6(1+\beta s_b) & 6(2+\beta s_b) & 0 \\ 3\beta(\beta p_1+2s_a) & 0 & 6(\beta p_1+2s_a) \end{bmatrix} \tag{4.63}$$

Obtaining the dimensionless stiffness $\mathbf{K}_{2n} = \mathbf{C}_{2n}^{-1}$ and using (4.45), the dimensional stiffness is obtained for out-of-plane motion as follows:

$$\mathbf{K}_2 = \begin{bmatrix} k_{2,zz} & k_{2,z\phi} & k_{2,z\eta} \\ & k_{2,\phi\phi} & k_{2,\phi\eta} \\ \text{sym.} & & k_{2,\eta\eta} \end{bmatrix} = \begin{bmatrix} \dfrac{6EI_{xa}}{a^3}K_{2n,zz} & \dfrac{6EI_{xa}}{a^2}K_{2n,z\phi} & \dfrac{6EI_{xa}}{a^2}K_{2n,z\eta} \\ & \dfrac{6EI_{xa}}{a}K_{2n,\phi\phi} & \dfrac{6EI_{xa}}{a}K_{2n,\phi\eta} \\ \text{sym.} & & \dfrac{6EI_{xa}}{a}K_{2n,\eta\eta} \end{bmatrix} \tag{4.64}$$

where

$$\frac{1}{K_{2n,zz}} = 2\left(2+\beta^3 p_1 + 3\beta s_b + 3\beta^2 s_a\right) - 6\frac{(1+\beta s_b)^2}{2+\beta s_b} - \frac{3}{2}\beta^2\left(\beta p_1 + 2s_a\right)$$

$$\frac{1}{K_{2n,z\phi}} = -2\left(2+\beta^3 p_1 + 3\beta s_b + 3\beta^2 s_a\right)\frac{2+\beta s_b}{1+\beta s_b} + 6(1+\beta s_b) + \frac{3}{2}\beta^2\frac{(\beta p_1 + 2s_a)(2+\beta s_b)}{1+\beta s_b}$$

$$\frac{1}{K_{2n,z\eta}} = -\frac{4}{\beta}\left(2+\beta^3 p_1 + 3\beta s_b + 3\beta^2 s_a\right) + \frac{12}{\beta}\frac{(1+\beta s_b)^2}{2+\beta s_b} + 3\beta(\beta p_1 + 2s_a)$$

$$\frac{1}{K_{2n,\phi\phi}} = -72\frac{(1+2\beta s_b)^2(\beta p_1 + 2s_a)}{4\left(2+\beta^3 p_1 + 3\beta s_b + 3\beta^2 s_a\right)(\beta p_1 + 2s_a) - 3\beta^2(\beta p_1 + 2s_a)^2} + 6(2+\beta s_b)$$

$$\frac{1}{K_{2n,\varphi\eta}} = -12\frac{1+\beta s_b}{\beta} + \frac{2+\beta s_b}{\beta(1+\beta s_b)}\left\{4\left(2+\beta^3 p_1 + 3\beta s_b + 3\beta^2 s_a\right) - 3\beta^2(\beta p_1 + 2s_a)\right\}$$

$$\frac{1}{K_{2n,\eta\eta}} = -\frac{9}{2}\beta^2\frac{(\beta p_1 + 2s_a)^2(2+\beta s_b)}{\left(2+\beta^3 p_1 + 3\beta s_b + 3\beta^2 s_a\right)(2+\beta s_b) - 3(1+\beta s_b)^2} + 6(\beta p_1 + 2s_a)$$

Stiffness of Connected Folded Beam

Two folded beams (Fig. 4.1d) may be connected to each other to support a movable structure. Let us consider a half structure of the suspended structure shown in Fig. 4.1d, which is illustrated in Fig. 4.6a. To set up a model for the stiffness analysis, we first consider the symmetry of the microstructure in Fig. 4.6a. Since the structure is symmetric about the x axis, force and moment distribution in the beams in Fig. 4.6a are also symmetric about the x axis. The movable structure is assumed to be rigid. As shown in Fig. 4.6a, there is one more symmetry line along the y axis. For the displacement δ_x, the forces and moments in the structure are not symmetric about the y axis. We then draw the free-body diagrams of the right and left halves of the folded beams in Fig. 4.6b. In the diagram, points D_1 and D denote the ends of the folded beams that are connected to different points of the rigid movable structures. Points C_1 and C are the points of the connected folded beam at which the connecting beam crosses the symmetric line (i.e., the y axis). As the movable structure moves by δ_x along the x axis in Fig. 4.6a, the forces and moments shown in Fig. 4.6b are generated at points C, D, and D_1 to balance the movable structures. The senses of the forces and moments will be determined from the force and moment equilibrium described below.

If we assume that the material of the structure in Fig. 4.6a are linear, all forces and moments in the free-body diagram of Fig. 4.6b must be proportional to the displacement δ_x. As a result, when the movable structure is displaced by $-\delta_x$ (the same magnitude of the displacement in the opposite direction), the

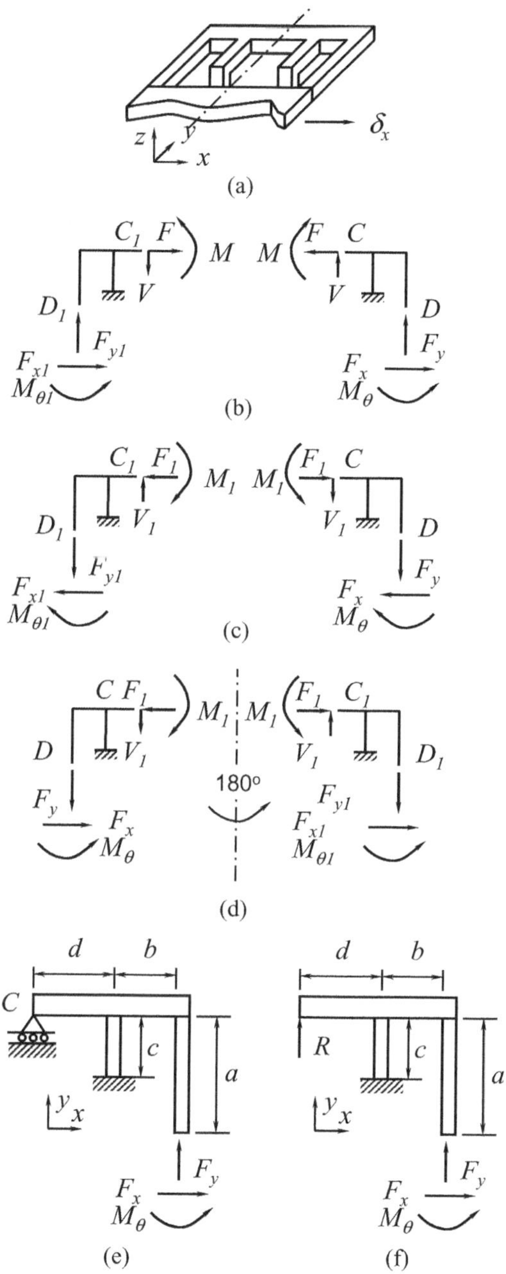

Figure 4.6 Folded beams connected to each other.

forces and moments are drawn in the opposite direction, as shown in Fig. 4.6c. To appeal to the symmetry argument, we rotate the free-body diagrams 180° about the symmetry line (the y axis), as shown in Fig. 4.6d. The resulting free-body diagram in Fig. 4.6d becomes a diagram in which the movable structures moves rightward by δ_x. Comparing Fig. 4.6b and d, the free-body diagrams for the displacement δ_x, we find the forces and moment at D_1 and D as follows:

$$F_{x1} = F_x \qquad F_{y1} = -F_y \qquad M_{\theta 1} = M_\theta$$

Since C_1 and C are the same point of the connecting beam, the forces and moments in opposite directions at C and C_1 must vanish for symmetry: $M = M_1 = 0$ and $F = F_1 = 0$. Furthermore, a comparison of the shear forces leads to $V_1 = V$. The shear force V at point C is proportional to the displacement δ_x of the movable structure along the x axis. Based on this information on the forces and moments, we construct the model shown in Fig. 4.6e, where point C can move on a roller and the displacement at C in the y direction remains zero. The force and moment for the model of Fig. 4.6e is illustrated in Fig. 4.6f, where the reaction R is applied in the y direction. We can repeat the preceding procedure described for the folded beam of Fig. 4.4 to obtain a set of linear equations that is expressed as follows:

$$\begin{Bmatrix} D_x \\ D_y \\ D_\theta \\ D_{y1} \end{Bmatrix} = \mathbf{C}_n \begin{Bmatrix} F_x \\ F_y \\ M_\theta/a \\ R \end{Bmatrix} \tag{4.65}$$

where

$$D_x = \frac{6EI_{za}}{a^3}\delta_x \qquad D_y = \frac{6EI_{za}}{a^3}\delta_y \qquad D_\theta = \frac{6EI_{za}}{a^2}\theta \qquad D_{y1} = \frac{6EI_{za}}{a^3}\delta_{y1}$$

$$\beta = \frac{b}{a} \qquad \gamma = \frac{c}{a} \qquad \xi = \frac{d}{a} \tag{4.66}$$

$$p = \frac{EI_{za}}{EI_{zb}} \qquad q = \frac{EI_{za}}{EI_{zc}} \qquad u = \frac{EI_{za}}{EI_{zd}}$$

$$\mathbf{C}_n = \begin{bmatrix} C_{n,xx} & C_{n,xy} & C_{n,x\theta} & C_{n,xy1} \\ C_{n,yx} & C_{n,yy} & C_{n,y\theta} & C_{n,yy1} \\ C_{n,\theta x} & C_{n,\theta y} & C_{n,\theta\theta} & C_{n,\theta y1} \\ C_{n,y1x} & C_{n,y1y} & C_{n,y1\theta} & C_{n,y1y1} \end{bmatrix}$$

$$= \begin{bmatrix} & \mathbf{C}_{1n} & & \begin{matrix} 3\xi q\left[1-(1-\gamma)^2\right] \\ 6\beta\gamma\xi q \\ 6\gamma\xi q \end{matrix} \\ 3\xi q\left[1-(1-\gamma)^2\right] & 6\beta\gamma\xi q & 6\gamma\xi q & 2\xi^2(3\gamma q + \xi u) \end{bmatrix} \tag{4.67}$$

In (4.66), δ_{y1} is the displacement of point C (the point of application of reaction R) in the y direction, which will be set equal to zero. In the preceding equation, $\mathbf{C}_{1n}$ is the compliance matrix defined in (4.34). The foregoing displacement–force equation can be solved numerically or analytically. If it is desired to obtain a closed-form stiffness k_{xx}, we may use Cramer's rule (Kreyszig, 1979). Since the microstructure of Fig. 4.6e imposes the boundary conditions $D_y = D_\theta = D_{y1} = 0$, the force to cause displacement δ_x can be written from (4.65) using Cramer's rule as follows:

$$F_x = \frac{\begin{vmatrix} D_x & C_{n,xy} & C_{n,x\theta} & C_{n,xy_1} \\ 0 & C_{n,yy} & C_{n,y\theta} & C_{n,yy_1} \\ 0 & C_{n,\theta y} & C_{n,\theta\theta} & C_{n,\theta y_1} \\ 0 & C_{n,y_1y} & C_{n,y_1\theta} & C_{n,y_1y_1} \end{vmatrix}}{\begin{vmatrix} C_{n,xx} & C_{n,xy} & C_{n,x\theta} & C_{n,xy_1} \\ C_{n,yx} & C_{n,yy} & C_{n,y\theta} & C_{n,yy_1} \\ C_{n,\theta x} & C_{n,\theta y} & C_{n,\theta\theta} & C_{n,\theta y_1} \\ C_{n,y_1x} & C_{n,y_1y} & C_{n,y_1\theta} & C_{n,y_1y_1} \end{vmatrix}} = D_x \frac{\begin{vmatrix} C_{n,yy} & C_{n,y\theta} & C_{n,yy_1} \\ C_{n,\theta y} & C_{n,\theta\theta} & C_{n,\theta y_1} \\ C_{n,y_1y} & C_{n,y_1\theta} & C_{n,y_1y_1} \end{vmatrix}}{\begin{vmatrix} C_{n,xx} & C_{n,xy} & C_{n,x\theta} & C_{n,xy_1} \\ C_{n,yx} & C_{n,yy} & C_{n,y\theta} & C_{n,yy_1} \\ C_{n,\theta x} & C_{n,\theta y} & C_{n,\theta\theta} & C_{n,\theta y_1} \\ C_{n,y_1x} & C_{n,y_1y} & C_{n,y_1\theta} & C_{n,y_1y_1} \end{vmatrix}} \tag{4.68}$$

where the denominator and numerator denote the determinants of the corresponding matrices. Substituting the definition of D_x, (4.66), into equation (4.68) leads to the stiffness k_{xx} as follows:

$$k_{xx} = \frac{F_x}{\delta_x} = \frac{6EI_{za}}{a^3} \frac{\begin{vmatrix} C_{n,yy} & C_{n,y\theta} & C_{n,yy_1} \\ C_{n,\theta y} & C_{n,\theta\theta} & C_{n,\theta y_1} \\ C_{n,y_1y} & C_{n,y_1\theta} & C_{n,y_1y_1} \end{vmatrix}}{\begin{vmatrix} C_{n,xx} & C_{n,xy} & C_{n,x\theta} & C_{n,xy_1} \\ C_{n,yx} & C_{n,yy} & C_{n,y\theta} & C_{n,yy_1} \\ C_{n,\theta x} & C_{n,\theta y} & C_{n,\theta\theta} & C_{n,\theta y_1} \\ C_{n,y_1x} & C_{n,y_1y} & C_{n,y_1\theta} & C_{n,y_1y_1} \end{vmatrix}} \tag{4.69}$$

Note that the stiffness above represents the stiffness of the folded beam in Fig. 4.6e. The effective stiffness of the connected folded beams (shown in Fig. 4.1d) is given by four times the preceding stiffness (i.e., $k_{\text{eff}} = 4 \times k_{xx}$.

If the flexural rigidities EI_{zb} and EI_{zd} of beams b and d are much greater than those of beams a and c (i.e., $p \ll 1$ and $u \ll 1$), the problem is greatly simplified. In this case, beams b and d can be thought of as rigid bars, so the stiffness k_{xx} is expressed as a combination of the stiffness of beams a and c, as shown in Fig. 4.7b and c. Let k_a and k_c be the stiffness of the stiffness of beams a and c, which is defined as

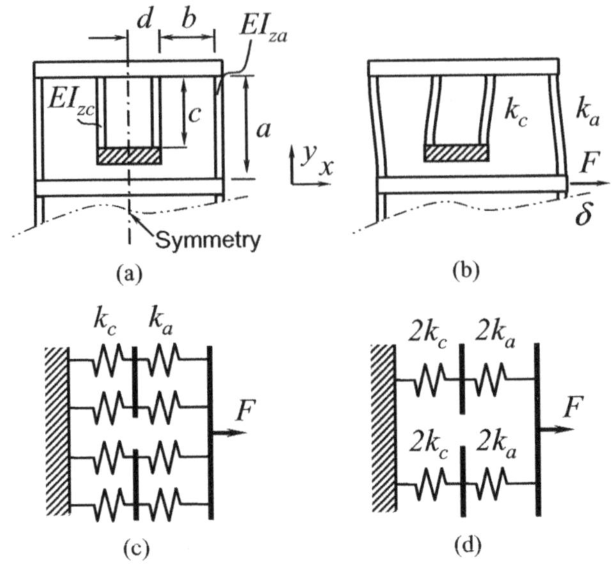

Figure 4.7 Stiffness of folded beams of different lengths.

$$k_a = \frac{12EI_{za}}{a^3}$$
$$k_c = \frac{12EI_{zc}}{c^3} \tag{4.70}$$

It is noted that the stiffnesses above are the stiffnesses of the straight beam k_{xx} in (4.14) or k in (3.62). Since beams b and d are rigid and are not angularly displaced due to the symmetry argument, the folded beams of Fig. 4.7a can be considered to be the connected springs shown in Fig. 4.7c. This spring connection is converted into the spring connections Fig. 4.7d if we use formula (4.1) for the springs in parallel. According to the effective spring formulas (4.1) and (4.2), the effective stiffness, k_{eff}, of a folded beam with rigid bars is then given by

$$k_{\text{eff}} = \frac{F}{\delta} = \frac{1}{1/(2k_c)+1/(2k_a)} + \frac{1}{1/(2k_c)+1/(2k_a)} = \frac{4}{1/k_a+1/k_c}$$
$$= \frac{4k_a k_c}{k_a+k_c} \tag{4.71}$$

If the lengths of beams a and c are the same and if beams b and d are considered as rigid bars and their flexural rigidities are identical, the effective stiffness k_{eff} is obtained by setting $k_a = k_c$ in equation (4.71) as follows:

$$k_{\text{eff}} = 2k \tag{4.72}$$

where

$$k = \frac{12EI_{za}}{a^3}$$

is the stiffness of beam a. For example, consider the case $E = 150\,\text{GPa}, a = 50\,\mu\text{m}$, $b = 20\,\mu\text{m}$, $c = 30\,\mu\text{m}$, $d = 14\,\mu\text{m}$, $t = 2\,\mu\text{m}$, $w_a = w_c = 2\,\mu\text{m}$, and $w_b = w_d = 4\,\mu\text{m}$. From these data, $\beta = 0.4$, $\gamma = 0.6$, $\xi = 0.28$, $p = u = 0.0156$, and $q = 1$ are obtained. Substituting these dimensionless parameters into (4.71) gives 63.16 N/m, while for the same data, four times the stiffness of (4.69) gives 62.76 N/m. As shown in this case, if beams b and d are sufficiently rigid (e.g., $p << 1$ and $u << 1$), the stiffness of (4.71) can be used to evaluate the stiffness of the structure with the connected folded beams (shown in Fig. 4.1d).

4.3 TRUSSES

In architecture and structural engineering, a *truss* is defined as a structure that consists of at least one triangular substructure made of slender beams. The truss, which acts as a basic structure in architecture and structural engineering, transmits loads (forces or moments) from one point to others via axial forces (i.e., compressive and tensile forces). A truss can be used in MEMS to form a relatively rigid structure (e.g., a movable structure) or to generate linear and angular displacements.

Let us begin by considering the truss (V-shaped beam) in Fig. 4.8, which consists of a lower beam of length l_b and two side beams of length l. The side beams then make an angle α with the lower beam as follows:

$$\alpha = \cos^{-1} \frac{l_b}{2l}$$

A substrate (e.g., silicon substrate) or a slender beam may serve as the lower beam. For a specific design, the triangular structure may be connected in a plane or three-dimensionally. In MEMS structures, the truss in Fig. 4.8a can act as a displacement or force generator that converts the strain in the beam into displacement or force. For example, if the truss in Fig. 4.8a is used for a thermal actuator, the axial strain, which may be generated by electrical current through the beam, causes a displacement of the vertex of the beam. The strain in the beams may be generated by various means (e.g., Joule heating, which elongates the beam).

Assuming that the lower beam (the substrate) is rigid, we consider the displacement of the vertex when a vertical force is applied to the vertex as shown in Fig. 4.8a. Since we shall use the force–displacement relation (4.12)

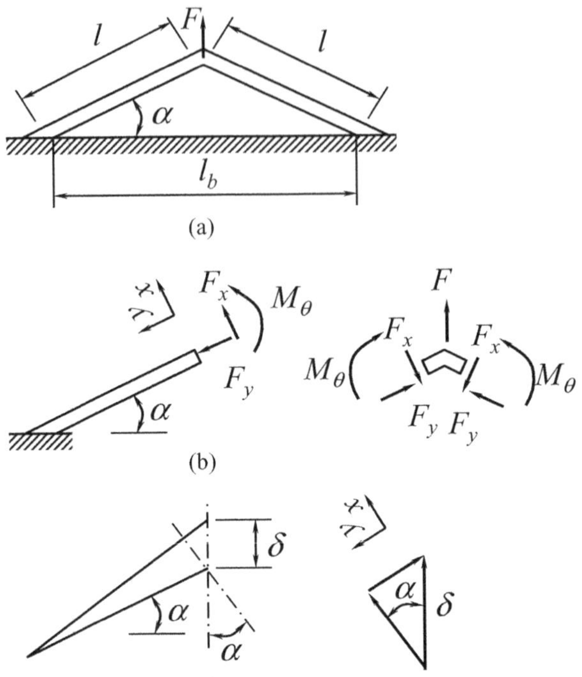

Figure 4.8 Deflection of a V-shaped beam.

of a straight beam, let the y axis coincide with the longitudinal axis of the straight beam, as shown in Fig. 4.8b. We shall use the definition of force, moment, and displacement shown in Fig. 4.3a to analyze the problem of Fig. 4.8. From the right free-body diagram of Fig. 4.8b, the force equilibrium equation in the vertical direction is written as

$$F - 2F_x \cos\alpha + 2F_y \sin\alpha = 0$$

When the vertex is displaced a small distance δ, in the vertical direction, a geometric relation is obtained from Fig. 4.8c:

$$\delta_x = \delta\cos\alpha \qquad \delta_y = -\delta\sin\alpha \qquad \theta = 0$$

The zero angle condition above is imposed by the symmetry argument (i.e., the structure is symmetric about the vertical line passing through the vertex). It is assumed that the geometric nonlinearity is negligible. The force–displacement relation of the straight beam for in-plane motion is given by (4.12), which is rewritten here:

$$\begin{Bmatrix} F_x \\ F_y \\ M_\theta \end{Bmatrix} = \begin{bmatrix} \frac{12EI_z}{l^3} & 0 & -\frac{6EI_z}{l^2} \\ 0 & \frac{EA}{l} & 0 \\ -\frac{6EI_z}{l^2} & 0 & \frac{4EI_z}{l} \end{bmatrix} \begin{Bmatrix} \delta_x \\ \delta_y \\ \theta \end{Bmatrix}$$

where E denotes the effective stiffness $\tilde{E}$, defined in (3.5) and (3.6). Substituting the preceding force–displacement relation into the preceding force equilibrium equation and using the geometric relation above gives the following expression, which relates the applied force F to the displacement δ:

$$\begin{aligned} F &= 2F_x \cos\alpha - 2F_y \sin\alpha \\ &= 2\left(\frac{12EI_z}{l^3}\delta\cos\alpha\right)\cos\alpha - 2\left[\frac{EA}{l}(-\delta\sin\alpha)\right]\sin\alpha \\ &= 2\left(\frac{12EI_z}{l^3}\cos^2\alpha + \frac{EA}{l}\sin^2\alpha\right)\delta \end{aligned}$$

The equation above gives the stiffness of a V-shaped beam as follows:

$$k = \frac{F}{\delta} = 2\left(\frac{12EI_z}{l^3}\cos^2\alpha + \frac{EA}{l}\sin^2\alpha\right) \tag{4.73}$$

Since the preceding stiffness is due to the two symmetric straight beams in Fig. 4.8, the stiffness k_1 of one inclined beam can be considered as half the preceding stiffness, which is given by

$$k_1 = \frac{F/2}{\delta} = \frac{12EI_z}{l^3}\cos^2\alpha + \frac{EA}{l}\sin^2\alpha \tag{4.74}$$

To examine the contribution of the bending moment in the straight beam to stiffness, the stiffness (4.73) is written in a different form:

$$k = \frac{F}{\delta} = 2\frac{EA}{l}\sin^2\alpha\left(1 + \frac{12I_z}{Al^2}\frac{1}{\tan^2\alpha}\right)$$

The preceding stiffness of a straight beam may be obtained by the stiffness transformation, which is dealt with in Section 4.4. Using the definition of the moment of inertia $I_z = tw^3/12$ and the cross-sectional area $A = wt$, $12I_z/Al^2$ is expressed as $(w/l)^2$ and the preceding stiffness is expressed as

$$k = 2\frac{EA}{l}\sin^2\alpha\left[1 + \left(\frac{w}{l}\right)^2\frac{1}{\tan^2\alpha}\right]$$

If $w/(l \tan\alpha) << 1$, the stiffness above is approximated as

$$k \cong 2\frac{EA}{l}\sin^2\alpha \tag{4.75}$$

which results from the axial force in the two straight beams in Fig. 4.8a. If the angle α approaches zero, the stiffness (4.73) becomes the following stiffness, which is due to the bending moment in the two straight beams:

$$k = \frac{24EI_z}{l^3} \tag{4.76}$$

The stiffness above can be thought of as the stiffness of two straight beams in series (i.e., $k = 2 \times 12EI_z/l^3$). From the derivations and discussion above, we can interpret a truss (the V-shaped beam in Fig. 4.8a) subjected to forces as follows: If the angle α is not zero, a truss consisting of slender beams can be analyzed without reference to the bending moment in the beam; and if the angle is zero, the stiffness is considered to be the stiffness due to the bending moment.

According to the preceding interpretation of a truss, the V-shaped beam is modeled as a simple linkage in Fig. 4.9a, whose vertices are replaced with pin joints. With the model we obtain the same stiffness, (4.75). The force equilibrium requirement (Fig. 4.9b) gives

$$F - 2F_1 \sin\alpha = 0$$

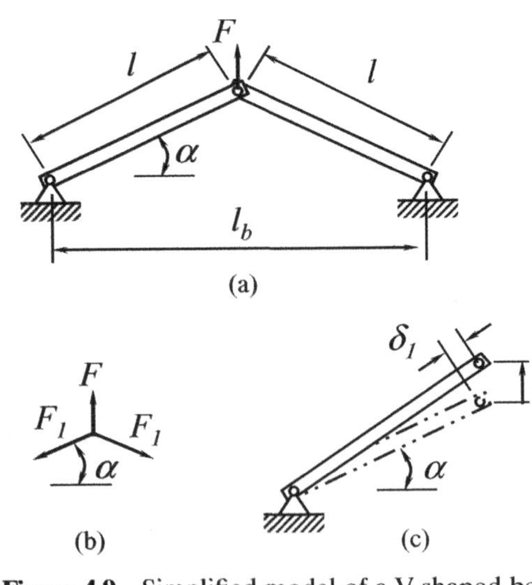

Figure 4.9 Simplified model of a V-shaped beam.

and the geometric relation from Fig. 4.9c leads to

$$\delta_1 = \delta \sin \alpha$$

Force–displacement will be obtained from the stress–strain equation (3.3):

$$\sigma = E\varepsilon$$

Substituting the definitions of the stress and area, $\sigma = F_1/A$ and $\varepsilon = \delta_1/l$, into the equation above, we have

$$F_1 = EA\frac{\delta_1}{l}$$

As we substitute the geometric relation and the force–deformation equation into the preceding force equilibrium equation, we obtain the same stiffness (4.75):

$$k = \frac{F}{\delta} = 2\frac{EA}{l}\sin^2 \alpha$$

where E represents the effective Young's modulus. As a result, if a truss is made of slender straight bars in a triangular shape and if the angles between the bars are not zero degrees, the truss can be replaced by a truss with pin joints. This type of replacement is illustrated in Fig. 4.10. In Fig. 4.10a, a microtruss made of slender straight beams of length l_1, l_2, and l_3 is subjected to forces F and P at the vertices in arbitrary directions in the truss plane. Since the truss is made of slender beams, the truss with rigid connections at vertices can be replaced by a truss with pin joints to analyze the truss. Note that this replacement holds for triangular trusses. If a nontriangular structure (e.g., a rectangle) is subjected to a force, the replacement of rigid joints with pin joints is not valid since the replacement can cause rotation of a beam about its pin joint.

The same argument holds for trusses of slender straight beams that undergo change in strain as shown in Fig. 4.11a. To evaluate the vertical displacement δ along the symmetry line of the truss in Fig. 4.11a, the original truss is replaced by one with pin joints, as shown in Fig. 4.11b. Let the average strain in the beam be denoted by ε. As shown in Fig. 4.11c, the point at the vertex is displaced by δ during the application of strain. From the geometry in Fig. 4.11c, we obtain a simple geometric relation:

$$l\varepsilon = \delta \sin \alpha$$

which gives the displacement–strain relation directly:

$$\delta = \frac{l}{\sin \alpha}\varepsilon \tag{4.77}$$

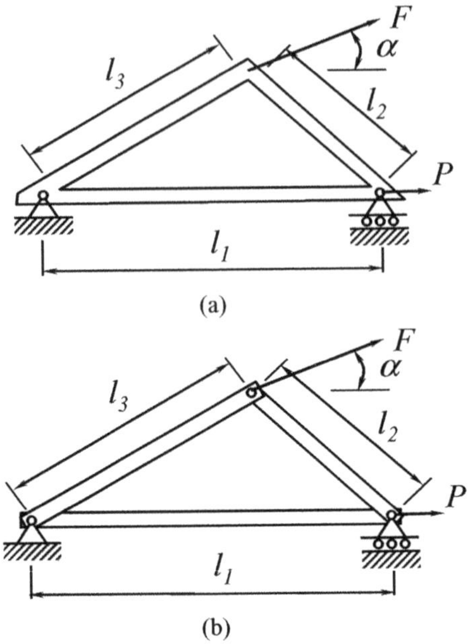

Figure 4.10 Triangular structure.

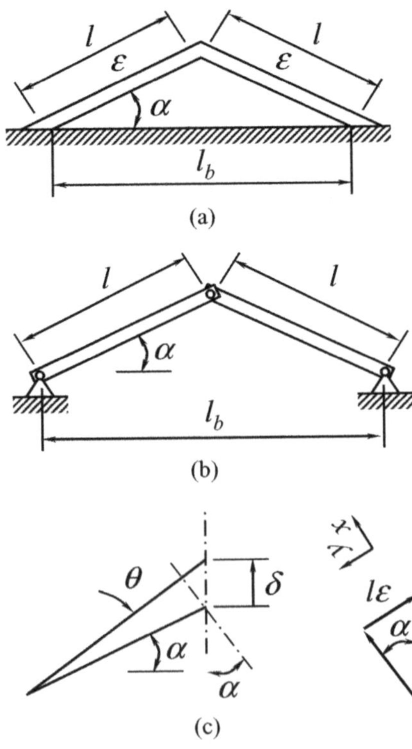

Figure 4.11 V-shaped beam under strain.

This relation holds if the angle α is not zero. If the angle is zero, the strain ε will generate a residual stress in the beam, which is discussed in Section 4.6.

4.4 STIFFNESS TRANSFORMATION

In the foregoing sections we derived the stiffness of element structures. As noted during the derivation of the stiffness of the element structures, the algebraic work to obtain the final form of the stiffness matrix is complicated and generally takes a long time. If we have a full set of stiffnesses of a flexure with respect to a coordinate system and if we wish to obtain the stiffness with respect to a different coordinate system, we can use a stiffness transformation described below. Since many MEMS structures, including movable structures and springs attached to movable structures, are in a plane (e.g., the x–y plane in Fig. 4.1), a stiffness transformation in the same plane is useful in obtaining the stiffness with respect to a different coordinate system.

Transformation of Stiffness of In-Plane Motion Springs

We can consider two transformations of stiffness related to in-plane motion: mirror and rotation of the coordinates. Combining the two transformations of the stiffness, we can obtain the transformed stiffness from the given stiffness matrix. We begin by considering a general coplanar flexure in Fig. 4.12a, which acts as a spring. The stiffness and force–displacement relation of the flexure shown in Fig. 4.12a may be obtained in the following form if we repeat the procedure described in the foregoing sections.

$$\begin{Bmatrix} F_x \\ F_y \\ M_\theta \end{Bmatrix} = \begin{bmatrix} k_{xx} & k_{xy} & k_{x\theta} \\ k_{yx} & k_{yy} & k_{y\theta} \\ k_{\theta x} & k_{\theta y} & k_{\theta\theta} \end{bmatrix} \begin{Bmatrix} \delta_x \\ \delta_y \\ \theta \end{Bmatrix} \tag{4.78}$$

If the flexure is mirrored about the vertical line (i.e., the y axis), we have the flexure in Fig. 4.12b, where new coordinates x' and y', and forces and moment corresponding to the new coordinates, are introduced for convenience. Comparing Fig. 4.12a and b gives the following relationships:

$$F_x = -F_{x'} \qquad F_y = -F_{y'} \qquad M_\theta = -M_{\theta'}$$

$$x = -x' \qquad y = y' \qquad \theta = -\theta'$$

Substituting the relations above into the force–displacement relation above, we find that

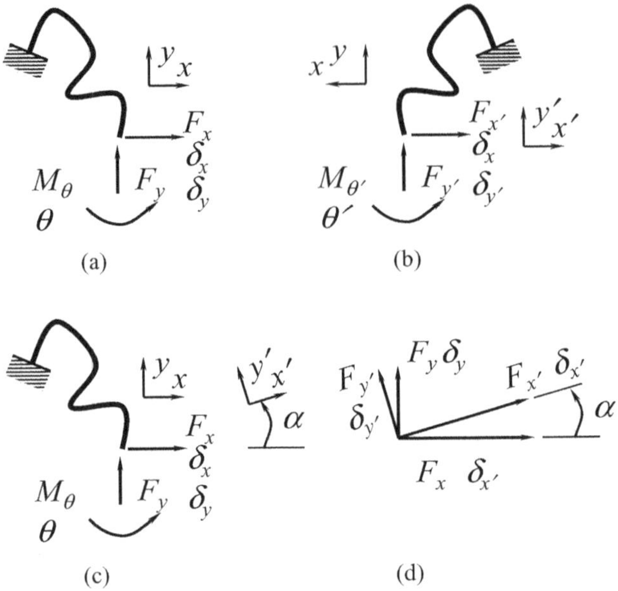

Figure 4.12 Stiffness transformation of a spring in a plane.

$$\begin{Bmatrix} -F_{x'} \\ F_{y'} \\ -M_{\theta'} \end{Bmatrix} = \begin{bmatrix} k_{xx} & k_{xy} & k_{x\theta} \\ k_{yx} & k_{yy} & k_{y\theta} \\ k_{\theta x} & k_{\theta y} & k_{\theta\theta} \end{bmatrix} \begin{Bmatrix} -\delta_{x'} \\ \delta_{y'} \\ -\theta' \end{Bmatrix}$$

Since the first equation of the expression above may be written as

$$F_{x'} = -[k_{xx}(-\delta_{x'}) + k_{xy}\delta_{y'} + k_{x\theta}(-\theta')] = k_{xx}\delta_{x'} - k_{xy}\delta_{y'} + k_{x\theta}\theta'$$

and others are expressed in similar form, the force–displacement relation of the mirrored flexure is given by

$$\begin{Bmatrix} F_{x'} \\ F_{y'} \\ M_{\theta'} \end{Bmatrix} = \begin{bmatrix} k_{xx} & -k_{xy} & k_{x\theta} \\ -k_{yx} & k_{yy} & -k_{y\theta} \\ k_{\theta x} & -k_{\theta y} & k_{\theta\theta} \end{bmatrix} \begin{Bmatrix} \delta_{x'} \\ \delta_{y'} \\ \theta' \end{Bmatrix} \tag{4.79}$$

Note from equation (4.79) that k_{xx}, k_{yy}, $k_{\theta\theta}$, $k_{x\theta}$, and $k_{\theta x}$ remain unchanged during the mirror operation while the others' signs are changed during the same operation.

Let us consider the rotation of the coordinate by an angle α in Fig. 4.12c, where the rotated coordinate system is labeled x' and y'. Referring to the rotation of the coordinates and forces in Fig. 4.12d, we may write the forces and moment in the x–y coordinates as functions of those in the x'–y' coordinates:

$$F_x = \cos\alpha F_{x'} - \sin\alpha F_{y'}$$
$$F_y = \sin\alpha F_{x'} + \cos\alpha F_{y'}$$
$$M_\theta = M_{\theta'}$$

The last equation is obtained from the fact that the moment M_θ remains unchanged for the rotation. For convenience, the preceding equations are expressed in matrix form as

$$\begin{Bmatrix} F_x \\ F_y \\ M_\theta \end{Bmatrix} = \mathbf{T}_1 \begin{Bmatrix} F_{x'} \\ F_{y'} \\ M_{\theta'} \end{Bmatrix} \tag{4.80}$$

where

$$\mathbf{T}_1 = \begin{bmatrix} \cos\alpha & -\sin\alpha & 0 \\ \sin\alpha & \cos\alpha & 0 \\ 0 & 0 & 1 \end{bmatrix} \tag{4.81}$$

is called the *transformation matrix*. The same transform holds for the displacement vector during the same rotation in Fig. 4.12d:

$$\begin{Bmatrix} \delta_x \\ \delta_y \\ \theta \end{Bmatrix} = \mathbf{T}_1 \begin{Bmatrix} \delta_{x'} \\ \delta_{y'} \\ \theta' \end{Bmatrix} \tag{4.82}$$

Since we wish to obtain the force–displacement relation in the new coordinate, we substitute the preceding two equations into the force–displacement relation (4.78) and find that

$$\mathbf{T}_1 \begin{Bmatrix} F_{x'} \\ F_{y'} \\ M_{\theta'} \end{Bmatrix} = \mathbf{K}_1 \mathbf{T}_1 \begin{Bmatrix} \delta_{x'} \\ \delta_{y'} \\ \theta' \end{Bmatrix}$$

where $\mathbf{K}_1$ denotes the stiffness matrix of (4.78). Upon multiplying the preceding equation by the inverse transformation matrix $\mathbf{T}_1^{-1}$, we find the force–displacement relation with respect to the new coordinates x'–y' as follows:

$$\begin{Bmatrix} F_{x'} \\ F_{y'} \\ M_{\theta'} \end{Bmatrix} = \mathbf{T}_1^{-1}\mathbf{K}_1\mathbf{T}_1 \begin{Bmatrix} \delta_{x'} \\ \delta_{y'} \\ \theta' \end{Bmatrix} \tag{4.83}$$

where

$$\mathbf{T}_1^{-1} = \begin{bmatrix} \cos\alpha & \sin\alpha & 0 \\ -\sin\alpha & \cos\alpha & 0 \\ 0 & 0 & 1 \end{bmatrix} \tag{4.84}$$

is the inverse of the transformation matrix $\mathbf{T}_1$, which is equivalent to the transpose of $\mathbf{T}_1$ (i.e., $\mathbf{T}_1^{-1} = \mathbf{T}_1^{\mathrm{T}}$ in this case). The transformed stiffness denoted by $\mathbf{K}_1'$ is defined as

$$\mathbf{K}_1' = \mathbf{T}_1^{-1}\mathbf{K}_1\mathbf{T}_1 \tag{4.85}$$

which may be written in the following form:

$$\mathbf{K}_1' = \begin{bmatrix} \cos\alpha & \sin\alpha & 0 \\ -\sin\alpha & \cos\alpha & 0 \\ 0 & 0 & 1 \end{bmatrix} \begin{bmatrix} k_{xx} & k_{xy} & k_{x\theta} \\ k_{yx} & k_{yy} & k_{y\theta} \\ k_{\theta x} & k_{\theta y} & k_{\theta\theta} \end{bmatrix} \begin{bmatrix} \cos\alpha & -\sin\alpha & 0 \\ \sin\alpha & \cos\alpha & 0 \\ 0 & 0 & 1 \end{bmatrix}$$

The transformed stiffness $\mathbf{K}_1'$ is then expressed as

$$\mathbf{K}_1' = \begin{bmatrix} k_{x'x'} & k_{x'y'} & k_{x'\theta'} \\ k_{y'x'} & k_{y'y'} & k_{y'\theta'} \\ k_{\theta'x'} & k_{\theta'y'} & k_{\theta'\theta'} \end{bmatrix}$$
$$= \begin{bmatrix} k_{xx}c^2 + k_{yy}s^2 + 2k_{xy}sc & (-k_{xx} + k_{yy})sc + k_{xy}(c^2 - s^2) & k_{x\theta}c + k_{y\theta}s \\ (-k_{xx} + k_{yy})sc + k_{xy}(c^2 - s^2) & k_{xx}s^2 + k_{yy}c^2 - 2k_{xy}sc & -k_{x\theta}s + k_{y\theta}c \\ k_{x\theta}c + k_{y\theta}s & -k_{x\theta}s + k_{y\theta}c & k_{\theta\theta} \end{bmatrix} \tag{4.86}$$

where $c = \cos\alpha$ and $s = \sin\alpha$.

Transformation of Stiffness of Out-of-Plane Motion Springs

To analyze the out-of-plane force–displacement relation, the following equation, which was described in the foregoing section, may be used:

$$\begin{Bmatrix} F_z \\ M_\phi \\ T \end{Bmatrix} = \mathbf{K}_2 \begin{Bmatrix} \delta_z \\ \phi \\ \eta \end{Bmatrix} \tag{4.87}$$

where

$$\mathbf{K}_2 = \begin{bmatrix} k_{zz} & k_{z\phi} & k_{z\eta} \\ k_{\phi x} & k_{\phi\phi} & k_{\phi\eta} \\ k_{\eta z} & k_{\eta\phi} & k_{\eta\eta} \end{bmatrix} \tag{4.88}$$

As described in the foregoing transformation, we shall obtain the transformed stiffness of $\mathbf{K}_2$ as the coordinates are mirrored in Fig. 4.13b and rotated in Fig. 4.13c. Let us first consider the mirrored flexure in Fig. 4.13b. Referring to Fig. 4.13a and b, we have the following relation:

$$F_z = -F_{z'} \qquad M_\phi = -M_{\phi'} \qquad T = T'$$

$$z = -z' \qquad \phi = -\phi' \qquad \eta = \eta'$$

Substituting the relations above into the foregoing force–displacement relation gives

$$\begin{Bmatrix} -F_{z'} \\ -M_{\phi'} \\ T \end{Bmatrix} = \begin{bmatrix} k_{zz} & k_{z\phi} & k_{z\eta} \\ k_{\phi x} & k_{\phi\phi} & k_{\phi\eta} \\ k_{\eta z} & k_{\eta\phi} & k_{\eta\eta} \end{bmatrix} \begin{Bmatrix} -\delta_{z'} \\ -\phi' \\ \eta' \end{Bmatrix}$$

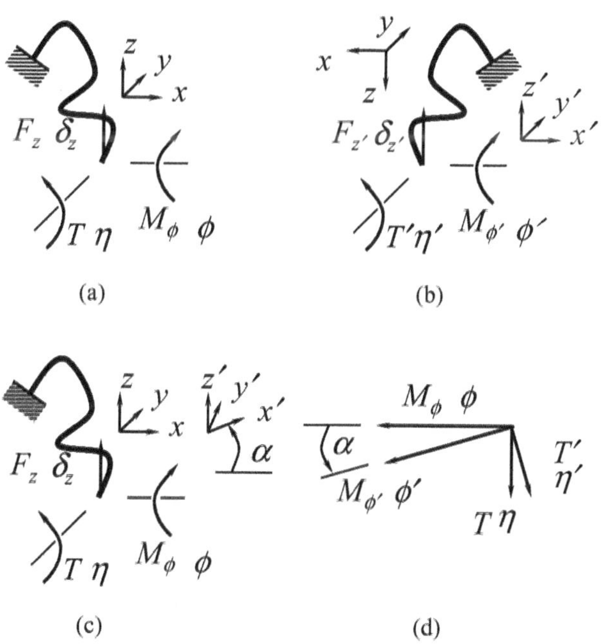

Figure 4.13 Transformation of stiffness associated with out-of-plane motion.

The equation above is rewritten to give the force–displacement relation of the mirrored flexure:

$$\begin{Bmatrix} F_{z'} \\ M_{\phi'} \\ T \end{Bmatrix} = \begin{bmatrix} k_{zz} & k_{z\phi} & -k_{z\eta} \\ k_{\phi x} & k_{\phi\phi} & -k_{\phi\eta} \\ -k_{\eta z} & -k_{\eta\phi} & k_{\eta\eta} \end{bmatrix} \begin{Bmatrix} \delta_{z'} \\ \phi' \\ \eta' \end{Bmatrix} \tag{4.89}$$

We consider the rotation of the coordinate by an angle α in the x–y plane in Fig. 4.13c, where the rotated coordinate system is labeled x' and y'. Referring to the rotation of the coordinates and corresponding force and moments in Fig. 4.13d, the force and moments in the x', y', and z' coordinates are expressed as functions of those with respect to the original coordinates as follows:

$$\begin{aligned} F_z &= F_{z'} \\ M_\phi &= \cos\alpha M_{\phi'} - \sin\alpha T' \\ T &= \sin\alpha M_{\phi'} + \cos\alpha T' \end{aligned}$$

The preceding equations are expressed in matrix form as

$$\begin{Bmatrix} F_z \\ M_\phi \\ T \end{Bmatrix} = \mathbf{T}_2 \begin{Bmatrix} F_{z'} \\ M_{\phi'} \\ T' \end{Bmatrix} \tag{4.90}$$

where

$$\mathbf{T}_2 = \begin{bmatrix} 1 & 0 & 0 \\ 0 & \cos\alpha & -\sin\alpha \\ 0 & \sin\alpha & \cos\alpha \end{bmatrix} \tag{4.91}$$

represents the transformation matrix for the out-of-plane motion. Similarly, the same transformation holds for the displacement vector during the same rotation in Fig. 4.13d:

$$\begin{Bmatrix} \delta_z \\ \phi \\ \eta \end{Bmatrix} = \mathbf{T}_2 \begin{Bmatrix} \delta_{z'} \\ \phi' \\ \eta' \end{Bmatrix} \tag{4.92}$$

Substituting equation (4.92) into the force–displacement relation (4.87) and multiplying the resulting equation by the inverse transformation matrix yields

$$\begin{Bmatrix} F_{z'} \\ M_{\phi'} \\ T' \end{Bmatrix} = \mathbf{T}_2^{-1}\mathbf{K}_2\mathbf{T}_2 \begin{Bmatrix} \delta_{z'} \\ \phi' \\ \eta' \end{Bmatrix} \tag{4.93}$$

where

$$\mathbf{T}_2^{-1} = \begin{bmatrix} 1 & 0 & 0 \\ 0 & \cos\alpha & \sin\alpha \\ 0 & -\sin\alpha & \cos\alpha \end{bmatrix} \tag{4.94}$$

is the inverse of the transformation matrix $\mathbf{T}_2$, which is equal to the transpose of $\mathbf{T}_2$. The transformed stiffness is denoted by $\mathbf{K}_2'$ as follows:

$$\mathbf{K}_2' = \mathbf{T}_2^{-1}\mathbf{K}_2\mathbf{T}_2 \tag{4.95}$$

The preceding equation may be expressed as

$$\mathbf{K}_2' = \begin{bmatrix} 1 & 0 & 0 \\ 0 & \cos\alpha & \sin\alpha \\ 0 & -\sin\alpha & \cos\alpha \end{bmatrix} \begin{bmatrix} k_{zz} & k_{z\phi} & k_{z\eta} \\ k_{\phi x} & k_{\phi\phi} & k_{\phi\eta} \\ k_{\eta z} & k_{\eta\phi} & k_{\eta\eta} \end{bmatrix} \begin{bmatrix} 1 & 0 & 0 \\ 0 & \cos\alpha & -\sin\alpha \\ 0 & \sin\alpha & \cos\alpha \end{bmatrix}$$

and is written in the form

$$\mathbf{K}_2' = \begin{bmatrix} k_{z'z'} & k_{z'\phi'} & k_{z'\eta'} \\ k_{\phi'z'} & k_{\phi'\phi'} & k_{\phi'\eta'} \\ k_{\eta'z'} & k_{\eta'\phi'} & k_{\eta'\eta'} \end{bmatrix}$$
$$= \begin{bmatrix} k_{zz} & k_{z\phi}c + k_{z\eta}s & -k_{z\phi}s + k_{z\eta}c \\ k_{z\phi}c + k_{z\eta}s & k_{\phi\phi}c^2 + k_{\eta\eta}s^2 + 2k_{\phi\eta}sc & (-k_{\phi\phi} + k_{\eta\eta})sc + k_{\phi\eta}(c^2 - s^2) \\ -k_{z\phi}s + k_{z\eta}c & (-k_{\phi\phi} + k_{\eta\eta})sc + k_{\phi\eta}(c^2 - s^2) & k_{\phi\phi}s^2 + k_{\eta\eta}c^2 - 2k_{\phi\eta}sc \end{bmatrix} \tag{4.96}$$

Transformation of a Compliance Matrix

We have obtained the stiffness transformation matrices with respect to the coordinate axes that are rotated by an angle in a plane. The same transformation holds for the compliance matrix. The transformed compliance matrix may be obtained by repeating the procedure for the stiffness transformation. We note that all transformations for the force and displacement hold for the compliance matrix, so we write down the transformed compliance matrices directly. Substituting c for k and $\mathbf{C}$ for $\mathbf{K}$ in the preceding transformed stiffness matrices, we have the following: a transformed compliance matrix after the mirror and rotations (Fig. 4.12b and c),

$$\begin{Bmatrix} \delta_{x'} \\ \delta_{y'} \\ \theta' \end{Bmatrix} = \mathbf{C}_1' \begin{Bmatrix} F_{x'} \\ F_{y'} \\ M_{\theta'} \end{Bmatrix} \tag{4.97}$$

where for the mirror operation,

$$\mathbf{C}_1' = \begin{bmatrix} c_{xx} & -c_{xy} & c_{x\theta} \\ -c_{yx} & c_{yy} & -c_{y\theta} \\ c_{\theta x} & -c_{\theta y} & c_{\theta\theta} \end{bmatrix} \tag{4.98}$$

and for the rotation operation,

$$\begin{Bmatrix} \delta_{x'} \\ \delta_{y'} \\ \theta' \end{Bmatrix} = \mathbf{C}_1' \begin{Bmatrix} F_{x'} \\ F_{y'} \\ M_{\theta'} \end{Bmatrix} \tag{4.99}$$

where

$$\mathbf{C}_1' = \begin{bmatrix} c_{x'x'} & c_{x'y'} & c_{x'\theta'} \\ c_{y'x'} & c_{y'y'} & c_{y'\theta'} \\ c_{\theta'x'} & c_{\theta'y'} & c_{\theta'\theta'} \end{bmatrix}$$
$$= \begin{bmatrix} c_{xx}c^2 + c_{yy}s^2 + 2c_{xy}sc & (-c_{xx} + c_{yy})sc + c_{xy}(c^2 - s^2) & c_{x\theta}c + c_{y\theta}s \\ (-c_{xx} + c_{yy})sc + c_{xy}(c^2 - s^2) & c_{xx}s^2 + c_{yy}c^2 - 2c_{xy}sc & -c_{x\theta}s + c_{y\theta}c \\ c_{x\theta}c + c_{y\theta}s & -c_{x\theta}s + c_{y\theta}c & c_{\theta\theta} \end{bmatrix} \tag{4.100}$$

where $s = \sin\theta$ and $c = \cos\theta$. In the preceding equations, $c_{xx}, \ldots, c_{\theta\theta}$ denote the components of the compliance matrix $\mathbf{C}_1$.

We also have a transformed compliance matrix after the mirror and rotations (Fig. 4.13b and c):

$$\begin{Bmatrix} \delta_{x'} \\ \phi' \\ T' \end{Bmatrix} = \mathbf{C}_2' \begin{Bmatrix} F_{z'} \\ M_{\phi'} \\ T \end{Bmatrix} \tag{4.101}$$

where for the mirror operation,

$$\mathbf{C}_2' = \begin{bmatrix} c_{zz} & c_{z\phi} & -c_{z\eta} \\ c_{\phi x} & c_{\phi\phi} & -c_{\phi\eta} \\ -c_{\eta z} & -c_{\eta\phi} & c_{\eta\eta} \end{bmatrix} \tag{4.102}$$

and for the rotation operation,

$$\mathbf{C}_2' = \begin{bmatrix} c_{z'z'} & c_{z'\phi'} & c_{z'\eta'} \\ c_{\phi'z'} & c_{\phi'\phi'} & c_{\phi'\eta'} \\ c_{\eta'z'} & c_{\eta'\phi'} & c_{\eta'\eta'} \end{bmatrix}$$

$$= \begin{bmatrix} c_{zz} & c_{z\phi}c + k_{z\eta}s & -c_{z\phi}s + c_{z\eta}c \\ c_{z\phi}c + c_{z\eta}s & c_{\phi\phi}c^2 + c_{\eta\eta}s^2 + 2c_{\phi\eta}sc & (-c_{\phi\phi} + c_{\eta\eta})sc + c_{\phi\eta}(c^2 - s^2) \\ -c_{z\phi}s + c_{z\eta}c & (-c_{\phi\phi} + c_{\eta\eta})sc + c_{\phi\eta}(c^2 - s^2) & c_{\phi\phi}s^2 + c_{\eta\eta}c^2 - 2c_{\phi\eta}sc \end{bmatrix} \quad (4.103)$$

In the preceding equations, $c_{zz}, \ldots, c_{\eta\eta}$ denote the components of the compliance matrix $\mathbf{C}_2$.

Stiffness of Springs with Respect to a Different Coordinate System

So far we have used the definitions of the force, moment, and linear and angular displacements illustrated in Fig. 4.3 to obtain positive components of compliance matrices for the planar springs in Fig. 4.1 and to easily obtain the corresponding stiffness matrices. These definitions of the parameters are not determined by the right-hand rule. Readers wishing to use a set of parameters following the right-hand rule can use the parameters, and the stiffness can simply be transformed for the new coordinate systems. The coordinate system used in Fig. 4.3 and a new coordinate system following the right-hand rule are illustrated in Fig. 4.14a and b, respectively. When we compare the systems, we find that all axes, forces, and moments are identical except the angles ϕ and η and the moments M_ϕ and M_η. Since all parameters are the same in the x–y plane, the stiffness $\mathbf{K}_1$ for in-plane motion in the new coordinate system is the same as that in the previous system; that is,

$$\mathbf{K}_1' = \mathbf{K}_1$$

However, the stiffness $\mathbf{K}_2$ is changed when we use a different coordinate system. Since $M_\phi = -M_{\phi'}$, $\phi = -\phi'$, $T_\phi = -T'$ and $\eta = -\eta'$, the force–displacement relation (4.21) may be written in the new coordinate system as follows:

$$\begin{Bmatrix} F_{z'} \\ -M_{\phi'} \\ -T' \end{Bmatrix} = \begin{bmatrix} k_{zz} & k_{z\phi} & k_{z\eta} \\ k_{\phi x} & k_{\phi\phi} & k_{\phi\eta} \\ k_{\eta z} & k_{\eta\phi} & k_{\eta\eta} \end{bmatrix} \begin{Bmatrix} \delta_{z'} \\ -\phi' \\ -\eta' \end{Bmatrix}$$

The equation above is rewritten as the force–displacement relation in the new coordinate system:

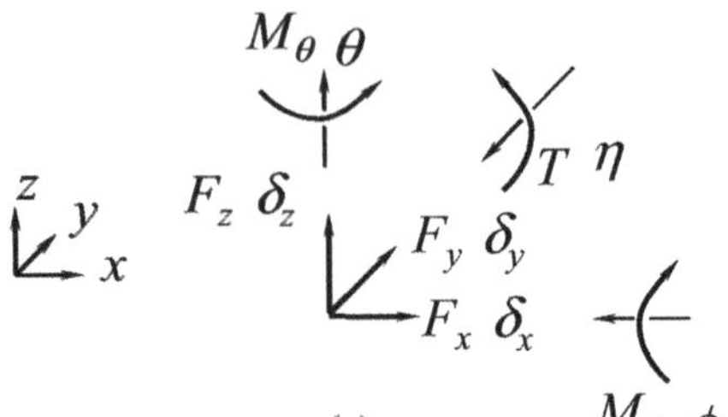

(a)

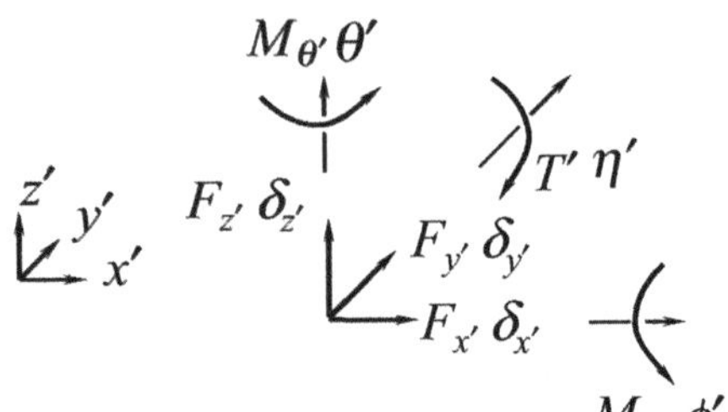

(b)

Figure 4.14 Change in the direction of the moment M_ϕ and T.

$$\begin{Bmatrix} F_{z'} \\ M_{\phi'} \\ T' \end{Bmatrix} = \begin{bmatrix} k_{zz} & -k_{z\phi} & -k_{z\eta} \\ -k_{\phi x} & k_{\phi\phi} & k_{\phi\eta} \\ -k_{\eta z} & k_{\eta\phi} & k_{\eta\eta} \end{bmatrix} \begin{Bmatrix} \delta_{z'} \\ \phi' \\ \eta' \end{Bmatrix}$$

From the equation above, the stiffness $\mathbf{K}_2$ for the out-of-plane motion is obtained as

$$\mathbf{K}_2' = \begin{bmatrix} k_{zz} & -k_{z\phi} & -k_{z\eta} \\ -k_{\phi x} & k_{\phi\phi} & k_{\phi\eta} \\ -k_{\eta z} & k_{\eta\phi} & k_{\eta\eta} \end{bmatrix} \tag{4.104}$$

Similarly, the compliance matrix $\mathbf{C}_1'$ is obtained as $\mathbf{C}_1' = \mathbf{C}_1$, and $\mathbf{C}_2'$ is obtained as follows:

$$\mathbf{C}_2' = \begin{bmatrix} c_{zz} & -c_{z\phi} & -c_{z\eta} \\ -c_{\phi x} & c_{\phi\phi} & c_{\phi\eta} \\ -c_{\eta z} & c_{\eta\phi} & c_{\eta\eta} \end{bmatrix}$$

Example 4.1 Using the typical structures described earlier in the chapter, we can make many variations whose configurations are changed or whose springs are tilted. The microstructure shown in Fig. 4.15 is a variation of the original structure of Fig. 4.1b, whose L-shaped beams make an angle of α with a vertical line. It is desired to evaluate the displacement δ when a force F is applied to the central rigid beam as shown in Fig. 4.15a.

Even though the springs are rotated by the angle α, a microstructure is symmetric with respect to the horizontal and vertical axes passing through the center of the structure. As a result, the central beam can move rightward while all other displacements and rotations of the central beam vanish due to the symmetry argument. Furthermore, referring to Fig. 4.15b, we need only $k_{x'x'}$ to obtain the overall stiffness in the rightward direction in Fig. 4.15a. We shall first evaluate the stiffness of the inclined spring in Fig. 4.15b and then obtain the overall stiffness by using the stiffness formula for the springs in parallel (Fig. 4.15c), which is described in (4.1). Since the stiffness of the L-shaped beam is given by (4.57), it is rewritten for further discussion:

$$\mathbf{K}_1 = \begin{bmatrix} \dfrac{3EI_{za}}{a^3}\dfrac{4+\beta p}{1+\beta p} & -\dfrac{9EI_{za}}{a^3}\dfrac{1}{\beta(1+\beta p)} & -\dfrac{3EI_{za}}{a^2}\dfrac{2+\beta p}{1+\beta p} \\ & \dfrac{3EI_{za}}{a^3}\dfrac{1+4\beta p}{\beta^3 p(1+\beta p)} & 3\dfrac{EI_{za}}{a^2}\dfrac{1}{\beta(1+\beta p)} \\ \text{sym.} & & \dfrac{EI_{za}}{a}\dfrac{4+3\beta p}{1+\beta p} \end{bmatrix} \tag{a}$$

In Fig. 4.15b we know that the x' and y' axes are rotated by $-\alpha$ with respect to the x and y axes. Substituting $-\alpha$ for α in (4.86), the transformed stiffness $k_{x'x'}$ is obtained as

$$k_{x'x'} = k_{xx}\cos^2\alpha + k_{yy}\sin^2\alpha - 2k_{xy}\sin\alpha\cos\alpha \tag{b}$$

Substitution of k_{xx}, k_{yy}, and k_{xy} of (a) into equation (b) leads to

$$\begin{aligned} k_{x'x'} &= \frac{3EI_{za}}{a^3}\frac{4+\beta p}{1+\beta p}\cos^2\alpha + \frac{3EI_{za}}{a^3}\frac{1+4\beta p}{\beta^3 p(1+\beta p)}\sin^2\alpha \\ &\quad - 2\left[-\frac{9EI_{za}}{a^3}\frac{1}{\beta(1+\beta p)}\right]\sin\alpha\cos\alpha \\ &= \frac{3EI_{za}}{a^3}\left[\frac{4+\beta p}{1+\beta p}\left(\cos^2\alpha + \frac{1}{\beta^3 p}\sin^2\alpha\right) + 6\frac{1}{\beta(1+\beta p)}\sin\alpha\cos\alpha\right] \end{aligned} \tag{c}$$

The right springs in Fig. 4.15 can be considered as the left springs mirrored about the vertical line, so the stiffness k_{xx} of the right spring is equal to that of the left spring from the mirror operation (4.79). Consequently, the overall

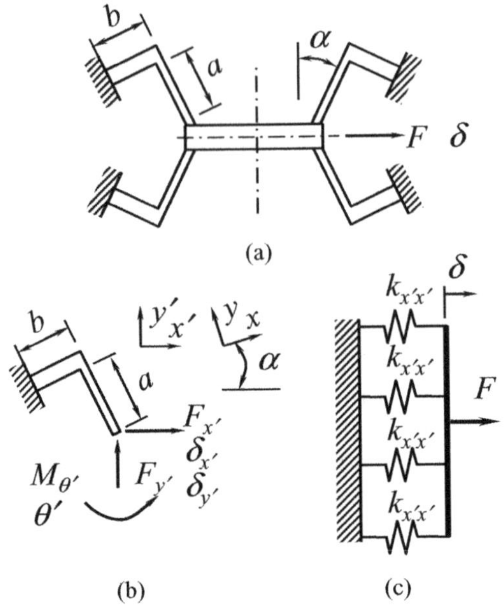

Figure 4.15 Stiffness of a structure with inclined L-shaped springs.

stiffness of the structure in the δ direction can be thought of as the effective stiffness of the four parallel springs in Fig. 4.15c, which is given by

$$k = \frac{F}{\delta} = 4k_{x'x'} = \frac{12EI_{za}}{a^3}\left[\frac{4+\beta p}{1+\beta p}\left(\cos^2\alpha + \frac{1}{\beta^3 p}\sin^2\alpha\right) + 6\frac{1}{\beta(1+\beta p)}\sin\alpha\cos\alpha\right] \tag{d}$$

If we substitute 0 for α in equation (d), we have the stiffness of the microstructure of $\alpha = 0$ in Fig. 4.15a as follows:

$$k\big|_{\alpha=0} = \frac{12EI_{za}}{a^3}\frac{4+\beta p}{1+\beta p} \tag{e}$$

If we assume further that the length b approaches zero [i.e., $\beta \to 0$ in equation (e)], the preceding stiffness becomes the overall stiffness of the microstructure of four straight beams (Fig. 4.1a):

$$k\big|_{\alpha=0} = 4\frac{12EI_{za}}{a^3} = \frac{48EI_{za}}{a^3} \tag{f}$$

The stiffness of other microstructures with inclined flexures can be obtained similarly.

4.5 STATIC BEHAVIOR OF PLANAR STRUCTURES

In the preceding sections we have defined common microsprings, derived their stiffness, and obtained corresponding transformed stiffness. The static behavior of the planar microstructures is related to the stiffness that has been expressed in matrix form in preceding sections. To clarify the static behavior of the microstructures, we combine the force and moment equilibrium equations discussed in Chapter 3, the geometry of displacements, and the force–displacement relations expressed using the stiffness matrices derived in preceding sections. In Example 4.1 we have shown the overall stiffness of the planar structure as force is applied in the horizontal direction. In this section we describe in detail how to obtain the overall stiffness of microstructures that consist of movable structures and flexures.

Stiffness of Planar Structures for In-Plane Displacements

Design of symmetric microstructures such as the structures in Fig. 4.1 is much easier than that of asymmetric microstructures since the symmetry cancels many displacements. Let us consider the structure with springs shown in Fig. 4.16a, which is symmetric about the dashed-dotted lines that pass through the center of the structures. We assume that the movable structures at the center are rigid and that the springs are made of planar flexures.

If the microstructure in Fig. 4.15a is subjected to a force F_1 in the x direction, the vertical and angular displacements of the movable structure vanish due to the symmetry, and the movable structure is displaced by δ_1. The displacement δ_1 may be obtained by the procedure described in Example 4.1. We illustrate a general procedure to obtain the linear and angular stiffnesses of the structure in Fig. 4.16a. The general procedure will be extended to obtain overall torsional stiffness of structures. If we take the local coordinate systems shown in the free-body diagram of Fig. 4.16b, the stiffness matrices of springs are identical with respect to the local coordinate systems, as noted in Section 4.3. For example, the forces and moment acting on the upper right spring labeled 1 in Fig. 4.16a are given by (4.12) and (4.15), which is rewritten as

$$\begin{Bmatrix} F_{x1} \\ F_{y1} \\ M_{\theta 1} \end{Bmatrix} = \begin{bmatrix} k_{xx} & k_{xy} & k_{x\theta} \\ k_{yx} & k_{yy} & k_{y\theta} \\ k_{\theta x} & k_{\theta y} & k_{\theta\theta} \end{bmatrix} \begin{Bmatrix} \delta_{x1} \\ \delta_{y1} \\ \theta_1 \end{Bmatrix}$$

where the force vector represents the forces in the x_1 and y_1 directions and the moment at the end of the upper right spring in Fig. 4.16a, and the displacement vectors denote the linear and angular displacements, corresponding to the force vector. The force–displacement relations acting on other springs (labeled 2, 3, and 4) are the same as the above, but the numbers in the subscripts in the preceding equation are replaced with 2, 3, and 4, depending on the spring

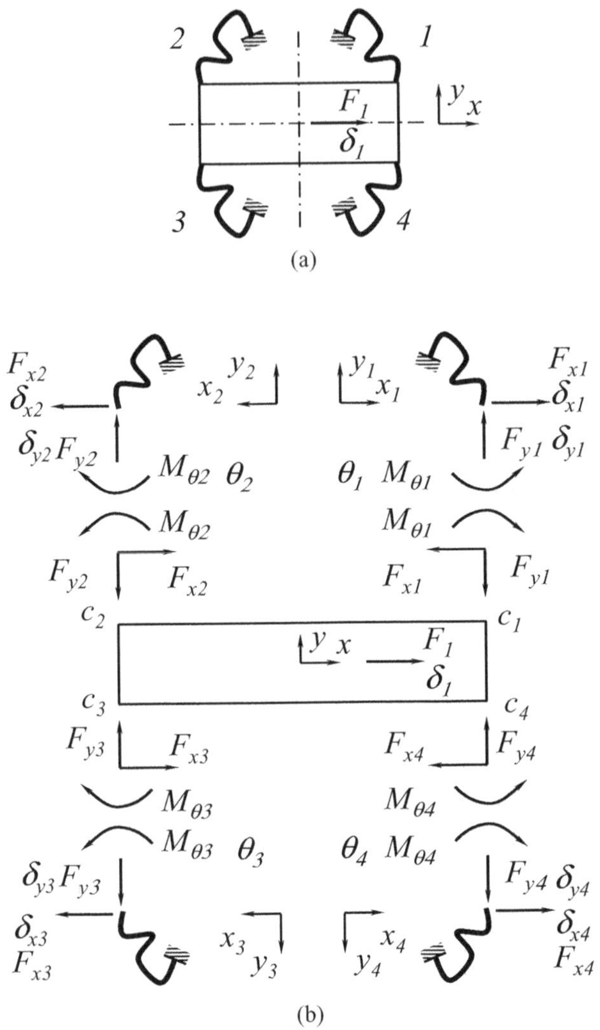

Figure 4.16 Stiffness of structures in coplanar motion.

locations. We have the following geometric conditions of the springs when the movable structure moves rightward by δ_1:

For spring 1: $\delta_{x1} = \delta_1$, $\delta_{y1} = 0$, $\theta_1 = 0$

For spring 2: $\delta_{x2} = -\delta_1$, $\delta_{y2} = 0$, $\theta_2 = 0$

For spring 3: $\delta_{x3} = -\delta_1$, $\delta_{y3} = 0$, $\theta_3 = 0$

For spring 4: $\delta_{x4} = \delta_1$, $\delta_{y4} = 0$, $\theta_4 = 0$

where the zero displacements in the y direction and the zero angle in the θ direction are imposed by the symmetry condition. Substituting the displacements above into the preceding force–displacement relation leads to:

For spring 1: $F_{x1} = k_{xx}\delta_1$, $F_{y1} = k_{yx}\delta_1$, $M_{\theta 1} = k_{\theta x}\delta_1$

For spring 2: $F_{x2} = -k_{xx}\delta_1$, $F_{y2} = -k_{yx}\delta_1$, $M_{\theta 2} = -k_{\theta x}\delta_1$

For spring 3: $F_{x3} = -k_{xx}\delta_1$, $F_{y3} = -k_{yx}\delta_1$, $M_{\theta 3} = -k_{\theta x}\delta_1$

For spring 4: $F_{x4} = k_{xx}\delta_1$, $F_{y4} = k_{yx}\delta_1$, $M_{\theta 4} = k_{\theta x}\delta_1$

Now we consider the force and moment equilibrium of the movable structure in Fig. 4.16b, which gives

$$\sum F_x = -F_{x1} + F_{x2} + F_{x3} - F_{x4} + F_1 = 0$$
$$\sum F_y = -F_{y1} - F_{y2} + F_{y3} + F_{y4} = 0$$
$$\sum M = -M_{\theta 1} + M_{\theta 2} - M_{\theta 3} + M_{\theta 4} = 0$$

The second and third equations above are satisfied automatically due to the symmetry argument (if we substitute the preceding forces and moment into the second and third equations above, we can verify it). Substituting the preceding relations between the forces and the displacement δ_1 and between the moment and the displacement δ_1 into the first equation of the preceding equilibrium equations, we find the displacements due to the force F_1 acting on the movable structure in the x direction:

$$\delta_1 = \frac{F_1}{4k_{xx}}$$

From the preceding equation we can define the stiffness of the structure in Fig. 4.16a in the x direction as follows:

$$k_{l1} = \frac{F_1}{\delta_1} = 4k_{xx} \tag{4.105}$$

If we replace F_1 acting at the center in the y direction in the free-body diagram of Fig. 4.16 by a force F_2, and if we repeat the foregoing procedure, we have the stiffness of the structure in the y direction as follows:

$$k_{l2} = \frac{F_2}{\delta_2} = 4k_{yy} \tag{4.106}$$

where δ_2 denotes the displacement in the y direction due to the force F_2.

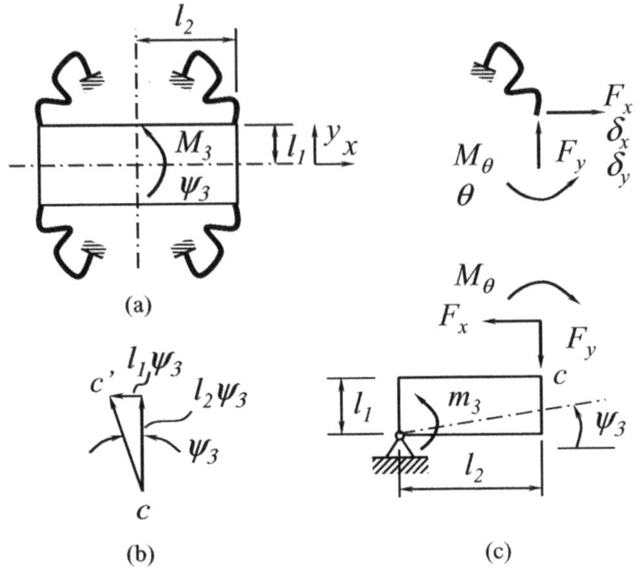

Figure 4.17 Stiffness of structures in coplanar motion.

We consider the same microstructure subjected to a moment M_3 in Fig. 4.17a. The movable structure is displaced by an in-plane angular displacement ψ_3 which is related to the applied moment M_3. Since the structure of Fig. 4.17a is symmetric about the x and y directions, the displacements in the x and y directions are zero and the movable structure is displaced by an angle ψ_3. Therefore, we consider the first quadrant (i.e., the upper right portion) of the movable structure in Fig. 4.17a, whose length and width are given by $l_2 \times l_1$. Free-body diagrams for the spring and the first quadrant, which is pined at the lower left corner, are shown in Fig. 4.17c, and the upper right corner of the movable structure is displaced from the initial position c to the new position c' during the small angular displacement ψ_3 as shown in Fig. 4.17b. In the free-body diagram, m_3 denotes one-fourth of the applied moment (M_3), and forces and moments are generated by the spring due to the angular displacement. From the moment equilibrium of the movable structure about the pinned corner in Fig. 4.17c, we have

$$m_3 + F_x l_1 - F_y l_2 - M_\theta = 0 \tag{4.107}$$

The following geometric relations are obtained from Fig. 4.17b:

$$\delta_x = -l_1 \psi_3 \qquad \delta_y = l_2 \psi_3 \qquad \theta = \psi_3 \tag{4.108}$$

where δ_x, δ_y, and θ denote the linear and angular displacements of the lower end of the spring as the movable structure is displaced by ψ_3. The force–displacement relation of the spring is given by

$$\begin{Bmatrix} F_x \\ F_y \\ M_\theta \end{Bmatrix} = \begin{bmatrix} k_{xx} & k_{xy} & k_{x\theta} \\ k_{yx} & k_{yy} & k_{y\theta} \\ k_{\theta x} & k_{\theta y} & k_{\theta\theta} \end{bmatrix} \begin{Bmatrix} \delta_x \\ \delta_y \\ \theta \end{Bmatrix} \tag{4.109}$$

Substituting (4.108) and (4.109) into the moment equilibrium equation (4.107) yields

$$\begin{aligned} m_3 &= -F_x l_1 + F_y l_2 + M_\theta \\ &= -\left[k_{xx}(-l_1\psi_3) + k_{xy}l_2\psi_3 + k_{x\theta}\psi_3\right]l_1 + \left[k_{yx}(-l_1\psi_3) + k_{yy}l_2\psi_3 + k_{y\theta}\psi_3\right]l_2 \\ &\quad + \left[k_{\theta x}(-l_1\psi_3) + k_{\theta y}l_2\psi_3 + k_{\theta\theta}\psi_3\right] \end{aligned}$$

From the foregoing equation, we obtain the torsional stiffness of a quadrant structure of Fig. 4.17a as follows:

$$\begin{aligned} k_{t3} &= \frac{m_3}{\psi_3} \\ &= k_{xx}l_1l_1 - k_{xy}l_1l_2 - k_{x\theta}l_1 - k_{yx}l_2l_1 + k_{yy}l_2l_2 + k_{y\theta}l_2 - k_{\theta x}l_1 + k_{\theta y}l_2 + k_{\theta\theta}\psi_3 \\ &= k_{xx}l_1^2 - 2k_{xy}l_1l_2 - 2k_{x\theta}l_1 + k_{yy}l_2^2 + 2k_{y\theta}l_2 + k_{\theta\theta} \end{aligned} \tag{4.110}$$

The last expression is obtained from the fact that the stiffness matrix is symmetric. Since the moment M_3 acting on the entire structure in Fig. 4.17a is four times m_3, the following moment equilibrium holds for the structure:

$$M_3 = 4m_3 = 4k_{t3}\psi_3$$

Consequently, for the angular displacement ψ_3, the torsional stiffness of the entire structure in Fig. 4.17a can be defined as

$$\begin{aligned} k_{T3} &= \frac{M_3}{\psi_3} = 4k_{t3} \\ &= 4\left(k_{xx}l_1^2 - 2k_{xy}l_1l_2 - 2k_{x\theta}l_1 + k_{yy}l_2^2 + 2k_{y\theta}l_2 + k_{\theta\theta}\right) \end{aligned} \tag{4.111}$$

Stiffness of Planar Structures Caused by Out-of-Plane Displacements

We have now obtained two linear stiffnesses, k_1 and k_2, and a torsional stiffness k_{t3}. Since the same planar structure has three more degrees of freedom (δ_3, ψ_1, ψ_2), as shown in Fig. 4.18a, the structure provides three more stiffnesses. Let us first consider the displacement δ_3 in the z direction when a force F_3 is

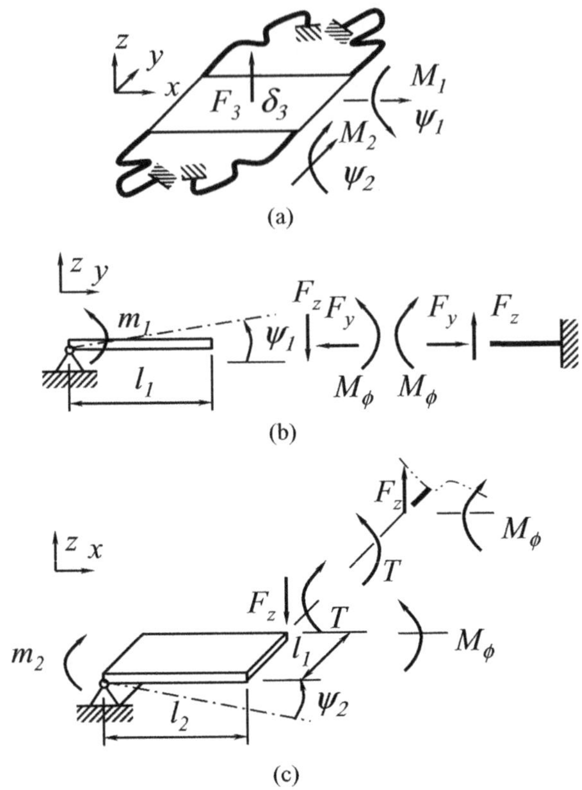

Figure 4.18 Linear and torsional stiffness of structures in out-of-plane motion.

applied at the center of the movable plate in Fig. 4.18a (the moments $M_1 = M_2 = 0$). The structure in Fig. 4.18a is symmetric about the two orthogonal lines (symmetry line) which pass through the center and which are parallel to the x and y axes, so angular displacements ψ_1 and ψ_2 must be zero. That is, the movable structure can move in the z direction without any angular displacement. We first rewrite the force–displacement relation (4.21) for general planar springs as follows:

$$\begin{Bmatrix} F_z \\ M_\phi \\ T \end{Bmatrix} = \begin{bmatrix} k_{zz} & k_{z\phi} & k_{z\eta} \\ k_{\phi z} & k_{\phi\phi} & k_{\phi\eta} \\ k_{\eta z} & k_{\eta\phi} & k_{\eta\eta} \end{bmatrix} \begin{Bmatrix} \delta_z \\ \phi \\ \eta \end{Bmatrix} \tag{4.112}$$

where the force vector on the left-hand side in equation (4.112) denotes the force and moments acting at the end of the upper right spring, and the displacement vector represents the linear and angular displacements of the end of the spring. Substituting the angles $\phi = 0$ and $\eta = 0$ into equation (4.112) gives $F_z = k_{zz}\delta_z$. We obtain the force equilibrium equation, $F_3 = 4F_z$, from a free-body diagram for the movable structure, which gives the stiffness of the structures with four identical springs as follows:

$$k_{l3} = \frac{F_3}{\delta_3} = 4k_{zz} \tag{4.113}$$

Two more torsional stiffnesses are obtained from the moment equilibrium of Fig. 4.18b and c. Let us suppose that the movable structure in Fig. 4.18a is subjected to the moment M_1 and that the other force, F_3, and moment, M_2, are zero. Under moment M_1, the movable structure is allowed to rotate by an angle ψ_1, and the other linear and angular displacements, δ_3 and ψ_2, remain zero, due to the symmetry. Thus, we can consider an upper right portion of the structure in Fig. 4.18a to analyze the stiffness that is related to the moment M_1. To obtain the stiffness, we need to draw free-body diagrams of the first quadrant of movable structure and the spring connected to the quadrant, which are shown in Fig. 4.18b. As shown in the free-body diagram, the first quadrant of the movable structure is thought of as a rigid beam (plate) that is pined at the left end and subjected to m_1 (one-fourth the applied moment M_1). The forces F_y, and F_z and the moment M_ϕ of the spring are exerted on the quadrant as shown in Fig. 4.18b. From the moment equilibrium about the left end of the quadrant in Fig. 4.18b, the moment equilibrium equation is given by

$$m_1 - F_z l_1 + F_y l \psi_1 + M_\phi = 0$$

where the force F_y is zero from (4.109) since the linear and angular displacements except ψ_1 are zero. When we consider the geometry of the pinned structure and the spring in Fig. 4.18b, the geometric conditions are given by

$$\delta_z = l_1\psi_1 \qquad \phi = -\psi_1 \qquad \eta = 0$$

To obtain the angular displacement ψ_1, we need the force–displacement relation for the out-of-plane motion, (4.112). Substitution of the force–displacement, (4.112), and the geometric conditions above into the preceding moment equilibrium equation leads to

$$\begin{aligned} m_1 &= F_z l_1 - M_\phi \\ &= \left[k_{zz}(l_1\psi_1) + k_{z\phi}(-\psi_1)\right] l_1 - \left[k_{\phi z}(l_1\psi_1) + k_{\phi\phi}(-\psi_1)\right] \end{aligned}$$

The equation above gives the stiffness corresponding to m_1 as follows:

$$\begin{aligned} k_{t1} &= \frac{m_1}{\psi_1} \\ &= k_{zz} l_1 l_1 - 2k_{z\phi} l_1 + k_{\phi\phi} \end{aligned} \tag{4.114}$$

The four springs in Fig. 4.18a resist the applied moment M_1, so the following moment equilibrium holds for the entire structure:

$$M_1 = 4m_1 = 4k_{t1}\psi_1$$

As a result, for the angular displacement ψ_1, the torsional stiffness of the entire structure in Fig. 4.18a may be written

$$\begin{aligned} k_{T1} &= \frac{M_1}{\psi_1} = 4k_{t1} \\ &= 4\left(k_{zz}l_1l_1 - 2k_{z\phi}l_1 + k_{\phi\phi}\right) \end{aligned} \tag{4.115}$$

For the moment M_2 acting along the y axis in Fig. 4.18a, we refer to the free-body diagram of the first quadrant of the structure in Fig. 4.18c and can repeat the procedure to find the stiffness associated with the moment M_2. The moment equilibrium about the pinned end of the movable structure in Fig. 4.18c gives

$$m_2 + F_z l_2 + T = 0$$

The geometric conditions from Fig. 4.18c are given by

$$\delta_z = -l_2\psi_2 \qquad \phi = 0 \qquad \eta = -\psi_2$$

Combining the preceding moment equilibrium equation and geometric conditions with (4.112), we find the moment–angular displacement relation:

$$\begin{aligned} m_2 &= -F_z l_2 - T \\ &= -\left[k_{zz}(-l_2\psi_2) + k_{z\eta}(-\psi_2)\right]l_2 - \left[k_{\eta z}(-l_2\psi_2) + k_{\eta\eta}(-\psi_2)\right] \end{aligned}$$

From the equation above, we obtain the stiffness k_{t2}, which is related to the moment m_2 and the angular displacement ψ_2:

$$\begin{aligned} k_{t2} &= \frac{m_2}{\psi_2} \\ &= k_{zz}l_2^2 + 2k_{z\eta}l_2 + k_{\eta\eta} \end{aligned} \tag{4.116}$$

Referring to Fig. 4.18a, in which the four springs support the movable structure, we find the stiffness of the overall structure as follows:

$$\begin{aligned} k_{T2} &= \frac{M_2}{\psi_2} = 4k_{t2} \\ &= 4\left(k_{zz}l_2^2 + 2k_{z\eta}l_2 + k_{\eta\eta}\right) \end{aligned} \tag{4.117}$$

4.6 RESIDUAL STRESS

Residual stress is the stress that remains in a body that is in equilibrium. It is usually caused by manufacturing processes and surface treatments (e.g., casting, forging, welding, CVD, electroplating).

Residual Stress in a Guitar

To further explain the residual stresses, we first consider in Fig. 4.19b a string of a guitar whose tension is adjusted by a tuning knob that is not shown in the figure. The string, of initial length l at a temperature T in Fig. 4.19a, is stretched by δ to generate the required tension N_0 in the string (Fig. 4.19b). The temperature T remains constant during the stretching. We first evaluate the stress due to the tension and then examine the effect of the temperature change on the stress. Referring to the stress–strain relation $\sigma = E\varepsilon$, and using the definition of strain $\varepsilon = \delta/l$, we have

$$\sigma_{0i} = E\frac{\delta}{l} \tag{4.118}$$

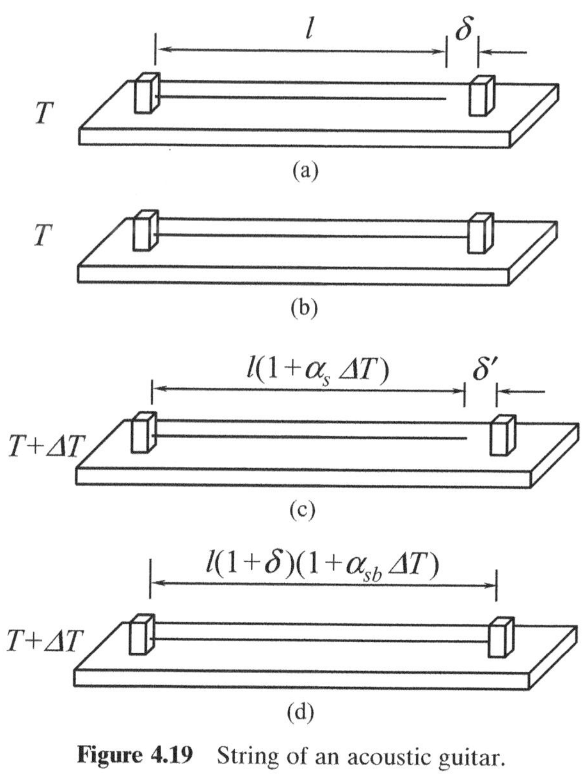

Figure 4.19 String of an acoustic guitar.

where σ_{0i} denotes an initial stress, which can be thought of as a residual stress at temperature T. Substituting the stress $\sigma_{0i} = N_{0i}/A$, where A is the cross-sectional area of the string, into equation (4.118) gives the initial tension N_{0i}:

$$N_{0i} = EA\frac{\delta}{l} \tag{4.119}$$

In Section 5.8 it will be shown that the preceding tension in a string is related to the natural frequency of the string.

Let us turn our attention to the effect of a temperature change on the stress in a string. The guitar is exposed to a temperature change ΔT (Fig. 4.19d) for a sufficient time to allow thermal expansion of the string and all other parts, so that

$$\begin{aligned}\delta' &= (l+\delta)(1+\alpha_{sb}\Delta T) - l(1+\alpha_s\Delta T)\\ &= \delta(1+\alpha_{sb}\Delta T) + l(\alpha_{sb}-\alpha_s)\Delta T\\ &\cong \delta + l(\alpha_{sb}-\alpha_s)\Delta T\end{aligned}$$

where α_{sb} and α_s denote the coefficients of thermal expansion of the substrate and the string, respectively. Substitution of the elongation above into the stress–strain relation $\sigma = E\varepsilon$, where E is Young's modulus, gives the residual stress σ_0 at the temperature $T + \Delta T$:

$$\sigma_0 = E\frac{\delta'}{l} = E\frac{\delta + l(\alpha_{sb}-\alpha_s)\Delta T}{l} = \sigma_{0i} + E(\alpha_{sb}-\alpha_s)\Delta T \tag{4.120}$$

and the residual stress difference:

$$\Delta\sigma_0 = \sigma_0 - \sigma_{0i} = E(\alpha_{sb}-\alpha_s)\Delta T$$

The axial force corresponding to the preceding stress is given by

$$N_0 = \sigma_0 A = N_{0i} + EA(\alpha_{sb}-\alpha_s)\Delta T \tag{4.121}$$

where N_{0i} represents the initial axial force (tension for the string). Note from the preceding equations that the stress in a stretched string remains constant at a temperature and that if the temperature is changed, the stress is also changed and is proportional to the temperature change ΔT. According to the definition of residual stress, the stress in a tensioned string can be defined as a residual stress since the system, including the string and substrate in Fig. 4.19b and d, is in equilibrium and no external forces are applied. The equations for the residual stress, (4.118) and (4.120), also hold for more general cases (e.g., the string is replaced by a straight beam; either the string or the substrate undergoes strain change). For these general cases, the residual stress–temperature relation (4.120) is generalized as follows:

$$\sigma_0 = \sigma_{0i} + E\varepsilon_r \tag{4.122}$$

where

$$\varepsilon_r = \varepsilon_{sb} - \varepsilon_s \tag{4.123}$$

is a relative strain and ε_{sb} and ε_s denote the average strains of the substrate and the structure suspended over the substrate, respectively. For the temperature change in Fig. 4.19b and d, the strains ε_{sb} and ε_s can be expressed as $\alpha_{sb}\Delta T$ and $\alpha_s\Delta$T, and then the general equation (4.122) for the straight structure becomes (4.120).

Residual Stress in Microbeams

Let us turn our attention to a fixed–fixed beam rectangular cross section which may be fabricated over a sacrificial layer using surface micromachining, described in Chapter 2. For convenience of explanation, we consider the two identical fixed–fixed beams shown in Fig. 4.19b, which may be fabricated on a substrate. One fixed–fixed beam is cut at the right end to obtain the cantilever beam in Fig. 4.19a. Thus, from a free body for a beam segment from the cantilever beam of Fig. 4.19a we know that the axial force of the straight cantilever beam is zero; that is, the residual stress is zero. However, the residual stress (axial stress) of the fixed–fixed beam may occur due to the fabrication process or the temperature change. In this case, as discussed in the guitar string problem, the residual stress can be considered as stress due to elongation of the cantilever beam of Fig. 4.19a to form the fixed–fixed beam in Fig. 4.9b. In this case, the stress–elongation relation (4.118) gives the residual stress in the fixed–fixed beam at temperature T as follows:

$$\sigma_{0i} = E\varepsilon \tag{4.124}$$

where E denotes the effective Young's modulus, defined in (3.5) and (3.6), and ε represents the strain, δ/l, in the fixed–fixed beam when we assume that the cantilever beam of Fig. 4.19a is deformed to form the fixed–fixed beam in Fig. 4.19b. If the fixed–fixed beam undergoes a change in relative strain (e.g., strain change due to a temperature change ΔT), the stress and strain relation (4.122) can be used to evaluate the stress in the beam,

$$\sigma_0 = \sigma_{0i} + E\varepsilon_r \tag{4.125}$$

where E denotes the effective Young's modulus, defined in (3.5) and (3.6), and

$$\varepsilon_r = \varepsilon_{sb} - \varepsilon_s$$

is the relative strain (ε_{sb} and ε_s denote the average strains of the substrate and the straight structure stressed and suspended over the substrate, respectively).

If the initial residual stress is much greater than the stress change (i.e., $\sigma_{0i} >> E\varepsilon_r$), the residual stress can be assumed to remain constant: $\sigma_0 = \sigma_{0i}$.

Residual Stress in a Structure with Rigid Portions

During the surface micromaching process decribed in Chapter 2, a structural layer on a sacrificial layer is deposited; if the residual stress exists in the structural layer, the residual stress is uniformly distributed in the layer plane (i.e., $\sigma_x = \sigma_y$). When the structural layer is cut into a desired structure (e.g., the structure in Fig. 4.20a) and released, the residual stress is redistributed depending on the structure stiffness and boundary conditions. It is noted that the structural layer and the desired structure before release can be considered as structures that are uniformly stretched or shrinked in every direction on the layer plane; this situation is similar to that of the thermally expanded structure at a uniform temperature.

Compared to the cantilever beam and fixed–fixed beam shown in Fig. 4.19, many microstructures fabricated on a substrate consist of one or more movable structures and flexures that serve as springs, as shown in Fig. 4.20a. In this case, the residual stress in the beam is different from (4.124), which is obtained from the string and beam without a rigid portion. We first briefly examine the residual stress in the structure in the longitudinal direction (i.e., the y direction) of the springs in Fig. 4.20a and then derive equations of the residual stress in the general planar structures over a substrate. Let us suppose that the structure in Fig. 4.20a is stressed due to a fabrication process (e.g., LPCVD film) or the dif-

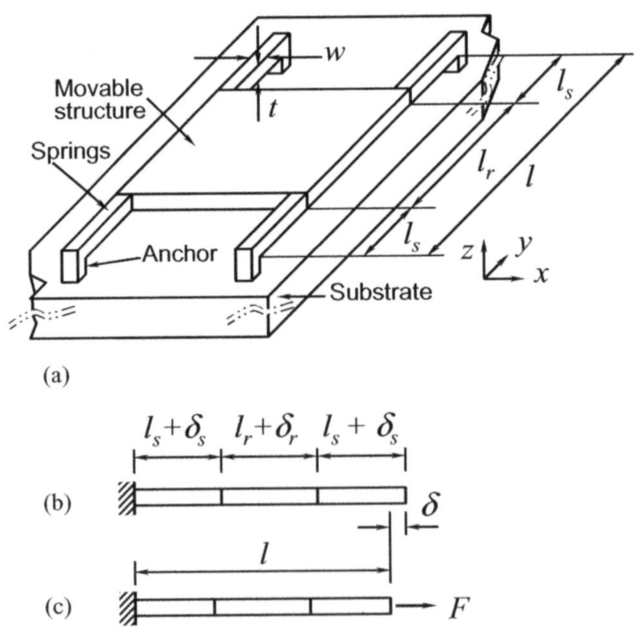

Figure 4.20 Microstructure under residual stress.

ference in the coefficients of thermal expansion. It is assumed that the movable structure is rigid since its cross-sectional area is much larger than that of springs.

Let the length of the springs and the movable plate be represented by l_s and l_r, respectively, as shown in Fig. 4.18a. Then the structure length in the y direction is $l = l_r + 2l_s$. Now we consider a structure in Fig. 4.20b (side view), which is the same structure as that in Fig. 4.20a except for the two upper anchors on the substrate. In this case, the structure elongates by the amount δ, as indicated in Fig. 4.20b (where the elongation is assumed positive). The elongation of the structure is then expressed as

$$\delta = \delta_s + \delta_r + \delta_s = (l_s + l_r + l_s)\varepsilon = l\varepsilon$$

where ε denotes the strain of the structure in the y direction. We exert a force F on the end of the springs, as shown in Fig. 4.18c, to obtain the same configuration and stress in Fig. 4.20a. Since the structure is compressed to make the same situation in Fig. 4.20a, the following strain of one spring is obtained from the fact that the two springs are contracted since the movable plate is rigid,

$$\varepsilon_y = \frac{-\delta/2}{l_s + \delta_s} \cong -\frac{\delta/2}{l_s} = -\frac{\delta}{2l_s}$$

The final expression is obtained since the beam length l_s is much larger than the displacement δ_s. Substituting the preceding two equations into the stress–strain relation, we have the stress in the straight beam in Fig. 4.20a as follows:

$$\sigma_y = E\varepsilon_y = -E\varepsilon\frac{l}{2l_s} \tag{4.126}$$

The axial force in the straight beam is then given by

$$F = \sigma_y A = -EA\varepsilon\frac{l}{2l_s} \tag{4.127}$$

where A denotes the cross-sectional area of the straight spring in Fig. 4.20a. Equation (4.127) shows that if a free structure such as a cantilever beam is compressed to form a structure whose ends are fixed at anchors, compressive stress is generated in the structure. The equation also states that if there are rigid portions in the structure, the flexible parts are more compressed and the compressive stress in the flexible parts is increased. Equation (4.127) also holds for stretching the beam to form a fixed–fixed structure. From the foregoing equation and the discussion above, we understand that if the compressed structure is released (i.e., the two upper anchors in Fig. 4.20a), the compressed structure will be expanded. This understanding of the relationship between the residual stress in the stressed structures and the strain of the corresponding released structure will be used to derive further expressions for residual stress in an arbitrarily shaped structure, described below.

Stiffness and Residual Stress of Planar Structures Due to In-Plane Motion

We study planar structures under residual stress, which has rotational symmetry, to derive an expression for their torsional stiffness and to understand the effect of the residual stress on the torsional displacements. Let us begin with the structure shown in Fig. 4.21a, which consists of a rectangle and two springs lying in the x–y plane and which has 180° rotational symmetry with respect to the center of the structure. The dimensions, moment M_3, and angular displacement ψ_3 of the structure are defined in Fig. 4.21a. We assume that the designed dimensions of Fig. 4.21a are changed to those of the stress-free structure in Fig. 4.21b and that to form the structure in Fig. 4.21a, the ends of two springs are displaced by δ_1 and δ_2, as shown in the free-body diagram of Fig. 4.21c. If we assume that the strain ε in Fig. 4.21b is positive, the following geometric relationships are obtained:

$$\delta_1 = -l_4\varepsilon \qquad (4.128)$$
$$\delta_2 = -(l_1 + l_3)\varepsilon$$

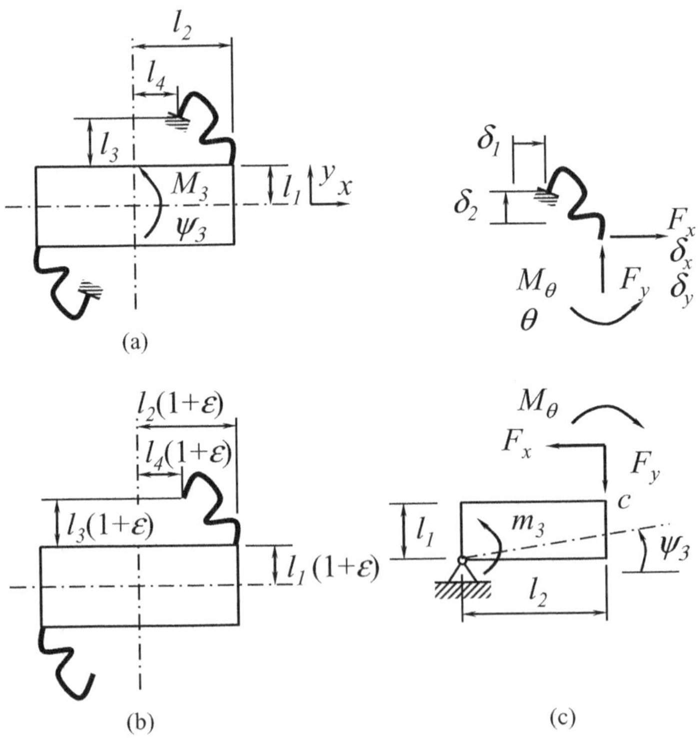

Figure 4.21 Structure with rotational symmetry.

Since the structure of Fig. 4.21a has rotational symmetry, we can consider only the structure shown in the free-body diagram in Fig. 4.21c to obtain the torsional stiffness when the moment M_3 is applied to a movable structure. Let m_3, F_x, F_y, and M_θ represent half the moment M_3 and the forces and moment acting on the springs. Moment equilibrium about the lower left end of the movable structure in Fig. 4.21c gives

$$m_3 + F_x l_1 - F_y l_2 - M_\theta = 0 \tag{4.129}$$

From the same free-body diagram, we have the geometric relations of the linear and angular displacements during the in-plane angular displacement ψ_3:

$$\delta_x = -l_1\psi_3 \qquad \delta_y = l_2\psi_3 \qquad \theta = \psi_3 \tag{4.130}$$

Using the force–displacement relation (4.12) and (4.15), we have the following force–displacement relation:

$$\begin{Bmatrix} F_x \\ F_y \\ M_\theta \end{Bmatrix} = \begin{bmatrix} k_{xx} & k_{xy} & k_{x\theta} \\ k_{yx} & k_{yy} & k_{y\theta} \\ k_{\theta x} & k_{\theta y} & k_{\theta\theta} \end{bmatrix} \begin{Bmatrix} \delta'_x \\ \delta'_y \\ \theta \end{Bmatrix} \tag{4.131}$$

where δ_x and δ_y represent the relative displacements of the lower end of the spring in the x and y directions in Fig. 4.21c. Substituting (4.130) into (4.131), we find the forces and moment of the spring as a function of the angular displacement ψ_3:

$$\begin{Bmatrix} F_x \\ F_y \\ M_\theta \end{Bmatrix} = \begin{bmatrix} k_{xx} & k_{xy} & k_{x\theta} \\ k_{yx} & k_{yy} & k_{y\theta} \\ k_{\theta x} & k_{\theta y} & k_{\theta\theta} \end{bmatrix} \begin{Bmatrix} -l_1\psi_3 - \delta_1 \\ l_2\psi_3 - \delta_2 \\ \psi_3 \end{Bmatrix} \tag{4.132}$$

Substitution of equation (4.132) into the moment equilibrium equation (4.129) leads to a relation between the applied moment m_3 and the angular displacement ψ_3, which is given by

$$\begin{aligned} m_3 &= -F_x l_1 + F_y l_2 + M_\theta \\ &= -\left[k_{xx}(-l_1\psi_3 - \delta_1) + k_{xy}(l_2\psi_3 - \delta_2) + k_{x\theta}\psi_3\right] l_1 \\ &\quad + \left[k_{yx}(-l_1\psi_3 - \delta_1) + k_{yy}(l_2\psi_3 - \delta_2) + k_{y\theta}\psi_3\right] l_2 \\ &\quad + \left[k_{\theta x}(-l_1\psi_3 - \delta_1) + k_{\theta y}(l_2\psi_3 - \delta_2) + k_{\theta\theta}\psi_3\right] \\ &= \left(k_{xx}l_1^2 - 2k_{xy}l_1 l_2 - 2k_{x\theta}l_1 + k_{yy}l_2^2 + 2k_{y\theta}l_2 + k_{\theta\theta}\right)\psi_3 \\ &\quad + \left(k_{xx}l_1 - k_{xy}l_2 - k_{x\theta}\right)\delta_1 + \left(k_{xy}l_1 - k_{yy}l_2 - k_{y\theta}\right)\delta_2 \end{aligned} \tag{4.133}$$

which is a function the angular displacement ψ_3 and the displacements δ_1 and δ_2 of the end of the spring. Taking the derivative of equation (4.133) with respect to ψ_3, we have the stiffness k_{t3} as follows:

$$\begin{aligned} k_{t3} &= \frac{\partial m_3}{\partial \psi_3} \\ &= k_{xx}l_1^2 - 2k_{xy}l_1l_2 - 2k_{x\theta}l_1 + k_{yy}l_2^2 + 2k_{y\theta}l_2 + k_{\theta\theta} \end{aligned} \tag{4.134}$$

It is noted that the stiffness k_{t3} is a function of the stiffness of the spring and the length l_1 and l_2 of the movable structure and is independent of δ_1 and δ_2, which results from the residual stress. In other words, the stiffness of linear structures under residual stress is not affected by residual stress but remains constant. With reference to (4.134), the moment (4.133) is expressed in a simple form as follows:

$$m_3 = k_{t3}\psi_3 + (k_{xx}l_1 - k_{xy}l_2 - k_{x\theta})\delta_1 + (k_{xy}l_1 - k_{yy}l_2 - k_{y\theta})\delta_2 \tag{4.135}$$

which is the moment to maintain the angular displacement ψ_3 of the movable structure while the upper end of the spring undergoes displacements δ_1 and δ_2 in the x and y directions, respectively.

Combining (4.128) and the preceding equation yields the moment m_3 as a function of the angular displacement ψ_3 and the strain ε:

$$m_3 = k_{t3}\psi_3 - [(k_{xx}l_1 - k_{xy}l_2 - k_{x\theta})l_4 + (k_{xy}l_1 - k_{yy}l_2 - k_{y\theta})(l_1 + l_3)]\varepsilon \tag{4.136}$$

It is worth noting that the preceding moment is required to maintain the angular displacement ψ_3, while the entire structure undergoes the strain ε, which may be caused by the residual stress or thermal expansion. If we set the moment in equation (4.136) equal to zero, we obtain the angular displacement ψ_3, the response to the strain ε:

$$\psi_3 = \frac{(k_{xx}l_1 - k_{xy}l_2 - k_{x\theta})l_4 + (k_{xy}l_1 - k_{yy}l_2 - k_{y\theta})(l_1 + l_3)}{k_{t3}}\varepsilon \tag{4.137}$$

As a result, for $m_3 = 0$ the structure in Fig. 4.21a is in self-equilibrium because no external force is exerted on it. Since the moment M_3 in Fig. 4.21a is twice the moment m_3 of (4.135), the moment M_3 is given by

$$\begin{aligned} M_3 &= 2m_3 \\ &= 2[k_{t3}\psi_3 + (k_{xx}l_1 - k_{xy}l_2 - k_{x\theta})\delta_1 + (k_{xy}l_1 - k_{yy}l_2 - k_{y\theta})\delta_2] \end{aligned} \tag{4.138}$$

If we use (4.136), we obtain the expression for the moment M_3 as a function of ψ_3 and ε:

$$
\begin{aligned}
M_3 &= 2m_3 \\
&= 2\left\{k_{t3}\psi_3 - \left[\left(k_{xx}l_1 - k_{xy}l_2 - k_{x\theta}\right)l_4 + \left(k_{xy}l_1 - k_{yy}l_2 - k_{y\theta}\right)(l_1 + l_3)\right]\varepsilon\right\}
\end{aligned}
\tag{4.139}
$$

Taking the derivative of equation (4.139) with respect to ψ_3 gives the torsional stiffness k_{T3} of the structure in Fig. 4.21:

$$
\begin{aligned}
k_{T3} &= \frac{\partial M_3}{\partial \psi_3} = 2k_{t3} \\
&= 2\left(k_{xx}l_1^2 - 2k_{xy}l_1l_2 - 2k_{x\theta}l_1 + k_{yy}l_2^2 + 2k_{y\theta}l_2 + k_{\theta\theta}\right)
\end{aligned}
\tag{4.140}
$$

which is independent of displacements δ_1 and δ_2.

The force–displacement relation (4.132) can be used to evaluate the stress in the springs. Substituting displacements δ_1 and δ_2 and the angular displacement ψ_3, which can be obtained from (4.138) as a function of m_3, into (4.132), we obtain the forces F_x and F_y and the moment M_θ at the end of the spring. Using the forces and moment in the spring, we can evaluate the stress due to the interaction of the movable structure and the spring. If we set the moment M_θ acting on the movable structure equal to zero, the stress in the springs will become the residual stress since the entire structure is in self-equilibrium. We calculate this further in the following example.

Example 4.2 Since residual stress plays an important role in some structures, such as a fixed–fixed beam, researchers wish to evaluate the residual stress in the structures. Structures in self-equilibrium can be used to measure residual stress. A structure in self-equilibrium is used, which consists of two identical slender beams and a relatively rigid beam (Fig. 4.22a). The angular displacement ψ_3 of the rigid beam is estimated by measuring the displacement of the end of the rigid beam as shown in Fig. 4.22b. It is desired to obtain a relation between the displacement δ of the bar shown in Fig. 4.22b and the residual

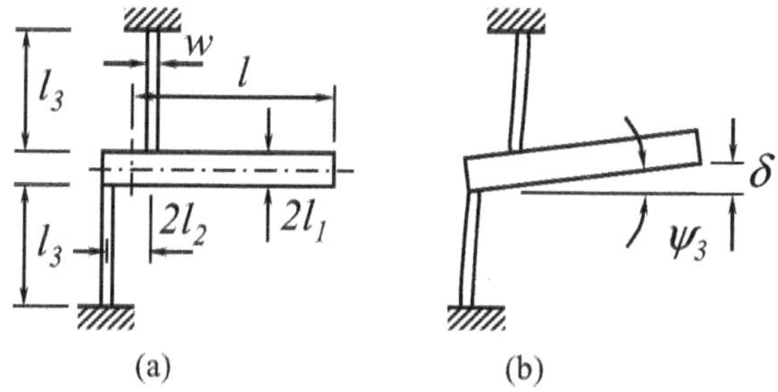

Figure 4.22 Residual stress-measuring structure.

stress in a fixed–fixed beam of length l_f which is placed near the structure of Fig. 4.22 on a substrate.

Our aim is to evaluate the residual stress in the fixed–fixed beam. Since it is difficult to measure the residual stress from a fixed–fixed beam directly, we use the self-equilibrium structure as a test structure. After removing the sacrificial layer under the structure (if we use surface micromachining), the structure whose dimensions are given in Fig. 4.22a is deformed into the shape of Fig. 4.22b, which includes information on the residual stress of the fixed–fixed beam. Let us begin by considering the stiffness (4.14) of the straight beam, which is rewritten here for convenience:

$$\mathbf{K}_1 = \begin{bmatrix} \frac{12EI_z}{l_3^3} & 0 & -\frac{6EI_z}{l_3^2} \\ 0 & \frac{EA}{l_3} & 0 \\ -\frac{6EI_z}{l_3^2} & 0 & \frac{4EI_z}{l_3} \end{bmatrix} \tag{a}$$

where E and I_z are the effective Young's modulus and the moment of inertia of the slender beam, defined as $tw^3/12$. When the structure of Fig. 4.22b is in equilibrium, the angular displacement ψ_3 is given by (4.137), which is given by

$$\psi_3 = \frac{(k_{xx}l_1 - k_{xy}l_2 - k_{x\theta})l_4 + (k_{xy}l_1 - k_{yy}l_2 - k_{y\theta})(l_1 + l_3)}{k_{t3}}\varepsilon \tag{b}$$

where $l_4 = l_2$ for the straight beam in Fig. 4.22. From (b) we have the following strain ε as a function of the angular displacement ψ_3:

$$\varepsilon = \frac{k_{t3}}{(k_{xx}l_1 - k_{xy}l_2 - k_{x\theta})l_4 + (k_{xy}l_1 - k_{yy}l_2 - k_{y\theta})(l_1 + l_3)}\psi_3 \tag{c}$$

Substituting (a) into (4.134) yields the stiffness of the structure that is defined in Fig. 4.21c:

$$k_{t3} = \frac{EA}{l_3}l_2^2\left[1 + \frac{12I_z}{Al_3^2}\left\{\left(\frac{l_1}{l_2}\right)^2 + \frac{l_1l_3}{l_2^2} + \frac{1}{3}\left(\frac{l_3}{l_2}\right)^2\right\}\right] \tag{d}$$

where A denotes the cross-sectional area of the slender beam, defined to be wt. Substituting (d) into (c) and using (a), we find the following strain ε, which appears in the freed structure as shown in Fig. 4.21b:

$$\varepsilon = -\psi_3 \frac{l_2}{l_1 + l_3} \frac{1 + \dfrac{12I_z}{Al_3^2}\left\{\left(\dfrac{l_1}{l_2}\right)^2 + \dfrac{l_1 l_3}{l_2^2} + \dfrac{1}{3}\left(\dfrac{l_3}{l_2}\right)^2\right\}}{1 - \dfrac{12I_z}{Al_3^2}\dfrac{l_1}{(l_1 + l_3)}\left(1 + \dfrac{l_3}{2l_1}\right)} \tag{e}$$

Note that the strain which appears when the fixed–fixed beam is freed (i.e., when the anchors of the fixed–fixed beam are removed) is a function of the measured angular displacement ψ_3 and the geometry of the structure of Fig. 4.22a. In equation (e), $12I_z/Al_3^2 = (w/l_3)^2$. If the right terms in the denominator and numerator in equation (e) are much smaller than unity, the equation reduces to

$$\varepsilon \cong -\psi_3 \frac{l_2}{l_1 + l_3} \tag{f}$$

From the geometry in Fig. 4.22b, a relation between the angular displacement ψ_3 and the measured displacement δ is expressed as

$$\psi_3 = \frac{\delta}{l} \tag{g}$$

Substituting (g) into (e), we find that

$$\varepsilon = -\frac{l_2}{l_1 + l_3} \frac{1 + \dfrac{12I_z}{Al_3^2}\left\{\left(\dfrac{l_1}{l_2}\right)^2 + \dfrac{l_1 l_3}{l_2^2} + \dfrac{1}{3}\left(\dfrac{l_3}{l_2}\right)^2\right\}}{1 - \dfrac{12I_z}{Al_3^2}\dfrac{l_1}{(l_1 + l_3)}\left(1 + \dfrac{l_3}{2l_1}\right)} \frac{\delta}{l} \tag{h}$$

When we substitute (g) into (f), if (f) holds we obtain the simplified strain as

$$\varepsilon \cong -\frac{l_2}{l_1 + l_3} \frac{\delta}{l} \tag{i}$$

We note from equations (h) and (i) that the negative displacement (i.e., $-\delta$) gives a positive strain. If the fixed–fixed beam is fabricated near the residual stress-measuring structure (Fig. 4.22) on a substrate, the residual stress σ_0 in the fixed–fixed beam (Fig. 4.19b) is obtained as

$$\sigma_0 = -E\varepsilon \tag{j}$$

where E is the effective Young's modulus defined in (3.5) and (3.6). In equation (j) the negative sign indicates that if the released structure expands, a compressive stress occurs in the structure anchored on the substrate. Readers may be interested in the stress in the spring beam (the straight beam) in Fig. 4.22b, which can be calculated by combining (4.128) and (4.132). Substituting (4.128) into (4.132) leads to

$$\begin{Bmatrix} F_x \\ F_y \\ M_\theta \end{Bmatrix} = \begin{bmatrix} k_{xx} & k_{xy} & k_{x\theta} \\ k_{yx} & k_{yy} & k_{y\theta} \\ k_{\theta x} & k_{\theta y} & k_{\theta\theta} \end{bmatrix} \begin{Bmatrix} -l_1\psi_3 + l_4\varepsilon \\ l_2\psi_3 + (l_1 + l_3)\varepsilon \\ \psi_3 \end{Bmatrix} \tag{k}$$

As seen in (b), the angular displacement ψ_3 is expressed as a function of the strain ε, so the forces and moment are proportional to the strain ε. The stresses on the cross section of the narrow straight beam can be obtained from equation (k). The forces F_x and F_y generate the shear and normal stresses, and the moment M_θ gives additional normal force. We can consider the normal stress due to the force F_y as the average normal force since the normal stress due to M_θ varies linearly perpendicular to the longitudinal direction of the narrow beam [the linear normal stress distribution is given by (3.41)]. Thus, the normal stress due to the axial force F_y may be taken as a characteristic stress that represents the stress distribution. The normal stress is given by

$$\begin{aligned} \sigma_y &= \frac{-F_y}{A} = \frac{1}{A}\{k_{yx}(-l_1\psi_3 + l_4\varepsilon) + k_{yy}[l_2\psi_3 + (l_1 + l_3)\varepsilon] + k_{y\theta}\psi_3\} \\ &= -\frac{1}{A}\{(-k_{yx}l_1 + k_{yy}l_2 + k_{y\theta})\psi_3 + [k_{yx}l_4 + k_{yy}(l_1 + l_3)]\varepsilon\} \end{aligned} \tag{l}$$

where the negative sign is added since the positive force F_y is compressive in Fig. 4.3 (recall that we defined the tensile stress as positive in Chapter 3). To evaluate the preceding stress in the straight beam in Fig. 4.22b, we substitute the stiffness (a) into the preceding stress and set l_4 equal to l_2, to obtain

$$\sigma_y = -\frac{E}{l_3}[l_2\psi_3 + (l_1 + l_3)\varepsilon] \tag{m}$$

Dividing equation (m) by (j) gives a stress ratio, a ratio of the axial stress in the test structure in Fig. 4.22 to that in the fixed–fixed beam, which is placed near the residual stress-measuring structure of Fig. 4.22a:

$$\frac{\sigma_y}{\sigma_0} = 1 + \frac{l_1}{l_3} + \frac{l_2}{l_3}\frac{\psi_3}{\varepsilon} \tag{n}$$

Substituting (e) into equation (n) gives

$$\frac{\sigma_y}{\sigma_0} = \left(1 + \frac{l_1}{l_3}\right)(1 - f_1) \tag{o}$$

where

$$f_1 = \frac{1 - \dfrac{12I_z}{Al_3^2}\dfrac{l_1}{(l_1+l_3)}\left(1+\dfrac{l_3}{2l_1}\right)}{1+\dfrac{12I_z}{Al_3^2}\left\{\left(\dfrac{l_1}{l_2}\right)^2 + \dfrac{l_1 l_3}{l_2^2} + \dfrac{1}{3}\left(\dfrac{l_3}{l_2}\right)^2\right\}} \tag{p}$$

If we design the structure where the right terms in the denominator and numerator in equation (p) are much smaller than unity, the stress ratio reduces to

$$\frac{\sigma_y}{\sigma_0} \approx 0 \tag{r}$$

Thus, the axial stress (it can also be considered residual stress) of the beam in the residual stress-measuring structure in Fig. 4.22b is much smaller than the axial stress (the residual stress of interest) in the fixed–fixed beam since the movable structure in Fig. 4.22a can rotate about a point to reduce the stress. The reader therefore should understand the difference between the axial stress (the residual stress of interest) in the fixed–fixed beam and the axial stress (a residual stress) of the residual stress-measuring structure.

In example 4.2 we developed the residual stress–displacement relation to evaluate the residual stress in a fixed–fixed beam and to calculate the axial stress of the residual stress-measuring structure. We have also understood that if structures are allowed to rotate about a point, the axial stress of the structure is greatly reduced. Using the same principle, we can explain that the axial stress (residual stress) in the self-equilibrium structure with the L-shaped springs shown in Fig. 4.23 will be very small. The movable plate in Fig. 4.23 can rotate about its center and the L-shaped beam can be also deformed easily (compared to the straight beam), so the axial force (residual stress) in the spring beam is negligible compared to that of a fixed–fixed beam fabricated near the structure in Fig. 4.23. Readers in doubt about this fact may repeat the procedure described in Example 4.2 with the stiffness of the L-shaped spring to verify the axial force-reducing effect. In the structure of Fig. 4.23, since the axial force (residual stress) along the beam can be negligible, the axial stress of the L-shaped springs (more generally, any folded and bent springs) can be considered to be virtually zero and the stiffness of the structure in Fig. 4.23 is not affected by the residual stress. The residual stress effect on the stiffness of a fixed–fixed beam was discussed in Section 3.7. Therefore, the effect of the residual stress caused by fabrication process or temperature change on the stiffness and residual stress of folded or bent flexures will not be considered further.

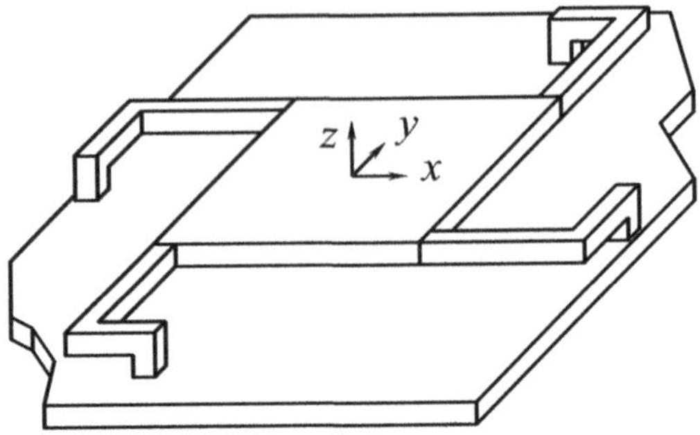

Figure 4.23 Plate with L-shaped springs.

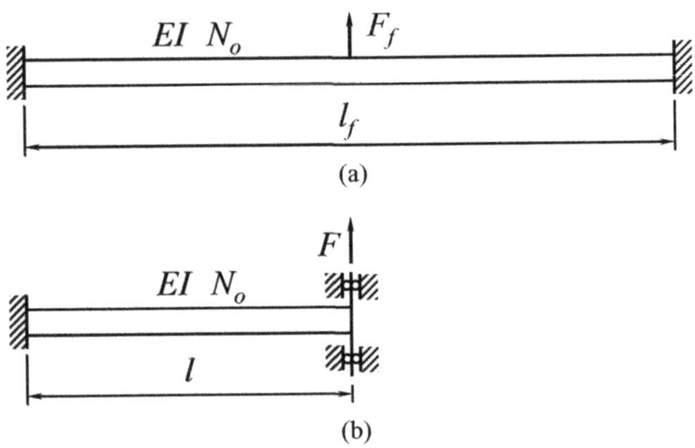

Figure 4.24 Fixed–fixed beam under a concentrated force.

4.7 CUBIC FORCE OF STRUCTURES

In preceding sections we have derived the compliance and stiffness matrices of flexures and have investigated the static behavior of planar structures. We have also understood that if the flexures are not straight, their residual stress is negligible compared to that in a fixed–fixed beam. In Section 3.7 we derived the linear and cubic stiffnesses of a fixed–fixed straight beam (Fig. 4.24a) whose geometric nonlinearity generates the cubic force. If the nonlinear force due to cubic stiffness (i.e., geometric nonlinearity) is much less than the linear force due to linear stiffness, the nonlinear force can be negligible. However, if the displacement perpendicular to the fixed–fixed beam is comparable to or greater than the width of the beam, the cubic force due to the geometric nonlinearity must be taken into account. Let us consider a fixed–fixed beam (Fig. 4.24a) of flexural rigidity EI, axial force N_0, and length l_f which is subjected to a force F_f concentrated at its center. Since the beam of Fig. 4.24a is symmetric about its center, we can use a beam of length l under a force F, where $l = l_f/2$ and $F = F_f/2$, whose right end is on rollers (Fig. 4.24b). The linear and cubic stiffness of the half beam in Fig. 4.24b was derived in Example 3.3.

Model for Cubic Force

To explore the cubic force and to derive further expressions associated with the geometric nonlinearity, we consider the bar actuator in Fig. 4.25a, which consists of two bars connected by a spring of linear stiffness k_s, a roller attached to the end of the lower bar, and a sleeve for guiding the relative bar motion in its axial direction. The upper end of the actuator is attached to a screw (not shown in the figure) to adjust the spring force, and the roller of the lower end can move laterally in a rail when a lateral force F is applied to the roller as shown in Fig. 4.25b. Let l_0, y, and l represent the initial length of the actuator where the force due to the spring in the sleeve is zero, the initial elongation of the actuator, and the length of the actuator when the actuator is elongated, respectively. When the actuator is subjected to a lateral force F, the lower end of the actuator moves by x and the actuator is then angularly displaced by θ, as shown in Fig. 4.25b. We shall analyze the restoring force of the actuator in the x direction and further derive the stored energy for use in subsequent chapters.

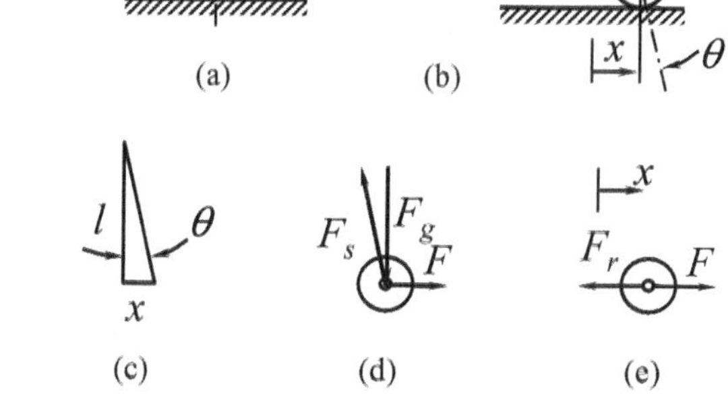

Figure 4.25 Simple model of a cubic force.

From the triangle in Fig. 4.25c, we find the elongation ζ of the spring along the axial direction of the actuator as follows:

$$\zeta = \sqrt{l^2 + x^2} - l_0 \cong l\left[1 + \frac{1}{2}\left(\frac{x}{l}\right)^2\right] - l_0 = y + \frac{1}{2}\frac{x^2}{l}$$

To obtain the right-hand term of the equation above for $x/l << 1$, the geometric relation $l = l_0 + y$ is used. The spring force in the axial direction of the actuator is then written

$$F_s = k_s\zeta = k_s\left(y + \frac{1}{2}\frac{x^2}{l}\right)$$

Referring to the free-body diagram of the roller in Fig. 4.25d, we find the force acting on the actuator in the x direction, which is given by

$$F = F_s \sin\theta = k\left(y + \frac{1}{2}\frac{x^2}{l}\right)\frac{x}{l} = k\frac{y}{l}x + \frac{k}{2l^2}x^3$$

If the preceding force is constant, and if the actuator subjected to the force is in equilibrium at x, we can define the force F as a restoring force F_r of the actuator at x since the magnitude of the restoring force is equal to that of the force applied in the x direction (i.e., $F_r = F$ in Fig. 4.25e, where the force components in the x direction are shown). Therefore, the cubic restoring force of the actuator in the x direction is expressed as

$$F_r = k_1x + k_3x^3 \tag{4.141}$$

where

$$k_1 = k_s\frac{y}{l} \tag{4.142}$$

$$k_3 = \frac{k_s}{2l^2} \tag{4.143}$$

are functions of the stiffness k_s of the spring and the length l and initial elongation y of the actuator. The preceding stiffnesses are defined as the linear and cubic stiffnesses of the actuator in the x direction, respectively. It is worth noting that the stiffnesses k_1 and k_3 are proportional to the spring stiffness k_s in the axial direction of the actuator and that k_1 varies from positive values to negative values as we adjust y in Fig. 4.25a.

Characteristics of Cubic Force

We now examine further the characteristics of the cubic force. In general, we can consider four cases of stiffness, based on the following combinations of stiffnesses: (1) $k_1 > 0$ and $k_3 > 0$; (2) $k_1 < 0$ and $k_3 > 0$; (3) $k_1 > 0$ and $k_3 < 0$; (4) $k_1 < 0$ and $k_3 < 0$. At this stage it is assumed that the linear cubic stiffnesses k_1 and k_3 can be adjusted independently. For the case of $k_1 > 0$ and $k_3 > 0$, the restoring force F_r is shown in Fig. 4.26a, where the force increases as the displacement x increases. If we adjust the initial elongation y from a positive value to a negative value, the force curve changes from Fig. 4.26a to Fig. 4.26b, in which the force curve is folded. For $k_1 > 0$ and $k_3 < 0$, the force curve is shown in Fig. 4.26c, where the curve is also folded. If k_1 in Fig. 4.26c is changed from a positive value to a negative value, the folding disappears and the force curve

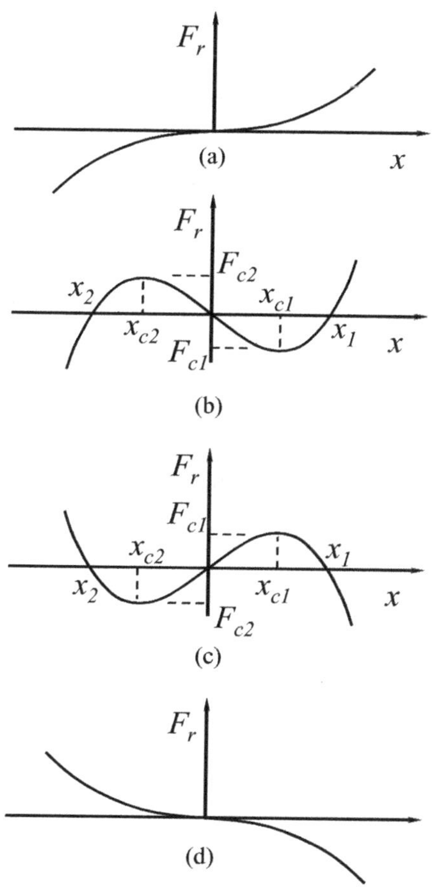

Figure 4.26 Cubic force.

is changed to the curve shown in Fig. 4.26d. We examine the force curves graphically. For $k_1 > 0$ and $k_3 > 0$, if no external force is exerted on the roller in Fig. 4.25a, the stable position is $x = 0$, which is verified from (4.141). The restoring force at the displacement x is determined by (4.141). If a force increment ΔF (i.e., the restoring force increment ΔF_r for equilibrium) is added to the roller in equilibrium in Fig. 4.25b, the roller will be further displaced by Δx. In this case, we define the stiffness k as follows:

$$k = \lim_{\Delta x \to 0} \frac{\Delta F_r}{\Delta x} = \left. \frac{\partial F_r}{\partial x} \right|_{x=x} = k_1 + 3k_3 x^2 \tag{4.144}$$

where x denotes the stable position. The stiffness above is equal to the slope in the force–displacement graphs in Fig. 4.26a–d. For $k_1 > 0, k_3 > 0$ in Fig. 4.26a, the preceding equation reduces to

$$k = k_1 + 3k_3 x^2 \big|_{x=0} = k_1$$

As shown in the equation above, the stiffness k at the stable position $x = 0$ is the linear stiffness k_1, which is positive in Fig. 4.26a. In this case, if the roller at $x = 0$ is displaced by a small increment Δx, the restoring ΔF_r will be

$$\Delta F_r = k\Delta x = k_1 \Delta x$$

The preceding restoring force of the actuator in equilibrium is always positive for $\Delta x > 0$ and negative for $\Delta x < 0$ (i.e., the restoring force always acts in the opposite direction of the displacement). Thus, the stable position is always zero (i.e., the origin in Fig. 4.24a) and the actuator at the position is always stable. In other words, if the stiffness k of an actuator in equilibrium at a position is positive, the actuator is said to be *stable* at the position. If the stiffness k is negative at an equilibrium position, the actuator is *unstable*. Unstable cases are described further below.

For $k_1 < 0$ and $k_3 > 0$, Fig. 4.26b shows three real roots, $x = 0, x_1$, and x_2, and for $k_1 > 0$ and $k_3 < 0$, Fig. 4.26c also gives three roots, $x = 0, x_1$, and x_2. We shall examine the stability of the actuator at the positions. The figures also show two critical positions, labeled x_{c1} and x_{c2}, that are also examined below. Setting F_r in (4.141) equal to zero, we find the three roots for Fig. 4.26b and c as follows:

$$\begin{aligned} x_0 &= 0 \\ x_{1,2} &= \pm\left(\frac{-k_1}{k_3}\right)^{1/2} \end{aligned} \tag{4.145}$$

where the subscripts 1 and 2 of x correspond to positive and negative signs, respectively. To determine their stability, we examine the stiffness at the positions. In Fig. 4.26b, the stiffness (i.e., the slope) at $x = x_1$ and x_2 are positive, so positions x_1 and x_2 are defined as stable positions. However, the position $x = 0$ in Fig. 4.26b is an unstable position since the slope of the force–displacement curve is negative at the position. As a result, If the roller of $k_1 < 0$ and $k_3 > 0$ is placed at $x = 0$, the roller will be moved to $x = x_1$ or x_2. Similarly, for $k_1 > 0$ and $k_3 < 0$ shown in Fig. 4.26c, the position $x = 0$ is the stable position since the stiffness is positive at the position; the other solutions, $x = x_1$ and x_2, are unstable solutions because the stiffnesses are negative at the position. For $k_1 < 0$ and $k_3 < 0$ shown in Fig. 4.26d, the position $x = 0$ is the solution of (4.141), but it is unstable because the stiffness at the position is negative. If the roller is released at $x = 0$, the roller will move away from $x = 0$.

In Fig. 4.26b and c, since the force–displacement curve is folded, each curve has two points of interest, which are labeled x_{c1} and x_{c2} in the graphs. At these positions, the slope (i.e., the stiffness k) is zero, so if the roller of the actuator is placed at these positions, it will become critical. Positions x_{c1} and x_{c2} can then be defined as critical positions. Let the roller of the actuator first be positioned at stable position x_2 in Fig. 4.26b and then gradually increase the force F from zero (the roller will move rightward). When the roller reaches the critical position x_{c2}, a small increase in the force results in instability (i.e., zero stiffness), so the roller will jump from the critical position x_{c2} to a new position at which the roller can be in equilibrium. If we place the roller at x_1 and if we push the roller slowly leftward, we experience the same phenomenon (i.e., the roller will move from x_{c1} to another equilibrium position). This system (in this case, the actuator) is then bistable at x_1 and x_2. For $k_1 > 0$ and $k_3 < 0$ shown in Fig. 4.26c, the system has a stable position, $x = 0$, and is stable for $x_{c2} < x < x_{c1}$. For the range of $x > x_{c1}$ and $x < x_{c1}$, the system is not stable.

Let us obtain the critical positions x_{c1} and x_{c2} in Fig. 4.26b and c. Setting the stiffness of (4.144) equal to zero leads to the critical positions:

$$x_{c1,c2} = \pm\left(\frac{-k_1}{3k_3}\right)^{1/2} \tag{4.146}$$

where the subscripts c_1 and c_2 correspond to positive and negative signs, respectively. Substituting (4.146) into the restoring force equation (4.141), we have the critical forces as follows:

$$\begin{aligned} F_{c1,c2} &= \pm\left[k_1\left(\frac{-k_1}{3k_3}\right)^{1/2} + k_3\left(\frac{-k_1}{3k_3}\right)^{3/2}\right] = \pm\left(\frac{-k_1}{3k_3}\right)^{1/2}\left[k_1 + k_3\left(\frac{-k_1}{3k_3}\right)\right] \\ &= \pm\frac{2k_1}{3}\left(\frac{-k_1}{3k_3}\right)^{1/2} = \frac{2k_1}{3}x_{c1,c2} \end{aligned} \tag{4.147}$$

Equation (4.147) holds for the folded force curves of Fig. 4.26b and c (i.e., $k_1 < 0$ and $k_3 > 0$; $k_1 > 0$ and $k_3 < 0$).

As described above, the force curve for $k_1 < 0$ and $k_3 > 0$ is folded as shown in Fig. 4.26b and has two stable positions, x_1 and x_2. If the linear stiffness k_1 is adjustable from a positive value to a negative value and if the cubic stiffness k_3 remains a positive value, the force curve can be adjusted between Fig. 4.26a and b. For example, the linear stiffness of the actuator in Fig. 4.25a can be adjusted between positive and negative values by controlling y as shown in Fig. 4.25a and (4.142). The linear stiffness k_1 of the fixed–fixed beam in Fig. 4.24a is also controlled if the axial force N_0 is adjusted by an axial force controller, which may use thermal expansion by Joule heating. When k_3 remains constant and k_1 is changed from a positive to a negative value, the stable position $x = 0$ in Fig. 4.26a is moved to one of the stable positions x_1 and x_2 in Fig. 4.26b. We obtain the stiffness at the stable positions in Fig. 4.26. Substituting (4.145) into (4.144), we find that

$$k_{x=x1,x2} = -2k_1 \tag{4.148}$$

For convenience, we define the following distances:

$$|x_1 - x_{c1}| = |x_2 - x_{c2}| = \frac{\sqrt{3}-1}{\sqrt{3}}\left(\frac{-k_1}{k_3}\right)^{1/2} \tag{4.149}$$

If the displacement of the roller from the stable positions in Fig. 4.26b is much less than that of (4.149), the restoring force can be thought of as a linear spring force and is represented by

$$F_r = k_{x=x1,x2}(x - x_{1,2}) = -2k_1(x - x_{1,2}) \tag{4.150}$$

However, if the displacement from the stable positions is comparable to $|x_1 - x_{c1}|$, the restoring force should be expressed as a cubic equation. To obtain the restoring force in this case, we consider the curve in Fig. 4.27. Let the displacement from the stable positions x_1 and x_2 be represented by δ. The displacement from the origin can then be written as $x = x_{1,2} + \delta$. When we substitute $x = x_{1,2} + \delta$ into the restoring force (4.141), we have

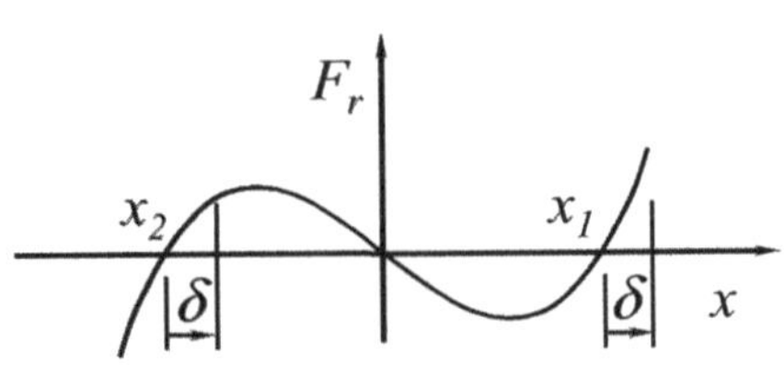

Figure 4.27 Force around a stable position.

$$
\begin{aligned}
F_r &= k_1(x_{1,2}+\delta)+k_3(x_{1,2}+\delta)^3 \\
&= k_1(x_{1,2}+\delta)+k_3(x_{1,2}^3+3x_{1,2}^2\delta+3x_{1,2}\delta^2+\delta^3) \\
&= (k_1x_{1,2}+k_3x_{1,2}^3)+(k_1+3k_3x_{1,2}^2)\delta+3k_3x_{1,2}\delta^2+k_3\delta^3
\end{aligned}
$$

Since the quantity in the first set of parentheses of the last expression is the force at the stable positions, the quantity vanishes. Thus, we write the preceding equation in a simple form,

$$
F_r = k_l\delta + k_q\delta^2 + k_c\delta^3 \tag{4.151}
$$

where

$$
\begin{aligned}
k_l &= k_1 + 3k_3x_{1,2}^2 \\
k_q &= 3k_3x_{1,2} \\
k_c &= k_3
\end{aligned}
\tag{4.152}
$$

denote the linear, quadratic, and cubic stiffness at the stable positions x_1 and x_2. Using the definitions x_1 and x_2 which are expressed in (4.145), the stiffnesses above are expressed as functions of k_1 and k_3:

$$
\begin{aligned}
k_l &= -2k_1 \\
k_q &= \pm 3(-k_1k_3)^{1/2} \\
k_c &= k_3
\end{aligned}
\tag{4.153}
$$

where the positive and negative signs for k_q correspond to the stable positions x_1 and x_2, respectively. It is noted that k_c is equal to k_3 even though the linear stiffness k_1 is changed from a positive to a negative value (i.e., the restoring force curve is changed from Fig. 4.26a to Fig. 4.26b). It is also noted from equations (4.153) that the quadratic stiffness is positive for the stable position x_1 but negative for the stable position x_2. We shall examine the direction of the restoring force due to the quadratic stiffness k_q. If the roller moves by δ from x_1, the restoring force contribution due to the quadratic stiffness k_q will be

$$
F_{r,kq} = k_q\delta^2 = 3(-k_1k_3)^{1/2}\,\delta^2
$$

which is always positive. It is noted that since the positive restoring force at x is defined as shown in Fig. 4.25e, the preceding restoring force $F_{r,kq}$ expressed in terms of δ from the stable position x_1 is always directed to the origin ($x = 0$). Similarly, for the second stable position, x_2, the restoring force contribution due to k_q is expressed as

$$F_{r,k_q} = -3(-k_1k_3)^{1/2}\,\delta^2$$

which is always negative. Then the preceding restoring force expressed in terms of δ from the stable position x_2 is always directed to the origin ($x = 0$). Therefore, if the actuator is positioned at one of the stable positions x_1 and x_2, and displaced by δ, the restoring force due to the quadratic stiffness is always directed to the origin, $x = 0$, while the restoring force due to the linear and cubic stiffness is directed to the stable position at which the actuator is positioned.

4.8 POTENTIAL ENERGY

In preceding sections we have considered the cubic force generated by a nonlinear actuator. In general, it is convenient to define a potential energy that is corresponding to the cubic force, $F_r = k_1x + k_3x^3$. To define the potential energy, we first consider the force equilibrium of the roller in the x direction in Fig. 4.25e, which is expressed as

$$-F_r\mathbf{i} + F\mathbf{i} = 0$$

where F, F_r, and $\mathbf{i}$ denote the applied force, the restoring force in equilibrium, and the unit vector in the x direction, respectively. Taking the dot product of the vector equation above with the displacement vector $dx\mathbf{i}$ gives

$$dU_p = F\mathbf{i}\cdot dx\mathbf{i} = F_r\mathbf{i}\cdot dx\mathbf{i}$$

where U_p is the potential energy. The preceding equation can be interpreted as meaning that if the actuator (or a system) is subjected to an applied force F and displaced in quasi-equilibrium, the work done by the applied force is stored in elastic members (the spring in Fig. 4.25) of the actuator. Integrating the preceding equation yields the potential energy,

$$U_p = \int_0^x F_r\,dx \tag{4.154}$$

More generally, the potential energy can be expressed as

$$U_p = -\int_A^B \mathbf{F}_r\cdot d\mathbf{r} \tag{4.155}$$

where $\mathbf{F}_r$, $\mathbf{r}$, A, and B represent the restoring force vector, the displacement vector, and the starting and ending points of the integration, respectively. For the actuator in Fig. 4.25a, the restoring force vector from Fig. 4.25e is defined as

$$\mathbf{F}_r = -F_r \mathbf{i}$$

Substituting the cubic force (4.141) into potential energy equation (4.155) gives the potential energy due to the cubic restoring force:

$$U_p = \int_0^x (k_1 x + k_3 x^3)\, dx = \frac{1}{2} k_1 x^2 + \frac{1}{4} k_3 x^4 \tag{4.156}$$

Since the general expression for potential energy is given by (4.155), we obtain $\mathbf{F}_r \cdot d\mathbf{r} = F_x dx + F_y dy + F_z dz = -dU_p$, where F_x, F_y, and F_z denote the force components in the x, y, and z directions. So we can obtain the force–potential relations as

$$F_x = -\frac{\partial U_p}{\partial x} \qquad F_y = -\frac{\partial U_p}{\partial y} \qquad F_z = -\frac{\partial U_p}{\partial z}$$

These are expressed in compact form as

$$\mathbf{F}_r = -\nabla U_p \tag{4.157}$$

where

$$\nabla = \mathbf{i}\frac{\partial}{\partial x} + \mathbf{j}\frac{\partial}{\partial y} + \mathbf{k}\frac{\partial}{\partial z}$$

is the gradient operator and **i**, **j**, and **k** denote the unit vectors for the rectangular coordinate system. The negative sign in the preceding force–energy relation denotes that the restoring force acts in the direction in which the potential energy decreases. We can verify that if we substitute the potential energy (4.156) into (4.157), we obtain the cubic force (4.141). The potential energy U_p is illustrated in Fig. 4.28a for $k_1 > 0$, and $k_3 > 0$, Fig. 4.28b for $k_1 < 0$ and $k_3 > 0$, Fig. 4.28c for $k_1 > 0$ and $k_3 < 0$, and Fig. 4.28d for $k_1 < 0$ and $k_3 < 0$. These potential energy curves correspond to the force curves shown in Fig. 4.26. In Fig. 4.28a–c, we define one or two potential wells, which are regions surrounding a local minimum of potential energy. In Fig. 4.28a and c, there is one potential well at the center of the graph. However, the potential energy curve in Fig. 4.26b (for $k_1 < 0$ and $k_3 > 0$) has two potential wells around x_1 and x_2. These potential wells are used in subsequent chapters to explain the dynamic behavior of mass.

Using the preceding definition of the potential energy, we shall express the potential energies that correspond to stable positions x_1 and x_2 and critical positions x_{c1} and x_{c2}. Substituting the stable positions x_1 and x_2, (4.145), into the potential energy (4.156) due to the cubic force, we find the potential energies corresponding to the stable positions as follows:

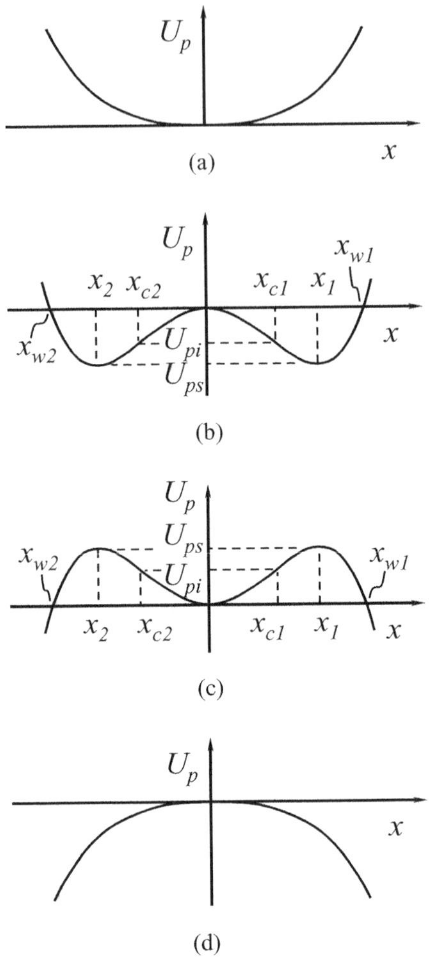

Figure 4.28 Potential energy due to a cubic force.

$$U_{ps} = -\frac{k_1^2}{4k_3} \tag{4.158}$$

Substituting the critical positions x_{c1} and x_{c2}, (4.146), which correspond to the infection points of the potential energy curve into the potential energy (4.156) leads to potential energies corresponding to the inflection points:

$$U_{pi} = \frac{-5k_1^2}{36k_3} \tag{4.159}$$

In addition to the stable positions and critical positions that were defined in the preceding section, we can further define zero potential energy positions,

x_{w1} and x_{w2}, shown in Fig. 4.28b and c, which are the positions at $U_p = 0$ and represent a characteristic width of the potential energy curve. Setting $U_p = 0$ in (4.156), the zero potential energy positions are given by

$$x_{w1,w2} = \pm\left(\frac{-2k_1}{k_3}\right)^{1/2} \tag{4.160}$$

If the potential energy U_p of a system is expressed in terms of the displacement x, the restoring force in the x direction is obtained from (4.157) as follows:

$$\mathbf{F}_r = -\nabla U_p = -\frac{\partial U_p}{\partial x}\mathbf{i} \tag{4.161}$$

where $\mathbf{i}$ denotes the unit vector in the x direction. For convenience, if the restoring force in the opposite direction of displacement is defined to be positive, as shown in Fig. 4.25e, the preceding restoring force is written as

$$F_r = \frac{\partial U_p}{\partial x} \tag{4.162}$$

The stiffness k is defined as the derivative of the preceding restoring force with respect to x:

$$k = \frac{\partial F_r}{\partial x} = \frac{\partial^2 U_p}{\partial x^2} \tag{4.163}$$

From the fact described earlier in this section that the restoring force is zero at stable and unstable positions and that the stiffness is positive at stable positions and negative at unstable positions, we obtain the following stable and unstable conditions: stable conditions,

$$\frac{\partial U_p}{\partial x} = 0 \qquad \frac{\partial^2 U_p}{\partial x^2} > 0 \tag{4.164}$$

and unstable conditions,

$$\frac{\partial U_p}{\partial x} = 0 \qquad \frac{\partial^2 U_p}{\partial x^2} < 0 \tag{4.165}$$

Since stiffness of zero or less results in system instability, the equation to obtain critical positions is given by

$$\frac{\partial^2 U_p}{\partial x^2} = 0 \tag{4.166}$$

From this condition we can obtain the stable positions of a system in equilibrium. The critical positions of the system are evaluated from the critical condition (4.166). For example, when we substitute potential energy equation (4.156) corresponding to the cubic restoring force (4.141) into the preceding five equations, we find the restoring force (4.141), stiffness (4.144), stable and unstable positions (4.145), and the critical positions (4.146) for $k_1k_3 < 0$.

Potential Energy of a System

The cubic force of nonlinear springs and the corresponding potential energy have been studied. If it is desired to find stable positions for the actuator roller in Fig. 4.25b, which is subjected to the force F, we can obtain them from the force equilibrium equation in the x direction in Fig. 4.25e. The force equilibrium of the roller gives

$$F = F_r$$

At this stage we assume that $k_1 < 0$ and $k_3 > 0$. For a constant force F, the force equilibrium above can be represented graphically as in Fig. 4.29a. For further derivation we define a restoring force, F_{rs}, of the system at the equilibrium positions (a–d in Fig. 4.29a) as follows:

$$F_{rs} = F_r - F = k_1 x + k_3 x^3 - F \tag{4.167}$$

where F_r and F denote the restoring force of the nonlinear spring and the applied force, respectively. As shown in Fig. 4.29b, the preceding restoring force curve is moved downward as the force increases. Setting $F_{rs} = 0$, we find the stable and unstable positions labeled in Fig. 4.29b. For example, if the force F is zero, the equilibrium equation gives three positions: $x = 0, x_1$, and x_2. When the force F increases, the curve moves downward and the solutions are changed from the previous positions to x_a, x_b, and x_c. As the force reaches the critical force ($F = F_{c2}$ in Fig. 4.29b), the curve provides one critical position, x_{c2}, and one stable solution, x_{j1}. The force F is beyond the critical force, and the curve gives only one solution, x_d. For $F_{c2} \leq F \leq F_{c1}$, setting the restoring force of (4.167) equal to zero gives the following positions in closed form (described in detail in Section 5.1):

$$\begin{aligned}
x_a &= 2\left(\frac{-k_1}{3k_3}\right)^{1/2} \cos\left[\frac{1}{3}(\phi + 2\pi)\right] \\
x_b &= 2\left(\frac{-k_1}{3k_3}\right)^{1/2} \cos\left[\frac{1}{3}(\phi + 4\pi)\right] \\
x_c &= 2\left(\frac{-k_1}{3k_3}\right)^{1/2} \cos\frac{\phi}{3}
\end{aligned} \tag{4.168}$$

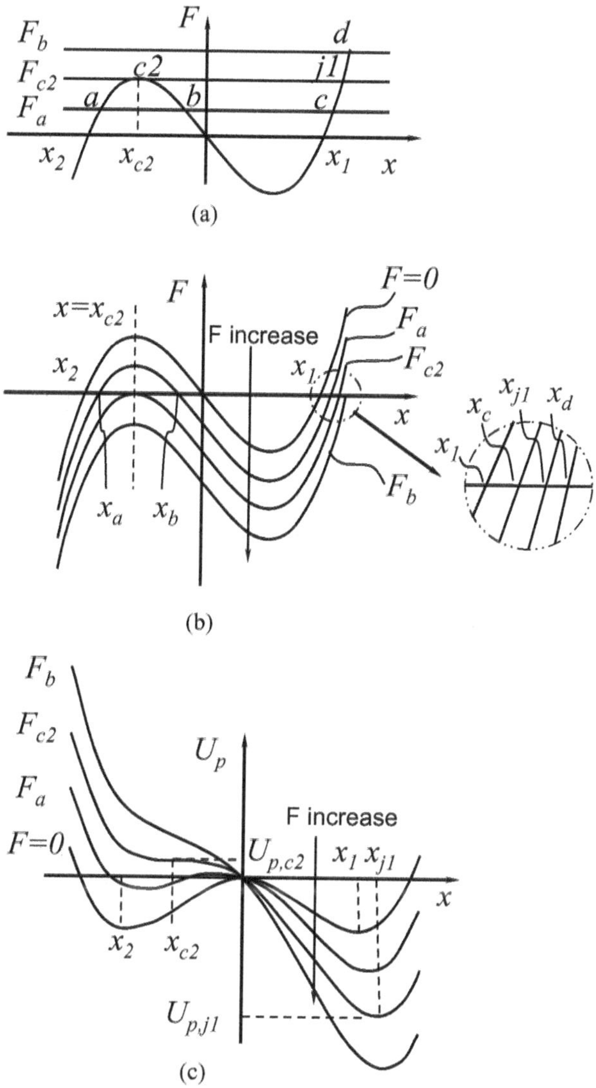

Figure 4.29 Restoring force and potential energy.

where

$$\phi = \cos^{-1}\left(\frac{27k_3^2 F}{2(-3k_1k_3)^{3/2}}\right) \tag{4.169}$$

For $F = 0$, equation (4.169) gives $\phi = 0$, and the positions, x_a, x_b, and x_c of (4.168) reduce to x_2, 0, and x_1, which are given in (4.145). It is also noted in (4.167) that x_a and x_c are the stable positions for $k_1 < 0$ and $k_3 > 0$ and that x_b is the unstable position. If $F > F_{c2}$ or $F < F_{c1}$, a single solution (e.g., x_d in Fig. 4.29b) is found, which is obtained analytically in Chapter 5.

If we assume that the force F is constant, we can define the stiffness of the system at the positions of (4.168) as follows:

$$k_s = \frac{\partial F_{rs}}{\partial x} = \frac{\partial F_r}{\partial x} = k_1 + 3k_3x^2 \tag{4.170}$$

which is the same as that of the spring defined in (4.144). As the force F reaches the critical force F_{c2}, the preceding stiffness becomes zero, and then the roller jumps from the critical position, x_{c2}, to the jump position, x_{j1}, in Fig. 4.29b. To find the jump position, we note from (4.169) that the auxiliary parameter ϕ is zero as the force F approaches F_{c2} and that ϕ is π as F approaches F_{c1}. Setting $\phi = 0$ in the third equation of (4.168) yields the jump position x_{j1}:

$$x_{j1} = x_c\big|_{F=F_{c2}} = 2\left(\frac{-k_1}{3k_3}\right)^{1/2} \tag{4.171}$$

We also find the following jump position x_{j2} as the force F approaches another critical force F_{c1}:

$$x_{j2} = x_a\big|_{F=F_{c1}} = -2\left(\frac{-k_1}{3k_3}\right)^{1/2} \tag{4.172}$$

Comparing the two equations above with the critical positions (4.146), we have the following relations: $x_{j1,j2} = 2x_{c1,c2}$. That is, the jump positions are two times the critical positions. If we assume that the force F is constant and conservative (e.g., the gravitational force), we can define the potential energy for the system as follows:

$$U_p = \int_0^x F_{rs}\,dx = \frac{k_1}{2}x^2 + \frac{k_3}{4}x^4 - Fx \tag{4.173}$$

The potential energy above is as shown in Fig. 4.29c when the force F increases from zero. Since the restoring force is given by $F_{rs} = \partial U_p/\partial x$, the minima of the potential energy curves of Fig. 4.29c represent the stable positions, which are given by the stable positions x_a and x_c in (4.168). Substituting $x = x_a$ and $x = x_c$ into equation (4.173), we find the minimum potential energies of the stable positions x_a and x_c:

$$U_{p,x_a} = \frac{k_1}{2}x_a^2 + \frac{k_3}{4}x_a^4 - Fx_a \tag{4.174}$$

$$U_{p,x_b} = \frac{k_1}{2}x_b^2 + \frac{k_3}{4}x_b^4 - Fx_b \tag{4.175}$$

Since in many cases the potential energies at the critical positions and the corresponding jump positions are important, we find the energies: for $F = F_{c2}$,

$$U_{p,x_{c2}} = \frac{k_1}{2}x_{c2}^2 + \frac{k_3}{4}x_{c2}^4 - F_{c2}x_{c2} = \frac{1}{12}\frac{k_1^2}{k_3} \tag{4.176}$$

$$U_{p,x_{j1}} = \frac{k_1}{2}x_{j1}^2 + \frac{k_3}{4}x_{j1}^4 - F_{c2}x_{j1} = -\frac{2}{3}\frac{k_1^2}{k_3} \tag{4.177}$$

and for $F = F_{c1}$,

$$U_{p,x_{c1}} = U_{p,x_{c2}} = \frac{1}{12}\frac{k_1^2}{k_3} \tag{4.178}$$

$$U_{p,x_{j2}} = U_{p,x_{j1}} = -\frac{2}{3}\frac{k_1^2}{k_3} \tag{4.179}$$

4.9 ANALOGY BETWEEN POTENTIAL ENERGIES

In the preceding two sections we studied the cubic force of a spring and obtained potential energy stored in a system. For easy understanding of the cubic force and the corresponding energy, we can develop the analogy of potential energy to gravitational potential energy with which we are familiar. Let us begin with the mass in Fig. 4.30a, which is supported by a wire. From force equilibrium, we have $F = mg$. Since the gravitational force mg is equal to the applied force F and tends to push the mass to a lower position, the gravitational force can be thought of as a restoring force. We then define the restoring force as follows:

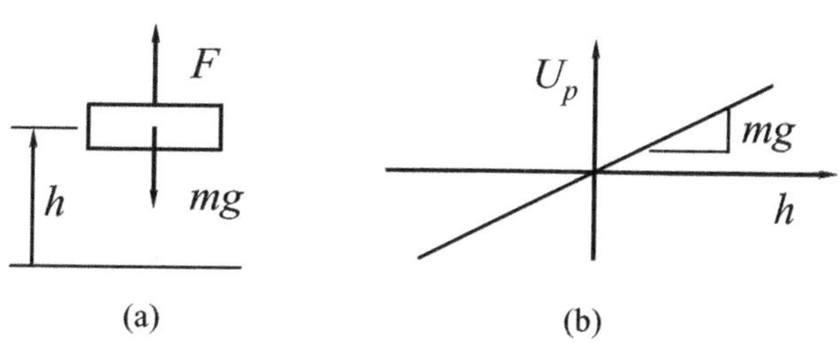

Figure 4.30 Potential energy due to a gravitational force.

$$F_r = mg$$

Integrating the restoring force above from 0 to h gives the gravitational energy as

$$U_p = -\int \mathbf{F}_r \cdot d\mathbf{r} = \int_0^h mg\, dh = mgh \tag{4.180}$$

which is shown in Fig. 4.30b. Now let us suppose that we apply a horizontal force to a mass m in a bowl as shown in Fig. 4.31a. The bowl surface y is defined as a function of the horizontal displacement x. We wish to represent the potential energy due to the cubic force of a nonlinear spring by the bowl surface function y. It is assumed that there is no friction between the mass and the bowl surface. From the free-body diagram in Fig. 4.31b, we find the horizontal and vertical force equilibrium equations as follows:

$$F = V \sin\theta \qquad V\cos\theta = mg$$

where V and θ denote the reaction force from the bowl and the angle made with the horizontal axis, respectively. The preceding equations give the horizontal force as

$$F = mg\tan\theta \tag{4.181}$$

The input energy due to the applied force F can be written

$$dU_p = F\, dx = mg\tan\theta\, dx$$

From the geometry in Fig. 4.31c, we have $\tan\theta = dy/dx$ and then the energy above may be expressed as

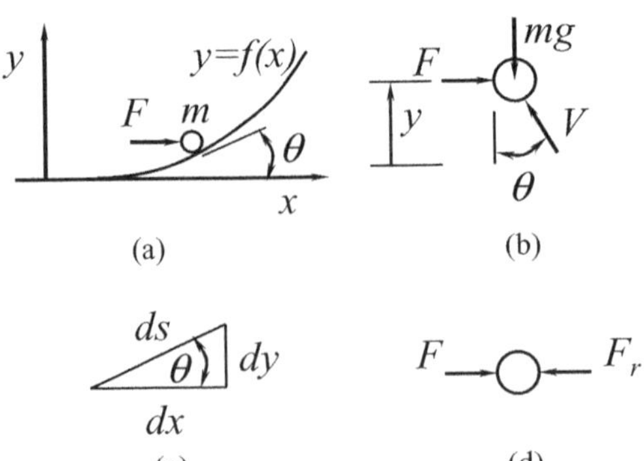

Figure 4.31 Analogy of potential energies.

$$dU_p = mgdy$$

Integrating the preceding energy equation from 0 to y yields the potential energy due to the work done by the force F:

$$U_p = \int_0^y mg\,dy = mgy \tag{4.182}$$

It is noted that equation (4.182) is the same as (4.180). This is due to the fact that the gravitational potential energy above is a point function whose value depends on the initial and final values (the vertical distance in the gravitational potential energy). Since it is desired to express the potential energy due to the cubic force as the bowl surface function y, we consider the gravitational potential energy as the potential energy due to the cubic force, and then we have from equation (4.182),

$$y = \frac{U_p}{mg} \tag{4.183}$$

Substituting the potential energy equation (4.173) into (4.183), we find the bowl surface function that represents the potential energy due to the cubic force:

$$y = \frac{U_p}{mg} = \frac{1}{mg}\left(\frac{k_1}{2}x^2 + \frac{k_3}{4}x^4 - Fx\right) \tag{4.184}$$

If the bowl surface has the shape of the function above, the potential energy is given by

$$U_p = mgy = mg\frac{1}{mg}\left(\frac{k_1}{2}x^2 + \frac{k_3}{4}x^4 - Fx\right) = \frac{k_1}{2}x^2 + \frac{k_3}{4}x^4 - Fx$$

The equation above is equal to (4.173), so we can represent the potential energy due to the cubic force by the bowl surface function y. Note that (4.183) holds for any arbitrary potential energy, and the potential energy can be represented graphically by (4.183). As shown in the force equilibrium in the horizontal direction in Fig. 4.31d, the applied force F is balanced by a restoring force F_r, and we can then consider the force F in (4.181) to be the restoring force,

$$F_r = F = mg\tan\theta = mg\frac{\partial y}{\partial x}$$

Substituting the bowl surface function (4.184) into the equation above gives the restoring force as

$$F_r = mg\frac{\partial y}{\partial x} = k_1x + k_3x^3 - F \tag{4.185}$$

which is equal to the restoring force of the system, (4.167). Therefore, the slope of the bowl surface function is proportional to the restoring force.

To facilitate understanding of the nonlinear actuator, we construct three bowl shapes (Fig. 4.32a–c) that show the analogy of the potential energy due to the cubic force on the bowl surface in the gravitational field: (a) $k_1 > 0$ and $k_3 > 0$, $F = 0$; (b) $k_1 < 0$ and $k_3 > 0$, $F = 0$; (c) $k_1 < 0$ and $k_3 > 0$, $0 < F < F_{c2}$, where F_{c2} is given by (4.147). When we place a steel ball at the lowest point in the bowl in Fig. 4.32a, the ball does not move and remains at the initial point. However, the linear stiffness k_1 becomes negative, the bowl surface is folded as shown in Fig. 4.32b, and the ball is forced to move one of the stable positions, x_1 or x_2. If the bowl shape of Fig. 4.32a is changed to that of Fig. 4.32c, the ball tends to move rightward and find a new stable position at x_b. As such, we can use the analogy between the bowl shape (curve) and the

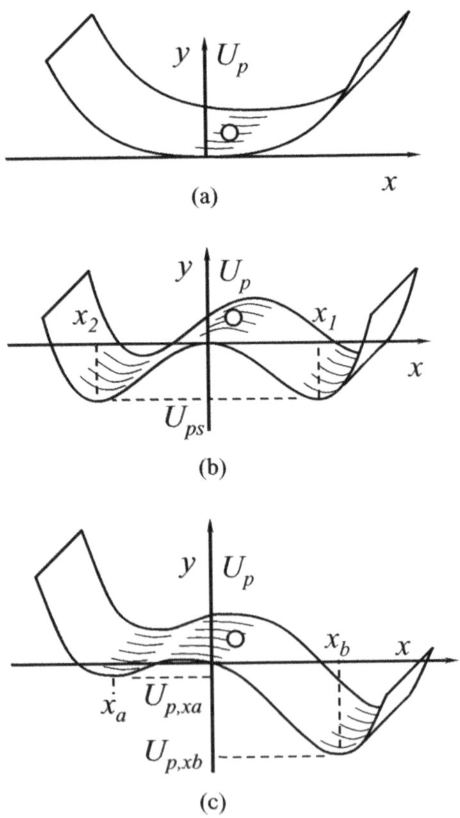

Figure 4.32 Analogy of potential energies.

potential energy due to the cubic restoring force in order to explain the behavior of a nonlinear actuator. We also know from our experience that a steel ball moving from a position in the bowl to a lower position will achieve a certain velocity and then vibrate in the bowl. This issue of a moving object is dealt with in a subsequent chapter.

Example 4.3 A straight beam subjected to an axial force in Fig. 4.33 represents a beam that is used in many MEMS devices. In the MEMS structure of Fig. 4.34, four identical straight beams of flexural rigidity EI and residual strain $-\varepsilon$ (i.e., a compressed beam) (Fig. 4.34) support a central rigid structure and are represented by the simple structure shown in Fig. 4.33. The strain along the beam may be adjusted to control the overall stiffness. We wish to obtain the overall stiffness in the x direction in terms of the residual strain. When the beam is buckled, it is desired to evaluate the critical positions and the corresponding critical force as functions of the strain.

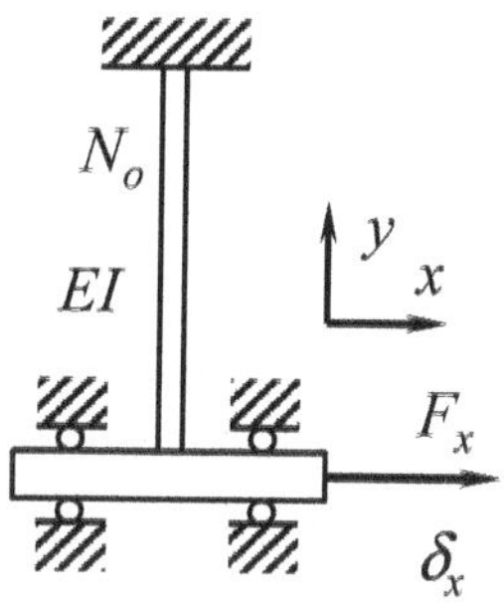

Figure 4.33 Model of a straight beam subjected to a residual force and geometric nonlinearity.

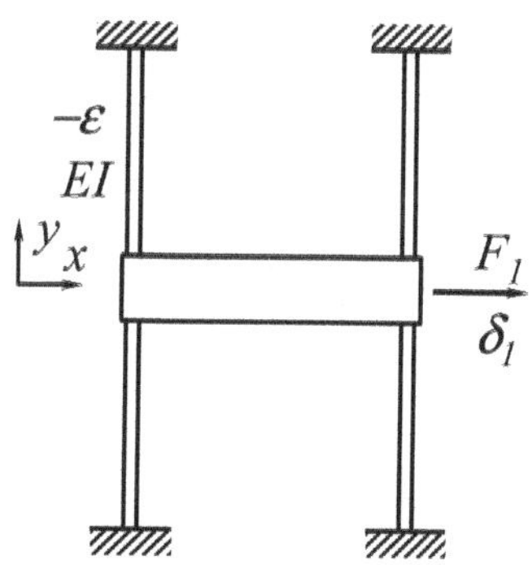

Figure 4.34 Microstructure consisting of four straight beams.

The linear and cubic stiffness of the beam were obtained in Example 3.3 and are given by

$$k_1 = 12\frac{EI}{l^3} + \frac{6}{5}\frac{N_0}{l} \tag{a}$$

$$k_3 = \frac{18}{25}\frac{EA}{l^3} \tag{b}$$

where E, I, l, A, and N_0 denote the effective Young's modulus defined in (3.5) and (3.6), the moment of inertia of the cross-sectional area of the beam, the length of the straight beam, the cross-sectional area, and the axial force in the beam, respectively. According to the stress–strain relation, the axial force is expressed as

$$N_0 = -EA\varepsilon \tag{c}$$

Substituting (c) into (a) gives the linear stiffness as a function of the strain as follows:

$$k_1 = 12\frac{EI}{l^3} - \frac{6}{5}\frac{EA\varepsilon}{l} = 12\frac{EI}{l^3}\left(1 - \frac{1}{10}\frac{Al^2}{I}\varepsilon\right) \tag{d}$$

The residual strain may be due to the fabrication process and may be adjusted by controlling a parameter (e.g., current through the beam generates heat to adjust the residual strain ε). Since the structure in Fig. 4.34 is symmetric about the orthogonal axes that pass through the center of the structure and is parallel to the x and y axes, the force applied to the structure along the x axis is given by

$$F_1 = 4F_x = k_l\delta_1 + k_c\delta_1^3 \tag{e}$$

where

$$k_l = 4k_1 = 48\frac{EI}{l^3}\left(1 - \frac{1}{10}\frac{Al^2}{I}\varepsilon\right) \tag{f}$$

$$k_c = 4k_3 = \frac{72}{25}\frac{EA}{l^3} \tag{g}$$

As shown in the equations above, the linear stiffness k_l of the restoring force is adjustable from a positive value to a negative value by controlling the strain ε while the cubic stiffness k_c remains constant. The restoring force varies from the force in Fig. 4.26a if $k_1 > 0$ and $k_3 > 0$ to the force in Fig. 4.26b if k_1 is changed from a positive value to a negative value and if k_3 remains constant. The corresponding potential energy curves are shown in Fig. 4.28a and b. From (f), the linear stiffness will be zero or negative if the following condition is satisfied:

$$\varepsilon \geq \frac{10I}{Al^2} \tag{h}$$

where the equals sign holds for the zero linear stiffness and the > (greater than) sign for the negative linear stiffness. Note that the preceding strain corresponds to the critical load (buckling load), defined in (3.81). For $k_1 < 0$ and $F = 0$, the stable positions x_1 and x_2 are given by (4.145):

$$x_{1,2} = \pm\left(\frac{-k_1}{k_3}\right)^{1/2} \tag{i}$$

and the critical positions x_{c1} and x_{c2} are determined by (4.146):

$$x_{c1,c2} = \pm\left(\frac{-k_1}{3k_3}\right)^{1/2} \tag{j}$$

Substituting (b) and (d) into (i) and (j), we have

$$x_{1,2} = \pm\left[-\frac{50}{3}\frac{I}{A}\left(1-\frac{1}{10}\frac{Al^2}{I}\varepsilon\right)\right]^{1/2} \tag{k}$$

and

$$x_{c1,c2} = \pm\left[-\frac{50}{9}\frac{I}{A}\left(1-\frac{1}{10}\frac{Al^2}{I}\varepsilon\right)\right]^{1/2} \tag{l}$$

Equations (k) and (l) hold if (h) is satisfied (i.e., the force is folded as shown in Fig. 4.28b). The restoring forces of the four straight beams at x_1 and x_2 are zero. The critical restoring force of the four straight beams is given by (4.147):

$$F_{c1,c2} = \pm\frac{2k_1}{3}\left(\frac{-k_1}{3k_3}\right)^{1/2} = \mp\frac{160\sqrt{2}}{3}\frac{EI^{3/2}}{l^3A^{1/2}}\left[-\left(1-\frac{1}{10}\frac{Al^2}{I}\varepsilon\right)\right]^{3/2} \tag{m}$$

where the negative and positive signs on the right-hand side of equation (m) correspond to x_{c1} and x_{c2}, respectively. The preceding forces are the forces generated by one beam. The critical force $F_{1,c}$ in Fig. 4.34 is given by four times the preceding forces. The jump positions are obtained from (4.171) and (4.172) (i.e., $x_{j1,j2} = 2x_{c1,c2}$). Equations (j)–(m) and the jumping position hold only if (h) is satisfied. If $k_1 > 0$ and $k_3 > 0$, the restoring force (e) should be used. The equations derived in this example will be used to obtain the static and dynamic responses of the strain control actuator.

PROBLEMS

4.1 Identical springs are connected in series and in parallel (Fig. P4.1). If the stiffness of each spring is k and if the right rigid bar is subjected to a lateral force F, find the displacement of the right bar.

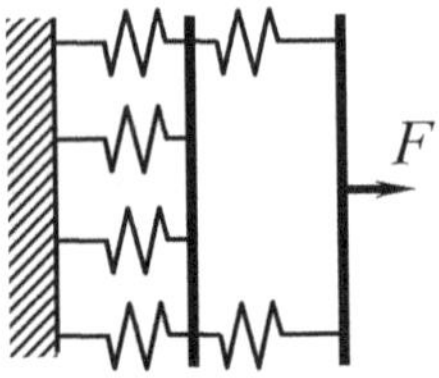

Figure P4.1

4.2 A free end of an in-plane serpentine spring is subjected to a horizontal force F_x, a vertical force F_y, and a moment M_θ to generate in-plane motion of the free end of the spring (Fig. P4.2). If a straight beam of length c is rigid and the serpentine (dimensions shown in the figure) of flexural rigidity is EI, construct the compliance matrix $\mathbf{C}_1$ and find the stiffness k_{xx}.

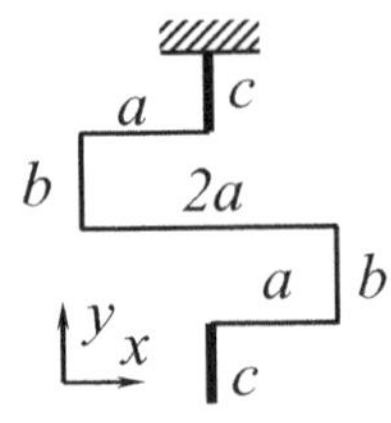

Figure P4.2

4.3 An arc of subtended angle α and flexural rigidity EI is subjected to a horizontal force F_x, a vertical force F_y, and a moment M_θ at its free end (Fig. P4.3). Construct the compliance matrix $\mathbf{C}_1$ and find the stiffness k_{xx} for $\alpha = 90°$.

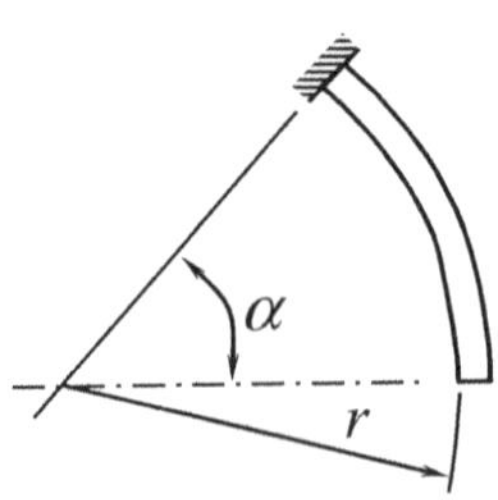

Figure P4.3

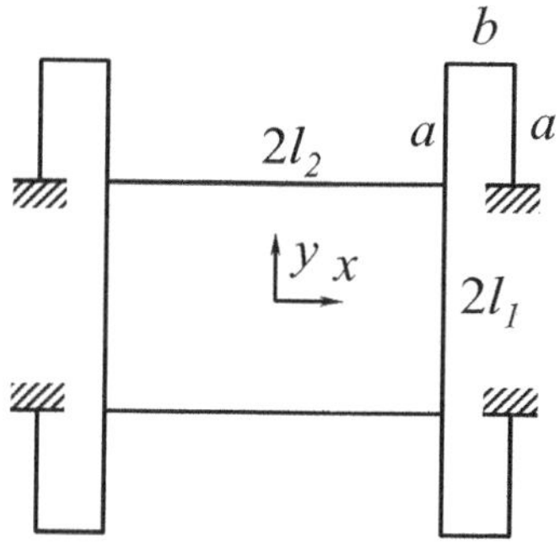

Figure P4.4

4.4 A movable rigid plate is suspended by four folded beams of flexural rigidity EI (Fig. P4.4). Find the displacements in the x and y directions as the forces F_x and F_y are applied to the center of the movable plate.

4.5 A rigid circular disk supported by four L-shaped springs of flexural rigidity EI has 90° rotational symmetry (Fig. P4.5). Determine the torsional stiffness of the structure.

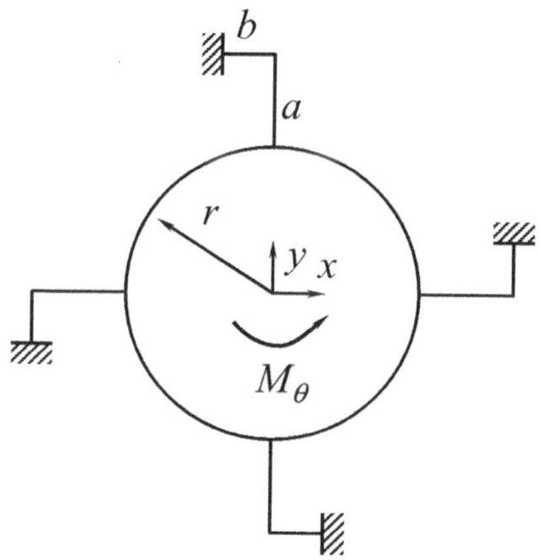

Figure P4.5

4.6 In an experiment using a residual stress-measuring structure, the residual stress of a slender fixed–fixed beam near the microstructure shown in Fig. 4.20 is obtained as σ_r. Find the stress in the straight beam in Fig. 4.20.

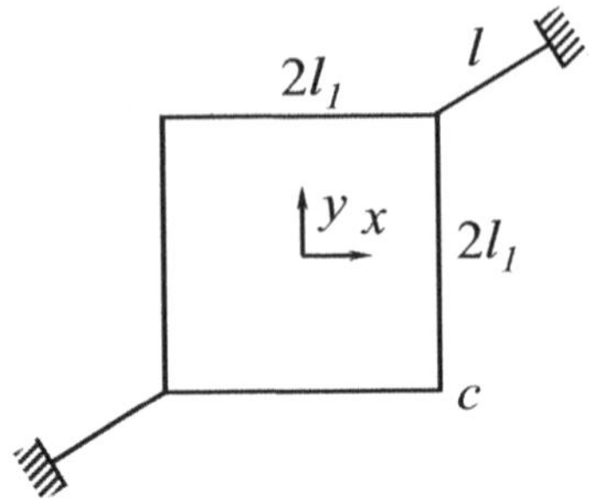

Figure P4.7

4.7 Two springs of flexural rigidity EI makes an angle of 30° with the horizontal line and support a rigid square plate of length $2l_1$ (Fig. P4.7). The structure and a slender fixed–fixed beam on a wafer are released when the surface micromachining has been completed. If the residual stress in the released fixed–fixed beam is measured as σ_r, find the displacement of the lower right corner of the square plate during the release period.

CHAPTER 5

DYNAMICS

In Chapters 3 and 4, a theory was developed to explain the static response of microstructures, and linear and cubic stiffnesses, which relate to the static response of commonly used microstructures, were expressed in closed form. In this chapter we describe displacement, velocity, and acceleration and discuss the governing equations for the dynamic behavior of microstructures. Combining these governing equations with the linear and cubic stiffnesses developed in Chapter 4, we derive closed-form solutions for the dynamic response of microstructures.

5.1 CUBIC EQUATION

In Chapters 3 and 4, the static response of nonlinear structures was expressed in terms of the linear stiffness k_1 and the cubic stiffness k_3. The dynamic responses of the structures are expressed as a function of the linear and cubic stiffnesses. The static and dynamic behavior of the structures is then expressed in cubic equations. To calculate the static and dynamic response, it is necessary to obtain the roots of a cubic equation.

In general, a cubic equation is expressed as

$$Lx^3 + A_1x^2 + B_1x + C_1 = 0 \tag{5.1}$$

Principles of Microelectromechanical Systems, By Ki Bang Lee

where x, L, A_1, B_1, and C_1 denote an unknown and its coefficients. Dividing equation (5.1) by the leading coefficient L, we have a simple form of the cubic equation as follows:

$$x^3 + ax^2 + bx + c = 0 \tag{5.2}$$

where a, b, and c represent new coefficients given by A_1/L, B_1/L, and C_1/L, respectively. To explicate equation (5.2), we consider two separate functions, $y_1 = x^3 + ax^2 + bx$ and $y_2 = -c$, which are the curve and the straight line in Fig. 5.1. Note that in many cases the curve y_1 is unchanged and the straight line y_2 moves upward or downward. For example, the static behavior of the actuator (shown in Fig. 4.25), which is subjected to a force F, is written as a cubic equation (4.141), where the force relates to the coefficient c of equation (5.2). Therefore, if the force is changed from one value to another, the curve y_2 in Fig. 5.1 moves upward and downward. The solutions of (5.2) may be defined as intersections of y_1 and y_2, which appear in Fig. 5.1. It is noted that the folded curve (Fig. 5.1b) can have three distinct real roots if the line y_2 is placed between y_u and y_l. Substituting $x = y - a/3$ into equation (5.2) we obtain the simplest form of the cubic equation:

$$y^3 + py + q = 0 \tag{5.3}$$

where

$$p = b - \frac{a^2}{3} \qquad q = c - \frac{ab}{3} + \frac{2a^3}{27} \tag{5.4}$$

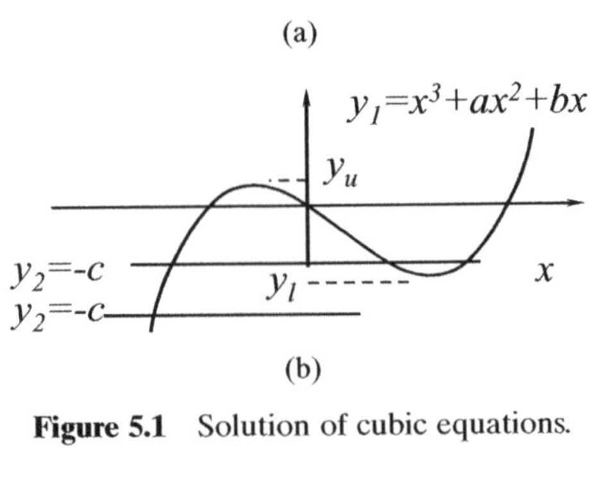

Figure 5.1 Solution of cubic equations.

Cubic equation (5.3) was solved by Niccolò Tartaglia, an Italian mathematician, but its solution was published by Gerolamo Cardano, another Italian mathematician. For convenience, a discriminant Δ is defined as

$$\Delta = 4p^3 + 27q^2 \tag{5.5}$$

Substituting (5.4) into equation (5.5), we have the discriminant in terms of the coefficients a, b, and c, given by

$$\Delta = 4a^3c - a^2b^2 - 18abc + 4b^3 + 27c^2 \tag{5.6}$$

For $\Delta > 0$, Cardano's formula is given as

$$\begin{aligned} y_1 &= \sqrt[3]{A} + \sqrt[3]{B} \\ y_2 &= -\frac{\sqrt[3]{A} + \sqrt[3]{B}}{2} + i\sqrt{3}\frac{\sqrt[3]{A} - \sqrt[3]{B}}{2} \\ y_3 &= -\frac{\sqrt[3]{A} + \sqrt[3]{B}}{2} - i\sqrt{3}\frac{\sqrt[3]{A} - \sqrt[3]{B}}{2} \end{aligned} \tag{5.7}$$

where

$$\begin{aligned} A &= -\frac{q}{2} + \sqrt{\frac{q^2}{4} + \frac{p^3}{27}} \\ B &= -\frac{q}{2} - \sqrt{\frac{q^2}{4} + \frac{p^3}{27}} \end{aligned} \tag{5.8}$$

The solution is known as *Cardano's formula* (Uspensky, 1948).

Next, we shall obtain convenient forms of the formula. If $A > 0$ and $B > 0$, the first equation of (5.7) gives a real root, but the others provide two complex roots. However, if A or B is negative, one of (5.7) gives a real root and the other two equations provide complex roots; in this case, the calculation becomes more complicated since $\sqrt[3]{-s}$, where s is a positive real number, has three complex roots. We can manipulate Cardano's formula (5.7) to make it simpler. Note that if A or B is negative, A and B in the set of the solutions (5.7) can be changed as follows:

$$\begin{aligned} &\text{If } A < 0: \ \sqrt[3]{A} = -\sqrt[3]{-A} \\ &\text{If } B < 0: \ \sqrt[3]{B} = -\sqrt[3]{-B} \end{aligned} \tag{5.9}$$

Even though these mathematical manipulations are performed on A and B, it is proved that the first equation of (5.7) still gives a real number and the others provide a pair of complex roots. Therefore, the roots are obtained as $x = y - a/3$. If we are only interested in the real root of a cubic equation ($\Delta > 0$), the real root is given by

$$
\begin{aligned}
x_1 &= y_1 - \frac{a}{3} = -\frac{a}{3} + \sqrt[3]{A} + \sqrt[3]{B} \\
x_1 &= y_1 - \frac{a}{3} = -\frac{a}{3} + \sqrt[3]{A} - \frac{p}{3}\frac{1}{\sqrt[3]{A}}
\end{aligned}
\tag{5.10}
$$

The second equation of (5.10) holds since $\sqrt[3]{B} = -p/3\sqrt[3]{A}$.

In many science and engineering problems, including MEMS and NEMS problems, jump phenomena of physical quantities are observed: for example, the collapse of a movable plate under electrostatic force to a stationary electrode, and the jumps in an oscillator (known as a Duffing oscillator) subjected to a cubic restoring force. In the remaining portions of this book, these jump phenomena are analyzed to obtain closed-form solutions for static and dynamic behavior. To describe jump phenomena, more convenient forms of the roots of the cubic equation may be used. As shown in Fig. 5.1b, two critical points, labeled y_u and y_l in Fig. 5.1b, can be defined if the curve is folded. When the straight line is between y_u and y_l in Fig. 5.1b, the discriminant Δ is less than zero and we can obtain a different form of (5.7), as described below. For $\Delta < 0$, we define the following quantity for convenience:

$$
\Delta_1 = \frac{q^2}{4} + \frac{p^3}{27} \tag{5.11}
$$

and we rewrite A and B of (5.8) as follows:

$$
A = a_1 + ib_1 = \sqrt{a_1^2 + b_1^2}\, e^{i\phi} \qquad B = a_1 - ib_1 = \sqrt{a_1^2 + b_1^2}\, e^{-i\phi} \tag{5.12}
$$

where

$$
a_1 = -\frac{q}{2} \qquad b_1 = \sqrt{-\Delta_1} \qquad \phi = \tan^{-1}\frac{b_1}{a_1} = \cos^{-1}\frac{a_1}{\sqrt{a_1^2 + b_1^2}} \tag{5.13}
$$

Substituting (5.12) into (5.7), noting the relationships

$$
-\frac{1}{2} + i\frac{\sqrt{3}}{2} = e^{i2\pi/3} \qquad -\frac{1}{2} - i\frac{\sqrt{3}}{2} = e^{-i2\pi/3}
$$

and recalling Euler's formula, $e^{iy} = \cos y + i\sin y$, a different form of (5.7) is given by

$$
\begin{aligned}
y_1 &= \left(a_1^2 + b_1^2\right)^{1/6} \left(e^{i\phi/3} + e^{-i\phi/3}\right) = 2\sqrt{\frac{-p}{3}} \cos\frac{\phi}{3} \\
y_2 &= \left(a_1^2 + b_1^2\right)^{1/6} \left(e^{i\phi/3} e^{i2\pi/3} + e^{-i\phi/3} e^{-i2\pi/3}\right) = 2\sqrt{\frac{-p}{3}} \cos\left[\frac{1}{3}(\phi + 2\pi)\right] \\
y_3 &= \left(a_1^2 + b_1^2\right)^{1/6} \left(e^{i\phi/3} e^{i4\pi/3} + e^{-i\phi/3} e^{-i4\pi/3}\right) = 2\sqrt{\frac{-p}{3}} \cos\left[\frac{1}{3}(\phi + 4\pi)\right]
\end{aligned} \tag{5.14}
$$

Substituting (5.4) into equations (5.14) and using $x = y - a/3$ gives three distinct real roots:

$$
\begin{aligned}
x_1 &= \frac{1}{3}\left(-a + 2\sqrt{a^2 - 3b} \cos\frac{\phi}{3}\right) \\
x_2 &= \frac{1}{3}\left\{-a + 2\sqrt{a^2 - 3b} \cos\left[\frac{1}{3}(\phi + 2\pi)\right]\right\} \\
x_3 &= \frac{1}{3}\left\{-a + 2\sqrt{a^2 - 3b} \cos\left[\frac{1}{3}(\phi + 4\pi)\right]\right\}
\end{aligned} \tag{5.15}
$$

where

$$
\phi = \cos^{-1} \frac{-a^3 + (9/2)ab - (27/2)c}{\left(a^2 - 3b\right)^{3/2}} \tag{5.16}
$$

It is noted that equation (5.16) holds for $\Delta \le 0$, where the equals sign gives a single root and a multiple root. It is also noted that for $\Delta = 0$, which corresponds to y_u and y_l in Fig. 5.1b, the argument of the inverse cosine in equation (5.16) becomes 1 or −1 and then the corresponding ϕ is 0 or π. These conditions may be used to find critical quantities, resulting in a critical situation of a system (e.g., the pull-in voltage of the parallel plate is described in Chapter 9). The preceding roots for $\Delta \le 0$ can be rewritten in terms of the coefficients of the more general cubic equation (5.1) as follows:

$$
\begin{aligned}
x_1 &= \frac{1}{3L}\left(-A_1 + 2\sqrt{A_1^2 - 3LB_1} \cos\frac{\phi}{3}\right) \\
x_2 &= \frac{1}{3L}\left\{-A_1 + 2\sqrt{A_1^2 - 3LB_1} \cos\left[\frac{1}{3}(\phi + 2\pi)\right]\right\} \\
x_3 &= \frac{1}{3L}\left\{-A_1 + 2\sqrt{A_1^2 - 3LB_1} \cos\left[\frac{1}{3}(\phi + 4\pi)\right]\right\}
\end{aligned} \tag{5.17}
$$

where

$$\phi = \cos^{-1} \frac{-A_1^3 + (9/2)LA_1B_1 - (27/2)L^2C_1}{\left(A_1^2 - 3LB_1\right)^{3/2}} \tag{5.18}$$

Example 5.1 A cubic equation is given by $-2x^3 + 4x^2 + 2x - 4 = 0$. We want to obtain the real roots of the equation.

The equation is factored into $-2(x^3 - 2x^2 - x + 2) = -2(x - 1)(x + 1)(x - 2) = 0$. We then obtain the roots as follows: $x = -1, 1, 2$. However, at this stage it is assumed that we do not know the roots, so we obtain the solutions by using the procedure described above. The equation is changed to $x^3 - 2x^2 - 1x + 2 = 0$ by dividing the equation by -2. We then have $a = -2$, $b = -1$, and $c = 2$. The discriminant is obtained from (5.6) as $\Delta = -36$. Since $\Delta < 0$, the equation will give three distinct roots. Equation (5.15) or (5.17) gives the roots as follows:

$$\phi = \cos^{-1} \frac{-10}{7^{3/2}} = 2.1412$$

$$x_1 = \frac{1}{3}\left(2 + 2\sqrt{7}\cos\frac{\phi}{3}\right) = \frac{1}{3}\left(2 + 2\sqrt{7}\cos\frac{2.1412}{3}\right) = 2$$

$$x_2 = \frac{1}{3}\left\{2 + 2\sqrt{7}\cos\left[\frac{1}{3}(\phi + 2\pi)\right]\right\} = -1$$

$$x_3 = \frac{1}{3}\left\{2 + 2\sqrt{7}\cos\left[\frac{1}{3}(\phi + 4\pi)\right]\right\} = 1$$

If a cubic restoring force of an actuator (e.g., Fig. 4.25) is in equilibrium with a constant external force, the force equilibrium equation is given by

$$k_1x + k_3x^3 - F = 0 \tag{5.19}$$

which provides the coefficients $a = 0$, $b = k_1/k_3$, and $c = F/k_3$. Setting the discriminant (5.6) equal to zero gives an equation for the critical force,

$$\Delta = 4\left(\frac{k_1}{k_3}\right)^3 + 27\left(\frac{F}{k_3}\right)^2 = 0$$

which yields the critical force of the actuator as follows:

$$F_{c1,c2} = \pm\frac{2k_1}{3}\left(\frac{-k_1}{3k_3}\right)^{1/2} \tag{5.20}$$

where the subscripts c_1 and c_2 correspond to positive and negative signs, respectively. Equation (5.20) is the same as (4.147) obtained from zero effective stiffness at the critical force. It is noted from equation (5.20) that if either the linear or cubic stiffness is negative, the critical force is obtained. In other words, if the force curve is folded as shown in Fig. 4.26b and c, the force equilibrium equation (5.19) yields the foregoing critical forces. It is also noted that if $k_1k_3 < 0$ and the force F is within the two critical forces, the discriminant Δ is less than zero. Otherwise, the discriminant Δ becomes a positive value. For $\Delta > 0$ (or $k_1k_3 > 0$), the single real root of equation (5.19) is obtained from (5.10) as follows:

$$x = \frac{1}{k_3}\left(\frac{k_1k_3}{3}\right)^{1/2}\left(S^{1/3} - \frac{1}{S^{1/3}}\right) \tag{5.21}$$

where

$$S = \phi_1 + \sqrt{\phi_1^2 + 1} \qquad \phi_1 = \sqrt{\frac{27k_3}{4k_1^3}}\,F$$

If $\Delta > 0$ and $k_1k_3 < 0$ (i.e., the force F is outside the range F_{c1} to F_{c2}), the root of the cubic equation (5.19) is expressed as

$$x = \left(\frac{-k_1}{3k_3}\right)^{1/2}\left(T^{1/3} + \frac{1}{T^{1/3}}\right) \tag{5.22}$$

where

$$T = \phi_2 + \sqrt{\phi_2^2 - 1} \qquad \phi_2 = \mathrm{sgn}(k_3)\sqrt{\frac{-27k_3}{4k_1^3}}\,F$$

and if $T < 0$, $T^{1/3} = -(-T)^{1/3}$. In the preceding equation, $\mathrm{sgn}(k_3)$ denotes the sign of k_3. For $\Delta \le 0$ (i.e., $k_1k_3 < 0$ and the force F is between the critical forces F_{c1} and F_{c2}), equation (5.17) yields a set of real roots as follows:

$$\begin{aligned}
x_1 &= \frac{2}{k_3}\left(\frac{-k_1k_3}{3}\right)^{1/2}\cos\frac{\phi}{3}\\
x_2 &= \frac{2}{k_3}\left(\frac{-k_1k_3}{3}\right)^{1/2}\cos\left[\frac{1}{3}(\phi + 2\pi)\right]\\
x_3 &= \frac{2}{k_3}\left(\frac{-k_1k_3}{3}\right)^{1/2}\cos\left[\frac{1}{3}(\phi + 4\pi)\right]
\end{aligned} \tag{5.23}$$

where

$$\phi = \cos^{-1}\frac{27k_3^2F}{2(-3k_1k_3)^{3/2}} \tag{5.24}$$

If $k_3 > 0, x_2 \le x_3 \le x_1$. If $k_3 \le 0, x_1 \le x_3 \le x_2$. If F is one of the critical forces, two of the preceding roots give an identical value (a multiple root).

5.2 DESCRIPTION OF MOTION

Structures for MEMS devices can move in more than one direction. For example, the movable structure of a typical resonator in Fig. 5.2 vibrates in the x direction over a substrate. The movable structure of the resonator has displacement, velocity, and acceleration with respect to the substrate. To describe the motion of the movable structure, we need to develop expressions for displacement, velocity, and acceleration. The displacement of a point p is given by a displacement vector $\mathbf{r}$ in Fig. 5.3, where $\mathbf{i}$, $\mathbf{j}$, and $\mathbf{k}$ denote the unit vectors (with respect to a rectangular coordinate system x, y, and z) which are unchanged in direction. Referring to Fig. 5.3, we can define the displacement at time t as follows:

$$\mathbf{r} = x\mathbf{i} + y\mathbf{j} + z\mathbf{k} \tag{5.25}$$

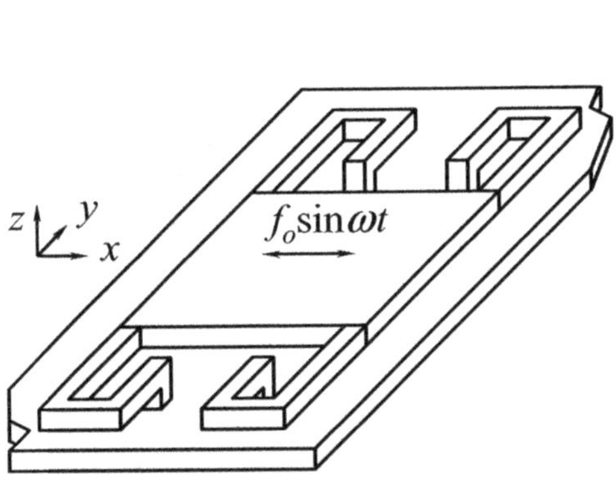

Figure 5.2 Typical resonator.

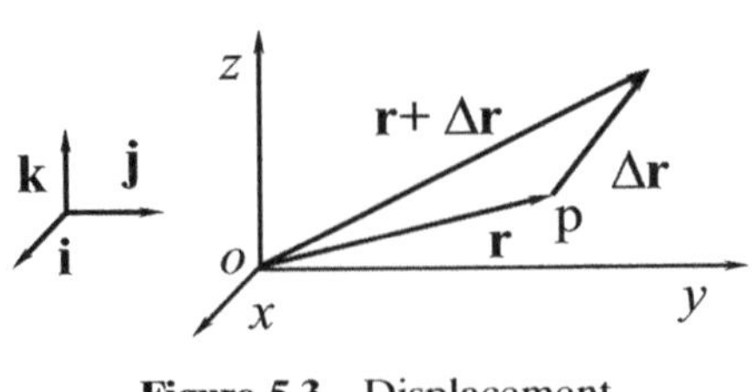

Figure 5.3 Displacement.

If point p moves from $\boldsymbol{r}$ to $\boldsymbol{r}+\Delta\boldsymbol{r}$ during time Δt, the displacement is given by

$$\mathbf{r}+\Delta\mathbf{r}=(x+\Delta x)\mathbf{i}+(y+\Delta y)\mathbf{j}+(z+\Delta z)\mathbf{k}$$

and we obtain the velocity vector $\mathbf{v}$ by taking the limit as Δt approaches zero:

$$\mathbf{v}=\lim_{\Delta t\to 0}\frac{\Delta\mathbf{r}}{\Delta t}=\frac{d\mathbf{r}}{dt}=\frac{dx}{dt}\mathbf{i}+\frac{dy}{dt}\mathbf{j}+\frac{dz}{dt}\mathbf{k} \tag{5.26}$$

Similarly, the acceleration of point p is obtained as

$$\mathbf{a}=\lim_{\Delta t\to 0}\frac{\Delta\mathbf{v}}{\Delta t}=\frac{d\mathbf{v}}{dt}=\frac{d^2x}{dt^2}\mathbf{i}+\frac{d^2y}{dt^2}\mathbf{j}+\frac{d^2z}{dt^2}\mathbf{k} \tag{5.27}$$

Note, from the definitions of velocity and acceleration above that the unit vectors are considered as constants since they are unchanged in magnitude and direction.

Let us consider a point on a disk that rotates at a constant angular velocity $\boldsymbol{\omega}=\omega\mathbf{k}$ with respect to a fixed *XYZ coordinate system* (also referred to as an *inertial frame of reference*) whose unit vectors are $\mathbf{i}', \mathbf{j}'$, and $\mathbf{k}'$, as shown in Fig. 5.4a. For a description of motion of the point, we consider a rotating coordinate system, fixed on the rotating disk, whose unit vectors are $\mathbf{i}, \mathbf{j}$, and $\mathbf{k}$. The point is denoted by a displacement $\mathbf{r}$ at time t in Fig. 5.4a. The point is angularly displaced during Δt as shown in Fig. 5.4b, where r, θ, $\Delta\theta$, and Δs denote the magnitude of the displacement vector $\mathbf{r}$, the angular displacement at t, the increase in the angular displacement during Δt, and the arc made by the point, respectively. Since the arc is given by $\Delta s = r\Delta\theta$, the velocity of the point is expressed as

$$\upsilon=\lim_{\Delta t\to 0}\frac{\Delta s}{\Delta t}=\lim_{\Delta t\to 0}\frac{r\Delta\theta}{\Delta t}=r\frac{d\theta}{dt}=r\dot{\theta}$$

where the dot denotes a derivative with respect to time. The velocity may be expressed in vector form as $\mathbf{v}=\upsilon\mathbf{j}=r\dot{\theta}\mathbf{j}$. If we express the angular velocity in vector form, $\boldsymbol{\omega}=\dot{\theta}\mathbf{k}$, the velocity can be expressed as $\mathbf{v}=\dot{\theta}r\mathbf{j}=\dot{\theta}\mathbf{k}\times r\mathbf{i}=\boldsymbol{\omega}\times\mathbf{r}$, where $\mathbf{r}=r\mathbf{i}$. In general, the velocity of a point on a rigid body rotating at an angular velocity $\boldsymbol{\omega}$ can be given by

$$\mathbf{v}=\frac{d\mathbf{r}}{dt}=\boldsymbol{\omega}\times\mathbf{r} \tag{5.28}$$

where $\mathbf{r}$ represents the displacement vector of the point in the body. Referring to Fig. 5.4c, which shows the velocity change, the acceleration of the point is obtained as

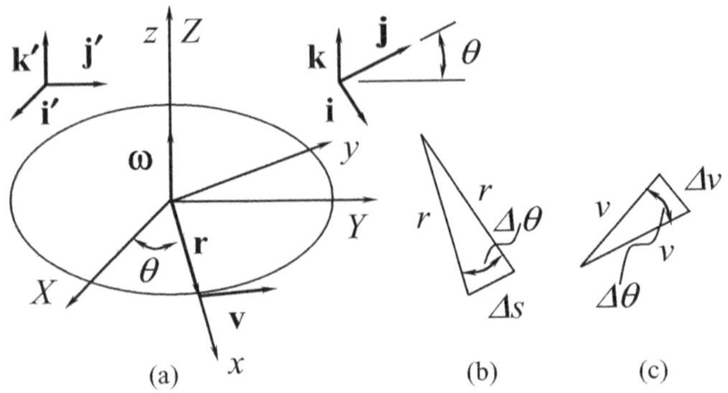

Figure 5.4 Velocity of a point on a rotating disk.

$$a = \lim_{\Delta t \to 0} \frac{\Delta v}{\Delta t} = \lim_{\Delta t \to 0} \frac{v \Delta \theta}{\Delta t} = v\dot{\theta} = r\dot{\theta}^2$$

and the acceleration **a** is directed to the center of the disk. The angular velocity of the point fixed on the disk is rewritten in vector form as

$$\mathbf{a} = \boldsymbol{\omega} \times \mathbf{v} = \boldsymbol{\omega} \times (\boldsymbol{\omega} \times \mathbf{r}) \tag{5.29}$$

With the displacement vector **r** of the point, equation (5.29) gives the acceleration of the position as $\mathbf{a} = \omega\mathbf{k} \times (\omega\mathbf{k} \times r\mathbf{i}) = -r\omega^2\mathbf{i}$, which is the same as the one obtained from Fig. 5.4c. For further derivation, we define the time derivative of the unit vectors. Substituting **i**, **j**, and **k** into (5.28), we find the time derivative of the unit vectors fixed on the rotating coordinate system:

$$\begin{aligned} \dot{\mathbf{i}} &= \boldsymbol{\omega} \times \mathbf{i} = \begin{vmatrix} \mathbf{i} & \mathbf{j} & \mathbf{k} \\ \omega_x & \omega_y & \omega_z \\ 1 & 0 & 0 \end{vmatrix} = \omega_z \mathbf{j} - \omega_y \mathbf{k} \\ \dot{\mathbf{j}} &= \omega_x \mathbf{k} - \omega_z \mathbf{i} \\ \dot{\mathbf{k}} &= \omega_y \mathbf{i} - \omega_x \mathbf{j} \end{aligned} \tag{5.30}$$

where ω_x, ω_y, and ω_z denote the components of the angular velocity $\boldsymbol{\omega}$ with respect to a fixed XYZ coordinate system.

Referring to Fig. 5.5, where the origin of a moving coordinate system xyz moves translationally and rotationally with respect to a fixed coordinate system XYZ, we shall derive more general relations for displacement, velocity, and acceleration of a point p that moves relative to the rotating coordinate system xyz. The unit vectors are represented by **i′**, **j′**, and **k′** for fixed coordi-

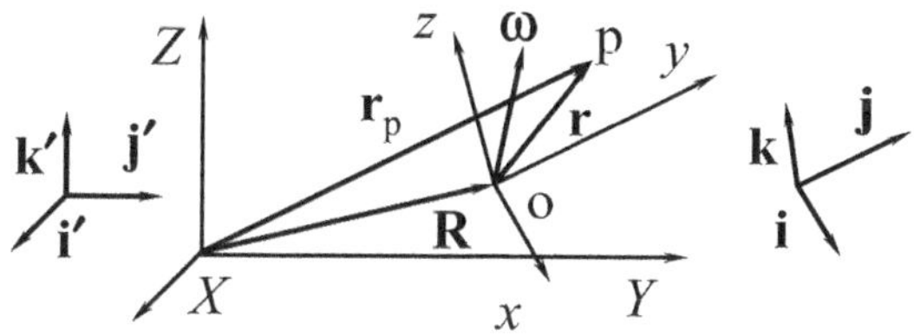

Figure 5.5 Motion with respect to a moving coordinate system.

nates and **i**, **j**, and **k** for rotating coordinates, respectively. Point p can be defined as the vector sum of the displacement vector **R** from the origin of XYZ to the origin of zyz and the displacement vector **r** extending from the origin of xyz to point p,

$$\mathbf{r}_p = \mathbf{R} + \mathbf{r} = \mathbf{R} + x\mathbf{i} + y\mathbf{j} + z\mathbf{k} \tag{5.31}$$

where $\mathbf{r}_p$ denotes the vector from the origin of XYZ to point p. Taking the time derivative of equation (5.31) and using (5.30) gives the velocity of point p:

$$\begin{aligned}\dot{\mathbf{r}}_p &= \dot{\mathbf{R}} + \dot{\mathbf{r}} = \dot{\mathbf{R}} + \dot{x}\mathbf{i} + \dot{y}\mathbf{j} + \dot{z}\mathbf{k} + x\dot{\mathbf{i}} + y\dot{\mathbf{j}} + z\dot{\mathbf{k}} \\ &= \dot{\mathbf{R}} + \dot{x}\mathbf{i} + \dot{y}\mathbf{j} + \dot{z}\mathbf{k} + \boldsymbol{\omega} \times (x\mathbf{i} + y\mathbf{j} + z\mathbf{k}) \\ &= \dot{\mathbf{R}} + \dot{\mathbf{r}}_r + \boldsymbol{\omega} \times \mathbf{r}\end{aligned} \tag{5.32}$$

whre $\dot{\mathbf{R}}$ and $\dot{\mathbf{r}}_r$, respectively, denote the velocity of the origin of xyz with respect to XYZ and the velocity of point p relative to the moving coordinates xyz (i.e., $\dot{\mathbf{r}}_r = \dot{x}\mathbf{i} + \dot{y}\mathbf{j} + \dot{z}\mathbf{k}$). Comparing the first and last lines of equation (5.32) gives the following relationship:

$$\dot{\mathbf{r}} = \dot{\mathbf{r}}_r + \boldsymbol{\omega} \times \mathbf{r}$$

which states that the velocity of point p with respect to the moving coordinates is the vector sum of velocities due to the translation and rotation relative to the moving coordinates. To obtain the acceleration $\ddot{\mathbf{r}}_p$ of point p, we take the time derivative of (5.32):

$$\begin{aligned}\ddot{\mathbf{r}}_p &= \ddot{\mathbf{R}} + \ddot{\mathbf{r}}_r + \boldsymbol{\omega} \times \dot{\mathbf{r}}_r + \dot{\boldsymbol{\omega}} \times \mathbf{r} + \boldsymbol{\omega} \times \dot{\mathbf{r}} \\ &= \ddot{\mathbf{R}} + \ddot{\mathbf{r}}_r + \boldsymbol{\omega} \times \dot{\mathbf{r}}_r + \dot{\boldsymbol{\omega}} \times \mathbf{r} + \boldsymbol{\omega} \times (\dot{\mathbf{r}}_r + \boldsymbol{\omega} \times \mathbf{r}) \\ &= \ddot{\mathbf{R}} + \boldsymbol{\omega} \times (\boldsymbol{\omega} \times \mathbf{r}) + \dot{\boldsymbol{\omega}} \times \mathbf{r} + \ddot{\mathbf{r}}_r + 2\boldsymbol{\omega} \times \dot{\mathbf{r}}_r\end{aligned} \tag{5.33}$$

where the first three terms represent accelerations due to translation and rotation of the xyz coordinate systems with respect to the fixed XYZ system, and the fourth and fifth terms are caused by the motion of point p relative to the xyz coordinate system. The fourth terms may be expressed as $\ddot{\mathbf{r}}_r = \ddot{x}\mathbf{i} + \ddot{y}\mathbf{j} + \ddot{z}\mathbf{k}$. In particular, the fifth term, $2\boldsymbol{\omega} \times \dot{\mathbf{r}}_r$, is called *Coriolis acceleration*.

5.3 GOVERNING EQUATIONS OF DYNAMICS

For a description of the motion of bodies at a velocity much less than the velocity of light, a variety of governing equations have been developed, but these equations were derived from one basic equation, *Newon's second law*, which is expressed mathematically as

$$\mathbf{F} = m\mathbf{a} \tag{5.34}$$

where m and $\mathbf{a}$ denote the mass and acceleration of the body, respectively. The governing equation (5.34) was formulated from experiments. It is noted that Newton's second law holds for a particle (i.e., in the case where the rotational effect are negligible). If the rotational effect cannot be neglected (i.e., the kinetic energy due to the rotational motion is not much smaller than that due to the translational motion), an additional governing equation should be used for a description of the motion. In MEMS, most movable structures move translationally or rotationally in a plane. Consider a rigid body that rotates at an angular velocity ω about its center of mass c and moves in a plane as shown in Fig. 5.6. In the figure, x' and y' denote the coordinates with respect to a fixed coordinate system, and x and y represent coordinates fixed on the body and measured from the center of mass of the body. The displacement vectors are defined as shown in the figure, and θ and dm denote an angle of the rotating coordinates and an infinitesimal mass at $\mathbf{r}$, respectively. Since the moving coordinates are fixed on the body and the relative velocity and acceleration are zero, and the acceleration equation (5.33) is reduced to

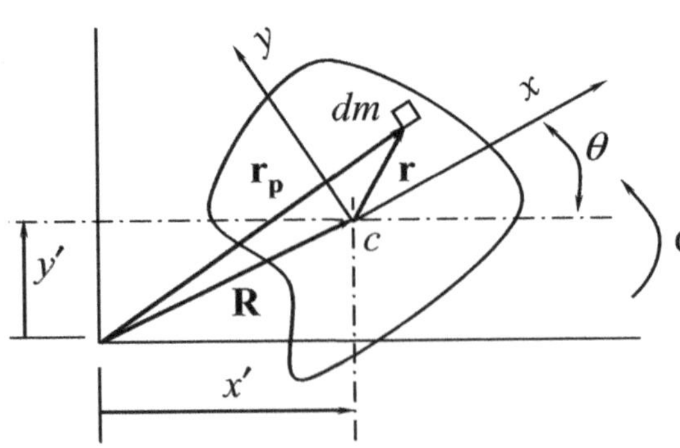

Figure 5.6 Rigid body in translational and rotational motion on a plane.

$$\ddot{\mathbf{r}}_p = \ddot{\mathbf{R}} + \boldsymbol{\omega} \times (\boldsymbol{\omega} \times \mathbf{r}) + \dot{\boldsymbol{\omega}} \times \mathbf{r}$$

Multiplying the preceding equation by the mass dm and integrating over mass, we have the force to accelerate the center of mass, which is given by

$$\begin{aligned}\mathbf{F} &= \int d\mathbf{F} = \int \ddot{\mathbf{r}}_p \, dm = \int_m \left[\ddot{\mathbf{R}} + \boldsymbol{\omega} \times (\boldsymbol{\omega} \times \mathbf{r}) + \dot{\boldsymbol{\omega}} \times \mathbf{r}\right] dm \\ &= \ddot{\mathbf{R}} \int_m dm + \boldsymbol{\omega} \times \left(\boldsymbol{\omega} \times \int_m \mathbf{r}\, dm\right) + \dot{\boldsymbol{\omega}} \times \left(\int_m \mathbf{r}\, dm\right) \\ &= m\mathbf{a} \end{aligned} \tag{5.35}$$

where

$$\int_m \mathbf{r}\, dm = 0$$

is used, which holds since the origin of the moving coordinate system coincides with the center of mass. Performing the cross product of the vector $\mathbf{r}\, dm$ and the preceding acceleration vector, and integrating over mass, we find the moment to rotate the body about the center of mass, which is expressed as

$$\begin{aligned}\mathbf{M} &= \int \mathbf{r} \times d\mathbf{F} = \int \mathbf{r} \times \ddot{\mathbf{r}}_p \, dm = \int_m \mathbf{r} \times \left[\ddot{\mathbf{R}} + \boldsymbol{\omega} \times (\boldsymbol{\omega} \times \mathbf{r}) + \dot{\boldsymbol{\omega}} \times \mathbf{r}\right] dm \\ &= \int_m \mathbf{r}\, dm \times \ddot{\mathbf{R}} + \int_m \mathbf{r} \times [\boldsymbol{\omega} \times (\boldsymbol{\omega} \times \mathbf{r})]\, dm + \int_m \mathbf{r} \times (\dot{\boldsymbol{\omega}} \times \mathbf{r})\, dm \end{aligned} \tag{5.36}$$

Note the following for the body in the plane:

$$\int_m \mathbf{r}\, dm = 0 \qquad \boldsymbol{\omega} = \omega \mathbf{k} \qquad \mathbf{r} = x\mathbf{i} + y\mathbf{j} \qquad \boldsymbol{\omega} \times (\boldsymbol{\omega} \times \mathbf{r}) = -\omega^2 \mathbf{r}$$

$$\mathbf{r} \times [\boldsymbol{\omega} \times (\boldsymbol{\omega} \times \mathbf{r})] = \mathbf{r} \times (-\omega^2 \mathbf{r}) = 0$$

$$\int_m \mathbf{r} \times (\dot{\boldsymbol{\omega}} \times \mathbf{r})\, dm = \int_m (x\mathbf{i} + y\mathbf{j}) \times [\dot{\omega}\mathbf{k} \times (x\mathbf{i} + y\mathbf{j})]\, dm = \int_m r^2 \, dm \dot{\omega} \mathbf{k}$$

where r is the magnitude of the displacement vector $\mathbf{r}$ relative to the moving coordinate system, we have the equation of motion of the body rotating in a plane:

$$\mathbf{M} = I_{zc} \dot{\omega} \mathbf{k} \tag{5.37}$$

where

$$I_{zc} = \int r^2 \, dm = \int (x^2 + y^2)\, dm \tag{5.38}$$

is defined as the moment of inertia of the body about the center of mass, and the subscripts z and c denote the axis of rotation and the center of mass. If we

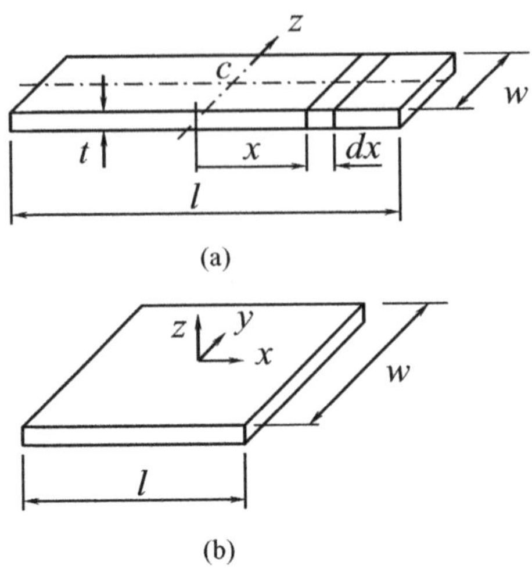

Figure 5.7 Beam and plate.

have a rigid slender beam ($t << l$ and $w << l$) and a plate of density ρ and thickness t (Fig. 5.7a and b), their moment of inertia about the z axis passing through the center of mass are obtained, respectively, as

$$I_{zc} = \int_{-l/2}^{l/2} x^2 \rho w t \, dx = \frac{ml^2}{12} \tag{5.39}$$

and

$$I_{zc} = \int_{-w/2}^{w/2} \int_{-l/2}^{l/2} (x^2 + y^2) \rho t \, dx \, dy = \frac{m}{12}(l^2 + w^2) \tag{5.40}$$

It should be noted that (5.37) holds only for a body that moves and rotates in a plane. If the body rotates an arbitrary angular velocity (i.e., $\boldsymbol{\omega} = \omega_x \mathbf{i} + \omega_y \mathbf{j} + \omega_z \mathbf{k}$), the moment equation (5.36) gives more terms, since $\mathbf{r} \times (\dot{\boldsymbol{\omega}} \times \mathbf{r})$ becomes more complicated. It should be noted from (5.35) and (5.37) that if forces are applied to a body on a plane, the center of body will be accelerated by the vector sum of the forces, and the body will be accelerated angularly by the moment due to the forces about its center of mass.

Example 5.2 Consider a man (Fig. 5.8) walking at a velocity $\dot{r}$ outward on a disk that rotates at a constant angular velocity ω about its center. Since the man walks along a guide, the man moves in the radial direction. It is desired to find forces acting on the man (1) when the man moves at a constant velocity, and (2) when the man moves sinusoidally relative to a position r_0.

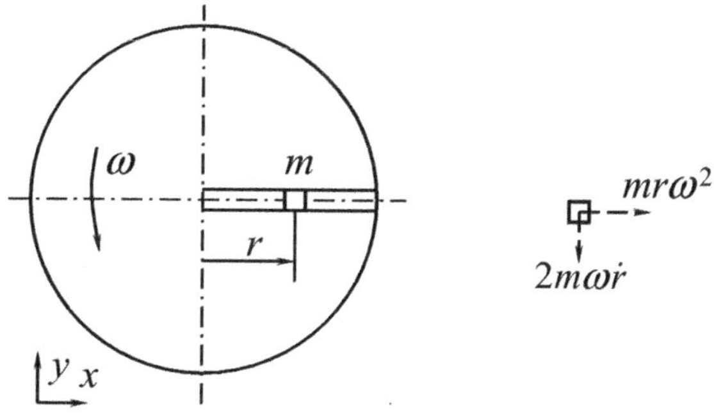

(a) (b)

Figure 5.8 Acceleration of a man walking on a rotating disk.

To evaluate the force acting on the man, we shall use (5.33) and (5.34). Let the angular velocity and the position of the man be denoted by $\boldsymbol{\omega} = \omega\mathbf{k}$ and $\mathbf{r}_r = r\mathbf{i}$, and we find the acceleration as

$$\begin{aligned}\ddot{\mathbf{r}}_p &= \omega\mathbf{k}\times(\omega\mathbf{k}\times r\mathbf{i}) + 2\omega\mathbf{k}\times\dot{r}\mathbf{i} \\ &= -r\omega^2\mathbf{i} + 2\omega\dot{r}\mathbf{j}\end{aligned} \tag{a}$$

where $\mathbf{i}$ and $\mathbf{j}$ denote the unit vectors with respect to the rotating coordinate systems fixed on the disk. The force exerted by the guide and the surface of the disk is given by

$$\mathbf{F} = m\mathbf{a} = m\ddot{\mathbf{r}}_p = -mr\omega^2\mathbf{i} + 2m\omega\dot{r}\,\mathbf{j} \tag{b}$$

which accelerates the man. The inertial force that the man feels is obtained as

$$\mathbf{F}_i = -\mathbf{F} = mr\omega^2\mathbf{i} - 2m\omega\dot{r}\mathbf{j} \tag{c}$$

where the first and second terms are known as the *centrifugal force* and the *Coriolis force*, respectively, and are illustrated in Fig. 5.8b.

If the disk and guide can sense the inertial force in the y direction (e.g., we install a sensor for this purpose), we can measure the Coriolis force F_c, defined as positive in the negative y direction, and in turn, obtain the angular acceleration as

$$\omega = \frac{F_c}{2mv} \tag{d}$$

If the man moves relative to a position r_0, the motion is approximated as $r = r_0 + r_a \sin\omega_a t$, where r_a, ω_a and t denote the amplitude of oscillation, the angular frequency, and time, respectively. Since the velocity and acceleration are given by $\dot{r} = r_a\omega_a \cos\omega_a t$ and $\ddot{r} = -r_a\omega_a^2 \sin\omega_a t = -r\omega_a^2$, the acceleration of the man is expressed as

$$\ddot{\mathbf{r}}_p = -\left(r_0\omega^2 + r_a\omega_a^2 \sin\omega_a t\right)\mathbf{i} + 2\omega\omega_a r_a \cos\omega_a t\mathbf{j} \tag{e}$$

The inertial force in the y direction is then given by

$$\mathbf{F}_i = -\mathbf{F} = m\left(r_0\omega^2 + r_a\omega_a^2 \sin\omega_a t\right)\mathbf{i} - 2m\omega\upsilon\mathbf{j} \tag{f}$$

where $\upsilon = \dot{r} = r_a\omega_a \cos\omega_a t$. If the sensor installed on the disk measures the inertial force in the negative y direction, the readout from the sensor will be

$$F_c = 2m\omega\upsilon = 2mr_a\omega\omega_a \cos\omega_a t \tag{g}$$

which is a sinusoidal force that is proportional to the angular velocity ω of the disk.

A measurement of the Coriolis force due to the vibrating structure in a rotating substrate, as described in Example 5.2, serves as the basic principle of vibratory gyroscopes. For example, the vibratory gyroscope shown in Fig. 5.9 consists of a movable plate that can vibrate in the x direction over a substrate, and springs one end of which is connected to the movable plate and whose other end is attached to the substrate. If the substrate rotates at an input angular velocity (angular rate) Ω while the movable plate vibrates at a velocity $\upsilon = \upsilon_0 \cos\omega t$, the movable plate is accelerated at an acceleration of $a = 2\Omega\upsilon$, and the inertial force (the Coriolis force), $F_c = 2m\Omega\upsilon$, is generated to press the movable plate in the negative z direction (Fig. 5.9). A motion detection sensor

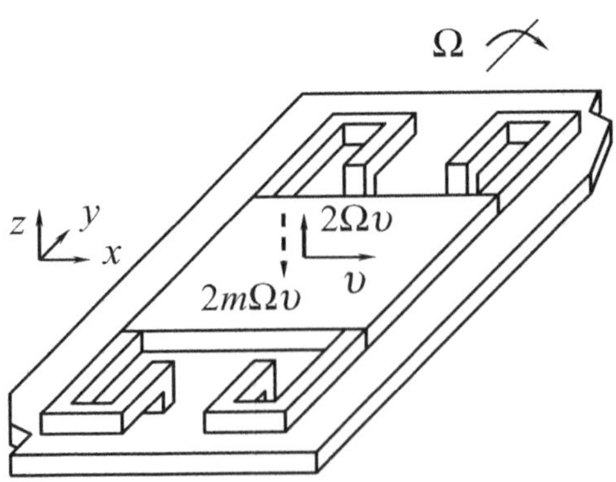

Figure 5.9 Coriolis force acting on a moving plate.

(e.g., a capacitance sensor to measure a capacitance change due to the motion of the movable plate) can measure the displacement of the movable plate. The Coriolis force can be written as

$$F_c = 2m\Omega v = 2mv_0\Omega\cos\omega t \tag{5.41}$$

where F_c is defined as positive when the force acts downward (i.e., in the negative z direction). The Coriolis force is used in Chapter 10 to obtain the dynamic response of a vibratory gyroscope.

5.4 ENERGY CONVERSION BETWEEN POTENTIAL AND KINETIC ENERGY

If a system has mass and stiffness, the system can vibrate mechanically since the mechanical vibration includes energy conversion between potential and kinetic energy. The mechanical vibration may begin with an initial displacement or may be caused by an applied force. Let's start by looking at one of the simplest cases: the energy conversion of a particle of mass m that slides down along an inclined plate under gravitational force. If there is no friction, and if the mass m starts from rest at position 1 in Fig. 5.10a, we know from experience that the mass will accelerate with time and that the velocity will increase. Since the velocity relates to the distance from position 1, we can obtain the velocity as a function of the distance. Taking the dot product on both sides of Newton's second law (5.34) with an increment of displacement $d\mathbf{r}$ and integrating, we have

$$\int_{r_1}^{r_2}\mathbf{F}\cdot d\mathbf{r} = \int_{r_1}^{r_2} m\frac{d^2\mathbf{r}}{dt^2}\cdot d\mathbf{r} = \int_{t_1}^{t_2}\frac{1}{2}\frac{d}{dt}\left(\frac{d\mathbf{r}}{dt}\cdot\frac{d\mathbf{r}}{dt}\right)dt = \int_{v_1^2}^{v_2^2}\frac{1}{2}m\,dv^2$$
$$= \frac{1}{2}mv_2^2 - \frac{1}{2}mv_1^2$$

When the kinetic energy is defined as

$$U_k = \frac{1}{2}mv^2 \tag{5.42}$$

the preceding integration may be expressed as

$$\int_{r_1}^{r_2}\mathbf{F}\cdot d\mathbf{r} = U_{k,2} - U_{k,1} \tag{5.43}$$

where the subscripts 1 and 2 denote positions 1 and 2. This equation states that work done on the particle by the force F is equal to an increase in the

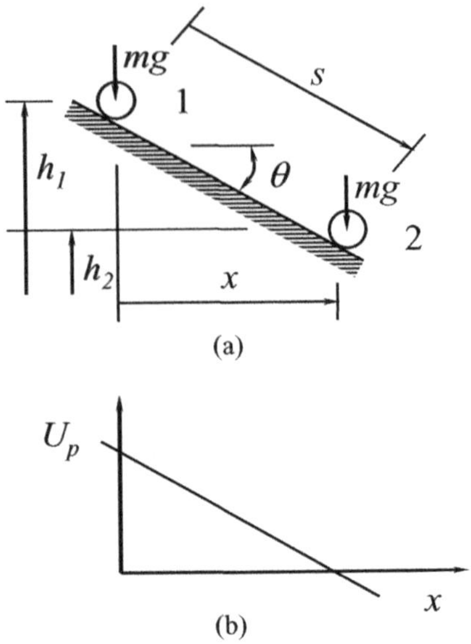

Figure 5.10 Particle moving on an inclined plane.

kinetic energy of a particle. In Fig. 5.10a, the force is given by $\mathbf{F} = mg(\sin\theta\mathbf{i} + \cos\theta\mathbf{j})$ and then the left side of equation (5.43) is expressed as

$$\int_{r_1}^{r_2}\mathbf{F}\cdot d\mathbf{r} = \int_0^s(-mg\,\mathbf{j})\cdot ds(\cos\theta\mathbf{i} - \sin\theta\mathbf{j}) = \int_0^s mg\sin\theta\, ds = mgs\sin\theta$$

where **i** and **j** denote the unit vectors in the horizontal and vertical directions. Since $\Delta h = h_1 - h_2 = s(\sin\theta)$, (5.43) is written as

$$mg\,\Delta h = \frac{1}{2}mv_2^2 - \frac{1}{2}mv_1^2 \tag{5.44}$$

If we define a restoring force as $\mathbf{F}_r = -mg\mathbf{j}$, the potential energy due to the gravity is given by (4.155):

$$U_p = -\int \mathbf{F}_r\cdot d\mathbf{r} = -\int_0^h -mg\mathbf{j}\cdot dh\mathbf{j} = mgh \tag{5.45}$$

where h is the vertical displacement of the mass. As a result, (5.43) is rewritten as

$$-\Delta U_p = \Delta U_k \tag{5.46}$$

where $\Delta U_p = mg(h_2 - h_1)$ and $\Delta U_k = \frac{1}{2}mv_2^2 - \frac{1}{2}mv_1^2$. Rearranging this equation gives

$$U = U_p + U_k = C \tag{5.47}$$

where U represents the total energy of the particle. Since the gravitational force is a conservative force, equations (5.46) and (5.47) hold. For any conservative forces (e.g., the electrostatic force of a charged particle), equations (5.46) and (5.47) also hold, a principle known as the *principle of the conservation of mechanical energy*. However, if the force applied to the particle is a nonconservative force (e.g., friction force), equation (5.47) for energy conservation does not hold, since energy is dissipated into heat. Figure 5.10b is a plot of the potential energy against the horizontal distance x.

Let us consider a mass m suspended by a spring of linear and cubic stiffnesses, k_1 and k_3, in Fig. 5.11a. While a constant force F is applied to the mass, we displace the mass from $x = 0$ by a distance x_i and then release the mass. The mass is then allowed to oscillate in the x direction. Since we assume that no friction force is applied to the system, the mechanical energy will be conserved. The curve of the potential energy corresponding to a cubic restoring force, discussed in Section 4.8, depends on the stiffnesses k_1 and k_3 and the constant force F. For example, the potential energy curve due to a cubic restoring force, $F_r = k_1x + k_3x^3 - F$, acting on the mass is shown in Fig. 5.11b for $k_1 > 0$, $k_3 > 0$, and $F = 0$, in Fig. 5.11c for $k_1 < 0$, $k_3 > 0$, and $F = 0$, and in Fig. 5.11d for $k_1 < 0$, $k_3 > 0$, and $F > 0$. If the mass m is placed at $x = x_i$, we know that the maximum velocity occurs at the lowest point of the curves (with reference to the principle of the conservation of mechanical energy). To obtain the velocity of the mass at x, we can write the equation of conservation of energy as

$$(U_p + U_k)_{x=x_i} = U_p + U_k$$

$$\frac{k_1}{2}x_i^2 + \frac{k_3}{4}x_i^4 - Fx_i = \frac{k_1}{2}x^2 + \frac{k_3}{4}x^4 - Fx + \frac{m}{2}\upsilon^2$$

Note from the equation above that the potential energy of the system includes the force effect. The preceding energy equation gives the velocity of the mass at x as follows:

$$\upsilon = \sqrt{\frac{1}{m}\left[k_1(x_i^2 - x^2) + \frac{k_3}{2}(x_i^4 - x^4) - 2F(x_i - x)\right]} \tag{5.48}$$

In Fig. 5.11b ($k_1 > 0$, $k_3 > 0$, and $F = 0$), the maximum velocity is achieved at $x = 0$, which is given by

$$\upsilon_{\max} = \sqrt{\frac{1}{m}\left(k_1 x_i^2 + \frac{k_3}{2}x_i^4\right)} = x_i\sqrt{\frac{k_1}{m}\left(1 + \frac{k_3}{2k_1}x_i^2\right)} \tag{5.49}$$

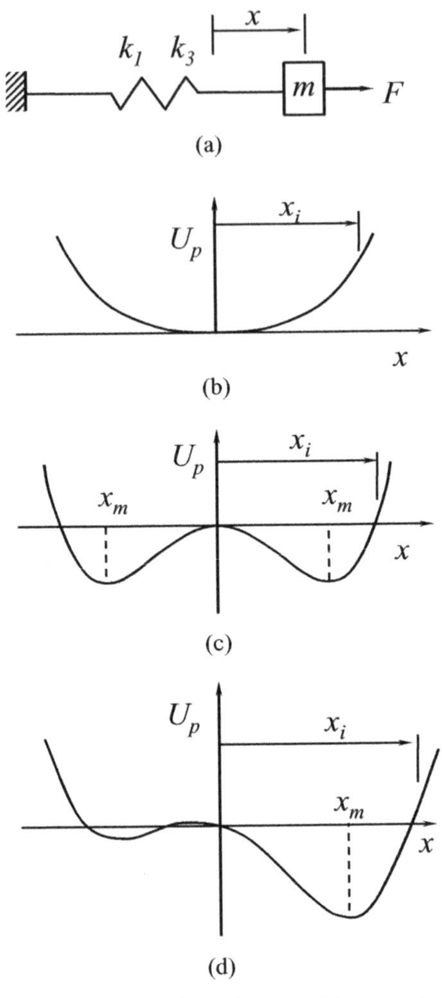

Figure 5.11 Mass subjected to a cubic restoring force.

For $k_1 < 0$ and $k_3 > 0$, the curve of the total potential energy depends on the magnitude and sign of the force F. If $F = 0$, the maximum velocity occurs at the two lowest points given by (4.145), labeled at $x = x_m$ in Fig. 5.11c. If the force F is between the critical forces, (4.147) is rewritten as:

$$F_{c1,c2} = \pm\frac{2k_1}{3}\left(\frac{-k_1}{3k_3}\right)^{1/2}$$

where the subscripts c_1 and c_2 correspond to the positive and negative signs. The maximum velocity is then given by

$$v_{\max} = \sqrt{\frac{1}{m}\left[k_1\left(x_i^2 - x_m^2\right) + \frac{k_3}{2}\left(x_i^4 - x_m^4\right) - 2F\left(x_i - x_m\right)\right]} \tag{5.50}$$

where x_m is the distance, (4.168), corresponding to the minimum potential energy, which is shown in Fig. 5.11d. The distance x_m is given by (4.168) as follows: If $0 \le F \le F_{c2}$,

$$x_m = 2\left(\frac{-k_1}{3k_3}\right)^{1/2} \cos\frac{\phi}{3} \tag{5.51}$$

and if $F_{c1} \le F < 0$,

$$x_m = 2\left(\frac{-k_1}{3k_3}\right)^{1/2} \cos\left[\frac{1}{3}(\phi + 2\pi)\right] \tag{5.52}$$

where

$$\phi = \cos^{-1} \frac{27k_3^2 F}{2(-3k_1k_3)^{3/2}}$$

For $0 \le F \le F_{c2}$, the potential energy and two minimum positions are shown in Fig. 5.11d. If $F = 0$, which is shown in Fig. 5.11c, the distance is reduced to $x_{\min} = (-k_1/3k_3)^{1/2}$. If $F > F_{c2}$ or $F < F_{c1}$, the distance for the minimum potential energy is obtained from (5.21) and is given by

$$x_m = \left(\frac{k_1}{3k_3}\right)^{1/2} \left[\left(\phi_1 + \sqrt{1+\phi_1^2}\right)^{1/3} - \left(-\phi_1 + \sqrt{1+\phi_1^2}\right)^{1/3}\right] \tag{5.53}$$

where

$$\phi_1 = \sqrt{\frac{27k_3}{4k_1^3}}\, F \tag{5.54}$$

It is noted that for $F > F_{c2}$ or $F < F_{c1}$, the potential energy has one well which is not symmetric about a vertical axis passing through its minimum; for example, if the force F becomes greater than F_{c2}, the left well in Fig. 5.11d disappears and the energy potential energy curve has a single well.

5.5 FREE VIBRATION OF UNDAMPED SYSTEMS

In section 5.4 we investigated the displacement–velocity relation (or relation between the potential and kinetic energy) in a conservative system. If we wish to find a dynamic response of a conservative system as a function of time, we need to solve an equation of motion which is formulated from Newton's

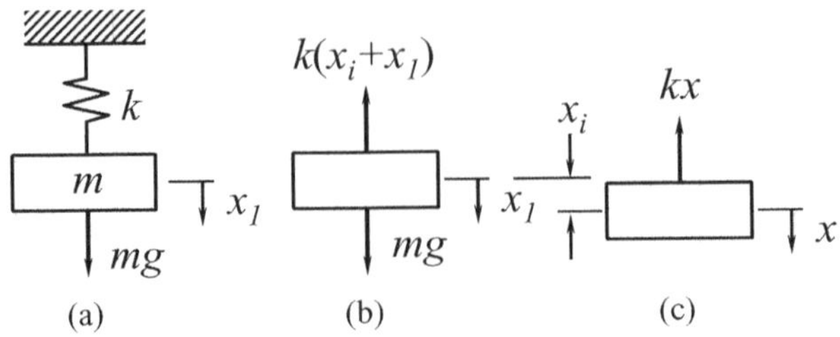

Figure 5.12 Mass–spring system.

second law or its derivatives. Let us consider the mass–spring system shown in Fig. 5.12a, which is a conservative system. Since a mass m connected to a spring of stiffness k is under gravity, the mass is displaced initially by x_i. If we displace the mass by an additional distance x_0 and release it, the mass will vibrate up and down. Newton's second law is applied to the free body in Fig. 5.12b to give the equation of motion for free vibration,

$$-kx_1 + mg = m\ddot{x}_1$$

where x_1 is the displacement measured from the position at which the spring is unstretched. Defining a new coordinate, $x = x_1 - x_i$, which represents displacement from the equilibrium displacement x_i, and subtracting the static equilibrium equation at $x = 0$, $-kx_i + mg = 0$, from the preceding equation, gives the equation of motion for the free vibration as

$$m\ddot{x} + kx = 0 \tag{5.55}$$

This is a differential equation that describes the displacement x of the mass relative to the equilibrium position x_i. To obtain the displacement in terms of time t, this equation should be solved with initial conditions. If the mass is released from rest at $x = x_0$, the initial conditions at time $t = 0$ are given by $x = x_0$ and $\dot{x} = 0$. Assuming a solution in the form $x = e^{\lambda t}$, we obtain the characteristic equation (5.55) as follows: $m\lambda^2 + k = 0$, which gives two roots,

$$\lambda = \pm i\sqrt{\frac{k}{m}}$$

where $i = \sqrt{-1}$. The solution of the preceding equation of motion is written as

$$x = A_1 e^{i\omega_n t} + B_1 e^{-i\omega_n t} \tag{5.56}$$

where A_1 and B_1 denote constants and

$$\omega_n = \sqrt{\frac{k}{m}} \tag{5.57}$$

is defined as the *natural angular frequency* (also known as the *resonant angular frequency*) of the mass–spring system. Recalling Euler's formula, $e^{i\theta t} = \cos\theta + i\sin\theta$, we can express the preceding solution in a different form,

$$x = A\cos\omega_n t + B\sin\omega_n t \tag{5.58}$$

where A and B denote constants. Applying the initial conditions at $t = 0, x = x_0$ and $\dot{x} = 0$, to the foregoing solution yields the constants $A = x_0$ and $B = 0$, and we obtain the solution as

$$x = x_0 \cos\omega_n t \tag{5.59}$$

This solution states that the mass m vibrates between x_0 and $-x_0$.

The potential and kinetic energies at x are obtained as

$$U_p = \frac{1}{2}kx^2 = \frac{1}{2}kx_0^2\cos^2\omega_n t$$

$$U_k = \frac{1}{2}mv^2 = \frac{1}{2}mx_0^2\omega_n^2\sin^2\omega_n t$$

Using (5.57), we obtain the total energy of the mass–spring system as follows:

$$\begin{aligned} U_t &= U_p + U_k = \frac{1}{2}kx_0^2\cos^2\omega_n t + \frac{1}{2}mx_0^2\omega_n^2\sin^2\omega_n t \\ &= \frac{1}{2}kx_0^2 = \frac{1}{2}mv_{\max}^2 \end{aligned}$$

where $v_{\max} = x_0\omega_n$. As described in Section 5.4, the mechanical energy of a mass–spring system remains constant since the energy does not dissipate. This energy conservation is shown in Fig. 5.13a, which shows graphically that the total energy remains constant during vibration. The displacement of the mass corresponding to the energy illustrated in Fig. 5.13a is shown in Fig. 5.13b and c, in which the displacement is plotted against time. The curve of Fig. 5.13a is the potential energy expressed as a function of the displacement from the equilibrium displacement x_i shown in Fig. 5.12c. Since the initial displacement x_i is proportional to the constant force mg, the potential energy curve is shifted from its original position to a new position as shown in Fig. 5.14, where x_1, x_i, x, $U_{p,1}$, $U_{p,1}$, and ΔU_p denote the displacement measured from the position at which the spring is unstretched, the initial displacement from x_1, the displacement from x_i, the potential energies, and the energy shift due to the constant

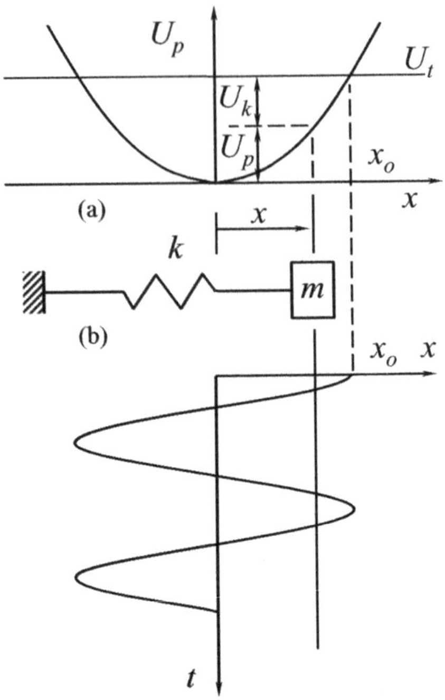

Figure 5.13 Free vibration of a mass–spring system.

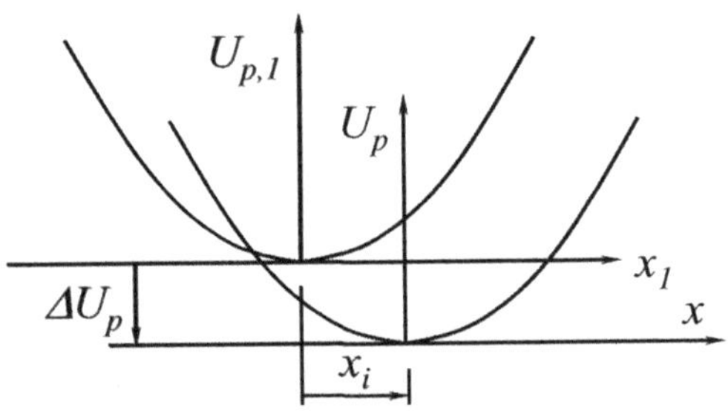

Figure 5.14 Shift of a potential energy curve.

force mg, respectively. Because the linear spring is used in the mass–spring system, the potential energy $U_{p,1}$ is given by (4.173) as follows:

$$
\begin{aligned}
U_{p,1} &= \left(\frac{k_1}{2}x_1^2 + \frac{k_3}{4}x_1^4 - Fx_1\right)_{k_3=0,F=mg} = \frac{k_1}{2}x_1^2 - mgx_1 \\
&= \frac{k_1}{2}\left[x_1^2 - \frac{2mg}{k_1}x_1 + \left(\frac{mg}{k_1}\right)^2\right] - \frac{k_1}{2}\left(\frac{mg}{k_1}\right)^2 = \frac{k_1}{2}\left(x_1 - \frac{mg}{k_1}\right)^2 - \frac{1}{2}\frac{(mg)^2}{k_1} \\
&= \frac{k_1}{2}(x_1 - x_i)^2 - \Delta U_p = U_p - \Delta U_p
\end{aligned}
$$

where $x_i = mg/k_1$, $U_p = k_1(x_1 - x_i)^2/2 = k_1x^2/2$, and $\Delta U_p = (mg)^2/2k_1$ are defined as the initial displacement, the potential energy, and the potential energy shift due to the constant force, mg, respectively. It is noted from the equation above that the potential energy is shifted by x_i and $-\Delta U_p$ when a mass–spring system with a linear spring is subjected to a constant force. In other words, the potential energy of a linear system can always be expressed in the quadratic form above and its curve shape is not distorted when a constant force is applied to the mass. Therefore, for linear systems such as the mass–spring system shown in Fig. 5.12 and a mass–spring–damper system with a linear spring (described in the following section), the equation of motion will be described in a form where the initial displacement due to the constant force does not appear as shown in (5.55).

The energy conservation of a conservative system may be used to find the angular frequency of the linear system. For example, to find the natural frequency of the mass–spring system in Fig. 5.12a, we can set the maximum potential energy equal to the maximum kinetic energy, that is,

$$\frac{1}{2}kx_0^2 = \frac{1}{2}mx_0^2\omega_n^2$$

This equation gives the same result as (5.57), $\omega_n = \sqrt{k/m}$. The natural frequency of the system is then written as

$$f_n = \frac{\omega_n}{2\pi} = \frac{1}{2\pi}\sqrt{\frac{k}{m}} \tag{5.60}$$

Note that the preceding method, using the maximum potential and kinetic energies to find the natural frequency, is valid only for a system in which the spring is linear. If a nonlinear force is not negligible and plays a role, the natural frequency evaluated by equation (5.60) will deviate from that measured from experiment or obtained from a numerical simulation.

5.6 VIBRATION OF DAMPED SYSTEMS

If a damping force, associated with energy dissipation, is involved in a vibration system, the total mechanical energy of the system is not conservative but is dissipated. For an analysis of this damped system, a damper is introduced to the mass–spring system described in Section 5.5. The system in Fig. 5.15a is then called a mass–spring–damper system, in which the damping force (drag force) is assumed to be proportional to the velocity of the mass; mathematically, the damping force F_d is expressed as

$$F_d = c\dot{x} \tag{5.61}$$

where c denotes the damping coefficient of the system (the damping force is discussed further in Chapter 6). Applying Newton's second law to the free body of Fig. 5.15b gives the equation of motion (linear differential equation) as

$$m\ddot{x} + c\dot{x} + kx = F \tag{5.62}$$

where the force F applied may be expressed as a sum of a constant force F_b and a sinusoidal force of amplitude F_0 and angular frequency ω as follows:

$$F = F_b + F_0 \sin \omega t \tag{5.63}$$

We shall first find the response x of the system during free vibration and then deal with the forced vibration.

Free Vibration of Damped Systems

When the mass in Fig. 5.15a is displaced to a position and then released, the mass may vibrate. In this case, the force term is dropped out from the equation of motion (5.62) and the differential equation will then be written as

$$m\ddot{x} + c\dot{x} + kx = 0 \tag{5.64}$$

We need to solve this equation for x with initial conditions. Assuming that $x = e^{\lambda t}$, we have a characteristic equation, $m\lambda^2 + c\lambda + k = 0$, which gives the roots as follows:

$$\lambda_{1,2} = -\frac{c}{2m} \pm \sqrt{\left(\frac{c}{2m}\right)^2 - \frac{k}{m}}$$

Depending on the discriminant D, which is defined as the quantity in the square root,

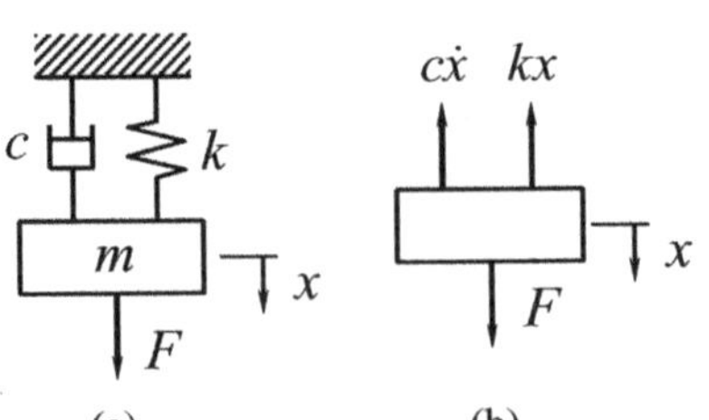

Figure 5.15 Mass–spring–damper system.

$$D=\left(\frac{c}{2m}\right)^2-\frac{k}{m} \tag{5.65}$$

three cases are encountered: an overdamped case for $D > 0$, a critically damped case for $D = 0$, and an underdamped case for $D < 0$. For the overdamped case ($D > 0$), the solution of the differential equation is given by

$$x = Ae^{\lambda_1 t} + Be^{-\lambda_2 t} \tag{5.66}$$

where A and B denote constants that can be determined from initial conditions. For the critically damped case, setting $D = 0$ gives the solution,

$$x = Ae^{-(c/2m)t} + Bte^{-(c/2m)t} \tag{5.67}$$

Note, from the preceding critically and overdamped cases that the mass does not vibrate even though an initial motion is given. This phenomenon is due to the energy dissipation, which is greater than or equal to a critical value. For a further description, it is convenient to define a critical damping coefficient. Setting the discriminant $D = 0$, we have the critical damping coefficient,

$$c_c = 2\sqrt{mk} \tag{5.68}$$

For the underdamped case, $D < 0$, the roots of the characteristic equation are given by

$$\lambda_{1,2} = -\frac{c}{2m} \pm i\sqrt{\frac{k}{m}-\left(\frac{c}{2m}\right)^2}$$

Defining a damping factor as

$$\zeta = \frac{c}{c_c} = \frac{c}{2\sqrt{mk}} \tag{5.69}$$

and using the definition of the natural angular frequency, $\omega_n = \sqrt{k/m}$, the preceding roots may be written as

$$\lambda_{1,2} = -\zeta\omega_n \pm i\omega_n\sqrt{1-\zeta^2}$$

and then the solution of the differential equation (5.64) is obtained as

$$x = e^{-\zeta\omega_n t}\left(A\cos\omega_d t + B\sin\omega_d t\right) \tag{5.70}$$

where

$$\omega_d = \omega_n \sqrt{1-\zeta^2} \tag{5.71}$$

is defined as the damped natural angular frequency and is less than the natural angular frequency ω_n. If the initial conditions are given by $x(0) = x_0$ and $\dot{x}(0) = 0$, the constants are found to be $A = x_0$ and $B = -\zeta x_0 / \sqrt{1-\zeta^2}$, and the foregoing solution is expressed as

$$x = x_0 e^{-\zeta\omega_n t}\left(\cos\omega_d t - \frac{\zeta}{\sqrt{1-\zeta^2}}\sin\omega_d t\right) \tag{5.72}$$

This response is shown in Fig. 5.16, and the potential energy U_p and the total energy U_t (the sum of the potential and kinetic energy) are illustrated in Fig. 5.17. As shown in both figures, the amplitude of the vibration decays with time

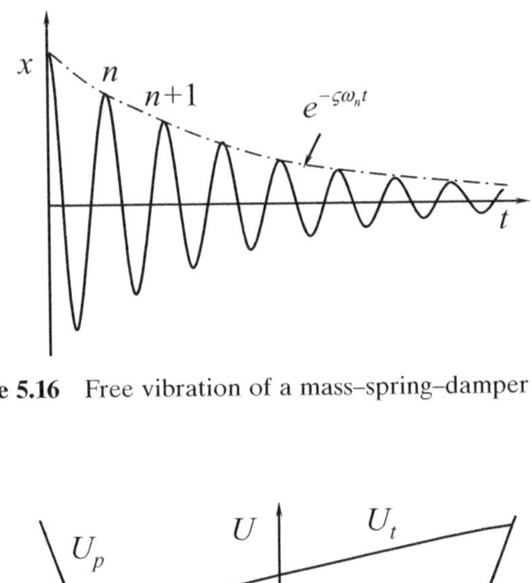

Figure 5.16 Free vibration of a mass–spring–damper system.

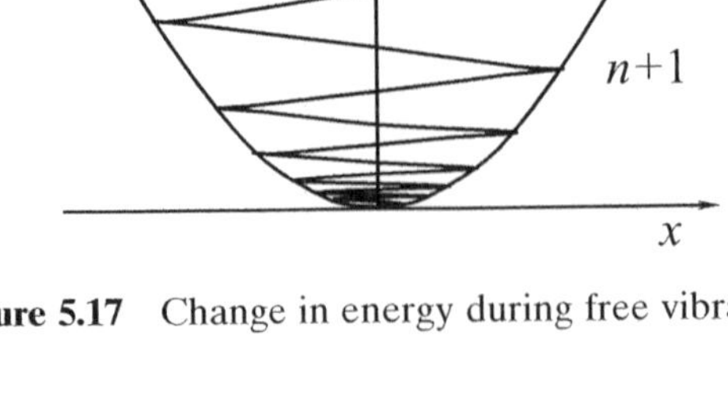

Figure 5.17 Change in energy during free vibration.

since the total energy dissipates gradually into heat. It is noted that the total energy at maximum and minimum amplitudes (labeled n and $n + 1$ in Figs. 5.16 and 5.17) is equal to the potential energy since the velocity at the amplitude is zero. As a result, the total energy decreases inside the potential energy curve as shown in Fig. 5.17.

Forced Vibration of Damped Systems

Let us turn our attention to the forced vibration of the mass–spring–damper system shown in Fig. 5.15. When a constant force F_b and a sinusoidal force of amplitude F_0 and angular frequency ω are applied to the mass, equation (5.62) is written

$$\ddot{x} + 2\zeta\omega_n\dot{x} + \omega_n^2 x = \frac{1}{m}(F_b + F_0 \sin \omega t) \tag{5.73}$$

The solution x is considered as a sum of the homogeneous solution x_h and the particular solution x_p (i.e., $x = x_p + x_h$). The homogeneous solution x_h is the solution of (5.64), which is given by (5.70). The particular solution is assumed to be $x_p = P + Q \sin \omega t + R \cos \omega t$, where P, Q, and R are constants to be determined. Substituting x_p into (5.73) and comparing both sides, we have the constants

$$P = \frac{F_b}{m\omega_n^2} = \frac{F_b}{k}$$

$$Q = \frac{F_0}{m} \frac{\omega_n^2 - \omega^2}{(\omega_n^2 - \omega^2)^2 + (2\zeta\omega_n\omega)^2}$$

$$R = -\frac{F_0}{m} \frac{2\zeta\omega_n\omega}{(\omega_n^2 - \omega^2)^2 + (2\zeta\omega_n\omega)^2}$$

We then obtain the particular solution,

$$x_p = \frac{F_b}{k} + \frac{F_0}{m} \frac{\omega_n^2 - \omega^2}{(\omega_n^2 - \omega^2)^2 + (2\zeta\omega_n\omega)^2} \sin \omega t - \frac{F_0}{m} \frac{2\zeta\omega_n\omega}{(\omega_n^2 - \omega^2)^2 + (2\zeta\omega_n\omega)^2} \cos \omega t$$

Using the identity

$$\begin{aligned} A\sin\theta - B\cos\theta &= \sqrt{A^2 + B^2}\left(\frac{A}{\sqrt{A^2 + B^2}}\sin\theta - \frac{B}{\sqrt{A^2 + B^2}}\cos\theta\right) \\ &= \sqrt{A^2 + B^2}\,(\cos\phi \sin\theta - \sin\phi \cos\theta) \\ &= \sqrt{A^2 + B^2}\,\sin(\theta - \phi) \end{aligned}$$

where $\phi = \tan^{-1}(B/A)$, the preceding solution is written as

$$x_p = \frac{F_b}{k} + \frac{F_0}{m}\frac{1}{\sqrt{(\omega_n^2-\omega^2)^2+(2\zeta\omega_n\omega)^2}}\sin(\omega t-\phi) \tag{5.74}$$

where

$$\phi = \tan^{-1}\frac{2\zeta\omega_n\omega}{\omega_n^2-\omega^2} \tag{5.75}$$

denotes the phase. Then the complete solution is expressed as the sum of the homogeneous and particular solutions, which is given by

$$\begin{aligned} x = {} & \frac{F_b}{k} + e^{-\zeta\omega_n t}(A\cos\omega_d t + B\sin\omega_d t) \\ & + \frac{F_0}{m}\frac{1}{\sqrt{(\omega_n^2-\omega^2)^2+(2\zeta\omega_n\omega)^2}}\sin(\omega t-\phi) \end{aligned} \tag{5.76}$$

where the constants A and B are determined from the initial conditions. From this solution, the first term on the right-hand side is the static displacement due to the constant force F_b, the second term is a transient response of the system that decays with time quickly as shown in Fig. 5.18 (calculated with initial conditions: $x(0) = x_0$ and $\dot{x}(0) = 0$), and the third term is the steady-state response of the system. The period during which the transient term lasts is represented by the time constant τ, which is defined as $\tau = 1/\zeta\omega_d$. For $\zeta << 1$, $\tau = 1/\zeta\omega_n$ because ω_d approaches ω_n. The mechanical energy is shown in Fig. 5.19, where the energy increases rapidly with time and reaches the steady state. In most MEMS, we are interested in the steady-state response, which represents the dynamic response of a mass–spring–damper system. The steady-state response is then given by

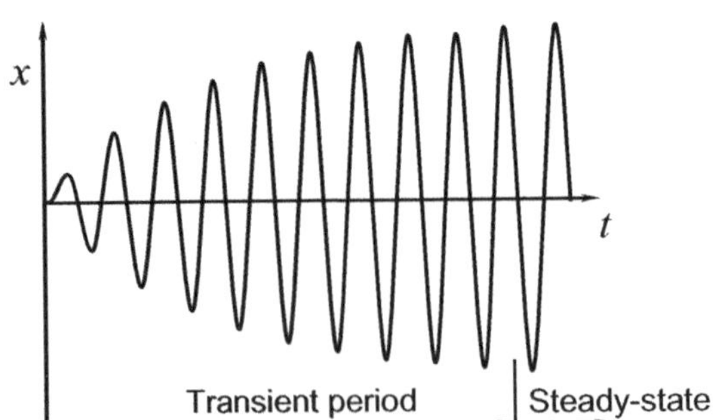

Figure 5.18 Transient response of an underdamped system.

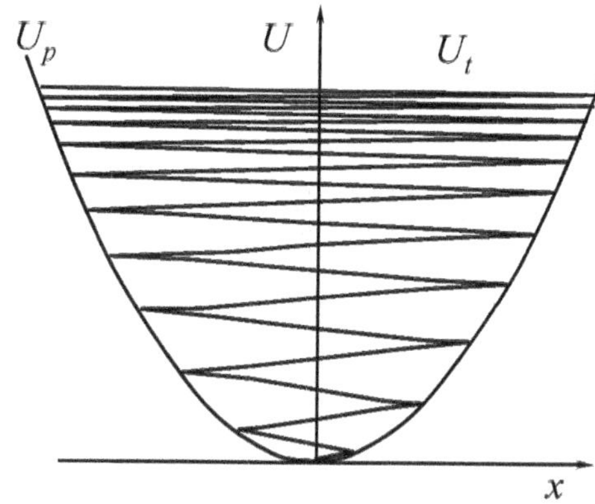

Figure 5.19 Change in energy during a transient period.

$$x = \frac{F_0}{m} \frac{1}{\sqrt{(\omega_n^2 - \omega^2)^2 + (2\zeta\omega_n\omega)^2}} \sin(\omega t - \phi) \tag{5.77}$$

If we are only interested in the steady-state response of a linear system, we can use a different method, based on superposition and complex numbers, which is relatively convenient for complicated system analysis. When a sinusoidal force of amplitude F_0 and angular velocity ω is applied to a mass–spring–damper system, the equation of motion is written

$$m\ddot{x} + c\dot{x} + kx = F_0 \sin \omega t \tag{5.78}$$

or

$$\ddot{x} + 2\zeta\omega_n\dot{x} + \omega_n^2 x = \frac{F_0}{m} \sin \omega t \tag{5.79}$$

where ζ and ω_n denote the damping factor, (5.69), and the natural angular frequency, (5.57), respectively. We can consider the following:

$$\ddot{x} + 2\zeta\omega_n\dot{x} + \omega_n^2 x = \frac{F_0}{m} \cos \omega t \tag{5.80}$$

which is a form in which the cosine is substituted for the sine in (5.79). Multiplying the preceding equation by i, defined as $\sqrt{-1}$, and adding the resulting equation to (5.79), we have

$$\ddot{x}(1+i) + 2\zeta\omega_n\dot{x}(1+i) + \omega_n^2 x(1+i) = \frac{F_0}{m}(\cos \omega t + i \sin \omega t)$$

Defining $x_1 = x(1 + i)$ and recalling the Euler formula, $e^{iy} = \cos y + i\sin y$, we have a simple differential equation which represents both (5.79) and (5.80):

$$\ddot{x}_1 + 2\zeta\omega_n\dot{x}_1 + \omega_n^2 x_1 = \frac{F_0}{m}e^{i\omega t} \tag{5.81}$$

Substituting $x_1 = x_0 e^{i\omega t}$ into equation (5.81), we have

$$x_0 = \frac{F_0}{m}\frac{1}{\omega_n^2 - \omega^2 + i2\zeta\omega_n\omega}$$

Noting the identity

$$\frac{1}{a+ib} = \frac{1}{\sqrt{a^2+b^2}\,e^{i\theta}} = \frac{1}{\sqrt{a^2+b^2}}e^{-i\theta}$$

where $\theta = \tan^{-1}(b/a)$, we have

$$x_0 = \frac{F_0}{m}\frac{1}{\sqrt{(\omega_n^2-\omega^2)^2 + (2\zeta\omega_n\omega)^2}}e^{-i\phi}$$

$$\phi = \tan^{-1}\frac{2\zeta\omega_n\omega}{\omega_n^2 - \omega^2} \tag{5.82}$$

and then the solution is expressed as

$$x_1 = \frac{F_0}{m}\frac{1}{\sqrt{(\omega_n^2-\omega^2)^2 + (2\zeta\omega_n\omega)^2}}e^{i(\omega t-\phi)} \tag{5.83}$$

Since $x_1 = x(1 + i)$, the solution above can be rewritten as

$$x(1+i) = \frac{F_0}{m}\frac{1}{\sqrt{(\omega_n^2-\omega^2)^2 + (2\zeta\omega_n\omega)^2}}\{[\cos(\omega t-\phi) + i\sin(\omega t-\phi)]\}$$

Comparing both sides, we find that

$$x = \mathrm{Im}(x_1) = \frac{F_0}{m}\frac{1}{\sqrt{(\omega_n^2-\omega^2)^2 + (2\zeta\omega_n\omega)^2}}\sin(\omega t-\phi) \tag{5.84}$$

$$x = \mathrm{Re}(x_1) = \frac{F_0}{m}\frac{1}{\sqrt{(\omega_n^2-\omega^2)^2 + (2\zeta\omega_n\omega)^2}}\cos(\omega t-\phi) \tag{5.85}$$

where

$$\phi = \tan^{-1} \frac{2\zeta\omega_n\omega}{\omega_n^2 - \omega^2} \tag{5.86}$$

is the phase. It is noted that the solution (5.83) provides two solutions, (5.84) and (5.85), which are the steady-state solutions of the equation of motion of mass–spring–damper system when the forces $F_0 \sin \omega t$ and $F_0 \cos \omega t$ are applied to the mass.

For convenience, we define the magnification factor M as follows:

$$M = \frac{x_0}{x_s} = \frac{x_0 k}{F_0} = \frac{1}{\sqrt{(1-\Omega^2)^2 + (2\zeta\Omega)^2}} \tag{5.87}$$

where

$$\Omega = \frac{\omega}{\omega_n} = \frac{f}{f_n} \tag{5.88}$$

is the frequency ratio and $x_s = F_0/m\omega_n^2 = F_0/k$ represents a static displacement corresponding to the force F_0. The steady-state response of the mass–stiffness–damper system is then expressed as

$$x = x_0 \sin(\omega t - \phi) \tag{5.89}$$

where x_0 is evaluated from (5.87) and ϕ is given by (5.86), which may be written in terms of Ω and ζ as

$$\phi = \tan^{-1} \frac{2\zeta\Omega}{1-\Omega^2}$$

Dissipated Energy

When the response of a damped system to a sinusoidal force reaches a steady state, an amount of energy is dissipated into heat during every cycle of oscillation, due to the damping force. Since energy dissipation is associated with the response, we shall evaluate the energy dissipation. The damping force is proportional to the velocity of mass and is then expressed as $F_d = c\upsilon$, where c and υ are the damping coefficient and the velocity, respectively. During one cycle of the steady-state response, the energy dissipated is obtained as

$$U_d = \int F_d \, dx = \int_0^\tau F_d \frac{dx}{dt} dt = \int_0^\tau F_d \upsilon \, dt = \int_0^\tau c\upsilon^2 \, dt \tag{5.90}$$

where τ denotes the period, defined as $2\pi/\omega$. Substituting (5.89) into equation (5.90) gives the energy dissipated during the period:

$$U_d = \pi c x_0^2 \omega \tag{5.91}$$

With the definitions of maximum velocity, $\upsilon_{\max} = x_0\omega$, and angular frequency, $\omega = 2\pi/\tau$, equation (5.91) gives the dissipated energy per cycle as

$$U_d = \frac{1}{2} c \upsilon_{\max}^2 \tau \tag{5.92}$$

Quality Factor

Earlier in this section we obtained the dynamic response of mass–stiffness–damper systems and evaluated the energy dissipated due to the damping force. The magnification factor (5.87) states that the steady-state response of a resonator is a function of the damping factor and frequency ratio. With reference to the damping force and the dissipated energy, we may use the damping coefficient, defined in (5.61), or the damping factor, (5.69). Since in most MEMS devices (e.g., a mechanical resonator) the damping coefficient and factor are typically small, we generally use the *quality factor*, which relates to the damping coefficient or energy dissipation and has a value greater than 1/2 for underdamped systems. The quality factor (also known as the *Q factor*) may be defined in three ways, described below. These three quality factors are interchangeable for linear vibrating systems that have a damping factor much below unity but give different numerical values for nonlinear systems.

The first definition of a quality factor is the ratio of the center frequency f_c (at which the amplitude of vibration is maximum) to the bandwidth, which is defined as $\Delta f = f_2 - f_1$, where f_2 and f_1 represent the *upper and lower half-power frequencies* (also known as the *upper and lower cutoff frequencies*), respectively, at which the energy in a system is half the maximum energy at resonance. The quality factor is then expressed as

$$Q = \frac{f_c}{\Delta f} \tag{5.93}$$

where f_c and Δf are as shown in Fig. 5.20. The square of the magnification factor (5.87) is written

$$M^2 = \frac{1}{(1-\Omega^2)^2 + (2\zeta\Omega)^2}$$

At the maximum multitude, the derivative of the equation above with respect to Ω is zero:

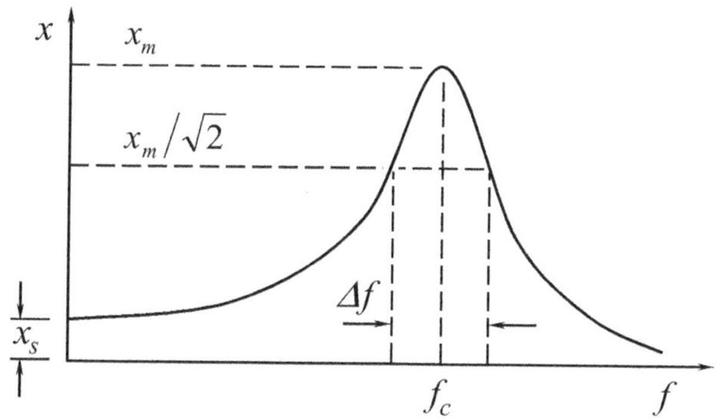

Figure 5.20 Frequency response of a one-DOF system.

$$\frac{\partial M^2}{\partial \Omega} = -\frac{-4r(1-\Omega^2)+8\zeta^2\Omega}{\left[(1-\Omega^2)^2+(2\zeta\Omega)^2\right]^2} = 0$$

whose numerator gives the frequency ratio Ω_c, which corresponds to the maximum amplitude,

$$\Omega_c = \sqrt{1-2\zeta^2}$$

or

$$f_c = f_n\sqrt{1-2\zeta^2} \tag{5.94}$$

where f_n denotes the natural frequency of the system. Note that the center frequency is lower than the natural frequency and close to the natural frequency for $\zeta << 1$. Substituting Ω_c into the magnification factor (5.87), we find that

$$M|_{\Omega=\Omega_c} = \frac{1}{2\zeta} \tag{5.95}$$

The preceding two equations are exact solutions for the center frequency f_c and the amplitude ratio at f_c. If $\zeta << 1$, the center frequency f_c can be written as

$$f_c \cong f_n \tag{5.96}$$

However, if the damping ratio is not much less than 1, f_c from (5.94) is significantly less than the natural frequency f_n. Using the definition of the half-power points at which the stored energy is half the maximum stored energy at f_c (it is noted that the energy stored in a linear system is proportional to the square of the amplitude), the magnification factor, (5.87), can be written as

$$\frac{1}{2}(M)^2_{\Omega=\Omega_c} = \frac{1}{\left(1-\Omega_h^2\right)^2 + \left(2\zeta\Omega_h\right)^2}$$

Substitution of (5.95) into the preceding equation yields

$$\Omega_h^2 = 1-2\zeta^2 \pm \sqrt{\left(1-2\zeta^2\right)^2 - 1 + 8\zeta^2} \cong 1 \pm 2\zeta$$

which gives the frequency ratio corresponding to the half-power points as follows:

$$\Omega_{h,2} \cong \sqrt{1+2\zeta} \cong 1+\zeta$$

$$\Omega_{h,1} \cong \sqrt{1-2\zeta} \cong 1-\zeta$$

The bandwidth is then obtained as

$$\Delta f = f_2 - f_1 = f_n\left(\Omega_{h,2} - \Omega_{h,1}\right) \cong 2\zeta f_n \tag{5.97}$$

Therefore, we find the quality factor from the definition (5.93),

$$Q = \frac{f_c}{\Delta f} \cong \frac{f_n}{\Delta f} \cong \frac{1}{2\zeta} \tag{5.98}$$

This is a relation between the quality factor and the damping factor, which holds for a linear one-DOF system with $\zeta << 1$. If the damping ratio is comparable to unity or if the vibrating system is nonlinear, the preceding relation does not hold.

Comparing (5.95) and (5.98), we find the second definition of the quality factor,

$$Q = M\big|_{\Omega=\Omega_c} = \left.\frac{x_0}{x_s}\right|_{\Omega=\Omega_c} \tag{5.99}$$

From the second definition of the quality factor, the maximum vibration amplitude x_m may be written as

$$x_m = x_0\big|_{\Omega=\Omega_c} = x_s Q = \frac{F_0}{k} Q \tag{5.100}$$

The third definition of the quality factor can be obtained from the ratio of the dissipated energy (5.91) to the stored energy. Using the definition of the damping factor (5.69) and the natural frequency (5.60) and the relation between the quality and damping factors (5.98), the ratio of the dissipated energy to the total mechanical energy (which is equal to the maximum potential or kinetic energy) at a frequency is given by

$$\frac{U_d}{U_p} = \frac{\pi c x_0^2 \omega}{k x_0^2/2} = \frac{2\pi\left(2\zeta\sqrt{mk}\right)\omega}{k} = 4\pi\zeta\omega\sqrt{\frac{m}{k}}$$
$$= 4\pi\zeta\frac{\omega}{\omega_n} = 2\pi\frac{1}{Q}\frac{\omega}{\omega_n}$$

At resonance ($\omega \cong \omega_n$ for $\zeta << 1$), the ratio is expressed as

$$\left.\frac{U_d}{U_p}\right|_{\omega=\omega_n} = 2\pi\frac{1}{Q}$$

which gives the third definition of the quality factor as

$$Q = 2\pi\left.\frac{U_p}{U_d}\right|_{\omega=\omega_n} \tag{5.101}$$

It is worth noting that for linear vibrating systems with a damping factor much less than unity, the three definitions of the quality factor [equations (5.93), (5.99) and (5.101)] give the same value of $Q \cong 1/(2\zeta)$, and that if a vibrating system is nonlinear or the damping factor ζ is not much less than unity, the three definitions of the quality factor give different numerical values.

5.7 MULTIDEGREE-OF-FREEDOM SYSTEMS

We have investigated a one-degree-of-freedom (DOF) vibrating system, in which only one coordinate is required to describe the motion of the system. In the real world, however, we encounter multidegree-of-freedom systems, which need more than one coordinate to describe their motions. In general, to describe the motion of n-DOF systems, we need n coordinates. Let us begin with the two-DOF vibrating system illustrated in Fig. 5.21, where two masses, m_1 and m_2, are supported by springs, labeled k_1 and k_2, and dampers, indicated as c_1 *and* c_2, which are attached to foundations. A spring of stiffness k_c and a damper of damping coefficient c_c connect the two masses to couple them mechanically. Assuming this system to be linear and applying Newton's second law to the free-body diagram of Fig. 5.21b, we find the equation of motion of the two masses in matrix form, given by

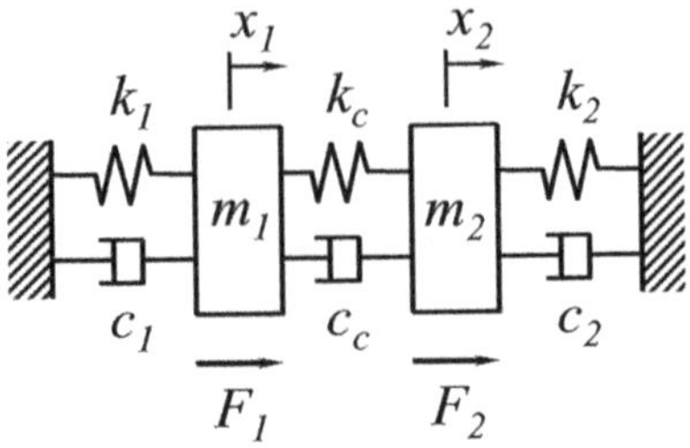

(a)

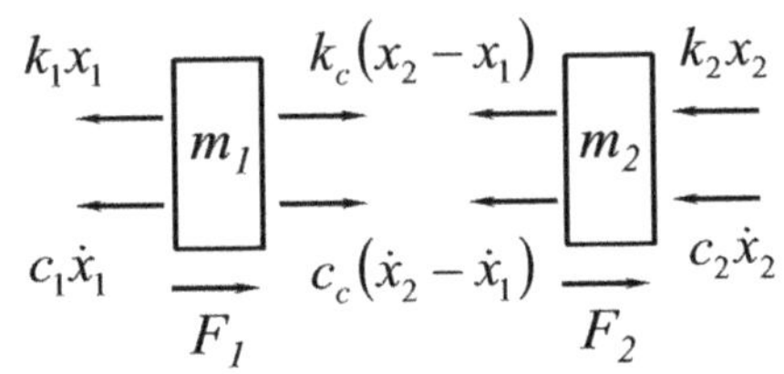

(b)

Figure 5.21 Coupled vibrating system.

$$\begin{bmatrix} m_1 & 0 \\ 0 & m_2 \end{bmatrix}\begin{bmatrix} \ddot{x}_1 \\ \ddot{x}_2 \end{bmatrix} + \begin{bmatrix} c_1 + c_c & -c_c \\ -c_c & c_2 + c_c \end{bmatrix}\begin{bmatrix} \dot{x}_1 \\ \dot{x}_2 \end{bmatrix} + \begin{bmatrix} k_1 + k_c & -k_c \\ -k_c & k_2 + k_c \end{bmatrix}\begin{bmatrix} x_1 \\ x_2 \end{bmatrix} = \begin{bmatrix} F_1 \\ F_2 \end{bmatrix} \tag{5.102}$$

where F_1 and F_2 denote the force acting on the masses, which can be either constant or time varying. When the forces F_1 and F_2 are given, we can repeat the analyses shown in Section 5.6 to obtain the responses x_1 and x_2. The analysis is more complicated than that for a one-DOF system. In many vibrating systems, including MEMS, the damped natural frequencies and the corresponding mode shape (e.g., $x_2/x_1 = 2$ in Fig. 5.21a) are of practical importance. The analysis used to obtain the natural frequencies and the corresponding mode shape (also a damping factor if needed) is called *modal analysis*. In this section we find the natural frequency and mode shape in relatively simple cases.

Setting the damping and force terms equal to zero in equation (5.102), we have the following differential equation for free vibration:

$$\begin{bmatrix} m_1 & 0 \\ 0 & m_2 \end{bmatrix}\begin{bmatrix} \ddot{x}_1 \\ \ddot{x}_2 \end{bmatrix} + \begin{bmatrix} k_1 + k_c & -k_c \\ -k_c & k_2 + k_c \end{bmatrix}\begin{bmatrix} x_1 \\ x_2 \end{bmatrix} = \begin{bmatrix} 0 \\ 0 \end{bmatrix} \tag{5.103}$$

To obtain the dynamic response, we assume that the response can be expressed in the following forms: $x_1 = X_1 \sin \omega t$ and $x_2 = X_2 \sin \omega t$, where X_1, X_2, ω, and t denote the amplitudes, the angular frequency, and time, respectively. Substituting the response into the preceding differential equation gives

$$\begin{bmatrix} k_1 + k_c - m_1\omega^2 & -k_c \\ -k_c & k_2 + k_c - m_2\omega^2 \end{bmatrix} \begin{bmatrix} X_1 \\ X_2 \end{bmatrix} = \begin{bmatrix} 0 \\ 0 \end{bmatrix} \tag{5.104}$$

This equation gives nontrivial solutions if the determinant of its square matrix is equal to zero:

$$\begin{vmatrix} k_1 + k_c - m_1\omega^2 & -k_c \\ -k_c & k_2 + k_c - m_2\omega^2 \end{vmatrix} = 0$$

which yields

$$\omega^4 - \left(\frac{k_1 + k_c}{m_1} + \frac{k_2 + k_c}{m_2} \right)\omega^2 + \frac{k_1 + k_c}{m_1}\frac{k_2 + k_c}{m_2} - \frac{k_c^2}{m_1 m_2} = 0$$

The polynomial above gives two distinct roots:

$$\omega_{1,2}^2 = \frac{1}{2}\left[\frac{k_1 + k_c}{m_1} + \frac{k_2 + k_c}{m_2} \mp \sqrt{\left(\frac{k_1 + k_c}{m_1} - \frac{k_2 + k_c}{m_2} \right)^2 + 4\frac{k_c^2}{m_1 m_2}} \right] \tag{5.105}$$

where subscripts 1 and 2 correspond to negative and positive signs, respectively. The natural frequency is then obtained as $f_{1,2} = \omega_{1,2}/2\pi$. The mode shapes corresponding to $\omega_{1,2}$ can be obtained from one of the equations (5.104):

$$\left(\frac{X_2}{X_1} \right)_{1,2} = \frac{1}{k_c}\left(k_1 + k_c - m_1\omega_{1,2}^2 \right)$$

This relation gives a relation between the amplitudes of vibration and then may be expressed in vector form as

$$\begin{bmatrix} X_1 \\ X_2 \end{bmatrix}_{1,2} = C \begin{bmatrix} 1 \\ \dfrac{k_1 + k_c - m_1\omega_{1,2}^2}{k_c} \end{bmatrix} \tag{5.106}$$

where C is a constant. The natural angular frequency (5.105) and mode shape (5.106) are functions of the masses and stiffness involved in the problem. Since the problem in Fig. 5.21a is a two-DOF problem, the two natural frequencies and two corresponding mode shapes are obtained as described above.

Let us examine simple cases of the solutions. If the masses are identical and the stiffnesses are the same (i.e., $k = k_1 = k_2 = k_c$, $m = m_1 = m_2$), we find the natural frequencies and mode shapes as follows:

$$\omega_1 = \sqrt{\frac{k}{m}} \qquad \begin{bmatrix} X_1 \\ X_2 \end{bmatrix}_1 = C\begin{bmatrix} 1 \\ 1 \end{bmatrix}$$

$$\omega_2 = \sqrt{\frac{k+2k_c}{m}} \qquad \begin{bmatrix} X_1 \\ X_2 \end{bmatrix}_2 = C\begin{bmatrix} 1 \\ -1 \end{bmatrix}$$

We may interpret these results by using the models in Fig. 5.22. The model shown in the left figure in Fig. 5.22a corresponds to ω_1, where the masses m_1 and m_2 are in phase with equal amplitude (i.e., two masses moves left or right at the same time). In this case, the coupling stiffness does not play a role since the spring remains undeformed. The vibrating system is considered as a one-DOF equivalent system, which is shown in the right figure in Fig. 5.22a. Corresponding to the second natural frequency ω_2, the situation is shown in the left figure in Fig. 5.22b. The masses vibrate out of phase with equal amplitudes, so the center of the system is considered as a symmetry line, indicated as a dashed–dotted line in Fig. 5.22b. It is interpreted that one of the masses is supported by the stiffness k and $2k_c$ (a half spring gives twice the stiffness of the original spring), as shown in the right figure in Fig. 5.22b, and that in turn the equivalent model gives the natural angular frequency in the preceding equation. It should be noted that the natural frequency (5.105) and the corresponding mode (5.106) are obtained without reference to the damping force or energy dissipation of the vibrating system. Even though an energy dissipation mechanism such as the damping force is involved in a real system, the natural frequency and mode shape, obtained from the modal analysis above, can be considered close to those of the real system. We examined a similar case for a one-DOF system in Section 5.6; if a one-DOF system is slightly damped (i.e., $\zeta << 1$ or $Q >> 1$), its damped natural frequency (resonant frequency) can be thought of as the natural frequency of the undamped one-DOF system, as shown in (5.71). The two-DOF system of Fig. 5.21 can be

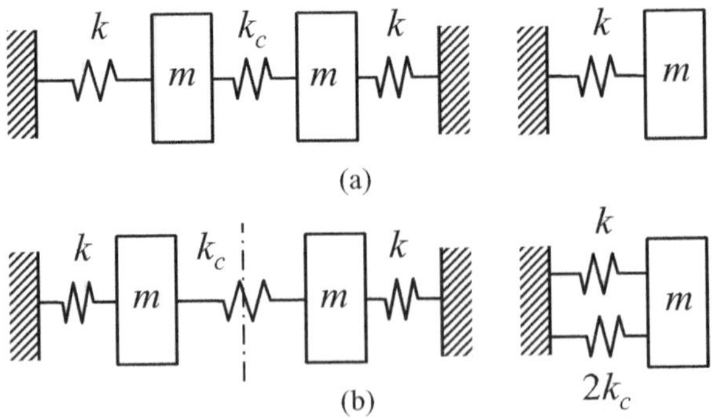

Figure 5.22 Mode shapes and equivalent models of coupled vibrating systems.

actuated by one or two sinusoidal forces (F_1 and F_2 in Fig. 5.21a) at the natural frequency (5.105).

Three-Dimensional Motions of a Symmetric Structure

If a structure is designed to be symmetric about the two orthogonal axes passing through the center of the structure, its motions in three linear and three angular directions will be decoupled. Let us consider the typical structure shown in Fig. 5.23, which may be used in MEMS design. The central plate, which is rigid compared to the four flexures supporting the plate, has six degrees-of freedom: x, y, z; ϕ, η, θ. We then write the equations of motion in these directions as follows:

$$\begin{aligned} m\ddot{x}+c_x\dot{x}+k_x x &= F_x \\ m\ddot{y}+c_y\dot{y}+k_y y &= F_y \\ m\ddot{z}+c_z\dot{z}+k_z z &= F_z \\ I_\phi\ddot{\phi}+c_\phi\dot{\phi}+k_\phi\phi &= M_\phi \\ I_\eta\ddot{\eta}+c_\eta\dot{\eta}+k_\eta\eta &= M_\eta \\ I_\theta\ddot{\theta}+c_\theta\dot{\theta}+k_\theta\theta &= M_\theta \end{aligned} \tag{5.107}$$

where m, I, F, and M represent the mass of the plate, the moment of inertia of the plate, and the force and moment acting on the plate center, respectively, and the subscripts denote the directions. Several types of flexures may be used, and the stiffnesses of typical flexures have been obtained in Chapter 4. Since none of the differential equations above is coupled, their solutions are obtained in the form of (5.76). For example, if we are only interested in the steady-state response to a sinusoidal force $F_z = F_{zo}\sin\omega t$, the response z can be obtained directly from (5.89) as follows:

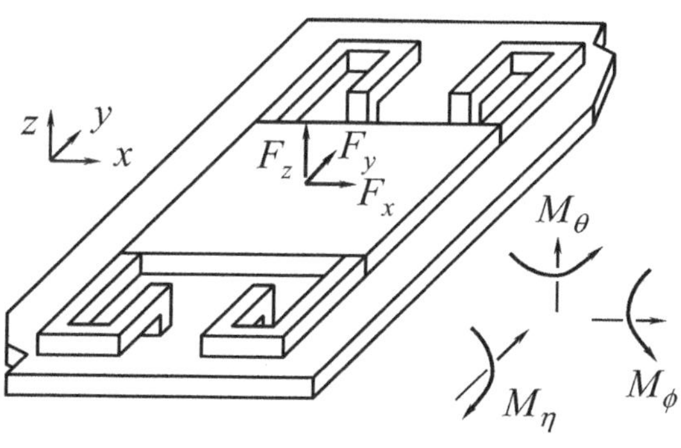

Figure 5.23 Three-dimensional motions of a symmetric structure.

$$z = z_0 \sin(\omega t - \psi) \tag{5.108}$$

where

$$z_0 = \frac{F_{z0}}{k_z} \frac{1}{\sqrt{(1-\Omega_z^2)^2 + (2\zeta_z \Omega_z)^2}}$$

$$\Omega_z = \frac{\omega}{\omega_{nz}} \qquad \omega_{nz} = \sqrt{\frac{k_z}{m}} \qquad \zeta_z = \frac{c_z}{2\sqrt{mk_z}} \qquad \psi = \tan^{-1}\frac{2\zeta_z\Omega_z}{1-\Omega_z^2}$$

The preceding problem, in which the structure is symmetric about its center, has been considered as decoupled, and its motion has been described by the solution of a one-DOF problem. However, if the symmetry is broken, the equations of motion are coupled and their solutions become complicated. Let us consider the structure in Fig. 5.24a, which is unsymmetric about a line (parallel to the x axis) that passes through the plate center but is symmetric about the y axis. The steady-state angular response of the plate to the moment M_η can be described by the solution in the same form as (5.108). With reference to the stiffness of a slender straight beam (4.14), the motion of the structure in the x direction is not coupled with that in the y direction and can then be described in the same form as that of a one-DOF model, as discussed. The motions in the z and ϕ directions (corresponding to M_ϕ) are coupled, so we need two equations of motion to describe them. To obtain the equations of motion, we refer to Fig. 5.24b. When two straight beams are deflected in the z and ϕ directions, the force and moment acting on the plate are given by (4.22):

$$\begin{Bmatrix} F_{z1} \\ M_{\phi 1} \end{Bmatrix} = 2\begin{bmatrix} k_{zz} & k_{z\phi} \\ k_{\phi z} & k_{\phi\phi} \end{bmatrix}\begin{Bmatrix} z_1 \\ \phi \end{Bmatrix} = 2\begin{bmatrix} \dfrac{12EI_x}{l^3} & -\dfrac{6EI_x}{l^2} \\ -\dfrac{6EI_x}{l^2} & \dfrac{4EI_x}{l} \end{bmatrix}\begin{Bmatrix} z_1 \\ \phi \end{Bmatrix}$$

where z_1 denotes the deflection of the right end of the beam in the z direction. For convenience, the preceding equation may be written as

$$F_{z1} = 2(k_{zz}z_1 + k_{z\phi}\phi)$$
$$M_{\phi 1} = 2(k_{\phi z}z_1 + k_{\phi\phi}\phi)$$

Applying Newton's second law to the plate shown in the right figure in Fig. 5.24b, we find that

$$F_z - F_{z1} - c_z\dot{z} = m\ddot{z}$$
$$F_{z1}l_c - M_{\phi 1} + M_\phi - c_\phi\dot{\phi} = I_\phi\ddot{\phi}$$

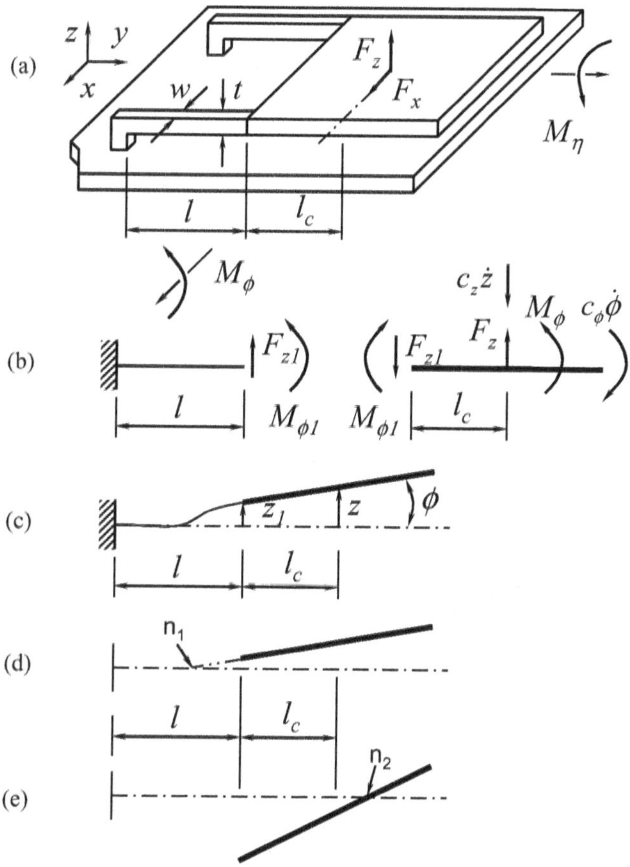

Figure 5.24 Modal analysis of a plate supported by springs.

Substituting the preceding force–deflection and moment–deflection relations into the equations above and using the geometric relation given by $z_1 = z - l_c\phi$ from Fig. 5.24c, we obtain two differential equations for the motion in the z and ϕ directions:

$$m\ddot{z} + c_z\dot{z} + 2k_{zz}z + 2(-k_{zz}l_c + k_{z\phi})\phi = F_z$$

$$I_\phi\ddot{\phi} + c_\phi\dot{\phi} + 2(-k_{zz}l_c + k_{z\phi})z + 2(k_{zz}l_c^2 - 2k_{z\phi}l_c + k_{\phi\phi})\phi = M_\phi$$

or in matrix form,

$$\begin{bmatrix} m & 0 \\ 0 & I_\phi \end{bmatrix}\begin{bmatrix} \ddot{z} \\ \ddot{\phi} \end{bmatrix} + \begin{bmatrix} c_z & 0 \\ 0 & c_\phi \end{bmatrix}\begin{bmatrix} \dot{z} \\ \dot{\phi} \end{bmatrix} + \begin{bmatrix} k_z & -k_c \\ -k_c & k_\phi \end{bmatrix}\begin{bmatrix} z \\ \phi \end{bmatrix} = \begin{bmatrix} F_z \\ M_\phi \end{bmatrix} \tag{5.109}$$

where

$$\begin{aligned}
k_z &= 2k_{zz} = \frac{24EI_x}{l^3} \\
k_c &= -2(-k_{zz}l_c + k_{z\phi}) = \frac{24EI_x}{l^3}l_c + \frac{12EI_x}{l^2} \\
&= \frac{12EI_x}{l^2}\left(2\frac{l_c}{l}+1\right) \\
k_\phi &= 2(k_{zz}l_c^2 - 2k_{z\phi}l_c + k_{\phi\phi}) = \frac{24EI_x}{l^3}l_c^2 - 4\left(-\frac{6EI_x}{l^2}\right)l_c + \frac{8EI_x}{l} \\
&= \frac{8EI_x}{l}\left[3\left(\frac{l_c}{l}\right)^2 + 3\frac{l_c}{l} + 1\right]
\end{aligned} \tag{5.110}$$

If the force and moment are given, the dynamic response z and ϕ can be obtained. To perform modal analysis, we let the damping and force terms be zero and obtain

$$\begin{bmatrix} m & 0 \\ 0 & I_{xc} \end{bmatrix}\begin{bmatrix} \ddot{z} \\ \ddot{\phi} \end{bmatrix} + \begin{bmatrix} k_z & -k_c \\ -k_c & k_\phi \end{bmatrix}\begin{bmatrix} z \\ \phi \end{bmatrix} = \begin{bmatrix} 0 \\ 0 \end{bmatrix} \tag{5.111}$$

Setting $z = z_0 \sin \omega t$ and $\phi = \phi_0 \sin \omega t$, we find that

$$\begin{bmatrix} k_z - m\omega^2 & -k_c \\ -k_c & k_\phi - I_\phi\omega^2 \end{bmatrix}\begin{bmatrix} z_0 \\ \phi_0 \end{bmatrix} = \begin{bmatrix} 0 \\ 0 \end{bmatrix} \tag{5.112}$$

Letting the determinant of the above be zero gives a polynomial for the natural frequency,

$$\omega^4 - \left(\frac{k_z}{m} + \frac{k_\phi}{I_\phi}\right)\omega^2 + \frac{k_z}{m}\frac{k_\phi}{I_\phi} - \frac{k_c^2}{mI_\phi} = 0 \tag{5.113}$$

which yields the natural angular frequencies

$$\omega_{1,2}^2 = \frac{1}{2}\left[\frac{k_z}{m} + \frac{k_\phi}{I_\phi} \mp \sqrt{\left(\frac{k_z}{m} - \frac{k_\phi}{I_\phi}\right)^2 + 4\frac{k_c^2}{mI_\phi}}\right] \tag{5.114}$$

where the negative and positive signs, respectively, correspond to ω_1 and ω_2, and the corresponding modes are obtained from (5.112) as follows:

$$\begin{bmatrix} z_0 \\ \phi_0 \end{bmatrix}_{1,2} = D\begin{bmatrix} 1 \\ \dfrac{k_z - m\omega_{1,2}^2}{k_c} \end{bmatrix} \tag{5.115}$$

where D is a constant. Let us examine the mode shape given by equation (5.115). The mode shape corresponding to the lower frequency ω_1 is illustrated in Fig. 5.24d, where z and ϕ are in phase. It is noted from the figure that there exists a nodal point n_1 which is a fixed point during vibration at ω_1 and exists outside the plate. The mode shape for ω_2 is shown in Fig. 5.24e, where z and ϕ are out of phase. We also observe a nodal point n_2 between both ends of the plate, which is fixed while the plate vibrates at a frequency of ω_2. The nodal points, n_1 and n_2, can be obtained from geometric relationships using (5.115).

5.8 CONTINUOUS SYSTEMS

If the mass and flexure of a system are distributed continuously, the system is said to be a *continuous system*. For example, a tensioned string and a fixed–fixed beam are continuous systems, thought of as bodies made of an infinite number of particles. As a result, an infinite number of coordinates are required to describe the particles, and the degrees of freedom of continuous systems will then be infinite.

String

If a slender beam is stretched and if the bending moment effect on the beam deflection is much less than the tension effect, the beam can be dealt with as a string. For example, as discussed in Section 3.6, if the dimensionless number, $l\sqrt{N/EI}$, of a fixed–fixed beam is much larger than 50, the bending moment contribution to the deflection is negligible compared to the tension contribution. Let us suppose that a string, tensioned between two supports as shown in Fig. 5.25a, is vibrating in a direction perpendicular to its longitudinal direction in Fig. 5.25b. Let x, y, and θ denote the coordinate along the string,

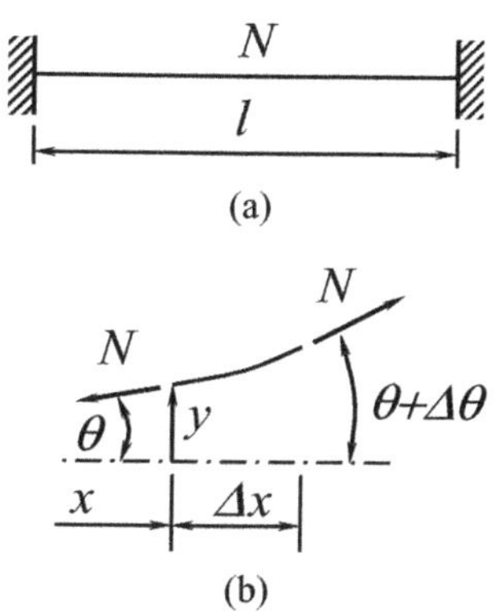

Figure 5.25 Vibration of a string.

the displacement of the string at x, and the angle of the sting, as shown in Fig. 5.25b, respectively. Applying Newton's law to a small element of the string of mass m_1 per unit length gives

$$N(\theta+\Delta\theta)-N\theta=m_1\Delta x\frac{\partial^2 y}{\partial t^2}$$

Noting that $\theta=\partial y/\partial x$, dividing the equation above by $m_1\,\Delta x$, letting $c=\sqrt{N/m_1}$, which represents the velocity of wave propagation, and taking the limit as $\Delta x\to 0$ gives the equation of motion, called a *wave equation*,

$$\frac{\partial^2 y}{\partial t^2}=c^2\frac{\partial^2 y}{\partial x^2} \tag{5.116}$$

The solution may be obtained by using the method of separation of variables. Assuming the solution to be

$$y=Y(x)G(t) \tag{5.117}$$

and substituting into the preceding wave equation, we have

$$\frac{1}{c^2}\frac{1}{G}\frac{d^2G}{dt^2}=\frac{1}{Y}\frac{d^2Y}{dx^2}$$

The left- and right-hand sides of the equation above are a function of time t, and the right-hand side is expressed in terms of the coordinate x, which suggests that the equation must be a constant. For convenience, we let the constant be $-\omega^2/c^2$. If we let the constant be arbitrary and conduct the procedure described below, we shall obtain the same constant. Setting the equation above equal to $-\omega^2/c^2$, we find two ordinary differential equations as follows:

$$\begin{aligned}&\frac{d^2G}{dt^2}+\omega^2G=0\\&\frac{d^2Y}{dx^2}+\frac{\omega^2}{c^2}Y=0\end{aligned} \tag{5.118}$$

The solutions to these equations are given by

$$G=A\cos\omega t+B\sin\omega t$$

$$Y=C_1\cos\left(\frac{\omega}{c}x\right)+D_1\sin\left(\frac{\omega}{c}x\right)$$

where A, B, C_1, and D_1 denote constants, and then the general solution for the string deflection is expressed as

$$y = \left[C_1 \cos\left(\frac{\omega}{c}x\right) + D_1 \sin\left(\frac{\omega}{c}x\right) \right] (A\cos\omega t + B\sin\omega t)$$

or, equivalently,

$$y = \left[C\cos\left(\frac{\omega}{c}x\right) + D\sin\left(\frac{\omega}{c}x\right) \right] \sin(\omega t - \phi) \tag{5.119}$$

where ϕ is a phase. The constants C and D can be determined from boundary conditions. If both ends of the string are fixed as shown in Fig. 5.25a, the boundary conditions are $y = 0$ at $x = 0$ and $x = 1$. Applying these conditions to the foregoing equation yields

$$C = 0 \qquad Y(l) = D\sin\left(\frac{\omega}{c}\right)x = 0$$

This equation gives the angular frequency at which the string can vibrate at maximum amplitude,

$$\omega_n = \frac{n\pi c}{l} \tag{5.120}$$

where $n = 1, 2, 3, \ldots$. The natural frequency is obtained as

$$f_n = \frac{\omega_n}{2\pi} = \frac{nc}{2l} \tag{5.121}$$

Substituting (5.120) into (5.119), we obtain the transverse deflection of the string as

$$y_n = D\sin\left(\frac{\omega_n}{c}x\right)\sin(\omega t - \phi) \tag{5.122}$$

This is the solution of the wave equation (5.116), and the wave (5.122) is called a *transverse wave* since the particles consisting of the string oscillate perpendicular to the direction of energy transfer (i.e., the longitudinal direction of the string). This equation can be interpreted as indicating that at the natural frequency (5.121), the string is allowed to vibrate at a maximum amplitude and that the function $\sin(\omega_n x/c)$ represents a mode shape at f_n. The first and second mode shapes are illustrated in Fig. 5.26. Both experiment and theory show that if the string is actuated either by a point force at a point (except nodal points) in the string or by distributed force over the length, the mode shapes at resonance are given by $\sin(\omega_n x/c)$, obtained from (5.122).

Figure 5.26 String vibration modes.

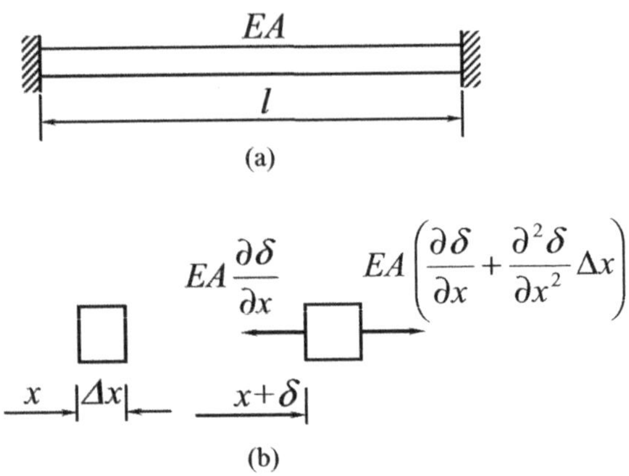

Figure 5.27 Longitudinal vibration of a beam.

Longitudinal Vibration of Rods

Since localized displacements due to stress in a slender rod propagate along its length, stress can travel from one point to another. This is called a *longitudinal wave* since the rod particles vibrate in the direction parallel to the wave propagation direction (i.e., the longitudinal direction). This longitudinal wave, which can propagate in three-dimensional space, exists as a solid, liquid, and gas and is also known as a *sound* or *acoustic wave*. Let us examine the longitudinal wave in the rod shown in Fig. 5.27a. Strain due to stress is generated at a point in the rod above in Fig. 5.27a and propagates along the rod. The small element of the rod, with Young's modulus E, cross-sectional area A, and density ρ, shown in the left figure in Fig. 5.27b, is displaced due to the stress propagated, as illustrated in the right figure in Fig. 5.27b. The force on the left face of the displaced element is $EA\partial\delta/\partial x$, where x is the coordinate along the rod length and the force on the right face is expressed as a Taylor series expansion of that on the left face. Application of Newton's second law to the right figure in Fig. 5.27b gives

$$EA\left(\frac{\partial\delta}{\partial x}+\frac{\partial^2\delta}{\partial x^2}dx\right)-EA\left(\frac{\partial\delta}{\partial x}\right)=\rho A\,dx\frac{\partial^2\delta}{\partial t^2}$$

Defining the velocity of wave propagation, $c=\sqrt{E/\rho}$, we obtain the wave equation for a slender rod of uniform cross section:

$$\frac{\partial^2\delta}{\partial t^2}=c^2\frac{\partial^2\delta}{\partial x^2} \qquad (5.123)$$

Equation (5.123) has the same form as (5.116), and with reference to (5.119), we then write its solution as follows:

$$\delta=\left[C\cos\left(\frac{\omega}{c}x\right)+D\sin\left(\frac{\omega}{c}x\right)\right]\sin(\omega t-\phi) \qquad (5.124)$$

If boundary conditions are given, the constants in equation (5.124) can be obtained. When both ends are fixed, as shown in Fig. 5.27a (i.e. $\delta=0$ at $x=0$ and $x=l$), we have

$$\delta=D\sin\left(\frac{\omega_n}{c}x\right)\sin(\omega t-\phi) \qquad (5.125)$$

where

$$\omega_n=\frac{n\pi c}{l} \qquad (5.126)$$

is the natural angular frequency and the natural frequency is then given by $f_n=\omega_n/2\pi=nc/2l$.

Modal Analysis of Straight, Uniform Beams

If the bending moment effect in a beam is not negligible and if we are interested in the transverse vibration of the beam, the full beam equation should be considered to obtain the natural frequency and its corresponding mode shape. Assuming that the flexural rigidity EI and the density ρ along the beam length are constants, we find the following equation of motion of the beam by adding the inertial and damping terms into the beam equation (3.54) or by applying Newton's second law to the small element of Fig. 5.28b:

$$\rho A\frac{\partial^2 y}{\partial t^2}+c\frac{\partial y}{\partial t}+\frac{\partial^2}{\partial x^2}\left(EI\frac{\partial^2 y}{\partial x^2}\right)-\left[N_0+\frac{1}{2}\frac{EA}{l}\int_0^l\left(\frac{\partial y}{\partial x}\right)^2dx\right]\frac{\partial^2 y}{\partial x^2}=q \qquad (5.127)$$

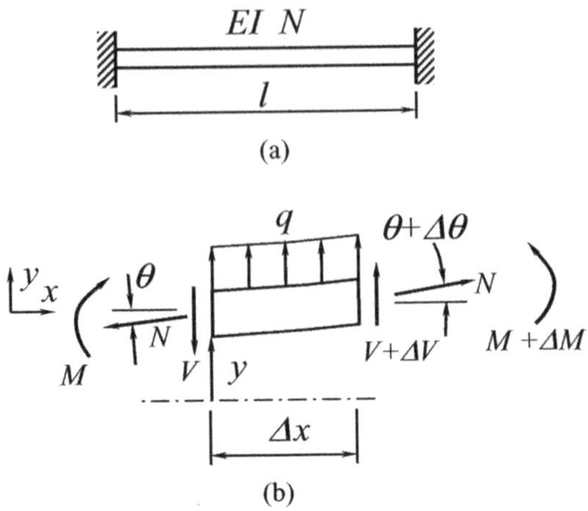

Figure 5.28 Vibrating beam under tension.

where ρ, A, y, t, c, E, N_0, and q denote the density, the cross-sectional area of the beam, the beam deflection (response), time, the damping coefficient per unit length, the effective Young's modulus defined by (3.5) and (3.6), the axial force (tension defined as positive), and the load per unit length, respectively. The load q can be either constant or time-varying. Let us first obtain the natural frequency and the corresponding mode shape of the linear beam system and then analyze the nonlinear differential equation above in the following section.

Setting the damping coefficient, the axial force, and the geometric nonlinear term equal to zero, we obtain the linear equation

$$\rho A \frac{\partial^2 y}{\partial t^2} + EI \frac{\partial^4 y}{\partial x^4} = 0 \tag{5.128}$$

Assuming a solution in the form

$$y = B\psi(x)\sin(\omega t - \phi) \tag{5.129}$$

where B, ψ, ω, and ϕ denote a constant, the mode shape, the angular frequency, and the phase, respectively, and substituting the assumed solution into the preceding equation of motion, we find an ordinary differential equation for the mode shape, which is given by

$$\frac{d^4\psi}{dx^4} - \beta^4\psi = 0 \tag{5.130}$$

where

$$\beta^4 = \frac{m_1\omega^2}{EI} \tag{5.131}$$

and $m_1 = \rho A$ is the mass per unit length. Substituting $\psi = e^{\lambda x}$ into differential equation (5.130) yields the characteristic equation, $\lambda^4 - \beta^4 = 0$, which gives four distinct roots: $\lambda = \pm\beta$; $\lambda = \pm\beta i$. We then express the mode shape as follows:

$$\psi = Q\cosh\beta x + R\sinh\beta x + S\cos\beta x + T\sin\beta x \tag{5.132}$$

The constants, Q, R, S, and T are determined from four boundary conditions.

If the beam is fixed at the both ends as shown in Fig. 5.28a, the boundary conditions are given by $\psi = 0$, $d\psi/dx = 0$ at $x = 0$ and $x = l$. The derivative of ψ is given by

$$\frac{d\psi}{dx} = Q\beta\sin\beta x + R\beta\cosh\beta x - S\beta\sin\beta x + T\beta\cos\beta x$$

Using the boundary conditions $\psi = 0$ and $d\psi/dx = 0$ at $x = 0$, we obtain S = –Q, T = –R. Applying the other two boundary conditions yields

$$\begin{bmatrix} \cosh\beta l - \cos\beta l & \sinh\beta l - \sin\beta l \\ \sinh\beta l + \sin\beta l & \cosh\beta l - \cos\beta l \end{bmatrix} \begin{bmatrix} Q \\ R \end{bmatrix} = 0 \tag{5.133}$$

For a nontrivial solution, the zero determinant of the matrix leads to

$$\cosh\beta l\cos\beta l = 1 \tag{5.134}$$

The dimensionless quantity βl satisfying equation (5.134) can be determined by a numerical method. Note that the quantity βl gives information on the natural frequency of the beam since β is expressed as a function of the frequency, as shown in (5.131). A relation between constants Q and R, which corresponds to βl of (5.134), can be obtained from (5.133) as follows:

$$R = -\frac{\cosh\beta l - \cos\beta l}{\sinh\beta l - \sin\beta l}Q$$

Substitution of the relation above into (5.132) gives the mode shape as

$$\psi = Q\left[\cosh\beta x - \cos\beta x - \frac{\cosh\beta l - \cos\beta l}{\sinh\beta l - \sin\beta l}(\sinh\beta x - \sin\beta x)\right] \tag{5.135}$$

where Q is a constant. Equation (5.134) may be written for large $\beta l >> 0$ as

$$\cos \beta l = \frac{1}{\cosh \beta l} \cong 0$$

which gives a relatively simple expression for βl:

$$\beta_n l \cong \frac{\pi}{2}(1+2n) \tag{5.136}$$

where n is a positive integer. From (5.131), the natural angular frequency is obtained as

$$\omega_n = \left(\frac{\beta_n^4 EI}{m_1}\right)^{1/2} = (\beta_n l)^2 \frac{1}{l^2}\sqrt{\frac{EI}{m_1}} \tag{5.137}$$

or the corresponding natural frequency is given by

$$f_n = \frac{\omega_n}{2\pi} = \frac{(\beta_n l)^2}{2\pi l^2}\sqrt{\frac{EI}{m_1}} \tag{5.138}$$

where $\beta_n l$ is obtained from Table 5.1. Similarly, the natural frequencies and corresponding mode shape of beams with different boundary conditions can be obtained.

Table 5.1 and Fig. 5.29 show βl and the mode shapes of fixed–fixed beams, cantilever beams (fixed–free beams), and free–free beams. From Table 5.1 and equation (5.138) we know that the natural frequencies of the beam vibration are not harmonic (i.e., $f_n \neq nf_1$). The first and second mode shapes of the three beams in Table 5.1 are sketched in Fig. 5.29a (fixed–fixed beam), Fig. 5.29b (cantilever beam), and Fig. 5.29c (free–free beam). Note that the mode shapes are normalized by their maximum values (i.e., the constant for the mode shape is selected to set its maximum equal to 1 [eg, Q of (5.135) for the first mode of the fixed–fixed beam is determined as 0.6297].

TABLE 5.1 Dimensionless Number βl for Uniform Beams

Beam	$\beta_1 l$	$\beta_2 l$	$\beta_3 l$	$\beta_n l$, $n > 3$
	4.7300	7.8532	10.9956	$(2n + 1)/2$
	1.8751	4.6940	7.8547	$(2n - 1)/2$
	4.7300	7.8532	10.9956	$(2n + 1)/2$

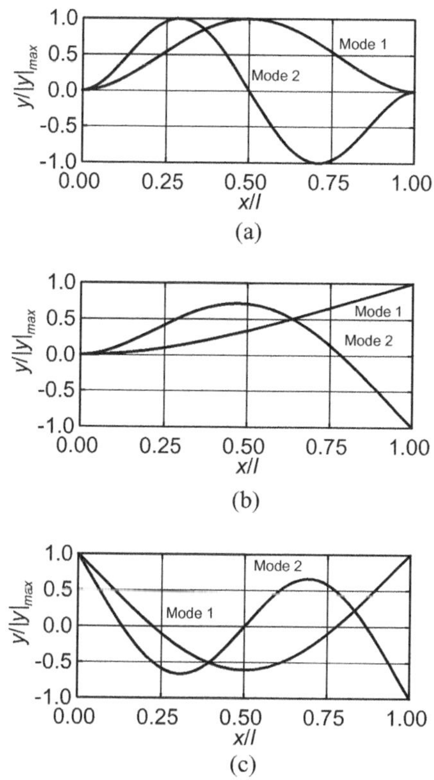

Figure 5.29 Mode shapes of vibrating beams.

5.9 EFFECTIVE MASS, DAMPING, AND STIFFNESS

Beams such as fixed–fixed beams are used for elements of MEMS. We shall obtain the effective mass, damping, and stiffness of beams and then evaluate the effective mass and stiffness of a spring beam.

One-DOF Model of Straight, Uniform Beams

Since the differential equation, (5.127), for a beam is nonlinear, we cannot find an exact solution of the equation. We will then obtain a one-DOF model for an approximate solution which is sufficiently accurate to evaluate the natural frequency. If the flexural rigidity, EI, remains constant along the longitudinal direction of the beam, the differential equation (5.127) is written as

$$\rho A \frac{\partial^2 y}{\partial t^2} + c\frac{\partial y}{\partial t} + EI\frac{\partial^4 y}{\partial x^4} - \left[N_0 + \frac{1}{2}\frac{EA}{l}\int_0^l \left(\frac{\partial y}{\partial x}\right)^2 dx\right]\frac{\partial^2 y}{\partial x^2} = q \qquad (5.139)$$

To obtain an approximate solution, we use Galerkin's method, discussed in Section 3.7. The force per length (distributed load), q, is time varying and is assumed to be expressed as

$$q = q_m \psi_m \sin \omega t \tag{5.140}$$

where q_m, ψ_m, and ω represent a maximum force per length over the beam, a force distribution function where force over the length is normalized by q_m, and the angular frequency of the force, respectively. The dynamic response (vibration amplitude, deflection) y at a point in the beam (e.g., midspan of a fixed–fixed beam) is of interest and is assumed to be

$$y_m = y_0 \sin(\omega t - \phi) \tag{5.141}$$

where y_0, ω, t, and ϕ denote the maximum vibration amplitude at the point, the angular frequency, the time, and the phase, respectively. The vibration amplitude along the beam is then written as

$$y = y_m \psi_1 = y_0 \psi_1 \sin(\omega t - \phi) \tag{5.142}$$

where ψ_1 is the mode shape of the vibrating beam, which is normalized by y_0. For convenience, as discussed in Section 3.7, a mode shape ψ is introduced for easy integration, as follows:

$$\psi_1 = \eta \psi \tag{5.143}$$

where η is a constant and ψ is a mode shape for easy integration. Substituting the vibration amplitude y (5.142), with its time derivatives substituted into (5.139):

$$\frac{\partial y}{\partial t} = y_0 \psi_1 \omega \cos(\omega t - \phi) = \psi_1 \dot{y}_m$$

$$\frac{\partial^2 y}{\partial t^2} = -y_0 \psi_1 \omega^2 \sin(\omega t - \phi) = \psi_1 \ddot{y}_m$$

multiplying the resulting equation by the mode shape ψ_1, and integrating gives an ordinary differential equation which describes the motion of the point of interest in the beam:

$$m_e \ddot{y}_m + c_e \dot{y}_m + k_1 y_m + k_3 y_m^3 = F_q \tag{5.144}$$

where

$$m_e = \int_0^l \rho A \psi_1^2 \, dx = \eta^2 l \int_0^1 \rho A \psi^2 \, dX = \eta^2 \rho A l \int_0^1 \psi^2 \, dX \tag{5.145}$$

$$c_e = \int_0^l c\psi_1^2\, dx = \eta^2 l \int_0^1 c\psi^2\, dX = \eta^2 c l \int_0^1 \psi^2\, dX \tag{5.146}$$

$$k_1 = \eta^2 \frac{EI}{l^3} \int_0^1 \left(\frac{d^2\psi}{dX^2} \right)^2 dX + \eta^2 \frac{N_0}{l} \int_0^1 \left(\frac{d\psi}{dX} \right)^2 dX \tag{5.147}$$

$$k_3 = \frac{1}{2} \eta^4 \frac{EA}{l^3} \left[\int_0^1 \left(\frac{d\psi}{dX} \right)^2 dX \right]^2 \tag{5.148}$$

$$F_q = \int_0^l q\psi_1\, dx = \eta l \left[\int_0^1 \psi_m \psi\, dX \right] q_m \sin \omega t = F_0 \sin \omega t \tag{5.149}$$

$$F_0 = \eta q_m l \int_0^1 \psi_m \psi\, dX \tag{5.150}$$

denote, respectively, the effective mass, a effective damping coefficient, the linear and cubic stiffness defined in (3.103) and (3.104), the sinusoidal force to actuate the beam, and the force amplitude. An equivalent model for the beam vibration is shown in Fig. 5.30, which consists of the effective mass and damping coefficient and the linear and cubic stiffnesses. If differential equation (5.144) is solved for y_m, the vibration amplitude of the other points of the beam is obtained by (5.142). Note from equation (5.147) that the linear stiffness k_1 includes the effects of the axial force N_0 as well as the flexural rigidity EI, and the solution y_m then depends on the axial force. Note also that for a cantilever beam (a fixed–free beam), the axial force N_0 and cubic stiffness k_3 vanish. If the cubic stiffness of the differential equation (5.144) becomes zero or is negligible, the natural frequency of the beam is given by

$$f_n = \frac{1}{2\pi} \sqrt{\frac{k_1}{m_e}} \tag{5.151}$$

Example 5.3 The fixed–fixed slender straight beam shown in Fig. 5.28 vibrates at the lowest natural frequency. The mass of the entire beam is m, and the damping coefficient per unit length is given as c. If the vibration mode shape is assumed as the mode shape, $\psi(X) = X^2(X - 1)^2$, where X is the coordinate along the beam normalized by the beam length, determine the elements of the equivalent vibrating system shown in Fig. 5.30 and obtain the response of the

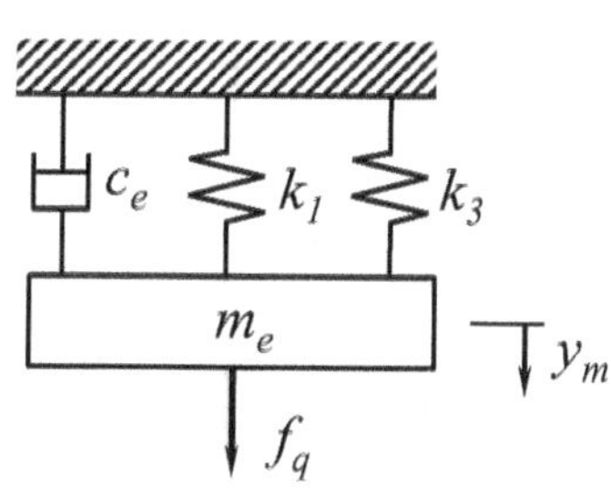

Figure 5.30 Equivalent model of beams.

beam to a load that is distributed uniformly over the beam length and varies sinusoidally with time.

If we select the vibration amplitude at the midspan as that of interest, the mode shape at the midspan (i.e. $X = x/l = 1/2$) yields $\psi = 1/16$, and in turn we have the normalized mode shape $\psi_1 = \eta\psi$, where $\eta = 16$. Substituting the mode shape $\psi = X^2(X-1)^2$ and the constant $\eta = 16$ into the preceding equations for the effective elements of the equivalent system, we have

$$\int_0^1 \psi^2\,dX = \int_0^1 \left[X^2(X-1)^2\right]^2 dX$$
$$= \int_0^1 \left(X^8 - 4X^7 + 6X^6 - 4X^6 + X^4\right)dX = \frac{1}{630} \tag{a}$$

$$m_e = \eta^2 \rho A l \int_0^1 \psi^2\,dX = 16^2 \frac{1}{630}\rho A l = \frac{128}{315} m \tag{b}$$

where $m = \rho A l$ is the mass of the entire beam. Similarly,

$$c_e = \eta^2 c l \int_0^1 \psi^2\,dX = \frac{128}{315} c_l \tag{c}$$

where $c_l = cl$ denotes the damping coefficient of the entire beam. The linear and cubic stiffness are given by (3.108) and (3.109) or (5.147) and (5.148):

$$k_1 = \eta^2 \frac{EI}{l^3}\frac{4}{5} + \eta^2 \frac{N_0}{l}\frac{2}{105} = \frac{1024}{5}\frac{EI}{l^3} + \frac{512}{105}\frac{N_0}{l} \tag{d}$$

$$k_3 = \eta^4 \frac{EA}{l^3}\left(\frac{2}{105}\right)^2 = \frac{262{,}144}{11{,}025}\frac{EA}{l^3} \tag{e}$$

With $q = q_m$ and $\psi_q = 1$ (if we assume that the sinusoidal force is a uniformly distributed force given by $q = q_m \sin \omega t$), the force amplitude F_0 (5.150) is given by

$$F_0 = \eta q_m l \int_0^l \psi_q \psi\,dx = \eta q_m l \frac{1}{30} = \frac{8}{15} q_m l \tag{f}$$

If $N_0 = 0$, the natural frequency for a small amplitude is given by

$$f_n = \frac{1}{2\pi}\sqrt{\frac{k_1}{m_e}} = \frac{3\sqrt{14}}{\pi}\frac{1}{l^2}\sqrt{\frac{EI}{m_1}} \cong 3.5730\frac{1}{l^2}\sqrt{\frac{EI}{m_1}} \tag{g}$$

The natural frequency of the beam is given by the approximate method, Galerkins method, using assumed mode shapes. The exact natural frequency of the fixed–fixed beam is written from (5.138) and Table 5.1 as follows:

$$f_n = \frac{(\beta_n l)^2}{2\pi l^2}\sqrt{\frac{EI}{m_1}} = \frac{(4.7300)^2}{2\pi}\frac{1}{l^2}\sqrt{\frac{EI}{m_1}} = 3.5608\frac{1}{l^2}\sqrt{\frac{EI}{m_1}} \tag{h}$$

The approximate resonant frequency (g) is in good agreement with the exact resonant frequency (h) with an error of 0.3%. Therefore, if the vibration amplitude is much smaller than the thickness of the beam, the expressions for f_n above can be used with sufficient accuracy. In this case, the cubic stiffness can be negligible and the steady-state response is given in the form (5.141), which is written

$$y_m = y_0 \sin(\omega t - \phi) \tag{i}$$

where the vibration of amplitude y_0 at the midspan of the beam is given by (5.87):

$$y_0 = \frac{F_0}{k_1}\frac{1}{\sqrt{(1-\Omega^2)^2 + (2\varsigma\Omega)^2}} \tag{j}$$

where F_0 and k_1 denote the force amplitude (f) and the linear stiffness (d) and

$$\Omega = \frac{\omega}{\omega_n} = \frac{f}{f_n} \tag{k}$$

$$\zeta = \frac{c_e}{2\sqrt{m_e k_1}} \tag{l}$$

$$\phi = \tan^{-1}\frac{2\zeta\Omega}{1-\Omega^2} \tag{m}$$

represent the frequency ratio, damping factor, and phase, respectively. Since equation (i) is the vibration of the midspan of the fixed–fixed beam, the dynamic response at a point of the beam is given by (5.142):

$$y = y_m \psi_1 = y_0 \psi_1 \sin(\omega t - \phi) \tag{n}$$

where y_0 is the vibration amplitude (j) and ψ_1 is the mode shape, which is expressed as

$$\psi_1 = \eta\psi = 16X^2(X-1)^2 = 16\left(\frac{x}{l}\right)^2\left(\frac{x}{l}-1\right)^2 \tag{o}$$

Noted that equation (i) is the response of a one-DOF model of the fixed–fixed beam and that if the frequency f is much higher than the natural frequency f_n, higher resonances (e.g., the second mode shown in Fig. 5.29a) can be found.

However, in general, the vibration amplitude of a higher mode is smaller than that of the first mode. For higher-the mode motion or natural frequency, suitable mode shapes may be used. If the displacement at the midspan of the fixed–fixed beam of Fig. 5.28 is comparable to the thickness of the beam, the effect of the cubic stiffness k_3 on the dynamic response must be taken into account. The nonlinear dynamic response, including the cubic stiffness effect, is dealt with in Section 5.11.

Similarly, for the cantilever beam, if we take the vibration of amplitude at the free end of the beam as the vibration of interest, we have the approximate mode shape $\psi(X) = X^2(X^2 - 4X + 6)$, where $X = x/l$ and the constant $\eta = 1/3$ (refer to Chapter 3 to obtain the mode shape), we obtain the following effective elements for the equivalent system:

$$m_e = \frac{104}{405}m \qquad c_e = \frac{104}{405}c_l \qquad k_1 = \frac{16}{5}\frac{EI}{l^3} \qquad k_3 = 0 \tag{5.152}$$

where m denotes the mass of the beam and $c_l = cl$. The cubic stiffness k_3 above is zero since there is no geometric nonlinearity along the beam. If a uniformly distributed force is applied sinusoidally to the cantilever beam, then $q = q_m$ and $\psi_q = 1$ and the force F_0 is given by

$$F_0 = \frac{2}{5}q_m l \tag{5.153}$$

The dynamic response of the cantilever beam can be found using equations (i)–(n) in Example 5.3.

Effective Mass and Stiffness of Flexures

As discussed in Chapter 4, various flexures are used as springs: for example, a straight beam, an L-shaped beam, and a folded beam. If the total mass of the flexures is comparable to the movable mass (e.g., a relatively rigid plate), the mass of the flexures should be taken into account.

Let us begin with a straight slender beam one end of which is attached to a support, with the other end fixed to a movable structure, as shown in Fig. 5.31a. The L-shaped beam in Fig. 5.31b, whose upper beam is much shorter than the vertical beam, and the string (its mode shape is a straight line) can be analyzed using this model, described below. When the lower end of the beam, of density ρ and cross-sectional area A, is displaced by δ as shown in Fig. 5.31c and d, the deflection y of the beam of Fig. 5.31a is given by (3.60):

$$y = \frac{F}{12EI}\left(3lx^2 - 2x^3\right)$$

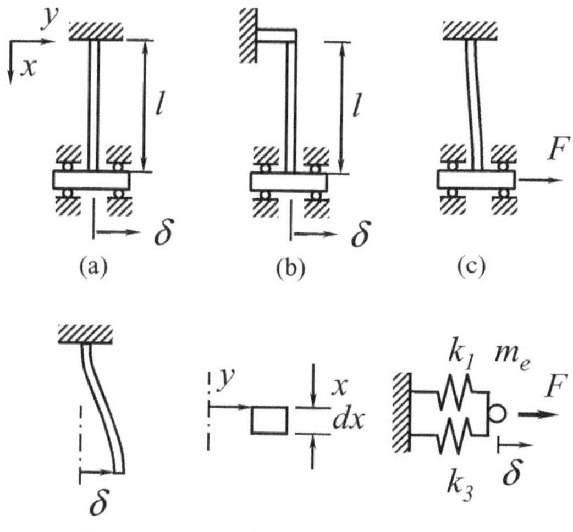

Figure 5.31 Effective mass and stiffness of beams.

where F, EI, l, and x denote the applied force, the flexural rigidity of the beam, the beam length, and the coordinate from the upper end of the beam, respectively. The maximum displacement δ and the stiffness k of the beam are obtained as

$$\delta = y\big|_{x=l} = \frac{Fl^3}{12EI} \qquad k = \frac{F}{\delta} = \frac{12EI}{l^3}$$

From the preceding beam deflection and its maximum deflection, we express the beam deflection y as a function of the mode shape ψ_1:

$$y = \delta\psi_1 \tag{5.154}$$

where

$$\psi_1 = \frac{1}{l^3}\left(3lx^2 - 2x^3\right) = 3X^2 - 2X^3 \tag{5.155}$$

where $X = x/l$. This is the mode shape that is normalized to unity at the lower end of the beam. For the convenience of integration, the preceding mode shape may be expressed as

$$\psi_1 = \eta\psi \tag{5.156}$$

where η is a constant. Since (5.155) is the most convenient form for integration, $\eta = 1$ and

$$\psi = \psi_1 = 3X^2 - 2X^3 \tag{5.157}$$

the displacement y, (5.154), may be written as a time-varying function,

$$y = \delta\psi_1 = \delta_0\psi_1 \sin\omega t \tag{5.158}$$

where

$$\delta = \delta_0 \sin\omega t \tag{5.159}$$

is the sinusoidal displacement of the lower end of the beam of Fig. 5.31d. To obtain the effective mass m_e of the straight beam, we use the energy method discussed earlier in the chapter. With the velocity and stored kinetic energy of the small element of Fig. 5.31e, which are given by

$$\dot{y} = \delta_0\psi_1\omega\cos\omega t = \psi_1\dot{\delta}$$

$$dU_k = \frac{1}{2}dm\dot{y}^2 = \frac{1}{2}\rho Al\frac{dx}{l}\left(\psi_1\dot{\delta}\right)^2 = \frac{1}{2}m\dot{\delta}^2\psi_1^2\,dX$$

we obtain the kinetic energy of the beam as follows:

$$U_k = \int dU_k = \frac{1}{2}m\dot{\delta}^2\int_0^1 \psi_1^2\,dX$$

or

$$U_k = \frac{1}{2}m_e\dot{\delta}^2 \tag{5.160}$$

where

$$m_e = m\int_0^1 \psi_1^2\,dX = m\eta^2\int_0^1 \psi^2\,dX \tag{5.161}$$

is the effective mass of the beam of mass m. It is noted that this effective mass obtained by the energy method is the same as that obtained using Galerkins method, which is given by (5.145). The effective mass of the straight beam in Fig. 5.31a is obtained by substituting the mode shape (5.155) into equation (5.161) as follows:

$$m_e = m\int_0^1 \left(3X^2 - 2X^3\right)^2 dX = \frac{13}{35}m \tag{5.162}$$

If the geometric nonlinearity of the straight beam is negligible (i.e., the displacement δ is much less than the beam width; more accurately, (u) of Example 3.3 is satisfied), the stiffness of the beam is given by (3.103) or (m) in Example 3.3, which is written

$$k_1 = \eta^2 \frac{EI}{l^3} \int_0^1 \left(\frac{d^2\psi}{dX^2} \right)^2 dX + \eta^2 \frac{N_0}{l} \int_0^1 \left(\frac{d\psi}{dX} \right)^2 dX \tag{5.163}$$

where N_0 is the initial axial force, defined by the tension as positive. Substituting the mode shape (5.157) and $\eta = 1$ into equation (5.163) gives the linear stiffness as

$$\begin{aligned} k_1 &= \frac{EI}{l^3} \int_0^1 [6(1-2X)]^2 \, dX + \frac{N_0}{l} \int_0^1 [6(X-X^2)]^2 \, dX \\ &= 12\frac{EI}{l^3} + \frac{6}{5}\frac{N_0}{l} \end{aligned} \tag{5.164}$$

If the maximum displacement δ_0 is comparable to the width of the beam, the geometric nonlinearity of the straight beam is not negligible and the cubic stiffness should be taken into account to evaluate the restoring force due to geometric nonlinearity. In this case, the cubic stiffness of the slender straight beam is given by (o) of Example 3.3:

$$k_3 = \frac{18}{25}\frac{EA}{l^3} \tag{5.165}$$

If the axial force N_0 is zero and the cubic stiffness above is negligible, the stiffness of the straight beam shown in Fig. 5.31a is given by

$$k = k_1 = \frac{F}{\delta} = \frac{12EI}{l^3} \tag{5.166}$$

A model of the straight beam, based on the preceding stiffnesses and effective mass, is illustrated in Fig. 5.31f, where the mass of the movable structure does not appear since the mass will be considered in the entire system of the movable structure and springs. If the cubic stiffness is negligible (e.g., the vibration amplitude is much less than the beam width), the cubic stiffness k_3 can be dropped from the model in Fig. 5.31f. If a very short beam is added to the vertical beam as shown in Fig. 5.31b, the axial stress N_0 and the cubic stiffness due to the geometric nonlinearity in the longer beam will be greatly reduced, with the stiffness in the beam of Fig. 5.31b approaching the stiffness equation (5.166).

The folded beam of Fig. 5.32a is used as a spring to support a movable structure and to generate a restoring force. The folded beam consists of two

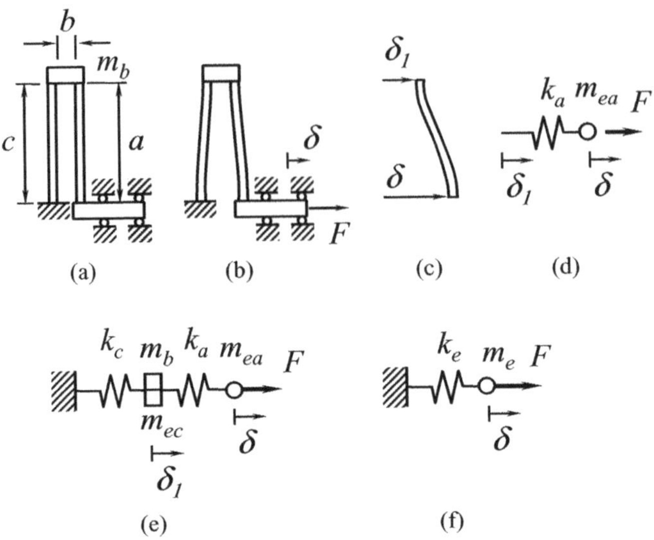

Figure 5.32 Effective mass and stiffness of folded beams.

identical beams of length a connected by a beam of length b and mass m_b. Since beam b is much shorter than beam a or the flexural rigidity EI of beam b is much greater than that of beam a, beam b can be assumed to be a rigid beam. When the movable structure connected to the lower end of beam a is displaced by δ due to the applied force F as shown in Fig. 31b, the rigid beam b is displaced by $\delta/2$ because the vertical line passing through the center of the beam b is the symmetry line. As a result, the rigid beam remains horizontal during displacement of the movable structure. Earlier we obtained the effective mass and stiffness of the straight beam c and will find the effective mass and stiffness of the entire folded beam. Let us consider the deflection of beam a in Fig. 5.32c, where the upper and lower ends of the beam are displaced by δ_1 and δ, respectively. We then express the displacement of the beam a at a point in terms of a mode shape ψ_1, as follows:

$$y = \delta_1 + (\delta - \delta_1)\psi_1 = [\delta_{10} + (\delta_0 - \delta_{10})\psi_1]\sin \omega t \tag{5.167}$$

where $\delta_1 = \delta_{10}\sin \omega t$, $\delta = \delta_0 \sin \omega t$, and ψ_1 is the mode shape, which depends on geometric constraints. Since the upper and lower ends of beam a do not undergo angular displacement, we can use the mode shape ψ_1 given by (5.155). The velocity and kinetic energy of the small beam element at x from the upper end of beam a are expressed as

$$\dot{y} = [\delta_{10} + (\delta_0 - \delta_{10})\psi_1]\omega \cos \omega t = \left[\dot{\delta}_1 + \left(\dot{\delta} - \dot{\delta}_1\right)\psi_1\right]$$

$$dU_k = \frac{1}{2}dm\dot{y}^2 = \frac{1}{2}\rho Aa\frac{dx}{a}\left[\dot{\delta}_1 + \left(\dot{\delta} - \dot{\delta}_1\right)\psi_1\right]^2$$
$$= \frac{1}{2}m\left[\dot{\delta}_1^2 + 2\dot{\delta}_1\left(\dot{\delta} - \dot{\delta}_1\right)\psi_1 + \left(\dot{\delta} - \dot{\delta}_1\right)^2\psi_1^2\right]dX$$

where $X = x/a$ and $m = \rho Aa$. The kinetic energy stored in beam a is then obtained as

$$U_k = \int dU_k = \frac{1}{2}m\int_0^1\left[\dot{\delta}_1^2 + 2\dot{\delta}_1\left(\dot{\delta} - \dot{\delta}_1\right)\psi_1 + \left(\dot{\delta} - \dot{\delta}_1\right)^2\psi_1^2\right]dX$$
$$= \frac{1}{2}m\dot{\delta}^2\left[\left(\frac{\dot{\delta}_1}{\dot{\delta}}\right)^2 + 2\frac{\dot{\delta}_1}{\dot{\delta}}\left(1 - \frac{\dot{\delta}_1}{\dot{\delta}}\right)\int_0^1\psi_1\,dX + \left(1 - \frac{\dot{\delta}_1}{\dot{\delta}}\right)^2\int_0^1\psi_1^2\,dX\right]$$
$$= \frac{1}{2}m\dot{\delta}^2\left[\left(\frac{\delta_1}{\delta}\right)^2 + 2\frac{\delta_1}{\delta}\left(1 - \frac{\delta_1}{\delta}\right)\int_0^1\psi_1\,dX + \left(1 - \frac{\delta_1}{\delta}\right)^2\int_0^1\psi_1^2\,dX\right] \quad (5.168)$$

We then find the effective mass m_e of beam *a as* follows:

$$m_{ea} = \frac{U_k}{\dot{\delta}^2/2} = m\left[\left(\frac{\delta_1}{\delta}\right)^2 + 2\frac{\delta_1}{\delta}\left(1 - \frac{\delta_1}{\delta}\right)\int_0^1\psi_1\,dX + \left(1 - \frac{\delta_1}{\delta}\right)^2\int_0^1\psi_1^2\,dX\right]$$
$$= m\left[\left(\frac{\delta_1}{\delta}\right)^2 + 2\frac{\delta_1}{\delta}\eta\left(1 - \frac{\delta_1}{\delta}\right)\int_0^1\psi\,dX + \left(1 - \frac{\delta_1}{\delta}\right)^2\eta^2\int_0^1\psi^2\,dX\right] \quad (5.169)$$

For $\delta_1 = 0$, the effective mass above is reduced to (5.161). Since equation (5.169) is a function of δ_1 and δ, we need one more equation to express the effective mass m_{ea} as a function of δ, which will be obtained from relations between geometry and stiffness. When the force F is applied to the beam as shown Fig. 32b, the displacement δ_1 is given by

$$\delta_1 = \frac{F}{k_c}$$

Since the effective stiffness of the folded beam is given by (4.2) as

$$\frac{1}{k_e} = \frac{1}{k_a} + \frac{1}{k_c}$$

the displacement δ is written

$$\delta = \frac{F}{k_e} = F\left(\frac{1}{k_a} + \frac{1}{k_c}\right)$$

The ratio δ_1/δ is written as a function of the stiffnesses of the beams,

$$\frac{\delta_1}{\delta} = \frac{k_a}{k_a + k_c} \qquad 1 - \frac{\delta_1}{\delta} = \frac{k_c}{k_a + k_c} \tag{5.170}$$

Substituting these relations and the mode shape ψ_1 into (5.169) gives the effective mass m_{ea}. For example, the stiffness of the beams a and c is given by

$$k_a = k_c = \frac{12EI}{a^3}$$

which gives $\delta_1/\delta = 1/2, 1 - \delta_1/\delta = 1/2$. We use the mode shape (5.155) and obtain the following:

$$\int_0^1 \psi_1 \, dX = \int_0^1 (3X^2 - 2X^3) dX = \frac{1}{2}$$

$$\int_0^1 \psi_1^2 \, dX = \int_0^1 (3X^2 - 2X^3)^2 \, dX = \frac{13}{35}$$

Substituting $\delta_1/\delta = 1/2, 1 - \delta_1/\delta = 1/2$ and the equations above into (5.169), we find the effective mass of beam a as follows:

$$m_{ea} = \frac{83}{140} m_a \tag{5.171}$$

where $m_a = \rho A a$. As described, the stiffness of beam a is $k_a = 12EI/a^3$. The equivalent model of beam a is shown in Fig. 5.32d. Let us obtain the effective mass due to beams b and c. The effective mass of beam c is given by (5.161) and then we can construct a model, shown in Fig. 5.32e, consisting of the effective mass and stiffness of the folded beam. Referring to the model, we express the kinetic energy stored in the entire folded beam as follows:

$$U_k = \frac{1}{2}(m_{ec} + m_b)\dot{\delta}_1^2 + \frac{1}{2} m_{ea} \dot{\delta}^2 = \frac{1}{2}\left[(m_{ec} + m_b)\left(\frac{\dot{\delta}_1}{\dot{\delta}}\right)^2 + m_{ea}\right]\dot{\delta}^2$$

which gives the effective mass of the folded beam,

$$m_e = (m_{ec} + m_b)\left(\frac{\dot{\delta}_1}{\dot{\delta}}\right)^2 + m_{ea} \tag{5.172}$$

The velocity ratio in equation (5.172) is given by (5.170):

$$\frac{\dot{\delta}_1}{\dot{\delta}} = \frac{\delta_{10}\omega\cos\omega t}{\delta_0\omega\cos\omega t} = \frac{\delta_1}{\delta} = \frac{k_a}{k_a + k_c}$$

Upon substitution of the preceding velocity ratio into the foregoing effective mass, we find the effective mass of the folded beam (Fig. 5.32a) as a function of the mass and stiffness of the beams, as follows:

$$m_e = (m_{ec} + m_b)\left(\frac{k_a}{k_a + k_c}\right)^2 + m_{ea} \tag{5.173}$$

where m_{ec}, m_{ea}, m_b, k_c, and k_a denote the effective mass of beams c and a, defined by (5.161) and (5.169), respectively, the mass of beam b, and the stiffnesses of beams c and a. For the folded beam of Fig. 5.32a, consisting of two identical beams and a rigid connecting beam, the mass of beams c and a are m_a and the effective masses are given by (5.162): $m_{ec} = 13m_c/35 = 13m_a/35$, and (5.171), $m_{ea} = 83m_a/140$. We then obtain the effective mass m_e of the folded beam as

$$m_e = \left(\frac{13}{35}m_a + m_b\right)\left(\frac{1}{2}\right)^2 + \frac{83}{140}m_a = \frac{1}{4}m_b + \frac{24}{35}m_a \tag{5.174}$$

Since beams a and c are identical and beam b is assumed to be a rigid beam, the stiffness of the folded beam can be written as

$$\frac{1}{k_e} = \frac{1}{k_a} + \frac{1}{k_c}$$

which gives the effective stiffness of the folded beam, k_e, as

$$k_e = \frac{k_a}{2} = \frac{6EI}{a^3} \tag{5.175}$$

The equivalent model of the folded beam may be constructed as in Fig. 5.32f, based on the preceding effective mass and stiffness. It is noted that the effective mass (5.174) and the effective stiffness (5.175) are obtained from the folded beam, which consists of two identical slender beams and a rigid connecting beam as shown in Fig. 5.32a. If a different folded beam (e.g., beam lengths a and c are different), we should evaluate the effective mass and stiffness by using different mode shapes.

Similarly, we can obtain the effective mass and stiffness of the connected folded beam illustrated in Fig. 5.33. It is assumed that a quantity, defined as EI/l^3, where EI and l denote the flexural rigidity and length of a beam, respectively, of the connecting beam is much greater than that of the other

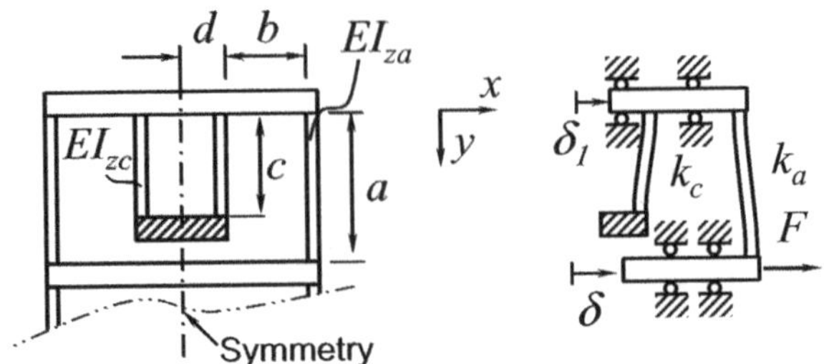

Figure 5.33 Effective mass and stiffness of connected folded beams.

slender beams. Under this assumption, the connecting beam located at the top of the structure in Fig. 5.33 can be considered to be a rigid beam, and then we construct a model of the right portion of the connected folded beam of Fig. 5.33a, which is illustrated in Fig. 5.33b, where the rigid connecting beam is considered to move almost horizontally. As a result, we can use the models in Fig. 5.32e and (f) as the equivalent models for the half portion of the connected folded beam of Fig. 5.33b. The effective mass of the half structure shown in Fig. 5.33b is then given by (5.173):

$$m_e = (m_{ec} + m_{bd})\left(\frac{k_a}{k_a + k_c}\right)^2 + m_{ea}$$

where

$$m_{ec} = \frac{13}{35}m_c \qquad m_{ea} = \frac{83}{140}m_a \qquad m_{bd} = m_b + m_d$$

are the effective masses of beams c and a, defined in (5.162) and (5.171), and the total mass of beams b and d. The effective stiffnesses of beams c and a a are expressed as

$$k_c = \frac{12EI_c}{c^3} \qquad k_a = \frac{12EI_a}{a^3}$$

Substituting the effective masses and stiffnesses of beams c and a into the preceding effective mass, we find the effective mass of the half portion of the connected folded beam shown in Fig. 5.33b:

$$m_e = \left(\frac{13}{35}m_c + m_{bd}\right)\left(\frac{k_a}{k_a + k_c}\right)^2 + \frac{83}{140}m_a \tag{5.176}$$

and the stiffness

$$\frac{1}{k_e}=\frac{1}{k_a}+\frac{1}{k_c} \quad \text{or} \quad k_e=\frac{k_a k_c}{k_a+k_c} \tag{5.177}$$

Example 5.4 We wish to obtain the effective mass and stiffness of the microstructure shown in Fig. 5.34a, commonly encountered in MEMS design. The structure consists of a central mass M supported by four folded beams. It is also necessary to evaluate the natural frequency of the structure.

Using the equivalent model of the folded beam of Fig. 5.32f, we construct the equivalent model of the microstructure of Fig. 5.34a, which is shown in Fig. 5.34b. Since the effective mass and stiffness of a folded beam are given by (5.174) and (5.175), the effective mass m_{eff} of the entire structure is written as

$$m_{\text{eff}} = M + 4m_e = M + 4\left(\frac{1}{4}m_b + \frac{24}{35}m_a\right) = M + m_b + \frac{96}{35}m_a \tag{a}$$

where m_a and m_b represent masses of the beams a and b, respectively. If the total masses of beams a and b are defined as $m_{bt} = 4m_b$ and $m_{at} = 8m_a$, respectively, equation (a) may be rewritten as follows:

$$m_{\text{eff}} = M + \frac{1}{4}m_{bt} + \frac{12}{35}m_{at} \tag{b}$$

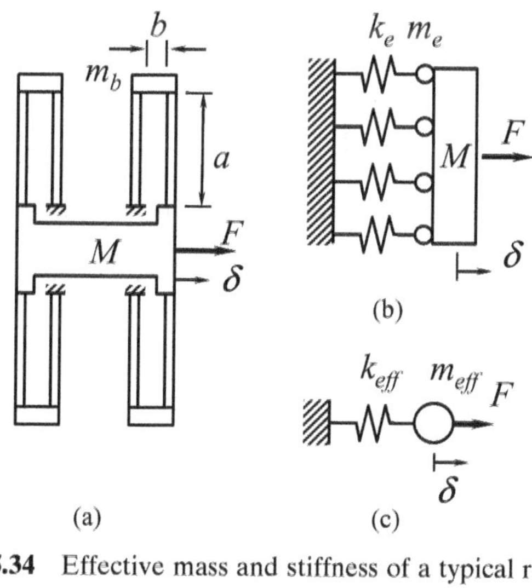

Figure 5.34 Effective mass and stiffness of a typical resonator.

The effective stiffness of the entire structure is obtained as

$$k_{\text{eff}} = 4k_e = 4\left(\frac{k_a}{2}\right) = 2k_a \tag{c}$$

where $k_a = 12EI_a/a^3$. The equivalent model of the entire structure of Fig. 5.34a is illustrated in Fig. 5.34c. Therefore, the natural frequency corresponding to the horizontal direction is given by

$$f_n = \frac{1}{2\pi}\sqrt{\frac{k_{\text{eff}}}{m_{\text{eff}}}} = \frac{1}{2\pi}\sqrt{\frac{2k_a}{M + m_b + (96/35)m_a}} \tag{d}$$

Note that if the total mass of the flexures (springs) is much less than the central mass (movable mass), the effective mass of the entire structure is close to the mass of the central mass. For example, if the effective mass of the flexures ($m_b + 96m_a/35$) is 4% of the central mass M, the natural frequency, (d), of the microstructure is 2% less than $f_n = \sqrt{2k_a/M}/2\pi$. In general, if the entire mass of flexures of microstructures is much less than that of the rigid movable mass (i.e., the entire area of the flexures on a mask for fabrication is much less than that of the relatively rigid movable structure), the contribution of the mass of the flexures to the natural frequency is negligible and the natural frequency of the microstructure then becomes

$$f_n \cong \frac{1}{2\pi}\sqrt{\frac{k_{\text{eff}}}{M}} \tag{e}$$

5.10 SYSTEMS WITH REPEATED STRUCTURES

Systems with two or more DOFs were analyzed in Section 5.7 to obtain the natural frequencies and corresponding mode shapes. n-DOF systems, in general, produce the determinant of an $n \times n$ matrix for the natural frequencies, which is difficult to solve in closed form (if $n >> 1$) and is usually solved by numerical methods. In many MEMS applications, such as RF filters, a series of identical structures are used to improve performance, such as bandwidth or sensitivity. Possible general configurations are illustrated in Fig. 5.35a–c, where m, k, k_c, ζ, and $x_1, x_2, \ldots, x_n$ represent the mass of the repeated structure, the stiffness to support the mass, the coupling stiffness, a physical coordinate, and the displacement of the masses corresponding to the mass labeled $j = 1, 2, \ldots, n$, respectively. The lengths l_1, l_2, and l_3 will be explained later. We shall obtain closed-form solutions for the natural

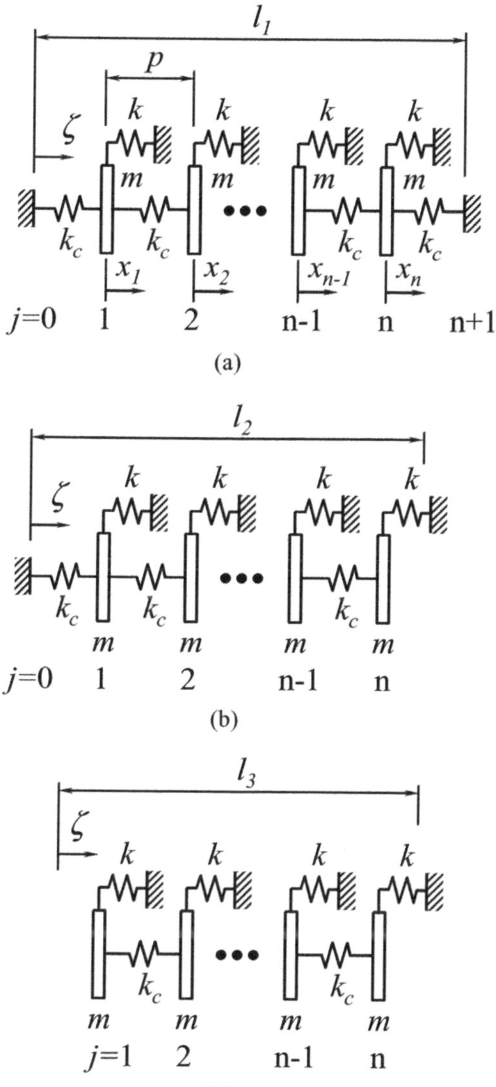

Figure 5.35 Series of resonators.

frequencies and the corresponding mode shapes of the n-DOF systems shown in Fig. 5.35.

Equations of Motion

Upon application of Newton's second law to the masses subjected to the spring forces in Fig. 5.35a, we find the equations of motion as follows:

$$\begin{aligned}
m\ddot{x}_1 + (k+2k_c)x_1 - k_c x_2 &= 0 \\
m\ddot{x}_2 - k_c x_1 + (k+2k_c)x_2 - k_c x_3 &= 0 \\
&\vdots \\
m\ddot{x}_{n-1} - k_c x_{n-2} + (k+2k_c)x_{n-1} - k_c x_n &= 0 \\
m\ddot{x}_n - k_c x_{n-1} + (k+2k_c)x_n &= 0
\end{aligned} \tag{5.178}$$

which may be expressed in matrix form as

$$\mathbf{m}\ddot{\mathbf{x}} + \mathbf{k}\mathbf{x} = \mathbf{0} \tag{5.179}$$

where

$$\mathbf{m} = \begin{bmatrix} m & & & & \\ & m & & \Phi & \\ & & \cdots & & \\ & \Phi & & m & \\ & & & & m \end{bmatrix}$$

$$\mathbf{k} = \begin{bmatrix} k+2k_c & -k_c & & & \\ -k_c & k+2k_c & -k_c & \Phi & \\ & -k_c & \cdots & -k_c & \\ & \Phi & -k_c & k+2k_c & -k_c \\ & & & -k_c & k+2k_c \end{bmatrix}$$

$$\mathbf{x} = \{x_1 \quad x_2 \quad \cdots \quad x_{n-1} \quad x_n\}^{\mathrm{T}}$$

$$\mathbf{0} = \{0 \quad 0 \quad \cdots \quad 0 \quad 0\}^{\mathrm{T}}$$

denote the mass matrix, the stiffness matrix, the displacement vector, and the zero vector. The preceding equation may be expressed in a generalized form of the equation of motion for the discrete systems shown in Fig. 5.35, as follows:

$$m\ddot{x}_j - k_c x_{j-1} + (k+2k_c)x_j - k_c x_{j+1} = 0 \tag{5.180}$$

where j is an integer to designate a mass in the systems. Substituting $j = 0, 1, \ldots, n+1$ into equation (5.180) and noting that $x_0 = 0$ and $x_{n+1} = 0$, we again obtain (5.178). To derive general solutions for the natural frequencies of the n-DOF system of Fig. 5.35a, we first review the vibrations in rod and string that were dealt with in Section 5.8. The vibrations of rod and string illustrated in Fig. 5.36 can be interpreted as a standing wave or stationary wave that is produced by the interference of two waves of the same amplitude and frequency which travel in opposite directions. In the standing wave, there are

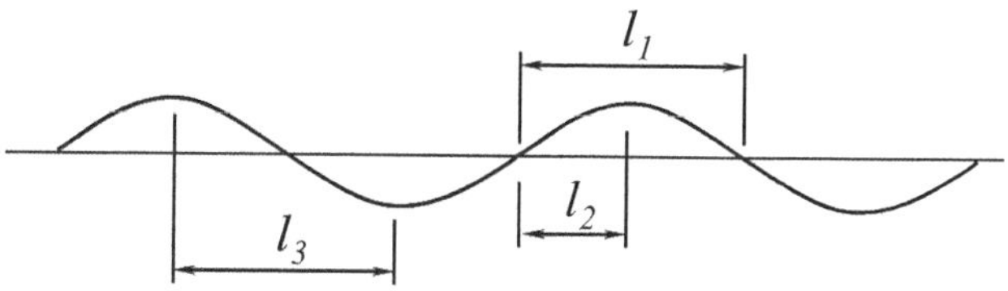

Figure 5.36 Standing wave.

nodes (nodal points) at which the two waves are canceled and the amplitude becomes zero. If the media, such as string and rod, have boundaries as shown in Figs. 5.25a and 5.27a, the propagating and reflected waves produce the standing wave. Therefore, the wave of Fig. 5.26 can be thought of as a standing wave. We shall compare the vibrations of the systems in Fig. 5.35a–c with the standing wave in Fig. 5.36. If the number of degrees of freedom of the systems in Fig. 5.35 is sufficiently large, the system can be assumed to be a continuous medium in which the wave travels. Under this assumption and the boundary boundaries shown in Fig. 5.35, we select three portions of the standing wave, labeled l_1, l_2, and l_3, from Fig. 5.36. The portions selected correspond to the lengths of the systems in Fig. 5.35a–c, also labeled l_1, l_2, and l_3. We assume the vibration of the masses as follows:

$$x_j = \psi \sin \omega t \tag{5.181}$$

where j, ω, and t denote, respectively, numbers to identify the mass, the angular frequency of the vibration, and time, and

$$\psi = (A \sin \beta\zeta_j + B \cos \beta\zeta_j) \tag{5.182}$$

is the mode shape, which is a function of the coordinate ζ_j and constant β associated with the frequency ω. Substituting the preceding displacement of masses into the equation of motion of the discrete system, (5.180), we find that

$$\begin{aligned}&[(k+2k_c)-m\omega^2](A\sin\beta\zeta_j+B\cos\beta\zeta_j)-k_c[A(\sin\beta\zeta_{j+1}+\sin\beta\zeta_{j-1})\\&\quad+B(\cos\beta\zeta_{j+1}+\cos\beta\zeta_{j-1})]=0\end{aligned} \tag{5.183}$$

From Fig. 5.35a–c, the relations $\zeta_{j+1} = \zeta_j + p$ and $\zeta_{j-1} = \zeta_j - p$ are observed. Substituting these relations into equation (5.183), recalling the trigonometric identities, $\sin(a + b) = \sin a \cos b + \cos a \sin b$ and $\cos(a + b) = \cos a \cos b - \sin a \sin b$, and noting that

$$\sin\beta\zeta_{j+1} + \sin\beta\zeta_{j-1} = 2\sin\beta\zeta_j \cos\beta p$$
$$\cos\beta\zeta_{j+1} + \cos\beta\zeta_{j-1} = 2\cos\beta\zeta_j \cos\beta p$$

gives a different form of the equation of motion:

$$(A\sin\beta\zeta_j + B\cos\beta\zeta_j)[k - m\omega^2 + 2k_c(1-\cos\beta p)] = 0 \tag{5.184}$$

Noting that the quantity in the left set of parentheses is the mode shape ψ given by (5.182), and using the identity $1 - \cos\beta p = 2\sin^2(\beta p/2)$, the preceding discrete equation of motion (5.183) is expressed as follows:

$$\psi\left(k - m\omega^2 + 4k_c\sin^2\frac{\beta p}{2}\right) = 0 \tag{5.185}$$

which gives two equations: $\psi = 0$ to find β, and $[k - m\omega^2 + 4k_c\sin^2(\beta p/2)] = 0$ to gives the natural frequency corresponding to β. If β is determined from boundary conditions, the natural angular frequencies is then given by

$$\omega_i = \sqrt{\frac{1}{m}\left(k + 4k_c\sin^2\frac{\beta_i p}{2}\right)} \tag{5.186}$$

Note that (5.185) is a fundamental equation to solve the problems in Fig. 5.35 or similar problems in which masses and springs are connected in series.

Case I We shall first obtain the natural frequencies and corresponding mode shapes of the system shown in Fig. 5.35a. The boundary conditions for Fig. 5.35a are as follows: $\psi = 0$ at $\zeta = 0$ and $\zeta = l_1 = (n+1)p$ (i.e., $\psi = 0$ $j = 0$ and $j = n + 1$). Applying these boundary conditions to the mode shape $\psi = (A\sin\beta\zeta_j + B\cos\beta\zeta_j)$, we obtain

$$B = 0 \quad \text{and} \quad A\sin\beta(n+1)p = 0$$

The latter equation gives the dimensionless number as follows:

$$\beta_i p = \frac{i\pi}{n+1} \tag{5.187}$$

where n and i denote the total number of the mass, and the ith mode shape and the range of i is $1 \le i \le n$. Substitution of the preceding dimensionless number $\beta_i p$ into (5.186) yields the ith natural frequency of the system in Fig. 5.35 as follows:

$$\omega_i = \sqrt{\frac{1}{m}\left\{k + 4k_c\left(\sin\frac{\pi i}{2(n+1)}\right)^2\right\}} \tag{5.188}$$

where $i = 1, 2, \ldots, n$. Substituting $B = 0$ and $\zeta_j = jp$ obtained from Fig. 5.35a into (5.182), we have the mode shape ψ_i, which corresponds to the preceding natural frequency,

$$\psi_i = A\sin\beta_i\zeta_j = A\sin\beta_i pj = A\sin\left(\frac{i\pi}{n+1}j\right) \tag{5.189}$$

where $j = 1, 2, \ldots, n$ denotes the number to designate the jth mass. The reader may verify that the preceding natural angular frequency and mode shape are solution of the equation of motion (5.180) by substituting them into the equation of motion (5.178) or (5.180). The mode shape vector may be written as

$$\{\psi\}_i = A\left[\sin\left(\frac{i\pi}{n+1}\right), \sin\left(\frac{i\pi}{n+1}2\right), \ldots, \sin\left(\frac{i\pi}{n+1}n\right)\right]^{\mathrm{T}} \tag{5.190}$$

where A and the superscript T denote a constant and the transpose of a vector. We shall use some examples to test the preceding solutions further. Let us first consider the system of Fig. 5.35a with one mass (i.e., $n=1$). When we substitute 1 for i, j, and n into (5.188) and (5.189), we have the following natural frequency and mode shape:

$$\omega_1 = \sqrt{\frac{1}{m}\left[k + 4k_c\left(\sin\frac{\pi}{4}\right)^2\right]} = \sqrt{\frac{k+2k_c}{m}}$$

$$\psi_1 = A\sin\left(\frac{i\pi}{n+1}j\right) = A\sin\frac{\pi}{2} = A$$

Since $k + 2k_c$ is the effective stiffness of the system with one mass in Fig. 5.35a, the preceding natural angular frequency is that of the one-mass system. The mode shape of the multimass system is interpreted as relative amplitudes of the masses, so the constant A can be understood as the amplitude of the single mass. For the two-mass system of Fig. 5.35a, we substitute 2 for n into (5.188), to obtain

$$\omega_i = \sqrt{\frac{1}{m}\left[k + 4k_c\left(\sin\frac{\pi i}{6}\right)^2\right]}$$

If, further, we substitute 1 and 2 for i into the preceding equation, we find that

$$\omega_1 = \sqrt{\frac{1}{m}\left[k + 4k_c\left(\sin\frac{\pi}{6}\right)^2\right]} = \sqrt{\frac{k+k_c}{m}}$$

$$\omega_2 = \sqrt{\frac{1}{m}\left[k + 4k_c\left(\sin\frac{\pi}{3}\right)^2\right]} = \sqrt{\frac{k+3k_c}{m}}$$

If we set $k = 0$, we obtain $\omega_1 = \sqrt{k_c/m}$ and $\omega_1 = \sqrt{3k_c/m}$, which are the natural angular frequencies of the two-mass system with $k_1 = k_2 = k_c$ that is illustrated in Fig. 5.21 and whose natural frequencies were discussed following (5.106). Similarly, we expand the mode shape (5.189) corresponding to ω_1 and ω_2 above as follows:

$$\{\psi\}_1 = A\begin{Bmatrix} \sin\frac{\pi}{3} \\ \sin\frac{2\pi}{3} \end{Bmatrix} = A_1\begin{Bmatrix} 1 \\ \sin\frac{2\pi}{3}\Big/\sin\frac{\pi}{3} \end{Bmatrix} = A_1\begin{Bmatrix} 1 \\ 1 \end{Bmatrix}$$

$$\{\psi\}_2 = A\begin{Bmatrix} \sin\frac{2\pi}{3} \\ \sin\frac{4\pi}{3} \end{Bmatrix} = A_2\begin{Bmatrix} 1 \\ \sin\frac{4\pi}{3}\Big/\sin\frac{2\pi}{3} \end{Bmatrix} = A_2\begin{Bmatrix} 1 \\ -1 \end{Bmatrix}$$

Since the constants A_1 and A_2 are arbitrary and the mode shape shows the relative positions of masses, the preceding two mode shapes corresponding to the first and second natural frequencies are the same as those shown following (5.106). Further verification shows that equations (5.188) and (5.189) are valid for all ranges of n (i.e., $1 \le n \le \infty$). The natural frequencies (5.188) and the first mode shape (5.189) of the six-mass system ($k = k_c$) of Fig. 5.35a is shown in Fig. 5.37a and b, respectively. From Fig. 5.37a, the natural frequency (circle), normalized by a characteristic frequency $f_c = \sqrt{k/m}/2\pi$, increases with the number i, which designates the mode number. If the number of mass increases (i.e., n increases), the natural frequencies of the systems are placed along the solid line in Fig. 5.36a. The first mode shape (circle in Fig. 5.37b), corresponding to the first resonant frequency, is in the solid line that represents $\sin(\pi\zeta/7)$, where ζ is real.

Case II The second case shown in Fig. 5.35b corresponds to the range l_2 of Fig. 5.36. The boundary conditions are given as follows: $\psi = 0$ at $\zeta = 0$ and $d\psi/d\zeta = 0$ at $\zeta = l_2 = (n + 1/2)p$. To understand the boundary condition at $\zeta = l_2$, we consider a system mirrored about $\zeta = l_2$, which becomes the system with $2n$ masses shown in Fig. 5.35a. The mode in Fig. 5.36 shows that the masses n and $n + 1$ both have the same displacement, and the length of the coupling stiffness between the masses n and $n + 1$ remains unchanged. As a result, the half portion of the $2n$-mass system is represented by the system shown in Fig. 5.35b. Applying the boundary conditions into the mode shape (5.182), we find that $B = 0$ and

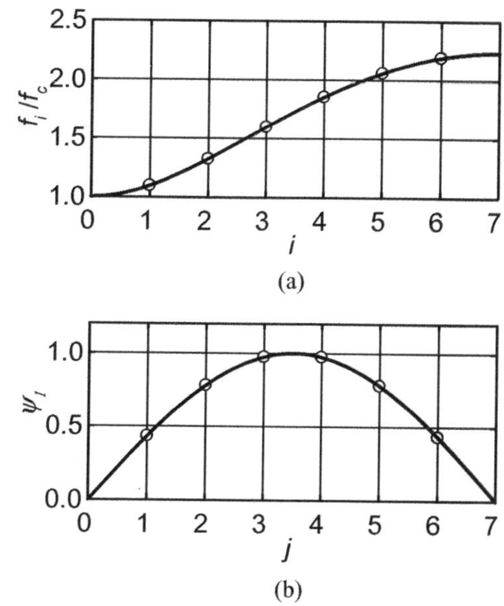

Figure 5.37 Natural frequencies and first mode shape of a six-mass system.

$$\left.\frac{d\psi}{d\zeta}\right|_{\zeta=(n+1/2)p} = A\beta\cos\left[\beta p\left(n+\frac{1}{2}\right)\right] = 0$$

which gives the dimensionless number

$$\beta_i p = \frac{\pi/2+\pi(i-1)}{n+1/2} = \frac{\pi(2i-1)}{2n+1}$$

Substitution of the equation above into (5.186) gives the natural angular frequency,

$$\omega_i = \sqrt{\frac{1}{m}\left\{k+4k_c\left[\sin\frac{\pi(2i-1)}{2(2n+1)}\right]^2\right\}} \tag{5.191}$$

where $i = 1, 2, \ldots, n$. Substituting $B = 0$, $\zeta_j = jp$, and $\beta_j p$ into the mode shape (5.182), we obtain

$$\psi_i = A\sin\beta_i\zeta_j = A\sin(\beta_i pj) = A\sin\left[\frac{\pi(2i-1)}{2n+1}j\right] \tag{5.192}$$

where $j = 1, 2, \ldots, n$ is the number that identifies the mass. Equations (5.191) and (5.192) hold for $1 \le n \le \infty$.

Case III The third case shown in Fig. 5.35c represents the range l_3 of the standing wave shown in Fig. 5.36. For this case, the boundary must satisfy $d\psi/d\zeta = 0$ at $\zeta = 0$ and $\zeta = l_3 = np$. Application of the boundary conditions to mode shape equation (5.182) gives $A = 0$ and

$$\left.\frac{d\psi}{d\zeta}\right|_{\zeta=np} = -B\beta \sin \beta np = 0$$

which provides the dimensionless number

$$\beta_i p = \frac{\pi(i-1)}{n}$$

Substituting the number into the natural frequency (5.186) gives

$$\omega_i = \sqrt{\frac{1}{m}\left\{k + 4k_c\left[\sin\frac{\pi(i-1)}{2n}\right]^2\right\}} \tag{5.193}$$

where $i = 1, 2, \ldots, n$. Referring to $\zeta_j = (j - \frac{1}{2})p$, the mode shape (5.182) is written

$$\psi_i = A \sin \beta_i \zeta_j = A \sin\left[\beta_i p\left(j - \frac{1}{2}\right)\right] = A \sin\left[\frac{\pi(i-1)}{2n}(2j-1)\right] \tag{5.194}$$

where $j = 1, 2, \ldots, n$. The natural frequency and corresponding mode shape above are valid for $1 \le n \le \infty$.

5.11 DUFFING'S EQUATION

In preceding sections of this chapter, dynamic responses of linear differential equations have been obtained. However, as described in Chapter 4, some systems have nonlinearity, which leads to nonlinear dynamic response. *Duffing's equation*, which includes the linear and cubic stiffnesses, is used to understand nonlinear systems (e.g., Figs.5.30 and 4.25) and is written

$$m\ddot{x} + c\dot{x} + k_1 x + k_3 x^3 = F_0 \cos \omega t_1 \tag{5.195}$$

where $m, c, x, k_1, k_3, F_0, \omega$, and t_1 denote the mass, damping coefficient, displacement, linear and cubic stiffnesses, force amplitude, angular frequency, and time, respectively. A sixth polynomial representing an equation for the steady-state response of equation (5.195) will first be obtained based on Duffing's work, and we shall then present an analytical closed-form solution for the steady-state response. For analytical work, we assume that the linear stiffness k_1 is positive and the cubic stiffness k_3 can be either positive or negative, and that the system represented by equation (5.195) is underdamped. We further assume that the response of the preceding differential equation is represented by a sinusoidal response with the angular frequency ω, and that higher harmonic response is negligible. Under these assumptions, the response to the sinusoidal force is then expressed as

$$x = x_0 \cos(\omega t_1 - \phi) \tag{5.196}$$

where x_0 and ϕ represent the amplitude and phase of the response, respectively. Since substitution of equation (5.196) into the foregoing differential equation produces complicated expressions, we introduce a simple transformation, which allows us to perform relatively simple mathematical operation as follows: $\omega t_1 - \phi = \omega t$ and thus $\omega t_1 = \omega t + \phi$. The equation of motion and the response are then written as

$$m\ddot{x} + c\dot{x} + k_1 x + k_3 x^3 = F_0 \cos(\omega t + \phi) \tag{5.197}$$

$$x = x_0 \cos \omega t \tag{5.198}$$

Before proceeding further, it is convenient to nondimensionalize equations (5.197) and (5.198) as follows (nondimensionalization was described in Section 1.4):

$$\frac{d^2 X}{dT^2} + 2\zeta \frac{dX}{dT} + X + k_r X^3 = \cos(\Omega T + \phi) \tag{5.199}$$

$$X = X_0 \cos \Omega T \tag{5.200}$$

where

$$\begin{gathered} X = \frac{x}{x_c} = \frac{k_1 x}{F_0} \qquad x_c = \frac{F_0}{k_1} \qquad X_0 = \frac{x_0}{x_c} = \frac{k_1 x_0}{F_0} \\ T = \frac{t}{t_c} = \frac{t}{\sqrt{m/k_1}} = t\sqrt{\frac{k_1}{m}} \qquad t_c = \sqrt{m/k_1} \qquad \zeta = \frac{c}{2\sqrt{mk_1}} \\ k_r = \frac{k_3 x_c^2}{k_1} = \frac{k_3}{k_1^3} F_0^2 \qquad \Omega = \omega t_c = \frac{\omega}{\sqrt{k_1/m}} \end{gathered} \tag{5.201}$$

denote the dimensionless displacement, a characteristic length, the amplitude, a dimensionless time, a characteristic time, the damping factor, a nonlinearity factor representing the nonlinearity, and a dimensionless angular frequency, respectively. Recalling the trigonometric identity, $\cos(\Omega T + \phi) = \cos\Omega T \cos\phi - \sin\Omega T \sin\phi$, the foregoing differential equation is written

$$\frac{d^2X}{dT^2} + 2\zeta\frac{dX}{dT} + X + k_r X^3 = \cos\Omega T\cos\phi - \sin\Omega T\sin\phi \tag{5.202}$$

Substituting the response (5.200) into equation (5.202), we find that

$$X_0\left(1-\Omega^2\right)\cos\Omega T - 2\zeta X_0\Omega\sin\Omega T + k_r X_0^3\cos^3\Omega T = \cos\Omega T\cos\phi - \sin\Omega T\sin\phi \tag{5.203}$$

which is a function of two unknowns: the amplitude X_0 and the phase ϕ. Since we require two independent equations to obtain the amplitude and phase, we multiply the preceding equation by $\sin\Omega T$ and $\cos\Omega T$, respectively and integrate the resulting equations from 0 to 2π (period) using the relations.

$$\int_0^{2\pi}\cos^2\Omega T\, d\Omega T = \int_0^{2\pi}\sin^2\Omega T\, d\Omega T = \pi$$

$$\int_0^{2\pi}\sin\Omega T\cos\Omega T\, d\Omega T = 0$$

$$\int_0^{2\pi}\sin\Omega T\cos^3\Omega T\, d\Omega T = 0$$

$$\int_0^{2\pi}\cos\Omega T\cos^3\Omega T\, d\Omega T = \frac{3}{4}\pi$$

The foregoing mathematical manipulation yields two equations:

$$X_0\left(1-\Omega^2\right) + k_r X_0^3\frac{3}{4} = \cos\phi$$

$$2\zeta X_0\Omega = \sin\phi$$

Substituting the preceding two equations into the identity $\cos^2\phi+\sin^2\phi=1$, we have

$$\left[X_0\left(1-\Omega^2\right) + k_r X_0^3\frac{3}{4}\right]^2 + (2\zeta X_0\Omega)^2 = 1 \tag{5.204}$$

which is a sixth-order polynomial of X_0. If $k_r = 0$ (i.e. $k_3 = 0$), equation (5.204) yields the magnification factor of (5.87) discussed in Section 5.6. Since there was no general solution of a sixth-order polynomial, numeral methods such as the Newton–Raphson method (Atkinson, 1978) have been used to solve the preceding equation for X_0.

Analytic Solution for Steady-State Response

We shall obtain an exact solution of the preceding equation in closed form. For the derivation of the exact solution, it is assumed that $k_r \neq 0$. Setting $Y = X_0^2$, expanding the preceding equation, and rearranging terms gives the following cubic equation:

$$Y^3 + \frac{8}{3k_r}(1-\Omega^2)Y^2 + \frac{16}{9k_r^2}\left[(1-\Omega^2)^2 + 4\zeta^2\Omega^2\right]Y - \frac{16}{9k_r^2} = 0 \quad (5.205)$$

or

$$Y^3 + aY^2 + bY + c = 0 \quad (5.206)$$

where

$$\begin{aligned} a &= \frac{8}{3k_r}(1-\Omega^2) \\ b &= \frac{16}{9k_r^2}\left[(1-\Omega^2)^2 + 4\zeta^2\Omega^2\right] \\ c &= -\frac{16}{9k_r^2} \end{aligned} \quad (5.207)$$

The general solution of the preceding equation was described in Section 5.1. We first evaluate the discriminant Δ, defined in (5.6):

$$\Delta = 4a^3c - a^2b^2 - 18abc + 4b^3 + 27c^2$$

If $\Delta > 0$, the cubic equation (5.205) gives one real root,

$$Y = -\frac{a}{3} + \sqrt[3]{A} - \frac{p}{3}\frac{1}{\sqrt[3]{A}} \quad (5.208)$$

where A and p are values defined in (5.8) and (5.4) which are functions of a, b, and c. If $\Delta \leq 0$, the cubic equation yields three distinct real roots (5.15), which are expressed as

$$\begin{aligned} Y_1 &= \frac{1}{3}\left(-a + 2\sqrt{a^2-3b}\cos\frac{\psi}{3}\right) \\ Y_2 &= \frac{1}{3}\left\{-a + 2\sqrt{a^2-3b}\cos\left[\frac{1}{3}(\psi+2\pi)\right]\right\} \\ Y_3 &= \frac{1}{3}\left\{-a + 2\sqrt{a^2-3b}\cos\left[\frac{1}{3}(\psi+4\pi)\right]\right\} \end{aligned} \quad (5.209)$$

where

$$\psi = \cos^{-1}\frac{-a^3+(9/2)ab-(27/2)c}{\left(a^2-3b\right)^{3/2}} \tag{5.210}$$

The dimensionless amplitude X_0 is then obtained as

$$X_0 = \sqrt{Y} \tag{5.211}$$

and the dimensional amplitude x_0 is given by the third equation of (5.201), which is written

$$x_0 = X_0\frac{F_0}{k_1} \tag{5.212}$$

It is noted that the dimensionless steady-state response X_0 is a function of the damping ratio ζ, the nonlinearity factor k_r, and the frequency ratio Ω. By comparing the preceding solution (5.211) with numerical solutions or experimental data, we find that if $\Delta < 0$, X_{01}, X_{02}, and X_{03}, which correspond to Y_1, Y_2, and Y_3 in (5.209), provide a stable solution for a larger amplitude, an unstable solution, and a stable solution for a smaller amplitude, and that a relation among the amplitudes is expressed by $X_{01} \geq X_{02} \geq X_{03}$. So far, we have obtained an exact solution for the steady-state response of a system whose behavior is expressed by Duffing's equation (5.195). The preceding steady-state response X_0 in Fig. 5.38 is plotted against the frequency ratio Ω, while

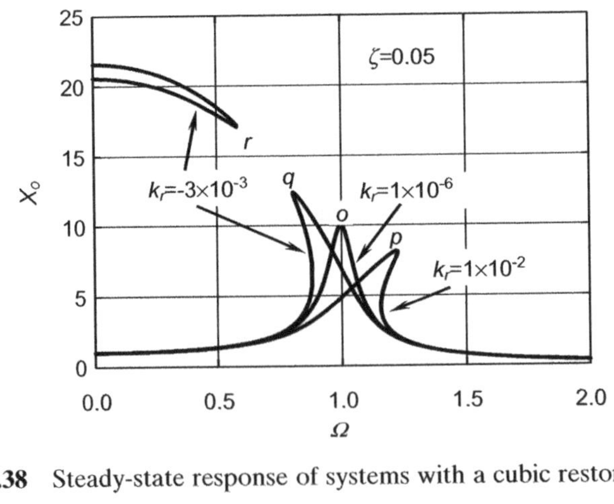

Figure 5.38 Steady-state response of systems with a cubic restoring force.

the damping factor remains at a value of 0.05 and the nonlinearity factor k_r is set at values of 1×10^{-2}, 1×10^{-6}, and -3×10^{-3}. The amplitude curves for $k_r > 0$ (hard spring) and $k_r < 0$ (soft spring) are inclined rightward and leftward, respectively, while the amplitude curve for $k_r = 1 \times 10^{-6}$ (relatively linear spring) is not distorted. It is also observed from Fig. 5.38 that for $k_r < 0$, (5.211) shows three solutions in the range from $\Omega = 0$ to a certain value of Ω (e.g., $\Omega = 0.58$ for the given ζ and k_r in Fig. 5.38). When the nonlinearity factor k_r increases further, the points of the curves, labeled r and q, are connected to each other and the resulting shape becomes a tube.

Experiment shows the jump phenomena shown in Fig. 5.39 when the frequency in Duffing's equation (5.195) for $k_1 > 0$ sweeps up or down. Consider a system with a hard spring ($k_1 > 0, k_3 > 0$) whose frequency response is shown in Fig. 5.39a. As the frequency sweeps up from 1 to 5 in Fig. 5.39a, the amplitude follows along the path 1–2–3–4–5, where the amplitude jumps from 3 to 4. If the frequency sweeps down from 5, the amplitude varies along the path 5–4–6–2–1, where the amplitude jumps from 6 to 2. It is noted from the figure that path 3–6, which corresponds to the unstable solution, X_{02}, corresponding to Y_2 in (5.209), is not obtained from experiments. Similarly, as the frequency sweeps up for a system with a soft spring (i.e., $k_1 > 0$, $k_3 < 0$) in Fig. 5.39b, the amplitude along 1–2–3–5–6 is achieved, and for a frequency sweep down from point 6, the amplitude along 6–5–4–2–1 is obtained. Jumps, 3–5 for frequency sweep up and 4–2 during frequency sweep down, are observed. The path 3–4 is not achieved from experiments because it is the unstable solution, X_{02}, which corresponds to Y_2 in (5.209).

Since the maximum amplitude and the corresponding frequency shown in Fig. 5.38 represents a characteristic of the system, we find them in closed form. We rearrange terms of (5.205) to give a quadratic equation with respect to $1 - \Omega^2$:

$$Y\left(1-\Omega^2\right)^2+\left(\frac{3k_r}{2}Y^2-4\zeta^2 Y\right)\left(1-\Omega^2\right)+\frac{9k_r^2}{16}Y^3+4\zeta^2 Y-1=0 \quad (5.213)$$

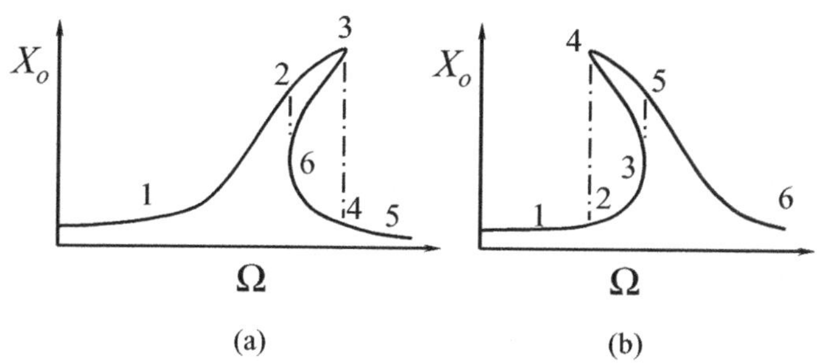

Figure 5.39 Jump phenomena of a nonlinear oscillator.

which leads to

$$1-\Omega^2 = \frac{1}{2Y}\left[-\left(\frac{3k_r}{2}Y^2 - 4\zeta^2 Y\right) \pm \sqrt{D_1}\right] \tag{5.214}$$

where D_1 is the discriminant of the quadratic equation, which is given by

$$D_1 = \left(\frac{3k_r}{2}Y^2 - 4\zeta^2 Y\right)^2 - 4Y\left(\frac{9k_r^2}{16}Y^3 + 4\zeta^2 Y - 1\right) \tag{5.215}$$

If D_1 becomes zero, (5.214) will give one solution, which corresponds to the maxima, labeled o, p, and q in Fig. 5.38. Setting $D_1 = 0$ and rearranging terms gives an equation for the maxima as

$$-4Y_m\left[3k_r\zeta^2 Y_m^2 + 4\zeta^2(1-\zeta^2)Y_m - 1\right] = 0$$

where Y_m denotes the dimensionless maximum amplitude squared. The preceding equation yields two distinct roots:

$$Y_m = \frac{2}{3}\frac{1-\zeta^2}{k_r}\left[-1+\sqrt{1+\frac{3k_r}{4\zeta^2(1-\zeta^2)}}\right]$$

$$Y_{m,u} = \frac{2}{3}\frac{1-\zeta^2}{k_r}\left[-1-\sqrt{1+\frac{3k_r}{4\zeta^2(1-\zeta^2)}}\right]$$

which are the dimensionless maximum amplitudes squared, corresponding to the lower and upper graphs in Fig. 5.38, respectively. The dimensionless maximum amplitude is expressed as a function of the damping factor ζ and the nonlinearity factor k_r:

$$X_{0m} = \sqrt{Y_m} = \sqrt{\frac{2}{3}\frac{1-\zeta^2}{k_r}\left[-1+\sqrt{1+\frac{3k_r}{4\zeta^2(1-\zeta^2)}}\right]} \tag{5.216}$$

$$X_{0m,u} = \sqrt{Y_{m,u}} = \sqrt{\frac{2}{3}\frac{1-\zeta^2}{k_r}\left[-1-\sqrt{1+\frac{3k_r}{4\zeta^2(1-\zeta^2)}}\right]} \tag{5.217}$$

Experimental data show that the lower amplitude is observed, so we takes X_{0m} as the maximum dimensionless amplitude, which corresponds to o, p, and q in Fig. 5.38. Substituting Y_m and $Y_{m,u}$ into (5.214) (note $D_1 = 0$) gives the frequency ratios

$$\Omega_m = \sqrt{1-2\zeta^2 + \frac{3}{4}k_r Y_m} \tag{5.218}$$

$$\Omega_{m,u} = \sqrt{1 - 2\zeta^2 + \frac{3}{4} k_r Y_{m,u}} \tag{5.219}$$

which denote the frequency ratio at the maximum amplitude, labeled o, p, and q in Fig. 5.38, and the frequency ratio corresponding to $X_{0m,u}$, respectively. Substituting Y_m and $Y_{m,u}$ into (5.218) gives the frequency ratio as a function of ζ and k_r:

$$\left(\frac{f}{f_n}\right)_m = \left(\frac{\omega}{\omega}\right)_m = \Omega_m = \sqrt{1 - 2\zeta^2 - \frac{1-\zeta^2}{2}\left[1 - \sqrt{1 + \frac{3k_r}{4\zeta^2(1-\zeta^2)}}\right]} \tag{5.220}$$

$$\left(\frac{f}{f_n}\right)_{m,u} = \left(\frac{\omega}{\omega}\right)_{m,u} = \Omega_{m,u} = \sqrt{1 - 2\zeta^2 - \frac{1-\zeta^2}{2}\left[1 + \sqrt{1 + \frac{3k_r}{4\zeta^2(1-\zeta^2)}}\right]} \tag{5.221}$$

Experimental response as well as data from numerical analyses show that the lower amplitude in Fig. 5.38 is achieved, so we consider (5.216) and (5.220) as the maximum amplitude and the corresponding frequency, respectively. It is noted that the preceding maximum amplitudes and corresponding frequency ratios are exact solutions of the steady-state amplitude equation (5.204). If the system represented by Duffing's equation is slightly damped (i.e., $\zeta << 1$), the preceding amplitude and frequency ratio are expressed as

$$X_{0m} \cong \sqrt{\frac{2}{3k_r}\left(-1 + \sqrt{1 + \frac{3k_r}{4\zeta^2}}\right)} \tag{5.222}$$

$$\left(\frac{f}{f_n}\right)_m \cong \sqrt{1 - 2\zeta^2 - \frac{1}{2}\left(1 - \sqrt{1 + \frac{3k_r}{4\zeta^2}}\right)} \tag{5.223}$$

If a condition, $3|k_r|/4\zeta^2 << 1$, is satisfied, equations (5.222) and (5.223) are further reduced to

$$X_{0m} \cong \frac{1}{2\zeta} \tag{5.224}$$

$$\left(\frac{f}{f_n}\right)_m \cong \sqrt{1 - 2\zeta^2 + \frac{3k_r}{16\zeta^2}} \tag{5.225}$$

Note from equations (5.224) and (5.225) expressed under the assumptions $k_1 > 0, \zeta << 1$, and $3|k_r|/4\zeta^2 << 1$, that (5.224) is equal to (5.95) for the maximum amplitude (magnification factor) of the slightly damped linear system; that is, if a resonator satisfies the conditions $k_1 > 0$, $\zeta << 1$, and $3|k_r|/4\zeta^2 << 1$, the resonator can be considered as a linear resonator, whose response is described in Section 5.6. Since the nonlinearity factor k_r is not a constant but is a function of the linear and cubic stiffnesses and the force amplitude, as shown in the seventh equation of (5.201), the resonant frequency ratio $(f/f_n)_m$ at the maximum amplitude depends on the force amplitude F_0: For a hard spring system, $k_1 > 0$ and $k_3 > 0$, the frequency at resonance increases with the force amplitude F_0; for a soft spring system, $k_1 > 0$ and $k_3 < 0$, the frequency decreases as the force amplitude increases.

Vibrating Systems with Negative Linear Stiffness and Positive Cubic Stiffness

Earlier in this section we derived a steady-state response of a vibrating system that has a positive linear stiffness and either positive or negative cubic stiffness. As discussed in Section 4.7, since a system with negative linear and cubic stiffnesses is unstable, we do not describe this system further. We can consider one more case: a system with a negative linear stiffness and a positive cubic stiffness. If the system subjected to a constant force F_b is at a stable position x_d in equilibrium as shown in Fig. 5.40a, the force equilibrium equation is given by setting the restoring force (4.167) equal to zero,

$$k_1 x_d + k_3 x_d^3 - F_b = 0 \tag{5.226}$$

where k_1 is negative. The stable position x_d is obtained in closed form, which was described in Section 5.1. When the mass is displaced by δ from the stable position, the restoring force is given by (4.167), which is rewritten as $F_r = k_1 x + k_3 x^3 - F_b$. The restoring force and corresponding potential energy are shown in Fig. 5.40a and b, respectively. Applying Newton's second law to the mass subjected to the restoring force and the damping force, which is assumed to be proportional to the velocity of the mass, we have the following equation of motion of the system:

$$m\ddot{x} + c\dot{x} + k_1 x + k_3 x^3 = F_b + F_0 \cos \omega t_1 \tag{5.227}$$

where the first and second terms on the right-hand side of equation (5.227) denote the constant force and a sinusoidal force. For $k_1 < 0$, the mass vibrates about its stable position x_d in Fig. 5.40, and $x = x_d + \delta$ is substituted into equation (5.227) to obtain

$$m\ddot{\delta} + c\dot{\delta} + k_1 (x_d + \delta) + k_3 (x_d + \delta)^3 = F_b + F_0 \cos \omega t_1$$

Subtracting the equilibrium equation (5.226) from the equation above, which is expanded in terms of x_d and δ, we have the following equation of motion of the system:

$$m\ddot{\delta} + c\dot{\delta} + k_l \delta + k_q \delta^2 + k_c \delta^3 = F_0 \cos \omega t_1 \tag{5.228}$$

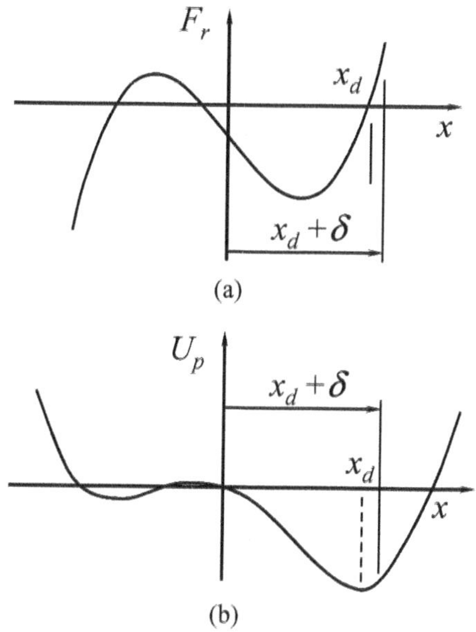

Figure 5.40 Restoring force and potential energy of a system with a negative linear stiffness.

where k_l, k_q, and k_c denote the linear, quadratic, and cubic stiffnesses at the stable positions and were discussed in Sec. 4.7. The stiffnesses are expressed as follows:

$$\begin{aligned} k_l &= k_1 + 3k_3x_d^2 \\ k_q &= 3k_3x_d \\ k_c &= k_3 \end{aligned} \tag{5.229}$$

It is noted that the differential equation above includes a quadratic stiffness which was not involved in (5.195). If the dynamic response δ is much smaller than the stable position x_d and if the linear force $k_l\delta$ is much larger than the quadratic and cubic forces, $kq\delta^2$ and $k_c\delta^3$, the preceding equation of motion can be expressed as

$$m\ddot{\delta} + c\dot{\delta} + k_l\delta = F_0 \cos \omega t_1 \tag{5.230}$$

The solution to equation (5.230) was obtained in Section 5.6, given by (5.89). The natural frequency and the damping of a lightly damped system may be expressed from equation (5.230) as follows:

$$f_n = \frac{1}{2\pi}\sqrt{\frac{k_l}{m}} \qquad \zeta = \frac{c}{2\sqrt{mk_l}} \tag{5.231}$$

and the quality factor is given by $Q = 1/2\zeta = \sqrt{mk_l}/c$. It is noted that even though the original linear stiffness k_1 is negative, the linear stiffness k_l at the equilibrium is positive, as given by the first equation of (5.229). For example, if the constant force F_b is zero, substituting the equilibrium position $x_d = \pm\sqrt{-k_1/k_3}$, which is given by (4.145), into (5.229) yields the stiffnesses, (4.153), which are rewritten for convenience:

$$\begin{aligned} k_l &= -2k_1 \\ k_q &= \pm 3(-k_1 k_3)^{1/2} \\ k_c &= k_3 \end{aligned} \tag{5.232}$$

where $-k_1$ is positive since k_1 is negative. The natural frequency (5.231) for $F_b = 0$ is reduced to

$$f_n = \frac{1}{2\pi}\sqrt{\frac{-2k_1}{m}} \qquad \zeta = \frac{c}{2\sqrt{-2mk_1}}$$

The equation of motion (5.230) is an approximation of (5.228), so its solution represents an approximation of the dynamic behavior of the system. As a result, the preceding natural frequency and the quality factor are approximate values. We examine this situation graphically using Fig. 5.40b. Since the potential energy graph in Fig. 5.40b is not symmetric about the vertical line passing through x_d, the vibration amplitude of the mass excited by a force becomes larger and larger when the mass has higher potential energy. If the total energy of the system reaches an energy sufficient to overcome the energy barrier formed between the two wells shown in Fig. 5.40b, the mass travels over the two wells and its travel range becomes wider. The motion of the mass traveling through two energy wells relates to a chaotic motion, so it is difficult to obtain a closed-form solution for the motion. However, if the mass oscillates in one of the wells in Fig. 5.40b with an amplitude much smaller than the width of the well, and equation (5.230) can give an approximate solution, the solution δ from (5.228) tends to be larger than the solution of (5.230) and the natural frequency become lower than that of (5.231). If the magnitude of the constant force F_b is larger than that of the critical force, defined in (5.20), one of the wells disappears and the stable position x_d can be estimated by (5.21). In this section we have described Duffing's equation, which represents the dynamic behavior of a system subjected to a cubic restoring force, and derived analytic solutions for steady-state response in closed form. To become familiar with the solutions derived, a dynamic response will be obtained in the following example.

Example 5.5 A typical configuration of a MEMS resonator with four straight beams is shown in Fig. 5.41. Slender straight beams of flexural rigidity EI and length l are subjected to an axial force (the tension is defined as positive) which is adjusted by Joule heating (which is dealt with in Chapter 9). We have the following data for the resonator: $EI = 2 \times 10^{-13}\,\mathrm{N \cdot m^2}$, Young's modulus $E = 150\,\mathrm{GPa}$, $l = 1 \times 10^{-4}\,\mathrm{m}$, cross-sectional area of the beam $A = 4 \times 10^{-12}\,\mathrm{m^2}$, mass $M = 1 \times 10^{-11}\,\mathrm{kg}$, and damping coefficient $c = 2 \times 10^{-7}\,\mathrm{N \cdot s/m}$. Assuming that the axial force $N_0 = 0$, we wish to find the vibration amplitude x_0 and the frequency at resonance when a sinusoidal force of force amplitude, F_0, of $1 \times 10^{-7}\,\mathrm{N}$ and $5 \times 10^{-7}\,\mathrm{N}$ is applied to the mass, respectively. Under the assumption $N_0 = -3 \times 10^{-4}\,\mathrm{N}$ and $F_0 = 1 \times 10^{-7}\,\mathrm{N}$, we shall find the stable position of the mass and the natural frequency for low vibration at the stable position.

The linear and cubic stiffnesses of the beam are given by equation (n) and (o) of Example 3.3, written

$$k_1 = 12\frac{EI}{l^3} + \frac{6}{5}\frac{N_0}{l} \tag{a}$$

$$k_3 = \frac{18}{25}\frac{EA}{l^3} \tag{b}$$

We assume that the spring beam mass is negligible. Substituting $N_0 = 0$ and the other data into equations (a) and (b) yields the stiffnesses of the four beams: $k_1 = 9.6\,\mathrm{N/m}$ and $k_3 = 1.73 \times 10^{12}\,\mathrm{N/m^3}$. The damping factor is obtained as $\zeta = c/2\sqrt{mk_1} = 0.0102$, and the nonlinearity factor is estimated as $k_r = k_3F_0^2/k_1^3 = 1.953 \times 10^{-5}$. Substituting the damping and nonlinearity factors into (5.220) and (5.216), we find the frequency ratio Ω_m and the dimensionless maximum amplitude X_{0m}: $\Omega_m = 1.017$, and $X_{0m} = 48.18$, which gives the maximum vibration amplitude $x_0 = X_{0m}F_0/k_1 = 5.02 \times 10^{-7}\,\mathrm{m}$ and the corresponding frequency $f_m = \Omega_m\sqrt{k_1/m}/2\pi = 1.59 \times 10^5\,\mathrm{Hz}$. The frequency at

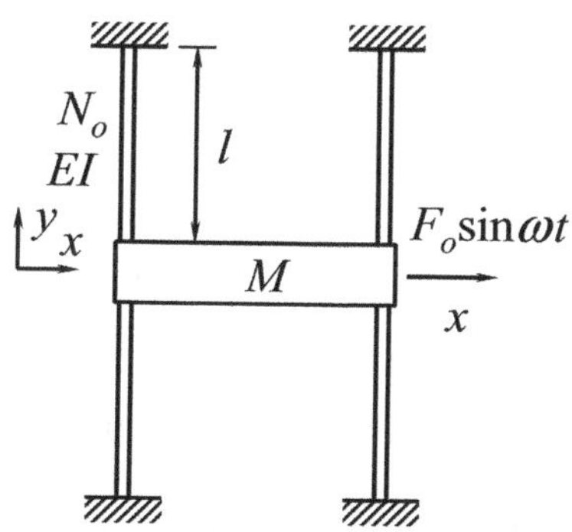

Figure 5.41 Typical resonator with straight beams.

resonance is close to the natural frequency $f_n = \sqrt{k_1/m}/2\pi = 1.56\times10^5$ Hz. As the force amplitude increases to 5×10^{-7} N, repeating the calculation procedure above gives the following: $k_r = 4.883 \times 10^{-4}$, $X_{0m} = 39.19$, and $\Omega_m = 1.250$. The amplitude x_0 and the frequency f_m are given by 2.04×10^{-6} m and 1.95×10^5 Hz. As shown in this calculation, the vibration amplitude and natural frequency increases with the force amplitude of the sinusoidal signal. When the axial force is set as $N_0 = -3 \times 10^{-4}$ N and the force amplitude is $F_0 = 1 \times 10^{-7}$ N, (a) and (b) yield $k_1 = -4.8$ N/m and $k_3 = 1.73 \times 10^{12}$ N/m^3. Since $k_1 < 0$, the straight beams are buckled and find stable positions that are given by (5.51) with the constant force $F_b = 0$, $x_m = \pm\sqrt{-k_1/k_3} = \pm 1.67\times10^{-6}$ m. At the stable positions, the linear stiffness is obtained from the first equation of (5.232), $k_l = -2k_1 = 9.6$ N/m, which leads to the natural frequency for small-amplitude vibration, $f_n = 1.56 \times 10^5$ Hz.

PROBLEMS

5.1 Find the roots of the following cubic equations: (a) $x^3 - x^2 + x - 1 = 0$; (b) $x^3 - x^2 + 2x - 2 = 0$; (c) $x^3 - x^2 + G = 0$.

5.2 A man walks along a guide at a constant velocity v on a disk in the vertical direction in Fig. P5.2 while the disk rotates at a constant angular velocity ω. Find the inertial force acting on the man at the position shown.

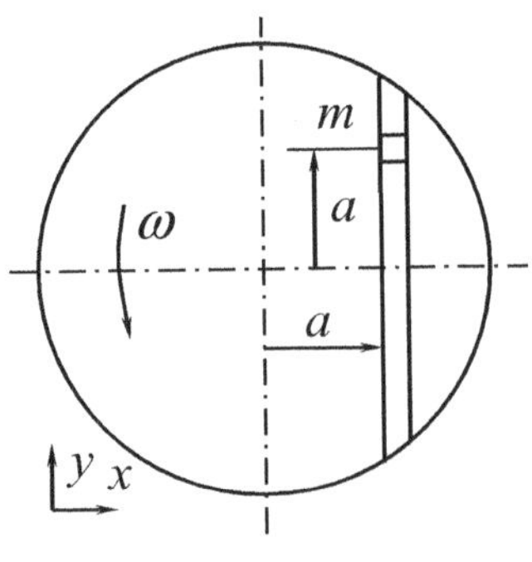

Figure P5.2

5.3 A small ball of mass m subjected to the gravitational force is in equilibrium on a rigid bowl whose wall is defined by a function $y = -x^3 + x$ (Fig. P5.3). Determine the equilibrium position of the ball. If we displace the mass by a distance leftward from the equilibrium position and release it, the mass will oscillate in the well. It is assumed that the kinetic energy due to the rotation of the ball and the friction and damping forces are negligible. Determine the initial displacement to confine the ball within the well.

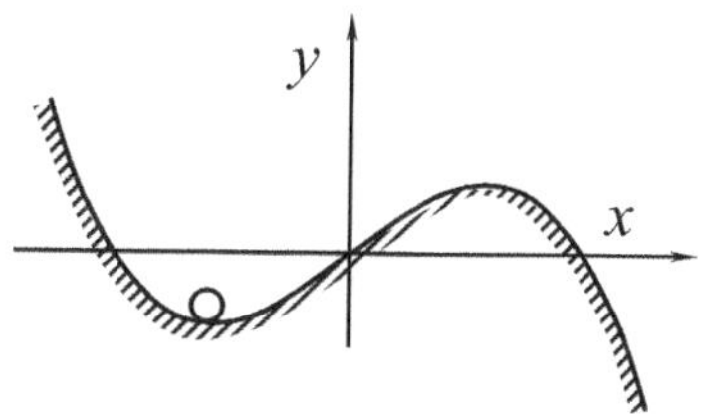

Figure P5.3

5.4 The equation of motion of a system is given by $\dot{x}+x=\sin\omega t$. With an initial condition $x=0$, find the transient and steady-state responses.

5.5 Determine the steady-state response of the differential equation $\ddot{x}+\dot{x}/10+x=\sin t+\cos t$.

5.6 Referring to the mass–spring system in Fig. P5.6 obtain equations for the natural frequencies and mode shapes.

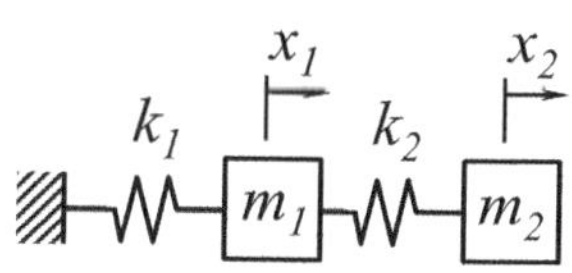

Figure P5.6

5.7 A plate of mass m, supported by a slender straight beam as shown in the Fig. P5.7, vibrates while sinusoidal force is applied to the center of the plate. Find the natural frequencies as a function of the mass m, flexural rigidity EI of the beam, and geometry.

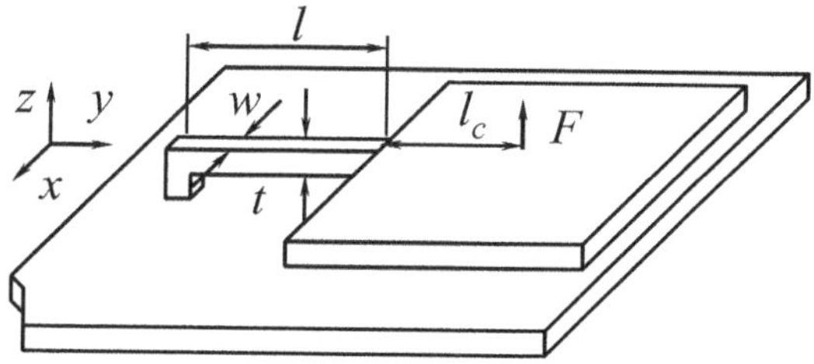

Figure P5.7

5.8 The microstructure shown in Fig. P5.8 vibrates in the horizontal direction. Beam b is considered rigid, and identical beams of mass m, length a, and flexural rigidity EI are used. Construct an equivalent model for a one-DOF system and find the natural frequency, as the vibration amplitude is much less than

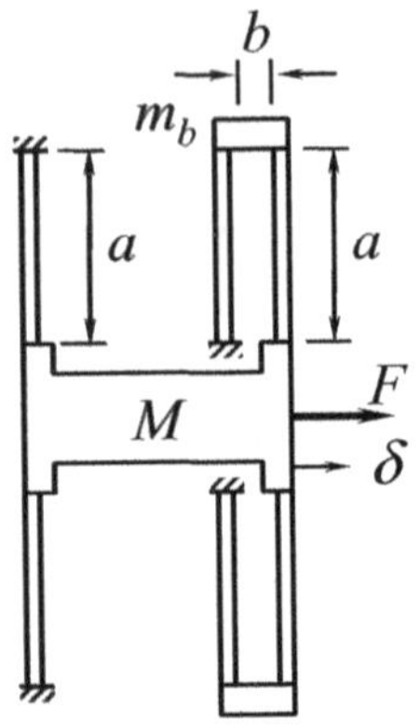

Figure P5.8

the thickness of the beam. An engineer wishes to use the same structure for a linear actuator that requires a displacement greater than three times the beam width. Define the potential problem that arises from the design, and suggest a solution to the problem.

5.9 Identical masses and springs are connected in series as shown in Fig. P5.9. Find an expression for the natural frequencies of the system. Evaluate the lowest natural frequency for $k = 1\,\text{N/m}$ and $m = 1 \times 10^{-10}\,\text{kg}$.

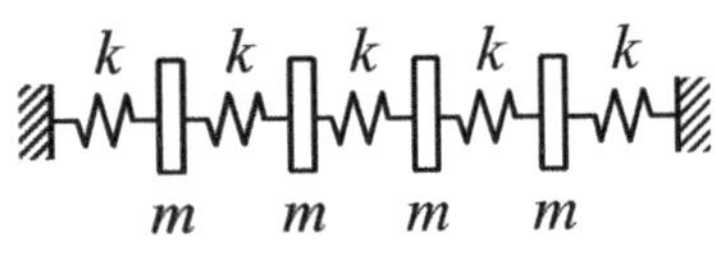

Figure P5.9

CHAPTER 6

FLUID DYNAMICS

As discussed in Chapter 5, the dynamic response of MEMS is affected by energy dissipation involved in the system (i.e., viscous damping due to surrounding gases or structural damping dissipates the mechanical energy into heat). If microstructures operate in high vacuum where the energy dissipation due to fluid is much less than that due to structural damping, the viscous damping may be neglected. However, if the pressure of surrounding gas is not extremely low, the damping force of the gas must be taken into account. In this chapter we describe governing equations for gas and liquid flows and derive a damping coefficient from the governing equations.

6.1 VISCOUS FLOW

Let us begin with a simple problem, illustrated in Fig. 6.1, to understand the viscous flow of fluid (liquid and gas). In Fig. 6.1, a plate of area A is placed on top of an oil film of thickness h, and a force F is applied to move the plate at a constant velocity of U in the x direction. Experiments show that the driving force F is proportional to the velocity and the area and inversely proportional to the thickness; mathematically, this relation may be expressed by $F \propto UA/h$ or, more conveniently, the expression is written as

$$F = \mu \frac{U}{h} A \tag{6.1}$$

Principles of Microelectromechanical Systems, By Ki Bang Lee

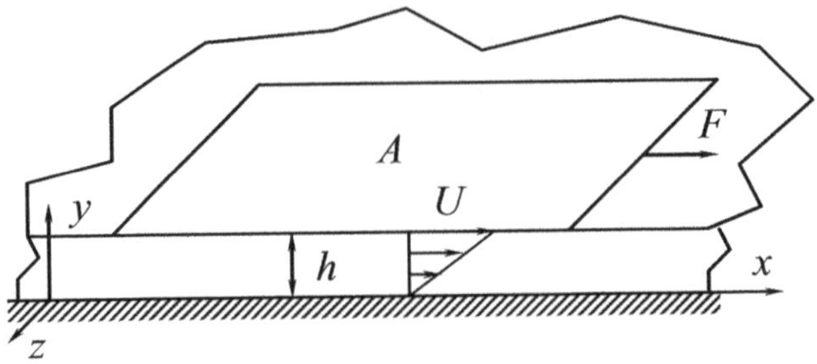

Figure 6.1 Viscous flow due to a moving plate on an oil film.

where μ represents the viscosity, which is a property of fluid. The shear stress acting on the plate is then given by

$$\tau = \frac{F}{A} = \mu \frac{U}{h}$$

which was given by Isaac Newton. If a fluid obeys the relation above, the fluid is called a *Newtonian fluid*. The equation above may be expressed in a general form to give the shear stress τ in the oil film,

$$\tau = \mu \frac{\partial u}{\partial y} \tag{6.2}$$

where u and y denote the velocity of the oil and the coordinate, as defined in Fig. 6.1. The drag force or damping force F_d is equal to the magnitude of the driving force but is directed in the opposite direction. The damping coefficient is then defined as

$$c = \frac{F_d}{U} = \frac{\mu A}{h} \tag{6.3}$$

To further understand the damping force due to fluid flowing around a microstructure, it is instructive to derive the governing equations of fluid: the continuity equation and the Navier–Stokes equation (momentum equation).

6.2 CONTINUITY EQUATION

While fluids flow, mass must be conserved. Let us consider the fluid in Fig. 6.2, which may be taken from the oil film in Fig. 6.1. The small volume $dx\ dy\ dz$ is fixed relative to a fixed coordinate system xyz and mass flows to and from the small volume as shown in Fig. 6.2, where ρ, u, v, w, and Δt denote the density of the fluid, velocities in the x, y, and z directions, and time increment. When

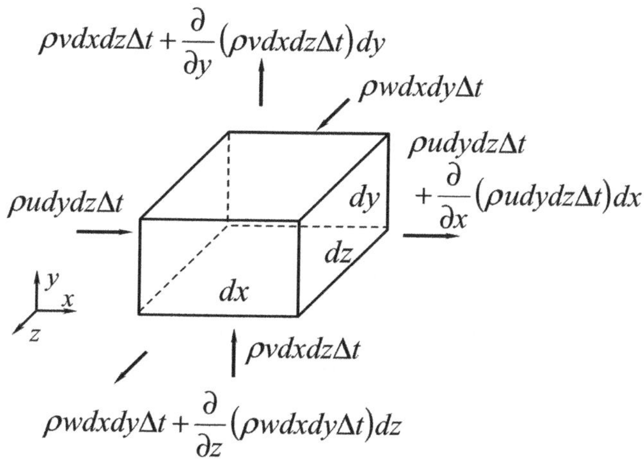

Figure 6.2 Mass flow through a small volume.

we consider the mass balance of the volume during time Δt, we obtain the following mass balance equation:

$$\left(\frac{\partial \rho u}{\partial x} + \frac{\partial \rho v}{\partial y} + \frac{\partial \rho w}{\partial z}\right) dx\, dy\, dz\, \Delta t = -\Delta(\rho\, dx\, dy\, dz)$$

where the negative sign indicates that the positive net mass flow-out causes decreases in the mass in the volume. Dividing the preceding equation by dx dy dz Δt, we have the continuity equation for compressible fluid:

$$\frac{\partial \rho u}{\partial x} + \frac{\partial \rho v}{\partial y} + \frac{\partial \rho w}{\partial z} = -\frac{\partial \rho}{\partial t} \tag{6.4}$$

This equation may be expressed in the simple form

$$\nabla \cdot \rho \mathbf{u} = -\frac{\partial \rho}{\partial t} \tag{6.5}$$

where

$$\nabla = \frac{\partial}{\partial x}\mathbf{i} + \frac{\partial}{\partial y}\mathbf{j} + \frac{\partial}{\partial z}\mathbf{k}$$

is the gradient operator and the vector $\mathbf{u} = u\mathbf{i} + v\mathbf{j} + w\mathbf{k}$ is the velocity vector, whose components are u, v, and w in the x, y, and z directions. If the density ρ is constant, the continuity equation is reduced to

$$\frac{\partial u}{\partial x}+\frac{\partial \upsilon}{\partial y}+\frac{\partial w}{\partial z}=0$$

or

$$\nabla \cdot \mathbf{u}=0 \tag{6.6}$$

6.3 NAVIER–STOKES EQUATION

Because the mass in the volume of Fig. 6.2 is accelerated or decelerated, Newton's second law applies to the flowing mass. Before proceeding with the dynamic problem, we first consider a derivative that represents a time derivative of property n carried by fluid. The property n depends on time t and the coordinates x, y, and z [i.e., $n = f(t,x,y,z)$]. Let dn, dt, dx, dy, and dz represent changes in the variables. The total differential of n is written

$$dn=\frac{\partial n}{\partial t}dt+\frac{\partial n}{\partial x}dx+\frac{\partial n}{\partial y}dy+\frac{\partial n}{\partial z}dz$$

and dividing the above by dt and noting the definition of velocities ($u = dx/dt$, $\upsilon = dy/dt$, and $w = dz/dt$) gives

$$\frac{dn}{dt}=\frac{\partial n}{\partial t}+u\frac{\partial n}{\partial x}+\upsilon\frac{\partial n}{\partial y}+w\frac{\partial n}{\partial z}$$

This is known as the *material derivative*. On the right-hand side of the preceding equation, the first term is called the *local derivative* and the others are called *convective derivatives* since they appear if the fluid velocities and spatial derivative of property both exist. The following notation is conventionally used for the derivative:

$$\begin{aligned}\frac{D}{Dt}&=\frac{\partial}{\partial t}+u\frac{\partial}{\partial x}+\upsilon\frac{\partial}{\partial y}+w\frac{\partial}{\partial z}\\&=\frac{\partial}{\partial t}+\mathbf{u}\cdot\nabla\end{aligned} \tag{6.7}$$

The material derivative is then expressed simply as follows:

$$\frac{Dn}{Dt}=\frac{\partial n}{\partial t}+\mathbf{u}\cdot\nabla n \tag{6.8}$$

Let us apply Newton's second law to the mass in the volume $dx\,dy\,dz$, which is fixed with respect to the coordinate system xyz in Fig. 6.3. Noting the momentum $m_x = u\,dm = \rho\,dx\,dy\,dz\,u$ in the x direction and applying Newton's second law to mass in the volume gives

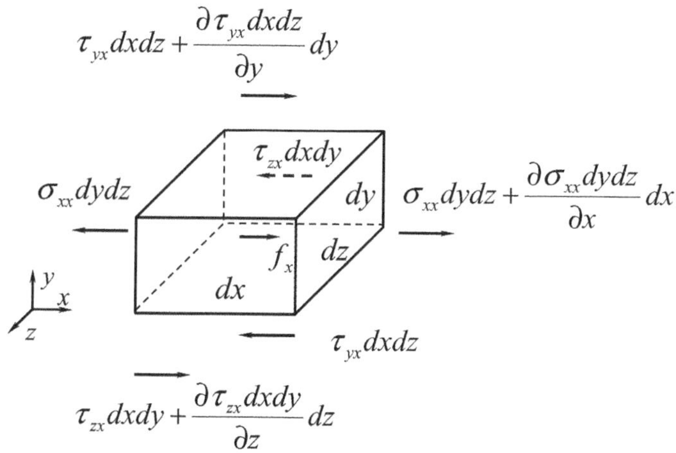

Figure 6.3 Forces acting on a fluid element.

$$\frac{Dm_x}{Dt} = \sum F_x$$

where the right-hand side represents the sum of x-component forces acting on the volume. In fluid dynamics, if the velocity of fluid is less than 0.3 times that of sound in the fluid, the fluid is considered incompressible. Assuming the fluid to be incompressible, referring to the forces in Fig. 6.3 and dividing the equation above by the volume $dx\ dy\ dz$, we have

$$\rho\frac{Du}{Dt} = \frac{\partial \sigma_{xx}}{\partial x} + \frac{\partial \tau_{xy}}{\partial y} + \frac{\partial \tau_{xz}}{\partial z} + f_x \tag{6.9}$$

where σ_{xx}, τ_{xy}, τ_{xz}, and f_x denote the normal stress, shear stresses, and body force, respectively. For the derivation above, the shear stress symmetry, $\tau_{yx} = \tau_{xy}$ and $\tau_{zx} = \tau_{xz}$, is used. In order to express equation (6.9) as a function of velocities, we need relations between the stresses and velocities. Stokes presented the following relations, which are general forms of the stress–velocity relation (6.2):

$$\begin{aligned}
\sigma_{xx} &= -p + 2\mu\frac{\partial u}{\partial x} \\
\tau_{xy} &= \mu\left(\frac{\partial u}{\partial y} + \frac{\partial v}{\partial x}\right) \\
\tau_{xz} &= \mu\left(\frac{\partial u}{\partial z} + \frac{\partial w}{\partial x}\right)
\end{aligned} \tag{6.10}$$

where p denotes the hydrostatic pressure. The original relations include the compressibility effect, but the preceding expressions are written for incompressible fluid (without the effect). Substituting the foregoing equations into (6.9) and recalling the continuity equation (6.6), we have the *Navier–Stokes equation* (also called the *momentum equation*) in the x direction:

$$
\begin{aligned}
\rho\frac{Du}{Dt} &= \frac{\partial}{\partial x}\left(-p+2\mu\frac{\partial u}{\partial x}\right)+\frac{\partial}{\partial y}\left[\mu\left(\frac{\partial u}{\partial y}+\frac{\partial v}{\partial x}\right)\right]+\frac{\partial}{\partial z}\left[\mu\left(\frac{\partial u}{\partial z}+\frac{\partial w}{\partial x}\right)\right]+f_x \\
&= -\frac{\partial p}{\partial x}+\mu\left(\frac{\partial^2 u}{\partial x^2}+\frac{\partial^2 u}{\partial y^2}+\frac{\partial^2 u}{\partial z^2}\right)+\mu\frac{\partial}{\partial x}\left(\frac{\partial u}{\partial x}+\frac{\partial v}{\partial y}+\frac{\partial w}{\partial z}\right)+f_x \\
&= -\frac{\partial p}{\partial x}+\mu\left(\frac{\partial^2 u}{\partial x^2}+\frac{\partial^2 u}{\partial y^2}+\frac{\partial^2 u}{\partial z^2}\right)+f_x
\end{aligned}
$$

In the equation above, the quantity in the second set of parentheses in the second line vanishes due to the continuity equation (6.6). Similarly, the Navier–Stokes equations in the other directions are obtained. A full set of Navier–Stokes equations for incompressible fluid is given in the Cartesian coordinate system (rectangular coordinate system) as follows:

$$
\begin{aligned}
\rho\left(\frac{\partial u}{\partial t}+u\frac{\partial u}{\partial x}+v\frac{\partial u}{\partial y}+w\frac{\partial u}{\partial z}\right) &= -\frac{\partial p}{\partial x}+\mu\left(\frac{\partial^2 u}{\partial x^2}+\frac{\partial^2 u}{\partial y^2}+\frac{\partial^2 u}{\partial z^2}\right)+f_x \\
\rho\left(\frac{\partial v}{\partial t}+u\frac{\partial v}{\partial x}+v\frac{\partial v}{\partial y}+w\frac{\partial v}{\partial z}\right) &= -\frac{\partial p}{\partial y}+\mu\left(\frac{\partial^2 v}{\partial x^2}+\frac{\partial^2 v}{\partial y^2}+\frac{\partial^2 v}{\partial z^2}\right)+f_y \qquad (6.11) \\
\rho\left(\frac{\partial w}{\partial t}+u\frac{\partial w}{\partial x}+v\frac{\partial w}{\partial y}+w\frac{\partial w}{\partial z}\right) &= -\frac{\partial p}{\partial z}+\mu\left(\frac{\partial^2 w}{\partial x^2}+\frac{\partial^2 w}{\partial y^2}+\frac{\partial^2 w}{\partial z^2}\right)+f_z
\end{aligned}
$$

If we multiply the equations above by the unit vectors **i**, **j**, and **k**, respectively, and add the resulting equations, we obtain the Navier–Stokes equation in a simple form,

$$\rho\frac{D\mathbf{u}}{Dt} = -\nabla p+\mu\nabla^2\mathbf{u}+\mathbf{f} \tag{6.12}$$

where ∇^2 and $\mathbf{f}$ represent the Laplacian (Laplace operator) defined by $\nabla^2 = \nabla\cdot\nabla = \partial^2/\partial x^2 + \partial^2/\partial y^2 + \partial^2/\partial z^2$ and the body force given by $\mathbf{f} = f_x\mathbf{i} + f_y\mathbf{j} + f_z\mathbf{k}$, respectively. In the microscale world, the body force is negligible, as discussed in Section 1.4, and we shall then use equation (6.12) without the body force term for further consideration. For the continuity and Navier–Stokes equations in different coordinate systems (e.g., a cylindrical coordinate system), the preceding continuity and Navier–Stokes equations in the rectangular system can be transformed into those in the different coordinate systems. If readers need the governing equations in different coordinate systems, they may refer to such textbooks as that of White (1974).

6.4 REYNOLDS EQUATION

The Reynolds equation is derived from the continuity and Navier–Stokes equations for thin fluid films which are squeezed between moving and stationary plates. To derive the Reynolds equation, the following are assumed:

1. The fluid is Newtonian [i.e., the fluid follows the relation (6.2)].
2. The thickness of the fluid film between plates is much less than the dimensions of plates (i.e., the length and width of the rectangular plate).
3. The pressure and density across the fluid film are considered to be constant.
4. The fluid inertia is negligible since the viscous force of the thin fluid film is much greater than the inertia force.
5. The body force (e.g., gravitational force) is neglected because the viscous force is dominant in the fluid film.
6. There is no slip at the fluid boundaries. In other words, the fluid velocity in contact with the plate is equal to that of the plate.

Assumption 6 will be changed later to modify the Reynolds equation. The Reynolds equation will be derived under the foregoing assumptions. If fluid problems do not satisfy assumptions 1 through 6, we cannot use Reynolds equation to solve the problems. Let us consider a fluid film, shown in Fig. 6.4, whose thickness is much less than the dimensions of plates (e.g., the length and width). In the figure, h and dh/dt denote the film thickness between plates and the velocity of the movable plate, respectively. Mass flows (denoted by $\dot{m}_x$ and $\dot{m}_y$) in the x and y directions are caused by pressure gradients and plate velocity. For a small column of the fluid in Fig. 6.4, the continuity equation may be written

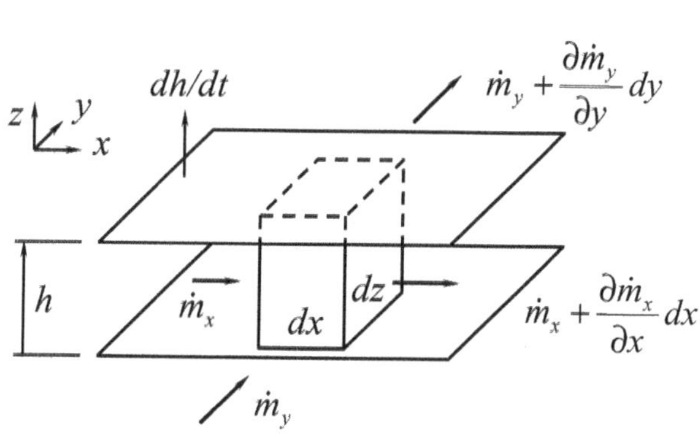

Figure 6.4 Mass flow through a small column between parallel plates.

$$\frac{\partial \dot{m}_x}{\partial x}dx + \frac{\partial \dot{m}_y}{\partial y}dy = -\frac{\partial m}{\partial t}$$

where m is the mass of the column, defined by $\rho\ dx\ dy\ h$, and the mass flows in the x and y directions are represented by

$$\dot{m}_x = \int_0^h \rho u\, dz\, dy = \rho\, dy \int_0^h u\, dz$$

$$\dot{m}_y = \int_0^h \rho \upsilon\, dx\, dz = \rho\, dx \int_0^h \upsilon\, dz$$

where u and υ are the fluid velocities in the x and y directions. Substituting the mass and mass flows into the foregoing continuity equation yields the continuity equation as a function of fluid velocities,

$$\frac{\partial}{\partial x}\left(\rho \int_0^h u\, dz\right) + \frac{\partial}{\partial z}\left(\rho \int_0^h \upsilon\, dz\right) = -\frac{\partial \rho h}{\partial t} \tag{6.13}$$

Substituting the velocities into equation (6.13), we shall obtain the Reynolds equation. The velocities can be obtained from the force balance of a fluid element shown in Fig. 6.5. Since the inertial force is neglected, the force equilibrium in the x direction gives

$$0 = \left(-\frac{\partial p}{\partial x} + \frac{\partial \tau}{\partial z}\right) dx\, dy\, dz$$

and substitution of $\tau = \mu\ \partial u/\partial z$ into the preceding equation leads to

$$\mu \frac{\partial^2 u}{\partial z^2} = \frac{\partial p}{\partial x} \tag{6.14}$$

Integrating twice over z and using the boundary conditions $u = 0$ at $z = 0$ and $z = h$, we find the velocity u as follows:

$$u = \frac{1}{2\mu}\frac{\partial p}{\partial x}(z^2 - hz) \tag{6.15}$$

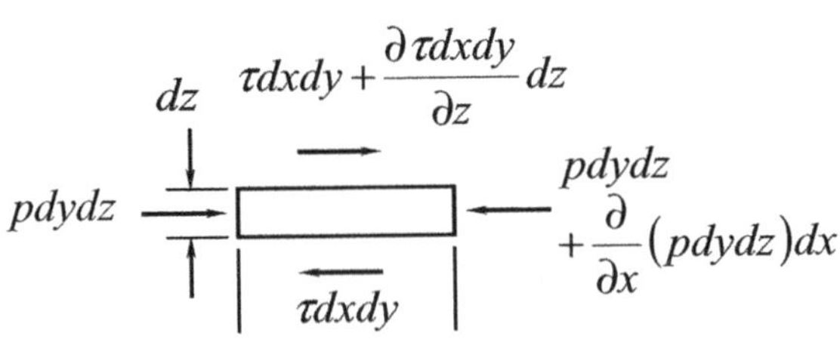

Figure 6.5 Forces acting on a small element.

Similarly, the velocity v is given by

$$v = \frac{1}{2\mu}\frac{\partial p}{\partial y}(z^2 - hz) \tag{6.16}$$

Integrating the preceding velocities from 0 to h gives the volumetric flow of the fluid per unit length in the x and y directions,

$$q_x = \int_0^h u\,dz = -\frac{1}{12\mu}\frac{\partial p}{\partial x}h^3$$

$$q_y = \int_0^h v\,dz = -\frac{1}{12\mu}\frac{\partial p}{\partial y}h^3$$

Substituting the volumetric flows above into the continuity equation (6.13) gives the Reynolds equation for compressible fluids:

$$\frac{\partial}{\partial x}\left(\frac{\rho}{\mu}\frac{\partial p}{\partial x}h^3\right) + \frac{\partial}{\partial y}\left(\frac{\rho}{\mu}\frac{\partial p}{\partial y}h^3\right) = 12\frac{\partial \rho h}{\partial t} \tag{6.17}$$

Assuming that $\rho = \text{constant}$ and $\mu = \text{constant}$, equation (6.17) becomes the Reynolds equation for incompressible fluids:

$$\frac{\partial}{\partial x}\left(h^3\frac{\partial p}{\partial x}\right) + \frac{\partial}{\partial y}\left(h^3\frac{\partial p}{\partial y}\right) = 12\mu\frac{\partial h}{\partial t} \tag{6.18}$$

If the film thickness h is uniform in the x and y directions and the film temperature is constant (i.e., $p/\rho = \text{constant}$ for ideal gas), we obtain a Reynolds equation for a compressible thin film of uniform thickness,

$$\frac{\partial}{\partial x}\left(p\frac{\partial p}{\partial x}\right) + \frac{\partial}{\partial y}\left(p\frac{\partial p}{\partial y}\right) = \frac{12\mu}{h^3}\frac{\partial ph}{\partial t} \tag{6.19}$$

In many cases it is very convenient to deal with dimensionless forms of governing equations. Let us make equation (6.19) dimensionless. We use the dimensionless parameters

$$X = \frac{x}{l_c} \qquad Y = \frac{y}{l_c} \qquad P = \frac{p}{p_c} \qquad H = \frac{h}{h_c} \qquad T = \frac{t}{t_c}$$

where l_c, p_c, h_c, and t_c denote the characteristic length, pressure, thickness, and time, respectively, and the capital letters represent the dimensionless parameters corresponding to the numerators. Using the dimensionless parameters above, we can nondimensionalize the preceding Reynolds equation as follows (the nondimensionalization procedure was discussed in Section 1.4):

$$\frac{\partial}{\partial X}\left(P\frac{\partial P}{\partial x}\right)+\frac{\partial}{\partial Y}\left(P\frac{\partial P}{\partial Y}\right)=\sigma\frac{\partial PH}{\partial T} \tag{6.20}$$

where

$$\sigma=\frac{12\mu l_c^2}{h_c^2 p_c t_c} \tag{6.21}$$

is the squeeze number. If the movable plate in a surrounding gas of pressure p_0 oscillates with a small amplitude at $h = h_0$ (i.e., $h = h_0 + h_1 \sin \omega t$, where h_1 is the vibration amplitude), we may take $h_c = h_0$, $t_c = 1/\omega$, and $p_c = p_0$ and find the squeeze number of the fluid film:

$$\sigma=\frac{12\mu l_c^2 \omega}{h_0^2 p_0} \tag{6.22}$$

The squeeze number plays an important role in determining whether or not the squeeze gas film is compressible. In general, if $\sigma << 1$, the film can be considered as incompressible. In this case, the Reynolds equation for $\rho =$ constant, (6.18), may be used to evaluate the pressure distribution and the damping force. If the density and viscosity remain constant and the film thickness is uniform, the Reynolds equation (6.17) reduces to a Poisson equation,

$$\frac{\partial^2 p}{\partial x^2}+\frac{\partial^2 p}{\partial y^2}=\frac{12\mu}{h^3}\frac{\partial h}{\partial t} \tag{6.23}$$

or in a simple form,

$$\nabla^2 p=-s \tag{6.24}$$

where s represents the source term of the Poisson equation and is given by

$$s=-\frac{12\mu}{h^3}\frac{\partial h}{\partial t}$$

So far, under assumptions 1 through 5, we have derived a Reynolds equation and discussed the fluid film compressibility. The foregoing Reynolds equations were derived under a no-slip condition of fluid particles on walls. However, if the film thickness of gas is less than a critical thickness or if the gas pressure decreases below a certain level, the gas particles slip on the walls. In this case, the preceding Reynolds equation is not applicable. Nevertheless, if the viscosity is replaced with the effective viscosity μ_e defined below, a Reynolds equation can be used to estimate the pressure or drag force. Researchers presented the effective viscosities. For example, the following was presented based on experimental data by Veijola et al. (1995):

$$\mu_e = \frac{\mu}{1 + 9.638 K_n^{1.159}} \tag{6.25}$$

where K_n is the Knusen number, defined by λ/L, where λ the mean free path of gas and L is the interplate gap.

6.5 COUETTE FLOW

Couette flow refers to a viscous laminar flow of fluid between parallel plates as shown in Fig. 6.6a. Let us consider a plate that is immersed in fluid (gas and liquid) and moves in the x direction while the plate is separated a distance h from a stationary plate (lower plate). A free-body diagram is shown in Fig. 6.6b, which illustrates the driving force, the drag force (shear force), and the normal force due to the pressure. The shear force drives the fluid between the plates to flow rightward. It is assumed that the plate length l is much greater than the gap h and that the effect of the upper flow to the shear force acting on the plate is neglected. Under the assumptions, the Navier–Stokes equation (6.11) is reduced to a simple equation,

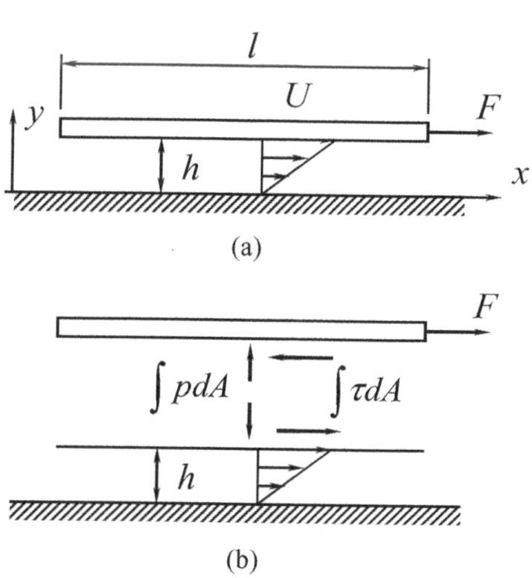

Figure 6.6 Flow driven by shear force.

$$-\frac{\partial p}{\partial x}+\mu\frac{\partial^2 u}{\partial y^2}=0$$

We assume further that the pressure gradient is negligible and then we solve the preceding equation with the boundary conditions $u = 0$ at $y = 0$ and $u = U$ at $y = h$ to give the following solution:

$$u=\frac{U}{h}y \tag{6.26}$$

The shear stress in the film and the drag force F_d are obtained as

$$\tau=\mu\frac{\partial u}{\partial y}\bigg|_{y=h}=\mu\frac{U}{h}$$
$$F_d=\tau A=\frac{\mu A}{h}U=cU \tag{6.27}$$

The damping coefficient due to Couette flow between parallel plates is then given by

$$c_{c0}=\frac{F_d}{U}=\frac{\mu A}{h} \tag{6.28}$$

The damping coefficient is proportional to the viscosity of the film and the plate area and inversely proportional to the gap h. If the plate length l is comparable to the gap h, equations (6.26) to (6.28) cause significant errors, which are discussed further in the remaining sections of this chapter. The damping force based on Couette flow has been used by researchers (e.g., Tang et al., 1989) for prediction of the quality factor of oscillating microstructures.

6.6 OSCILLATING PLATE IN A FLUID

As discussed in Section 6.5, in Couette flow, the upper plate moves at a constant velocity to generate the shear flow between parallel plates. For real microstructures, the damping force (drag force) due to Couette flow causes significant errors since only the damping forces between the moving plate and the substrate are considered. For better prediction of the damping force, Cho et al. (1994) included the shear force due to the inertia effect of gas near the oscillating structures. Let us suppose that an infinite plate oscillates in the x direction in Fig. 6.7a. This problem is known as *Stokes' second problem*. A similar case is shown in Fig. 6.7b, where the infinite plate immersed in a fluid oscillates in the x direction while the gap between the plate and a substrate

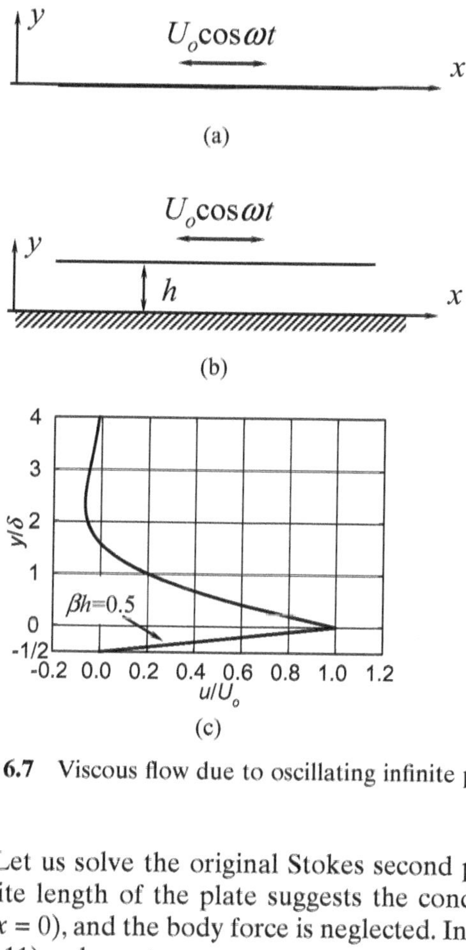

Figure 6.7 Viscous flow due to oscillating infinite plates.

remains constant. Let us solve the original Stokes second problem shown in Fig. 6.7a. The infinite length of the plate suggests the conditions ($\upsilon = w = 0$, $\partial u/\partial x = 0$, and $\partial p/\partial x = 0$), and the body force is neglected. In this case, Navier–Stokes equation (6.11) reduces to

$$\frac{\partial u}{\partial t} = \nu \frac{\partial^2 u}{\partial y^2} \tag{6.29}$$

where ν denotes the kinematic viscosity, defined by μ/ρ. This equation states that the shear force balances the inertial force in fluid and the fluid particles are driven by the shear stress of the plate. The displacement of the plate and its corresponding velocity are assumed to be $x_{y=0} = x_0 \sin \omega t$ and $u_{y=0} = x_0 \omega \cos \omega t = U_0 \cos \omega t$, where x_0, U_0, ω, and t denote the amplitudes of the displacement and velocity of the plate, the angular frequency, and the time, respectively. To solve the preceding linear partial differential equation, we shall use the complex number method, which was described in Section 5.6.

Case I Let us first consider the problem in Fig. 6.7a, where an infinite plate oscillates in a fluid. We shall replace $u = U\cos\omega t$ in the fluid with $u = Ue^{i\omega t}$ and solve the resulting equation. The velocity in the fluid will be given by the real part of the complex velocity (i.e., $u = \mathrm{Re}[u]$). Substituting the complex velocity into equation (6.29) yields an ordinary differential equation:

$$\frac{\partial^2 U}{\partial y^2} = i\frac{\omega}{\nu}U$$

Setting $U = e^{\lambda y}$ and noting that $\sqrt{i} = \pm(1+i)/\sqrt{2}$, the characteristic equation gives two roots,

$$\lambda = \sqrt{i\frac{\omega}{\nu}} = \pm\sqrt{\frac{\omega}{2\nu}}(1+i) = \pm\beta(1+i)$$

where

$$\beta = \sqrt{\frac{\omega}{2\nu}} \tag{6.30}$$

The velocity may be written as

$$u = Ue^{j\omega t} = \left(A_1 e^{\beta(1+i)y} + B_1 e^{-\beta(1+i)y}\right)e^{j\omega t} \tag{6.31}$$

Applying the boundary conditions $U = U_0$ at $y = 0$ and $U = 0$ at $y = \infty$, we have the complex velocity of the fluid as

$$u = U_0 e^{-\beta(1+i)y} e^{j\omega t} = U_0 e^{-\beta y} e^{j(\omega t - \beta y)}$$

The velocity is obtained as the real part of the equation above,

$$u = U_0 e^{-\beta y}\cos(\omega t - \beta y) \tag{6.32}$$

The shear stress in the fluid at y is then given by

$$\begin{aligned}\tau = \mu\frac{\partial u}{\partial y} &= -\mu U_0 \beta e^{-\beta y}[-\cos(\omega t - \beta y) + \sin(\omega t - \beta y)] \\ &= \sqrt{2}\,\mu U_0 \beta e^{-\beta y}\sin\left(\omega t - \beta y - \frac{\pi}{4}\right)\end{aligned} \tag{6.33}$$

If the length of the plate is finite and the plate area is denoted by A, the drag force is obtained from the free-body diagram of the plate as follows:

$$\begin{aligned} F_d &= -\tau|_{y=0} A = -\sqrt{2}\,\mu A\beta U_0 \sin\left(\omega t - \frac{\pi}{4}\right) \\ &= -U_0 A\sqrt{\rho\mu\omega}\sin\left(\omega t - \frac{\pi}{4}\right) \end{aligned} \tag{6.34}$$

where the negative sign denotes that the drag force directed to the negative x direction is defined as positive (refer to Fig. 3.9 for a definition of positive shear stress). The drag force is expressed in the form

$$\begin{aligned} F_d &= \mu A\beta U_0(\cos\omega t - \sin\omega t) \\ &= c_{st}\dot{x} + k_{st}x \end{aligned} \tag{6.35}$$

where

$$c_{st} = \mu A\beta = \sqrt{\frac{\rho\mu}{2}}A\omega^{1/2} \tag{6.36}$$

$$\begin{aligned} k_{st} &= -\mu A\beta\omega = -\sqrt{\frac{\rho\mu}{2}}A\omega^{3/2} \\ x &= x_0 \sin\omega t \qquad \dot{x} = u_0 = x_0\omega\cos\omega t \end{aligned} \tag{6.37}$$

denote the damping coefficient and stiffness due to the shear stress and the displacement, and the velocity of the plate, respectively. The damping coefficient and stiffness are proportional to the frequency. Note that the stiffness k_{st} is negative and that its magnitude increases with the angular frequency. If a spring of stiffness k is connected to the oscillating plate, the effective stiffness due to the spring stiffness and the stiffness of the flow will decrease from k as the frequency increases. The drag force above is due to one side of the plate. If both sides of a plate are immersed in a fluid, the drag force will be doubled.

Case II The infinite plate shown in Fig. 6.7b, immersed in a fluid, oscillates in the x direction while the gap between the plate and a substrate remains constant. The upper flow was solved as described earlier in this section. The lower flow between the plate and the substrate can be solved in the same manner. Imposing the boundary conditions ($u = 0$ at $y = 0$, $U = U_0$ at $y = h$) on solution (6.31) gives the following constants and the velocity u:

$$A_1 = \frac{U_0}{e^{\beta(1+i)h} - e^{-\beta(1+i)h}} \qquad B_1 = -A_1 = -\frac{U_0}{e^{\beta(1+i)h} - e^{-\beta(1+i)h}}$$

$$u = U_0\frac{e^{\beta(1+i)y} - e^{-\beta(1+i)y}}{e^{\beta(1+i)h} - e^{-\beta(1+i)h}}e^{i\omega t}$$

Recalling the identity $\sinh a = (e^a - e^{-a})/2$, we rewrite the preceding velocity in a different form,

$$u = U_0 \frac{\sinh[\beta(1+i)y]}{\sinh[\beta(1+i)h]} e^{i\omega t} \tag{6.38}$$

We wish to express equation (6.38) in a simple form. The following identities are used to simplify (6.38):

$$\begin{aligned} \sinh[\beta(1+i)y] &= \sinh\beta y \cos\beta y + i\cosh\beta y \sin\beta y \\ &= a_y e^{i\phi_y} \end{aligned} \tag{6.39}$$

where

$$\begin{aligned} a_y &= \sqrt{(\sinh\beta y \cos\beta y)^2 + (\cosh\beta y \sin\beta y)^2} \\ &= \sqrt{\cosh^2\beta y - \cos^2\beta y} = \sqrt{(\cosh 2\beta y - \cos 2\beta y)/2} \\ \phi_y &= \tan^{-1}\frac{\cosh\beta y \sin\beta y}{\sinh\beta y \cos\beta y} = \tan^{-1}\frac{\tan\beta y}{\tanh\beta y} \end{aligned} \tag{6.40}$$

The following are used to obtain the final expression for a_y above:

$$\sinh^2 a = \cosh^2 a - 1 \qquad \sin^2 a = 1 - \cos^2 a$$

$$\cosh^2 a = \frac{1+\cosh 2a}{2} \qquad \cos^2 a = \frac{1+\cos 2a}{2}$$

Similarly,

$$\sinh[\beta(1+i)h] = a_h e^{i\phi_h} \tag{6.41}$$

where

$$\begin{aligned} a_h &= \sqrt{\frac{\cosh 2\beta h - \cos 2\beta h}{2}} \\ \phi_h &= \tan^{-1}\frac{\tan\beta h}{\tanh\beta h} \end{aligned} \tag{6.42}$$

Upon substituting the preceding expressions into (6.38), we have

$$\begin{aligned} u &= U_0 \frac{\sinh[\beta(1+i)y]}{\sinh[\beta(1+i)h]} e^{j\omega t} = U_0 \frac{a_y}{a_h} e^{i(\omega t + \phi_y - \phi_h)} \\ &= U_0 \sqrt{\frac{\cosh 2\beta y - \cos 2\beta y}{\cosh 2\beta h - \cos 2\beta h}} e^{i(\omega t + \phi_y - \phi_h)} \end{aligned}$$

The velocity of fluid u between the plate and the substrate is then given as the real part of the preceding equation:

$$u = \mathrm{Re}(u) = U_0\sqrt{\frac{\cosh 2\beta y - \cos 2\beta y}{\cosh 2\beta h - \cos 2\beta h}}\cos(\omega t + \phi_y - \phi_h) \tag{6.43}$$

where ϕ_y and ϕ_h denote the phases defined by (6.40) and (6.42). The shear stress acting on the lower side of the plate is obtained as

$$\tau = \mu\frac{\partial u}{\partial y}\bigg|_{y=h} = \mu U_0\left(\beta\frac{\sinh 2\beta h + \sin 2\beta h}{\cosh 2\beta h - \cos 2\beta h}\cos\omega t - \sin\omega t\frac{\partial \phi_y}{\partial y}\bigg|_{y=h}\right)$$

where the derivative of the phase ϕ_y is given by

$$\begin{aligned}
\frac{\partial \phi_y}{\partial y}\bigg|_{y=h} &= \beta\frac{\partial \phi_y}{\partial \beta y}\bigg|_{y=h} \\
&= \beta\frac{\sec^2 \beta y \tanh \beta y - \tan \beta y \sec h^2 \beta y}{\left[1+(\tan\beta y/\tanh\beta y)^2\right]\tanh^2 \beta y}\bigg|_{y=h} \\
&= \beta\frac{\sinh\beta y\cosh\beta y - \sin\beta y\cos\beta y}{\cosh^2\beta y - \cos^2\beta y}\bigg|_{y=h} \\
&= \beta\frac{\sinh 2\beta h - \sin 2\beta h}{\cosh 2\beta h - \cos 2\beta h}
\end{aligned}$$

The shear stress beneath the plate is expressed as

$$\tau = \mu U_0\beta\left(\frac{\sinh 2\beta h + \sin 2\beta h}{\cosh 2\beta h - \cos 2\beta h}\cos\omega t - \frac{\sinh 2\beta h - \sin 2\beta h}{\cosh 2\beta h - \cos 2\beta h}\sin\omega t\right) \tag{6.44}$$

If $h \to \infty$, the shear stress above reduces to (6.33) at $y = 0$, but their signs are opposite. The drag force acting on the lower face of the plate is given by

$$\begin{aligned}
F_d &= \tau|_{y=h} A \\
&= U_0 A\mu\beta\left(\frac{\sinh 2\beta h + \sin 2\beta h}{\cosh 2\beta h - \cos 2\beta h}\cos\omega t - \frac{\sinh 2\beta h - \sin 2\beta h}{\cosh 2\beta h - \cos 2\beta h}\sin\omega t\right) \\
&= c_{st}\dot{x} + k_{st}x
\end{aligned} \tag{6.45}$$

where

$$c_{st} = \frac{\mu A}{h}\beta h\frac{\sinh 2\beta h + \sin 2\beta h}{\cosh 2\beta h - \cos 2\beta h} \tag{6.46}$$

$$k_{st} = -\frac{\mu A}{h}\omega\beta h \frac{\sinh 2\beta h - \sin 2\beta h}{\cosh 2\beta h - \cos 2\beta h} \tag{6.47}$$

$$x = x_0 \sin \omega t \qquad \dot{x} = u_0 = x_0 \omega \cos \omega t$$

represent the damping coefficient and stiffness due to the viscous flow of the fluid and the displacement and velocity of the plate, respectively. If $h \to \infty$, the damping coefficient and stiffness above go to (6.36) and (6.37). For $\beta h = 0.5$ and $\omega t = 0°$, the velocity distribution of the upper and lower flows is illustrated in Fig. 6.7c. It is noted from the figure that the velocity of the lower flow is almost linear (i.e., its velocity is close to that of Couette flow). Further comparison shows that if $\beta h << 1$, the viscous flow between an oscillating plate and a substrate can be considered as Couette flow. This is due to the fact that the inertial force is much smaller than the viscous force in the flow.

We next obtain simplified expressions for the damping coefficient and stiffness under the condition $\beta h \leq 1$. Substituting the following series of functions,

$$\sinh a = a + \frac{a^3}{3!} + \frac{a^5}{5!} + \frac{a^7}{7!} + \cdots \qquad \cosh a = 1 + \frac{a^2}{2!} + \frac{a^4}{4!} + \frac{a^6}{6!} + \cdots$$

$$\sin a = a - \frac{a^3}{3!} + \frac{a^5}{5!} - \frac{a^7}{7!} + - \cdots \qquad \cos a = 1 - \frac{a^2}{2!} + \frac{a^4}{4!} - \frac{a^6}{6!} + - \cdots$$

we have approximate expressions for c_{st}:

$$\begin{aligned} c_{st} &= \frac{\mu A}{h}\left[1 + \frac{4}{45}(\beta h)^4 + O\left((\beta h)^8\right)\right] \\ &\cong \frac{\mu A}{h}\left[1 + \frac{4}{45}(\beta h)^4\right] \end{aligned} \tag{6.48}$$

which is in excellent agreement with (6.46) within an error of 0.3% at $\beta h = 1$. This expression shows that if $\beta h \to 0$, the preceding damping coefficient approaches the expression based on Couette flow, $c_{c0} = \mu A/h$. Similarly, for $\beta h \leq 1$, the stiffness due to Stokes' flow is expressed as

$$\begin{aligned} k_{st} &= \frac{\mu A}{h}\omega\left\{\frac{2}{3}(\beta h)^2\left[1 - \frac{2}{15}(\beta h)^2 + O\left((\beta h)^4\right)\right]\right\} \\ &\cong \frac{2}{3}\frac{\mu A}{h}\omega(\beta h)^2\left[1 - \frac{2}{15}(\beta h)^2\right] \end{aligned} \tag{6.49}$$

which approaches 0 as $\beta h \to 0$. So far, the damping coefficient and stiffness due to an oscillating infinite plate have been obtained. However, real microstructures oscillating parallel to substrates are finite, so (6.46) may underestimate the damping coefficient. This topic is discussed in the following example and in subsequent sections.

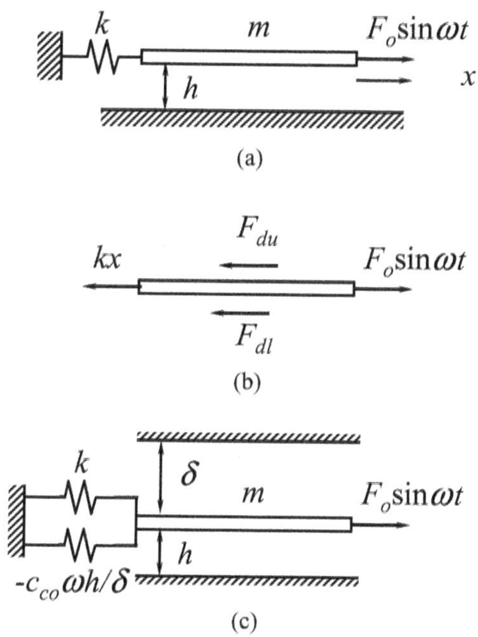

Figure 6.8 Vibration of a plate of large area.

Example 6.1 We consider a thin and large plate (Fig. 6.8a) oscillating in air. The plate, separated by h from a substrate, is supported by a spring of stiffness of k and oscillates in the x direction as a sinusoidal force is applied. Find the steady-state response of the plate and make an equivalent model of the vibration system if $\beta h << 1$.

We assume that the plate is large enough to be considered as an infinite plate. Applying Newton's second law to the free-body diagram of Fig. 6.8b gives

$$m\ddot{x} + F_{du} + F_{dl} + kx = F_0 \sin \omega t$$

and substituting the drag forces, (6.35) and (6.45), of the upper and lower flows into the equation above yields the equation of motion,

$$m\ddot{x} + c\dot{x} + k_{\text{eff}}x = F_0 \sin \omega t \tag{a}$$

where

$$\begin{aligned} c &= c_{\text{st},u} + c_{\text{st},l} \\ &= \frac{\mu A}{h}\left(\beta h + \beta h \frac{\sinh 2\beta h + \sin 2\beta h}{\cosh 2\beta h - \cos 2\beta h}\right) \end{aligned} \tag{b}$$

$$\begin{aligned} k_{\text{eff}} &= k - \mu A \beta \omega - \frac{\mu A}{h} \omega \beta h \frac{\sinh 2\beta h - \sin 2\beta h}{\cosh 2\beta h - \cos 2\beta h} \\ &= k - \frac{\mu A}{h} \omega \left(\beta h + \beta h \frac{\sinh 2\beta h - \sin 2\beta h}{\cosh 2\beta h - \cos 2\beta h} \right) \end{aligned} \tag{c}$$

denote the damping coefficient and an effective stiffness. Since c and k_{eff} depend on the frequency and are independent of the displacement and velocity, the steady-state response of the plate (if the system is under damped) is given by (5.89), which is rewritten here:

$$x = x_0 \sin(\omega t - \phi) \tag{d}$$

where

$$x_0 = \frac{F_0}{k_{\text{eff}}} \frac{1}{\sqrt{(1-\Omega^2)^2 + (2\zeta\Omega)^2}} \tag{e}$$

where

$$\Omega = \frac{\omega}{\omega_n} = \frac{f}{f_n} \qquad \phi = \tan^{-1} \frac{2\zeta\Omega}{1-\Omega^2} \qquad \omega_n = \sqrt{\frac{k_{\text{eff}}}{m}} \qquad \zeta = \frac{c}{2\sqrt{m k_{\text{eff}}}}$$

The quality factor is then given by

$$Q \cong \frac{1}{2\zeta} = \frac{\sqrt{m k_{\text{eff}}}}{c} \tag{f}$$

For $bh << 1$, c and k_{eff} can be written in simpler forms:

$$c \cong c_{c0}(1 + \beta h) = c_{c0}\left(1 + \frac{h}{\delta}\right) \tag{g}$$

$$k_{\text{eff}} \cong k - c_{c0}\omega\beta h = k - c_{c0}\frac{h}{\delta}\omega \tag{h}$$

where c_{c0} is the damping coefficient based on Couette flow, which is defined as $\mu A/h$, and $\delta = 1/\beta = \sqrt{2\nu/\omega}$ is interpreted as the penetration depth at which the amplitude of velocity (or displacement) in fluid falls to $1/e$ (about 37%) of that of the plate velocity (or displacement). If $\zeta << 1$ (i.e., is slightly damped), the quality factor is simplified further as follows:

$$Q = \frac{\sqrt{mk_{\text{eff}}}}{c}\bigg|_{\omega=\omega_{nc}} \cong \frac{\sqrt{m(k - c_{c0}(h/\delta)\omega_{nc})}}{c_{c0}(1+h/\delta)} = Q_{c0}\frac{\sqrt{1-(c_{c0}/k)(h/\delta)\omega_{nc}}}{1+h/\delta}$$
$$\cong Q_{c0}\left[1-\frac{h}{\delta}\left(1+\frac{c_{c0}\omega_{nc}}{k}\right)\right] = Q_{c0}\left[1-\frac{h}{\delta}\left(1+\frac{1}{Q_{c0}}\right)\right] \tag{i}$$

where ω_{nc} is the angular frequency given by $\sqrt{k/m}$ and Q_{c0} denotes the quality factor based on Couette flow and the stiffness k, which is given by

$$Q_{c0} = \frac{\sqrt{mk}}{c_{c0}} \tag{j}$$

Based on equations (g) and (h), we can construct a simple model for $\beta h << 1$, which is illustrated in Fig. 6.8c. The vibrating system of Fig. 6.8a is converted to an equivalent system which consists of two dampers based on Couette flow and two springs: the original spring of stiffness k and another spring of negative stiffness $-c_{c0}\omega h/\delta$. At room temperature, the kinematic viscosity is given by $\nu = 1.46 \times 10^{-5}\,\text{m}^2/\text{s}$ and the penetration depth $\delta = \sqrt{2\nu/\omega}$ is estimated as $22\,\mu\text{m}$ at $f = 10\,\text{kHz}$ and $6.8\,\mu\text{m}$ at $f = 100\,\text{kHz}$. If a large plate, separated a distance $2\,\mu\text{m}$ from a substrate, oscillates at a frequency of $10\,\text{kHz}$ in air, the ratio $h/\delta = 1/11$ is much less than 1 and the contribution of the inertial effect of the upper flow becomes small. If the plate has perforations or gratings or if the plate is short in length, the inertial effect of the fluid will become smaller or even negligible. Further discussion on this topic follows in a subsequent section.

6.7 CREEPING FLOW

In the preceding two sections we described the dampers based on Couette flow and Stokes' second problem (fluid motion due to an oscillating plate immersed in fluid). Since the damping coefficient due to Couette flow was obtained from the drag force (shear force) of a fluid between plates, the drag force due to viscous flow over the plate was neglected. If the length of microstructures is comparable to the gap between plates, the Navier–Stokes equation to be solved becomes much complicated, and then the drag force based on Couette flow deviates from that from simulation or experiment. As described in Section 6.6, the drag force based on Stokes's second problem was obtained under the assumption that an infinite plate oscillates and the inertia force of fluid then balances the shear force generated by the velocity parallel to the plate. However, if the plate length is finite (e.g., the length is five times the plate thickness), the inertial force is much less than the viscous forces, due to the velocity components parallel to and perpendicular to the plate. Then the drag forces based on Couette flow and the Stokes' second problem tend to be

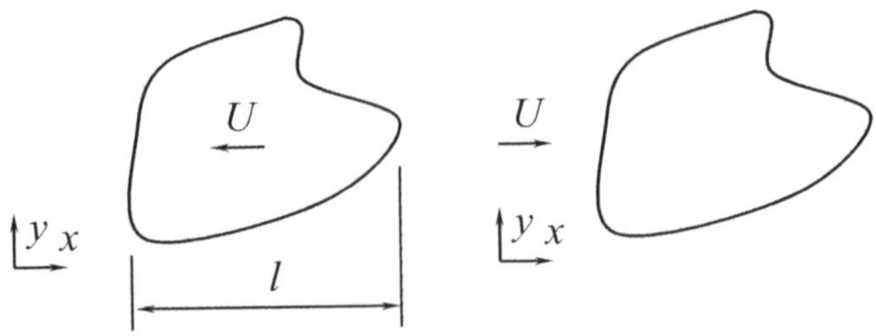

Figure 6.9 Creeping flow near an object.

underestimated for microstructures (i.e., the quality factor tends to be overestimated for microstructures).

Let us begin with an arbitrarily shaped structure in Fig. 6.9 whose characteristic length is given by l. In Fig. 6.9a, the structure immersed in a fluid moves leftward with a velocity U. The same structure in Fig. 6.9b is fixed in a free stream of velocity U. In both cases, the velocity of the fluid is U relative to the structure, and identical viscous and pressure forces are applied to the structures. Thus, we can solve the second problem (Fig. 6.9b) instead of the first (Fig. 6.9a) since the second is much easier to solve. If the velocity of fluid is less than a certain value, the inertial force of the fluid will be much less than the viscous forces in fluid. The Reynolds number, R_n, is a dimensionless number that is defined as the ratio of inertial to viscous forces. The Reynolds number is expressed as

$$R_n = \frac{\rho U l}{\mu} \tag{6.50}$$

As discussed in Section 1.4, in the microscale world, body forces such as the inertial and gravitational forces become much smaller than surface forces such as shear forces due to viscous flow. Thus, the Reynolds number in MEMS tends to be much smaller than unity and the fluid creeps around structures. Neglecting all body forces, we can write the continuity equation (6.6) and Navier–Stokes equations (6.11) for two-dimensional flow:

$$\frac{\partial u}{\partial x} + \frac{\partial v}{\partial y} = 0 \tag{6.51}$$

$$0 = -\frac{\partial p}{\partial x} + \mu\left(\frac{\partial^2 u}{\partial x^2} + \frac{\partial^2 u}{\partial y^2}\right) \tag{6.52}$$

$$0 = -\frac{\partial p}{\partial y} + \mu\left(\frac{\partial^2 v}{\partial x^2} + \frac{\partial^2 v}{\partial y^2}\right) \tag{6.53}$$

These are functions of the pressure p and the velocities u and υ. To express the equations above in a simple form, we introduce the stream functions ψ:

$$u = \frac{\partial \psi}{\partial y}$$
$$\upsilon = -\frac{\partial \psi}{\partial x} \tag{6.54}$$

which satisfy the continuity equation (6.51). Taking the derivative of (6.52) and (6.53) with respect to y and x, respectively, subtracting the resulting equations, and using the preceding definition of the stream function leads us to the basic equation for creeping flow, also known as Stokes' flow:

$$\frac{\partial^4 \psi}{\partial x^4} + 2\frac{\partial^4 \psi}{\partial x^2\, \partial y^2} + \frac{\partial^4 \psi}{\partial y^4} = 0$$

or simply

$$\nabla^4 \psi = 0 \tag{6.55}$$

where

$$\nabla^4 = \nabla^2 \nabla^2 = \frac{\partial^4}{\partial x^4} + 2\frac{\partial^4}{\partial x^2 \partial y^2} + \frac{\partial^4}{\partial y^4}$$

is the biharmonic operator. The preceding linear biharmonic equation for creeping flow is valid for $R_n << 1$. After solving equation (6.55) for the stream function, the velocities u and υ are obtained from (6.54) and the pressure distribution is given by (6.52) and (6.53).

For creeping flow, the force (due to pressure and shear stress) exerted on structures immersed in fluids is, in general, expressed as

$$\frac{F}{\mu U l} = \text{const} \tag{6.56}$$

Solutions for creeping flow due to simple shapes are found in existing papers or texts (e.g., White, 1974). For example, the thin disk normal to and parallel to the free stream shown in Fig. 6.10a and b is subjected to the following drag forces, respectively:

$$F = \frac{32}{3}\mu U a \tag{6.57}$$

$$F = 16\mu U a \tag{6.58}$$

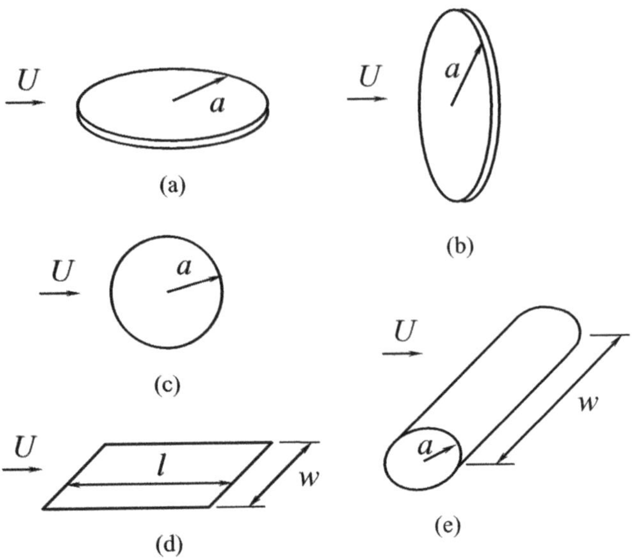

Figure 6.10 Creeping flow near various shapes.

where a denotes the radius of the disk. Even though the disks are exposed to the free stream in the different situations (Fig. 6.10a and b), the drag forces, (6.57) and (6.58), acting on the disk are not much different. A sphere (Fig. 6.10c) in the free stream undergoes the drag force given by

$$F = 6\mu U a \tag{6.59}$$

As shown in the preceding three equations, the viscous forces are proportional to the free-stream velocity. Some solutions for creeping flow involve the inertia effect of fluid, which extend the range of application. For example, if a rectangular plate is exposed parallel to the free stream as shown in Fig. 6.10d, the force applied to the plate is given by

$$F = 4\pi\mu U w \frac{1}{3.2 - \ln R_n} \tag{6.60}$$

where w is the width perpendicular to the free stream and R_n is Reynolds number, which is defined by $R_n = \rho U l/\mu$. Equation (6.60) is valid for $R_n \le 3$. For the circular beam of radius a and length w shown in Fig. 6.10e, the viscous force is expressed as

$$F = 4\pi\mu U w \frac{1}{2 - \ln R_n} \tag{6.61}$$

where $R_n = 2\rho U a/\mu$. Equation (6.61) holds for $R_n \le 1$.

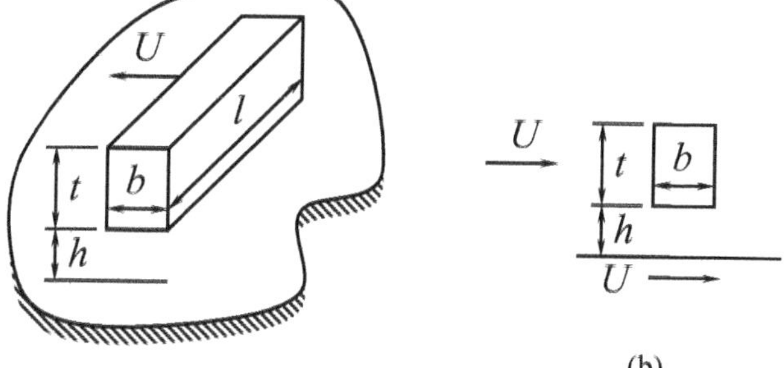

Figure 6.11 Rectangular beam moving over a substrate.

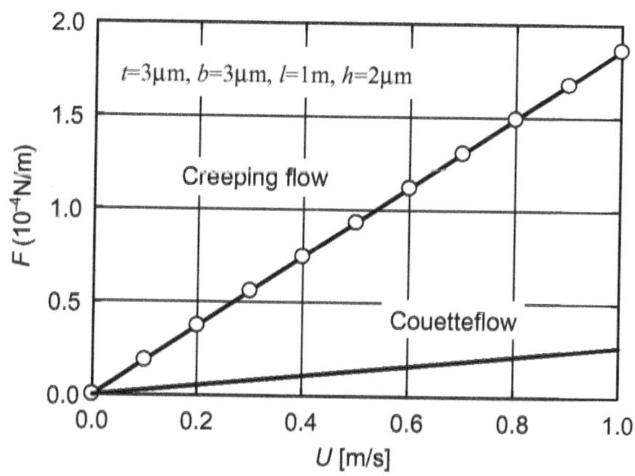

Figure 6.12 Viscous force acting on a rectangular beam.

In MEMS, structures move or oscillate over substrates in the direction parallel to the substrates. In Fig. 6.11a, a rectangular beam of width b, thickness t, and length l moves leftward at a velocity U, while the gap between the beam and substrate remains constant. For an analysis of the viscous flow and evaluation of the force exerted on the beam, the model of Fig. 6.11b can be used, where the beam is fixed in a free stream of velocity U and the infinite lower plate moves rightward. In this case it is very difficult to solve the flow near the beam analytically. We thus solve the viscous flow and the force acting on the beam numerically. If the length l is much larger than the other dimensions, t, b, and h, we can solve the biharmonic equation (6.55) and obtain the force on the beam as shown in Fig. 6.12, in which the data $b = 3\,\mu\text{m}$, $t = 3\,\mu\text{m}$, $l = 1\,\text{m}$, and $h = 2\,\mu\text{m}$ are used for the simulation. Commercially available software may be used to obtain the force. As discussed previously, the force (open circle) based on creeping flow is proportional to the free-stream velocity U, and the viscous force based on Couette flow between the beam and the substrate is also proportional to the velocity. However, the force calculated from the creeping flow is 6.9 times that from Couette flow. This huge discrepancy is due to contributions to the drag force from the pressure and shear stress. Therefore, if dimensions t and h of the rectangular structure moving over a substrate are comparable to the width b, the viscous force acting on the structure must be estimated by the creeping flow. Applying a curve-fitting technique to a set of simulation data, we obtain the viscous force acting on the rectangular beam (b, t, and h are much less than l) as follows:

$$F = \alpha\mu U l \tag{6.62}$$

where α is a dimensionless number, given by

$$\alpha = 6.80 + 0.70b + 1.20t - 1.03h$$

and b, t, and h are in micrometers. The preceding equation is valid for the ranges $2\,\mu\text{m} \leq b \leq 5\,\mu\text{m}$, $2\,\mu\text{m} \leq t \leq 5\,\mu\text{m}$, and $1\,\mu\text{m} \leq h \leq 5\,\mu\text{m}$. For $b >> h$ and $b >> t$, further numerical analysis shows that the force subjected to the structure in Fig. 6.11 approaches that due to Couette flow (i.e., the shear force under the structure becomes much greater than the other forces).

Example 6.2 Four identical rectangular beams of width b, thickness t, and length l are separated by a gap h from a substrate and used as springs (Fig. 6.13), with a central plate oscillating in the x direction with a small amplitude less then the beam width. If $b = 3\,\mu\text{m}$, $t = 3\,\mu\text{m}$, $l = 200\,\mu\text{m}$, $h = 2\,\mu\text{m}$, and the viscosity $\mu = 1.79 \times 10^{-5}\,\text{N·s/m}^2$, find the effective damping coefficient due to the four beams, which can be used for a one-DOF vibration system.

The deflection of the beam along the length is given by the mode shape discussed in Example 3.3, which is written

$$x = x_m\psi_1 = x_m\eta\psi \tag{a}$$

where x_m denotes the deflection of the central plate, η is 1 and the mode shape ψ is expressed in terms of $Y = y/l$ as follows:

$$\psi = 3Y^2 - 2Y^3 \tag{b}$$

To obtain the damping force due to the surrounding fluid, we use the effective damping coefficient (5.146), which is rewritten as

$$c_e = \eta^2 c_{ul} l \int_0^1 \psi^2 dY \tag{c}$$

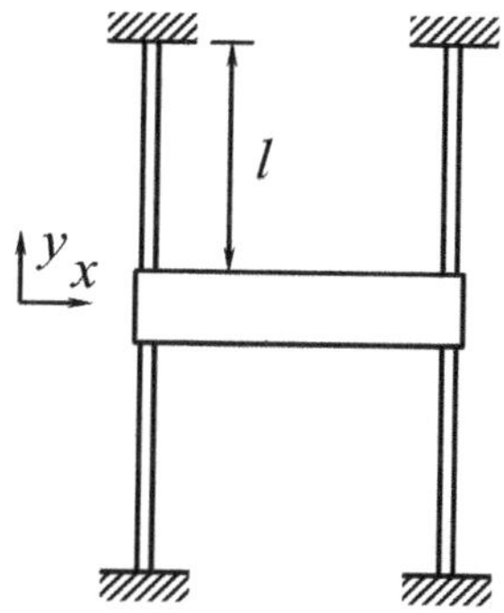

Figure 6.13 Squeeze film under a rectangular plate.

where c_{ul} is the damping coefficient per unit length of the beam, which is found from (6.62) as follows:

$$c_{ul} = \frac{F}{Ul} = \alpha\mu \tag{d}$$

The damping force of one beam is then given by

$$F_d = c_e u \tag{e}$$

where u denotes the velocity of the central mass. Substitution of (b) into (c) leads to the effective damping coefficient of one beam,

$$c_e = \eta^2 c_{ul} l \int_0^1 \psi^2 dY = c_{ul} l \int_0^1 Y^4 (3-2Y)^2 \, dY = \frac{13}{35} c_{ul} l \tag{f}$$

Substituting $b = 3\,\mu\text{m}$, $t = 3\,\mu\text{m}$, and $h = 2\,\mu\text{m}$ into α following (6.62) gives $\alpha = 10.44$, and in turn (f) yields $c_e = 1.39 \times 10^{-8}$ N·s/m. The damping force F due to the four beams is then given by

$$F = 4F_d = 4c_e u \tag{g}$$

Therefore, the effective damping coefficient due to the four beams is obtained as

$$c_{4\text{beam}} = \frac{F}{u} = 4c_e \tag{h}$$

The damping coefficient $c_{4\text{beam}}$ is estimated as 5.56×10^{-8} N·s/m. If the damping coefficient of the central plate is given by c_c, the damping coefficient of the structure in 6.13 is obtained as $c = c_c + c_{4\text{beam}}$.

It is noted from (5.145) and (5.146) that there is an analogy between the effective mass and the damping coefficient. For convenience, the equations are rewritten as

$$m_e = \eta^2 \rho A l \int_0^1 \psi^2 dY \qquad c_e = \eta^2 c_{ul} l \int_0^1 \psi^2 dY$$

The damping coefficient can be obtained by replacing the mass $\rho A l$ by the damping coefficient $c_{ul}l$, and vice versa. We can verify this by comparing the effective mass (5.162) and equation (f). Therefore, all equations for effective mass (e.g., equations described in Section 5.9) can be converted directly into those for a effective damping coefficient when we use the analogy.

6.8 SQUEEZE FILM

Squeeze film of fluid (in particular, gas) is used widely in MEMS devices. If a structure moves or oscillates against a substrate or electrode, the fluid between the structure and the substrate is squeezed out, and in turn the force (pressure) of the fluid is exerted on the structure. When the squeeze fluid is thin compared to the dimensions of the structure, the pressure and damping coefficient due to the squeeze film can be analyzed using Reynolds equation, which was described in Section 6.4.

Squeeze Damping Due to Strips

Let us consider a plate that moves up with a velocity of dh/dt from a substrate as shown in Fig. 6.14. The width and length of the plate are denoted by b and l, respectively, and the gap between the plate and substrate is uniform and represented by h. Let us begin with a simple case in which the width b is much less than the length l and the density of fluid can be considered as a constant.

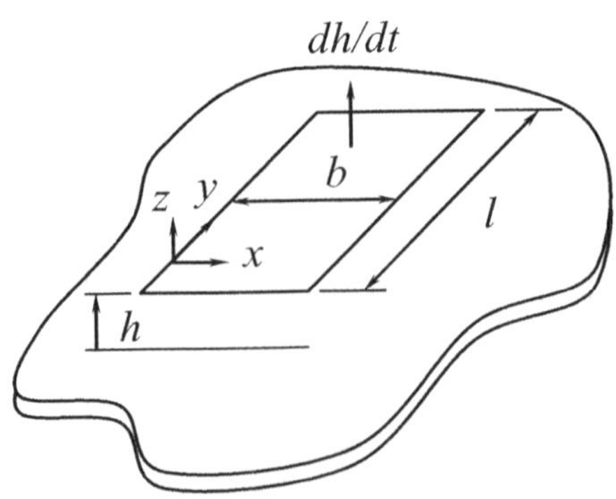

Figure 6.14 Squeeze film under a rectangular plate.

In this case, Reynolds equation (6.23) may be used, which is rewritten for further analysis as

$$\frac{\partial^2 p}{\partial x^2}+\frac{\partial^2 p}{\partial y^2}=\frac{12\mu}{h^3}\frac{\partial h}{\partial t}$$

Let the characteristic pressure and length in the x and y directions be p_0, b, and l. We shall compare the orders of magnitude of the first and second terms on the left-hand side of the equation above. For this purpose, we define dimensionless numbers $X = x/b$, $Y = y/l$, and $P = p/p_0$ and then compare the terms:

$$\frac{\partial^2 p/\partial y^2}{\partial^2 p/\partial x^2}=\frac{p_0/l^2}{p_0/b^2}\left[\frac{\partial^2 P/\partial Y^2}{\partial^2 P/\partial X^2}\right]\approx\left(\frac{b}{l}\right)^2$$

where the order of magnitude of the dimensionless quantity in brackets is unity. As shown in the preceding comparison, if $b/l \ll 1$, the second term in the preceding differential equation becomes much less than the first. In other words, if the plate length l is much greater than the width b, the change in pressure in the lengthwise direction is much less than that in the widthwise direction. Dropping the second term, we can write the differential equation as

$$\frac{\partial^2 p}{\partial x^2}=-s \tag{6.63}$$

where

$$s=-\frac{12\mu}{h^3}\frac{\partial h}{\partial t}$$

is the source term. We solve the foregoing equation with the boundary conditions ($p = 0$ at $x = 0$ and b) to give the pressure distribution and the force due to the fluid film, as follows:

$$\begin{aligned}p&=-\frac{sb^2}{2}\left[\left(\frac{x}{b}\right)^2-\frac{x}{b}\right]\\F&=\int_0^b pl\,dx=-l\frac{sb^2}{2}\int_0^b\left[\left(\frac{x}{b}\right)^2-\frac{x}{b}\right]dx\\&=\frac{slb^3}{12}\end{aligned} \tag{6.64}$$

Since the damping force in the opposite moving direction is defined as positive, the following damping force is obtained:

$$F_d = -F = \mu l \left(\frac{b}{h} \right)^3 \frac{\partial h}{\partial t} \tag{6.65}$$

The damping coefficient due to the fluid film of a strip (long plate) is then given by

$$c = \frac{F_d}{\partial h / \partial t} = \mu l \left(\frac{b}{h} \right)^3 \tag{6.66}$$

It is noted that the damping force is proportional to the length l of the strip and the cube of the ratio of width to film thickness, b/h.

Example 6.3 A fixed–fixed beam of width b and length l in Fig. 6.15a is separated by a gap h from a substrate. The beam in a surrounding gas of viscosity μ oscillates at an angular frequency with an amplitude small compared to the gap and the beam thickness. The vibration mode shape is assumed as the normalized mode shape, $\psi_1(X) = 16X^2(X - 1)^2$, where X is the coordinate along the beam normalized by the beam length. If $h << b$, $b << l$ and if the displacement of the midspan, y_m, is used as a characteristic displacement for a one-DOF model, derive the effective damping coefficient c_e for the one-DOF model.

Since $b << l$, the pressure gradient along the length l is negligible compared to that over the width b. Furthermore, since $h << b$, we can use a one-dimensional Reynolds equation to obtain the pressure beneath the plate. The damping coefficient per unit length is obtained as

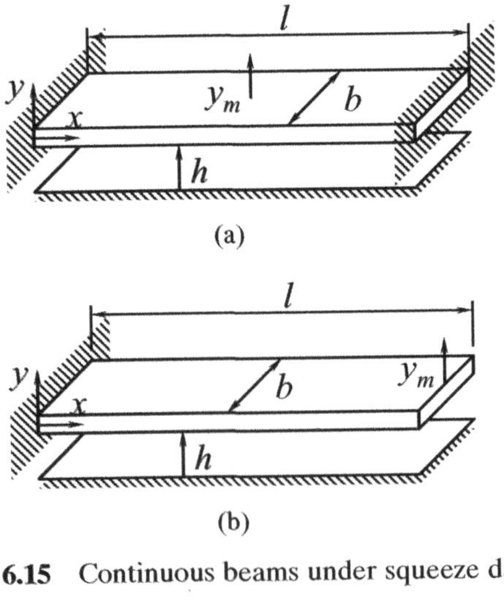

Figure 6.15 Continuous beams under squeeze damping.

$$c_{ul} = \frac{c_l}{l} = \mu\left(\frac{b}{h}\right)^3 \tag{a}$$

where c_l denotes the damping coefficient of the beam of length l, which is given by (6.66). If the beam sinusoidally oscillates, the vibration amplitude along the beam can be expressed as follows:

$$y = y_m\psi_1 = y_o\psi_1 \sin\omega t \tag{b}$$

where

$$\psi_1 = \eta\psi \tag{c}$$

Here η is a constant and ψ is a mode shape for easy integration. The effective damping coefficient, c_e, of continuous beams is given by (5.146), which is rewritten as

$$c_e = \eta^2 c_{ul} l \int_0^1 \psi^2 dX \tag{d}$$

Substituting the mode shape $\psi(X) = X^2(X-1)^2$ and the constant $\eta = 16$ gives the effective damping coefficient for the fixed–fixed beam,

$$c_e = \frac{128}{315} c_l \tag{e}$$

where $c_l = c_{ul}l$ denotes the damping coefficient. Substitution of (a) into equation (e) leads to the damping coefficient of a narrow rectangular fixed–fixed beam,

$$c_e = \frac{128}{315}\mu\left(\frac{b}{h}\right)^3 l \tag{f}$$

Since the damping coefficient is given in Example 6.3, we can obtain the quality factor, which is defined by $Q = \sqrt{m_e k_e}/c_e$ (if slightly damped), where m_e and k_e denote the effective mass and stiffness. Substituting (b) and (d) of Example 5.3 in Chapter 5, Section 5.9, and (f) of Example 6.3 into the quality factor equation, we have the quality factor of a fixed–fixed beam as follows:

$$Q = \frac{\sqrt{m_e k_e}}{c_e} = \sqrt{42}\frac{h^3}{\mu}\left(\frac{t}{bl}\right)^2 \sqrt{\rho_b \tilde{E}\left[1 + \frac{2}{7}\frac{\sigma_0}{\tilde{E}}\left(\frac{l}{t}\right)^2\right]} \tag{6.67}$$

where ρ_b, $\tilde{E}$, t, and σ_0 denote the density of the beam, the effective Young's modulus, and the thickness of the fixed–fixed beam and the residual stress (tensile stress defined as positive) in the beam, respectively. Similarly, for the cantilever beam of Fig. 6.15b, the effective damping coefficient c_e is given by (5.152), as $c_e = 104c_l/405$, where c_l represents the damping coefficient given by (a). Substituting equation (a) in Example 6.3 into equation (6.67), we find the effective damping coefficient of the cantilever beam:

$$c_e = \frac{104}{405}\mu\left(\frac{b}{h}\right)^3 l \tag{6.68}$$

Substituting (5.152) and equation (6.68) into the quality factor equation, we have the quality factor of a cantilever beam:

$$Q = 3\sqrt{\frac{3}{26}\frac{h^3}{\mu}\left(\frac{t}{bl}\right)^2}\sqrt{\rho_b\tilde{E}} \tag{6.69}$$

Note that the preceding quality factors are those corresponding to the lowest natural frequencies of the beams. If mode shapes for higher frequencies are given, the effective mass, stiffness, damping coefficient, and quality factor can be obtained in the same manner.

Squeeze Damping Due to Rectangular Plates

If the width of the plate is comparable to the length and the density is assumed to be constant, a two-dimensional Reynolds equation must be solved. We shall solve the following two-dimensional Reynolds equation (6.24) with the boundary conditions (zero pressure at the plate edges) to give the pressure distribution under the plate:

$$\frac{\partial^2 p}{\partial x^2} + \frac{\partial^2 p}{\partial y^2} = -s$$

where

$$s = -\frac{12\mu}{h^3}\frac{\partial h}{\partial t}$$

is the source term. The preceding Poisson equation with the simple boundary condition can be solved. We assume the following pressure distribution, which satisfies the boundary conditions:

$$p = \sum_{m=1}^{\infty}\sum_{n=1}^{\infty} a_{mn} \sin\frac{m\pi x}{b}\sin\frac{n\pi y}{l} \tag{6.70}$$

Substituting the pressure distribution above into the preceding differential equation, multiplying both sides of the resulting equation by $\sin(m\pi x/b)$ $\sin(m\pi y/l)$, and integrating over the area yields the coefficient:

$$\begin{aligned} a_{mn} &= s\frac{16}{\pi^2 mn}\frac{1}{(m\pi/b)^2+(n\pi/l)^2} \\ &= sb^2\frac{16}{\pi^4 mn}\frac{1}{m^2+(b/l)^2 n^2} \end{aligned}$$

where m and n are odd numbers (note that $a_{mn} = 0$ if m and n are even). The force due to the fluid film is obtained as follows:

$$F = \int_0^l\int_0^b p\,dx\,dy = \frac{64}{\pi^6}sb^3 l\sum_{m=\text{odd}}^{\infty}\sum_{n=\text{odd}}^{\infty}\frac{1}{m^2n^2\left[m^2+(b/l)^2 n^2\right]}$$

and substituting the source term into the equation above gives the damping force as

$$F_d = -F = \frac{768}{\pi^6}\sum_{m=\text{odd}}^{\infty}\sum_{n=\text{odd}}^{\infty}\frac{1}{m^2n^2\left\{m^2+(b/l)^2 n^2\right\}}\mu l\left(\frac{b}{h}\right)^3\frac{\partial h}{\partial t} \tag{6.71}$$

Using the definition of the damping coefficient, we have

$$c = \frac{F_d}{\partial h/\partial t} = \alpha\mu l\left(\frac{b}{h}\right)^3 \tag{6.72}$$

where α is a dimensionless coefficient that is given by

$$\alpha = \frac{768}{\pi^6}\sum_{m=\text{odd}}^{\infty}\sum_{n=\text{odd}}^{\infty}\frac{1}{m^2n^2\left[m^2+(b/l)^2 n^2\right]} \tag{6.73}$$

which serves as a correction factor for (6.66), as the length l is finite. Since the preceding α is expressed in infinite series, many terms must be added for sufficient accuracy, and we may use a computer to evaluate the preceding function. For, $b/l \le 1$, using a curve fitting technique and the function (6.73), we obtain a closed-form expression for α, which allows us to evaluate α as follows:

$$\alpha = 1 - 0.68\left(\frac{b}{l}\right) + 0.1\left(\frac{b}{l}\right)^2 \tag{6.74}$$

which is in good agreement with (6.73) within an error of 0.94% over the full range of $b/l = 0$ to 1. For $b/l > 0$, b and l switch places and then we can still use equation (6.74). In most MEMS devices using a plate, the squeeze number, defined as $\sigma = 12\mu b^2\omega/h_0^2 p_0$, is much less than 1 and then the preceding damping coefficients for the strip and plate can be used.

If the squeeze number is comparable to 1 or greater than 1, the linearized form of Reynolds equation (6.20) should be solved. Since the plate oscillates at the stable height h_0 and the pressure varies from the surrounding gas pressure p_0, the dimensionless film thickness (gap) and pressure may be expressed as follows:

$$H = 1 + \tilde{H} \qquad P = 1 + \tilde{P}$$

where $\tilde{H}$ and $\tilde{P}$ denote the dimensionless fluctuations of the gap and pressure, which are expressed as $\tilde{H} = H_1 \cos T$ and $\tilde{P} = P_1 \cos(T - \phi) = q \cos T + r \sin T$, where H_1, P_1, T, ϕ, q, and r represent the dimensionless amplitude for height and pressure, the dimensionless time defined by ωt, and the phase and dimensionless amplitudes, respectively. Substituting the foregoing quantities into Reynolds equation (6.20) and neglecting high-order terms leads to the linearized Reynolds equation,

$$\frac{\partial^2 \tilde{P}}{\partial X^2} + \frac{\partial^2 \tilde{P}}{\partial Y^2} - \sigma \frac{\partial \tilde{P}}{\partial T} = \sigma \frac{\partial \tilde{H}}{\partial T} \tag{6.75}$$

Substituting $\tilde{H} = H_1 \cos T$ and and $\tilde{P} = q \cos T + r \sin T$ into equation (6.75), solving for $\tilde{P}$, and integrating $\tilde{P}$ over the area gives the damping force, which is given by Blech (1983). The damping coefficient is given by

$$c = \frac{F_d}{\partial h/\partial t} = \alpha \mu l \left(\frac{b}{h}\right)^3 \tag{6.76}$$

where

$$\alpha = \frac{768}{\pi^6} \sum_{m=\text{odd}}^{\infty} \sum_{n=\text{odd}}^{\infty} \frac{m^2 + (b/l)^2 n^2}{m^2 n^2 \left\{\left[m^2 + (b/l)^2 n^2\right]^2 + \sigma^2/\pi^4\right\}} \tag{6.77}$$

$$\sigma = \frac{12\mu b^2 \omega}{h_0^2 p_0} \tag{6.78}$$

If $\sigma \to 0$, the preceding coefficient α reduces to (6.73), which was obtained without consideration of the squeeze number σ. As a result, for $\sigma << 1$, we can use (6.72) and (6.74) for evaluation of the damping coefficient due to the squeeze fluid (gas and liquid) film between the rectangular plate (or strip) and a substrate.

Squeeze Damping Due to Perforated Plates

As discussed in Section 5.6, if we wish a small force at a particular frequency to actuate a resonator, the energy dissipation should be minimized (i.e., the damping factor ζ is as small as possible, or equivalently, the quality factor Q is taken large). A large plate moving perpendicular to a substrate, fabricated by microfabrication (in particular, surface micromachining), is perforated both to adjust the damping force due to the fluid film and to etch away sacrificial layers. A perforated plate, separated a gap h from a substrate and moving with a velocity dh/dt upward, is shown in Fig. 6.16a, where square perforations of size a and pitch p_s are made on the plate as shown in Fig. 6.16b. We can estimate the damping force due to the fluid film between the plate and substrate by solving a Reynolds equation with suitable boundary conditions. We assume that the density of fluid remains constant. If the plate oscillates with a small amplitude at an angular frequency ω at a gap h_0, we use the following squeeze number:

$$\sigma = \frac{12\mu(p_s - a)^2\,\omega}{h_0^2\,p_0} \tag{6.79}$$

where the dimension, $p_s - a$ is used as the characteristic length since it represents the dimension associated with the force due to the fluid film. If $\sigma << 1$, the density of fluid can be considered as constant and Reynolds equation (6.24) is then solved for the pressure. However, since complicated boundary conditions ($p = 0$ at the hole, and the gradient of p is 0 along the symmetry line between perforations) are imposed on the problem, it is difficult to solve the Reynolds equation analytically with the boundary conditions. We then solve the Reynolds equation numerically, which is a Poisson equation (6.24).

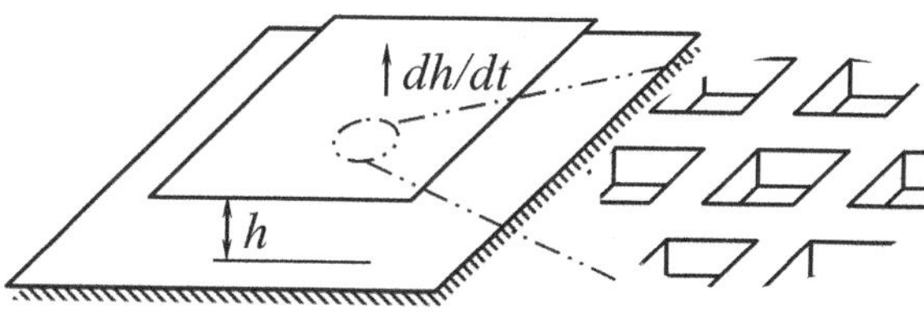

(a)

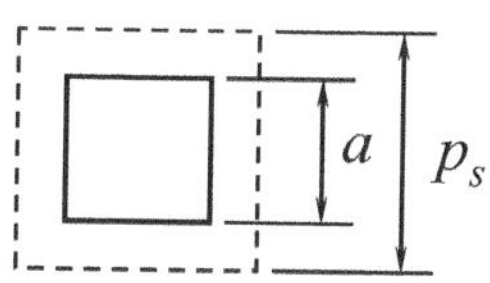

(b)

Figure 6.16 Squeeze damping of a perforated plate.

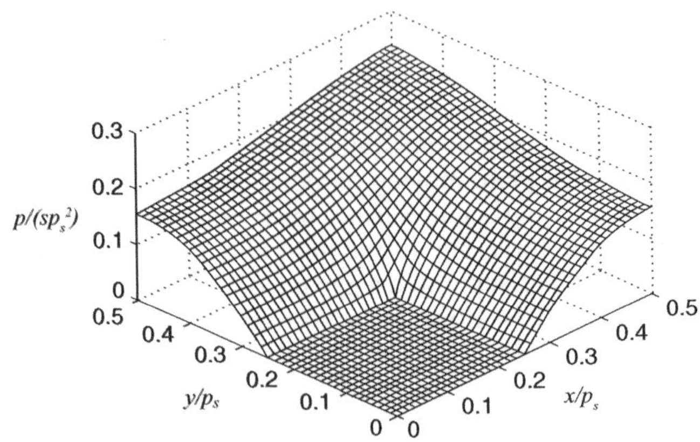

Figure 6.17 Pressure distribution of a perforated plate.

Readers may write a program or use commercially available software (e.g., electrical field solver, heat transfer solver) to solve the preceding Poisson equation for pressure, as shown in Fig. 6.17. The pressure distribution in Fig. 6.17 is obtained by the finite difference method (FDM) described in existing publications [e.g., the book by Anderson et al. (1984)]. On integrating the pressure over the area, we find the force F_d due to the fluid film and in turn obtain the damping coefficient, which is given by

$$c = \frac{F_d}{\partial h / \partial t} = \alpha \mu \frac{p_s^4}{h^3} \tag{6.80}$$

where α denotes a dimensionless number which is shown in Fig. 6.18. The coefficient α in the figure may be expressed in simple form as

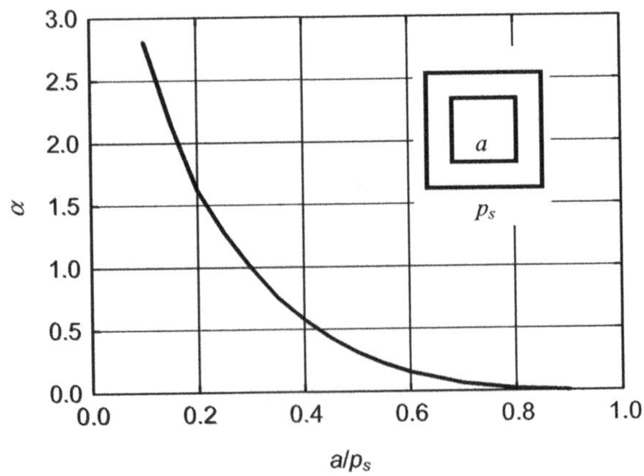

Figure 6.18 Force factor of a perforation cell.

$$\alpha = 4.4334e^{-5\gamma} + 0.4467\gamma^2 - 0.6352\gamma + 0.1469 \tag{6.81}$$

where $\gamma = a/p_s$ represents the ratio of hole size to pitch, respectively. This equation is valid in the range $\gamma = 0.1$ to 1.

PROBLEMS

6.1 Nondimensionalize the first equation of (6.11) and find a dimensionless number that represents the state of flow.

6.2 Find the dissipated energy of the oscillating plate in Fig. 6.8a and obtain the quality factor by using the dissipated and stored energy.

6.3 A central plate of area $100 \times 30\,\mu\text{m}^2$ and thickness $3\,\mu\text{m}$ in Fig. 6.13 oscillates with a small amplitude at a frequency of 85 kHz in the x direction in air at room temperature. The plate is separated by a distance $2\,\mu\text{m}$ and suspended by the four straight beams described in Example 6.2. Assuming that the damping forces of the plate and the beams can be described by Stokes' second problem and the creeping flow, respectively, find the overall damping coefficient due to the central plate and the beams.

6.4 A central mass of damping coefficient c_M is supported by four folded beams, each of which consists of two identical slender beams (damping coefficient per unit length c_{ua}) and a rigid beam (damping coefficient c_b). Determine the effective damping coefficient defined by $c_e = F_d/(dx/dt)$, where F_d and dx/dt denote the total damping force exerted on the central mass and the velocity, respectively.

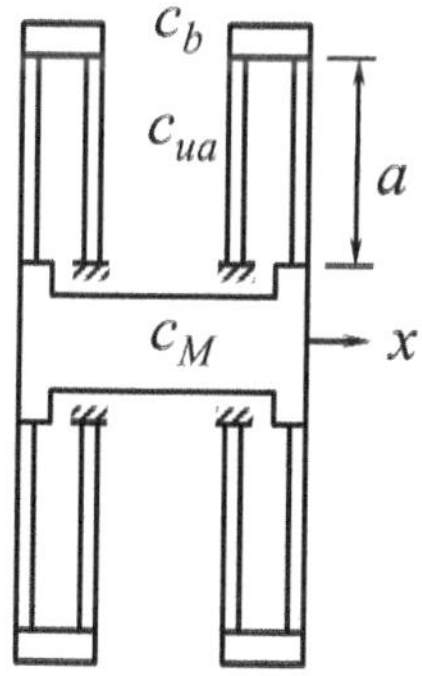

Figure P6.4

6.5 A circular plate of radius R is pushed against an oil film of viscosity μ on a horizontal substrate and moves with a velocity of w to substrate. The Laplacian for the cylindrical coordinate system is given by

$$\nabla^2 = \frac{1}{r}\frac{\partial}{\partial r}\left(\frac{1}{r}\frac{\partial}{\partial r}\right) + \frac{1}{r^2}\frac{\partial^2}{\partial \theta^2}$$

where r and θ represent the radius and angle of the cylindrical coordinate system. Derive the force exerted by the plate and the damping coefficient as functions of the plate radius, viscosity, and oil film thickness.

6.6 A square plate of area $20 \times 20\,\mu\text{m}^2$ is spaced by a gap $2\,\mu\text{m}$ from a substrate in air at room temperature and then oscillates with a small amplitude at a frequency of 40 kHz. Determine the damping coefficient due to the air film.

6.7 A plate of area $198 \times 198\,\mu\text{m}^2$ is perforated to reduce the damping coefficient and oscillates with a small amplitude at a frequency of 10 kHz in air at room temperature. If the plate is separated by $2\,\mu\text{m}$ from a substrate and square perforations of size $6\,\mu\text{m}$ and pitch $12\,\mu\text{m}$ are made on the plate, evaluate the damping coefficient due to the air.

CHAPTER 7

ELECTROMAGNETICS

In preceding chapters we described the microfabrication and static and dynamic behavior of structures. To design MEMS that function correctly, we need to drive or actuate structures in desired directions and need to detect or measure the movement of the structures. In the macroscale world we live in, electromagnetic motors (also known as electric motors) and engines have been used widely to convert electric and chemical energies into mechanical energy to drive mechanical parts. For sensing physical quantities such as force and displacement, various measuring mechanisms are used (e.g., capacitance detection). In this chapter we describe basic laws for electric circuits, forces, and moments based on electromagnetics, most of which can also be used to detect the motion of a body.

7.1 BASIC ELEMENTS OF ELECTRIC CIRCUITS

Passive elements, defined as elements that cannot generate energy, will be described that are related to mechanical–electrical energy conversion. Active elements such as batteries and amplifiers are capable of delivering energy to electrical circuits connected to them. Many documents and textbooks cover the characteristics and design of active elements, and the reader may refer to them to understand the active elements [e.g., Millman and Grabel (1987)]. Typical passive elements are the resistor, the capacitor, and the inductor, which are shown in Fig. 7.1. A resistor of resistance R in Fig. 7.1a is connected to a

Principles of Microelectromechanical Systems, By Ki Bang Lee

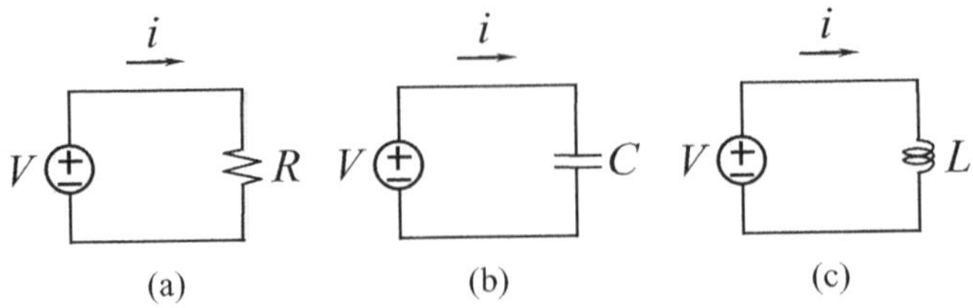

Figure 7.1 Basic elements of electrical circuits.

voltage source of voltage V, and a electric current i flows through the resistor. The voltage from a voltage source, known as the *electromotive force* (emf), produces electric current in an electric circuit. If the voltage V and resistance R are known, the current i is determined by *Ohm's law*,

$$i = \frac{V}{R} \tag{7.1}$$

where i, V, and R are in amperes (A), volts (V), and ohms (Ω), respectively. When a capacitor of capacitance C is connected to a voltage source in Fig. 7.1b, positive and negative electrical charges are, respectively, accumulated on the conductors of the capacitor. The accumulated charge Q is proportional to both the capacitance and the voltage, so is expressed in the form

$$Q = CV \tag{7.2}$$

where Q is in coulombs (C) and C is in farads (F). Since the current i is defined as the net charge per unit time flowing through a conductor, the current is expressed as

$$i = \frac{dQ}{dt} = C\frac{dV}{dt} \tag{7.3}$$

where t is time. If the current is given, the voltage across the capacitor is obtained as

$$V = \frac{1}{C}\int_0^t i\,dt + V(0) \tag{7.4}$$

where $V(0)$ is an initial voltage across the capacitor at $t = 0$. In Fig. 7.1c an inductor (e.g., a multiturn coil) is connected to a voltage source and the voltage and current relation is given by

$$V = L\frac{di}{dt} \tag{7.5}$$

where L is the inductance of the inductor in henrys (H). When the voltage is given, the current through the inductor is expressed as follows:

$$i = \frac{1}{L}\int_0^t V\,dt + i(0) \tag{7.6}$$

where $i(0)$ is an initial current. These are the basic definitions of the resistance, capacitance, and inductance of passive elements.

All three elements are related to the energy, U, which is supplied by the active elements. Since power is defined as energy per unit time,

$$P = \frac{dU}{dt} \tag{7.7}$$

the power is equal to a product of the voltage V and the current i since

$$P = \frac{dU}{dQ}\frac{dQ}{dt} = Vi \tag{7.8}$$

where the voltage V from the voltage source is defined to be the energy per unit charge (dU/dQ). From equations (7.7) and (7.8), we have the following electric energy, which is supplied by active elements from 0 to t:

$$U = \int_0^t Vi\,dt \tag{7.9}$$

The electric energy above is converted into heat in the resistor or stored in the capacitor and inductor. When the voltage V remains constant [i.e., direct-current (dc) voltage], equation (7.9) reduces to the simple form

$$U = Vit \tag{7.10}$$

We can apply to the resistor an alternating-current (ac) voltage with a dc voltage (bias voltage), $V = V_b + V_a \sin \omega t$, where V_b, V_a, and ω denote the amplitudes of the dc and ac voltages and the angular frequency of the ac voltage, respectively. Substituting the voltage into (7.9) and integrating the resulting equation from 0 to τ (a period defined by $2\pi/\omega$) gives

$$\begin{aligned}
U_\tau &= \int_0^\tau \frac{(V_b + V_a \sin \omega t)^2}{R}\,dt = \int_0^\tau \frac{1}{R}(V_b^2 + 2V_bV_a \sin \omega t + V_a^2 \sin^2 \omega t)\,dt \\
&= \frac{1}{\omega R}\int_0^{2\pi}\left[V_b^2 + 2V_bV_a \sin \omega t + V_a^2 \frac{1}{2}(1 - \cos 2\omega t)\right]d\omega t \\
&= \frac{\tau}{R}\left(V_b^2 + \frac{1}{2}V_a^2\right)
\end{aligned} \tag{7.11}$$

which is the electric energy supplied by the voltage source during the period τ and is equal to the energy converted into heat by the resistor. We then have the average energy consumed in the resistor during the period:

$$P_{\text{avg}} = \frac{U_\tau}{\tau} = \frac{V_b^2}{R} + \frac{1}{2}\frac{V_a^2}{R} \tag{7.12}$$

which is known as *Joule heat*. The first term on the right-hand side of (7.12) is the contribution of the dc voltage V_b. The power due to the dc voltage may be expressed as

$$P_{\text{dc}} = V_b i = \frac{V_b^2}{R} = Ri^2 \tag{7.13}$$

In order to consider the contribution of the ac voltage to the power, we introduce the root-mean-square (rms) voltage and current, defined by

$$V_{\text{rms}} = \frac{V_a}{\sqrt{2}} \qquad i_{\text{rms}} = \frac{i_a}{\sqrt{2}} \tag{7.14}$$

where i_a denotes the amplitude of the ac current. With the definition above, the ac contribution to the power can be expressed in a simple form:

$$P_{\text{ac}} = V_{\text{rms}} i_{\text{rms}} = \frac{(V_{\text{rms}})^2}{R} = R(i_{\text{rms}})^2 \tag{7.15}$$

We can then rewrite (7.12) in a more convenient form:

$$P_{\text{avg}} = P_{\text{dc}} + P_{\text{ac}} = \frac{1}{R}\left(V_b^2 + V_{\text{rms}}^2\right) = R\left(i_b^2 + i_{\text{rms}}^2\right) \tag{7.16}$$

where i_b denotes the current due to dc voltage. Note that the quantities in parentheses in equation (7.16) may be defined as the rms values of the voltage and current:

$$V_{\text{rms}} = \sqrt{V_b^2 + V_{\text{rms}}^2} \qquad i_{\text{rms}} = \sqrt{i^2 + i_{\text{rms}}^2} \tag{7.17}$$

A time-varying signal of period p can be expressed by a Fourier series,

$$f = f_{\text{dc}} + \sum_{j=1}^{\infty}\left(f_{s,j}\sin\omega_j t + f_{c,j}\cos\omega_j t\right) \tag{7.18}$$

where f_{dc}, ω_j, $f_{s,j}$, and $f_{c,j}$ represent the dc component, the angular frequency given by $2j\pi/p$, and the coefficients of the sine and cosine functions at ω_j, respectively. Extending the idea of the rms value shown in (7.16), we obtain the following rms value of the preceding time-varying signal:

$$f_{\mathrm{rms}} = \sqrt{f_{\mathrm{dc}}^2 + \frac{1}{2}\left(f_{c,1}^2 + f_{c,2}^2 + \cdots\right) + \frac{1}{2}\left(f_{s,1}^2 + f_{s,2}^2 + \cdots\right)}$$
$$= \sqrt{f_{\mathrm{dc}}^2 + \frac{1}{2}\sum_{j=1}^{\infty}\left(f_{c,j}^2 + f_{s,j}^2\right)} \tag{7.19}$$

Equation (7.19) is expressed equivalently as

$$f_{\mathrm{rms}} = \sqrt{\frac{1}{p}\int_0^p f^2 dt} \tag{7.20}$$

As shown in the preceding equations, note that the rms value of a physical quantity, which is time varying with a period, represents an average of the quantity during the period and that the square of the rms value is proportional to the power. In Fig. 7.1a, the voltage source supplies the resistor with the energy (7.12) and all the energy is dissipated into heat that may be used to heat up a body or fluid (e.g., air in a room).

If we replace the resistor in Fig. 7.1a with the capacitor in Fig. 7.1b and the inductor in Fig. 7.1c, the energy from the voltage source does not dissipate, but is stored in the elements. The following energies are stored in the capacitor and the inductor:

$$U = \int_0^t Vi\,dt = \int_0^t VC\frac{dV}{dt}dt = \int_0^V CV\,dV = \frac{1}{2}CV^2 \tag{7.21}$$

$$U = \int_0^t Vi\,dt = \int_0^t L\frac{di}{dt}i\,dt = \int_0^i Li\,di = \frac{1}{2}Li^2 \tag{7.22}$$

The preceding two energies are stored in the three-dimensional space around the elements and we discuss them in the remaining sections of the chapter.

For some MEMS (e.g., a thermal actuator), we need to heat up an object, such as the prismatic beam shown in Fig. 7.2, where a voltage source applies a voltage V across the beam length to allow an electric current i to flow through the beam. Experiments show that the resistance of the beam is proportional to the length l and inversely proportional to the area A. If the beam is homogeneous and made of an isotropic material, the resistance of the beam can then be given by

$$R = \rho_r \frac{l}{A} \tag{7.23}$$

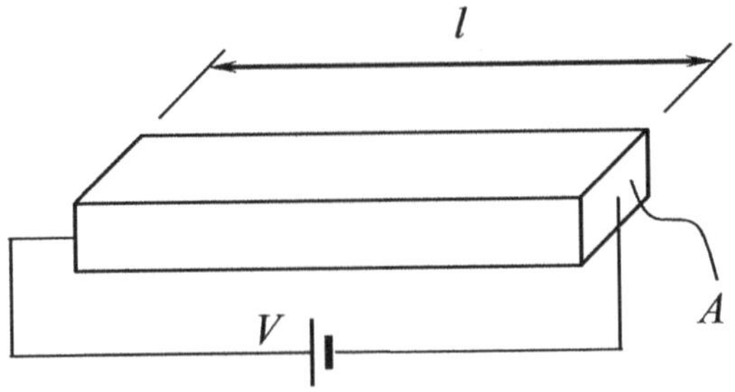

Figure 7.2 Joule heating.

where ρ_r is the resistivity in $\Omega \cdot$m. The heat generated per unit length per unit time is obtained from (7.16) as follows:

$$\dot{Q}_{\text{heat}} = \frac{P}{l} = \frac{V^2}{Rl} \tag{7.24}$$

Substituting (7.23) into equation (7.24) gives

$$\dot{Q}_{\text{heat}} = \frac{AV^2}{\rho_r l^2}$$

If the size of the beam increases n times and if the same material and voltage are used, the generated heat per unit time remains constant. The heat generated from objects (e.g., the beam of Fig. 7.2) results in thermal expansion of the objects. The thermally controlled objects may serve as thermal actuators, which are covered in subsequent chapters.

7.2 KIRCHHOFF'S CIRCUIT LAWS

Electrical circuits consist of a combination of active and passive components for a specific function or purpose. The relations between voltage and current in a circuit may be expressed by Kirchhoff's current and voltage laws. Before proceeding with Kirchhoff's laws, we consider simple cases with which we are familiar. Two small rivers in Fig. 7.3a are merged into a large river. If the mass flow rates of the small rivers are represented by $\dot{m}_1$ and $\dot{m}_2$, we know from our experience that the mass flow rate of the large river is given by $\dot{m}_3 = \dot{m}_1 + \dot{m}_2$, which states that at a junction, the sum of mass flow rates flowing in the junction is equal to the sum of mass flow rates flowing away from the junction. Now let us consider the case shown in Fig. 7.3b, where a pipe is connected to a water tank. We know that the mass flow rate $\dot{m}$ is proportional to the water level z and that this relation may be expressed in the following form: $z = R\dot{m}$,

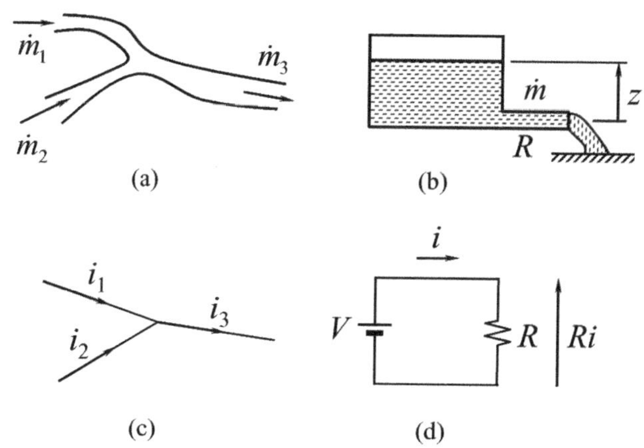

Figure 7.3 Analogy between fluidic and electrical systems.

where R denotes the flow resistance of the pipe. Since the pressure near the entrance of the pipe is proportional to the water level, this relation can be interpreted as meaning that the pressure applied across the fluid resistance (the pipe) is equal to the pressure drop through the resistance.

These phenomena in fluidic systems are analogous to the phenomena between electrical current and voltage in electrical circuits. For example, at the junction of the electrical circuit shown in Fig. 7.3c, the sum of current (electric current) flowing in the junction is equal to the sum of current flowing away from the junction; that is,

$$i_1 + i_2 - i_3 = 0$$

If we define the positive current as the current flowing in a junction and the negative current as the current flowing away from the junction, *Kirchhoff's current law* may be expressed by

$$\sum_j i_j = 0 \tag{7.25}$$

where i_j denotes the jth current. Since the voltage V from a voltage source in Fig. 7.3d is equal to the voltage drop Ri across the resistance (electric resistance), Kirchhoff's voltage law states that the sum of the voltage rise (e.g., the voltage of a battery) in the closed loop of a circuit is equal to the sum of the voltage drops in the loop. That is, *Kirchhoff's voltage law* can be expressed in the form

$$\sum_j V_j = 0 \tag{7.26}$$

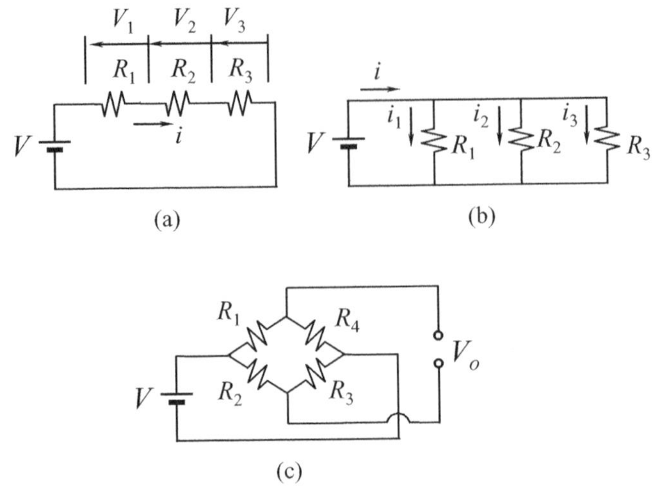

Figure 7.4 Series and parallel resistances.

where V_j represents the jth voltage rise or drop. For the circuit in Fig. 7.3d, equation (7.26) is written as $V - Ri = 0$.

Using the current and voltage laws for electric circuits, we can determine the voltage and current of the circuit. Let us analyze typical circuits in Fig. 7.4, which we frequently encounter. For a series connection of resistors in Fig. 7.4a, Kirchhoff's voltage law gives $V - R_1 i - R_2 i - R_3 i = 0$, which in turn yields $i = V/(R_1 + R_2 + R_3)$. If we are interested in the voltage across R_3, the voltage is obtained as

$$V_3 = R_3 i = \frac{R_3}{R_1 + R_2 + R_3} V$$

The other voltage drops can be obtained similarly. Suppose that resistors are connected in parallel, as shown in Fig. 7.4b. For the parallel resistors, Kirchhoff's current law gives $i - i_1 - i_2 - i_3 = 0$, which can be combined with Ohm's law to provide

$$i = \left(\frac{1}{R_1} + \frac{1}{R_2} + \frac{1}{R_3} \right) V$$

The equivalent resistances of series and parallel resistors, such as Fig. 7.4a and b are defined as $R = V/i$ and written, respectively, as follows:

$$R = R_1 + R_2 + \cdots + R_n \tag{7.27}$$

$$\frac{1}{R} = \frac{1}{R_1} + \frac{1}{R_2} + \cdots + \frac{1}{R_n} \tag{7.28}$$

Kirchhoff's laws and the equivalent resistances are very useful in analyzing electric circuits. As an example, let us analyze the output voltage of the Wheatstone bridge shown in Fig. 7.4c. We consider the two current flows through the resistors: R_1 and R_4; R_2 and R_3. The voltage V_1 at the upper point between R_1 and R_4 is given by

$$V_1 = \frac{R_4}{R_1 + R_4} V$$

and the voltage V_2 at the lower point between R_2 and R_3 is written as

$$V_2 = \frac{R_3}{R_2 + R_3} V$$

The output voltage V_o is obtained as a function of the input voltage V and the four resistances,

$$V_o = V_1 - V_2 = \left(\frac{R_4}{R_1 + R_4} - \frac{R_3}{R_2 + R_3} \right) V = \frac{R_2 R_4 - R_1 R_3}{(R_2 + R_3)(R_1 + R_4)} V \qquad (7.29)$$

If the resistances are varied, the output voltage V_o deviates from the initial value. A measuring device measures this voltage change, and we can then obtain the resistance change.

Kirchhoff's laws can also be used to analyze more complicated circuits. Some common circuits are shown in Fig. 7.5. Applying Kirchhoff's voltage law to the circuit in Fig. 7.5a and recalling the relations (7.1), (7.3), and (7.5) between the voltage and current, we have

$$L\frac{di}{dt} + Ri + \frac{1}{C}\int_0^t i\,dt = V - V(0)$$

Differentiating the equation above with respect to time, we find the following second-order differential equation:

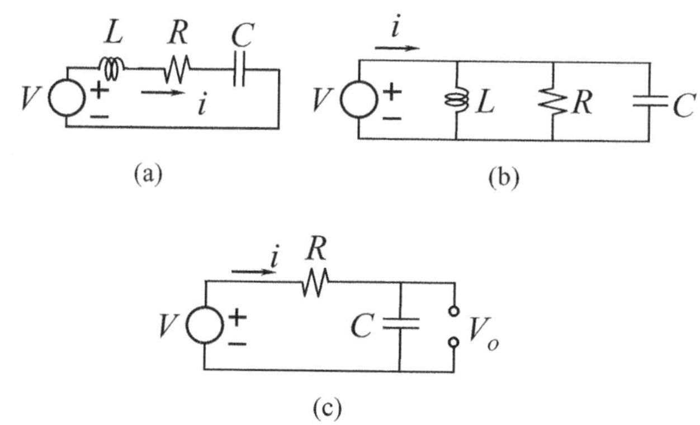

Figure 7.5 Circuits with an inductor, a resistor, and a capacitor.

$$L\frac{d^2i}{dt^2}+R\frac{di}{dt}+\frac{1}{C}i=\frac{dV}{dt} \tag{7.30}$$

We shall obtain the steady-state response to an ac voltage $V = V_a \cos \omega t$, where V_a denotes the amplitude of the ac voltage. Since equation (7.30) is linear, the response i can be expressed as

$$i = i_a \cos(\omega t - \phi) \tag{7.31}$$

where i_a and ϕ represent the amplitude and phase of the current i, respectively. Since we are interested in the steady-state response, we shall use the complex number method, described in Section 5.6. Note that all variables and constants in (7.30) are real numbers. In most electric circuits, the time constant representing the period of transient response is very short, so it is convenient to use the complex number method to obtain the steady-state response of electric circuits, as we used it in Section 5.6 to obtain the steady-state response of linear systems. If in the preceding differential equations we replace the applied voltage $V = V_a \cos \omega t$ with $V = V_a e^{j\omega t}$, where $j=\sqrt{-1}$, the response (7.31) can be written in complex number form:

$$i = i_a e^{j(\omega t-\phi)} = i_{a1} e^{j\omega t} \tag{7.32}$$

where $i_{a1} = i_a e^{-\phi}$ is a complex number that we determine later. If we obtain the solution i in complex number form, the response will be obtained as $i = \mathrm{Re}(i)$. If the ac voltage is given by an ac voltage $V = V_a \sin \omega t$, the response will be $i = \mathrm{Im}(i)$. Substituting the preceding current for the complex number into (7.30) yields

$$\left[(j\omega)^2 L + j\omega R + \frac{1}{C}\right] i_{a1} = j\omega V_a$$

and dividing the equation above by $j\omega$ gives

$$\left(j\omega L + R + \frac{1}{j\omega C}\right) i_{a1} = V_a$$

We then have the current in the complex number:

$$\begin{aligned} i_{a1} &= \frac{1}{j\omega L + R + 1/j\omega C} V_a = \frac{1}{R + j(\omega L - 1/\omega C)} V_a \\ &= \frac{1}{\sqrt{R^2 + (\omega L - 1/\omega C)^2}} V_a e^{-j\phi} \end{aligned} \tag{7.33}$$

where

$$\phi = \tan^{-1}\frac{\omega L - 1/\omega C}{R}$$

The current in the complex number is given by (7.32):

$$i = \frac{1}{\sqrt{R^2 + (\omega L - 1/\omega C)^2}} V_a e^{j(\omega t - \phi)} \tag{7.34}$$

which in turn yields the current i in the circuit in Fig. 7.5a:

$$i = \mathrm{Re}(i) = \frac{1}{\sqrt{R^2 + (\omega L - 1/\omega C)^2}} V_a \cos(\omega t - \phi) \tag{7.35}$$

As noted in Fig. 7.6 and (7.30)–(7.35), for convenience, physical variables (e.g., the current and voltage) and the governing equation, (7.30), in the real domain are converted to corresponding variables and the governing equation in the complex domain. Then the governing equations are solved in the complex domain for complex variables, and the complex variables are converted back to real variables. This strategy is very powerful if circuits (or in general, any linear systems) become complicated.

For easy evaluation of the steady-state response of linear electric circuits to a sinusoidal voltage or current, the impedances of the resistor, capacitor, and inductor are defined as follows:

$$Z_R = R \qquad Z_C = \frac{1}{j\omega C} \qquad Z_L = j\omega L \tag{7.36}$$

where Z represents the impedance, defined by $Z = V/i$, where V and i are the complex voltage and current across the passive element. The impedance above

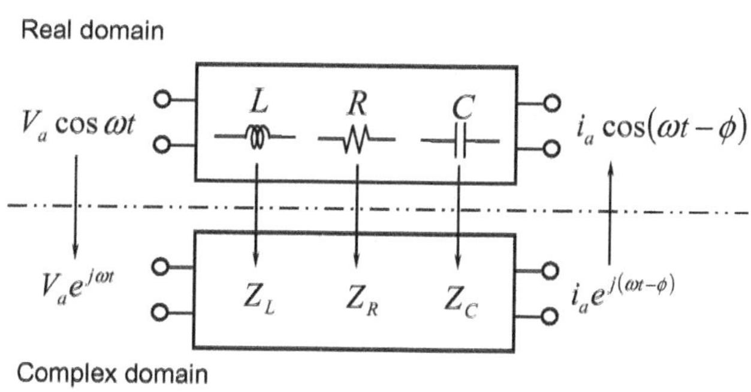

Figure 7.6 Electric circuits in real and complex domains.

can be interpreted as a measure of opposition to an ac voltage in the complex domain. A comparison between problems in the real and complex domains is shown in Fig. 7.6, where the symbols Z_L, Z_R, and Z_C represent the impedance caused by the inductor, resistor, and capacitor for modeling the electric system in complex domains. When we use the preceding impedance of the basic elements, we can use Kirchhoff's current and voltage laws to obtain a variable of interest directly, without considering the detailed differential equation. As an example, we shall again solve the preceding problem, shown in Fig. 7.5a, with using impedance. The voltage $V_a \cos \omega t$ in the real domain is considered as $V_a e^{j\omega t}$ in the complex domain. We then apply Kirchhoff's voltage law to the circuit in Fig. 7.5a to give

$$V - V_L - V_R - V_C = 0$$

Substituting the complex voltages $V_L = Z_L i$, $V_R = Ri$, and $V_C = Z_C i$ into the equation above, we find the complex current as follows:

$$i = \frac{V}{Z}$$

where Z is given by $Z = Z_L + Z_R + Z_C = j\omega L + R + 1/j\omega C$. Substitution of the complex voltage, $V = V_a e^{\omega t}$, into the equation above gives the complex current as

$$i = \frac{1}{|Z| e^{j\phi}} V_a e^{j\omega t} = \frac{1}{|Z|} V_a e^{j(\omega t - \phi)}$$

Substituting the impedance into the preceding equation yields

$$i = \frac{1}{\sqrt{R^2 + (\omega L - 1/\omega C)^2}} V_a e^{j(\omega t - \phi)}$$

where

$$\phi = \tan^{-1} \frac{\omega L - 1/\omega C}{R}$$

The current in the real domain is obtained as the real part of the complex current,

$$i = \mathrm{Re}(i) = \frac{1}{\sqrt{R^2 + (\omega L - 1/\omega C)^2}} V_a \cos(\omega t - \phi)$$

The foregoing equation is the same as (7.35), and we can then use the strategy to solve the problem that is illustrated in Fig. 7.6. In general, the impedance

Z for a linear circuit (or a linearized circuit at an operating point) in the complex domain of Fig. 7.6 is defined in the complex domain as follows:

$$Z = \frac{V}{i} = R + jX = |Z|e^{j\phi} \tag{7.37}$$

where $V, i, R,$ and X denote the complex voltage and current and the resistance and reactance, respectively. The resistance R dissipates the electrical energy into heat and the reactance X stores the electrical energy in space.

Let us apply the strategy shown in Fig. 7.6 to analyze the circuit in Fig. 7.5b. We wish to find the current i, the response to a sinusoidal voltage, in the real domain of Fig. 7.5b. Referring to Kirchhoff's current law and the definition of the impedance of the passive elements (7.36),

$$i = \frac{1}{Z}V$$

where

$$\frac{1}{Z} = \frac{1}{Z_L} + \frac{1}{Z_R} + \frac{1}{Z_C} = \frac{1}{j\omega L} + \frac{1}{R} + \frac{1}{1/j\omega C} = \frac{1}{R} + j\left(\omega C - \frac{1}{\omega L}\right)$$

is the impedance, which can also be obtained from (7.28) if we substitute Z for R, Z_L for R_1, R for R_2, and Z_C for R_3, respectively. Substituting $V = V_a e^{j\omega t}$ into the foregoing relation between the complex current and voltage, we have

$$i = \left[\frac{1}{R} + j\left(\omega C - \frac{1}{\omega L}\right)\right] V_a e^{j\omega t} = \sqrt{\frac{1}{R^2} + \left(\omega C - \frac{1}{\omega L}\right)^2}\, V_a e^{j(\omega t - \phi)}$$

where $\phi = -\tan^{-1}(\omega RC - R/\omega L) = \tan^{-1}[R/(\omega L) - \omega RC]$. We then obtain the following current i in the real domain:

$$\begin{aligned} i &= \mathrm{Re}\left[\sqrt{\frac{1}{R^2} + \left(\omega C - \frac{1}{\omega L}\right)^2}\, V_a e^{j(\omega t - \phi)}\right] \\ &= \sqrt{\frac{1}{R^2} + \left(\omega C - \frac{1}{\omega L}\right)^2}\, V_a \cos(\omega t - \phi) = i_a \cos(\omega t - \phi) \end{aligned} \tag{7.38}$$

which is the response to the voltage $V_a \cos \omega t$. If the input voltage is $V_a \sin \omega t$, the response will be $i = i_a \sin(\omega t - \phi)$. From equation (7.38) we have the ratio of the amplitudes of the current and voltage:

$$\frac{i_a}{V_a} = \sqrt{\frac{1}{R^2} + \left(\omega C - \frac{1}{\omega L}\right)^2} \tag{7.39}$$

If we are interested in the ratio of the amplitudes, i_a/V_a, we straightforwardly obtain the ratio from (7.37), given by

$$\frac{i_a}{V_a} = \frac{\left|i_a e^{j(\omega t-\phi)}\right|}{\left|V_a e^{j\omega t}\right|} = \left|\frac{i}{V}\right| = |Z| = \sqrt{\frac{1}{R^2} + \left(\omega C - \frac{1}{\omega L}\right)^2}$$

which is the same as (7.39). The low-pass filter shown in Fig. 7.5c can be used to filter high-frequency signals or to detect an envelope of a modulated signal. We shall use the impedance to obtain the output signal V_o in Fig. 7.5c. Assuming the input voltage to be $V_a \cos \omega t$ in the real domain, we have the output voltage V_o in the complex domain:

$$\begin{aligned} V_o &= \frac{Z_C}{Z_R + Z_C} V = \frac{1/j\omega C}{R + 1/j\omega C} V_a e^{j\omega t} = \frac{1}{1 + j\omega RC} V_a e^{j\omega t} \\ &= \frac{1}{\sqrt{1+(\omega RC)^2}} V_a e^{j(\omega t - \phi)} \end{aligned}$$

where $\phi = \tan^{-1}(\omega RC)$. The output in the real domain is then obtained as

$$V_o = \mathrm{Re}(V_o) = \frac{1}{\sqrt{1+(\omega RC)^2}} V_a \cos(\omega t - \phi) \tag{7.40}$$

It is noted that the amplitude of the output voltage V_o at $\omega = 1/RC$ is $V_a/\sqrt{2}$, and the corresponding angular frequency at this point is known as the *cutoff frequency*,

$$\omega_c = \frac{1}{RC} \tag{7.41}$$

which represents the bandwidth of the low-pass filter.

7.3 ELECTROSTATICS

In the microscale world, the electrostatic force, governed by Coulomb's law, becomes more important since the force is proportional to the area of an object. We begin with Coulomb's law and describe electrostatic force and moment acting on an object that is electrically charged. Let us consider two identical spheres in a vacuum, as shown in Fig. 7.7a, which are positively charged and supported under gravity by strings. Since the force equilibrium in Fig. 7.7b gives the electrostatic force $F = mg \tan \theta$, where mg denotes the gravitational force acting on the mass, the distance between the spheres depends on the electrostatic force. Experiments show that the electrostatic

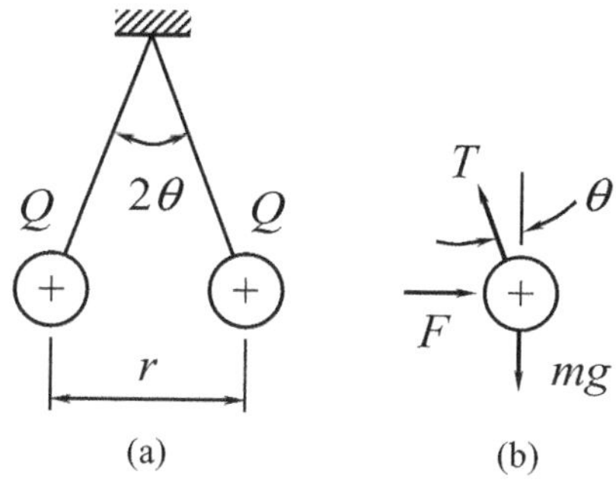

Figure 7.7 Force between two charged spheres.

force F is proportional to the amount of charge of the spheres and inversely proportional to the square of the distance between the spheres (i.e., $F \propto QQ/r^2$). The electrostatic force can be written in the form

$$F = \frac{1}{4\pi\varepsilon_0}\frac{Q^2}{r^2} \tag{7.42}$$

where ε_0 is the permittivity of vacuum, which is 8.854×10^{-12}F/m. If the spheres of charges, Q_1 and Q_2, are placed in a fluid (gas or liquid) of permittivity $\varepsilon = \varepsilon_0\varepsilon_r$, where ε_r denotes the relative permittivity of the fluid, the electrostatic force exerted to the right sphere may be written in vector form,

$$\mathbf{F} = \frac{1}{4\pi\varepsilon}\frac{Q_1Q_2}{r^2}\frac{\mathbf{r}}{r} \tag{7.43}$$

where $\mathbf{r}$ is the displacement vector from the left sphere to the right sphere and $\mathbf{r}/r$ then denotes a unit vector showing the direction of application of the force. Equation (7.43), known as *Coulomb's law*, serves as a basic equation for electrostatics.

Let us consider the electrostatic force generated by positive and negative charges. In Fig. 7.8a, a small sphere of positive charge Q is located at a point in space, and a particle of positive charge q is placed at a distance r from the center of the sphere. In this configuration, the sphere exerts an electrostatic force on the particle according to Coulomb's law, as follows:

$$\mathbf{F} = \frac{1}{4\pi\varepsilon_0}\frac{Qq}{r^2}\frac{\mathbf{r}}{r}$$

If the particle is placed on the dashed-dotted circle in Fig. 7.8a, the amplitudes of the electrostatic force acting on the particle are the same and the force is directed in the radial direction, as shown in Fig. 7.8a. In order words, a positive charge exerts a repulsive force on another positive charge. As we divide the preceding equation by q, we obtain the *electric field strength* (also known as the *electric field intensity*) **E**, which is defined as the electrostatic force acting on a unit positive charge:

$$\mathbf{E} = \frac{\mathbf{F}}{q} = \frac{1}{4\pi\varepsilon_0}\frac{Q}{r^2}\frac{\mathbf{r}}{r} \tag{7.44}$$

If the sphere is charged with a negative charge, $-Q$, the electrostatic force, exerted on the particle of positive charge q, will be

$$\mathbf{E}_1 = \frac{\mathbf{F}_1}{q} = -\frac{1}{4\pi\varepsilon_0}\frac{Q}{r^2}\frac{\mathbf{r}}{r}$$

which is illustrated in Fig. 7.8b. Since a force is a vector quantity, the electric field strength **E** is also a vector, as shown in (7.44). Suppose that the two charges Q_1 and Q_2 are first placed in space to form electric field strength and that we place a third one, Q, as shown in Fig. 7.8c. In this case, the third one experiences the forces $\mathbf{F}_1$ and $\mathbf{F}_2$ generated by Q_1 and Q_2, and we then write the force acting on Q as a sum of vectors as follows:

$$\mathbf{F} = \mathbf{F}_1 + \mathbf{F}_2 = \frac{1}{4\pi\varepsilon_0}\frac{Q_1 Q}{r_1^2}\mathbf{n}_1 + \frac{1}{4\pi\varepsilon_0}\frac{Q_2 Q}{r_2^2}\mathbf{n}_2$$

where $\mathbf{n}_1$ and $\mathbf{n}_2$ denote the unit vectors for the force $\mathbf{F}_1$ and $\mathbf{F}_2$, respectively. Dividing the foregoing force by Q gives the electric field strength **E** formed by Q_1 and Q_2:

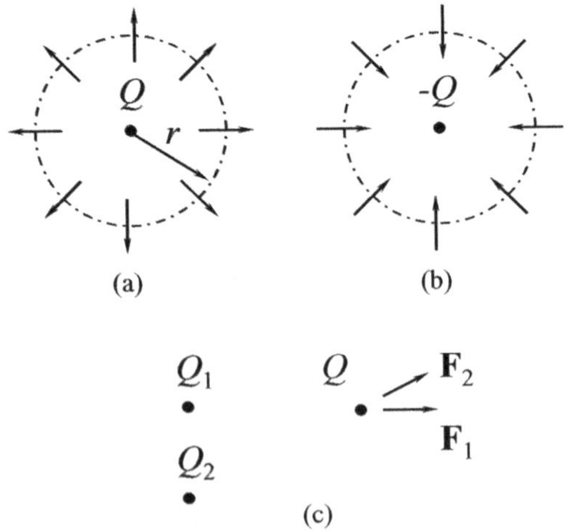

Figure 7.8 Forces due to charged particles.

$$\mathbf{E} = \frac{\mathbf{F}}{Q} = \frac{1}{4\pi\varepsilon_0}\frac{Q_1}{r_1^2}\mathbf{n}_1 + \frac{1}{4\pi\varepsilon_0}\frac{Q_2}{r_2^2}\mathbf{n}_2 = \mathbf{E}_1 + \mathbf{E}_2$$

which states that the electric field strength **E** is a vector, and **E**, due to more than one charge, can be obtained by the vector sum. When N charged particles are involved to generate an electric field strength **E**, the electric field strength **E** is obtained as

$$\mathbf{E} = \sum_{j=1}^{N} \mathbf{E}_j \tag{7.45}$$

Electric Potential Energy

We shall consider the electric potential energy that is associated with Coulomb force. Let us suppose (Fig. 7.9a) that a machine takes one positive charge Q_a and moves it from point 1 to point 2, closer to another positive charge, Q. During the process, a repulsive force is generated by the charges, and the machine then does work on the charge Q_a to overcome the repulsive force. Referring to the free-body diagram in Fig. 7.9b, we have $F_s = F$ or $\mathbf{F} = -\mathbf{F}_s = F\mathbf{i}$, where **F**, $\mathbf{F}_s$, and **i** denote the force exerted by the left charge Q on Q_a, the force exerted by the machine on Q_a, and the unit vector, respectively. Coulomb's law gives the force F as follows:

$$\mathbf{F} = \frac{1}{4\pi\varepsilon_0}\frac{QQ_a}{r^2}\mathbf{i} = EQ_a\mathbf{i}$$

where

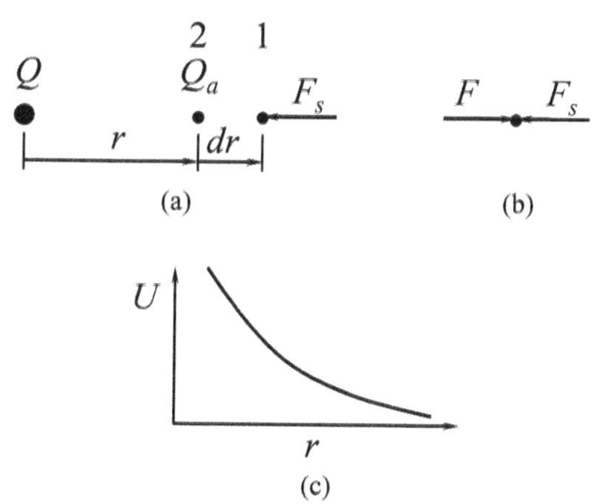

Figure 7.9 Electric potential energy.

$$E = \frac{1}{4\pi\varepsilon_0}\frac{Q}{r^2}$$

is the electric field strength generated by the charge Q. During the infinitesimal displacement, $-d\mathbf{r}$, from position 1 to position 2, the electric potential energy dU stored in space is equal to the work, $dW = -\mathbf{F}_s\cdot(-d\mathbf{r})$, done by the machine. The energy balance, $dU = dW = -\mathbf{F}_s\cdot(-d\mathbf{r}) = \mathbf{F}\cdot(-d\mathbf{r})$, may be written as follows:

$$dU = \mathbf{F}\cdot(-d\mathbf{r}) = \frac{1}{4\pi\varepsilon_0}\frac{QQ_a}{r^2}\mathbf{i}\cdot(-dr\,\mathbf{i}) = -\frac{1}{4\pi\varepsilon_0}\frac{QQ_a}{r^2}dr$$

Integrating the preceding equation from an infinite distance to a distance r gives the electric potential energy,

$$U = -\int_{\infty}^{r}\frac{1}{4\pi\varepsilon_0}\frac{QQ_a}{r^2}dr = \frac{1}{4\pi\varepsilon_0}\frac{QQ_a}{r} \tag{7.46}$$

which is plotted against the distance r in Fig. 7.9c. It is noted that the potential energy stored in space is used to overcome the repulsive force due to the same types of charges. If one charge is positive and the other is negative, the interactive force acting on the charges will be attractive according to Coulomb's law (7.43) and the electric potential energy will be negative. As the charge moves from point 1 to point 2 in Fig. 7.9a, the change in the potential energy is given by

$$U_{12} = U_2 - U_1 = \frac{QQ_a}{4\pi\varepsilon_0}\left(\frac{1}{r_2} - \frac{1}{r_1}\right) \tag{7.47}$$

Since the preceding energy balance for the infinitesimal displacement is rearranged as

$$dU = \mathbf{F}\cdot(-d\mathbf{r}) = Q_a\frac{1}{4\pi\varepsilon_0}\frac{Q}{r^2}\mathbf{i}\cdot(-dr\,\mathbf{i}) = -Q_a\mathbf{E}\cdot d\mathbf{r}$$

the difference in the potential energy per unit charge is obtained as

$$V_{12} = \frac{U_{12}}{Q_a} = -\int_1^2 \mathbf{E}\cdot d\mathbf{r} \tag{7.48}$$

where V_{12} is defined as the *voltage difference* (also known as the *voltage*), which is a scalar quantity, and $\mathbf{E}$ is the electric field strength generated by the charge Q, which is given by

$$\mathbf{E} = \frac{1}{4\pi\varepsilon_0}\frac{Q}{r^2}\frac{\mathbf{r}}{r}$$

We then obtain the voltage difference V_{12}, which represents the energy needed to move a unit positive charge from point 1 to point 2:

$$V_{12} = -\int_1^2 \mathbf{E}\cdot d\mathbf{r} = -\int_1^2 \frac{1}{4\pi\varepsilon_0}\frac{Q}{r^2}dr = \frac{Q}{4\pi\varepsilon_0}\left(\frac{1}{r_2}-\frac{1}{r_1}\right) \tag{7.49}$$

We can also obtain this equation by dividing (7.47) by the charge Q_a.

The relations between the electric potential energy U and the electric field strength $\mathbf{E}$ due to one or two point charges have been considered in the preceding paragraphs. If more charges are involved to generate the electric field, the electric field strength $\mathbf{E}$ is given by the superposition of the electric field, (7.45). Let us obtain a relation between the voltage V and the electric field strength $\mathbf{E}$. If an electric field is formed in which equipotential lines (i.e., constant voltage lines) are as shown in Fig. 7.10a, the relation between the voltage V (electric potential energy per unit positive charge) and the electric field strength $\mathbf{E}$ may be written from (7.48) as follows:

$$\begin{aligned} dV &= -\mathbf{E}\cdot d\mathbf{r} = -(E_x\mathbf{i}+E_y\mathbf{j}+E_z\mathbf{k})\cdot(dx\,\mathbf{i}+dy\,\mathbf{j}+dz\,\mathbf{k}) \\ &= -(E_x\,dx+E_y\,dy+E_z\,dz) \end{aligned}$$

where E_x, E_y, E_z, dx, dy, and dz denote the components of the vector $\mathbf{E}$ in the x, y, and z directions and the components of the displacement $d\mathbf{r}$, respectively.

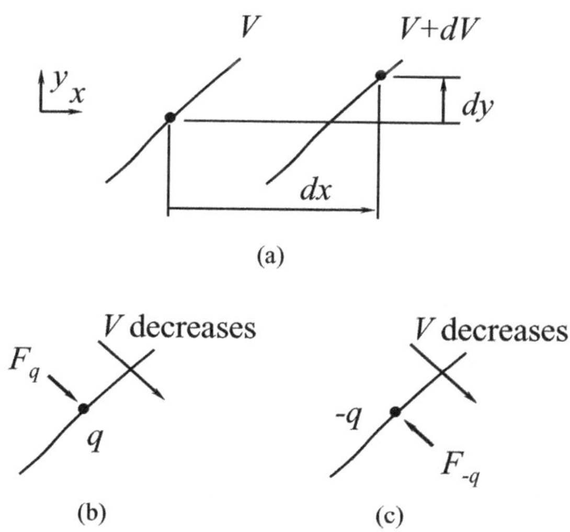

Figure 7.10 Force exerted on a charged particle in an electrical field.

If dy and dz are zero, the preceding equation reads $dV = -E_x dx$, which in turn yields the x component of the electric field strength **E**:

$$E_x = -\frac{\partial V}{\partial x}$$

Similarly,

$$E_y = -\frac{\partial V}{\partial y} \qquad E_z = -\frac{\partial V}{\partial z}$$

The electric field strength is then obtained as a function of the voltage V:

$$\begin{aligned}\mathbf{E} &= E_x\mathbf{i} + E_y\mathbf{j} + E_z\mathbf{k} = -\left(\frac{\partial V}{\partial x}\mathbf{i} + \frac{\partial V}{\partial y}\mathbf{j} + \frac{\partial V}{\partial z}\mathbf{k}\right) \\ &= -\nabla V \end{aligned} \tag{7.50}$$

where ∇ denotes the gradient operator. It is noted that the **E** is applied in the direction in which the voltage (or electric potential) decreases; that is, if a positive charge q in Fig. 7.10b is placed in the electric field, the charge is subjected to an electric force $\mathbf{F}_q = \mathbf{E}q = -q\nabla V$, which is applied in the direction in which the voltage decreases as shown in Fig. 7.10b. If a negative charge, $-q$, is placed in the same point in the field, an electric force $\mathbf{F}_{-q} = \mathbf{E}(-q) = q\nabla V$ is exerted on the charge as shown in Fig. 7.10c, where the negative charge tends to move in the direction in which the voltage V increases.

Laplace Equation

The electrostatic force acting between conductive bodies, which is electrically charged, is governed by the Laplace equation. It is instructive to derive the Laplace equation for the electrostatic problems. First we introduce electric flux due to the electric field and then obtain Laplace's equation. Before proceeding with electric problems, let us first consider the spring in Fig. 7.11a, from which water flows out. If flow from the spring is in the steady state, the mass flow rate (mass per unit time) remains constant. From our experience we know that the mass flow rate passing through the circle of radius r in Fig. 7.11a is equal to that passing through an enclosing curve; this relation, known as the *continuity equation*, was dealt with in Section 6.2. Now let us turn our attention to the positive electric charge Q in space shown in Fig. 7.11a. Like the water spring, we assume that an electric flux flows out from the charge. Due to the spherical symmetry, the electric flux as well as the electric field are considered to flow radially outward from the charge. We then assume that the electric flux density **D**, which is defined as the electric flux per unit area, is proportional to the electric field strength **E**; that is, the relation between **D** and **E** is expressed mathematically as

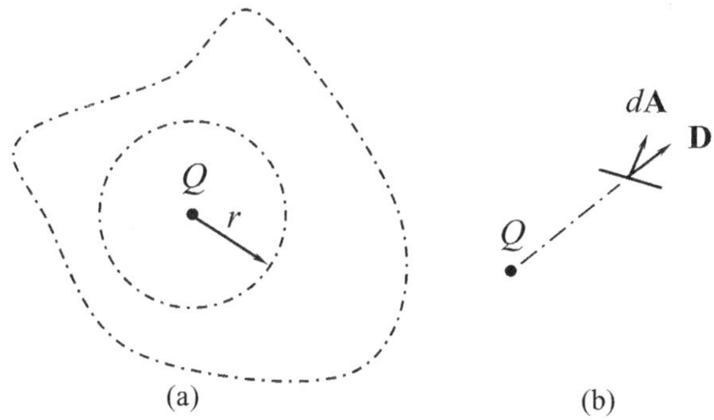

Figure 7.11 Electric flux from an electrically charged body.

$$\mathbf{D} = \alpha \mathbf{E}$$

where α is a constant. When we consider the sphere of radius r shown in Fig. 7.11a, the total flux ψ is given by

$$\psi = D \cdot 4\pi r^2 = \alpha E 4\pi r^2 = \alpha \frac{1}{4\pi\varepsilon_0} \frac{Q}{r^2} 4\pi r^2 = \frac{\alpha}{\varepsilon_0} Q$$

where D and E denote the magnitudes of $\mathbf{D}$ and $\mathbf{E}$. When we define the total flux ψ to be the amount of charge, Q (i.e., $\psi = Q$), the constant must be equal to the permittivity ε_0, and we then obtain the relation between $\mathbf{D}$ and $\mathbf{E}$ as follows:

$$\mathbf{D} = \varepsilon_0 \mathbf{E} \tag{7.51}$$

If we draw an arbitrary shape in Fig. 7.11a that encloses the charge Q completely, the total flux ψ is expressed as follows:

$$\psi = Q = \oint_A \mathbf{D} \cdot d\mathbf{A}$$

where $d\mathbf{A}$ is an infinitesimal area on the arbitrary shape outward from the charge as shown in Fig. 7.11b. Even though shapes of arbitrary size are drawn, the total flux is conserved and equal to the amount of charge. When we place two charges, Q_1 and Q_2, in the space in Fig. 7.12, the total flux passing through the left surface A_1 in Fig. 7.12 is Q_1, and that passing through the right surface A_2 is Q_2. If we consider a merged surface that encloses the charges Q_1 and Q_2 as shown in Fig. 7.12, the total flux will be the sum of Q_1 and Q_2, or, mathematically,

$$\oint_A \mathbf{D} \cdot d\mathbf{A} = \oint_{A_1} \mathbf{D} \cdot d\mathbf{A} + \oint_{A_2} \mathbf{D} \cdot d\mathbf{A} = Q_1 + Q_2$$

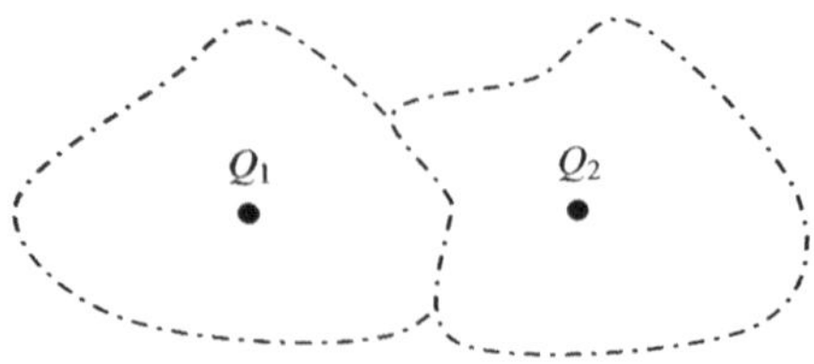

Figure 7.12 Electric flux from two charged bodies.

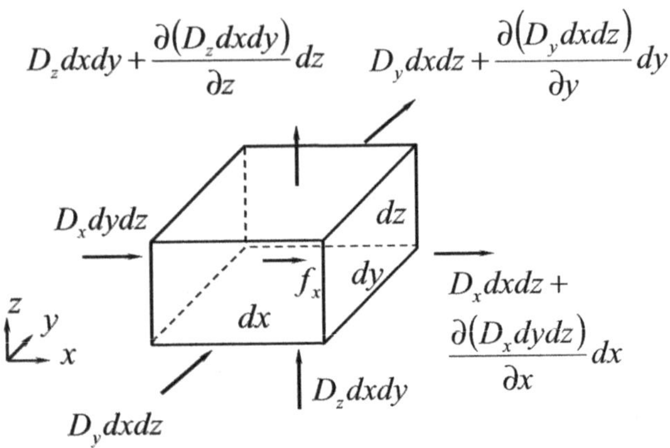

Figure 7.13 Electric flux through a rectangular volume.

where A denotes the surface of the merged shape. In general, if charges (Q_1, $Q_2, \cdots, Q_n$) are located in an enclosing surface A, the total electric flux is equal to the sum of the charges involved and is expressed as follows:

$$\oint_A \mathbf{D} \cdot d\mathbf{A} = \sum_j Q_j \tag{7.52}$$

which is known as *Gauss' law*. Equation (7.52) is the integral form of Gauss's law, which can be converted to the differential form of Gauss's law if we apply the preceding Gauss's law to the small volume of space shown in Fig. 7.13. The small volume is filled with a material of charge density ρ_c and the electric flux flows in and out of the volume. Referring to Fig. 7.13, Gauss's law may be written

$$\left(\frac{\partial D_x}{\partial x} + \frac{\partial D_y}{\partial y} + \frac{\partial D_z}{\partial z} \right) dx\, dy\, dx = \rho_c\, dx\, dy\, dz$$

which is divided by the volume $dxdydz$ to give the differential form of Gauss's law:

$$\nabla \cdot \mathbf{D} = \rho_c \tag{7.53}$$

Integrating equation (7.53) over volume yields

$$\int_{V_1} \nabla \cdot \mathbf{D} \, dV_1 = \int_{v} \rho_c \, dV_1$$

where V_1 is the volume enclosed by the surface A. The right-hand side of the preceding equation is the total charge in the volume and is equal to that of (7.52). This conclusion leads us to *Gauss' theorem* (also known as the *divergence theorem*),

$$\int_{V_1} \nabla \cdot \mathbf{D} \, dV_1 = \oint_A \mathbf{D} \cdot d\mathbf{A} \tag{7.54}$$

where **D** denote the electric flux density. If **D** is interpreted as a flux of a physical quantity, Gauss' theorem can be used to convert volume integrals of the flux to corresponding surface integrals, and vice versa. Substituting (7.50) into (7.51) and combining the resulting equation with Gauss' law (7.53) gives Poisson's equation,

$$\nabla^2 V = -\frac{\rho_c}{\varepsilon_0} \tag{7.55}$$

where $\nabla^2 = \nabla \cdot \nabla$ is the Laplacian (Laplace operator), defined as $\partial^2/\partial x^2 + \partial^2/\partial y^2 + \partial^2/\partial z^2$ in the Cartesian coordinate system (the rectangular coordinate system). If the charge density ρ_c is specified in space, the voltage distribution V can be obtained by solving equation (7.55). When a voltage V_1 is applied between conductive bodies such as the metal plates in Fig. 7.14a, the theory of electromagnetics shows that the charges are distributed on the surface of the conductive bodies and no charge is in the free space (or any insulating material of permittivity ε) between the bodies. In this situation, Poisson's equation reduces to Laplace's equation,

$$\nabla^2 V = 0 \tag{7.56}$$

If we wish to find the voltage distribution V between conductive bodies (e.g., Fig. 7.14), Laplace's equation should be solved with boundary conditions.

7.4 FORCE AND MOMENT DUE TO AN ELECTRIC FIELD

The working principles of many MEMS are based on the electrostatic force and moment due to Coulomb's force, which has been described in preceding sections. Let us begin with conductive plates in Fig. 7.14a, where an upper plate is spaced

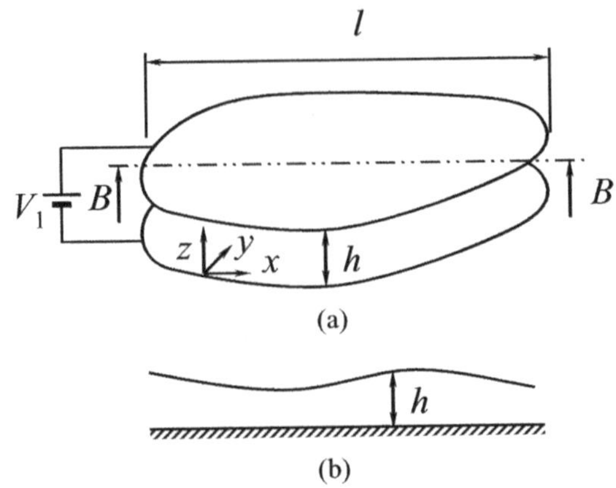

Figure 7.14 Plates with a small gap.

by h from a lower plate. The thin gap h between the plates is not uniform, as shown in Fig. 7.14b. Laplace's equation in Cartesian coordinates reads

$$\frac{\partial^2 V}{\partial x^2}+\frac{\partial^2 V}{\partial y^2}+\frac{\partial^2 V}{\partial z^2}=0 \tag{7.57}$$

To compare the orders of magnitude of the terms on the left-hand side of equation (7.57), we shall nondimensionalize the equation. We assume that the gap h is much less than the length of the plate and that the width is comparable to the length. Taking the dimensionless parameters $X = x/l$, $Y = y/l$, $Z = z/h$, and $V_n = V/V_c$, we have the dimensionless form of the Laplace equation:

$$\frac{h^2}{l^2}\frac{\partial^2 V_n}{\partial X^2}+\frac{h^2}{l^2}\frac{\partial^2 V_n}{\partial Y^2}+\frac{\partial^2 V_n}{\partial Z^2}=0$$

Since the gap is much less than the length and width, the first and second terms of the equation above can be neglected and the third terms remains. That is, if the gap between plates is much less than the other dimensions of the plates, Laplace's equation reduces to

$$\frac{d^2 V}{dz^2}\cong 0 \tag{7.58}$$

The following boundary conditions are imposed on the problem shown in Fig. 7.14a: $V = 0$ at $x = 0$ and $V = V_1$ at $x = h$. Equation (7.58) and the imposed boundary conditions lead us to the voltage distribution between the plates,

$$V = V_1\frac{z}{h} \tag{7.59}$$

The electric field strength **E** is then obtained from (7.50):

$$\mathbf{E}=-\left(\frac{\partial V}{\partial x}\mathbf{i}+\frac{\partial V}{\partial y}\mathbf{j}+\frac{\partial V}{\partial z}\mathbf{k}\right)\cong-\frac{V_1}{h}\mathbf{k} \tag{7.60}$$

which shows that **E** is directed from higher to lower voltage. It is noted that the preceding equation is valid since $(\partial V/\partial x)/(\partial V/\partial z) = h/l$ and $(\partial V/\partial y)/(\partial V/\partial z) \approx h/l$ (the x and y components of **E** are neglected for $h/l \ll 1$). Let us obtain the capacitance due to the plates shown in Fig. 7.14a. Substitution of (7.51) into (7.52) gives the total positive charge of the upper plate:

$$Q=\oint_A \mathbf{D}\cdot d\mathbf{A}=\oint_A \varepsilon_0\mathbf{E}\cdot d\mathbf{A} \tag{7.61}$$

The same amount of negative charge is induced on the lower plate. Most charges are accumulated on the lower face of the upper plate and on the upper face of the lower plate, so equation (7.61) may be written

$$Q\cong\int_A \varepsilon_0\left(-\frac{V}{h}\mathbf{k}\right)\cdot(-dA\mathbf{k})=\varepsilon_0 V\int_A \frac{1}{h}dA \tag{7.62}$$

where the right-hand side is obtained under the assumption of constant permittivity. The integral of equation (7.62) is taken over the lower face of the upper plate (i.e., the area facing the other plate). Using the definition of capacitance, $C = Q/V$, we find the capacitance of the plates, one of which is spaced by the gap from the other,

$$C=\frac{Q}{V}=\varepsilon_0\int_A \frac{1}{h}dA \tag{7.63}$$

As an example, for the parallel plates in Fig. 7.15, in which $h \ll l$ and $h \ll w$, the electric field strength **E** and the capacitance C are given, respectively, by

$$\mathbf{E}=-\frac{V}{h}\mathbf{k} \tag{7.64}$$

$$C=\frac{\varepsilon_0 A}{h} \tag{7.65}$$

Electric Potential Energy in Capacitors

Let us consider electric potential energy stored in a capacitor of capacitance C (Fig. 7.16b). Before proceeding with the capacitor, we first consider the analogous case in Fig. 7.16a, where we take a cup of water from the big bucket on the right and pour it into the big bucket on the left. Since the capacity of the cup is much less than that of the buckets, the water levels of the buckets

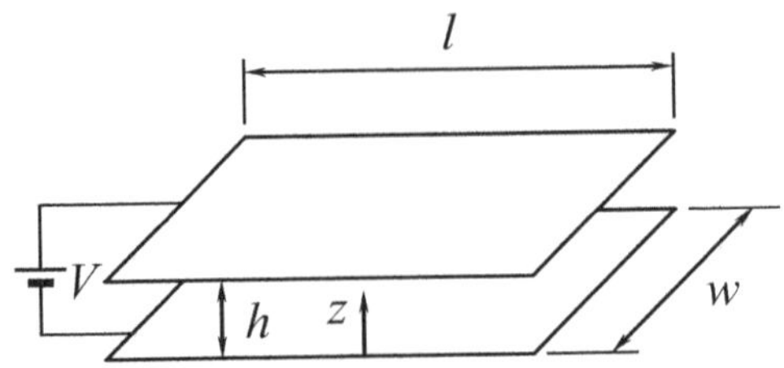

Figure 7.15 Parallel plates with a small gap.

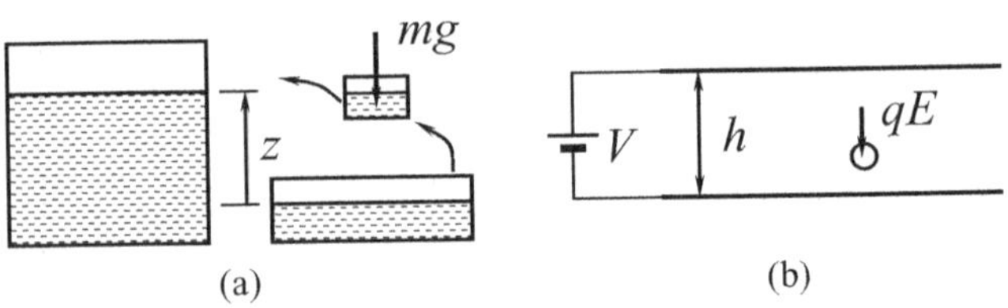

Figure 7.16 Small amount of energy added to a water tank and electric capacitor.

remain almost unchanged. If the difference in the water levels is given by z as shown in Fig. 7.16a, the work to raise the water of mass m by z is mgz, which is the gravitational potential energy stored in the water transferred. Let us turn our attention to the parallel-plate capacitor case in Fig. 7.16b, where plates spaced by a gap h are connected to a voltage source. If we take a charge q and move it from the lower plate to the upper plate, we must do an amount of work that is equal to the increase in electric potential energy. When we move the charge by dz from z, the work done is equal to the increase in the electrical potential energy, given by

$$dU = -\mathbf{F}\cdot d\mathbf{r} = -q\mathbf{E}\cdot d\mathbf{r} = -q\left(-\frac{V}{h}\mathbf{k}\right)\cdot dz\,\mathbf{k} = q\frac{V}{h}dz$$

where z is the coordinate from the lower plate and $\mathbf{F}$ is the force acting on the charge, which is given by $q\mathbf{E}$. Integrating the preceding equation from zero to h, we have the electric potential energy that is stored in the capacitor during the moving process:

$$U = \int_0^h q\frac{V}{h}dz = qV \tag{7.66}$$

This states that the potential energy is a function of the voltage difference and the amount of charge. The gap or distance between plates is not involved in the potential energy.

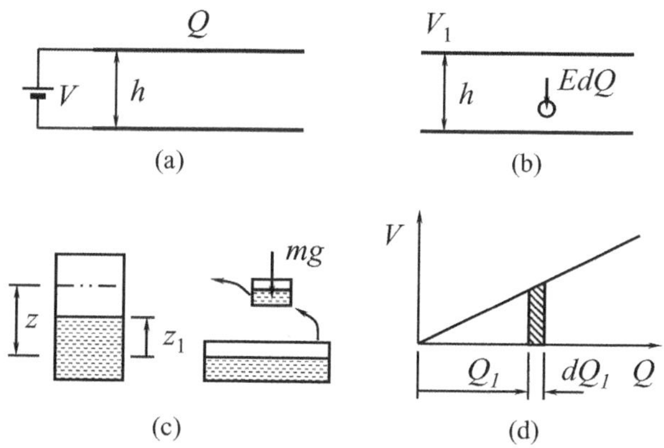

Figure 7.17 Stored energy in a water tank and electric capacitor.

We shall evaluate the energy (electric potential energy) U which is stored in a capacitor of capacitance C in Fig. 7.17a. We connect a voltage source to the capacitor and then disconnect the source from the capacitor. During this process, the capacitor is fully charged with $Q = CV$. Let us calculate the energy stored in the capacitor. We first consider an empty capacitor of capacitance C and move charges from the lower plate to the upper plate until the voltage reaches the desired voltage V. This process is very similar to that of water pumping, which is shown in Fig. 7.17c. The water level of the left bathtub of limited capacity is equal to that of the right bathtub with relatively infinite capacity (e.g., lake) at the initial stage. When we pump water, the water level z_1 becomes higher and higher. As a result, the work for pumping is proportional to the height z_1. The total work (or equivalently, the potential energy) of the pumped water is $mg(z/2) = mgz/2$, where $z/2$ denotes the height of the center of mass of the added water measured from the water source. Similarly, during the charge pumping process in Fig. 7.17b, the voltage across the plates and the amount of charge increase as shown in Fig. 7.17d. The increase in energy is obtained from (7.66) as follows:

$$dU = V\,dQ_1$$

where Q_1 represents the charge, which is a dummy variable introduced for convenience. The preceding equation gives the energy stored in the capacitor as

$$U = \int_0^Q V\,dQ \tag{7.67}$$

If the capacitance is linear (i.e., $Q = CV$ holds), equation (7.67) yields the energy stored in linear capacitors as follows:

$$U = \frac{1}{2}\frac{Q^2}{C} = \frac{1}{2}QV = \frac{1}{2}CV^2 \tag{7.68}$$

According to the theory of electromagnetics, the preceding energy is stored in the space near electrically charged conductive bodies and is, in general, expressed in the form

$$U = \frac{1}{2}\int_v \mathbf{E}\cdot\mathbf{D}\,dv \tag{7.69}$$

where the integral is taken over the volume v. Substituting $\mathbf{D} = \varepsilon\mathbf{E}$ into equation (7.69) and performing integration over the volume between the plates in Fig. 7.17, we have the following, which is the same as (7.68):

$$U = \frac{1}{2}\int_v \mathbf{E}\cdot\mathbf{D}\,dv = \frac{1}{2}\left(-\frac{V}{h}\mathbf{k}\right)\cdot\varepsilon\left(-\frac{V}{h}\mathbf{k}\right)Ah = \frac{1}{2}\frac{A}{h}V^2 = \frac{1}{2}CV^2$$

where C is the capacitance of the capacitor, which is defined to be $\varepsilon A/h$.

Force and Moment Acting on a Body with a Constant Charge

The force and moment acting on bodies with a constant charge shown in Fig. 7.18b are analogous to those acting on the bucket with water of mass m under

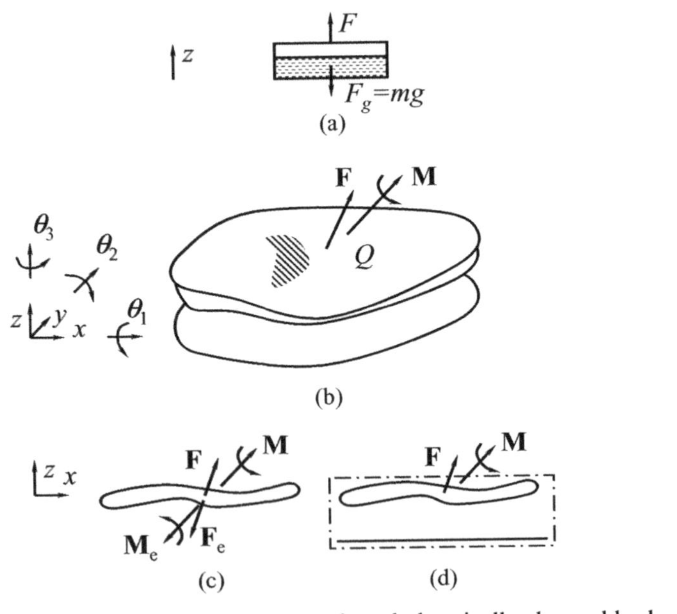

Figure 7.18 Displacement of a water tank and electrically charged body.

gravity in Fig. 7.18a. The force F to support the water bucket is in equilibrium with the gravitational force, F_g, acting on the bucket (i.e., $\mathbf{F} = -\mathbf{F}_g = mg\mathbf{k}$). Similarly, in Fig. 7.18b, a conductive body is close to another conductive body and is subjected to electrostatic force and moment generated by the electric field. The body is then in equilibrium, as shown in Fig. 7.18c; that is, $\mathbf{F} = -\mathbf{F}_e$ and $\mathbf{M} = -\mathbf{M}_e$, where $\mathbf{F}$, $\mathbf{M}$, $\mathbf{F}_e$, and $\mathbf{M}_e$ denote the force and moment exerted by a support (not shown in the figure) and the force and moment due to the electric field. We wish to find the forces and moments acting on the body. Let us consider the system enclosed in the dashed-dotted rectangle in Fig. 7.18d. The external force and moment across the boundary of the system are those due to the support. In this situation we can move the body by small displacement $d\mathbf{r}$ and angle $d\boldsymbol{\theta}$ by changing the position and angle of the support. The work done by the support is equal to the stored energy, which is expressed as

$$dU = \mathbf{F}\cdot d\mathbf{r} + \mathbf{M}\cdot d\boldsymbol{\theta}$$

Since $\mathbf{F} = -\mathbf{F}_e$ and $\mathbf{M} = -\mathbf{M}_e$, the preceding equation may be written as

$$\begin{aligned} dU &= -\mathbf{F}_e\cdot d\mathbf{r} - \mathbf{M}_e\cdot d\boldsymbol{\theta} \\ &= -(F_{ex}\mathbf{i} + F_{ey}\mathbf{j} + F_{ez}\mathbf{k})\cdot(dx\,\mathbf{i} + dy\,\mathbf{j} + dz\,\mathbf{k}) \\ &\quad -(M_{e1}\mathbf{i} + M_{e2}\mathbf{j} + M_{e3}\mathbf{k})\cdot(d\theta_1\,\mathbf{i} + d\theta_2\,\mathbf{j} + d\theta_3\,\mathbf{k}) \\ &= -(F_{ex}\,dx + F_{ey}\,dy + F_{ez}\,dz) - (M_{e1}\,d\theta_1 + M_{e2}\,d\theta_2 + M_{e3}\,d\theta_3) \end{aligned}$$

If dx is allowed and others are zero, the preceding equation is reduced to $dU = -F_{ex}dx$, which in turn gives the electrostatic force in the x direction as

$$F_{ex} = -\frac{\partial U}{\partial x}$$

Similarly, the electrostatic forces in the y and z directions are given by

$$F_{ey} = -\frac{\partial U}{\partial y} \qquad F_{ez} = -\frac{\partial U}{\partial z}$$

The electrostatic force due to the electric field is then given in vector form by

$$\begin{aligned} \mathbf{F}_e &= F_{ex}\mathbf{i} + F_{ey}\mathbf{j} + F_{ez}\mathbf{k} = -\left(\frac{\partial U}{\partial x}\mathbf{i} + \frac{\partial U}{\partial y}\mathbf{j} + \frac{\partial U}{\partial z}\mathbf{k}\right) \\ &= -\nabla U \end{aligned} \tag{7.70}$$

where ∇ is the gradient operator. Similarly, the electrostatic moment due to the electric field is written as

$$\mathbf{M}_e = M_{e1}\mathbf{i} + M_{e2}\mathbf{j} + M_{e3}\mathbf{k} = -\left(\frac{\partial U}{\partial \theta_1}\mathbf{i} + \frac{\partial U}{\partial \theta_2}\mathbf{j} + \frac{\partial U}{\partial \theta_3}\mathbf{k}\right)$$
$$= -\nabla_\theta U \tag{7.71}$$

where ∇_θ is defined as an angular gradient operator,

$$\nabla_\theta = \frac{\partial}{\partial \theta_1}\mathbf{i} + \frac{\partial}{\partial \theta_2}\mathbf{j} + \frac{\partial}{\partial \theta_3}\mathbf{k}$$

It is noted that the electrostatic force and moment are applied in the direction in which the energy stored in the electric field (i.e., the energy stored in the capacitor) decreases. The force and moment exerted by the support on the body are then obtained as

$$\mathbf{F} = -\mathbf{F}_e = \nabla U \tag{7.72}$$

$$\mathbf{M} = -\mathbf{M}_e = \nabla_\theta U \tag{7.73}$$

Force and Moment Acting on a Body Connected to a Voltage Source

Let us reconsider the electrostatic force and moment acting on the same body while a voltage source is connected to the body as shown in Fig. 7.19b. We first consider an analogous case in Fig. 7.19a, where a small water bucket is connected to a large water source (e.g., a lake). We know from our experience that the water levels of the bucket and water source are the same. As we raise the bucket, some amount of water flows out, so the water levels remain the same. In the case in Fig. 7.19b where the voltage source is analogous to the large water source, the electric current flows out when the upper body moves away from the lower body. While we can move the upper body a small distance $d\mathbf{r}$ and angle $d\boldsymbol{\theta}$, the increment in the stored energy is equal to the sum of the work done by the external force and moment and the charge flow, which may written as

$$dU = \mathbf{F} \cdot d\mathbf{r} + \mathbf{M} \cdot d\boldsymbol{\theta} + V dQ$$

where VdQ is the work done by the voltage source (the sense of positive charge flow is taken to be that in which the energy stored in the capacitor increases). Substituting $\mathbf{F} = -\mathbf{F}_e$ and $\mathbf{M} = -\mathbf{M}_e$ into the preceding equation and rearranging terms gives

$$dU - VdQ = -\mathbf{F}_e \cdot d\mathbf{r} - \mathbf{M}_e \cdot d\boldsymbol{\theta}$$

For convenience, we introduce the complementary energy U^*, which corresponds to the upper area in Fig. 7.19c and is given mathematically by

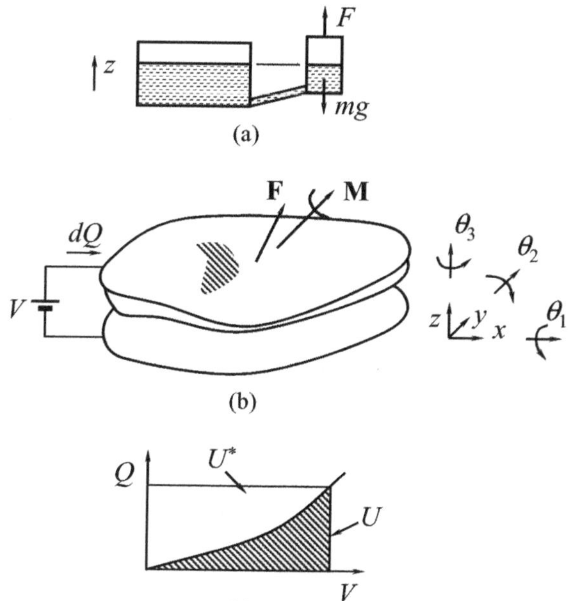

Figure 7.19 Water tank connected to a water source and conductor connected to a voltage source.

$$U^* = \int_0^V Q\,dV = QV - \int_0^Q V\,dQ = QV - U \tag{7.74}$$

where U is the energy stored in the capacitor, which is given by (7.67). For the constant voltage V, combining the preceding two equations yields

$$-dU^* = -\mathbf{F}_e \cdot d\mathbf{r} - \mathbf{M}_e \cdot d\boldsymbol{\theta}$$

Using the same process that was described in the preceding paragraphs, we find the electrostatic force and moment, $\mathbf{F}_e$ and $\mathbf{M}_e$, due to the electric field:

$$\begin{aligned}\mathbf{F}_e &= F_{ex}\mathbf{i} + F_{ey}\mathbf{j} + F_{ez}\mathbf{k} = \frac{\partial U^*}{\partial x}\mathbf{i} + \frac{\partial U^*}{\partial y}\mathbf{j} + \frac{\partial U^*}{\partial z}\mathbf{k} \\ &= \nabla U^*\end{aligned} \tag{7.75}$$

$$\begin{aligned}\mathbf{M}_e &= M_{e1}\mathbf{i} + M_{e2}\mathbf{j} + M_{e3}\mathbf{k} = \frac{\partial U^*}{\partial \theta_1}\mathbf{i} + \frac{\partial U^*}{\partial \theta_2}\mathbf{j} + \frac{\partial U^*}{\partial \theta_3}\mathbf{k} \\ &= \nabla_\theta U^*\end{aligned} \tag{7.76}$$

The force exerted by support is then obtained from the force and moment equilibrium as follows:

$$\mathbf{F} = -\mathbf{F}_e = -\nabla U^* \tag{7.77}$$

$$\mathbf{M} = -\mathbf{M}_e = -\nabla_\theta U \tag{7.78}$$

In most low-voltage applications, the permittivity is considered as a constant, and in turn the capacitor is linear; that is, the relation $Q = CV$ holds and the complementary energy U^* is equal to the energy U stored in the capacitor. Under this condition, U is substituted for U^* in equations (7.77) and (7.78), so the electrostatic force and moment are expressed as functions of U:

$$\mathbf{F}_e = \nabla U \tag{7.79}$$

$$\mathbf{M}_e = \nabla_\theta U \tag{7.80}$$

The force and moment exerted by the support are obtained as

$$\mathbf{F} = -\mathbf{F}_e = -\nabla U \tag{7.81}$$

$$\mathbf{M} = -\mathbf{M}_e = -\nabla_\theta U \tag{7.82}$$

It is convenient to express the preceding equations for the electrostatic force and moment in terms of the capacitance. If the total charge Q is conserved, as shown in Fig. 7.18b, and if the capacitor is linear, the energy equation $U = CV^2/2 = Q^2/2C$ holds, and the electrostatic force and moment, (7.70) and (7.71), may be written as

$$\mathbf{F}_e = -\nabla U = \frac{Q^2}{2C^2}\nabla C \tag{7.83}$$

$$\mathbf{M}_e = -\nabla_\theta U = \frac{Q^2}{2C^2}\nabla_\theta C \tag{7.84}$$

Similarly, for the linear capacitor with a voltage source shown in Fig. 7.19b, the electrostatic force and moment, (7.79) and (7.80), are written as

$$\mathbf{F}_e = \frac{V^2}{2}\nabla C \tag{7.85}$$

$$\mathbf{M}_e = \frac{V^2}{2}\nabla_\theta C \tag{7.86}$$

Equations (7.83) and (7.84) for the constant charge Q are actually the same as (7.85) and (7.86) since $V = Q/C$. The foregoing four equations state that the electrostatic force and moment are applied in the direction in which the capacitance C increases.

7.5 ELECTROSTATIC FORCES AND MOMENTS ACTING ON VARIOUS OBJECTS

In Section 7.4, we derived general forms of the electrostatic force and moment acting on objects. In this section we find the electrostatic force and moment of objects of various shapes, which will be used for electrostatic sensors and actuators.

Parallel Plates

Parallel plates (Fig. 7.20) are widely used for capacitors, sensors, and actuators. The movable plate of parallel plates moves without angular motion. We consider the two cases shown in Fig. 7.20a and b: the charge Q is conserved during motion; the voltage across the plates remains unchanged. The capacitance of the parallel plates is given by (7.65), which is rewritten for convenience:

$$C = \frac{Q}{V} = \frac{\varepsilon_0 A}{h}$$

If the positive charge Q is on the upper plate (movable plate) and the negative charge $-Q$ is on the lower plate (stationary plate), the electrostatic force acting on the movable plate is obtained from (7.83) as

$$\mathbf{F}_e = \frac{Q^2}{2C^2}\nabla C = \frac{Q^2}{2C^2}\frac{\partial}{\partial h}\left(\frac{\varepsilon_0 A}{h}\right)\mathbf{k} = -\frac{Q^2}{2C^2}\frac{\varepsilon_0 A}{h^2}\mathbf{k} = -\frac{Q^2}{2\varepsilon_0 A}\mathbf{k} \tag{7.87}$$

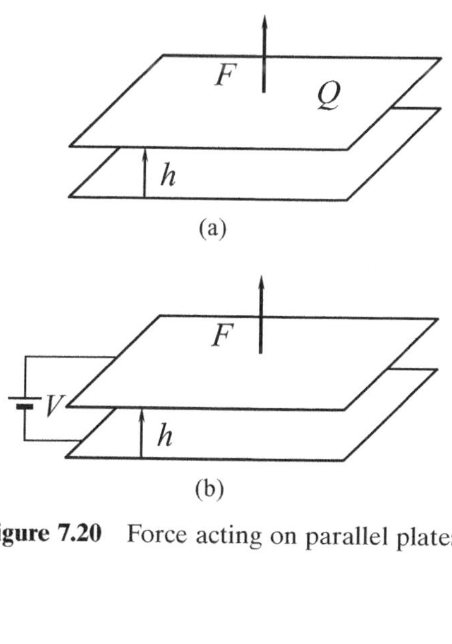

Figure 7.20 Force acting on parallel plates.

where **k** is the unit vector in the h direction in Fig. 7.20a. The force **F**, maintaining equilibrium, is then expressed as

$$\mathbf{F} = -\mathbf{F}_e = \frac{Q^2}{2\varepsilon_0 A}\mathbf{k}$$

Note that if the movable plate of parallel plates moves without any angular motion, the electrostatic force is independent of the gap h and remains constant.

If the plates are connected to a voltage source as shown in Fig. 7.20b, the electrostatic force acting on the movable plate is given by (7.85) as follows:

$$\mathbf{F}_e = \frac{V^2}{2}\nabla C = \frac{V^2}{2}\frac{\partial}{\partial h}\left(\frac{\varepsilon_0 A}{h}\right)\mathbf{k} = -\frac{\varepsilon_0 A V^2}{2h^2}\mathbf{k} \tag{7.88}$$

The electrostatic force $\mathbf{F}_e$ is inversely proportional to the square of the gap h. The force **F**, maintaining equilibrium, is obtained as

$$\mathbf{F} = -\mathbf{F}_e = \frac{\varepsilon_0 A V^2}{2h^2}\mathbf{k}$$

The electrostatic forces, (7.87) and 7.88), hold for capacitors in which the interplate gap h is much less than the other dimensions (e.g., length and width). If one of the other dimensions is comparable to the gap h, the preceding two equations result in significant errors.

Parallel and Series Capacitors

To evaluate the electrostatic force acting on strips (i.e., narrow plates), the equivalent capacitance can be used. If capacitors are connected in series as shown in Fig. 7.21a or in parallel as shown in Fig. 7.21b, the equivalent capacitance in Fig. 7.21c is obtained. For the series capacitors in Fig. 7.21a, the same amount of charge is accumulated in capacitor 1 of capacitance C_1 and in capacitor 2 of capacitance C_2 (i.e., $Q = Q_1 = Q_2$). Kirchhoff's voltage law gives the voltage relation $V = V_1 + V_2$. Substituting the voltage–charge relation, $V_1 = Q/C_1$, $V_2 = Q/C_2$, and $V = Q/C$ into the preceding voltage relation, we have the equivalent capacitance of series capacitors,

$$\frac{1}{C} = \frac{1}{C_1} + \frac{1}{C_2} \tag{7.89}$$

Similarly, for the parallel connection of the capacitor in Fig. 7.21b, the voltage across the capacitors is the same and the total charge on the capacitor is given by $Q = Q_1 + Q_2$. Substitution of the charge–voltage relations $V_1 = Q/C_1$,

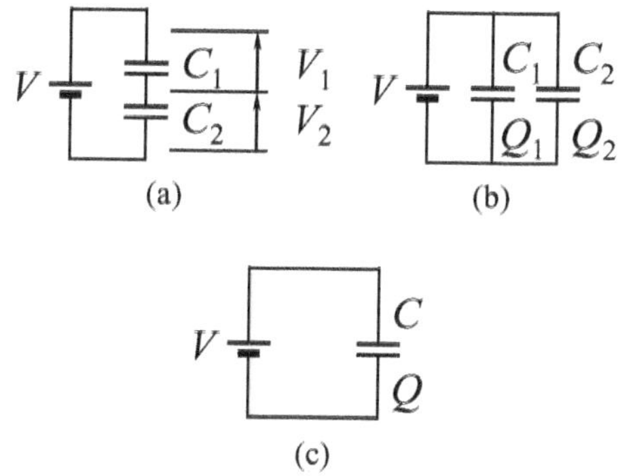

Figure 7.21 Series and parallel connections of capacitors.

$V_2 = Q/C_2$, and $V = Q/C$ into the preceding charge relation, we find the equivalent capacitance of parallel capacitors as

$$C = C_1 + C_2 \tag{7.90}$$

These equivalent capacitances are useful to evaluate the charge that is accumulated in capacitors and to calculate the voltage across a capacitor.

Strip Capacitors

If the thickness of two identical strips in Fig. 7.22a is much less than the gap h and the strip width w, the capacitance of the parallel strips can be obtained by *Palmer's formula*, (Palmer, 1927):

$$C = \varepsilon \frac{wl}{h}\left(1 + \frac{h}{\pi w} + \frac{h}{\pi w}\ln\frac{2\pi w}{h}\right) \tag{7.91}$$

Substituting the capacitance above into (7.85), we find the electrostatic force acting on the upper strip:

$$\mathbf{F}_e = \frac{V^2}{2}\nabla C = \frac{V^2}{2}\frac{\partial C}{\partial h}\mathbf{k} = -\frac{1}{2}\frac{\varepsilon wl}{h^2}\left(1 + \frac{1}{\pi}\frac{h}{w}\right)V^2\mathbf{k} \tag{7.92}$$

where the second term in the parentheses represents the fringe-field effect, which plays a role if the width w is comparable to the gap h. In Fig. 7.22b, the upper strip is spaced by a gap h from an infinite lower plate. The capacitance between the strip and the infinite plate can be obtained by considering series capacitors, which is illustrated in Fig. 7.22c. The capacitor structure is symmetric about the infinite plate. The upper and lower plates are connected to the voltage $+V_1$ and $-V_1$, respectively, and the infinite plate is connected to

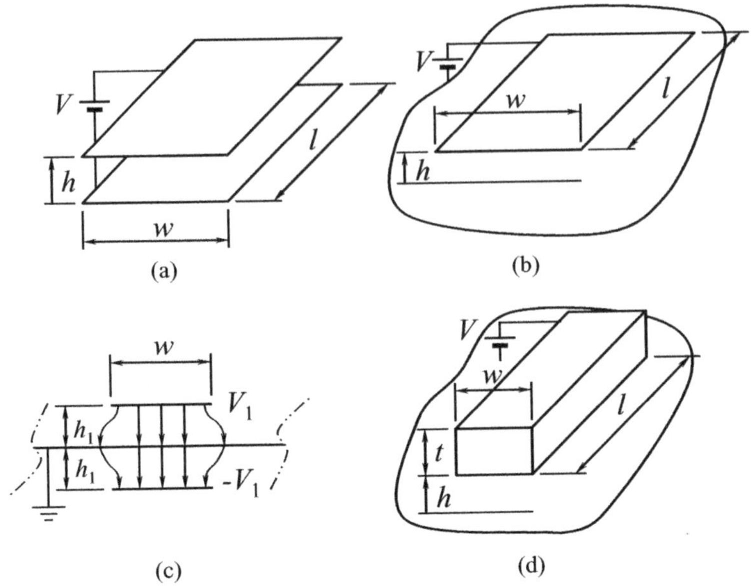

Figure 7.22 Strips and beam.

ground. The electric field strength **E** is directed from the upper strip to the lower strip as shown in Fig. 7.22c, and its magnitude is symmetric about the infinite plate. Setting $h = 2h_1$ and $V = V_1 - (-V_1) = 2V_1$, the capacitance between the upper and lower strips in Fig. 7.22c is equal to that of the strips shown in Fig. 7.22a (i.e., the capacitance of Fig. 7.22c can be considered as a series connection of two small capacitors). Let the upper and lower capacitances be represented by C_1 and use (7.89) to find the capacitance C_1, which is given by$C_1 = 2C$, where C denotes the capacitance between the upper and lower strips. Substituting $C = C_1/2$ and $h = 2h_1$ into (7.91) and rearranging terms, we obtain the capacitance between the upper (or lower) strip and the infinite plate in Fig. 7.22c:

$$C_1 = \varepsilon \frac{wl}{h_1}\left(1 + \frac{2h_1}{\pi w} + \frac{2h_1}{\pi w}\ln\frac{\pi w}{h_1}\right)$$

which is the capacitance of the structure in Fig. 7.22b. For consistency, we substitute C for C_1 and h for h_1 into the equation above to give the capacitance in Fig. 22b:

$$C = \varepsilon \frac{wl}{h}\left(1 + \frac{2h}{\pi w} + \frac{2h}{\pi w}\ln\frac{\pi w}{h}\right) \tag{7.93}$$

Upon substitution of the preceding capacitance into (7.85), we have the electrostatic force $\mathbf{F}_e$ acting on the upper strip:

$$\mathbf{F}_e = -\frac{1}{2}\frac{\varepsilon w l}{h^2}\left(1+\frac{2}{\pi}\frac{h}{w}\right)V^2\mathbf{k} \tag{7.94}$$

Note that if we compare equation (7.94) with (7.88), the quantity in the parentheses in (7.94) can be considered as a correction factor, and the contribution of the fringe field is evaluated as $2h/\pi w \cong 0.637h/w$.

Thick Beam Capacitors

If the thickness of a strip is comparable to the width and gap, the rectangular beam in Fig. 7.22d should be considered to evaluate the capacitance and electrostatic force. For evaluation of the capacitance of the beam, the analysis is more difficult and the solution becomes more complicated. For easy evaluation of the capacitance, expressions given by curve-fitting techniques can be used. For example, Meijs and Fokkema (1984) proposed an empirical expression for the capacitance of a long narrow beam over an infinite plate (Fig. 7.22d), which is given by

$$C = \varepsilon l\left[\frac{w}{h}+0.77+1.06\left(\frac{w}{h}\right)^{0.25}+1.06\left(\frac{t}{h}\right)^{0.5}\right] \tag{7.95}$$

where $w \ll l$ and $h \ll l$. The capacitance above is in good agreement with simulated capacitance within an error of 2% for $w/h \geq 1$ and $0.1 \leq t/h \leq 4$. In the range $w/h \geq 0.3$ and $4 \leq t/h \leq 10$, the maximum deviation of the foregoing capacitance from the simulated capacitance is 6%. Substituting the capacitance above into (7.85), we have the electrostatic force $\mathbf{F}_e$ acting on the rectangular beam:

$$\mathbf{F}_e = -\frac{1}{2}\frac{\varepsilon w l}{h^2}\left[1+0.265\left(\frac{h}{w}\right)^{0.75}+0.530\left(\frac{t}{w}\right)^{0.5}\left(\frac{h}{w}\right)^{0.5}\right]V^2\mathbf{k} \tag{7.96}$$

Capacitance of Cylindrical Beams

If we use carbon nanotubes as a structural material, we should obtain the electrostatic force acting on the tubes. The capacitance between two parallel cylindrical beams in Fig. 7.23a is obtained by the superposition of the electrical potentials due to two wires, whose derivation can be found from textbooks of electromagnetics (e.g., Demarest, 1998). The capacitance between parallel cylindrical beams in Fig. 7.23a reads

$$C = \frac{\pi\varepsilon l}{\cosh^{-1}(D/2a)} \tag{7.97}$$

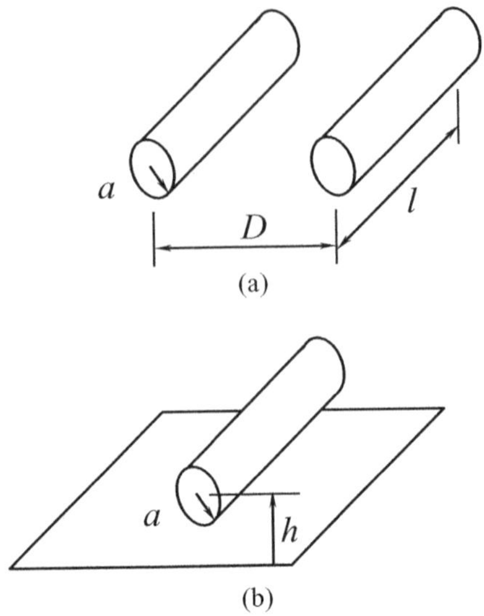

Figure 7.23 Capacitance of circular beams.

which holds for $a/l \ll 1$. When we use (7.89) for a series connection of capacitors, the capacitance between a cylindrical beam and an infinite plate (Fig. 7.23b) is then obtained as

$$C = \frac{2\pi\varepsilon l}{\cosh^{-1}(h/a)} \tag{7.98}$$

Referring to the following equation for the derivative of the inverse hyperbolic cosine function,

$$\frac{d}{dx}\cosh^{-1} x = \frac{1}{\sqrt{x^2 - 1}}$$

the derivative of the preceding capacitance with respect to the gap h is obtained as

$$\frac{\partial C}{\partial h} = -\frac{2\pi\varepsilon l}{a}\frac{1}{\left[\cosh^{-1}(h/a)\right]^2 \sqrt{(h/a)^2 - 1}}$$

The preceding derivative is substituted into the force–capacitance relation (7.85) in order to give the following electrostatic force acting on the cylindrical beam in Fig. 7.23b:

$$\mathbf{F}_e = \frac{1}{2}\frac{\partial C}{\partial h}V^2\mathbf{k} = -\frac{\pi\varepsilon l}{a}\frac{1}{\left[\cosh^{-1}(h/a)\right]^2\sqrt{(h/a)^2-1}}V^2\mathbf{k} \tag{7.99}$$

Inclined Plate

For torsional sensors and actuators, the inclined plate in Fig. 7.24a is used where the upper plate makes an angle of θ with a lower plate on a substrate. The upper plate overlaps the lower plate from $x = a$ to $x = b$ as shown in Fig. 7.24a. The free-body diagram for the upper plate is shown in Fig. 7.23b, where the forces (F_s and F_{sx}) and moment (M_s) exerted by the support balance the electrostatic force (F_e) and moment (M_e). It is noted that the electrostatic force F_e is applied perpendicular to the upper plate surface since the electric field $\mathbf{E}$ is applied perpendicular to the equipotential line (i.e., the upper plate

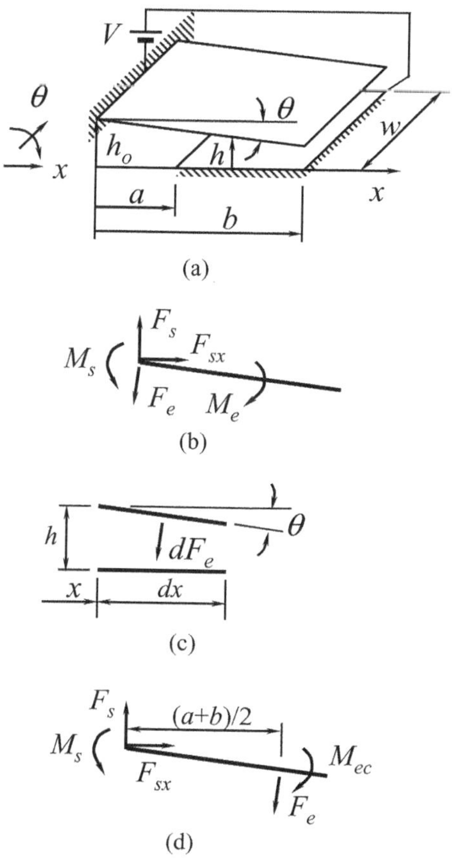

Figure 7.24 Forces and moments acting on an inclined plate.

surface), as shown in (7.50). Considering the force equilibrium in the x and y directions in Fig. 7.24b gives $F_{sx} = F_e \sin\theta \le F_e h_0/b$ and $F_s = F_e \cos\theta \approx F_e$ for $\theta << 1$. Since $F_{sx}/F_s \le h_0/b$ and $h_0/b << 1$, the force reaction F_{sx} in the x direction due to the electrostatic force can be neglected and the electrostatic force F_e is balanced by the force exerted by the support on the upper plate. From the geometry of Fig. 7.24a and c, the interplate gap is expressed as $h = h_0 - x\theta$. From a small element in Fig. 7.24c, the upper plate can be thought of as a small element whose capacitance is obtained as

$$dC = \frac{\varepsilon dA}{h} = \frac{\varepsilon w dx}{h_0 - x\theta}$$

Integrating the equation above from $x = a$ to $x = b$ yields the capacitance between the inclined and lower plates,

$$C = \int dC = \int_a^b \frac{\varepsilon w\, dx}{h_0 - x\theta} = \frac{\varepsilon w}{\theta} \ln\frac{h_0 - a\theta}{h_0 - b\theta} \tag{7.100}$$

The electrostatic force acting on the small element in Fig. 7.24c is then obtained from (7.88):

$$dF_e = \frac{\varepsilon w\, dx}{2h^2} V^2 = \frac{\varepsilon w V^2}{2(h_0 - x\theta)^2} dx$$

Integrating the equation above over the overlapped length, we have the electrostatic force F_e:

$$F_e = \int dF_e = \int_a^b \frac{\varepsilon w V^2}{2(h_0 - x\theta)^2} dx = \frac{\varepsilon w V^2}{2\theta}\left(\frac{1}{h_0 - b\theta} - \frac{1}{h_0 - a\theta}\right) \tag{7.101}$$

Similarly, the electrostatic moment M_e is given by

$$\begin{aligned} M_e &= \int dM_e = \int x\, dF_e = \int_a^b \frac{\varepsilon w V^2 x}{2(h_0 - x\theta)^2} dx \\ &= \frac{\varepsilon w V^2}{2\theta^2}\left(\frac{h_0}{h_0 - b\theta} - \frac{h_0}{h_0 - a\theta} + \ln\frac{h_0 - b\theta}{h_0 - a\theta}\right) \end{aligned} \tag{7.102}$$

Equations (7.101) and (7.102) can also be obtained from the electrostatic force–capacitance and electrostatic moment–capacitance relations, (7.85) and (7.86). Substituting the capacitance of the inclined plate, (7.100), into (7.85) gives the electrostatic force F_e:

$$\mathbf{F}_e = \frac{V^2}{2}\nabla C = \frac{\varepsilon w V^2}{2\theta}\mathbf{k}\frac{\partial}{\partial h_0}\left(\ln\frac{h_0 - a\theta}{h_0 - b\theta}\right)$$
$$= -\frac{\varepsilon w V^2}{2\theta}\left(\frac{1}{h_0 - b\theta} - \frac{1}{h_0 - a\theta}\right)\mathbf{k} \qquad (7.103)$$

where **k** denotes the unit vector in the h direction. Similarly,

$$\mathbf{M}_e = \frac{V^2}{2}\nabla_\theta C = \frac{\varepsilon w V^2}{2}\frac{\partial}{\partial\theta}\left(\frac{1}{\theta}\ln\frac{h_0 - a\theta}{h_0 - b\theta}\right)\mathbf{j}$$
$$= \frac{\varepsilon w V^2}{2\theta^2}\left(\frac{h_0}{h_0 - b\theta} - \frac{h_0}{h_0 - a\theta} + \ln\frac{h_0 - b\theta}{h_0 - a\theta}\right)\mathbf{j} \qquad (7.104)$$

where **j** represents the unit vector in the y direction in which the angle θ increases. Equations (7.103) and (7.104) are the same as equations (7.101) and (7.102). It is of interest to examine the electrostatic moment M_e above. Introducing a new coordinate, $x = (a + b)/2 + x_1$, where $(a + b)/2$ and x_1 denote the distance to the center of the overlapped length and the distance from the center, we write the electrostatic moment (7.102) as

$$M_e = \int x dF_e = \int_{-(b-a)/2}^{(b-a)/2} \frac{\varepsilon w V^2[(a+b)/2 + x_1]}{2[h_0 - (a+b)\theta/2 - x_1\theta]^2}dx_1$$
$$= \frac{a+b}{2}\varepsilon w V^2 \int_{-(b-a)/2}^{(b-a)/2} \frac{1}{[h_0 - (a+b)\theta/2 - x_1\theta]^2}dx_1$$
$$+ \frac{\varepsilon w V^2}{2}\int_{-(b-a)/2}^{(b-a)/2} \frac{x_1}{2[h_0 - (a+b)\theta/2 - x_1\theta]^2}dx_1$$
$$= F_e\frac{a+b}{2} + \frac{\varepsilon w V^2}{2\theta^2}\left[\left(h_0 - \frac{a+b}{2}\theta\right)\left(\frac{1}{h_0 - b\theta} - \frac{1}{h_0 - a\theta}\right) + \ln\frac{h_0 - b\theta}{h_0 - a\theta}\right] \qquad (7.105)$$

which can be written as $M_e = F_e(a + b)/2 + M_{ec}$, where $F_e(a + b)/2$ and M_{ec}, respectively, denote the moment about $x = 0$ due to F_e and the electrostatic moment about the center of the overlapping length, $x = (a + b)/2$, as shown in Fig. 7.24d. Equation (7.105) may be expressed in dimensionless form,

$$\frac{M_e}{F_e(a+b)/2} = 1 + \gamma \qquad (7.106)$$

where

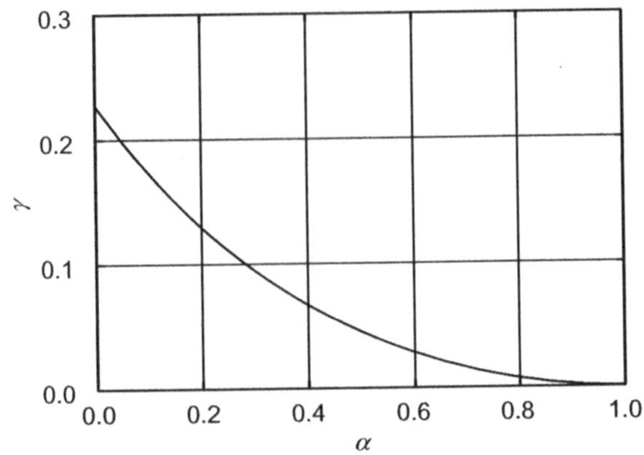

Figure 7.25 Forces and moments acting on an inclined plate at $\Theta = 0.5$.

$$\gamma = \frac{M_{ec}}{F_e(a+b)/2}$$

$$= \frac{1}{(a+b)\theta/2}\left[\left(h_0 - \frac{a+b}{2}\theta\right) + \frac{1}{\dfrac{1}{h_0 - b\theta} - \dfrac{1}{h_0 - a\theta}} \ln\frac{h_0 - b\theta}{h_0 - a\theta}\right]$$

$$= \frac{2}{(1+\alpha)\Theta}\left[1 - \frac{1}{2}(1+\alpha)\Theta + \frac{(1-\Theta)(1-\alpha\Theta)}{(1-\alpha)\Theta}\ln\frac{1-\Theta}{1-\alpha\Theta}\right] \qquad (7.107)$$

$$\Theta = \frac{b\theta}{h_0} \qquad \alpha = \frac{a}{b}$$

denote the ratio of the electrostatic moment (shown in Fig. 7.24d) about $x = (a + b)/2$ to the moment due to the electrostatic force applied at $x = (a + b)/2$, the dimensionless angle, and the ratio of the start point to the end point of the lower plate. The ratio γ is plotted against the ratio α in Fig. 7.25 when the angle Θ remains at 0.5. The ratio γ is 0.045 at $\alpha = 0.5$ and 0.01 at 0.75; in other words, the electrostatic moment M_e in (7.105) is 4.5% at $a/b = 0.5$ and 1% at $a/b = 0.75$ greater than the moment due to the electrostatic force about $(a + b)/2$. Therefore, if $a/b \geq 0.5$, the electrostatic moment M_e about $x = 0$ in Fig. 7.24b can be approximated as

$$M_e \cong \frac{F_e(a+b)}{2} \qquad (7.108)$$

Variable Capacitors

The capacitance of the capacitors described earlier in this section varies nonlinearly with the displacement of capacitor electrodes. For easy control of

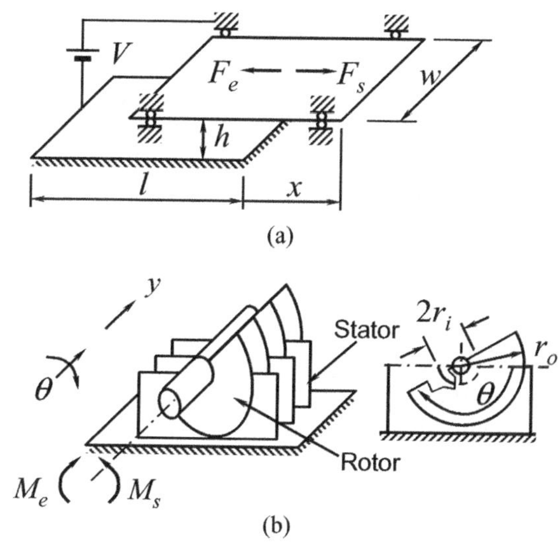

Figure 7.26 Force and moment acting on linear variable capacitors.

capacitance, linear capacitors have been used in electronic devices. For example, AM (amplitude modulation) radio can use a rotary-type variable capacitor (Fig. 7.26b) to select a desired radio frequency. Let us consider the simple linear variable capacitor illustrated in Fig. 7.26a, where a movable plate of width w can be moved with respect to a stationary plate while the interplate gap h remains constant. For $x >> h$, the capacitance of overlapping area, $A = w(l - x)$, is given by (7.65):

$$C = \frac{\varepsilon A}{h} = \frac{\varepsilon w(l-x)}{h} \tag{7.109}$$

If a voltage V is applied between the plates, an electrostatic force $\mathbf{F}_e$ due to the electric field is generated and a force $\mathbf{F}_s$ exerted by a support (not shown in the figure) balances $\mathbf{F}_e$. This electrostatic force acting on the movable plate can be obtained by using (7.85). Substituting the preceding capacitance into (7.85), we find the electrostatic force:

$$\mathbf{F}_e = -\frac{\varepsilon w}{2h} V^2 \mathbf{i} \tag{7.110}$$

where $\mathbf{i}$ denotes the unit vector in the x direction. It is noted that the capacitance, proportional to the displacement, gives the constant electrostatic force as the movable plate moves or remains stationary. The support force $\mathbf{F}_s$ in the x direction is given by $\mathbf{F}_s = -\mathbf{F}_e$. Furthermore, the electrostatic force $\varepsilon AV^2/2h^2$, given by (7.88), is applied to the plate in the direction perpendicular to the plate. If the displacement x is zero, the electrostatic force $\mathbf{F}_e$ given by (7.110) vanishes since the capacitance in Fig. 7.26a is symmetric about $x = 0$ (i.e., the derivative of capacitance at $x = 0$ is zero).

Similarly, we can obtain the electrostatic moment $\mathbf{M}_e$ and support moment $\mathbf{M}_s$ acting on the rotor of a rotary variable capacitor (Fig. 7.26b). The capacitance of the overlapping area A is written

$$C = 2n\frac{\varepsilon}{h}A = 2n\frac{\varepsilon}{h}\left[\pi\left(r_o^2 - r_i^2\right)\frac{\theta}{2\pi}\right] = n\frac{\varepsilon\left(r_o^2 - r_i^2\right)}{h}\theta \tag{7.111}$$

where h, n, r_o, r_i, and θ denote the interplate gap, the number of pairs of blades, the radius of the rotor, the radius of the concave portion of the stator, and the overlapping angle, respectively. Substituting the capacitance above into the moment–capacitance relation (7.86) yields the electrostatic moment:

$$\mathbf{M}_e = \frac{1}{2}\frac{\partial C}{\partial \theta}V^2\mathbf{j} = n\frac{\varepsilon\left(r_o^2 - r_i^2\right)}{2h}V^2\mathbf{j} \tag{7.112}$$

The supporting moment, $\mathbf{M}_s = -\mathbf{M}_e$, is generated by support (e.g., bearings) to maintain equilibrium. Since the friction force of the bearings can balance the electrostatic force, the rotary capacitor used in electronic devices remains in equilibrium.

Interdigitated Combs

A micro version of a linear capacitor (Fig. 7.27), was proposed by Tang et al. (1989). To take advantage of planar structures, the structures of Fig. 7.26 are arranged in a plane and interdigitated as illustrated in Fig. 7.27. A movable comb remains stationary or moves, while the fingers of the movable comb are interdigitated with those of a stationary comb. Since a voltage V is applied between the combs, the electrostatic force, (7.110), is generated as shown in Fig. 7.27. Referring to the dimensions of Fig. 7.27, the capacitance between combs is defined as

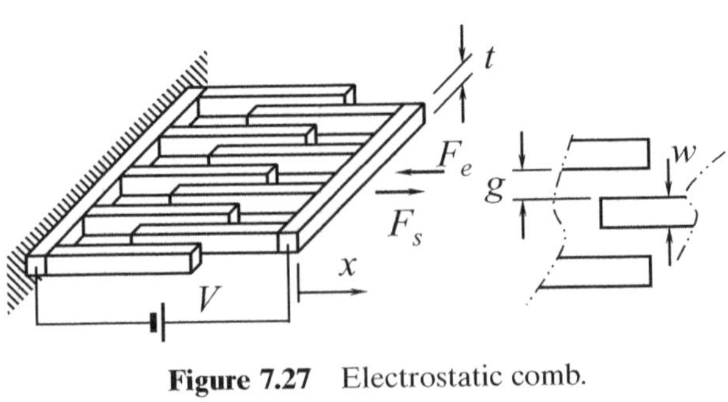

Figure 7.27 Electrostatic comb.

$$C = \frac{\varepsilon A}{g} = 2n\frac{\varepsilon t(l-x)}{g} \tag{7.113}$$

where n, t, l, x, and g represent the number of pairs of comb fingers, the thickness of the comb structures, the finger length, the displacement of the movable comb rightward, and the interfinger gap, respectively. The electrostatic force is then obtained as

$$\mathbf{F}_e = \frac{V^2}{2}\nabla C = \frac{1}{2}V^2\frac{\partial C}{\partial x}\mathbf{i} = -n\frac{\varepsilon t}{g}V^2\mathbf{i} \tag{7.114}$$

where $\mathbf{i}$ denotes a unit vector in the x direction. The electrostatic force remains constant if the applied voltage is unchanged. It should be noted that the electrostatic force above acting on the comb is independent of the finger width w in the right figure in Fig. 7.27.

Tailored Electrostatic Force

For a constant voltage, the electrostatic force between the interdigitated combs remains constant as one comb moves against the other in the direction of the comb finger. Electrostatic forces or moments of other structures described in the preceding paragraphs vary nonlinearly as the movable structures are displaced. In some cases we need to tailor the electrostatic force or moment for a specific requirement or purpose. For example, if we wish to generate a linear electrostatic force that is proportional to the displacement of a movable structure, we may tailor movable and stationary structures. Let us begin with the general shapes of the structures, which are illustrated in Fig. 7.28. A movable structure is supported by a support that is not shown in the figure, and a voltage source of voltage V is connected to the movable structure and a stationary structure. The movable structure is subjected both to the electrostatic force and moment and to the force and moment due to the support. Since the electrostatic force and moment are described by the force–capacitance and

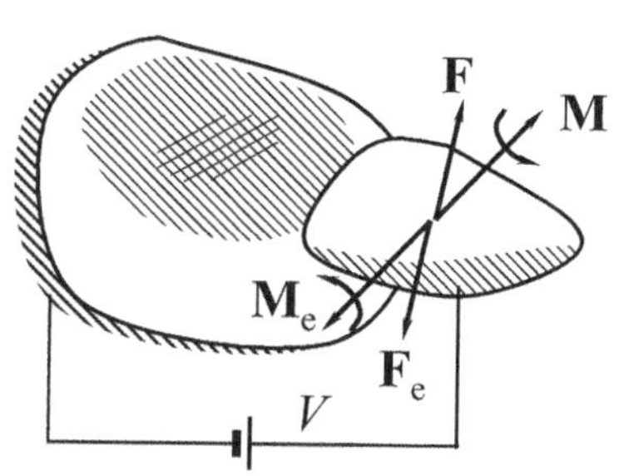

Figure 7.28 Movement of a body from a stationary body.

moment–capacitance relations, (7.85) and (7.86), tailoring the electrostatic force $\mathbf{F}_e$ and moment $\mathbf{M}_e$ reduces to designing the capacitance. For further derivation, equations (7.85) and (7.86) are rewritten as

$$\mathbf{F}_e = \frac{V^2}{2}\nabla C = \frac{1}{2}V^2\left(\frac{\partial C}{\partial x}\mathbf{i} + \frac{\partial C}{\partial y}\mathbf{j} + \frac{\partial C}{\partial z}\mathbf{k}\right)$$

$$\mathbf{M}_e = \frac{V^2}{2}\nabla_\theta C = \frac{1}{2}V^2\left(\frac{\partial C}{\partial \theta_1}\mathbf{i} + \frac{\partial C}{\partial \theta_2}\mathbf{j} + \frac{\partial C}{\partial \theta_3}\mathbf{k}\right)$$

where C is a function of the coordinates x, y, z, θ_1, θ_2, and θ_3. If we constrain the movable structure to move in only one direction, the tailored capacitance can easily be obtained. Let us consider one of simple cases, which is illustrated in Fig. 7.29a, where the movable comb with straight fingers is allowed to move in the x direction. The finger shape of the stationary comb will be tailored to generate a linear electrostatic force. In this case, the electrostatic force is expressed as

$$\mathbf{F}_e = \frac{1}{2}\frac{dC}{dx}V^2\mathbf{i} = (a_1 x + b_1)\mathbf{i}$$

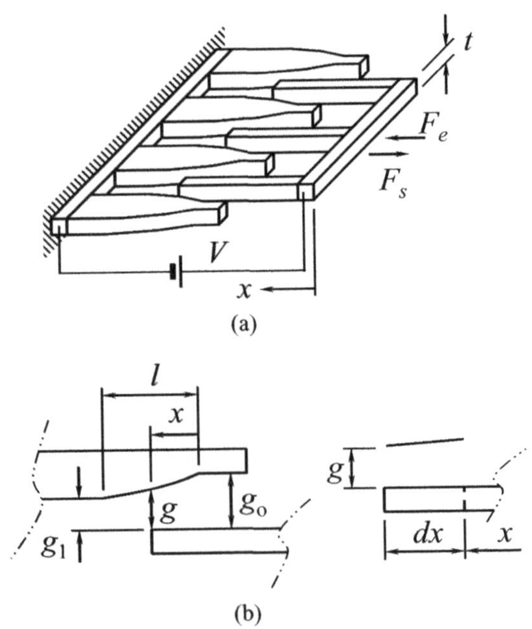

Figure 7.29 Electrode shape to generate a linear electrostatic force.

where a_1 and b_1 denote functions of the voltage and geometry to be determined. Since the voltage V remains constant, the foregoing equation reduces to

$$\frac{dC}{dx} = ax + b \tag{7.115}$$

where a and b are constants. Assuming that the structures are coplanar and that the gap, g, of the stationary comb fingers varies from g_0 to g_1 as shown in Fig. 7.29b, we find the capacitance change due to an infinitesimal displacement, dx, of the movable finger:

$$dC = \frac{\varepsilon dA}{g} = \frac{\varepsilon t}{g}dx \quad \text{or} \quad \frac{dC}{dx} = \frac{\varepsilon t}{g}$$

Substituting the capacitance change above into (7.115), we have the interfinger gap,

$$g = \frac{\varepsilon t}{ax + b}$$

Imposing the boundary conditions $g = g_0$ at $x = 0$ and $g = g_1$ at $x = l$ on the solution above, we obtain the interfinger gap:

$$g = g_0 \frac{1}{[(g_0/g_1) - 1](x/l) + 1} \tag{7.116}$$

The capacitance change rate is then given by

$$\frac{dC}{dx} = \frac{\varepsilon t}{g} = \frac{\varepsilon t}{g_0}\left[\left(\frac{g_0}{g_1} - 1\right)\frac{x}{l} + 1\right]$$

and we have the electrostatic force, $\mathbf{F}_e$, which is proportional to the displacement x:

$$\mathbf{F}_e = \frac{1}{2}\frac{\varepsilon t}{g_0}\left[\left(\frac{g_0}{g_1} - 1\right)\frac{x}{l} + 1\right]V^2\mathbf{i} \tag{7.117}$$

In Fig. 7.30, the electrostatic force and simulated electrostatic force above, obtained from a commercial software (program name: Maxwell by Ansoft), are plotted against the displacement x. For $x = 2$ to $9\,\mu$m, the preceding analytic electrostatic force is very close to the simulated force. Furthermore, the derivative of the analytic electrostatic forces, dF_e/dx, is almost the same as that from simulation at an operating point $x = 5\,\mu$m. This linear electrostatic force can

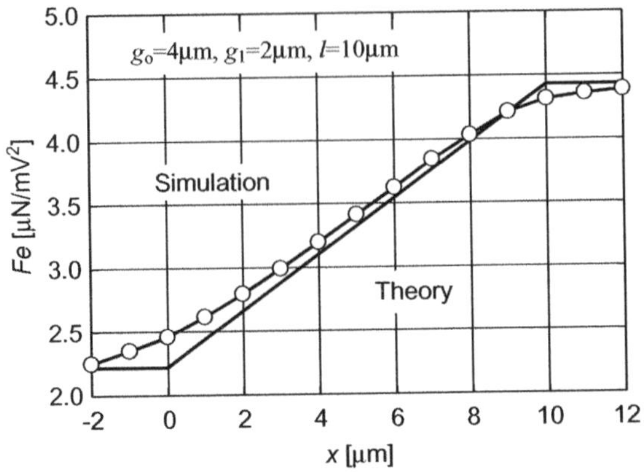

Figure 7.30 Theoretical and simulated electrostatic forces.

be used to control both the effective stiffness and the natural frequency of a structure.

7.6 ELECTROMAGNETIC FORCE

Some MEMS devices use electromagnetic force, which is usually generated by a current through a coil. As a magnetic material such as iron is exposed to a magnetic field (magnetic field strength **H**) as shown in Fig. 7.31, a magnetic flux flows in the direction of the magnetic field and the magnetic flux density **B** is proportional to the magnetic field strength; mathematically,

$$\mathbf{B} = \mu \mathbf{H} \tag{7.118}$$

where μ denotes the permeability of the material. The permeability can be written as $\mu = \mu_0 \mu_r$, where μ_0 is the permeability of free space, given by $4\pi \times 10^{-7}$ H/m, and μ_r is the relative permeability of the material. As the magnetic field strength **H** in a magnetic circuit (Fig. 7.32a) is generated by an N-turn coil, *Ampere's law* can be used to obtain **H**:

$$\oint \mathbf{H} \cdot d\mathbf{l} = \sum i = Ni \tag{7.119}$$

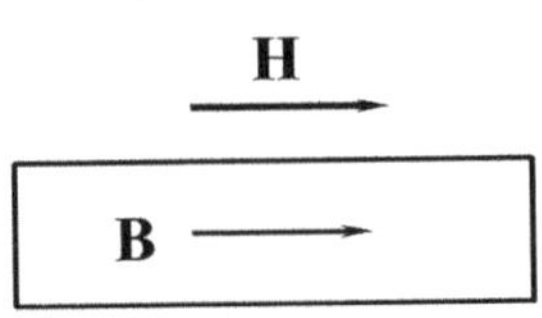

Figure 7.31 Magnetic flux density due to magnetic field strength.

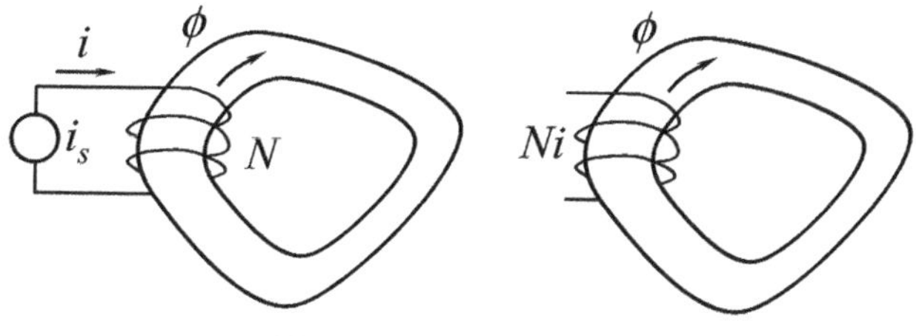

Figure 7.32 Magnetic circuit.

where **l**, N, and i denote the length (vector) along the magnetic circuit, the number of coil turns, and the current through the coil, respectively. From equation (7.119), if the cross-sectional area of the core of the circuit remains constant, the magnitude of the magnetic field strength **H** is obtained as

$$H = \frac{Ni}{l} \tag{7.120}$$

where l denotes the length of the magnet circuit in Fig. 7.32a. The magnetic flux is given by

$$\phi = BA = \mu HA \tag{7.121}$$

where B is the magnitude of the magnetic flux density given by (7.118) and A is the cross-sectional area of the core. The magnetic circuit in Fig. 7.32a is usually illustrated as in Fig. 7.32b, where the current source is replaced by Ni. Substituting (7.120) into equation (7.121), we have

$$\phi = \mu \frac{Ni}{l} A = \frac{\mu A}{l} Ni = \frac{Ni}{R} \tag{7.122}$$

where R denotes the reluctance in ampere-turns per weber (At/Wb), which is defined as

$$R = \frac{l}{\mu A} \tag{7.123}$$

Ni, known as the *magnetomotive force* (mmf), produces magnetic flux in a magnetic circuit. If the reluctance in Fig. 7.32b is represented by a series of reluctances, $R_1, R_2, \ldots, R_n$, the reluctance R is expressed as

$$R = R_1 + R_2 + \cdots + R_n \tag{7.124}$$

When the magnetomotive force Ni is applied to a magnetic circuit, the magnetic flux ϕ given by (7.122) flows in the circuit. As a result, the following magnetic energy is stored in the circuit:

$$U = \int_0^{\phi} Ni\, d\phi$$

This equation can be expressed in a more general form. According to the theory of electromagnetics (electromagnetism), if a material is exposed to a magnetic field, the following energy is stored in the material and the free space near the material:

$$U = \frac{1}{2}\int_v \mathbf{B}\cdot\mathbf{H}\, dv \tag{7.125}$$

where the integral is taken over the volume v. Note that equation (7.125) is in the same form as (7.69) for the electric energy stored in a material exposed to an electric field. If the core (labeled 1) of the magnetic circuit shown in Fig. 7.33a has an air gap h (labeled 2), the air gap acts as a reluctance. Let us analyze the magnetic circuit shown in Fig. 7.33a. Assuming that the magnetic flux ϕ flows in the circuit without leakage, we have two reluctances, R_1 and R_2, due to the core and the air gap:

$$R_1 = \frac{l}{\mu A} \qquad R_2 = \frac{g}{\mu_0 A}$$

we then find the total reluctance R, which is expressed as a sum of the reluctances:

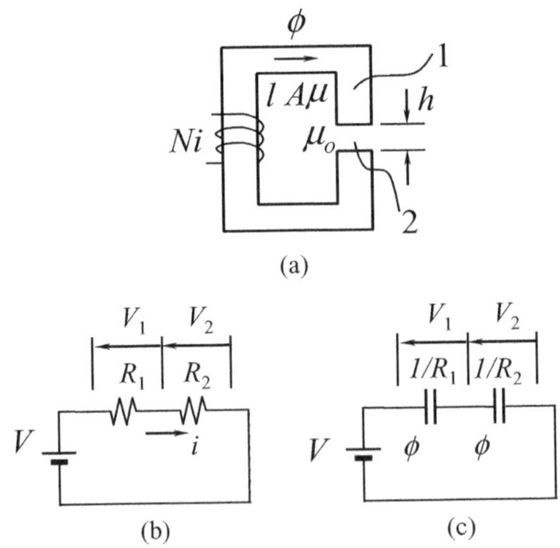

Figure 7.33 Magnetic circuit of a core with a gap.

$$R = R_1 + R_2 = \frac{1}{A}\left(\frac{l}{\mu} + \frac{g}{\mu_0}\right)$$

Equation (7.122) gives the magnetic flux through the circuit,

$$\phi = \frac{Ni}{R}$$

Then the energy stored in the magnetic circuit is obtained from (7.125):

$$\begin{aligned} U &= \frac{1}{2}\int_v \mathbf{B}\cdot\mathbf{H}\,dv = \frac{1}{2}\left[\left(\frac{\phi}{A}\right)^2 \frac{1}{\mu} Al + \left(\frac{\phi}{A}\right)^2 \frac{1}{\mu_0} Ag\right] \\ &= \frac{R\phi^2}{2} = \frac{(Ni)^2}{2R} = \frac{\phi(Ni)}{2} \end{aligned} \tag{7.126}$$

For convenience, we define the complementary energy for the magnetic field problem, which is in the same form as that for the electric field problem, (7.74), and is given by

$$U^* = \phi(Ni) - U \tag{7.127}$$

If magnetic materials are not saturated and remain linear (i.e., the permeabilities remain unchanged), the complementary energy above is equal to the energy (7.126):

$$U^* = U = \frac{(Ni)^2}{2R} = \frac{R\phi^2}{2} = \frac{\phi(Ni)}{2} \tag{7.128}$$

Comparing the energy, (7.126), stored in a magnetic circuit, with the energy stored in a inductor, $U = Li^2/2$, which is given by (7.22), we find the inductance L of the magnetic circuit as follows:

$$L = \frac{N^2}{R} \tag{7.129}$$

The magnetic circuit in Fig. 7.33a is analogous to the electric circuit in Fig. 7.33b. That is, the magnetomotive force Ni is analogous to the electromotive force V, the reluctances R_1 and R_2 to the resistances R_1 and R_2, and the magnetic flux ϕ to the electric current i.

Electromagnetic Force and Moment Acting on a Body

As a material exposed to a magnetic field, energy is stored in materials exposed to a magnetic field. The energy U and compliementary energy U^* were shown

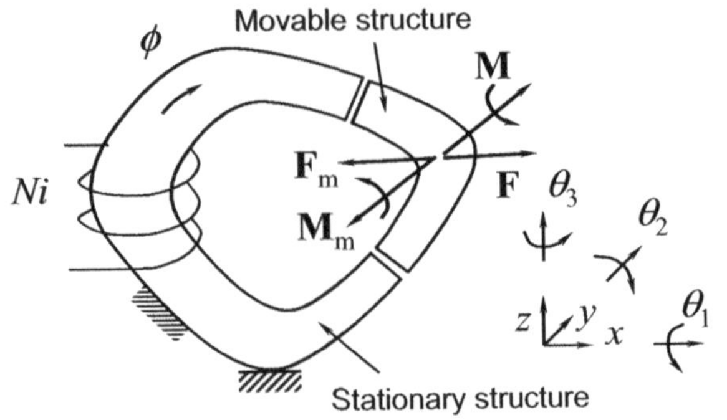

Figure 7.34 Electromagnetic force acting on a movable structure in a magnetic circuit.

in (7.126) and (7.127), respectively. If we compare the energies for the magnetic field with those for the electric field, (7.68) and (7.74), we find another analogy between the electric and magnetic energies; that is, the magnetic flux ϕ is analogous to the charge Q, the magnetomotive force Ni to the electromotive force V, and the reluctance R to the inverse of the capacitance $1/C$. The analogy is shown in Fig. 7.33c. Let us consider a magnetic circuit in Fig. 7.34, which can be obtained by cutting the magnetic circuit in Fig. 7.32b at two points. The movable structure, separated from the stationary structure, is supported by a support (not shown in the figure) to maintain equilibrium. Air gaps are allowed in the circuit as shown in Fig. 7.34, and the force and moment, **F** and **M**, exerted by the support balance the electromagnetic force and moment, $\mathbf{F}_m$ and $\mathbf{M}_m$. Since the analogy, described above and in Fig. 7.33c, holds between the electric and magnetic circuits, the electromagnetic force $\mathbf{F}_m$ and moment $\mathbf{M}_m$ can be obtained by variable substitution. If the magnetic flux ϕ remains constant, we find the electromagnetic force, which is given by (7.70):

$$\begin{aligned}\mathbf{F}_m &= F_{mx}\mathbf{i} + F_{my}\mathbf{j} + F_{mz}\mathbf{k} = -\left(\frac{\partial U}{\partial x}\mathbf{i} + \frac{\partial U}{\partial y}\mathbf{j} + \frac{\partial U}{\partial z}\mathbf{k}\right) \\ &= -\nabla U \end{aligned} \tag{7.130}$$

and the electromagnetic moment $\mathbf{M}_m$ is obtained from (7.71):

$$\begin{aligned}\mathbf{M}_m &= M_{m1}\mathbf{i} + M_{m2}\mathbf{j} + M_{m3}\mathbf{k} = -\left(\frac{\partial U}{\partial \theta_1}\mathbf{i} + \frac{\partial U}{\partial \theta_2}\mathbf{j} + \frac{\partial U}{\partial \theta_3}\mathbf{k}\right) \\ &= -\nabla_\theta U \end{aligned} \tag{7.131}$$

where U, ∇, and ∇_θ represent the magnetic energy defined by (7.125) or (7.126), the gradient operator, and the angular gradient operator defined following (7.71). Referring to the energy (7.126), we express the preceding equations as functions of reluctances, as follows:

$$\mathbf{F}_m = -\nabla U = -\nabla \frac{R\phi^2}{2} = -\frac{\phi^2}{2}\nabla R \tag{7.132}$$

$$\mathbf{M}_m = -\nabla_\theta U = -\nabla_\theta \frac{R\phi^2}{2} = -\frac{\phi^2}{2}\nabla_\theta R \tag{7.133}$$

If the magnetomotive force Ni remains constant, the magnetic force and moment acting on the movable structure are given by (7.75) and (7.76):

$$\begin{aligned}\mathbf{F}_m &= F_{mx}\mathbf{i} + F_{my}\mathbf{j} + F_{mz}\mathbf{k} = \frac{\partial U^*}{\partial x}\mathbf{i} + \frac{\partial U^*}{\partial y}\mathbf{j} + \frac{\partial U^*}{\partial z}\mathbf{k} \\ &= \nabla U^* \end{aligned} \tag{7.134}$$

$$\begin{aligned}\mathbf{M}_m &= M_{m1}\mathbf{i} + M_{m2}\mathbf{j} + M_{m3}\mathbf{k} = \frac{\partial U^*}{\partial \theta_1}\mathbf{i} + \frac{\partial U^*}{\partial \theta_2}\mathbf{j} + \frac{\partial U^*}{\partial \theta_3}\mathbf{k} \\ &= \nabla_\theta U^* \end{aligned} \tag{7.135}$$

The preceding four equations for the force and moment are obtained from the analogy between the electric and magnetic circuits shown in Fig. 7.33c. These equations can also be derived from the magnetic circuit; the derivation with the magnetic circuit is similar to that described in Section 7.4. If the material behaves as a linear material, the energy U stored in the magnetic circuit is equal to the corresponding complementary energy U^*, and in turn, equations (7.134) and (7.135) for the constant magnetomotive force Ni is reduced to

$$\mathbf{F}_m = \nabla U \tag{7.136}$$

$$\mathbf{M}_m = \nabla_\theta U \tag{7.137}$$

With reference to the complementary energy U^*, (7.128), equations (7.136) and (7.137) for the constant magnetomotive force Ni can be written as

$$\mathbf{F}_m = \nabla U = \nabla \frac{(Ni)^2}{2R} = -\frac{(Ni)^2}{2R^2}\nabla R \tag{7.138}$$

$$\mathbf{M}_m = \nabla_\theta U = \nabla_\theta \frac{(Ni)^2}{2R} = -\frac{(Ni)^2}{2R^2}\nabla_\theta R \tag{7.139}$$

Note from (7.132), (7.133), (7.138), and (7.139) that the electromagnetic force and moment exerted on a part in a electromagnetic circuit are applied in the direction in which the reluctance R decreases or in which the inductance L

increases since the reluctance is inversely proportional to the inductance, as shown in (7.129). This phenomenon is similar to that in electrostatics; as discussed in Section 7.5, the attractive electrostatic force and moment are applied in the direction in which the capacitance increases.

Electromagnetic Relay

As a typical application of an electromagnet, let us consider the electromagnetic relays illustrated in Figs. 7.35a and 7.36a, where the dimensions and areas are also shown. The relays in these figures consist of a core with an N-turn coil, an armature (movable bar), and an air gap h. For analysis of the relays,

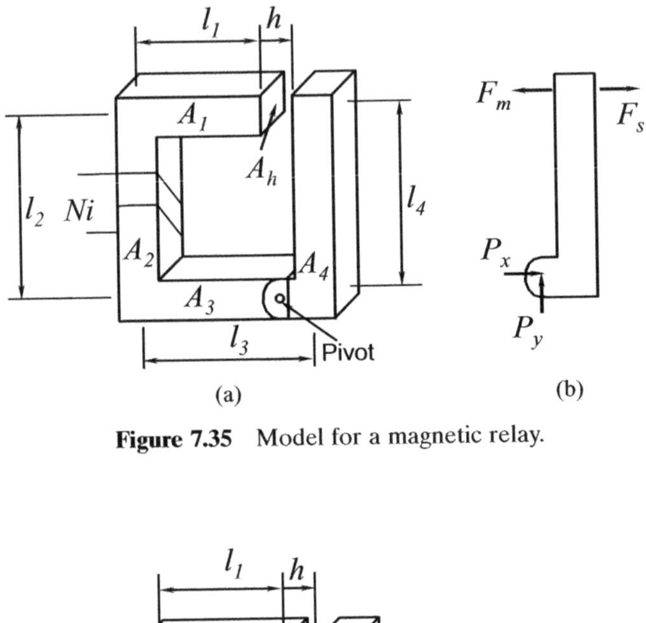

Figure 7.35 Model for a magnetic relay.

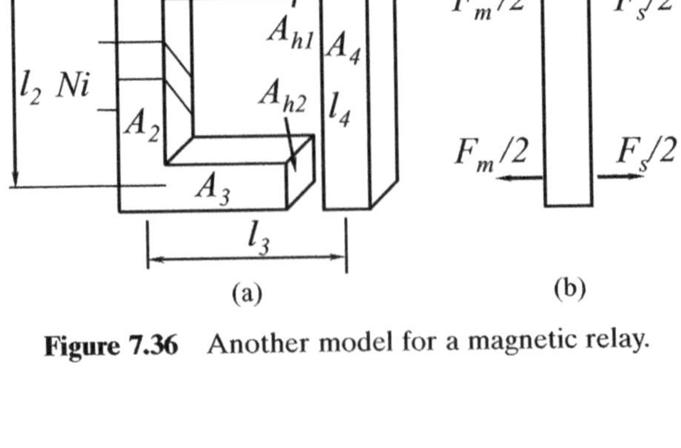

Figure 7.36 Another model for a magnetic relay.

we first express the reluctances as functions of the gap h, permeabilities, and geometries. The reluctance R of the magnetic circuit of the relays may be written

$$R = R_c + R_h \tag{7.140}$$

where

$$R_c = \sum_{j=1}^{m} \frac{l_j}{\mu_j A_j} \tag{7.141}$$

$$R_h = \sum_{j=1}^{n} \frac{h}{\mu_0 A_{hj}} \tag{7.142}$$

denote the reluctance of the cores and the armatures and the reluctance of the air gaps, respectively. For example, equation (7.140) for the relay in Figs. 7.35a (with one air gap) reads

$$R = R_c + R_h = R_c + \frac{h}{\mu_0 A_h} = R_c + r_d h \tag{7.143}$$

where

$$R_c = \frac{l_1}{\mu A_1} + \frac{l_2}{\mu A_2} + \frac{l_3}{\mu A_3} + \frac{l_4}{\mu A_4} \tag{7.144}$$

$$r_d = \frac{1}{\mu_0 A_h} \tag{7.145}$$

Similarly, if the armature of the relay of Fig. 7.36a (with two air gaps) is constrained to move without angular displacement, the reluctance is expressed in the same form:

$$R = R_c + r_d h$$

where

$$R_c = \frac{l_1}{\mu A_1} + \frac{l_2}{\mu A_2} + \frac{l_3}{\mu A_3} + \frac{l_4}{\mu A_4}$$

$$r_d = \frac{1}{\mu_0 A_{h1}} + \frac{1}{\mu_0 A_{h2}} \tag{7.146}$$

If the electric current is adjusted for the magnetic flux ϕ to remain constant during an infinitesimal displacement dh in Fig. 7.35a, we obtain the electromagnetic force $\mathbf{F}_m$, shown in the free-body diagram of Fig. 7.35b, by substituting (7.143) into (7.132),

$$\mathbf{F}_m = -\frac{\phi^2}{2}\nabla R = -\frac{\phi^2}{2}\nabla(R_c + r_d h) = -\frac{\phi^2}{2} r_d \mathbf{i} \tag{7.147}$$

where r_d and $\mathbf{i}$ denote the reluctance change rate, defined in (7.145), and the unit vector in the direction in which h increases, respectively. It is noted form equation (7.147) that if the magnetic flux remains constant, the electromagnetic force acting on the portions of the armature that face the air gap is independent of the gap h. The support force is obtained from the moment balance about the pivot point in Fig. 7.35b, $\mathbf{F}_s = -\mathbf{F}_m$. Equation (7.147) also holds for the magnetic relay shown in Fig. 7.36, but the reluctance change rate r_d is that defined by (7.146).

In most electromagnetic applications, such as the relays in Figs. 7.35a and 7.36a, electric current change rate is much greater than the velocity or angular rate of movable structures such as the armatures in Figs. 7.35a and 7.36a. As a result, the magnetomotive force Ni can be considered to be constant. For a constant Ni we can substitute (7.143) into (7.138) to obtain the magnetic force $\mathbf{F}_m$ acting on the armature in Fig. 7.35a:

$$\begin{aligned} \mathbf{F}_m &= -\frac{(Ni)^2}{2R^2}\nabla R = -\frac{(Ni)^2}{2(R_c + r_d h)^2}\nabla(R_c + r_d h) \\ &= -\frac{(Ni)^2}{2(R_c + r_d h)^2} r_d \mathbf{i} \end{aligned} \tag{7.148}$$

where R_c is defined by (7.144) and r_d denotes the reluctance change rate, which is given by (7.145) and (7.146) for the relays of Figs. 7.35a and 7.36a, respectively. The electromagnetic force $\mathbf{F}_m$ decreases as the gap h increases. As such, we can use the equations developed in this section to obtain the force and moment acting on objects that are parts of the electromagnetic circuit.

7.7 FORCE ACTING ON A MOVING CHARGE IN ELECTRIC AND MAGNETIC FIELDS

If a particle of electric charge q in Fig. 7.37 moves at a velocity $\mathbf{v}$ in an electric field (electric field strength $\mathbf{E}$) and a magnetic field (magnetic flux density $\mathbf{B}$), the particle undergoes a Lorentz force, which is given by

$$\mathbf{F} = q(\mathbf{E} + \mathbf{v} \times \mathbf{B}) \tag{7.149}$$

An electric current flowing through the conductor in Fig. 7.38 can be considered as a charge flow rate, so the conductor exposed to a magnetic field also

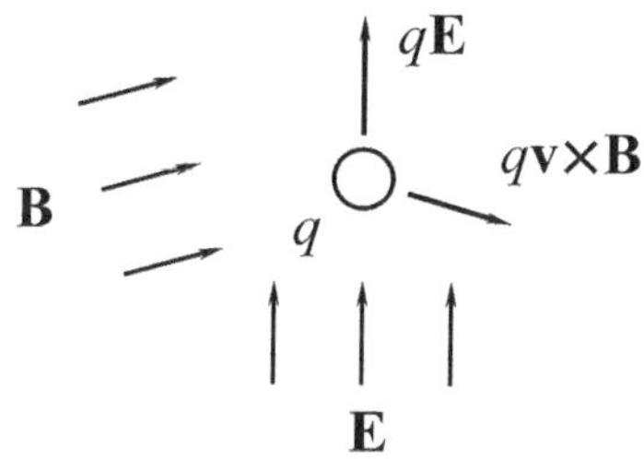

Figure 7.37 Moving electric charge in electric and magnetic fields.

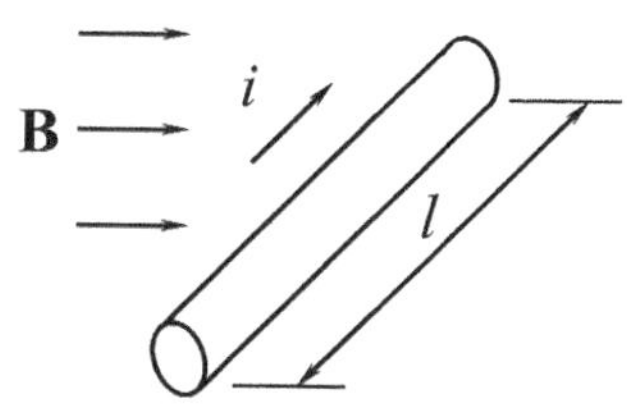

Figure 7.38 Force acting on a conductor carrying current in a magnetic field.

experiences a Lorentz force. As a charge in the electric current i flows during a time t, the charge velocity is obtained as $\mathbf{v} = \mathbf{l}/t$, where $\mathbf{l}$ denotes a vector form of the conductor length (which corresponds to displacement of the electric charge during time t), showing the charge flow. Rearranging (7.149), we find the force acting on the conductor as follows:

$$\mathbf{F} = q\mathbf{v} \times \mathbf{B} = q\frac{\mathbf{l}}{t} \times \mathbf{B} = \frac{q}{t}\mathbf{l} \times \mathbf{B} = i\mathbf{l} \times \mathbf{B} \tag{7.150}$$

which is also known *Fleming's left-hand rule.*

Moment Acting on Planar Coils Exposed to a Magnetic Field

Some MEMS devices employ planar coils to generate an electromagnetic force or moment. Typical configurations for planar coils are illustrated in Figs. 7.39a and 7.40, which can be used for torsional actuators. We shall obtain the forces and moment acting on the planar coils. A series of square coils are formed on a substrate in Fig. 7.39a, and electric current i flows from a to b. Let $\mathbf{l} = l\mathbf{j}$ and $\mathbf{B} = B\mathbf{i}$ in Fig. 7.39b represent the length vector of the coil segment and the magnetic flux density. The Lorentz force $\mathbf{F}_1$, (7.150), acting on the coil segment in Fig. 7.39b, reads

$$\mathbf{F}_1 = il\mathbf{j} \times B\mathbf{i} = -ilB\mathbf{k} \tag{7.151}$$

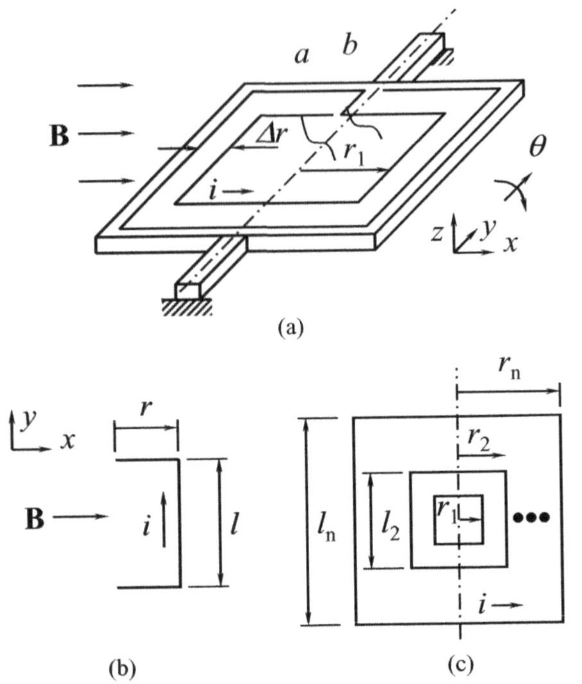

Figure 7.39 Multiturn rectangular planar coil on a plate.

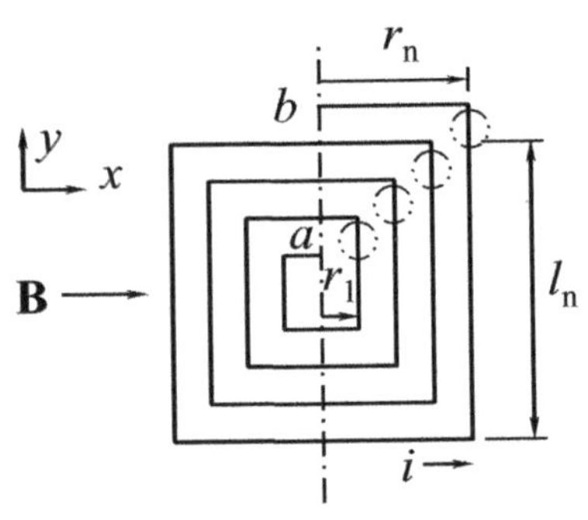

Figure 7.40 Rectangular spiral planar coil.

and the moment $\mathbf{M}_1$ due to the force above is given by

$$\mathbf{M}_1 = \mathbf{r} \times \mathbf{F} = r\mathbf{i} \times (-ilB\mathbf{k}) = rliB\mathbf{j} \tag{7.152}$$

where $\mathbf{r} = r\mathbf{i}$ is a vector from the center to the coil of length l in Fig. 7.39b. The magnetic moment $\mathbf{M}_m$ can be obtained from a combination of the preceding moments. Defining the following geometric ratios for the equivalent model shown in Fig. 7.39c,

$$\alpha = \frac{\Delta r}{r_1} \qquad \Delta r = \frac{r_n - r_1}{n-1} \qquad \beta = \frac{\Delta l}{l_1} \qquad \Delta l = \frac{l_n - l_1}{n-1} \tag{7.153}$$

we express the moment as follows:

$$\begin{aligned}
\mathbf{M}_m &= \sum_j \mathbf{M}_j = \mathbf{r} \times \mathbf{F} \\
&= 2iBr_1l_1\mathbf{j} + 2iB(r_1 + \Delta r)(l_1 + \Delta l)\,\mathbf{j} + \cdots + 2iB[r_1 + (n-1)\Delta r][l_1 + (n-1)\Delta l]\,\mathbf{j} \\
&= 2iBr_1l_1\mathbf{j}\{1 + (1+\alpha)(1+\beta) + \cdots + [1 + (n-1)\alpha][1 + (n-1)\beta]\} \\
&= 2iBr_1l_1\mathbf{j}\left\{n + (\alpha+\beta)[1 + 2 + \cdots + (n-1)] + \alpha\beta\left[1^2 + 2^2 + \cdots + (n-1)^2\right]\right\}
\end{aligned}$$

Introducing the identities

$$1 + 2 + \cdots + n = \frac{n(n+1)}{2}$$

$$1^2 + 2^2 + \cdots + n^2 = \frac{n(n+1)(2n+1)}{6}$$

where n denotes a natural number, the equation above is reduced to

$$\mathbf{M}_m = 2nr_1l_1iB\left[1 + (\alpha+\beta)\frac{n-1}{2} + \alpha\beta\frac{(n-1)(2n-1)}{6}\right]\mathbf{j} \tag{7.154}$$

where n, r_1, l_1, i, and B represent the number of coil turns, the distance of the first coil turn from the coil center as shown in Fig. 7.39a and c, the length of the first coil in the y direction shown in Fig. 7.39b, the current through the coil, and the magnetic flux density, respectively. The dimensionless numbers α and β are given by (7.153). If $n = 1$ (i.e., a single-turn coil), the preceding equation is reduced to $\mathbf{M}_m = 2r_1l_1iB\mathbf{j}$, which is twice that of the half-turn of the coil, (7.152). The rectangular spiral shown in Fig. 7.40 can be used for a planar coil. Since the spiral coil is considered as the rectangular coil with extended portions that is circled in Fig. 7.40, the magnetic moment, M_{ms}, of the spiral coil can be thought of as a sum of the moment of the rectangular coil, (7.154), and an additional moment due to the extended portions. Using (7.152), we obtain the additional moment as follows:

$$\begin{aligned}
\Delta\mathbf{M}_m &= iBr_1\Delta l\mathbf{j} + iB(r_1 + \Delta r)\Delta l\mathbf{j} + \cdots + iB[r_1 + (n-1)\Delta r]\Delta l\mathbf{j} \\
&= iBr_1l_1\mathbf{j}\{\beta + (1+\alpha)\beta + \cdots + [1 + (n-1)\alpha]\beta\} \\
&= iBr_1l_1\mathbf{j}\{n\beta + \alpha\beta[1 + 2 + \cdots + (n-1)]\} \\
&= iBr_1l_1\mathbf{j}\left[n\beta + \alpha\beta\frac{n(n-1)}{2}\right] \\
&= nr_1\beta l_1iB\left(1 + \alpha\frac{n-1}{2}\right)\mathbf{j}
\end{aligned} \tag{7.155}$$

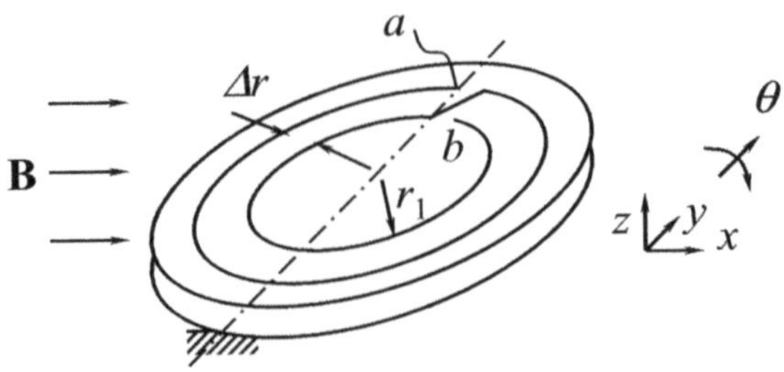

Figure 7.41 Multiturn circular planar coil on a plate.

The magnetic moment of the rectangular spiral coil shown in Fig. 7.40 is then expressed as

$$\begin{aligned}\mathbf{M}_{ms} &= \mathbf{M}_m + \Delta\mathbf{M}_m \\ &= 2nr_1l_1iB\left[1+(\alpha+\beta)\frac{n-1}{2}+\alpha\beta\frac{(n-1)(2n-1)}{6}+\frac{\beta}{2}\left(1+\alpha\frac{n-1}{2}\right)\right]\mathbf{j}\end{aligned} \quad (7.156)$$

Similarly, the circular coil on a substrate shown in Fig. 7.41, is subjected to a magnetic moment which is given by

$$\mathbf{M}_m = n\pi r_1^2 iB\left[1+\gamma(n-1)+\gamma^2\frac{(n-1)(2n-1)}{6}\right]\mathbf{j} \quad (7.157)$$

where

$$\gamma = \frac{\Delta r}{r_1} \qquad \Delta r = \frac{r_n - r_1}{n-1}$$

The derivation of the equation above is left to the reader as an exercise (Problem 7.7).

7.8 PIEZORESISTANCE

As described in Section 7.1, since a resistor generates heat as the electric current flows across a heater, the thermal expansion of the heater can be used for a thermally driven actuator. However, even though we apply force to the resistance or distort the resistance, the resistor does not generate a voltage or electrical charge. Nevertheless, application of force or distortion changes the resistance of the resistor, which can be measured if the resistor is connected to an electrical circuit for measuring resistance (e.g., a Wheatstone bridge), an effect known as the *piezoresistive effect*. To understand the piezoresistive effect, let us begin with the resistance of the beam of rectangular cross section

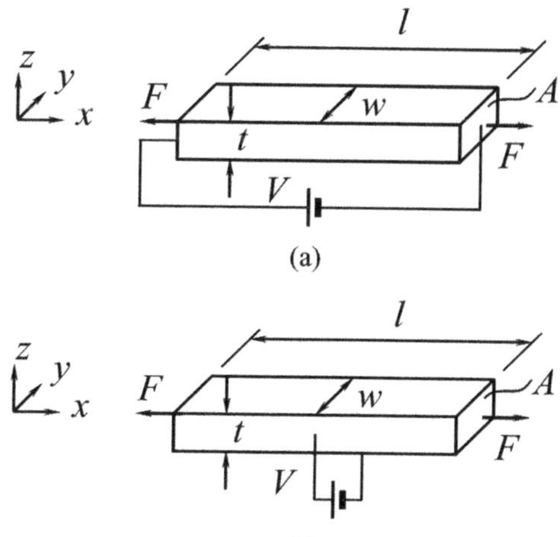

Figure 7.42 Piezoresistance.

illustrated in Fig. 7.42, which is the simplest case of piezoresistance. The rectangular beam of width w, thickness t, and length l is stretched by a tensile force F while a voltage is applied across both ends of the beam (i.e., the direction of the electric field coincides with that of stress). The resistance is given by (7.23), which is rewritten for convenience:

$$R = \rho_r \frac{l}{A} = \rho_r \frac{l}{wt} \tag{7.158}$$

where ρ_r and A denote the resistivity and cross-sectional area of the beam, respectively. As the force results in an elongation according to the stress–strain relation discussed in Section 3.2, the length is elongated by Δl, the width and thickness are contracted, and the resistance of the beam is then increased by ΔR. If Δw and Δt represent increases in the width and thickness during the stretching process, respectively, the resistance is expressed mathematically by

$$\begin{aligned} R + \Delta R &= (\rho_r + \Delta\rho_r)\frac{l + \Delta l}{(w + \Delta w)(t + \Delta t)} = \frac{\rho_r l}{wt}\frac{(1 + \Delta\rho_r/\rho_r)(1 + \Delta l/l)}{(1 + \Delta w/w)(1 + \Delta t/t)} \\ &\cong \frac{\rho_r l}{wt}\left(1 + \frac{\Delta\rho_r}{\rho_r}\right)\left(1 + \frac{\Delta l}{l}\right)\left(1 - \frac{\Delta w}{w}\right)\left(1 - \frac{\Delta t}{t}\right) \\ &\cong \frac{\rho_r l}{wt}\left(1 + \frac{\Delta\rho_r}{\rho_r} + \frac{\Delta l}{l} - \frac{\Delta w}{w} - \frac{\Delta t}{t}\right) \end{aligned}$$

Dividing the equation above by the initial resistance $R = \rho_r l/wt$, we find that

$$\frac{\Delta R}{R}=\frac{\Delta\rho_r}{\rho_r}+\frac{\Delta l}{l}-\frac{\Delta w}{w}-\frac{\Delta t}{t} \tag{7.159}$$

which relates the fractional change in resistance to the fractional change in resistivity and the strains in the three orthogonal directions. To relate $\Delta w/w$ and $\Delta t/t$ to $\Delta l/l$, we use the relation (3.4) between the longitudinal strain and the other strains in the plane normal to the axial direction, which gives

$$\frac{\Delta w}{w}=-\nu\frac{\Delta l}{l} \qquad \frac{\Delta t}{t}=-\nu\frac{\Delta l}{l}$$

where ν denotes Poisson's ratio. Substituting the relations above into (7.159), we obtain the following equation:

$$\frac{\Delta R/R}{\varepsilon_x}=1+2\nu+\frac{\Delta\rho_r/\rho_r}{\varepsilon_x} \tag{7.160}$$

where the dimensionless parameters in the preceding equation are defined as the gage factor, $(\Delta R/R)/\varepsilon_x$; the longitudinal strain, $\varepsilon_x = \Delta l/l$; and the ratio of the fractional change in resistivity to the longitudinal strain, $(\Delta\rho_r/\rho_r)/\varepsilon_x$. From an experiment, $(\Delta\rho_r/\rho_r)/\sigma_x$, where σ_x is the longitudinal stress defined by F/A, can be obtained, which is defined as the longitudinal piezoresistance coefficient,

$$\frac{\Delta\rho_r/\rho_r}{\sigma_x}=\pi_l \tag{7.161}$$

Recalling the stress–strain relation, $\sigma_x = E\varepsilon_x$, we have

$$\frac{\Delta\rho_r/\rho_r}{\varepsilon_x}=\pi_l E \tag{7.162}$$

If we substitute the definitions above into (7.160), we obtain the gage factor as a function of the longitudinal piezoresistance coefficient:

$$\frac{\Delta R/R}{\varepsilon_x}=1+2\nu+\pi_l E \tag{7.163}$$

It is of interest to examine the terms in equation (7.163). The number 1 and 2ν represent the resistance changes due to the length elongation and area contraction of the beam. The term $\pi_l E$ corresponds to the resistance change due to the stress (strain) in the beam. For metals, the term $1 + 2\nu$ is dominant and the third term can be neglected. However, for a semiconductor, the order of magnitude of the third term, $\pi_l E$, is much greater than that of the others

($v \leq 0.5$ for all materials), so the preceding gage factor for semiconductors can be written as

$$\frac{\Delta R/R}{\varepsilon_x} \cong \frac{\Delta\rho_r/\rho_r}{\varepsilon_x} = \pi_l E \tag{7.164}$$

which may be written as

$$\frac{\Delta\rho_r/\rho_r}{\sigma_x} = \pi_l \tag{7.165}$$

Similarly, we can measure the resistance of the beam in the y direction while the stress σ_x is applied in the x direction, as shown in Fig. 7.42b (i.e., the direction of electric field is perpendicular to that of stress). The transverse piezoresistance coefficient π_t is defined as

$$\frac{\Delta\rho_r/\rho_r}{\sigma_x} = \pi_t \tag{7.166}$$

Piezoresistance in Plane Stress

As shown in the preceding equations, the fractional change in resistivity $\Delta\rho_r/\rho_r$ (or resistance change) depends on the longitudinal piezoresistance coefficient π_l and the longitudinal stress σ_x. The fractional change in resistivity is, in general, a function of three-dimensional stresses, so more piezoresistance coefficients can be defined. As another example, consider a piezoresistor that is in plane stress as illustrated in Fig. 7.43, where σ_x and σ_y denote the longitudinal and transverse stresses, respectively. If we measure the resistance change between the ends of the beam while σ_t is applied and σ_l remains zero, another fractional change in resistivity can be defined in the same form as (7.166):

$$\frac{(\Delta\rho_r/\rho_r)_l}{\sigma_y} = \pi_t \tag{7.167}$$

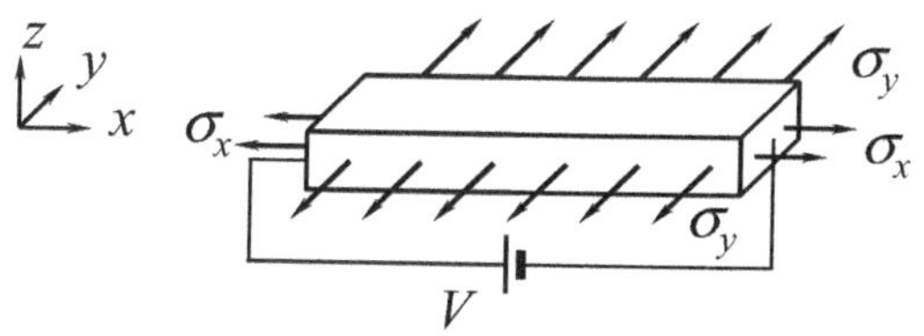

Figure 7.43 Piezoresistance under plane stress.

Assuming that the piezoresistance is linear so that the principle of superposition holds, we combine (7.165) and (7.167) to give

$$\left(\frac{\Delta\rho_r}{\rho_r}\right)_l = \pi_l\sigma_x + \pi_t\sigma_y \tag{7.168}$$

The piezoresistance coefficients π_l and π_t can be measured from a material. If the rectangular beam in Fig. 7.42 becomes thinner and wider ($w \geq 5t$), the stress σ_y is generated since the strain ε_y can be considered as zero, as discussed in Chapter 3. The first and second equations of (3.16) are rewritten to obtain the stress σ_y:

$$\varepsilon_x = \frac{1}{E}[\sigma_x - \nu(\sigma_y + \sigma_z)]$$

$$\varepsilon_y = \frac{1}{E}[\sigma_y - \nu(\sigma_z + \sigma_x)]$$

Setting $\varepsilon_y = 0$ and $\sigma_z = 0$, we have the stress σ_x and σ_y:

$$\sigma_y = \nu\sigma_x \qquad \sigma_x = \frac{E}{1-\nu^2}\varepsilon_x \tag{7.169}$$

Substituting the stresses into (7.168), we find the relation between the fractional change in resistivity and the longitudinal strain:

$$\left(\frac{\Delta\rho_r}{\rho_r}\right)_l = (\pi_l + \nu\pi_t)\sigma_x = E\frac{\pi_l + \nu\pi_t}{1-\nu^2}\varepsilon_x \tag{7.170}$$

Therefore, the gage factor for the wide beam under uniaxial loading is given by

$$\frac{(\Delta R/R)_l}{\varepsilon_x} = \frac{(\Delta\rho_r/\rho_r)_l}{\varepsilon_x} = E\frac{\pi_l + \nu\pi_t}{1-\nu^2} \tag{7.171}$$

$E/(1-\nu^2)$ is defined as the effective stiffness $\tilde{E}$ in (3.6). If we further define an effective longitudinal piezoresistance coefficient π_{el} for a wide beam under uniaxial loading as

$$\pi_{el} = \pi_l + \nu\pi_t \tag{7.172}$$

the equation (7.171) can be expressed in the simple form

$$\frac{(\Delta R/R)_l}{\varepsilon_x} = \pi_{el}\tilde{E} \tag{7.173}$$

which is in the same form as (7.164). We can also define the transverse gage factor for a wide beam under uniaxial loading. Let us consider a wider beam, shown in Fig. 7.42b, that becomes thinner and wider ($w \geq 5t$). The stress relations (7.169) still hold for the beam in Fig. 7.42. If the piezoresistance coefficients are symmetric with respect to a 90° rotation about the z axis [e.g., the (100) silicon wafer], the fractional change in resistivity in the y direction in Fig. 7.42b is written by applying the principle of superposition as follows:

$$\left(\frac{\Delta\rho_r}{\rho_r}\right)_t = \pi_t\sigma_x + \pi_l\sigma_y \tag{7.174}$$

Upon substitution of (7.169) into equation (7.174), we obtain the relation between the fractional change in resistivity and the strain:

$$\left(\frac{\Delta\rho_r}{\rho_r}\right)_t = (\pi_t + \nu\pi_l)\sigma_x = E\frac{\pi_t + \nu\pi_l}{1-\nu^2}\varepsilon_x \tag{7.175}$$

which gives the transverse gage factor for a wide beam under uniaxial loading:

$$\frac{(\Delta R/R)_t}{\varepsilon_x} = \frac{(\Delta\rho_r/\rho_r)_t}{\varepsilon_x} = E\frac{\pi_t + \nu\pi_l}{1-\nu^2} \tag{7.176}$$

As for the longitudinal gage factor, we define an effective transverse piezoresistance coefficient π_{et} for a wide beam under uniaxial loading as follows:

$$\pi_{et} = \pi_t + \nu\pi_l \tag{7.177}$$

Therefore, the transverse gage factor (7.176) can be expressed in terms of the effective longitudinal piezoresistance coefficient π_{et} and the effective Young's modulus $\tilde{E}$:

$$\frac{(\Delta R/R)_t}{\varepsilon_x} = \pi_{et}\tilde{E} \tag{7.178}$$

Piezoresistance Coefficients

As described in the foregoing paragraphs, to obtain a gage factor for slender or wide beams, we first evaluate the longitudinal and transverse piezoresistance coefficients. The coefficients can be obtained from an experiment such as Fig. 7.42a and b or may be evaluated from published data. For example, Smith (1954) reported the piezoresistance coefficients of silicon which are listed in Table 7.1. In the table, the coefficients, π_{11}, π_{12}, and π_{44} are measured values with respect to the principal axes of the silicon structure. If we use the

TABLE 7.1 Piezoresistance Coefficients of Silicon (10^{-11} Pa^{-1})

Doping	Resistivity (Ω·cm)	π_{11}	π_{12}	π_{44}
n-type	11.7	−102.2	53.4	−13.6
p-type	7.8	6.6	−1.12	138.1

(100) silicon wafer and the [100] axis of the silicon wafer coincides with the x axis in Fig. 7.42a and b, the longitudinal and transverse piezoresistance coefficients are given by

$$\pi_l = \pi_{11} \qquad \pi_t = \pi_{12} \tag{7.179}$$

However, if we use the (100) silicon wafer and the [110] axis is aligned with the x axis in Fig. 7.42a and b, the coefficient π_{44}, which represents a piezoresistive effect due to shear stress in silicon, plays a role. The longitudinal and transverse piezoresistance coefficients can then be obtained by a coordinate transform (Tufte, 1963),

$$\pi_l = \frac{1}{2}(\pi_{11} + \pi_{12} + \pi_{44}) \qquad \pi_t = \frac{1}{2}(\pi_{11} + \pi_{12} - \pi_{44}) \tag{7.180}$$

Using π_l and π_t of (7.179) and (7.180), the effective piezoresistance coefficients π_{el} and π_{et}, (7.172) and (7.177), are obtained, and the longitudinal and transverse gage factors, (7.173) and (7.178), can be evaluated for the piezoresistor configuration in Fig. 7.42a and b.

Example 7.1 Two beams of width w_b and thickness t_b are made of bulk silicon, as shown in Fig. 7.44a and b, where the longitudinal direction of the beam coincides with the [110] direction of the (100) silicon wafer. The two beams are subjected to a transverse force F and a moment M at their ends, and their cross sections are approximated as rectangles ($w_b >> 5t_b$, $l_b >> w_b$). Identical p-type piezoresistors ($t << t_b$) are formed on the beams at l_b from the ends of the beams, as shown in Fig. 7.44a and b. If the resistance of the resistors is measured as shown in the lower figures in Fig. 7.44a and b, find the resistance change as a function of the force and moment.

The basic configurations in Fig. 7.44a and b are used to measure force, moment, acceleration (inertial force), pressure, and so on. We shall first obtain the strain at the point where the resistors are placed, and then obtain the relation between the resistance and the strain due to the applied force and moment. If we cut the beam at the resistor site and isolate the right portion of the beam, the moment equilibrium at the resistor site gives the bending moment,

$$M_r = Fl_b + M \tag{a}$$

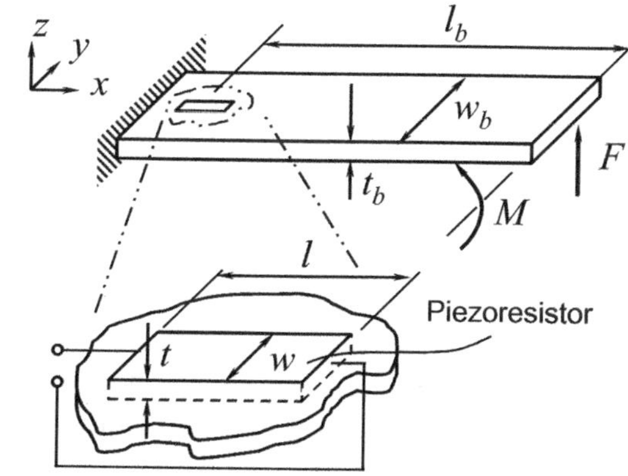

(a)

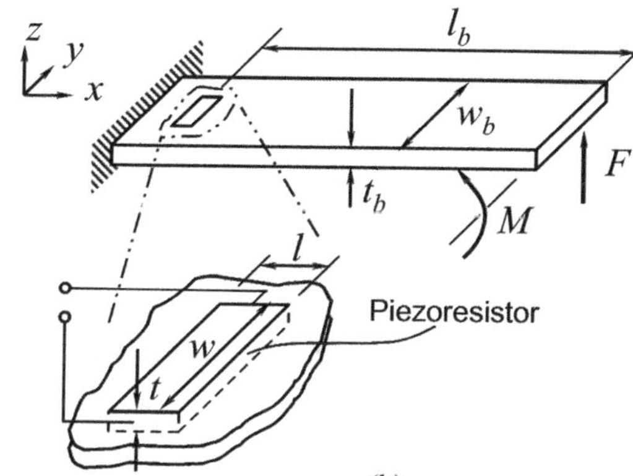

(b)

Figure 7.44 Cantilever beam with piezoresistors.

The stress at the site is evaluated by the stress–moment relation, (3.41), as follows:

$$\sigma_x = -\frac{M_r}{I} y \bigg|_{y=t_b/2} = -\frac{M_r}{2I} t_b \qquad \text{(b)}$$

where the moment of inertia of the rectangular cross section is given by (3.44):

$$I = \frac{w_b t_b^3}{12} \qquad \text{(c)}$$

Since $w_b >> 5t_b$, the following effective Young's modulus $\tilde{E}$ is used for evaluation of the longitudinal strain ε_x,

$$\tilde{E} = \frac{E}{1-\nu^2} \tag{d}$$

The strain ε_x is written as

$$\varepsilon_x = \frac{\sigma_x}{\tilde{E}} \tag{e}$$

Since the wide beam ($w_b >> 5t_b$) is subjected to the axial stress (b), the site is in plane stress (i.e., the stresses σ_x and σ_y defined by (7.169) are both generated on the surface). The longitudinal directions of the beams are aligned with the [110] direction of the wafer, so we use (7.180) for the evaluation of π_l and π_t, which is rewritten here for convenience:

$$\pi_l = \frac{1}{2}(\pi_{11} + \pi_{12} + \pi_{44}) \qquad \pi_t = \frac{1}{2}(\pi_{11} + \pi_{12} - \pi_{44}) \tag{f}$$

Since the x and y directions are the [110] direction of the (110) wafer, the longitudinal and transverse gage factors are obtained from (7.173) and (7.178), respectively:

$$\frac{(\Delta R/R)_l}{\varepsilon_x} = \pi_{el}\tilde{E} \qquad \text{where } \pi_{el} = \pi_l + \nu\pi_t \tag{g}$$

$$\frac{(\Delta R/R)_t}{\varepsilon_x} = \pi_{et}\tilde{E} \qquad \text{where } \pi_{et} = \pi_t + \nu\pi_l \tag{h}$$

Combining (b), (e), (g), and (h), we have the fractional change in the resistances of Fig. 7.44a and b, respectively:

$$(\Delta R/R)_l = \pi_{el}\tilde{E}\varepsilon_x = \pi_{el}\sigma_x = -\pi_{el}\frac{M_r}{2I}t_b \tag{i}$$

$$(\Delta R/R)_t = -\pi_{et}\frac{M_r}{2I}t_b \tag{j}$$

Substituting (a) into (i) and (j) yields the resistance change as functions of force and moment which applied at the end of beam:

$$(\Delta R/R)_l = -\pi_{el}\frac{t_b}{2I}(Fl_b + M) \tag{k}$$

$$(\Delta R/R)_t = -\pi_{et}\frac{t_b}{2I}(Fl_b + M) \tag{l}$$

where π_{el} and π_{et} are the effective piezoresistance coefficients defined by (g) and (h). If $w_b \approx t_b$, π_{el} and π_{et} can be replaced by π_l and π_t in (k) and (l), respectively. For a p-type resistor, equation (f) and the data of Table 7.1 gives $\pi_l = 71.79 \times 10^{-11}\,\mathrm{Pa}^{-1}$ and $\pi_t = -66.31 \times 10^{-11}\,\mathrm{Pa}^{-1}$. Assuming that $\nu = 0.22$ for the bulk silicon, equations (g) and (h) yield $\pi_{el} = 57.20 \times 10^{-11}\,\mathrm{Pa}^{-1}$ and $\pi_{et} = -50.52 \times 10^{-11}\,\mathrm{Pa}^{-1}$. Note that the magnitudes of π_{el} and π_{et} are less than that of π_l and π_t since the strain of the wide beam decreases due to the transverse stress, σ_y, which is generated in the wide beam ($w_b \geq 5t_b$). A ratio of the changes in the resistors may be defined as

$$\frac{(\Delta R/R)_t}{(\Delta R/R)_l} = \frac{\pi_{et}}{\pi_{el}} \tag{m}$$

Substituting the effective piezoresistance coefficients $\pi_{el} = 57.20 \times 10^{-11}\,\mathrm{Pa}^{-1}$ and $\pi_{et} = -50.52 \times 10^{-11}\,\mathrm{Pa}^{-1}$ into equation (m), the ratio is obtained as

$$\frac{(\Delta R/R)_t}{(\Delta R/R)_l} = -0.88$$

7.9 PIEZOELECTRICITY

Some solid materials, known as *piezoelectric materials*, generate an electric charge when they are deformed. This is known as the *piezoelectric effect*. This phenomenon also occurs in reverse; that is, if the materials are exposed to an electric field, they are deformed, called the *reverse piezoelectric effect*. The piezoelectric effect is used for sensors and the reverse effect, for actuators. A general configuration used in MEMS devices is illustrated in Fig. 7.45a. A thin piezoelectric film of thickness t, length l, and width w has electrodes on its top and bottom surfaces. If an input voltage V_i is applied across the film as shown in Fig. 7.45b, the shape of the film is deformed. As a force or stress is applied to the film in Fig. 7.45c, the film first generates an electric charge Q and then provides voltage V_o to the outside since the film acts as a capacitor, $V_o = Q/C$, where C is the capacitance of the film. As the force or stress acting on the film (or any piezoelectric material) remains constant, the charge generated, Q, leaks away though the film resistance itself or an outside resistance. If enough time passes, the charge will disappear and the output voltage in Fig. 7.45c will be zero. This charge leakage in piezoelectric materials is confined in its use to sensors dealing with a time-varying signal (e.g., microphones). If the piezoelectric film in Fig. 7.45b is used as an actuator, the film works under ac as well as dc voltage, since charges are supplied from a voltage source.

Let us analyze the response of the piezoelectric film in Fig. 7.45a to the applied voltage or stress. In general, the six stress components, three normal

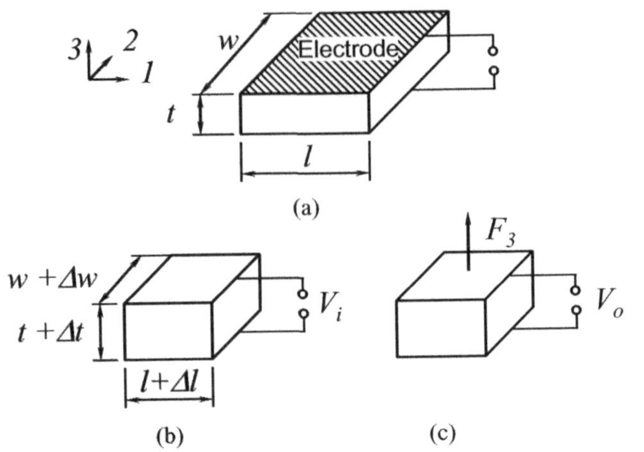

Figure 7.45 Piezoelectricity.

and three shear stresses, are related to the charge per unit area on three opposite faces of a rectangular parallelepiped (if we use Cartesian coordinates). We assume that the 3-direction in Fig. 7.45a is the polarization direction of an artificial piezoelectric material (i.e., a high voltage is applied along the 3-direction to make the material piezoelectric). If we are only interested in relations between a normal strain (or stress) and the applied charge (or voltage) parallel to the 3-direction, the relations are greatly simplified. As we apply a voltage V_i across the piezoelectric film as shown in Fig. 7.45b, where the film is on a frictionless surface, strains in the 1-, 2-, and 3-directions are developed and can be expressed as follows:

$$\begin{aligned} \varepsilon_1 &= d_{31}E_3 \\ \varepsilon_2 &= d_{32}E_3 = d_{31}E_3 \\ \varepsilon_3 &= d_{33}E_3 \end{aligned} \tag{7.181}$$

where ε_1, ε_2, ε_3, d_{31}, d_{33}, and E_3 denote the normal strains in the 1-, 2-, and 3-directions, defined respectively, as $\Delta l/l$, $\Delta w/w$, and $\Delta t/t$, the piezoelectric constants depending on materials and the electric field strength in the 3-direction, defined as V_i/t, respectively. It is noted that the preceding strain is like that due to the thermal expansion in Section 3.2; that is, if a piezoelectric material without constraint is exposed to a uniform electric field, the material expands or contracts without internal stress. Combining the foregoing equations with the stress–strain relationships (Hooke's law), we can obtain the stress or force acting on the material with constraint. It is also noted from the preceding equations that if the direction of E_3 is reversed, the signs of the strains are also reversed. Using the definitions of the strains and electric field strength, equation (7.181) may be written for the film as

$$\Delta l = d_{31}\frac{l}{t}V_i$$
$$\Delta w = d_{31}\frac{w}{t}V_i \qquad (7.182)$$
$$\Delta t = d_{33}V_i$$

Note that the change in thickness is proportional to the applied voltage V_i but independent of thickness. When the film is exposed to a stress parallel to the 3-direction, as shown in Fig. 7.45c, the electric field strength across the thickness of the film is obtained as

$$E_3 = g_{33}\sigma_3 \qquad (7.183)$$

The coefficient g_{33} is given by d_{33}/ε, where ε denotes the permittivity of the piezoelectric film. Referring to Fig. 7.45c, the voltage across the film is obtained and may be written as

$$V_0 = g_{33}\frac{t}{wl}F_3 \qquad (7.184)$$

where F_3 is the force acting on the film in the 3-direction. As the film is exposed to a stress in the 1-direction (i.e., the film is stretched by a force), the electric field strength across the thickness of the film is expressed as

$$E_3 = g_{31}\sigma_1 \qquad (7.185)$$

The coefficient g_{31} is given by d_{31}/ε, where ε denotes the permittivity of the piezoelectric film. The expression above may be written

$$V_0 = g_{31}\frac{t}{wl}F_1 \qquad (7.186)$$

where V_0 is the voltage across the film and F_1 is the force acting on the film in the 1-direction.

PROBLEMS

7.1 A sinusoidal voltage $V = V_a \cos \omega t$ is applied to the circuit shown in Fig. P7.1. Find the steady-state current i and voltage V_o.

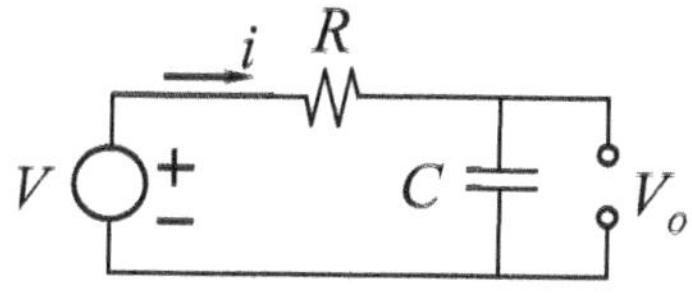

Figure P7.1

7.2 A particle of positive charge q is located near the lower face of the upper plate in vacuum in Fig. 7.16b. If the voltage V across the plates remains constant, determine the velocity of the particle when the particle reaches the lower plate.

7.3 Two identical beams face each other and a voltage is applied between the beams shown in Fig. P7.3. Obtain the capacitance between the beams and the electrostatic force acting on the upper beam.

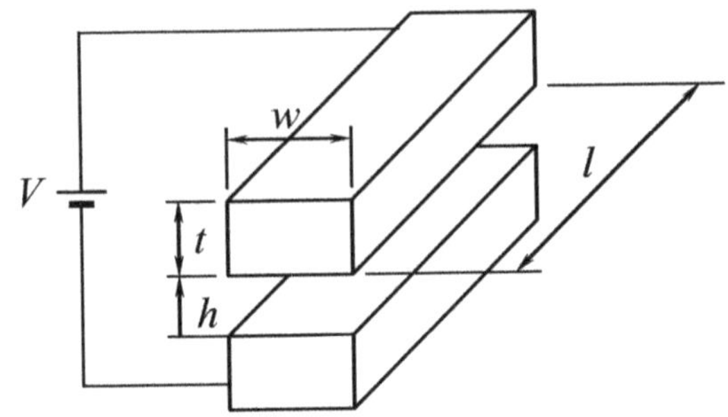

Figure P7.3

7.4 The movable comb of the electrostatic combs of overlap length l in Fig. 7.27 moves in the direction perpendicular to the x direction and in the plane in which the combs are placed. Assuming that the electrostatic force can be approximated as that of a parallel plate of interfinger gap g, find an expression for the electrostatic force in the direction as a function of the displacement and geometry of the comb.

7.5 Referring to Fig. 7.34, derive expressions (7.130), (7.131), (7.134), and (7.135) for the electromagnetic force $\mathbf{F}_m$ and moment $\mathbf{M}_m$ (not from analogy).

7.6 The magnetomotive force Ni (Fig. 7.35a) remains constant. Find the magnetic moment $\mathbf{M}_m$ acting on the armature about the pivot point and express the moment in terms of the magnetic force $\mathbf{F}_m$, (7.148), and the length l_4.

7.7 Derive the magnetic moment $\mathbf{M}_m$, (7.157), acting on the circular coil in Fig. 7.41.

7.8 We can measure the applied force and moment simultaneously if we add one more piezoresistor on the cantilever beam shown in Fig. 7.44a. Two beams of width w_b and thickness t_b are made of bulk silicon as shown in Fig. 7.44a, where the longitudinal direction of the beam coincides with the [110] direction of the (100) silicon wafer. The slender beam is subjected to a transverse force F and a moment M at the free end, and the cross-section of the beam are approximated as rectangle (w_b>>5t_b, l_b>>w_b). Identical p-type piezoresistors are formed on the beams at l_b from the free end of the beam (as shown in Fig. 7.44a) and at $l_b/2$ from the free end (not shown in Fig. 7.44a). The thickness of the resistors is negligible compared to that of the beam. If the resistance of the resistors is measured as shown in the lower figure of Fig. 7.44a and if a resistance measuring device measures the change in each resister, find the applied force F and moment M from the measured resistance changes.

CHAPTER 8

PIEZOELECTRIC AND THERMAL ACTUATORS

In previous chapters we described the principles for micro- and nanosystems: the statics and dynamics of flexible and rigid elements, fluid dynamics, and electromagnetics. Since MEMS are systems in which mechanical and electrical components are combined for specific purposes (e.g., actuating mechanical components, and sensing physical quantities), the mechanical and electrical behavior of the combined components must be investigated to design actuators and sensors and to conduct further research on MEMS. In MEMS, most actuators used for actuating mechanical components are also used for sensing physical quantities. We describe piezoelectric and thermal actuators in this chapter, and electrostatic and electromagnetic actuators are covered in Chapter 9.

8.1 COMPOSITE BEAMS

In Chapters 3 and 4 we described the static and dynamic behavior of beams made of uniform material. In some MEMS, composite elements, which are made of more than one material, are also used (e.g., bimetals actuated by temperature change and piezoelectric cantilever beams driven by an electric field). To analyze actuators using composite beams, we must find relations between the driving forces (e.g., the temperature for bimetals) and the responses (e.g., deflection). Let us begin by considering the straight composite beam shown in Fig. 8.1a, which consists of two different materials of different

Principles of Microelectromechanical Systems, By Ki Bang Lee

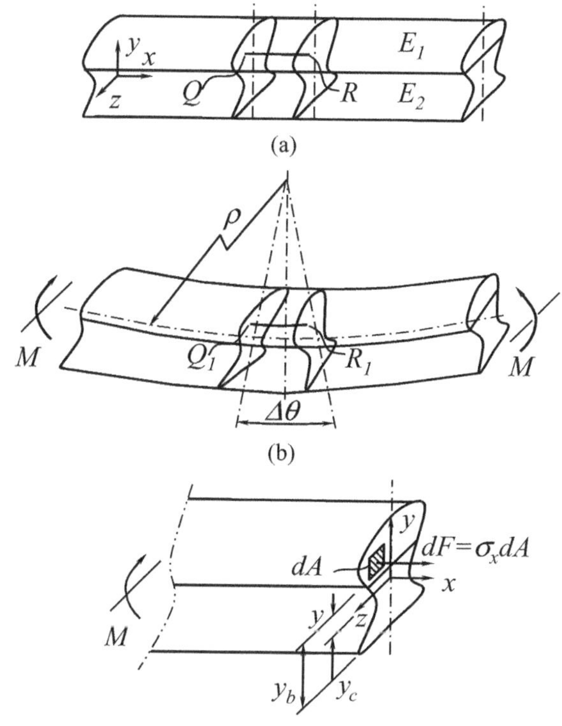

Figure 8.1 Composite beam.

Young's moduli, E_1 and E_2, and whose cross section is symmetric about the y axis. When a bending moment M is applied to the end of the beam as shown in Fig. 8.1b, the straight line QR, connecting two points on cross sections in Fig. 8.1a, is deformed into the curve Q_1R_1 in Fig. 8.1b. Due to the symmetry argument that was used in Chapter 3 to derive the moment–curvature relationship of the uniform material, the cross section in the beam remains plane (in which the points Q_1 and R_1 lie) after the moment is applied. Since the cross section of the beam is subjected to the moment M, the beam is deformed uniformly so that the beam makes the radius of curvature ρ indicated in Fig. 8.1b.

Moment–Curvature Relationships

We extend the theory developed in Chapter 3 to describe the mechanical behavior of the composite beam in Fig. 8.1. Using the fact that the internal moment due to the stress in the beam balances the applied moment M, we can obtain the radius of curvature and the stress distribution in the beam. Let y, Δx, and $\Delta\theta$ represent the coordinate measured from the neutral surface (where the normal stress is zero), the length of line QR in Fig. 8.1a, and the angle that is made by the planes, as shown in Fig. 8.1b. Because all quantities such as moment are independent of a 180° rotation about the vertical line in Fig. 8.1b, the cross sections in Fig. 8.1b remain plane, so (3.35)–(3.37), which were obtained for the uniform beam, still hold for the composite beam and are rewritten here for convenience:

$$\varepsilon_x = \frac{(\rho - y)\Delta\theta - \Delta x}{\rho\Delta\theta} = -\frac{y}{\rho} \tag{8.1}$$

$$\frac{1}{\rho} = \lim_{\Delta s \to 0} \frac{\Delta\theta}{\Delta s} = \frac{d\theta}{ds} \tag{8.2}$$

$$\varepsilon_x = -\frac{d\theta}{ds} y \tag{8.3}$$

where ε_x and $\Delta\theta$ denote the strain in the beam at y and the length of the arc, which is defined as $\Delta s = \rho\ \Delta\theta$, respectively. The strain equation above is obtained by substituting (8.2) into (8.1). Applying the force equilibrium condition to the free body in Fig. 8.1c gives the following equation:

$$\begin{aligned}\sum F_x &= \int_A \sigma_x dA = \int_A \tilde{E}\varepsilon_x dA = \int_A \tilde{E}\left(-\frac{d\theta}{ds} y\right) dA \\ &= -\frac{d\theta}{ds}\int_A \tilde{E} y\, dA = 0\end{aligned}$$

or

$$\int_A \tilde{E} y dA = 0$$

where σ_x, $\tilde{E}$, and A denote the normal stress along the x axis, the effective Young's modulus defined in (3.5) and (3.6), and the cross-sectional area of the beam, respectively. For convenience we define the coordinate y as $y = y_b - y_c$, where y_b is the distance from the bottom of the beam and y_c denotes the coordinate of the neutral surface from the bottom of the beam. Substituting $y = y_b - y_c$ into the preceding force equilibrium equation gives the coordinate of the neutral surface:

$$y_c = \frac{\int_A \tilde{E} y_b\, dA}{\int_A \tilde{E}\, dA} \tag{8.4}$$

Applying the moment equilibrium condition $\sum M = 0$ to the freebody in Fig. 8.1c and using (8.1)–(8.3), we have the following equation, which relates the applied moment M and the curvature ρ:

$$M = -\int_A y\sigma_x \, dA = -\int_A y\tilde{E}\varepsilon_x \, dA = -\int_A y\tilde{E}\left(-\frac{d\theta}{ds}y\right)dA$$
$$= \frac{d\theta}{ds}\int_A \tilde{E}y^2 \, dA = \frac{d\theta}{ds}\left(\tilde{E}I\right)_{\text{eff}} = \frac{\left(\tilde{E}I\right)_{\text{eff}}}{\rho} \tag{8.5}$$

where

$$\left(\tilde{E}I\right)_{\text{eff}} = \int_A \tilde{E}y^2 \, dA \tag{8.6}$$

is defined as the effective flexural rigidity. As a result, the radius of curvature ρ is obtained as

$$\rho = \frac{\left(\tilde{E}I\right)_{\text{eff}}}{M} \tag{8.7}$$

Substituting $d\theta/ds \cong d^2y/dx^2$ into (8.5) yields the moment–curvature relationship:

$$\left(\tilde{E}I\right)_{\text{eff}} \frac{d^2y}{dx^2} = M \tag{8.8}$$

Equation (8.8) can be solved for y if the boundary conditions are given. The strain and the corresponding stress at a point in the composite beam are then obtained by combining (8.1) and (8.7):

$$\varepsilon_x = -\frac{y}{\rho} = -\frac{M}{\left(\tilde{E}I\right)_{\text{eff}}} y \tag{8.9}$$

$$\sigma_x = \tilde{E}\varepsilon_x = -\frac{\tilde{E}}{\left(\tilde{E}I\right)_{\text{eff}}} My \tag{8.10}$$

where $\tilde{E}$ is Young's modulus at a point of interest (i.e., $\tilde{E}_1$ for material 1 and $\tilde{E}_2$ for material 2 in Fig. 8.1a).

Example 8.1 A slender straight composite beam which consists of the two beams of rectangular cross section shown in Fig. 8.2a, is used in a MEMS application. The composite beam can be made by bonding together two layers or by depositing an upper layer on a lower layer. It is desired to obtain the effective flexural rigidity $\left(\tilde{E}I\right)_{\text{eff}}$ and the radius of curvature ρ as functions of Young's moduli and geometry when a moment M is applied at the ends of the beam, as shown in Fig. 8.1b.

We use (8.4) to obtain the coordinate of the neutral surface:

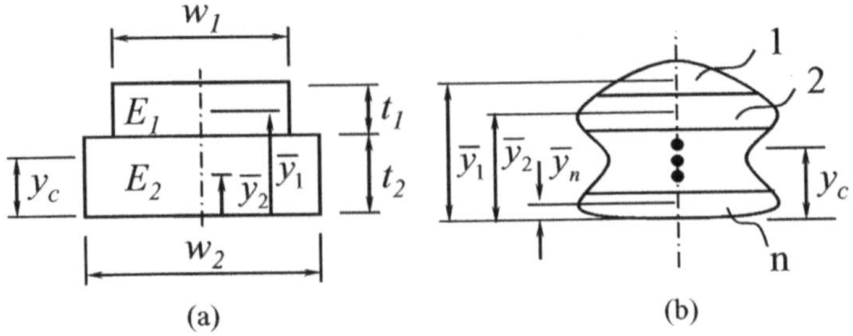

Figure 8.2 Cross sections of composite beams.

$$y_c = \frac{\int_A \tilde{E} y_b \, dA}{\int_A \tilde{E} \, dA} = \frac{\tilde{E}_1 A_1 \bar{y}_1 + \tilde{E}_2 A_2 \bar{y}_2}{\tilde{E}_1 A_1 + \tilde{E}_2 A_2} \tag{a}$$

where

$$\bar{y}_1 = \frac{\int_{A_1} \tilde{E}_1 y_b \, dA}{\int_{A_1} \tilde{E}_1 \, dA} = \frac{\int_{A_1} y_b \, dA}{A_1} \quad \text{and} \quad \bar{y}_2 = \frac{\int_{A_2} \tilde{E}_2 y_b \, dA}{\int_{A_2} \tilde{E}_2 \, dA} = \frac{\int_{A_2} y_b \, dA}{A_2} \tag{b}$$

represent the centroids of the areas, which are measured from the bottom of the beam. For rectangles, equations (a) and (b) become

$$\bar{y}_1 = \frac{t_1}{2} + t_2 \quad \text{and} \quad \bar{y}_2 = \frac{t_2}{2} \tag{c}$$

where t_1 and t_2 are the thickness of the rectangles. Substituting (c) into (a) gives the coordinate of the neutral surface,

$$y_c = \frac{\tilde{E}_1 w_1 t_1 (t_1/2 + t_2) + \tilde{E}_2 w_2 t_2 \, t_2/2}{\tilde{E}_1 w_1 t_1 + \tilde{E}_2 w_2 t_2} = \frac{t_2}{2} \frac{1 + pqr(2+p)}{1 + pqr} \tag{d}$$

where

$$p = \frac{t_1}{t_2} \qquad q = \frac{w_1}{w_2} \qquad r = \frac{\tilde{E}_1}{\tilde{E}_2} \tag{e}$$

are the ratios of the thicknesses, widths, and Young's moduli of the layers. The effective flexural rigidity $(\tilde{E}I)_{\text{eff}}$ is given by (8.6):

$$\begin{aligned}(\tilde{E}I)_{\text{eff}} &= \int_A \tilde{E} y^2 \, dA = \tilde{E}_1 \int_{A_1} y^2 \, dA + \tilde{E}_2 \int_{A_2} y^2 \, dA \\ &= \tilde{E}_1 I_{c1} + \tilde{E}_2 I_{c2}\end{aligned} \tag{f}$$

where

$$I_{c1} = \int_{A_1} y^2 \, dA \qquad I_{c2} = \int_{A_2} y^2 \, dA \tag{g}$$

are the moments of inertia of the upper and lower areas with respect to y_c. Referring to Fig. 8.2a, the coordinate y may be written as $y = y_b - y_c = (y_b - \bar{y}_1) + (\bar{y}_1 - y_c)$, and thus the moments of inertia of the upper area are given by

$$\begin{aligned} I_{c1} &= \int_{A_1} [(y_b - \bar{y}_1) + (\bar{y}_1 - y_c)]^2 \, dA \\ &= \int_{A_1} \left[(y_b - \bar{y}_1)^2 + 2(y_b - \bar{y}_1)(\bar{y}_1 - y_c) + (\bar{y}_1 - y_c)^2 \right] dA \\ &= I_{01} + A_1 (\bar{y}_1 - y_c)^2 \end{aligned} \tag{h}$$

where A_1 is the area of the upper area, defined by $A_1 = w_1 t_1$, and I_{01} is the moment of inertia of A_1 with respect to $\bar{y}_1$, which is given by

$$I_{01} = \int_{A_1} (y_b - \bar{y}_1)^2 \, dA = \int_{-t_1/2}^{t_1/2} y_{01}^2 w_1 \, dy_{01} = \frac{w_1 t_1^3}{12} \tag{i}$$

In equation (i) an auxiliary variable, $y_{01} = y_b - \bar{y}_1$, is introduced to perform the integration. Similarly, the moment of inertia, I_{c2}, is obtained as

$$I_{c2} = \int_{A_2} y^2 \, dA = I_{02} + A_2 (\bar{y}_2 - y_c)^2 \tag{j}$$

where A_2 is the area of the lower area, defined by $A_2 = w_2 t_2$, and I_{02} is the moment of inertia of A_2 with respect to $\bar{y}_2$:

$$I_{02} = \int_{A_1} (y_b - \bar{y}_2)^2 \, dA = \frac{w_2 t_2^3}{12} \tag{k}$$

Substituting (h) and (j) into (f), we obtain effective flexural rigidity $\left(\tilde{E}I\right)_{\text{eff}}$ as a function of the ratios p, q, and r:

$$\begin{aligned} \left(\tilde{E}I\right)_{\text{eff}} &= \tilde{E}_1 \left[\frac{w_1 t_1^3}{12} + w_1 t_1 \left(\frac{t_1}{2} + t_2 - y_c \right)^2 \right] \\ &\quad + \tilde{E}_2 \left[\frac{w_2 t_2^3}{12} + w_2 t_2 \left(y_c - \frac{t_2}{2} \right)^2 \right] \\ &= \tilde{E}_2 \frac{w_2 t_2^3}{12} \left\{ r \left[p^3 q + 3pq \left(2 + p - \frac{2y_c}{t_2} \right)^2 \right] + 1 + 3 \left(\frac{2y_c}{t_2} - 1 \right)^2 \right\} \end{aligned} \tag{l}$$

To simplify equation (l), we use (d) to obtain the following relations:

$$2+p-\frac{2y_c}{t_2}=2+p-\frac{1+pqr(2+p)}{1+pqr}=\frac{1+p}{1+pqr}$$

$$\frac{2y_c}{t_2}-1=\frac{1+pqr(2+p)}{1+pqr}-1=\frac{pqr(1+p)}{1+pqr}$$

Upon substitution of the equations above into (l), we find that

$$\begin{aligned}\left(\tilde{E}I\right)_{\text{eff}} &= \tilde{E}_2\frac{w_2t_2^3}{12}\left\{r\left[p^3q+3pq\left(\frac{1+p}{1+pqr}\right)^2\right]+1+3\left[\frac{pqr(1+p)}{1+pqr}\right]^2\right\}\\ &= \tilde{E}_2\frac{w_2t_2^3}{12}\frac{1}{1+pqr}\left[(1+pqr)(1+p^3qr)+3pqr(1+p)^2\right] \qquad \text{(m)}\end{aligned}$$

Since $\tilde{E}_2w_2t_2^3/12$ can be written as $\tilde{E}_2I_2$, the dimensionless effective flexural rigidity $\left(\tilde{E}I\right)_{\text{eff}}/\tilde{E}_2I_2$ is expressed in terms of the ratios p, q, and r as follows:

$$\frac{\left(\tilde{E}I\right)_{\text{eff}}}{\tilde{E}_2I_2}=\frac{1}{1+pqr}\left[(1+pqr)(1+p^3qr)+3pqr(1+p)^2\right] \qquad \text{(n)}$$

The $\left(\tilde{E}I\right)_{\text{eff}}$ above will be used to further develop the static and dynamic responses of the composite beam to external force and moment.

If the cross section of a slender beam is made of n-layered materials of different mechanical properties as shown in Fig. 8.2b, repeating the procedure to obtain (8.1) to (8.10) yields the following set of equations, which holds for a beam of cross section symmetric about the y axis (the vertical dashed-dotted line in Fig. 8.2b):

$$y_c=\frac{\int_A \tilde{E}y_b\,dA}{\int_A \tilde{E}\,dA}=\frac{\sum_{j=1}^{n}\tilde{E}_jA_j\bar{y}_j}{\sum_{j=1}^{n}\tilde{E}_jA_j}=\frac{\tilde{E}_1A_1\bar{y}_1+\tilde{E}_2A_2\bar{y}_2+\cdots+\tilde{E}_nA_n\bar{y}_n}{\tilde{E}_1A_1+\tilde{E}_2A_2+\cdots+\tilde{E}_nA_n} \qquad (8.11)$$

$$\begin{aligned}\left(\tilde{E}I\right)_{\text{eff}} &= \int_A \tilde{E}y^2\,dA=\tilde{E}_1\int_{A_1}y^2\,dA+\tilde{E}_2\int_{A_2}y^2\,dA+\cdots+\tilde{E}_n\int_{A_n}y^2\,dA\\ &= \tilde{E}_1I_{c1}+\tilde{E}_2I_{c2}+\cdots+\tilde{E}_nI_{cn}=\sum_{j=1}^{n}\tilde{E}_jI_{cj} \qquad (8.12)\end{aligned}$$

$$I_{c1}=\int_{A_1}y^2\,dA,\quad \ldots,\quad I_{cj}=\int_{A_j}y^2\,dA,\quad \ldots,\quad I_{cn}=\int_{A_n}y^2\,dA \qquad (8.13)$$

$$\rho = \frac{(\tilde{E}I)_{\text{eff}}}{M} \tag{8.14}$$

$$(\tilde{E}I)_{\text{eff}} \frac{d^2y}{dx^2} = M \tag{8.15}$$

Deflection of Composite Beams

Because we have the effective flexural rigidity and the moment–curvature relationships, we can analyze the composite beam. Let us apply moment–curvature relationship (8.8) to obtain the deflection of composite beams which are subjected to force and moment. As discussed in Chapter 3, the beam deflection is easily found by applying Castigliano's theorem, which also holds for composite beam problems since the energy due to the moment can be expressed as

$$U = \int_0^l \frac{M^2}{2(\tilde{E}I)_{\text{eff}}} dx \tag{8.16}$$

whose derivation is similar to that of (3.131); that is, equation (8.16) is derived from (8.8). As an example, we apply Castigliano's theorem to a cantilever beam made of composite material in Fig. 8.3. The upper layer of Young's modulus E_1 is partially layered on the lower layer of Young's modulus E_2, and the end of the beam is subjected to both a force F and a moment M. We shall find the displacement δ in the vertical direction and the angular displacement θ, illustrated in Fig. 8.3. We first obtain the moment M_x at x from the free end of the beam:

$$M_x = Fx + M \tag{8.17}$$

Substituting the moment above into the preceding energy equation gives the elastic energy stored in the beam:

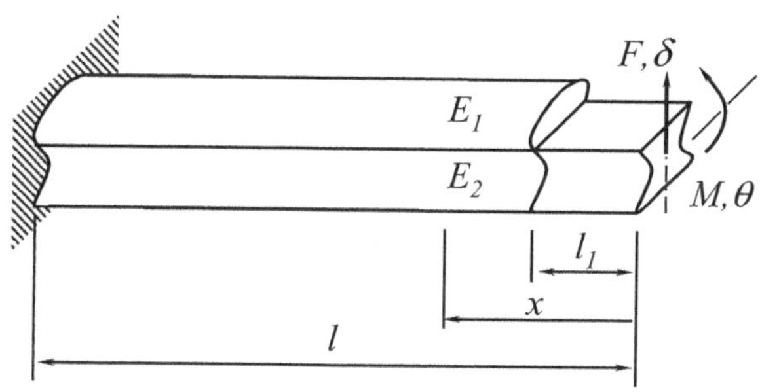

Figure 8.3 Deflection of a composite cantilever beam.

$$U = \int_0^l \frac{M^2}{2EI} dx = \int_0^l \frac{(Fx+M)^2}{2EI} dx$$
$$= \int_0^{l_1} \frac{(Fx+M)^2}{2E_2I_2} dx + \int_{l_1}^l \frac{(Fx+M)^2}{2(EI)_{\text{eff}}} dx \tag{8.18}$$

Under the assumption that the beam is linear, the complementary energy U^*, due to the applied force and moment, is equal to the elastic energy shown in equation (8.18). Applying Castigliano's theorem, (3.125) and (3.126), to equation (8.18), we have the deflection δ and angular displacement θ of the end:

$$\delta = \frac{\partial U^*}{\partial F} = \frac{\partial U}{\partial F} = \int_0^{l_1} \frac{Fx^2 + Mx}{E_2I_2} dx + \int_{l_1}^l \frac{Fx^2 + Mx}{(EI)_{\text{eff}}} dx$$
$$= \frac{1}{3}\left(\frac{l_1^3}{E_2I_2} + \frac{l^3 - l_1^3}{(EI)_{\text{eff}}}\right)F + \frac{1}{2}\left(\frac{l_1^2}{E_2I_2} + \frac{l^2 - l_1^2}{(EI)_{\text{eff}}}\right)M \tag{8.19}$$

$$\theta = \frac{\partial U^*}{\partial M} = \frac{\partial U}{\partial M} = \int_0^{l_1} \frac{Fx + M}{E_2I_2} dx + \int_{l_1}^l \frac{Fx + M}{(EI)_{\text{eff}}} dx$$
$$= \frac{1}{2}\left(\frac{l_1^2}{E_2I_2} + \frac{l^2 - l_1^2}{(EI)_{\text{eff}}}\right)F + \left(\frac{l_1}{E_2I_2} + \frac{l - l_1}{(EI)_{\text{eff}}}\right)M \tag{8.20}$$

If the lower layer is fully covered with the upper layer (i.e., $l_1 = 0$), equations (8.19) and (8.20) reduce to

$$\delta = \frac{l^3}{3(EI)_{\text{eff}}} F + \frac{l^2}{2(EI)_{\text{eff}}} M \tag{8.21}$$

$$\theta = \frac{l^2}{2(EI)_{\text{eff}}} F + \frac{l}{(EI)_{\text{eff}}} M \tag{8.22}$$

Similarly, beams under different boundary conditions are also analyzed by using the moment–curvature relationships, (8.8), or by applying Castigliano's theorem. In subsequent paragraphs, the mechanical behavior of composite beams due to the strain change in layers is investigated using the equations developed here.

Mechanical Behavior Due to Strain Change

In the preceding paragraphs, the effective flexural rigidity, radius of curvature, and moment–curvature relationship of composite beams were obtained. In MEMS as well as conventional large-scale machines, the composite beams are used for various purposes. For example, in an electric iron, when a bimetallic strip, which consists of two metals of different coefficients of thermal expan-

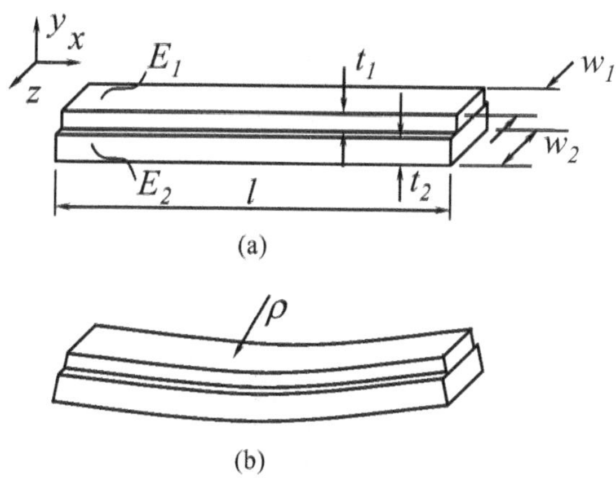

Figure 8.4 Composite material actuator.

sion (Fig. 8.4a), is exposed to a change in temperature, the bimetallic strip is deformed as shown in Fig. 8.4b and then serves as a temperature-sensitive switch. Piezoelectric actuators, also illustrated in the structure in Fig. 8.4a, is deformed when their piezoelectric layers are exposed to an electric field. The bimetallic and piezoelectric actuators can be considered as a composite beam and are driven by strain change in the layers.

To analyze the behavior of the composite beam, we define the dimensions in Fig. 8.4a, where E, w, t, and l denote Young' modulus, width, thickness, and length of a layer, respectively, and the subscripts 1 and 2, respectively, represent the upper and lower layers. The radius of curvature of the deformed beam is represented by ρ as indicated in Fig. 3.4b. For an analysis of the mechanical behavior of the composite beam, we develop the steps illustrated in Fig. 8.5. We first consider expansion of separate layers in Fig. 8.5a and then obtain the radius of curvature ρ shown in Fig. 8.5d by applying the principle of superposition. The two layers, which will be bonded to form a composite beam, are the same in length, but their lengths are different when they are exposed to any source that causes length change (i.e., strain change). For bimetallic and piezoelectric actuators, the sources are temperature and electric field, respectively. When the layers are separately exposed to a source whose distribution is uniform in space, the layer elongates in the longitudinal direction as shown in Fig. 8.5a. Let δ_1, δ_2, and δ represent elongations of layers 1 and 2 and the relative elongation defined by $\delta_2 - \delta_1$, respectively. At this stage we assume that the two layers elongate as shown in Fig. 8.5a, and we further assume that tensile forces F_1 and F_2 are applied to the ends of the layers as illustrated in Fig. 8.5b to form the straight composite beam in Fig. 8.5c, where the left end of the beam is fixed on a rigid wall and the right end is on a roller which moves freely in the longitudinal direction of the beam. It is noted that the right endpoints of the upper and lower layers are displaced by the same elongation δ_f

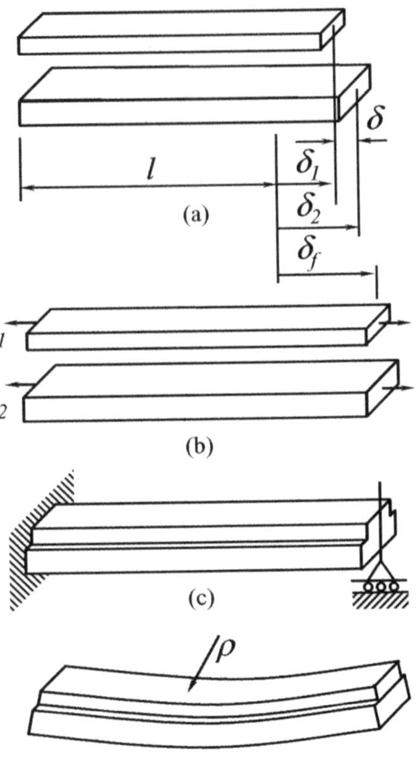

Figure 8.5 Procedure to obtain the deformed shape of a composite beam.

shown in Fig. 8.5b. When the roller is removed, the composite beam subjected to the strain change in its layers is curled into the curved beam shown in Fig. 8.5d, whose radius of curvature is ρ.

We shall follow the steps necessary to obtain the longitudinal strain and stress distributions in the layers and the radius of curvature ρ in Fig. 8.5d. The free-body diagram of Fig. 8.6a illustrates the forces acting on the ends of the straight beam after the stretched layers are bonded in Fig. 8.5c to form a composite beam. The forces and moment acting on the roller are illustrated in the free-body diagram of Fig. 8.6b. From the force equilibrium of Fig. 8.6b, $\Sigma F_x = -F_1 - F_2 = 0$, we have $F_1 = -F_2$. The moment equilibrium condition $\Sigma M_x = 0$ gives

$$M = -F_1 \frac{t_1 + t_2}{2} \tag{8.23}$$

We then take an element from the beam to illustrate the force and moment in Fig. 8.6c. The dimensions and parameters, shown in Figs. 8.2a and 8.4a, are

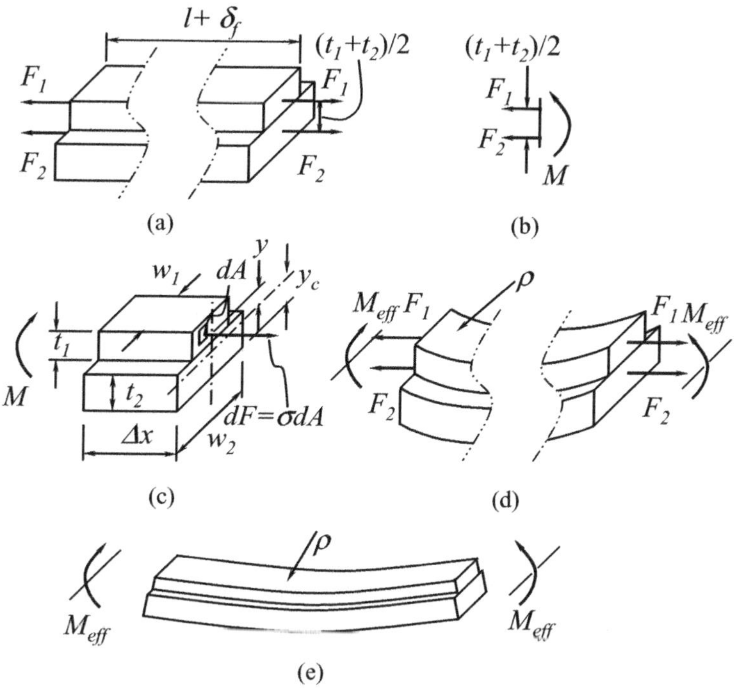

Figure 8.6 Composite beam in equilibrium.

used for consistency. The longitudinal stresses in the straight layers of Fig. 8.6a and b are given by

$$\sigma_{x1} = \tilde{E}_1\varepsilon_{x1} = \tilde{E}_1\frac{\delta_f - \delta_1}{l} = \tilde{E}_1(\varepsilon_f - \varepsilon_1) \tag{8.24}$$

$$\sigma_{x2} = \tilde{E}_2\varepsilon_{x2} = \tilde{E}_2\frac{\delta_f - \delta_2}{l} = \tilde{E}_2(\varepsilon_f - \varepsilon_2) \tag{8.25}$$

where σ_{x1} and σ_{x2} denote the stresses in layers 1 and 2, respectively, and ε_{x1} and ε_{x2} represent the strains corresponding to the σ_{x1} and σ_{x2} respectively. From the force equilibrium of the free body in Fig. 8.6c, we have $F_1 + F_2 = 0$ or

$$\sigma_{x1}A_1 + \sigma_{x2}A_2 = 0 \tag{8.26}$$

where A_1 and A_2 denote the cross-sectional areas of the upper and lower layers, respectively. Substituting (8.24) and (8.25) into equation (8.26) gives the strain of the straight beam in Figs. 8.5b and 8.6a as follows:

$$\varepsilon_f = \frac{\tilde{E}_1A_1\varepsilon_1 + \tilde{E}_2A_2\varepsilon_2}{\tilde{E}_1A_1 + \tilde{E}_2A_2} \tag{8.27}$$

Substitution of the preceding strain into (8.24) and (8.25) yields the stresses in the upper and lower layers:

$$\sigma_{x1} = \tilde{E}_1\left(\frac{\tilde{E}_1 A_1 \varepsilon_1 + \tilde{E}_2 A_2 \varepsilon_2}{\tilde{E}_1 A_1 + \tilde{E}_2 A_2} - \varepsilon_1\right) = \tilde{E}_1 \frac{\tilde{E}_2 A_2 (\varepsilon_2 - \varepsilon_1)}{\tilde{E}_1 A_1 + \tilde{E}_2 A_2} \tag{8.28}$$

$$\sigma_{x2} = \tilde{E}_2\left(\frac{\tilde{E}_1 A_1 \varepsilon_1 + \tilde{E}_2 A_2 \varepsilon_2}{\tilde{E}_1 A_1 + \tilde{E}_2 A_2} - \varepsilon_2\right) = -\tilde{E}_2 \frac{\tilde{E}_1 A_1 (\varepsilon_2 - \varepsilon_1)}{\tilde{E}_1 A_1 + \tilde{E}_2 A_2} \tag{8.29}$$

For convenience, setting $A_1 = w_1 t_1$ and $A_2 = w_2 t_2$, defining a relative strain,

$$\varepsilon = \frac{\delta_2 - \delta_1}{l} = \varepsilon_2 - \varepsilon_1 \tag{8.30}$$

and recalling the following ratios, which were used in Example 8.1,

$$p = \frac{t_1}{t_2} \qquad q = \frac{w_1}{w_2} \qquad r = \frac{\tilde{E}_1}{\tilde{E}_2} \tag{8.31}$$

the longitudinal stresses are obtained as functions of the effective Young's modulus of the lower layer and the relative strain and the ratios

$$\sigma_{x1} = \tilde{E}_2 \frac{r\varepsilon}{1 + pqr} \tag{8.32}$$

$$\sigma_{x2} = -\tilde{E}_2 \frac{pqr\varepsilon}{1 + pqr} \tag{8.33}$$

It is noted that the preceding stresses, σ_{x1} and σ_{x2}, represent the uniform axial stresses due to the forces F_1 and F_2 shown in Figs. 8.5b and 8.6a. For a positive ε (i.e., the elongation of the lower layer is greater than that of the upper layer), σ_{x1} and σ_{x2} become positive and negative, respectively.

Applying the moment equilibrium condition $\sum M = 0$ to Fig. 8.6c, we find the moment M, which is applied by the roller in Fig. 8.5c to keep the composite beam straight:

$$M = -\int_A y\,dF = -\int_A y\sigma_x\,dA$$

Substituting $y = y_b - y_c$ into the preceding equation gives

$$\begin{aligned} M &= -\left[\int_{A_1} \sigma_{x1}(y_b - y_c)\,dA + \int_{A_2} \sigma_{x2}(y_b - y_c)\,dA\right] \\ &= -[\sigma_{x1} A_1 (\bar{y}_1 - y_c) + \sigma_{x2} A_2 (\bar{y}_2 - y_c)] \end{aligned}$$

Since $F_1 = -F_2$ (i.e., $\sigma_{x1}A_1 = -\sigma_{x2}A_2$), $A_1 = w_1t_1$, and $A_2 = w_2t_2$ for the rectangular layers, the preceding moment is expressed as

$$\begin{aligned} M &= -[-\sigma_{x2}A_{21}(\bar{y}_1 - y_c) + \sigma_{x2}A_2(\bar{y}_2 - y_c)] \\ &= \sigma_{x2}A_2(\bar{y}_1 - \bar{y}_2) = \sigma_{x2}A_2\frac{t_1 + t_2}{2} = \frac{1}{2}\sigma_{x2}w_2t_2(t_1 + t_2) \\ &= \frac{1}{2}\sigma_{x2}w_2t_2^2(1 + p) \end{aligned}$$

Combining the preceding equation and (8.33) yields the following moment M:

$$M = -\frac{1}{2}\tilde{E}_2w_2t_2^2\frac{pqr(1+p)}{1+pqr}\varepsilon \tag{8.34}$$

The moment M above is caused by the forces F_1 and F_2, which are applied by the roller in Fig. 8.5c to keep the beam straight. The moment M can also be obtained by combining (8.23) and (8.32). When we remove the roller in Fig. 8.5c, the composite beam is curled into the curved beam in Fig. 8.5d since the moment due to the roller vanishes. It is interpreted that the moment M due to the couple of forces in Fig. 8.6c is balanced by a moment when the composite beam is released. Let M_{eff} represent the moment, which can be considered as an effective moment. From the moment balance equilibrium in Fig. 8.6d, we obtain the effective moment acting on curling the beam:

$$M_{\text{eff}} = -M = \frac{1}{2}\tilde{E}_2w_2t_2^2\frac{pqr(1+p)}{1+pqr}\varepsilon \tag{8.35}$$

It is noted that the effective moment M_{eff} is a function of the relative strain ε and the effective Young's modulus and geometry of the beam. Substituting equation (8.5) into (8.8) and solving the resulting equation with boundary conditions, we find the deflection of the composite beam. The radius of curvature ρ is obtained by substituting the effective flexural rigidity $(\tilde{E}I)_{\text{eff}}$, which is given by (m) in Example 8.1, and the preceding effective moment M_{eff} into (8.7):

$$\begin{aligned} \rho &= \frac{(\tilde{E}I)_{\text{eff}}}{M_{\text{eff}}} = \frac{t_2}{6\varepsilon}\frac{(1+pqr)(1+p^3qr) + 3pqr(1+p)^2}{pqr(1+p)} \\ &= \frac{t_2}{6\varepsilon}\frac{3(1+p)^2 + (1+pqr)(p^2 + 1/pqr)}{1+p} \end{aligned} \tag{8.36}$$

Since the total beam thickness may be expressed as $t = t_1 + t_2 = t_2(1 + p)$, the thickness of the lower layer is given by

$$t_2 = \frac{t}{1+p} \tag{8.37}$$

Substituting equation (8.37) into (8.36), we obtain the radius of curvature due to the change in the internal strain in a simple form:

$$\begin{aligned}\rho &= \frac{t}{6\varepsilon}\frac{3(1+p)^2+(1+pqr)(p^2+1/pqr)}{(1+p)^2} \\ &= \frac{t}{2\varepsilon}\left[1+\frac{1+pqr}{3(1+p)^2}\left(p^2+\frac{1}{pqr}\right)\right]\end{aligned} \tag{8.38}$$

which is a function of the relative strain ε defined in (8.30), the total thickness t, and the ratios p, q, and r, which are defined by (8.31). Note that if we apply the effective moment M_{eff} (8.35) to a stress-free beam like the bcam in Fig. 8.4a, the radius of curvature ρ of the deformed beam in Fig. 8.6e is equal to that given by equation (8.38). Equation (8.38) will be used to derive further static and dynamic behavior of the composite beam. If $p = q = r = 1$ (i.e., $t_1 = t_2$, $w_1 = w_2$, and $\tilde{E}_1 = \tilde{E}_2$), equation (8.38) is reduced to $\rho = 2t/3\varepsilon$. If $p \leq 0.01$, the radius of curvature is approximated as $\rho \approx (1 + 4pqr)t/6pqr\varepsilon$.

We shall obtain the strain and stress in the composite beam by applying the principle of superposition as shown in Fig. 8.6d. The strains of the uniformly stretched beams in Fig. 8.5b are given by $\varepsilon_{x1} = \sigma_{x1}/\tilde{E}_1$ and $\varepsilon_{x2} = \sigma_{x2}/\tilde{E}_2$, where σ_{x1} and σ_{x2} denote the uniform stresses, (8.32) and (8.33), respectively. When the effective moment M_{eff} is applied to the stress-free straight composite beam in Fig. 8.6e, the strain developed in the beam is given by (8.9), $\varepsilon_x = -y/\rho$, where ρ denote the radius of curvature of the deformed beam, which is given by (8.38). The total strain ε_{xt} in the released beam in Fig. 8.5d is then obtained as a sum of the uniform strain in Fig. 8.5.b and the linearly varying strain,

$$\varepsilon_{xt,j} = \varepsilon_{xj} + \varepsilon_x = \frac{\sigma_{xj}}{\tilde{E}_j} - \frac{y}{\rho} \tag{8.39}$$

where j is taken from 1 to 2, depending on the layer of interest. Setting $j = 1$, substituting (8.32), and recalling that $r = \tilde{E}_1/\tilde{E}_2$, we have the total strain in the upper layer:

$$\varepsilon_{xt,1} = \left[\frac{1}{1+pqr} - \frac{2}{1+\dfrac{1+pqr}{3(1+p)^2}\left(p^2+\dfrac{1}{pqr}\right)}\frac{y}{t}\right]\varepsilon \tag{8.40}$$

where y is the coordinate from the neutral surface. Similarly, setting $j = 2$ and substituting (8.33) into (8.39) gives the total strain in the lower layer:

$$\varepsilon_{xt,2} = \left[\frac{pqr}{1+pqr} - \frac{2}{1+\dfrac{1+pqr}{3(1+p)^2}\left(p^2+\dfrac{1}{pqr}\right)} \frac{y}{t} \right] \varepsilon \tag{8.41}$$

In each layer, the strain varies linearly and is proportional to the relative strain ε, defined in (8.30). Then the stresses $\sigma_{xt,1}$ and $\sigma_{xt,2}$ corresponding to the preceding strains are expressed as

$$\sigma_{xt,1} = \tilde{E}_1 \varepsilon_{xt,1} \qquad \sigma_{xt,2} = \tilde{E}_2 \varepsilon_{xt,2} \tag{8.42}$$

Using the preceding strain and stress equations, we evaluate the strain and stress at a point of interest in the beam. For example, if a strain gage is attached on the top of the composite beam, the strain on the top is obtained from (8.40). The coordinate y at the top of the beam is expressed as

$$\begin{aligned} y\big|_{top} &= t_1 + t_2 - y_c = t_2\left(1 + p - \frac{y_c}{t_2}\right) \\ &= \frac{t}{1+p}\left[1 + p - \frac{1}{2}\frac{1+pqr(2+p)}{1+pqr}\right] = t\left[1 - \frac{1}{2}\frac{1+pqr(2+p)}{(1+p)(1+pqr)}\right] \end{aligned} \tag{8.43}$$

where (d) of Example 8.1 is used for y_c. Combining (8.40) and equation (8.43), we have the strain on the top,

$$\varepsilon_{xt,1}\big|_{\text{top}} = \left\{ \frac{1}{1+pqr} - \frac{1}{1+\dfrac{1+pqr}{3(1+p)^2}\left(p^2+\dfrac{1}{pqr}\right)} \left[2 - \frac{1+pqr(2+p)}{(1+p)(1+pqr)}\right] \right\} \varepsilon \tag{8.44}$$

We then obtain the corresponding stress on the top of the beam:

$$\sigma_{xt,1}\big|_{\text{top}} = \tilde{E}_1 \varepsilon_{xt,1}\big|_{\text{top}} \tag{8.45}$$

The stress distributions before and after releasing the right end of the beam in Fig. 8.5c are illustrated in Fig. 8.7a and b, respectively. The stresses in the straight beam of Fig. 8.7a are given by (8.32) and (8.33). In the released beam in Fig. 8.7b, the stress distribution is obtained by superposition of the stress in Fig. 8.7a and the stress given by (8.42). It is noted that the stress is linear in the upper or lower layer but is discontinuous at the boundary of the two layers.

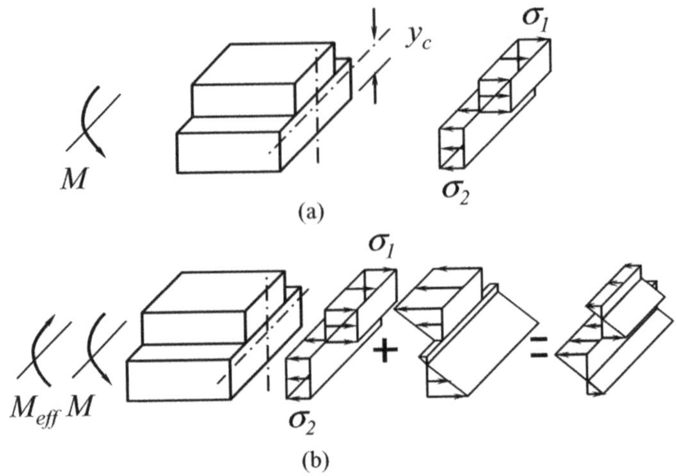

Figure 8.7 Stress distribution on the cross section of a composite beam.

However, the stress distribution satisfies the force and moment equilibrium conditions, so the released beam of Fig. 8.5d is in equilibrium. It is also noted that the stress distribution shown in Fig. 8.7b is that in a cross section far from the ends of the slender composite beam. The stress in cross section is zero at the ends of the beam and is changed to the stress shown in Fig. 8.7b within the order of magnitude of the total thickness of the beam.

Example 8.2 The left end of a composite beam subjected to the relative strain is fixed to a rigid wall as shown in Fig. 8.8a. We wish to find the beam deflection in terms of x.

We find the deflection in three different ways. From the geometry in Fig. 8.8b, the exact beam deflection is expressed as

$$y = \rho - \sqrt{\rho^2 - x^2} = \rho\left[1 - \sqrt{1 - \left(\frac{x}{\rho}\right)^2}\right] \tag{a}$$

where the radius of curvature ρ is given by (8.38):

$$\rho = \frac{t}{2\varepsilon}\left[1 + \frac{1 + pqr}{3(1+p)^2}\left(p^2 + \frac{1}{pqr}\right)\right] \tag{b}$$

Since $x << \rho$, equation (b) is reduced to an approximate solution,

$$y \cong \rho\left\{1 - \left[1 - \frac{1}{2}\left(\frac{x}{\rho}\right)^2\right]\right\} = \frac{x^2}{2\rho} \tag{c}$$

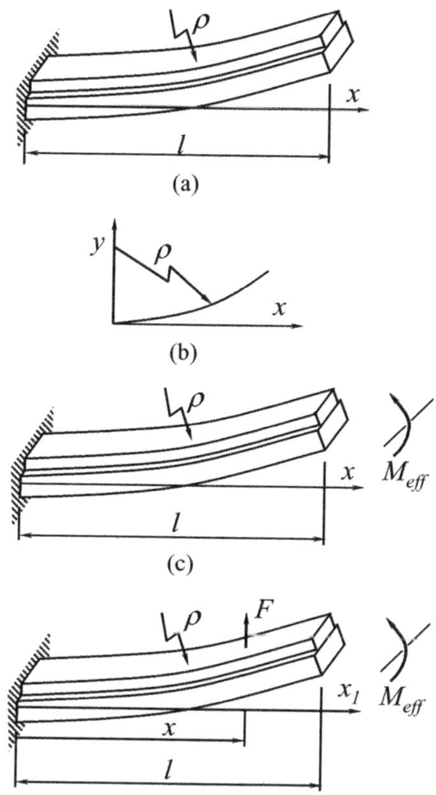

Figure 8.8 Deflection of a composite beam.

which holds only for $l << \rho$. If the beam length l is comparable to or greater than ρ, approximate solution (c) gives significant errors, so in this case, (a) must be used to evaluate the beam deflection. Substituting (b) into (c) yields

$$y = \frac{1}{1 + \frac{1+pqr}{3(1+p)^2}\left(p^2 + \frac{1}{pqr}\right)} \frac{x^2}{t} \varepsilon \tag{d}$$

The solution (c) is also obtained by solving the moment–curvature relationship, (8.8), which is given by

$$\left(\tilde{E}I\right)_{\text{eff}} \frac{d^2 y}{dx^2} = M_{\text{eff}} \tag{e}$$

where M_{eff} denotes the effective moment, (8.35), due to the relative strain in the layer. Since the radius of curvature is given by $\rho = (\tilde{E}I)_{\text{eff}} / M_{\text{eff}}$ in Fig. 8.8c, equation (e) may be written as

$$\frac{d^2 y}{dx^2} = \frac{1}{\rho} \tag{f}$$

Solving equation (f) with the boundary conditions $y = 0$ and $dy/dx = 0$ at $x = 0$, we find the following solution, which is the same as (c):

$$y = \frac{x^2}{2\rho} \tag{g}$$

We can use the energy method to obtain the beam deflection at x from the left end of the beam in Fig. 8.8d. To obtain the beam deflection at x, we place a dummy force F at x. For $x \leq x_1 \leq l$, where x_1 is a dummy variable, the bending moment at x is given by $M_{x1} = M_{\text{eff}}$ In the range of $0 \leq x_1 \leq x$, the moment is expressed as $M_{x1} = M_{\text{eff}} + F(x - x_1)$. We then express the energy stored in the beam in terms of the force and moment:

$$U = \int_0^l \frac{1}{2} \frac{M_{x1}^2}{(EI)_{\text{eff}}} dx_1 = \int_x^l \frac{1}{2} \frac{M_{\text{eff}}^2}{(EI)_{\text{eff}}} dx_1 + \int_0^x \frac{1}{2} \frac{[M_{\text{eff}} + F(x - x_1)]^2}{(EI)_{\text{eff}}} dx_1 \tag{h}$$

Applying Castigliano's theorem, we find the deflection at x from the fixed end of the beam:

$$y = \left.\frac{\partial U}{\partial F}\right|_{F=0} = \int_0^x \frac{M_{\text{eff}}(x - x_1)}{(EI)_{\text{eff}}} dx_1 = \frac{M_{\text{eff}} x^2}{2(EI)_{\text{eff}}} = \frac{x^2}{2\rho} \tag{i}$$

which is the same as (c) and (g). The angular displacement of the beam at x is given by

$$\theta = \frac{\partial y}{\partial x} = \frac{x}{\rho} \tag{j}$$

As shown in Example 8.2, the effective moment M_{eff} due to the strain change in a composite beam can be applied to the straight beam with zero strain to obtain the beam deflection. We thus use the effective moment and Castigliano's theorem (or any other energy method) to analyze composite deflections.

If a composite beam consists of multilayered material as shown in Fig. 8.9a, we can obtain the radius of curvature and the effective moment by repeating the procedure described above. Referring to the right figure in Fig. 8.9a, the strains of the separated layers are obtained as

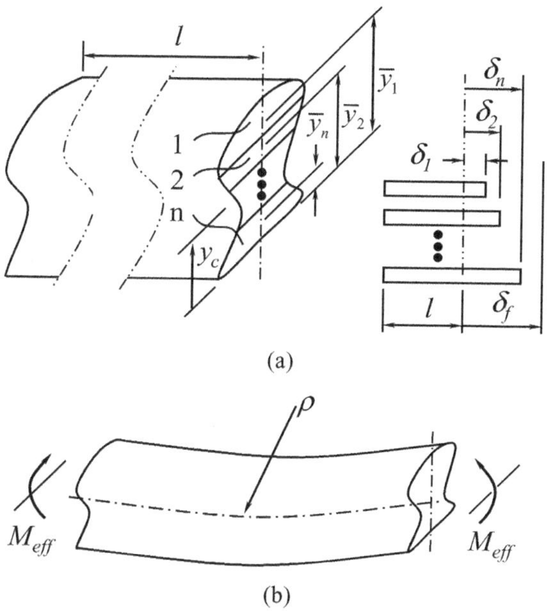

Figure 8.9 Deflection of beam of an arbitrarily shaped cross section.

$$\varepsilon_1 = \frac{\delta_1}{l}, \quad \varepsilon_2 = \frac{\delta_2}{l}, \quad \ldots, \quad \varepsilon_n = \frac{\delta_n}{l} \tag{8.46}$$

The force equilibrium condition in the longitudinal direction gives

$$\begin{aligned}\sum F_x &= \int_A \sigma_x \, dA \\ &= \tilde{E}_1(\varepsilon_f - \varepsilon_1)A_1 + \tilde{E}_2(\varepsilon_f - \varepsilon_2)A_2 + \cdots + \tilde{E}_n(\varepsilon_f - \varepsilon_n)A_n \\ &= \sum_{j=1}^{n} \tilde{E}_j(\varepsilon_f - \varepsilon_j)A_j = 0\end{aligned}$$

which yields the strain of the straight beam constrained by the roller:

$$\varepsilon_f = \frac{\sum_{j=1}^{n} \tilde{E}_j A_j \varepsilon_j}{\sum_{j=1}^{n} \tilde{E}_j A_j} \tag{8.47}$$

Applying the moment equilibrium condition, we obtain the effective moment M_{eff}, which is the moment used to make the same radius of curvature of the released beam:

$$
\begin{aligned}
M_{eff} &= \int_A \sigma_x \, dA \\
&= \sigma_{x1} A_1 (\bar{y}_1 - y_c) + \sigma_{x2} A_2 (\bar{y}_2 - y_c) + \cdots + \sigma_{xn} A_n (\bar{y}_n - y_c) \\
&= \sum_{j=1}^{n} \tilde{E}_j (\varepsilon_f - \varepsilon_j) A_j (\bar{y}_j - y_c)
\end{aligned}
\tag{8.48}
$$

We thus obtain the radius of curvature ρ by substituting effective moment (8.48) into (8.14) as follows:

$$
\rho = \frac{(\tilde{E}I)_{\text{eff}}}{M_{\text{eff}}} = \frac{\sum_{j=1}^{n} \tilde{E}_j I_{cj}}{\sum_{j=1}^{n} \tilde{E}_j (\varepsilon_f - \varepsilon_j) A_j (\bar{y}_j - y_c)} \tag{8.49}
$$

where I_{cj} denotes the moment of inertia of the cross section of the nth layer with respect to the neutral surface, which is defined in (8.13). Since we have the radius of curvature (8.49), we solve the moment–curvature equation (8.15) with boundary conditions to obtain the deflection of the composite beam.

8.2 PIEZOELECTRIC ACTUATORS

Piezoelectric materials, described in Section 7.9, serve as actuators driven by the electric field. In Fig. 8.10, piezoelectric actuators are illustrated that use the piezoelectric coefficient d_{33} (i.e., the actuators elongate in the direction of the applied field). A single layer of piezoelectric constant d_{33}, Young's

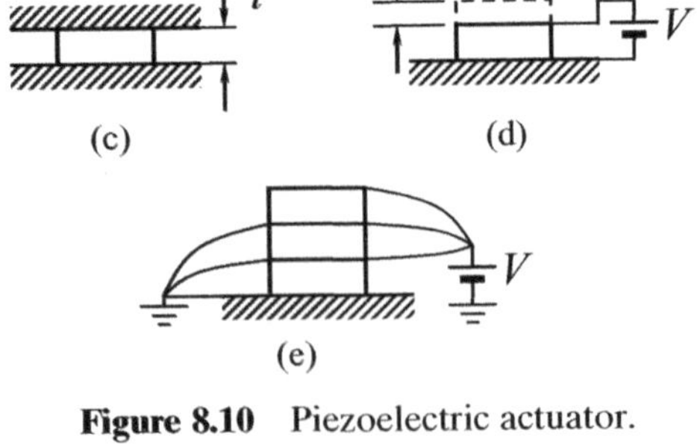

Figure 8.10 Piezoelectric actuator.

modulus E, area A, and thickness t is placed on a plane as shown in Fig. 8.10a. The top and bottom of the layer have thin electrodes used to apply electricity. When a voltage V is applied across the layer in Fig. 8.10b, the layer elongates by δ in the direction of the electric field:

$$\delta = \Delta t = d_{33}V \tag{8.50}$$

If the piezoelectric actuator is constrained by the rigid wall shown in Fig. 8.10c, the actuator subjected to the voltage V does not elongate but applies a force on the rigid wall. Let us obtain the force applied on the wall. The force of the constrained actuator can be analyzed by applying a force F to the elongated layer as shown in Fig. 8.10d. If the actuator under the voltage V elongates by δ, the force F is applied to return the top of the actuator to the initial position. Using the stress–strain relation $\sigma = E\varepsilon$, where $\sigma = F/A$ and $\varepsilon = -\delta/t$, we find the force generated by the actuator as follows:

$$F = -EA\frac{\delta}{t} = -EA\frac{d_{33}}{t}V \tag{8.51}$$

where the negative sign represents the compressive force (i.e., the actuator pushes the wall if it elongates). If the polarity of the voltage is reversed, the actuator pulls the wall. It is noted that EA/t is defined as the stiffness of the actuator and that the force generated by the actuator is proportional to EA but is inversely proportional to the thickness. The elongation (8.50) shows that elongation of the unconstrained piezoelectric actuator, illustrated in Fig. 8.10a, is proportional to the voltage and is independent of the thickness. Therefore, to achieve a large elongation, we can stack thin layers to form a multilayered actuator, as shown in Fig. 8.10e, where the bottoms of the layers are connected to the ground and the tops are connected to the voltage. The total elongation δ_t of the n-layered actuator is obtained as the sum of the elongation of each layer:

$$\delta_t = \delta_1 + \delta_2 + \cdots + \delta_n = nd_{33}V \tag{8.52}$$

It is noted that the total elongation increases by a factor of n. If the n-layered piezoelectric actuator in Fig. 8.10e is constrained by a rigid wall like that in Fig. 8.10c, the force generated by the actuator is given by

$$F = -EA\frac{\delta_t}{nt} = -EA\frac{d_{33}}{t}V \tag{8.53}$$

As expressed in equation (8.53), the force of the n-layered actuator is equal to that of a single-layer actuator, given by (8.51), since both the elongation and thickness increase by a factor of n.

Piezoelectric Cantilever Beam Actuator

If one or two of the composite layers of the cantilever beam in Fig. 8.11 are piezoelectric layers, the composite cantilever beam acts as an actuator, which is deflected in the direction perpendicular to the longitudinal direction. For convenience of explanation, we imagine Fig. 8.11a, in which the upper layer is a piezoelectric layer of piezoelectric constant d_{31} and the lower layer is a nonpiezoelectric layers such as silicon. It is assumed that the top and bottom of the upper layer have conductive layers whose thickness is much thinner than those of the upper and lower layers. The flexural rigidity and stiffness due to the conductive layers are then neglected. As the separated upper layer is connected to a voltage source of voltage V (electric field $E = -V/t_1$, i.e., the field is directed in the negative y direction), the upper layer contracts by δ_1, as shown in Fig. 8.11b. The contraction of the upper layer is given by (7.181) as follows:

$$\delta_1 = -d_{31}\frac{l}{t_1}V \tag{8.54}$$

where the negative sign denotes the contraction of the beam since d_{31} is assumed positive. The elongation of the lower layer is $\delta_2 = 0$, since the layer is nonpiezoelectric. The relative strain is defined from (8.30):

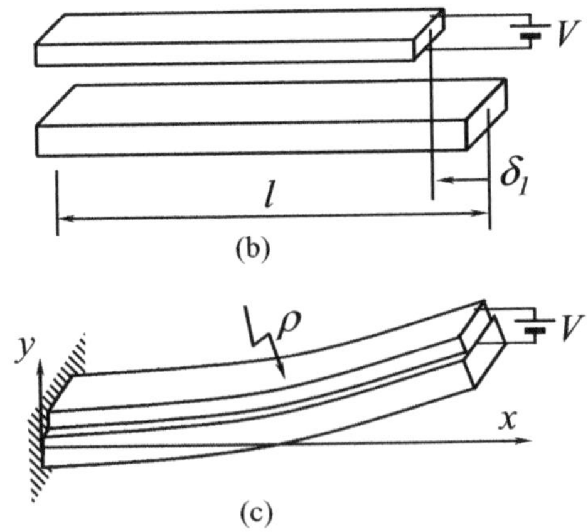

Figure 8.11 Piezoelectric cantilever beam actuator.

$$\varepsilon = \frac{\delta_2 - \delta_1}{l} = \frac{d_{31}}{t_1} V \tag{8.55}$$

The radius of curvature and the deflection of the cantilever beam are then obtained from the theory developed in Section 8.1. Since $t = t_1 + t_2 = t_2(1 + p)$ and $t_1 = t_1/t_2\, t_2 = [p/(1+p)]t$, where p is the thickness ratio $p = t_1/t_2$, the foregoing relative strain is expressed as

$$\varepsilon = \frac{1+p}{p}\frac{d_{31}}{t} V \tag{8.56}$$

Substituting the preceding relative strain ε into (8.38), we find the radius of curvature ρ at the applied voltage V:

$$\begin{aligned}\rho &= \frac{t}{2\varepsilon}\left[1 + \frac{1+pqr}{3(1+p)^2}\left(p^2 + \frac{1}{pqr}\right)\right] \\ &= \frac{t^2}{2d_{31}V}\frac{p}{1+p}\left[1 + \frac{1+pqr}{3(1+p)^2}\left(p^2 + \frac{1}{pqr}\right)\right]\end{aligned} \tag{8.57}$$

where p, q, and r denote the ratios defined by (8.31). Therefore, the deflection of the beam at x from the fixed end of the beam is obtained from (c) in Example 8.2:

$$y = \frac{x^2}{2\rho} = \frac{1+p}{p\left[1 + \frac{1+pqr}{3(1+p)^2}\left(p^2 + \frac{1}{pqr}\right)\right]}\left(\frac{x}{t}\right)^2 d_{31}V \tag{8.58}$$

If we are interested in the angular displacement θ at x, θ is given by dy/dx. Setting $x = l$, we have the maximum deflection y_l at the free end of the beam:

$$y_l = \frac{1+p}{p\left[1 + \frac{1+pqr}{3(1+p)^2}\left(p^2 + \frac{1}{pqr}\right)\right]}\left(\frac{l}{t}\right)^2 d_{31}V \tag{8.59}$$

When l/ρ is not much less than unity, the beam deflection y should be evaluated using (a) of Example 8.2. Note from the above equation (8.59) that the deflection at the free end is proportional to the square of the ratio of the beam length to the total thickness and is linearly proportional to the applied voltage. As a result, a small change in voltage causes a large deflection of the free end, and then the piezoelectric cantilever beam actuator acts as a linear actuator with a large gain. So far, we have obtained the deflection of the cantilever beam actuator, which consists of the upper piezoelectric layer and the lower nonpiezoelectric layer. In some cases we may need to use more than one piezoelectric layer to form a piezoelectric actuator. To obtain the response of this actuator to the voltage, we can use the theory developed in Section 8.1. For example, let us consider the actuator in Fig. 8.11, whose upper and lower layers are made of piezoelectric materials. In this case, the relative strain ε, (8.30), is written

$$\varepsilon = \frac{\delta_2 - \delta_1}{l} = -\frac{d_{31,2}}{t_2}V_2 + \frac{d_{31,1}}{t_1}V_1 = \frac{1+p}{t}\left(-d_{31,2}V_2 + \frac{1}{p}d_{31,1}V_1\right) \tag{8.60}$$

where $\delta_{31,2}$, $\delta_{31,1}$, V_2, and V_1 denote the piezoelectric constants δ_{31} of the lower and upper layers and the voltages applied to the lower and upper layers, respectively. Therefore, the radius of curvature ρ and the deflection y of the actuator are obtained from (8.38) as

$$\begin{aligned}\rho &= \frac{t}{2\varepsilon}\left[1 + \frac{1+pqr}{3(1+p)^2}\left(p^2 + \frac{1}{pqr}\right)\right] \\ &= \frac{t^2}{2(1+p)\left(-d_{31,2}V_2 + \frac{1}{p}d_{31,1}V_1\right)}\left[1 + \frac{1+pqr}{3(1+p)^2}\left(p^2 + \frac{1}{pqr}\right)\right]\end{aligned} \tag{8.61}$$

$$y = \frac{x^2}{2\rho} = \frac{1+p}{p\left[1 + \frac{1+pqr}{3(1+p)^2}\left(p^2 + \frac{1}{pqr}\right)\right]}\left(\frac{x}{t}\right)^2\left(-pd_{31,2}V_2 + d_{31,1}V_1\right) \tag{8.62}$$

Setting $\delta_{31,1} = 0$ or $V_2 = 0$, equation (8.62) reduces to (8.58). If $V_2 = -V_1$, and $\delta_{31,2} = \delta_{31,1}$, the driving terms in equation (8.62) reduce to $2\delta_{31,1}\, V_1$.

The beam deflection in (8.58) is the static response of the piezoelectric cantilever beam actuators to a constant voltage or a voltage whose frequency is much lower than the first natural frequency of the actuator. If the frequency of the voltage approaches or is greater than the first natural frequency, the vibration amplitude of the beam deviates from the foregoing beam deflections. The natural frequency of the cantilever actuator, given by (5.138), can be written as

$$f_n = \frac{(\beta_n l)^2}{2\pi l^2}\sqrt{\frac{\left(\tilde{E}I\right)_{\text{eff}}}{m_1}} \tag{8.63}$$

where n, $\beta_n l$, $\left(\tilde{E}I\right)_{\text{eff}}$, and m_1 denote positive integers, the dimensionless number listed in Table 5.1, the effective flexural rigidity given by (m) in Example 8.1, and the mass per unit beam length, respectively. The first and second mode shapes are illustrated in Fig. 5.29b, and more mode shapes are given by (5.135).

Piezoelectric Cantilever Beam Actuator Connected to a Flexible Structure

If the piezoelectric cantilever beam actuator described in the preceding paragraphs is connected to a flexible structure for a design purpose, the deflection or response of the actuator deviates from the static response, (8.58) and (8.62). Let us obtain the response of this type of actuator. We first find the response to the voltage of the actuator without flexible structure and then obtain the response of the actuator. Consider the piezoelectric actuator shown in Fig. 8.12a, whose upper piezoelectric layer is partially deposited on the lower non-piezoelectric layer. When a voltage V is applied across the upper layer, as shown in Fig. 8.12a, the relative strain ε, which is given by (8.55), is generated in the beam and the beam is then deflected. If the flexible structure is connected to the right end of the cantilever beam in Fig. 8.12a, we can construct an equivalent system, illustrated in Fig. 8.12b, which is subjected to the effective moment M_{eff}, (8.35), force F, and a moment M. The beam length is given by l, and the length of the undeposited portion is l_1. The displacement and angular displacement at the right end of the beam are represented by δ and θ, respectively.

We shall use Castigliano's theorem to obtain the response of the actuator. Since the moment at x from the right end is given by $M_x = Fx + M$ for $0 \le x \le l_1$ and $M_x = Fx + M + M_{\text{eff}}$ for $l_1 \le x \le l$, the elastic energy stored in the beam may be expressed as

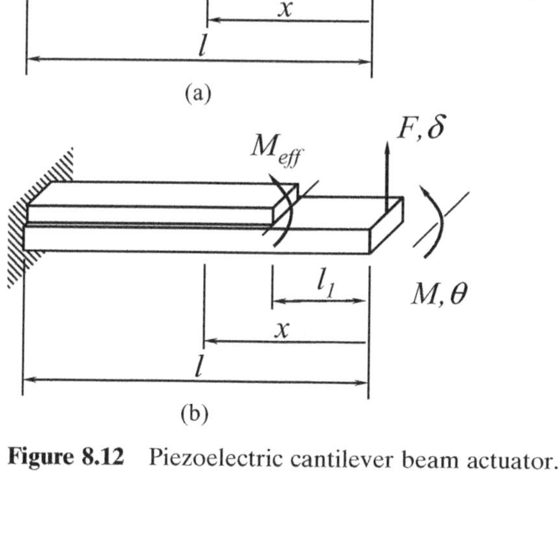

Figure 8.12 Piezoelectric cantilever beam actuator.

$$U = \int_0^l \frac{1}{2} \frac{M_x^2}{(EI)_{\text{eff}}} dx_1 = \int_0^{l_1} \frac{1}{2} \frac{(Fx+M)^2}{E_2 I_2} dx + \int_{l_1}^{l} \frac{1}{2} \frac{(Fx+M+M_{\text{eff}})^2}{(EI)_{\text{eff}}} dx \quad (8.64)$$

and we find that

$$\begin{aligned} \delta = \frac{\partial U}{\partial F} &= \int_0^{l_1} \frac{Fx^2 + Mx}{E_2 I_2} dx + \int_{l_1}^{l} \frac{Fx^2 + (M + M_{\text{eff}})x}{(EI)_{\text{eff}}} dx \\ &= \frac{1}{3}\left[\frac{l_1^3}{E_2 I_2} + \frac{l^3 - l_1^3}{(EI)_{\text{eff}}}\right] F + \frac{1}{2}\left[\frac{l_1^2}{E_2 I_2} + \frac{l^2 - l_1^2}{(EI)_{\text{eff}}}\right] M + \frac{1}{2}\frac{l^2 - l_1^2}{(EI)_{\text{eff}}} M_{\text{eff}} \quad (8.65) \end{aligned}$$

$$\begin{aligned} \theta = \frac{\partial U}{\partial M} &= \int_0^{l_1} \frac{Fx + M}{E_2 I_2} dx + \int_{l_1}^{l} \frac{Fx + M + M_{\text{eff}}}{(EI)_{\text{eff}}} dx \\ &= \frac{1}{2}\left[\frac{l_1^2}{E_2 I_2} + \frac{l^2 - l_1^2}{(EI)_{\text{eff}}}\right] F + \left[\frac{l_1}{E_2 I_2} + \frac{l - l_1}{(EI)_{\text{eff}}}\right] M + \frac{l - l_1}{(EI)_{\text{eff}}} M_{\text{eff}} \quad (8.66) \end{aligned}$$

In the preceding equation, $M_{\text{eff}}/(\tilde{E}I)_{\text{eff}}$ can be considered as the driving term due to the relative strain ε in the beam. For convenience, the following radius of curvature, which was introduced in (8.38), is used again to simplify equation (8.66):

$$\rho = \frac{(\tilde{E}I)_{\text{eff}}}{M_{\text{eff}}} = \frac{t}{2\varepsilon}\left[1 + \frac{1 + pqr}{3(1+p)^2}\left(p^2 + \frac{1}{pqr}\right)\right] \quad (8.67)$$

where t denotes the total thickness given by $t = t_1 + t_2$, and p, q, and r represent the ratios given by (8.31).

The piezoelectric cantilever beam is connected to a flexure whose stiffness can be evaluated using the theory described in Chapter 4. In Fig. 8.13a, the C-shaped structure is used for the flexible structure or spring for convenience of explanation. However, in general, a flexible structure of any shape can be attached to the end of the beam. Let the linear and torsional stiffnesses of the added structure be represented by k and k_t, respectively. Using the actuator shown in Fig. 8.12 and the stiffnesses due to the flexible structure, we set up a model of the actuator, which is illustrated in Fig. 8.13b. Referring to Fig. 8.13c, the force and moment due to the flexible structure are given by $F = -k\delta$ and $M = -k_t\theta$. Substituting $(\tilde{E}I)_{\text{eff}}/M_{\text{eff}} = \rho, F = -k\delta$ and $M = -k_t\theta$ into (8.65) and (8.66), we have the following equation, which will give the response of the actuator to the applied voltage:

$$\begin{bmatrix} a & b \\ c & d \end{bmatrix} \begin{Bmatrix} \delta/l \\ \theta \end{Bmatrix} = \begin{Bmatrix} e \\ f \end{Bmatrix} \quad (8.68)$$

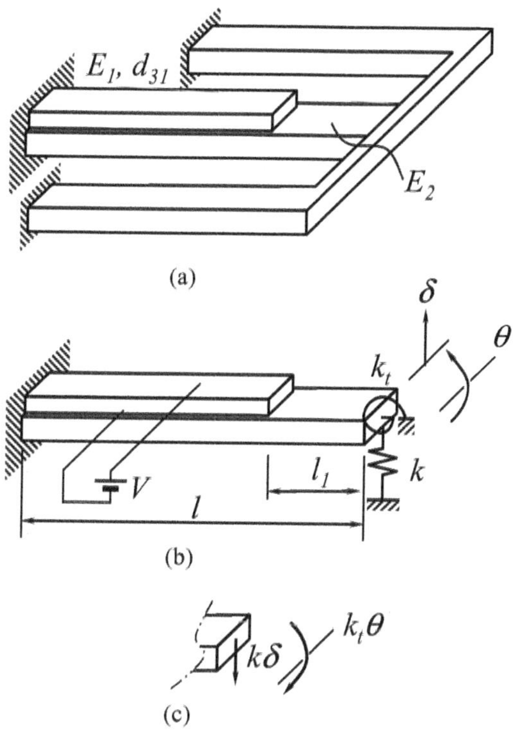

Figure 8.13 Piezoelectric cantilever beam actuator connected to flexures.

where

$$
\begin{aligned}
a &= 1+\frac{kl^3}{3E_2I_2}\left\{\left(\frac{l_1}{l}\right)^3+\frac{E_2I_2}{(EI)_{\text{eff}}}\left[1-\left(\frac{l_1}{l}\right)^3\right]\right\}\\
b &= \frac{1}{2}\frac{k_tl}{E_2I_2}\left\{\left(\frac{l_1}{l}\right)^2+\frac{E_2I_2}{(EI)_{\text{eff}}}\left[1-\left(\frac{l_1}{l}\right)^2\right]\right\}\\
c &= \frac{kl^3}{2E_2I_2}\left\{\left(\frac{l_1}{l}\right)^2+\frac{E_2I_2}{(EI)_{\text{eff}}}\left[1-\left(\frac{l_1}{l}\right)^2\right]\right\}\\
d &= 1+\frac{k_tl}{E_2I_2}\left[\frac{l_1}{l}+\frac{E_2I_2}{(EI)_{\text{eff}}}\left(1-\frac{l_1}{l}\right)\right]\\
e &= \frac{l^2-l_1^2}{2l\rho} \qquad f=\frac{l-l_1}{\rho}
\end{aligned}
\tag{8.69}
$$

The ratio $E_2I_2/(EI)_{\text{eff}}$ for equations (8.69) is obtained from (n) of Example 8.1:

$$\frac{\tilde{E}_2 I_2}{\left(\tilde{E}I\right)_{\text{eff}}} = \frac{1+pqr}{(1+pqr)(1+p^3qr)+3pqr(1+p)^2} \tag{8.70}$$

Equation (8.68) is solved to give the response:

$$\begin{Bmatrix} \delta/l \\ \theta \end{Bmatrix} = \begin{bmatrix} a & b \\ c & d \end{bmatrix}^{-1} \begin{Bmatrix} e \\ f \end{Bmatrix} = \frac{\begin{bmatrix} d & -b \\ -c & a \end{bmatrix}}{ad-bc} \begin{Bmatrix} e \\ f \end{Bmatrix} = \frac{1}{ad-bc} \begin{Bmatrix} de-bf \\ -ce+af \end{Bmatrix}$$

Substituting the driving term, e and f of (8.69), the foregoing equation can be written in the form

$$\delta = l\frac{d(1+l_1/l)-2b}{ad-bc}\frac{l-l_1}{2\rho} \tag{8.71}$$

$$\theta = \frac{-c(1+l_1/l)+2a}{ad-bc}\frac{l-l_1}{2\rho} \tag{8.72}$$

Upon substitution of (8.67) into equations (8.71) and (8.72), we find the following deflection and angular displacement of the actuator when the relative strain ε in the beam occurs.

$$\delta = l\frac{l-l_1}{t}\frac{d(1+l_1/l)-2b}{ad-bc}\frac{1}{1+\dfrac{1+pqr}{3(1+p)^2}\left(p^2+\dfrac{1}{pqr}\right)}\varepsilon \tag{8.73}$$

$$\theta = \frac{l-l_1}{t}\frac{-c(1+l_1/l)+2a}{ad-bc}\frac{1}{1+\dfrac{1+pqr}{3(1+p)^2}\left(p^2+\dfrac{1}{pqr}\right)}\varepsilon \tag{8.74}$$

where a, b, c, and d are the constants defined in (8.69), and p, q, and r denote the ratios given by $p = t_1/t_2$, $q = w_1/w_2$, and $r = \tilde{E}_1/\tilde{E}_2$, respectively. Note from the preceding equations that the relative strain in the composite beam drives the beam to give the linear and angular displacement in Fig. 8.13b. Substituting the strain–voltage relationship, (8.55), into equations (8.73) and (8.74) and recalling $t_1 = pt/(1+p)$ gives

$$\delta = \phi_\delta V \tag{8.75}$$

where

$$\phi_\delta = l\frac{l-l_1}{t^2}\frac{d(1+l_1/l)-2b}{ad-bc}\frac{1+p}{p\left[1+\dfrac{1+pqr}{3(1+p)^2}\left(p^2+\dfrac{1}{pqr}\right)\right]}d_{31} \tag{8.76}$$

can be defined as a constant that is a function of the effective Young's moduli and geometry of the actuator. The angular displacement is obtained similarly:

$$\theta = \phi_\theta V \tag{8.77}$$

where

$$\phi_\theta = \frac{l-l_1}{t^2}\frac{-c(1+l_1/l)+2a}{ad-bc}\frac{1+p}{p\left[1+\frac{1+pqr}{3(1+p)^2}\left(p^2+\frac{1}{pqr}\right)\right]}d_{31}$$

denotes another constant of the actuator.

8.3 THERMAL ACTUATORS

Thermal expansion can be used to actuate mechanical structures in specific directions. A well-known thermal actuator, the bimetallic actuator, which can serve as a temperature sensor in the electric iron, is shown in Fig. 8.14a. A bimetallic actuator consists of two layers of different coefficients of thermal expansion, as shown in Fig. 8.14a. Since a bimetallic actuator is driven by the difference between thermal expansions of two layers, the theory developed in the preceding sections is applicable to the bimetallic problem. Let us examine a bimetallic actuator exposed to two different temperatures, T_a and T_b. Let α_1 and α_2 represent the coefficients of thermal expansion of the upper and lower layers, respectively. We assume at this stage that α_2 is greater than α_1. Initially, the bimetallic actuator remains straight and is exposed to a temperature T_a (e.g., it is placed in an oven). If the temperature is increased from T_a to T_b, the thermal expansion of the lower layer is greater than that of the upper layer, and the bimetallic actuator is curled upward as shown in Fig. 8.14c and d. Let us consider separated beams which are exposed to the same temperature change. For the temperature change, the elongations of the upper and lower layers are given by $\delta_1 = \alpha_1(T_b - T_a)$ and $\delta_2 = \alpha_2(T_b - T_a)$, so the relative stain ε is obtained as

$$\varepsilon = \frac{\delta_2 - \delta_1}{l} = (\alpha_2 - \alpha_1)(T_b - T_a) \tag{8.78}$$

which causes the displacement of the bimetallic actuator in the transverse direction. Substituting the foregoing relative strain into (8.38) and (d) of Example 8.2, we obtain the radius of curvature ρ indicated in Figs. 8.14c and d and the beam deflection in Fig. 8.14d:

$$\rho = \frac{t}{2(\alpha_2 - \alpha_1)(T_b - T_a)}\left[1+\frac{1+pqr}{3(1+p)^2}\left(p^2+\frac{1}{pqr}\right)\right] \tag{8.79}$$

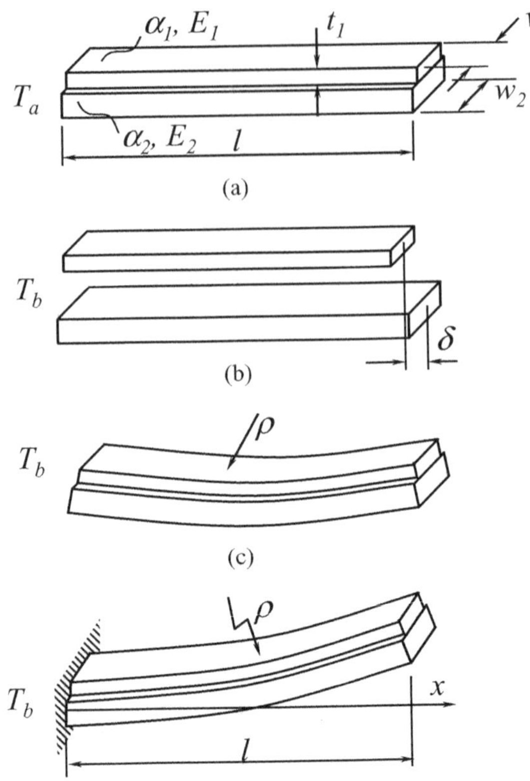

Figure 8.14 Bimetallic actuator.

$$y = \frac{1}{1 + \frac{1 + pqr}{3(1+p)^2}\left(p^2 + \frac{1}{pqr}\right)} \frac{x^2}{t} (\alpha_2 - \alpha_1)(T_b - T_a) \qquad (8.80)$$

where $p = t_1/t_2$, $q = w_1/w_2$, and $r = \tilde{E}_1/\tilde{E}_2$. If $p = q = r = 1$ (i.e., $t_1 = t_2$, $w_1 = w_2$, and $\tilde{E}_1 = \tilde{E}_2$), the radius of curvature becomes $\rho = 2t/[3(\alpha_2 - \alpha_1)(T_b - T_a)]$, and the beam deflection is given by $y = [3(\alpha_2 - \alpha_1)(T_b - T_a)x^2]/4t$. The angular displacement θ at x is then obtained as

$$\theta = \frac{\partial y}{\partial x} = \frac{2}{1 + \frac{1 + pqr}{3(1+p)^2}\left(p^2 + \frac{1}{pqr}\right)} \frac{x}{t} (\alpha_2 - \alpha_1)(T_b - T_a) \qquad (8.81)$$

The end of the bimetallic actuator in Fig. 8.15a can be connected to a flexible structure for a design purpose. If the upper layer partially covers the lower layer and if the flexible structure is considered as a linear spring of stiffness

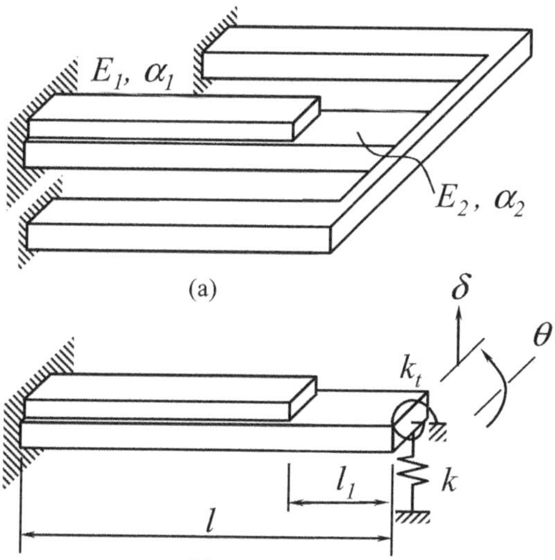

Figure 8.15 Bimetallic actuator connected to flexures.

k and a torsional spring of stiffness k_t, we can set up the model shown in Fig. 8.15b, which represents the mechanical behavior of the bimetallic actuator with flexible structure shown in Fig. 8.15a. It is worth noting that the model of the bimetallic actuator in Fig. 8.15b is identical to that of the piezoelectric cantilever beam actuator, except that the source to generate the relative strain ε in the beam is changed from voltage to temperature. As a result, the responses δ and θ of the bimetallic actuator in Fig. 8.15 to the temperature change $(T_b - T_a)$ are easily obtained by substituting the relative strain ε, (8.78), into (8.73) and (8.74), as follows:

$$\delta = \phi_1 (T_b - T_a) \tag{8.82}$$

where

$$\phi_1 = l \frac{l - l_1}{t} \frac{d(1 + l_1/l) - 2b}{ad - bc} \frac{1}{1 + \dfrac{1 + pqr}{3(1+p)^2}\left(p^2 + \dfrac{1}{pqr}\right)} (\alpha_2 - \alpha_1) \tag{8.83}$$

$$\theta = \phi_2 (T_b - T_a) \tag{8.84}$$

where

$$\phi_2 = \frac{l - l_1}{t} \frac{-c(1 + l_1/l) + 2a}{ad - bc} \frac{1}{1 + \dfrac{1 + pqr}{3(1+p)^2}\left(p^2 + \dfrac{1}{pqr}\right)} (\alpha_2 - \alpha_1) \tag{8.85}$$

In (8.83) and (8.85), a, b, c, and d are constants defined by

$$a = 1 + \frac{kl^3}{3E_2I_2}\left\{\left(\frac{l_1}{l}\right)^3 + \frac{E_2I_2}{(EI)_{\text{eff}}}\left[1-\left(\frac{l_1}{l}\right)^3\right]\right\}$$

$$b = \frac{1}{2}\frac{k_t l}{E_2I_2}\left\{\left(\frac{l_1}{l}\right)^2 + \frac{E_2I_2}{(EI)_{\text{eff}}}\left[1-\left(\frac{l_1}{l}\right)^2\right]\right\}$$

$$c = \frac{kl^3}{2E_2I_2}\left\{\left(\frac{l_1}{l}\right)^2 + \frac{E_2I_2}{(EI)_{\text{eff}}}\left[1-\left(\frac{l_1}{l}\right)^2\right]\right\}$$

$$d = 1 + \frac{k_t l}{E_2I_2}\left[\frac{l_1}{l} + \frac{E_2I_2}{(EI)_{\text{eff}}}\left(1-\frac{l_1}{l}\right)\right]$$

and p, q, and r denote the ratios given by $p = t_1/t_2$, $q = w_1/w_2$, and $r = \tilde{E}_1/\tilde{E}_2$. So far, we have obtained bimetallic actuators which are placed in uniform temperature fields (i.e., in an oven). If temperature varies along the length of a bimetallic actuator, the responses of the actuator differ from those derived in this paragraph. In the succeeding paragraphs we consider the effect of temperature variation on the behavior of a bimetallic actuator.

Heat Transfer from Microstructures

The mechanical behavior of components in MEMS, utilizing thermal expansion, is influenced by heat transfer since the temperature distribution in the components depends on heat transfer. Heat transfer includes conduction, convection, and radiation: conduction, in which heat (thermal energy) is transferred by temperature gradient from one atom to another; convection, where heat is transferred via a bulk, macroscopic motion of a fluid (liquid or gas); and radiation, in which the energy of material is converted to electromagnetic waves and propagates in space. Convection may be categorized into two groups: forced and natural. *Forced convection* involves a bulk fluid motion, which may be caused by a fan or natural wind, whereas *natural convection* is caused by the buoyancy force of a thermally expanded fluid. Thus, natural convection does not play a role if there is no fluid near a solid (i.e., in vacuum) or if the solid and fluid are not exposed to gravitation. Since microstructures are usually placed in a small package, in general, forced convection is not used in MEMS. Natural convection in MEMS is also negligible because the order of magnitude of the buoyancy force is much less than that of a viscous force. As a result, we can rule out the effect of convection in the analysis of most MEMS. Let us examine radiation in microstructures. The energy radiated from a structure is expressed as σAT^4 (if the structure is an ideal body, known as a *blackbody*), where σ, A, and T denote the Stefan–Boltzmann constant with a value of $5.669 \times 10^{-8} \text{W/m}^2 \cdot \text{K}^{-4}$, the structure area, and the structure temperature in kelvin, respectively. For a temperature that is much less than

1500°C, heat transfer due to radiation is, in general, much less than that due to conduction, which is described in detail in subsequent paragraphs. We are now left with conduction, which plays a dominant role in heat transfer in MEMS.

Conduction Heat Transfer from Slender Beams

Before analyzing conduction heat transfer from a complicated microstructure, it is convenient to develop formulaes for the conduction heat transfer of simple and slender beams used in MEMS. Figure 8.16 illustrates cross sections of rectangular beams [(a) and (b)] and a circular beam (c). To evaluate the heat transferred from the beams, we use the analogy that exists between electrostatic and thermal problems. Figure 8.16a illustrates a rectangular beam of width w and thickness t which is over an infinite plate and immersed in a fluid of permittivity ε or in vacuum. The voltage (electric potential) of the beam and plate are represented by V and V_s, respectively. An identical structure in Fig. 8.16b is placed in a temperature field instead of an electric field. The temperatures of the beam and plate are T and T_s, respectively, and the structure is immersed in a fluid of thermal conductivity k_a. For the electric problem shown in Fig. 8.16a, the electric field is governed by Poisson's equation, given by (7.55), which is rewritten here for further discussion:

$$\nabla^2 V = -\frac{\rho_c}{\varepsilon} \tag{8.86}$$

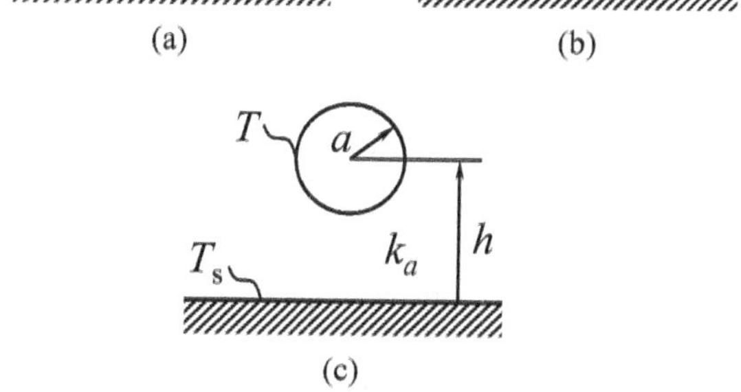

Figure 8.16 Analogy between electrostatic and thermal problems.

where ρ_c and ε denote the charge density and permittivity of the material between the beam and plate, respectively. To solve the steady-state heat conduction problem shown in Fig. 8.16b, we use the following Poisson's equation for heat conduction, which is obtained from an energy balance of a small element of the fluid between the beam and the plate:

$$\nabla^2 T = -\frac{\dot{Q}_t}{k_a} \tag{8.87}$$

where $\dot{Q}_t$ represents heat (per unit volume per unit time) generated in the fluid surrounding the structure. If $\rho_c = 0$ and $\dot{Q}_t = 0$, equations (8.86) and (8.87) reduce to Laplace's equations:

$$\nabla^2 V = 0 \tag{8.88}$$

$$\nabla^2 T = 0 \tag{8.89}$$

In Fig. 8.6a, the electric charge due to the voltage difference $V - V_s$, is accumulated on the surface of the beam and plate; dQ represents the charge on the surface area dA. The accumulated charge Q per unit length can be obtained by combining (7.61) and (7.50):

$$Q = \oint_A \mathbf{D} \cdot d\mathbf{A} = \oint_A \varepsilon_0 \mathbf{E} \cdot d\mathbf{A} = -\varepsilon_0 \oint_A \nabla V \cdot d\mathbf{A} \tag{8.90}$$

which may be expressed using the following definition of capacitance:

$$Q = C(V - V_s) = C\Delta V \tag{8.91}$$

where ΔV is the overall voltage difference, defined to be $V - V_s$. Let us examine the thermal problem in Fig. 8.16b. Since the heat transfer rate per unit area is given by $-k_a \nabla_T$, which is a general form of (3.20), the heat transfer rate Q_t transferred from the beam is expressed as

$$Q_t = -k_a \oint_A \nabla T \cdot d\mathbf{A} \tag{8.92}$$

which is also expressed in terms of the overall temperature difference, $T - T_s$:

$$Q_t = k_a S(T - T_s) = k_a S\,\Delta T \tag{8.93}$$

where S denotes the shape factor and ΔT is the overall temperature difference, defined to be $T - T_s$. Comparing (8.86) to (8.87) and (8.91) to (8.93), we find the following analogy between the electric problem in Fig. 8.16a and the thermal problem in Fig. 8.16b:

$$V \leftrightarrow T \qquad \varepsilon \leftrightarrow k_a \qquad \rho_c \leftrightarrow \dot{Q}_t \qquad Q \leftrightarrow Q_t \qquad C \leftrightarrow k_a S \tag{8.94}$$

where the sign $\leftrightarrow$ represents analogy between the left and right variables. Using this analogy, we can easily obtain the shape factor that is needed for evaluation of the conduction heat transfer rate Q_t, as given by (8.93). According to the analogy (8.94), the shape factor S is written

$$S = \frac{C|_{\varepsilon \to k_a}}{k_a} \tag{8.95}$$

where the numerator means that the permittivity ε in the equation for capacitances is replaced with thermal conductivity k_a. For example, the shape factor of a large thin plate (i.e., a parallel plate, $t << w$, $h << w$, and $t << h$ in Fig. 8.16b) is obtained from the capacitance of the parallel plate, (7.65), which is rewritten here for further derivation: $C = \varepsilon_0 A/h$. Using (8.95), the shape factor of the parallel plate is obtained as

$$S = \frac{C|_{\varepsilon \to k_a}}{k_a} = \frac{k_a A}{h}\frac{1}{k_a} = \frac{A}{h} = \frac{wl}{h} \tag{8.96}$$

where l is the length of the beam. The heat transfer rate Q_t from the upper plate to the lower plate is given by (8.93):

$$Q_t = k_a S \Delta T = k_a \frac{A}{h} \Delta T \tag{8.97}$$

This equation is also obtained from (3.20). For the large thin plate of Fig. 8.16b, where the heat moves from the upper plate to the lower one, the heat transfer equation (3.20) is written to give

$$Q_t = -k_a A \frac{\partial T}{\partial z} = -k_a A \frac{-\Delta T}{h} = k_a \frac{A}{h} \Delta T$$

The equation above is the same as (8.97). It is noted that all other capacitances between conductors are converted to the corresponding shape factor for the thermal problem.

Let us obtain more shape factors by using the analogy discussed above. If the thickness of the upper structure over an infinite plate in Fig. 8.16 is thin but the width w is comparable to the gap h, we can use (7.93) to provide the shape factor of the structure, which is given by

$$S = \frac{wl}{h}\left(1 + \frac{2h}{\pi w} + \frac{2h}{\pi w} \ln \frac{\pi w}{h}\right) \tag{8.98}$$

If the thickness t and width w are comparable to the gap h (i.e., the upper structure is a thick beam), (7.95) is used to obtain the shape factor of the structure, given by

$$S = l\left[\frac{w}{h} + 0.77 + 1.06\left(\frac{w}{h}\right)^{0.25} + 1.06\left(\frac{t}{h}\right)^{0.5}\right] \tag{8.99}$$

The shape above is in good agreement with the simulated capacitance, within an error of 2%, for $w/h \geq 1$ and $0.1 \leq t/h \leq 4$. In the range $w/h \geq 0.3$ and $4 \leq t/h \leq 10$, the maximum deviation of the foregoing shape factor from the simulated shape factor is 6%. If the upper beam is a cylindrical beam over an infinite plate, the shape factor is obtained from (7.98) as follows:

$$S = \frac{2\pi l}{\cosh^{-1}(h/a)} \tag{8.100}$$

Noted that the capacitance between two conductors, which is obtained in Section 7.5, is used to evaluate the electrostatic force between the conductors (structures), as shown in Section 7.5, and also to evaluate the conduction heat transfer between the same structures.

Heat Transfer from Slender Beams

Thermal actuators in MEMS generally use heat generated in a slender beam to actuate mechanical components. Temperature distribution along the beam is essential to analyze the response of thermal actuators. Let us consider the fixed–fixed beam in Fig. 8.17a, through which electric current flows to generate heat. T_0 and T_l represent temperatures at the left and right ends of the beam, respectively, and the substrate temperature is given by T_s. To obtain the temperature distribution along the beam, we set up a model (Fig. 8.17b) and use the following steady-state conduction equation:

$$k_s \frac{d^2T}{dx^2} - \frac{k_a S_1}{A}(T - T_s) + \frac{\dot{Q}_t}{A} = 0 \tag{8.101}$$

where S_1 is the shape factor per unit length, defined as $S_1 = S/l$, which is a dimensionless number, and $\dot{Q}_t$ is the heat generated per unit beam length per unit time. It is noted that this equation is obtained by repeating the procedure from (3.20) to (3.21) when the heat transfer rate from the beam of length Δx to the plate is given by (8.93) [i.e., $k_a S_1 \Delta x(T - T_s)$]. The solution of equation (8.101) will be expressed in a complicated form. However, if the order of magnitude of the second term is much less than that of the first term, equation (8.101) will be expressed in a simple form. In order to determine whether we solve the original equation, we define a dimensionless number, which repre-

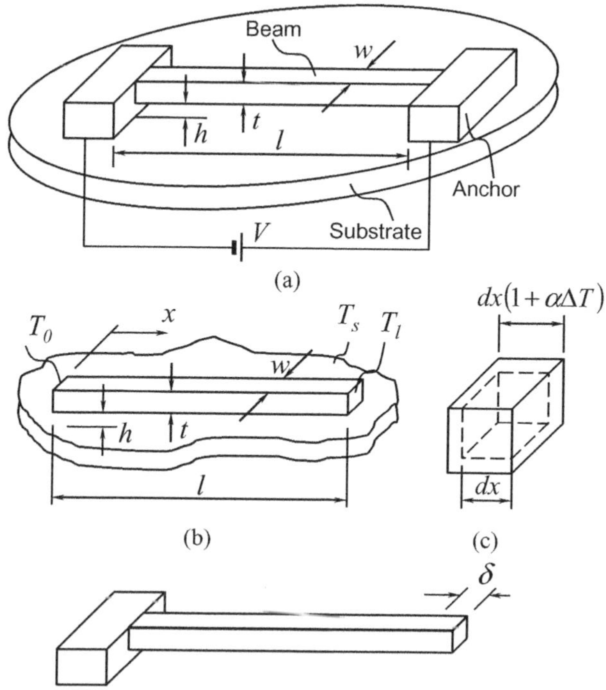

Figure 8.17 Joule-heated beam.

sents a comparison of the order of magnitude of the heat transfer through the surrounding fluid to that along the beam. Using the dimensional analysis described in Section 1.4, we introduce a dimensionless number,

$$\frac{\left[\dfrac{k_a S_1}{A}(T-T_s)\right]}{\left[k_s\dfrac{d^2T}{dx^2}\right]}=\frac{\dfrac{k_a S_1}{A}T_c}{k_s\dfrac{T_c}{l^2}}=\frac{k_a S_1 l^2}{k_s A}$$

where the brackets are used as an operator to extract the order of magnitude of quantities in the brackets, and T_c and l are used as the characteristic temperature difference and length, respectively. In subsequent analysis, the square root of the number above will appear, so we then define the dimensionless number as

$$ml=\sqrt{\frac{k_a S_1}{k_s A}}\,l \tag{8.102}$$

If $ml << 1$, the second term in (8.101) is much less the first term, so the conduction equation is approximated as

$$\frac{d^2T}{dx^2} = -\frac{\dot{Q}_t}{k_s A} \tag{8.103}$$

This case includes the following: The structure in Fig. 8.17a is placed in a vacuum chamber; the gap h is much greater than the thickness t and width w. If $T_0 = T_l = T_s$ and $ml = 0.36$, the solution of equation (8.103) deviates from that of the original equation (8.101) by only 1%. Therefore, if $ml \leq 0.36$, we can consider the solution of equation (8.103) as an approximate solution of (8.101).

We can solve the foregoing equation with the boundary conditions T_0 at $x = 0$ and T_l at $x = l$ to give the temperature distribution as follows:

$$T = \frac{\dot{Q}_t l^2}{2k_s A}\left[\frac{x}{l} - \left(\frac{x}{l}\right)^2\right] + (T_l - T_0)\frac{x}{l} + T_0 \tag{8.104}$$

If $T_0 = T_l = T_s$ as shown in Fig. 8.17a, equation (8.104) reduces to

$$T = \frac{\dot{Q}_t l^2}{2k_s A}\left[\frac{x}{l} - \left(\frac{x}{l}\right)^2\right] + T_0 \tag{8.105}$$

If the left end of the beam is fixed on a wall and the right end is free (i.e., a cantilever beam) or if the right end of the beam in Fig. 8.17 is in contact with an insulator, we can use the boundary conditions $T = T_0$ at $x = 0$ and $dT/dx = 0$ at $x = l$ to provide the following temperature along the beam:

$$T = \frac{\dot{Q}_t l^2}{2k_s A}\left[2\frac{x}{l} - \left(\frac{x}{l}\right)^2\right] + T_0 \tag{8.106}$$

The temperature at $x = l$ is obtained as

$$T_l = \frac{\dot{Q}_t l^2}{2k_s A} + T_0 \tag{8.107}$$

If $ml > 0.36$, we should solve (8.101) to obtain a more accurate solution of the problem. Let us solve (8.101) for T when the boundary conditions T_0 at $x = 0$ and $T = T_l$ at $x = l$ are imposed on the problem. For convenience, let us set $\theta = T - T_s$ and write (8.101) in the following form:

$$\frac{d^2\theta}{dx^2} - m^2\theta = -p \tag{8.108}$$

where

$$m = \sqrt{\frac{k_a S_1}{k_s A}} \qquad p = \frac{\dot{Q}_t}{k_s A} \tag{8.109}$$

represent a parameter in m^{-1} (associated with the dimensionless number ml already discussed) and another parameter in °C/m^2, respectively. The boundary conditions are changed to $\theta_0 = T_0 - T_s$ at $x = 0$ and $\theta_1 = T_l - T_0$ at $x = l$. Since the homogeneous solution of $(d^2\theta/dx^2) - m^2\theta = 0$ is given by $\theta_h = c_1 \cosh mx + c_2 \sinh mx$ and the particular solution is $\theta_p = p/m^2$, the general solution of (8.108) can be written

$$\theta = \theta_h + \theta_p = \frac{p}{m^2} + c_1 \cosh mx + c_2 \sinh mx \tag{8.110}$$

Imposing the boundary conditions θ_0 at $x = 0$ and θ_l at $x = l$, we find that

$$\begin{aligned}\theta = \theta_0 &\left(\cosh mx - \cosh ml \frac{\sinh mx}{\sinh ml} \right) + \theta_l \frac{\sinh mx}{\sinh ml} \\ &+ \frac{p}{m^2}\left(1 - \cosh mx + \frac{\cosh ml - 1}{\sinh ml} \sinh mx \right)\end{aligned}$$

Substitution of $\theta = T - T_s$, $\theta_0 = T_0 - T_s$, and $\theta_l = T_l - T_s$ into the foregoing equation yields the temperature distribution of the slender beam of uniform cross section as follows:

$$\begin{aligned}T = T_s &+ (T_0 - T_s)\left(\cosh mx - \frac{\sinh mx}{\tanh ml} \right) + (T_l - T_s)\frac{\sinh mx}{\sinh ml} \\ &+ \frac{p}{m^2}\left[1 - \cosh mx - (1 - \cosh ml)\frac{\sinh mx}{\sinh ml} \right]\end{aligned} \tag{8.111}$$

In the limit as $m \to 0$, equation (8.111) reduces to (8.104); that is,

$$\begin{aligned}T = \lim_{m \to 0} &\left\{ T_s + (T_0 - T_s)\left(\cosh mx - \frac{\sinh mx}{\sinh ml} \cosh ml \right) + (T_l - T_s)\frac{\sinh mx}{\sinh ml} \right. \\ &\left. + pl^2 \left[\frac{1 - \cosh mx}{(ml)^2} - \frac{1 - \cosh ml}{(ml)^2} \frac{\sinh mx}{\sinh ml} \right] \right\} \\ &= T_s + (T_0 - T_s)\left(1 - \frac{x}{l} \right) + (T_l - T_s)\frac{x}{l} + pl^2 \left(-\frac{1}{2}\frac{x^2}{l^2} + \frac{1}{2}\frac{x}{l} \right) \\ &= T_0 + (T_l - T_0)\frac{x}{l} + \frac{pl^2}{2}\left(\frac{x}{l} - \frac{x^2}{l^2} \right)\end{aligned}$$

Substituting $p = \dot{Q}_t / k_s A$ into the equation above yields (8.104). If the temperature of the left end of the beam is considered to be equal to that of the substrate (i.e., $T_0 = T_s$), (8.111) is reduced to

$$T = T_0 + (T_l - T_0)\frac{\sinh mx}{\sinh ml} + \frac{p}{m^2}\left[1 - \cosh mx - (1 - \cosh ml)\frac{\sinh mx}{\sinh ml}\right] \tag{8.112}$$

For $T_0 = T_l = T_s = 20°\text{C}$, $l = 100\,\mu\text{m}$, and $p/m^2 = 2000°\text{C}$ (p/m^2 is not a physical temperature at a point in the beam but has the dimensions of temperature), the temperature distribution of a fixed–fixed beam is shown in Fig. 8.18, in which the temperature is lowered as ml increases from zero. For further analysis, we need to know the heat transfer rate along the beam. Since the derivative of the temperature (8.111) with respect to x is expressed as

$$\begin{aligned}\frac{\partial T}{\partial x} = m\Bigg\{&(T_0 - T_s)\left(\sinh mx - \frac{\cosh mx}{\tanh ml}\right) + (T_l - T_s)\frac{\cosh mx}{\sinh ml}\\ &+ \frac{p}{m^2}\left[-\sinh mx + (\cosh ml - 1)\frac{\cosh mx}{\sinh ml}\right]\Bigg\}\end{aligned}$$

the heat transfer rate q_0 at the left end of the beam is given by

$$\begin{aligned}q_0 &= -k_s A \left.\frac{\partial T}{\partial x}\right|_{x=0}\\ &= -k_s m A\left[-(T_0 - T_s)\frac{1}{\tanh ml} + (T_l - T_s)\frac{1}{\sinh ml} + \frac{p}{m^2}\frac{\cosh ml - 1}{\sinh ml}\right]\end{aligned} \tag{8.113}$$

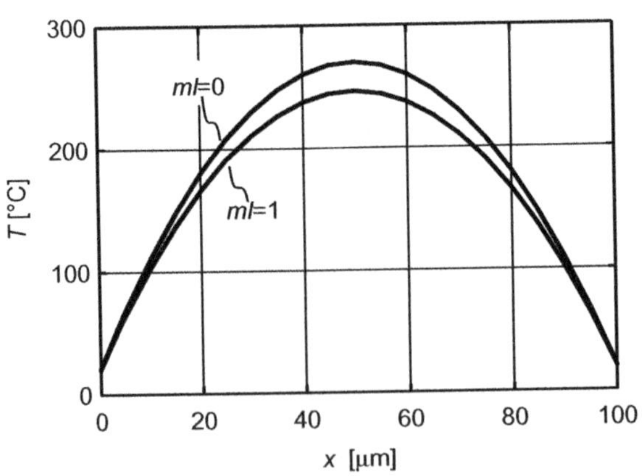

Figure 8.18 Temperature distribution.

From equation (8.113), the positive q_0 is defined as the heat flowing in the x direction in Fig. 8.17b. Similarly, at the right end of the beam, the heat transfer rate q_l is obtained as

$$\begin{aligned} q_l &= -k_s A \left.\frac{\partial T}{\partial x}\right|_{x=l} \\ &= -k_s m A\left[-(T_0 - T_s)\frac{1}{\sinh ml} + (T_l - T_s)\frac{1}{\tanh ml} + \frac{p}{m^2}\frac{1-\cosh ml}{\sinh ml}\right] \end{aligned} \tag{8.114}$$

Thermal Strain and Stress

When an elastic body under constraints is heated, the body is expanded and experiences thermal stress. Let us evaluate the thermal stress of the fixed–fixed beam heated by Joule heating, as shown in Fig. 8.17a. An element of length dx taken from the beam is expanded by the heating, as shown in Fig. 8.17c. If the initial temperature of the beam is equal to that of the substrate, and if the heat transfer from the beam to the substrate via the surrounding fluid is negligible (i.e., $ml << 1$), the temperature distribution of (8.104) is used to give the thermal expansion $d\delta$ of the element of length dx:

$$\begin{aligned} d\delta &= \alpha(T - T_s)dx \\ &= \alpha\left\{\frac{\dot{Q}_t l^2}{2k_s A}\left[\frac{x}{l} - \left(\frac{x}{l}\right)^2\right] + (T_l - T_0)\frac{x}{l} + T_0 - T_s\right\}dx \end{aligned}$$

where α is the coefficient of the thermal expansion. Integrating over the length of the beam yields the thermal expansion δ of the beam, as shown in Fig. 8.17d:

$$\begin{aligned} \delta &= \int_0^l \alpha\left\{\frac{\dot{Q}_t l^2}{2k_s A}\left[\frac{x}{l} - \left(\frac{x}{l}\right)^2\right] + (T_l - T_0)\frac{x}{l} + T_0 - T_s\right\}dx \\ &= \alpha l\left(\frac{\dot{Q}_t l^2}{12k_s A} + \frac{T_l - T_0}{2} + T_0 - T_s\right) \end{aligned} \tag{8.115}$$

and the corresponding strain ε, which is obtained as

$$\varepsilon = \frac{\delta}{l} = \alpha\left(\frac{\dot{Q}_t l^2}{12k_s A} + \frac{T_l - T_0}{2} + T_0 - T_s\right) \tag{8.116}$$

If $T_l = T_0 = T_s$, the preceding strain equation reduces to

$$\varepsilon = \frac{\alpha \dot{Q}_t l^2}{12k_s A} \tag{8.117}$$

As a result, the thermal stress in the fixed–fixed beam in Fig. 8.17a is obtained as

$$\sigma = \tilde{E}(-\varepsilon) = -\tilde{E}\alpha\left(\frac{\dot{Q}_t l^2}{12k_s A} + \frac{T_l - T_0}{2} + T_0 - T_s\right) \tag{8.118}$$

where the negative sign indicates that Joule heating of the fixed–fixed beam results in a compressive stress in the beam. If the right end is in contact with an insulator, (8.106) is used to give the following expansion of the beam:

$$\delta = \int_0^l \alpha(T - T_s)\,dx = \int_0^l \alpha\frac{\dot{Q}_t l^2}{2k_s A}\left[2\frac{x}{l} - \left(\frac{x}{l}\right)^2\right]dx$$
$$= \alpha\frac{\dot{Q}_t l^3}{3k_s A} \tag{8.119}$$

$$\varepsilon = \frac{\delta}{l} = \alpha\frac{\dot{Q}_t l^2}{3k_s A} \tag{8.120}$$

If the insulator in contact with the right end of the beam is rigid, the thermal stress caused in the beam is given by

$$\sigma = \tilde{E}(-\varepsilon) = -\tilde{E}\alpha\frac{\dot{Q}_t l^2}{3k_s A} \tag{8.121}$$

If the heat transfer through the surrounding fluid is not negligible, (8.111) should be used to evaluate the thermal expansion and stress of the beam. In this case, the thermal expansion of the beam with a free end is given by

$$\delta = \int_0^l \alpha(T - T_s)\,dx = \alpha l\left\{(T_0 + T_l - 2T_s)\frac{1}{ml}\frac{\cosh ml - 1}{\sinh ml}\right.$$
$$\left. + \frac{pl^2}{(ml)^2}\left[1 - \frac{1}{ml}\sinh ml + \frac{1}{ml}\frac{(\cosh ml - 1)^2}{\sinh ml}\right]\right\} \tag{8.122}$$

and the corresponding strain is expressed as

$$\varepsilon = \frac{\delta}{l} = \alpha\left\{(T_0 + T_l - 2T_s)\frac{1}{ml}\frac{\cosh ml - 1}{\sinh ml}\right.$$
$$\left. + \frac{pl^2}{(ml)^2}\left[1 - \frac{1}{ml}\sinh ml + \frac{1}{ml}\frac{(\cosh ml - 1)^2}{\sinh ml}\right]\right\} \tag{8.123}$$

Then the thermal stress in the fixed–fixed beam, whose ml is not much less than unity, is given by

$$\sigma = \tilde{E}(-\varepsilon) = -\tilde{E}\alpha\left\{(T_0 + T_l - 2T_s)\frac{1}{ml}\frac{\cosh ml - 1}{\sinh ml}\right.$$
$$\left. + \frac{pl^2}{(ml)^2}\left[1 - \frac{1}{ml}\sinh ml + \frac{1}{ml}\frac{(\cosh ml - 1)^2}{\sinh ml}\right]\right\} \qquad (8.124)$$

In the limit as $ml \to 0$, this equation goes to (8.118). For $T_l = T_0 = T_s$ and $ml \leq 0.32$, (8.118) is in good agreement with equation (8.124) within an error of 1%. As a result, if $T_l = T_0 = T_s$ and $ml \leq 0.32$, (8.118) can be used to evaluate the thermal stress in fixed–fixed beams.

Buckling Modes of Fixed–Fixed Beams

In the preceding paragraphs we described the thermal strain and stress in slender straight beams. If the normal stress due to any cause (e.g., the thermal stress discussed in the preceding paragraphs or the residual stress) is developed in a fixed–fixed beam, the stiffnesses in the y and z directions of the beam in Fig. 8.19 decrease with decreasing stress and become zero at the critical stress corresponding to the critical load P_{cr}, given by (3.80). We then obtain the critical stress at which the buckling phenomenon occurs:

$$\sigma_{cr} = -\frac{P_{cr}}{A} = -\frac{4\pi^2\tilde{E}I}{Al^2} \qquad (8.125)$$

If we use Galerkin's method to obtain the critical stress, the following approximate critical stress is obtained from (n) of Example 3.3 or (3.81):

$$\sigma_{cr} = -\frac{10\tilde{E}I}{A(l/2)^2} = -\frac{40\tilde{E}I}{Al^2} \qquad (8.126)$$

When the stress in the beam is reduced from an initial stress (or residual stress) to the critical stress above, the beam does not deviate from its initial position and remains straight. However, if the stress in the beam becomes less than the preceding critical stress, the beam deviates from its initial position and is deflected in the transverse direction as shown in Fig. 8.19a and b, depending on the width-to-thickness ratio. If $w < t$ in Fig. 7.19a, the beam is deflected in the positive or negative y direction. We cannot determine the direction of deflection of the beam since the beam is symmetric before buckling; that is we can only say that the beam in Fig. 8.19, subjected to a stress lower than the critical stress (8.125), will be buckled in the positive or negative y direction. However, if $w > t$ in Fig. 7.19b, the beam, subjected to a compressive stress beyond the critical stress, is deflected in the positive or negative z direction.

Even though we cannot determine the direction of deflection of the beam, it is of interest to evaluate how much the beam is deflected from

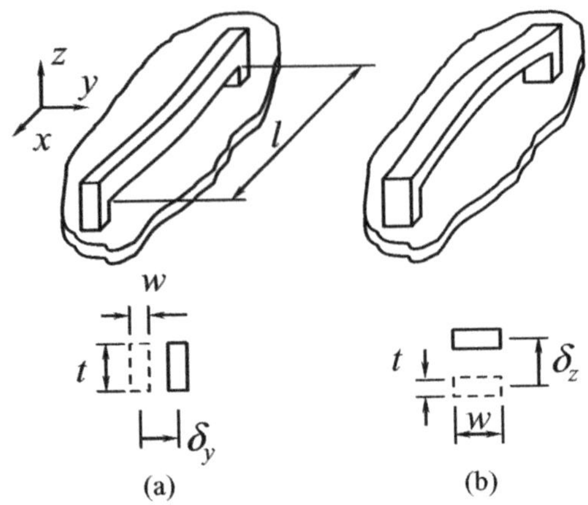

Figure 8.19 Buckling modes of fixed–fixed beams.

its initial position. To do this, we use the linear and cubic stiffness derived in Chapter 3. Let us suppose that the beam shown in Fig. 8.19a is subjected to a force F_y applied to the midspan in the y direction. We express the force in the form

$$F_y = 2(k_1\delta_y + k_3\delta_y^3) \tag{8.127}$$

where δ_y denotes the beam deflection of the midspan and k_1 and k_3 represent the stiffnesses, given by (n) and (o) in Example 3.3, which are rewritten here for convenience:

$$k_1 = 12\frac{\tilde{E}I_z}{l_1^3} + \frac{6}{5}\frac{N_0}{l_1}$$

$$k_3 = \frac{18}{25}\frac{\tilde{E}A}{l_1^3}$$

where $\tilde{E}, I_z, l_1, N_0$, and A represent the effective Young's modulus of the beam, the moment of inertia defined as $tw^3/12$, the half-length of the beam, the axial force, and the cross-sectional area of the beam, respectively. Since no force is exerted on the beam at the midspan, we can set the preceding force equal to zero to give the deflection δ_y of the midspan of the beam: $\delta_y = \pm\sqrt{-k_1/k_3}$. Substituting the linear and cubic stiffnesses above into the preceding deflection equation, we find that

$$\delta_y = \pm\sqrt{\frac{-k_1}{k_3}} = \pm\sqrt{-\frac{50}{3}\frac{I_z}{A}\left(1 + \frac{1}{10}\frac{N_0 l_1^2}{\tilde{E}I_z}\right)} \tag{8.128}$$

Substituting $I_z = tw^3/12$, $A = wt$, and $l_1 = l/2$, we obtain the deflection δ_y of the midspan of the buckled beam of $w/t < 1$, shown in Fig. 8.19a:

$$\delta_y = \pm w\sqrt{-\frac{25}{18}\left[1+\frac{3}{10}\frac{\sigma_0}{\tilde{E}}\left(\frac{l}{w}\right)^2\right]} \tag{8.129}$$

where σ_0 denotes the stress in the beam. If a beam with residual stress σ_r is heated by Joule heating, the total stress σ_0 is expressed as $\sigma_0 = \sigma_r + \sigma$, where σ represents the thermal stress, given by (8.118) or (8.124). Note that equations (8.128) and (8.129) hold for buckled beams; that is,

$$\sigma_0 \le -\frac{10}{3}\tilde{E}\left(\frac{w}{l}\right)^2 \tag{8.130}$$

where the term on the right-hand side represents the critical stress (8.126). For the buckling mode in Fig. 8.19b, the deflection δ_z of the buckled beam is obtained by switching δ_z and δ_y and substituting t for w in (8.128), as follows:

$$\delta_z = \pm\sqrt{-\frac{50}{3}\frac{I_y}{A}\left(1+\frac{1}{10}\frac{N_0 l_1^2}{\tilde{E}I_y}\right)} \tag{8.131}$$

Substituting $I_y = wt^3/12$, $A = wt$, and $l_1 = l/2$, we obtain the deflection δ_z of the midspan of the buckled beam of $w/t > 1$ shown in Fig. 8.19b:

$$\delta_z = \pm t\sqrt{-\frac{25}{18}\left[1+\frac{3}{10}\frac{\sigma_0}{\tilde{E}}\left(\frac{l}{t}\right)^2\right]} \tag{8.132}$$

Equation (8.132) holds for

$$\sigma_0 \le -\frac{10}{3}\tilde{E}\left(\frac{t}{l}\right)^2 \tag{8.133}$$

If the stress in the beam is caused by Joule heating, (8.118) for $m = 0$ and (8.124) for $m > 0$ can be used to evaluate the deflection of the buckled beam, as shown in Figs. 8.19a and b. The deflections of midspan (8.129) and (8.132) are plotted in Fig. 8.20, where $\delta/d = \delta_y/w$ and $X = 3\sigma_0 l^2/(10\tilde{E}w^2)$ for $t > w$ and $\delta/d = \delta_z/t$ and $X = 3\sigma_0 l^2/(10\tilde{E}t^2)$ for $t < w$. Note in Fig. 8.20 that the sensitivity of the midspan deflection to the stress in the beam at the critical stress σ_{cr} becomes infinite (i.e., $\partial\delta_y/\delta\sigma_0 = \pm\infty$ for $w/t < 1$ and $\partial\delta_z/\partial\sigma_0 = \pm\infty$ for $w/t > 1$. We should note that the deflection above is limited by the melting temperature of the beam; that is, the temperature of the beam approaches the melting temperature and the theory developed in this section cannot be applicable.

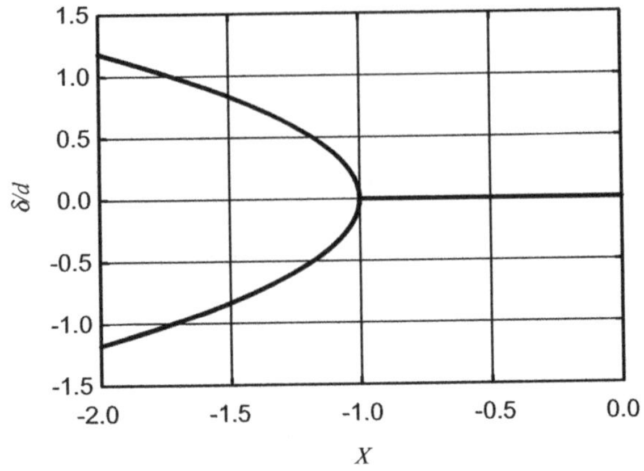

Figure 8.20 Deflection of midspan of buckled fixed–fixed beams.

Example 8.3 A fixed–fixed beam is heated by Joule heat as shown in Fig. 8.17a. The related data are: the resistance of the beam $R = 500\,\Omega$, length $l = 150\,\mu\text{m}$, width $w = 3\,\mu\text{m}$, thickness $t = 4\,\mu\text{m}$, gap $h = 2\,\mu\text{m}$, Young's modulus $E = 150\,\text{GPa}$, coefficient of thermal expansion $\alpha = 2.6 \times 10^{-6}\,^{\circ}\text{C}^{-1}$, thermal conductivity of beam $k_s = 20\,\text{W/m}\cdot{}^{\circ}\text{C}$, thermal conductivity of surrounding gas $k_a = 0.026\,\text{W/m}\cdot{}^{\circ}\text{C}$, residual stress $\sigma_0 = -10.0\,\text{MN/m}^2$, and room temperature 20°C. It is desired to obtain the maximum temperature and thermal stress in the beam when voltages of 2 and 4V are applied across the beam, respectively. If the beam is buckled, we wish to find the beam deflection at the midspan.

For the applied voltage of 2V, the heat generated per unit length is obtained as

$$\dot{Q}_t = \frac{V^2}{Rl} = \frac{2^2}{500 \times 150 \times 10^{-6}} = 53.33\ \text{W/m} \tag{a}$$

The cross-sectional area is given by $A = tw = 6 \times 10^{-12}\,\text{m}^2$. The dimensional number ml (8.102) is used to make a decision as to whether the heat transfer through the surrounding gas can be neglected. To calculate ml, we first need to obtain the shape factor per unit length S_1. For a rectangular beam in the range $w/h \geq 1$ and $0.1 \leq t/h \leq 4$, (8.99) is used to give S_1:

$$S_1 = \frac{S}{l} = \frac{w}{h} + 0.77 + 1.06\left(\frac{w}{h}\right)^{0.25} + 1.06\left(\frac{t}{h}\right)^{0.5} \tag{b}$$

Substituting the geometric data into equation (b) yields $S_1 = 4.9421$. We then use (8.109) to obtain m and p, which are rewritten here for convenience:

$$m = \sqrt{\frac{k_a S_1}{k_s A}} \qquad p = \frac{\dot{Q}_t}{k_s A} \tag{c}$$

Upon substitution of (b) and other data into the preceding equations, we have $m = 2.3139 \times 10^4 \text{m}^{-1}$ and $p = 2.2222 \times 10^{11}\,°\text{C/m}^2$, and also obtain $ml = 2.3139 \times 10^{-4} \times 150 \times 10^{-6} = 3.4708$. Since $ml > 0.32$, the heat through the surrounding gas plays an important role and cannot be neglected. For the evaluation of the maximum temperature and thermal stress in the beam, we use (8.111) and (8.124), rewritten below:

$$T = T_s + (T_0 - T_s)\left(\cosh mx - \frac{\sinh mx}{\tanh ml}\right) + (T_l - T_s)\frac{\sinh mx}{\sinh ml} + \frac{p}{m^2}\left[1 - \cosh mx - (1 - \cosh ml)\frac{\sinh mx}{\sinh ml}\right] \tag{d}$$

$$\sigma = -\tilde{E}\alpha\left\{(T_0 + T_l - 2T_s)\frac{1}{ml}\frac{\cosh ml - 1}{\sinh ml} + \frac{pl^2}{(ml)^2}\left[1 - \frac{1}{ml}\sinh ml + \frac{1}{ml}\frac{(\cosh ml - 1)^2}{\sinh ml}\right]\right\} \tag{e}$$

If $ml < 0.32$, (8.104) and (8.118) can be used for the determination of the temperature and stress. It is assumed from Fig. 8.17a that since $l >> h$, the temperatures at the substrate and anchors are equal (i.e., $T_0 = T_1 = T_s$). Since $t/w = 4/3$, $\tilde{E} = E = 150\,\text{GPa}$, which is given by (3.5). Substituting $x = l/2$ and other data into equations (d) and (e) gives $T_{\max} = 293.10°\text{C}$ and $\sigma = -7.4221 \times 10^7 \text{N/m}^2$. Because the thermal stress is added to the residual stress, the total stress σ_0 is expressed as $\sigma_0 = \sigma_r + \sigma = -1.0 \times 10^7 - 7.4221 \times 10^7 = -8.4221 \times 10^7 \text{N/m}^2$. For the beam in the range of $t > w$, (8.130) is used to obtain the stress at which the beam is buckled:

$$\sigma_0 \le -\frac{10}{3}\tilde{E}\left(\frac{w}{l}\right)^2 \tag{f}$$

where the right-hand side of equation (f) corresponds to the critical stress σ_{cr}. Using the data of the present example, the critical stress is obtained as $\sigma_{cr} = -2.0000 \times 10^8 \text{N/m}^2$. Since $\sigma > \sigma_{cr}$, the beam remains straight and the deflection of the midspan is $\delta_y = 0$. When the applied voltage increases to $V = 4\text{V}$, the preceding procedure is repeated to give $m = 2.3139 \times 10^4 \text{m}^{-1}$, $p = 8.8889 \times 10^{11}\,°\text{C/m}^2$, $T_{\max} = 1112.4°\text{C}$, and $\sigma_0 = -3.0689 \times 10^8 \text{N/m}^2$.

Since $\sigma < \sigma_{cr}$ and $w/t < 1$, we use (8.129) to give the deflection δ_y of the midspan:

$$\delta_y = \pm w\sqrt{-\frac{25}{18}\left[1 + \frac{3}{10}\frac{\sigma_0}{\tilde{E}}\left(\frac{l}{w}\right)^2\right]} \tag{g}$$

Substituting the data into equation (g), we find that $\delta_y = \pm 2.58 \times 10^{-6}\,\text{m} = \pm 2.58\,\mu\text{m}$. The deflection of the midspan is illustrated as a function of the stress in Fig. 8.20, in which δ/d and X represent δ_y/w and $3\sigma_0 l^2/10\tilde{E}w^2$, respectively.

V-Shaped Beam

In previous paragraphs we derived the response of a fixed–fixed beam to temperature change. The deflection of midspan of the beam remains zero as long as the stress in the beam is greater than the critical stress (i.e., the magnitude of the compressible stress is less than that of the critical stress). This means that much electric energy should be consumed to generate the motion or displacement of the beam. To avoid the waste of Joule heat, a V-shaped beam is used, whose mechanical behavior was studied in Section 4.3. Let us consider a V-shaped beam of half-length l in Fig. 8.21a, which is spaced by a gap h from a substrate and whose two ends are fixed to the substrate. The beam is designed to satisfy the condition $l \sin\theta >> w$, where θ and w denote the angle in Fig. 8.21 and the width of the beam, respectively. When a voltage V is applied through the V-shaped beam, the beam is expanded and the vertex of the beam is displaced by δ, as shown in Fig. 8.21a. The V-shaped beam, anchored on the substrate, can be considered as the pin-jointed truss shown in Fig. 8.21b, which was dealt with in Section 4.3. As a result, we shall use the deflection–strain relation, (4.77), to obtain the deflection of the vertex as follows:

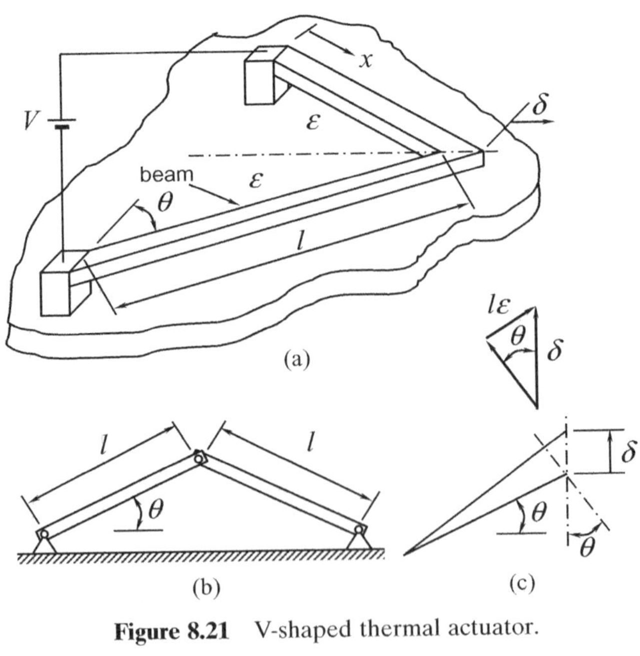

Figure 8.21 V-shaped thermal actuator.

$$\delta = \frac{l}{\sin\theta}\varepsilon \tag{8.134}$$

where ε denotes the strain of the beam of length l. If the strain ε is given, we can obtain the displacement δ from equation (8.134). As a thermal system, a V-shaped beam can be thought of as a fixed–fixed beam of length $2l$ (Fig. 8.17a), and the temperature distribution along the beam is obtained by substituting $2l$ for l in (8.111) as follows:

$$\begin{aligned} T = T_s + (T_0 - T_s)\left(\cosh mx - \frac{\sinh mx}{\tanh 2ml}\right) + (T_l - T_s)\frac{\sinh mx}{\sinh 2ml} \\ + \frac{p}{m^2}\left[1 - \cosh mx - (1 - \cosh 2ml)\frac{\sinh mx}{\sinh 2ml}\right] \end{aligned} \tag{8.135}$$

where x is the coordinate along the beam as shown in Fig. 8.21 and its range is given by $0 \le x \le 2l$. In equation (8.135), m and p are given by (8.109) and the heat generated from the beam is $\dot{Q}_t = V^2/2Rl$, where R is the total resistance between the ends of the V-shaped beam. The heated beam expands by εl as shown in Fig. 8.21c, and the vertex is displaced by δ along the symmetric line. Assuming that $T_l = T_0 = T_s$ and substituting $2l$ for l in (8.123), we find the strain ε of the beam of length l as follows:

$$\begin{aligned} \varepsilon &= \frac{\delta_1}{l} = \frac{\delta_2/2}{l} \\ &= \alpha\frac{pl^2}{(ml)^2}\left[1 - \frac{1}{2ml}\sinh 2ml + \frac{1}{2ml}\frac{(\cosh 2ml - 1)^2}{\sinh 2ml}\right] \end{aligned} \tag{8.136}$$

where δ_1 denotes the longitudinal elongation of the beam of length l and δ_2 represents the elongation of the straight beam of length $2l$, which is obtained by substituting $2l$ for l in (8.122). The preceding strain is equal to the strain of the beam of length $2l$, subjected to temperature (8.135). Substituting equation (8.136) into (8.134) and using $p = \dot{Q}_t/k_sA$, we find the displacement δ of the V-shaped beam in Fig. 8.21a as follows:

$$\begin{aligned} \delta &= \frac{\alpha}{\sin\theta}\frac{pl^3}{(ml)^2}\left[1 - \frac{1}{2ml}\sinh 2ml + \frac{1}{2ml}\frac{(\cosh 2ml - 1)^2}{\sinh 2ml}\right] \\ &= \frac{\alpha\dot{Q}_t l^3}{k_sA\sin\theta}\frac{1}{(ml)^2}\left[1 - \frac{1}{2ml}\sinh 2ml + \frac{1}{2ml}\frac{(\cosh 2ml - 1)^2}{\sinh 2ml}\right] \end{aligned} \tag{8.137}$$

which is valid for $\sin\theta >> w/l$. If the heat transfer via surrounding fluids is neglected (i.e., $ml \to 0$), equation (8.137) is reduced to

$$\delta = \frac{\alpha\dot{Q}_t l^3}{3k_sA\sin\theta} \tag{8.138}$$

The V-shaped beam in Fig. 8.21 has the stiffness (4.75), which is rewritten were for convenience:

$$k = \frac{F}{\delta} = 2\frac{EA}{l}\sin^2\theta \tag{8.139}$$

where F and A denotes the applied force in the δ direction and the cross-sectional area of the beam, respectively. Since the strain of the half beam of length l is given by ε in the longitudinal direction of the half beam, the V-shaped beam of strain ε in Fig. 8.21a may be considered to generate an effective force F_{eff}, corresponding to the displacement at the vertex:

$$\begin{aligned} F_{\text{eff}} &= k\delta \\ &= 2\sin\theta \frac{EA\alpha \dot{Q}_t l^2}{k_s A}\frac{1}{(ml)^2}\left[1 - \frac{1}{2ml}\sinh 2ml + \frac{1}{2ml}\frac{(\cosh 2ml - 1)^2}{\sinh 2ml}\right] \end{aligned} \tag{8.140}$$

That is, if the force above is applied to the V-shaped beam without voltage, the corresponding displacement is equal to the displacement δ, (8.137), due to Joule heating.

If the vertex of the beam is in contact with a rigid insulator, the thermally actuated V-shaped beam applies the following force F_w to the insulator:

$$\begin{aligned} F_w &= F_{\text{eff}} \\ &= 2\sin\theta \frac{EA\alpha \dot{Q}_t l^2}{k_s A}\frac{1}{(ml)^2}\left[1 - \frac{1}{2ml}\sinh 2ml + \frac{1}{2ml}\frac{(\cosh 2ml - 1)^2}{\sinh 2ml}\right] \end{aligned} \tag{8.141}$$

Bimetallic Actuator with Varying Temperature

In the description associated with Figs. 8.14 and 8.15, we obtained the response of the bimetallic actuator, in which temperature is distributed uniformly. When a bimetallic actuator is exposed to a temperature field that is not uniform in space, the radius of curvature ρ is not uniform along the actuator but varies. If the temperature distribution along the actuator is given, we can obtain the response of the bimetallic actuator to the temperature change. Let us imagine a bimetallic actuator whose material properties and dimensions are shown in Figs. 8.22a and b. The upper layer, isolated electrically from the lower layer, is patterned as shown in Fig. 8.22b to generate heat when a voltage is applied between the ends of the layer. We assume that the heat transfer through surrounding fluid is negligible compared to that along the beam. We also assume that since the thickness and width of the beam are much less than the length, the temperature varies only in the longitudinal direction of the beam. For a given temperature at x in Fig. 8.22a, the radius of curvature of the composite beam is given by (8.38)

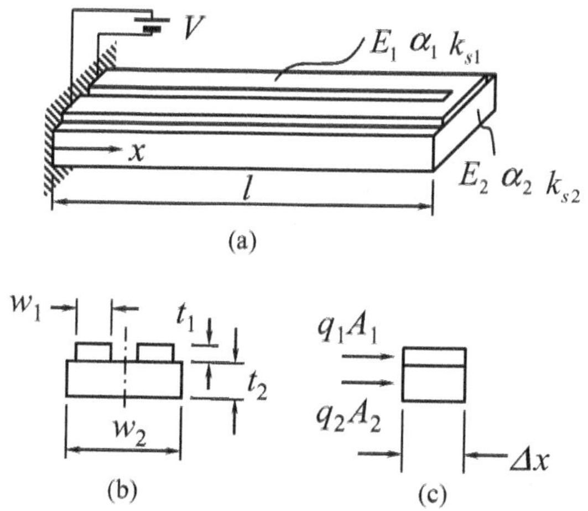

Figure 8.22 Thermal actuator with a thin-film heater.

$$\rho = \frac{t}{2\varepsilon}\left[1 + \frac{1+pqr}{3(1+p)^2}\left(p^2 + \frac{1}{pqr}\right)\right] \tag{8.142}$$

where t denotes the total thickness given by $t = t_1 + t_2$, ε stands for the relative strain, and p, q, and r denote the ratios given by (8.31), which can be written for the geometry in Fig. 8.22b as follows:

$$p = \frac{t_1}{t_2} \qquad q = \frac{2w_1}{w_2} \qquad r = \frac{\tilde{E}_1}{\tilde{E}_2} \tag{8.143}$$

where $2w_1$ is the total width of the upper layer. If the temperature at a point in the bimetallic actuator is changed from the initial temperature T_0 to a temperature T, the relative strain σ at a point is given by (8.78):

$$\varepsilon = (\alpha_2 - \alpha_1)(T - T_0) \tag{8.144}$$

and the following moment–curvature, (e) in Example 8.2, is used to obtain the deflection of the actuator:

$$\frac{d^2y}{dx^2} = \frac{M_{\text{eff}}}{\left(\tilde{E}I\right)_{\text{eff}}} = \frac{1}{\rho} \tag{8.145}$$

Upon substitution of (8.142) into equation (8.145), we have

$$\frac{d^2y}{dx^2} = \frac{1}{t}\frac{2}{1 + \frac{1+pqr}{3(1+p)^2}\left(p^2 + \frac{1}{pqr}\right)}\varepsilon \tag{8.146}$$

which states that the curvature $1/\rho = d^2y/dx^2$ is proportional to the relative strain ε. Combining (8.144) and (8.146), we find the temperature–curvature equation of the bimetallic actuator shown in Fig. 8.22 as follows:

$$\frac{d^2y}{dx^2} = \frac{\alpha_2 - \alpha_1}{t} \frac{2}{1 + \frac{1+pqr}{3(1+p)^2}\left(p^2 + \frac{1}{pqr}\right)}(T - T_0) \tag{8.147}$$

If the temperature difference, $T - T_0$, is given as a function of the coordinate x, the preceding differential equation can be solved. Since we assume that the heat transfer through surrounding gas is neglected, the heat flow from the right end of the actuator in Fig. 8.22a can be considered as zero; $dT/dx = 0$ at the right end. To perform the preceding integration, we then use the temperature distribution of (8.106), which is rewritten as follows:

$$T = \frac{\dot{Q}_t l^2}{2k_s A}\left[2\frac{x}{l} - \left(\frac{x}{l}\right)^2\right] + T_0 \tag{8.148}$$

where $\dot{Q}_t = V^2/2Rl$ is the heat generated per unit length and R is the resistance between the ends of the upper layer. In the foregoing temperature distribution, the thermal conductivity k_s plays an important role, and then we must express the thermal conductivity in terms of k_{s1} and k_{s2}. To obtain k_s, we first define the cross-sectional areas as follows: $A_1 = 2w_1t_1$, $A_2 = w_2t_2$, and $A = A_1 + A_2 = 2w_1t_1 + w_2t_2$. Let us represent the total flow rate and the heat flow rates through the upper and lower layers by q, q_1, and q_2, respectively. The total heat flow rate q in the x direction in Fig. 8.22c can be written as $q = q_1 + q_2$, where the subscripts 1 and 2 denote the upper and lower layers, respectively. The total heat flow rate is then expressed as a sum of the upper and lower heat flow rate:

$$-k_s \frac{\Delta T}{\Delta x} A = -k_{s1} \frac{\Delta T}{\Delta x} A_1 + \left(-k_{s2} \frac{\Delta T}{\Delta x}\right) A_2$$

where Δx, ΔT, k_s, k_{s1}, and k_{s2} denote the length of the element in Fig. 8.22c, the temperature difference across Δx, the thermal conductivity of the actuator, and the thermal conductivities of the upper and lower layers, respectively. Rearranging the terms in the preceding equation, the thermal conductivity k_s is obtained as

$$k_s = \frac{k_{s1}A_1 + k_{s2}A_2}{A} \tag{8.149}$$

which may be considered as an effective thermal conductivity of the beam along the longitudinal direction.

Substituting (8.148) into (8.147), we have

$$\frac{d^2y}{dx^2} = \frac{(\alpha_2-\alpha_1)\dot{Q}_t l^2}{tk_s A} \frac{1}{1+\frac{1+pqr}{3(1+p)^2}\left(p^2+\frac{1}{pqr}\right)} \left[2\frac{x}{l}-\left(\frac{x}{l}\right)^2\right]$$

$$= D_1\left[2\frac{x}{l}-\left(\frac{x}{l}\right)^2\right] \tag{8.150}$$

where

$$D_1 = \frac{(\alpha_2-\alpha_1)\dot{Q}_t l^2}{tk_s A} \frac{1}{1+\frac{1+pqr}{3(1+p)^2}\left(p^2+\frac{1}{pqr}\right)} \tag{8.151}$$

We integrate the foregoing differential equation twice and use the boundary conditions $y = 0$ and $dy/dx = 0$ at $x = 0$ to give the deflection of the actuator,

$$y = \frac{D_1 l^2}{12}\left(\frac{x}{l}\right)^3\left(4-\frac{x}{l}\right) \tag{8.152}$$

Substitution of (8.151) into equation (8.152) yields the deflection y of the bimetallic actuators:

$$y = \frac{(\alpha_2-\alpha_1)\dot{Q}_t l^4}{12tk_s A} \frac{1}{1+\frac{1+pqr}{3(1+p)^2}\left(p^2+\frac{1}{pqr}\right)} \left(\frac{x}{l}\right)^3\left(4-\frac{x}{l}\right) \tag{8.153}$$

where $\alpha_2 - \alpha_1$ represents the difference of coefficients of thermal expansion of the upper and lower layers, $\dot{Q}_t$ denotes the heat generated per unit length given by $\dot{Q}_t = V^2/2Rl$, l stands for the length of the bimetallic actuator, t represents the total thickness given by $t_1 + t_2$, k_s denotes the effective thermal conductivity (8.149), and p, q, and r represent the ratios given by (8.143). We rewrite the foregoing equation in a simpler form,

$$y = \frac{D}{3}\left(\frac{x}{l}\right)^3\left(4-\frac{x}{l}\right)V^2 \tag{8.154}$$

where

$$D = \frac{(\alpha_2-\alpha_1)l^3}{8Rtk_s A} \frac{1}{1+\frac{1+pqr}{3(1+p)^2}\left(p^2+\frac{1}{pqr}\right)} \tag{8.155}$$

is a constant of the bimetallic actuator in m/V^2. The maximum deflection is obtained at the free end of the actuator:

$$y_l = y|_{x=l} = DV^2 \tag{8.156}$$

Joule-Heated Structures

For various purposes, elastic beams attached to a relatively large structure may be heated by Joule heating. Typical configurations are illustrated in Fig. 8.23a and b. The central planar structure in Fig. 8.23a is supported by four slender straight beams of length l, and the structure in Fig. 8.23b is suspended by four L-shaped beams of length l; in general, other flexures can be used (e.g., the flexures discussed in Chapter 4). First we obtain the temperature distribution of the structures and next, obtain the stress in the beams and stiffness of the structures. Because the structures in Fig. 8.23 are symmetric about the two lines passing through its center, we can set up a model (Fig. 8.24a) that represents a quarter of the structures in Fig. 8.23. In Fig. 8.24a, the plate of area A_{p1} is suspended by a beam of length l attached on a substrate of temperature T_0. The initial temperature of the structure is equal to that of the substrate,

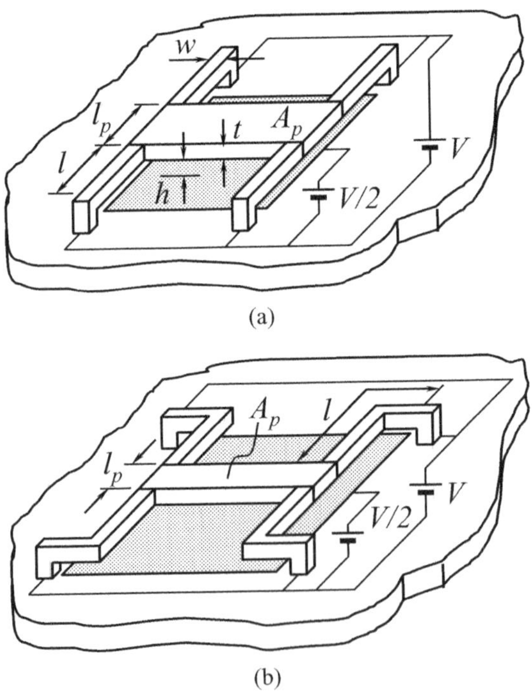

Figure 8.23 Structures heated by Joule heating.

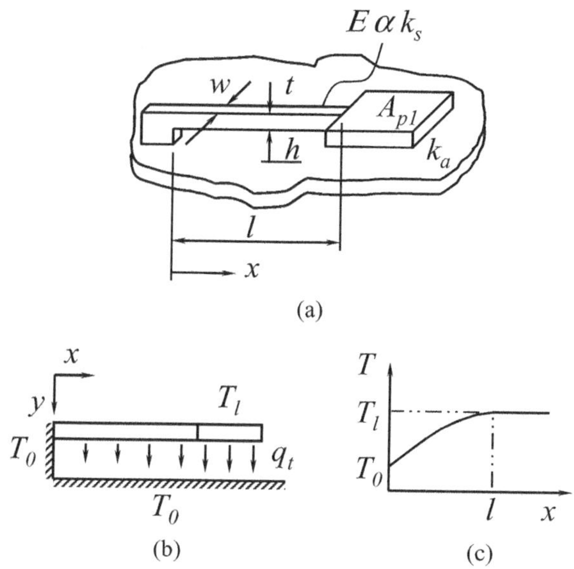

Figure 8.24 Equivalent model of structures heated by Joule heating.

and the structure is heated by Joule heating. The temperature is illustrated in Fig. 8.24b and c, where T_0, T_l, and q_t denote the temperatures at the left and the right ends of the beam and heat transfer rate from the structure. Since the cross-sectional area of the plate is much greater than that of the beam, the following is assumed: (1) the electrical resistance of the beam is much larger than that of the plate; (2) the heat generated by Joule heating in the beam is much greater than that in the plate; (3) the temperature in the plate remains constant while that in the beam varies with the longitudinal direction. We assume further that the heat transferred from the sides of the plate is neglected since area A_{p1} of the plate is much greater than that of the edges. Under this assumption, the temperature distribution is shown in Fig. 8.24c.

Referring to the geometric dimensions and material properties shown in Fig. 8.24, we obtain the temperature distribution of the beam. Since the heat transfer rate from the right end of the beam is equal to that transferred from the plate to the substrate, the energy balance is written

$$q_l = -k_a \frac{\partial T}{\partial y} A_{p1} \tag{8.157}$$

where q_l represents the heat transfer rate from the right end of the beam, and A_{p1} denotes the cross-sectional area, wt, of the beam and the plate area, given by $A_p/4$, respectively. The temperature distribution of the beam was obtained in (8.112) and the heat transfer rate q_l is obtained from (8.114) as follows:

$$q_l = -k_s m A\left[(T_l - T_0)\frac{1}{\tanh ml} + \frac{p}{m^2}\frac{1-\cosh ml}{\sinh ml}\right] \tag{8.158}$$

where $A = wt$. Substituting equation (8.158) and $\partial T/\partial y = (T_0 - T_l)/h$ into (8.157) and using the definition of m, (8.109), we find the temperature T_l as follows:

$$\begin{aligned} T_l &= T_0 + \frac{p}{m^2}\frac{\cosh ml - 1}{\cosh ml}\frac{1}{1+\dfrac{\tanh ml}{ml}\dfrac{k_a A_{p1}}{k_s A}\dfrac{l}{h}} \\ &= T_0 + \frac{p}{m^2}\frac{\cosh ml - 1}{\cosh ml}\frac{1}{1+(A_{p1}/lh)ml\tanh ml} \end{aligned} \tag{8.159}$$

In the limit as ml approaches zero, equation (8.159) is reduced to the following equation:

$$\begin{aligned} T_l &= \lim_{m\to 0}\left[T_0 + pl^2\frac{\cosh ml - 1}{(ml)^2}\frac{1}{\cosh ml}\frac{1}{1+(ml\tanh ml)(A_{p1}/lh)}\right] \\ &= T_0 + \frac{pl^2}{2} = T_0 + \frac{\dot{Q}_t l^2}{2k_s A} \end{aligned} \tag{8.160}$$

where $\dot{Q}_t$ is the heat generated per unit length. To obtain the last expression, the definition of p, (8.109), is used. It is noted that equation (8.160) is the same as (8.107), which may be considered as the temperature distribution of the cantilever beam in vacuum.

We use the preceding temperature distribution along the beams to find the thermal stress in the beam. Let us remove the upper anchors in Fig. 8.23a (while the temperature distribution is maintained) to allow the structures to expand, and then the thermal expansion δ of one beam in the longitudinal direction of the beams is obtained by substituting $T_s = T_0$ into (8.122):

$$\begin{aligned} \delta_l = \alpha l\Bigg\{&(T_l - T_0)\frac{1}{ml}\frac{\cosh ml - 1}{\sinh ml} \\ &+ \frac{pl^2}{(ml)^2}\left[1 - \frac{1}{ml}\sinh ml + \frac{1}{ml}\frac{(\cosh ml - 1)^2}{\sinh ml}\right]\Bigg\} \end{aligned} \tag{8.161}$$

The total beam expansion δ_t in the longitudinal direction is expressed as the sum of expansions due to the two beams and the plate:

$$\begin{aligned} \delta_t = 2\delta + \delta_p = \alpha\Bigg[2l\Bigg\{&(T_l - T_0)\frac{1}{ml}\frac{\cosh ml - 1}{\sinh ml} \\ &+ \frac{pl^2}{(ml)^2}\left[1 - \frac{1}{ml}\sinh ml + \frac{1}{ml}\frac{(\cosh ml - 1)^2}{\sinh ml}\right]\Bigg\} + l_p(T_l - T_0)\Bigg] \end{aligned} \tag{8.162}$$

To return the upper end of the beam to the initial position, the following external stress should be applied to the end of the beam:

$$\sigma = \tilde{E}\varepsilon = \tilde{E}\left(-\frac{\delta_t}{2l}\right) = -\tilde{E}\frac{\delta_t}{2l} \tag{8.163}$$

where $\tilde{E}$ and l denote the effective Young's modulus defined by (3.5) and (3.6) and the length of the straight beam shown in Fig. 8.23. Since the plate is considered as a body whose stiffness is much greater than that of the beam, the strain ε in equation (8.163) is given by $-\delta_t/2l$ (i.e., the central plate is thought of as a rigid body). The preceding stress is considered as the thermal stress of the structure in Fig. 8.23a, and we then obtain the thermal stress by substituting (8.162) into equation (8.163) as follows:

$$\sigma = -\tilde{E}\alpha\left[\left\{(T_l - T_0)\frac{1}{ml}\frac{\cosh ml - 1}{\sinh ml} + \frac{pl^2}{(ml)^2}\left[1 - \frac{1}{ml}\sinh ml + \frac{1}{ml}\frac{(\cosh ml - 1)^2}{\sinh ml}\right]\right\} + \frac{l_p}{2l}(T_l - T_0)\right] \tag{8.164}$$

where T_l is given by (8.159). If the data in Fig. 8.23a are given, we can evaluate the stress in the beam from (8.164). If ml approaches zero, equation (8.164) is reduced to the following simple form:

$$\begin{aligned}\sigma &= -\tilde{E}\alpha\left[(T_l - T_0)\frac{1}{2} + pl^2\frac{1}{12} + \frac{l_p}{2l}(T_l - T_0)\right]\\ &= -\tilde{E}\alpha\left[\frac{pl^2}{12} + \frac{T_l - T_0}{2}\left(1 + \frac{l_p}{l}\right)\right]\end{aligned} \tag{8.165}$$

Substituting (8.160) and the second equation of (8.109) into equation (8.165), we find that

$$\sigma = -\tilde{E}\frac{\alpha\dot{Q}_t l^2}{12k_s A}\left[1 + 3\left(1 + \frac{l_p}{l}\right)\right] = -\tilde{E}\frac{\alpha\dot{Q}_t l^2}{3k_s A}\left(1 + \frac{3l_p}{4l}\right) \tag{8.166}$$

which is the thermal stress in the beams as ml approaches zero. If the residual stress σ_r exists before Joule heating, the total stress is obtained as $\sigma_0 = \sigma_r + \sigma$, where σ denotes the thermal stress given by (8.164) or (8.166). It is noted that the thermal stress (8.166) is considered as the maximum compressible stress in the range $0 \leq ml \leq \infty$.

The linear and cubic stiffnesses of the structure with four straight beams in Fig. 8.23a is obtained from (n) and (o) in Example 3.3:

$$k_1 = 4\left(12\frac{\tilde{E}I}{l^3} + \frac{6}{5}\frac{N_0}{l}\right) = 48\frac{\tilde{E}I}{l^3} + \frac{24}{5}\frac{N_0}{l} \tag{8.167}$$

$$k_3 = 4\frac{18}{25}\frac{\tilde{E}A}{l^3} = \frac{72}{25}\frac{\tilde{E}A}{l^3} \tag{8.168}$$

where I, A, and N_0 denote the moment of inertia, the cross-sectional area of the beam given by wt, and the axial force defined as $\sigma_0 A$, where σ_0 is the stress in the beam, respectively. The moment of inertia is given by $I = w^3t/12$ for motion parallel to the substrate and $I = t^3w/12$ for motion perpendicular to the substrate. The resorting force of the structure in Fig. 8.23a is expressed as

$$F_r = k_1\delta + k_3\delta^3 \tag{8.169}$$

where δ is the displacement of the central plate. It is noted that since the stress (or axial force) in the beam is controlled by the applied voltage, the preceding restoring force is adjusted by the voltage. When the linear stiffness k_1 is equal to zero, the buckling phenomenon occurs; that is,

$$N_0 = -10\frac{\tilde{E}I}{l^2}$$

or

$$\sigma_0 = -10\frac{\tilde{E}I}{Al^2} \tag{8.170}$$

It is noted that the beams are buckled in the direction in which the moment inertia is lower than that in the other direction: parallel to the substrate for $w < t$; perpendicular to the substrate for $t < w$. After the beam has buckled, the structure responses are obtained by using the theory developed in Chapters 4 and 5. For a structure with L-shaped beams (or any bent beams) in Fig. 8.23b, the residual stress is negligible, as discussed in Chapter 4 since the L-shaped beam is flexible in the three orthogonal directions (compared to the straight beam in the longitudinal direction). As a result, the stiffness of the structure in Fig. 8.23b is not affected by Joule heating and can be evaluated by using the theory developed in Chapter 4.

Residual Stress Varying in Cross Section

As described so far, the residual stress is added to the other stresses, such as the thermal stress to obtain the total stress. If the residual stress is uniform in cross section, evaluation of the total stress is simple. However, as the residual stress varies in the cross section of a beam, the residual stress must be evaluated to perform further calculation. The residual stress also plays a role when a cantilever beam with a stress gradient in the cross section is released; the cantilever released is curled from its initial position, as shown in Fig. 8.25c. The residual stress distribution is assumed to be linear since the process condition remains constant (e.g., the gases and temperature during the deposition of structural layer during surface micromachining are not changed). We assumed further that the cross section of the beam is symmetric about the y axis shown in Fig. 8.25a. Under these assumptions, the stress, σ_x, of the fixed–fixed beam in Fig. 8.25a (or films deposited on a substrate) can be expressed as

$$\sigma_x = \sigma_0 + sy \tag{8.171}$$

where σ_0, s, and y denote a constant stress, a stress gradient, and the coordinate y, as shown in Fig. 8.25a, respectively. The beam then applies the following force to the rigid wall:

$$F = \int_A \sigma_x \, dA = \int_A (\sigma_0 + sy) \, dA = \sigma_0 A + s \int_A y \, dA$$

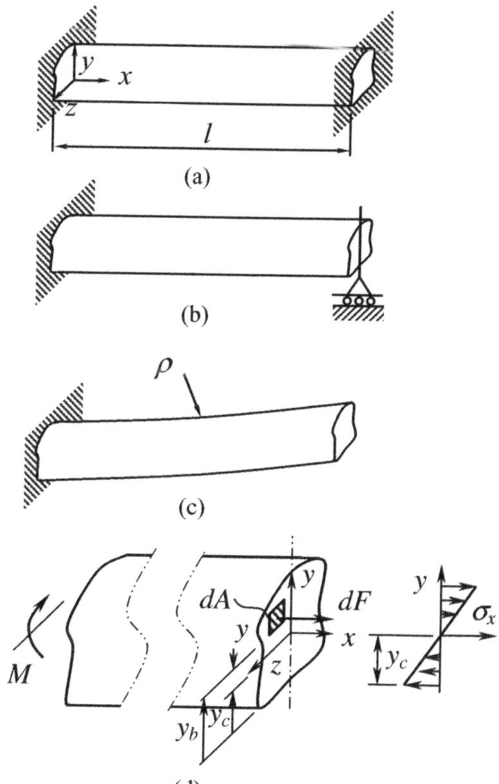

Figure 8.25 Residual stress.

If y is the coordinate from the centroid of cross section, then the integral on the right-hand side in the preceding equation vanishes and the preceding force becomes

$$F = \sigma_0 A \tag{8.172}$$

which is applied to the beam and walls. We now replace the right rigid wall with the roller shown in Fig. 8.25b. Since the roller cannot resist a force in the x direction, the right end of the beam expands or contracts in the x direction until the preceding force becomes zero. This expansion (or contraction) of the beam can be interpreted as meaning that a force of the same magnitude but opposite direction is applied to the right end. We then obtain the beam expansion in the longitudinal direction:

$$\delta = \frac{-(F)l}{\tilde{E}A} = -\frac{\sigma_0}{\tilde{E}} l \tag{8.173}$$

which states that if the stress σ_0 is compressive (i.e., negative), the beam will expand. Even though the beam is released in the longitudinal direction as shown in Fig. 8.25b, the beam is still subjected to a moment due to the stress gradient s. Referring to the free-body diagram of the beam in Fig. 8.25d, we shall obtain the moment due to the roller. From the force equilibrium condition

$$\sum M = M + \int_A y(\sigma_x - \sigma_0)\,dA = 0$$

we obtain the moment as

$$M = -\int_A y(\sigma_x - \sigma_0)\,dA = -s\int_A y^2 dA$$

In the preceding equation, the coordinate y is defined as $y = y_b - y_c$, where y_b is the coordinate in Fig. 8.25d and the location of the centroid from the bottom of the beam, which is given by

$$y_c = \frac{\int_A y_b\,dA}{A}$$

The foregoing moment may be written in a simple form:

$$M = -sI \tag{8.174}$$

where I represents the moment of inertia, given by

$$I = \int_A y^2 dA \tag{8.175}$$

When the roller in Fig. 8.25b is removed, the moment M acting on the beam vanishes and the beam is curled as shown in Fig. 8.25c. We can then obtain the effective moment M_{eff}, which is considered to result in the beam deformation in Fig. 8.25c as follows:

$$M_{\text{eff}} = -M = sI \tag{8.176}$$

Since we have the effective moment M_{eff}, we can obtain the radius of curvature ρ corresponding to the moment from the moment–curvature equation, (3.40), $\rho = \tilde{E}I/M_{\text{eff}}$:

$$\rho = \frac{\tilde{E}I}{sI} = \frac{\tilde{E}}{s} \tag{8.177}$$

It is worth noting that when a straight beam with linear residual stress is released as shown in Fig. 8.25c, the residual stress, (8.171), vanishes; σ_0 in (8.171) vanishes in Fig. 8.25b, and the remaining residual stress sy is canceled away in Fig. 8.25c since the moment acting on the beam becomes zero (i.e., $M_{\text{eff}} - M = 0$). Therefore, the stress of the released beam is evaluated as $\sigma = sy + \sigma_{\text{eff}}$, where σ_{eff} is given by (3.41), $\sigma_{\text{eff}} = -M_{\text{eff}}y/I$. Combining the preceding stress and (8.176), we find the stress in the beam which is released in Fig. 8.25c:

$$\sigma = sy - \frac{sI}{I}y = 0$$

In other words, if the residual stress is distributed linearly in the fixed–fixed beam in Fig. 8.25a, the stress in the beam becomes zero and is curled upon release of the beam.

Since $1/\rho = dy^2/dx^2$ for large ρ, (8.177) may be written as

$$\frac{d^2y}{dx^2} = \frac{s}{\tilde{E}} \tag{8.178}$$

which is a curvature–moment relationship, corresponding to (3.40). When we solve equation (8.178) with the boundary conditions for the cantilever in Fig. 8.25c: $y = 0$ and $dy/dx = 0$ at $x = 0$, we have the deflection y:

$$y = \frac{s}{2\tilde{E}}x^2 \tag{8.179}$$

which gives the maximum deflection y_l at $x = l$ as follows:

$$y_l = \frac{s}{2\tilde{E}} l^2 \tag{8.180}$$

If the maximum deflection is measured, the stress gradient s in the unreleased fixed–fixed beam is obtained as

$$s = \frac{2\tilde{E}}{l^2} y_l \tag{8.181}$$

When the radius curvature is measured, the stress gradient s is evaluated from (8.177):

$$s = \frac{\tilde{E}}{\rho} \tag{8.182}$$

PROBLEMS

8.1 The cantilever beam in the Fig. P8.1 is subject to a force at its free end; the related data are: $t_1 = 0.5\,\mu\text{m}$, $t_2 = 2\,\mu\text{m}$, $w_1 = w_2 = 5\,\mu\text{m}$, $l = 200\,\mu\text{m}$, $E_1 = 110\,\text{GPa}$, $E_2 = 160\,\text{GPa}$, $\nu_1 = 0.25$, and $\nu_2 = 0.27$, where ν denotes Poisson's ratio. A force of $4\,\mu\text{N}$ is applied to the end of the beam in the y direction. Find the effective flexural rigidity of the beam and the maximum deflection.

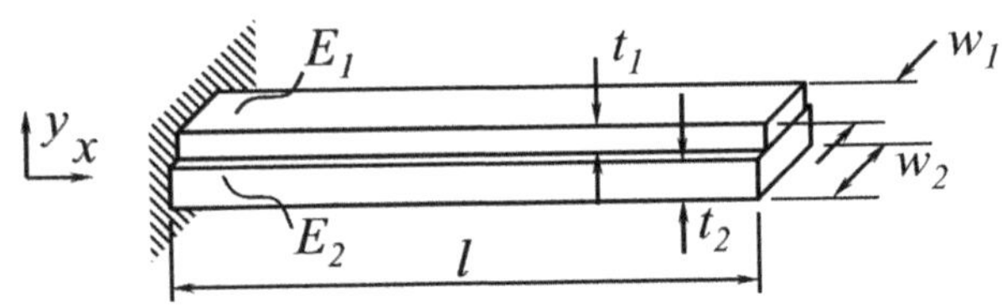

Figure P8.1

8.2 The upper layer of the cantilever beam in Problem 8.1 is a piezoelectric material whose d_{31} is $-5.43 \times 10^{-12}\,\text{m/V}$. If a voltage of 2 V is applied between the top and bottom of the upper layer, determine the deflection at the free end.

8.3 The coefficients of thermal expansion of the upper and lower layers in Problem 8.1 are given by $4.5 \times 10^{-6}\,{}^\circ\text{C}^{-1}$ and $2.6 \times 10^{-6}\,{}^\circ\text{C}^{-1}$, respectively. Find the deflection of the free end when the beam undergoes a temperature change of 100°C.

8.4 If the beam of Problems 8.2 and 8.3 experiences the voltage and temperature changes at the same time, determine the deflection of the free end of the beam.

8.5 At a distance of 20 μm from the fixed end of the beam in Problem 8.1, a piezoresistor will be formed. If a force is applied to the free end in the y direction, find a relation between the strain at the piezoresistor site and the force applied.

8.6 A bimetallic actuator with a tip is placed in an oven and is arranged as shown in Fig. P8.6. The displacement of the tip is given by $y = \beta \Delta T$, where β and ΔT denote a constant and the temperature change. At ΔT_1, the tip touches the upper cantilever beam, which is made of a uniform material and remains straight as the temperature increases by ΔT. The flexural rigidities of both cantilevers are EI. If the entice structure undergoes an additional temperature change of $2\Delta T$, find the deflection of the free end of the upper beam.

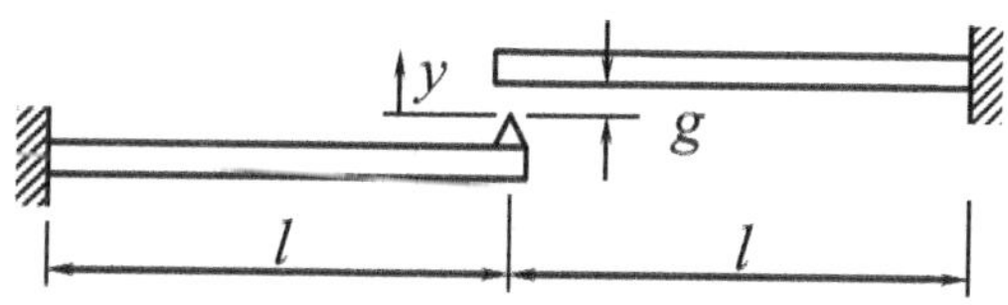

Figure P8.6

8.7 A V-shaped thermal actuator in (Fig. 8.21a) is fabricated. The related data are the resistance of the beam $R = 500\,\Omega$, length $l = 100\,\mu$m, width $w = 4\,\mu$m, thickness $t = 5\,\mu$m, gap $h = 2\,\mu$m, angle $\theta = 20°$, Young's modulus $E = 150$ GPa, and coefficient of thermal expansion $\alpha = 2.6 \times 10^{-6}\,°C^{-1}$, thermal conductivity of beam $k_s = 20$ W/m·°C, thermal conductivity of surrounding gas $k_a = 0.026$ W/m·°C, and room temperature 20°C. Find the displacement δ of the vertex of the beam actuator in the form $\delta = \gamma V^2$, where γ and V denote a constant and the voltage applied between the ends of the beam.

8.8 A surface-micromachined cantilever beam of length 100 μm, width 3 μm and thickness 2 μm is released. The free end of the beam is displaced upward by 10 μm during the release process. If Young's modulus of the beam material is 120 GPa, evaluate the stress gradient in the beam before and after release.

CHAPTER 9

ELECTROSTATIC AND ELECTROMAGNETIC ACTUATORS

Electrostatic actuators utilizing electrostatic attractive force have been used widely in MEMS because their fabrication processes are simple compared to those of other actuators and are compatible with that of existing integrated circuits; that is, an electrostatic actuator can be fabricated with electronic circuits for operating the actuator using the micromachining technologies discussed in Chapter 2. Electromagnetic actuators employing the electromagnetic force are also used. In this chapter we describe a variety of electrostatic and electromagnetic actuators and derive closed-form solutions for their static and dynamic responses.

9.1 ELECTROSTATIC ACTUATORS

Typical electrostatic actuators are shown in Fig. 9.1, in which the structural layers are anchored on substrates and allowed to move in specific directions. In the 1960s, Nathanson et al. demonstrated the cantilever beam with an electronic circuit on a substrate which is illustrated in Fig. 9.1a. The cantilever beam was driven by ac voltage with dc bias and served as a filter or an amplifier for an electronic device. Since the demonstration of the electrostatic cantilever beam actuator with an operating circuit, research on other types of electrostatic actuators as shown in Fig. 9.1b–e has been conducted. A fixed–fixed beam both ends of which are fixed on a substrate (Fig. 9.1b) are actuated electrostatically by applying a time-varying voltage between the beam and a

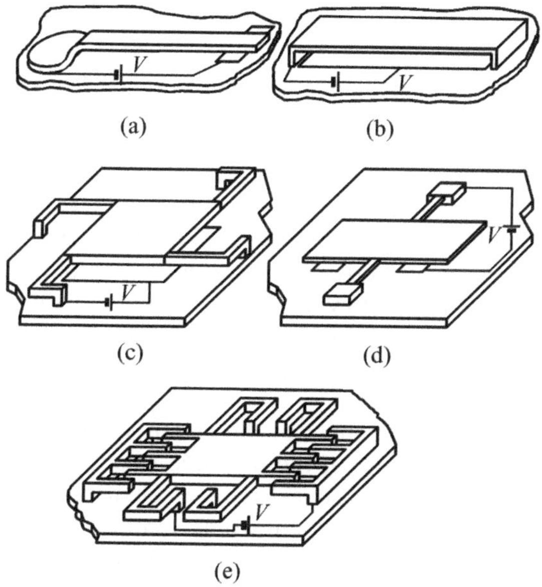

Figure 9.1 Electrostatic actuators.

lower electrode placed on the substrate. The beams of the actuators shown in Fig. 9.1a and b are continuously deformed by the electrostatic force, and thus the deflection and angular displacement of a point of the beam varies along the beam. Then analysis of the electrostatically driven beam actuators becomes difficult since the mechanical behavior and electrical behavior of the actuators in Fig. 9.1a and b are coupled with each other (i.e., the beam actuators became highly nonlinear).

The parallel-plate actuator in Fig. 9.1c, which has 90° rotational symmetry about the center of the plate, acts as an electrostatic actuator. A movable plate of the parallel-plate actuator is initially suspended by flexures over a stationary plate sitting on a substrate and displaced downward when a voltage is applied between the plates. Since the structure is symmetric (90° rotational symmetry in the case of Fig. 9.1c), the movable plate is displaced without any angular displacement, so the electrical and mechanical behavior of the parallel plate shown in Fig. 9.1c is simpler than that of the beam actuators in Fig. 9.1a and b. For this reason, the parallel-plate actuator in Fig. 9.1c can be considered as a model of highly complicated electrostatic actuators (Fig. 9.1a–d) if certain conditions (discussed later in the chapter) are satisfied. The torsional actuator in Fig. 9.1d, whose plate is suspended by two straight beams, is displaced angularly as the applied voltage increases. Since the torsional beam is compliant in the direction perpendicular to the substrate, the torsional plate is also displaced downward. When the electrostatic actuators shown in Fig. 9.1a–d are released during fabrication, their movable structures remain parallel to the substrates. For this reason, some researchers call them parallel-plate actuators or parallel plates. However, to explain the behavior of the actuators, we refer to them differently: cantilever beam actuator for (a), fixed–fixed beam actuator for (b), parallel-plate actuator for (c), and torsional actuator for (d). To stress the inherent characteristics of these actuators, the term *electrostatic* or *gap-closing* may be placed before the names. For example, the torsional actuator in Fig. 9.1d can be called an electrostatic gap-closing torsional actuator. All of the electrostatic actuators in Fig. 9.1a–d employ the electrostatic attractive force given by Coulomb's law, are usually operated at lower voltage (<10V in many cases), and show the jump phenomenon which occurs at a critical voltage (known as pull-in voltage) and causes mechanical contact of the suspended structure to the lower electrode or the substrate.

To avoid the mechanical contact of the electrostatic gap-closing actuators and to extend the travel range of the movable structure, Tang et al, demonstrated a comb structure, which consists of a set of interdigitated combs as shown in Fig. 9.1e; the structure is known as a *comb drive actuator* or *comb drive*. When a voltage is applied between the comb structures, the electrostatic attractive force is generated and causes the movable structure to move. A comb drive actuator does not show nonlinearity such as the jump phenomenon mentioned earlier. However, its electrostatic force is smaller than that of the electrostatic gap-closing actuators shown in Fig. 9.1a–d. As a result, the comb drive actuator, in general, needs high voltage to generate motion of the movable structure. To decrease the operating voltage, the comb drive actuator is usually operated in a low-pressure gas (or in vacuum) or at resonance. In subsequent sections we first describe a comb drive actuator and then illustrate the electrostatic gap-closing actuators since analysis of the comb drive actuator is much simpler than that of others.

9.2 COMB DRIVE ACTUATOR

A typical configuration for the comb drive actuator is illustrated in Fig. 9.2a. A movable plate with electrostatic combs is suspended by flexures and actuated in the direction parallel to the comb fingers when ac voltage with dc bias is applied between the combs. Another set of combs on the left-hand side is used as a capacitance sensor, which measures the movement of the movable plate. Using the theory described in preceding chapters, we can set up the model shown in Fig. 9.2b, where a mass m is supported by a spring of stiffness k and a damper of damping coefficient c. The electrostatic force for driving the mass is given by (7.114), whose magnitude is rewritten here for convenience:

$$F_{er} = n\frac{\varepsilon t}{g}V^2 \tag{9.1}$$

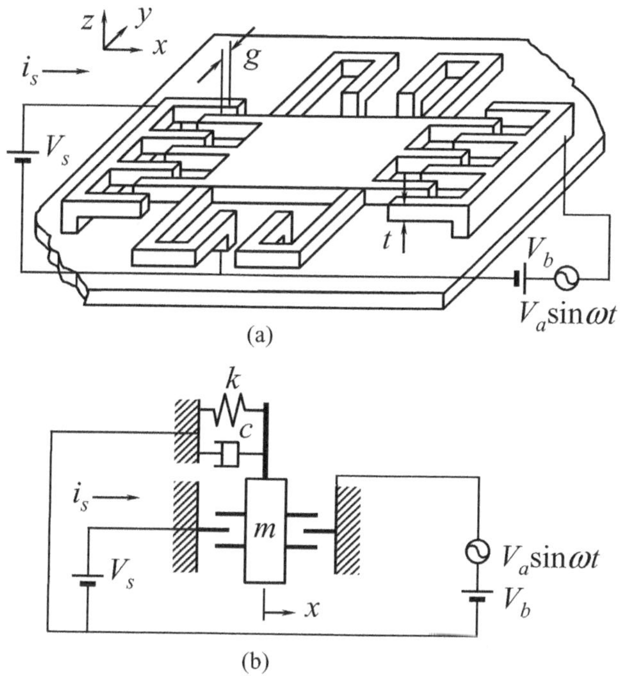

Figure 9.2 Electrostatic comb drive actuator.

where n, ε, t, g, and V represent the number of pairs of comb fingers, the permittivity, the thickness of the comb structures, the interfinger gap, and the actuation voltage, respectively. Similarly, the electrostatic force due to the sensing voltage V_s is written as

$$F_{el} = n\frac{\varepsilon t}{g}V_s^2 \tag{9.2}$$

Note from the preceding equations that the electrostatic forces are applied in the direction in which the capacitance of interdigitated combs increases (i.e., F_{er} and F_{el} acting on the mass are directed in the positive and negative x directions in Fig. 9.2b, respectively). Applying Newton's second law to the model in Fig. 9.2b gives the equation of motion of the movable structure:

$$m\ddot{x} + c\dot{x} + kx = F_e \tag{9.3}$$

where the dot represents the time derivative, x denotes the displacement of the mass, and F_e stands for the electrostatic force due to the voltage V and the sensing voltage V_s. If the voltage V is given by a sum of dc bias and ac voltage,

$$V = V_b + V_a \sin \omega t \tag{9.4}$$

where V_b, V_a, and ω denote the amplitudes of dc and ac voltage and the angular frequency, respectively, and if the sensing voltage remains constant, the time-varying force is obtained as

$$\begin{aligned} F_e = F_{er} - F_{el} &= n\frac{\varepsilon t}{g}(V_b + V_a \sin \omega t)^2 - n\frac{\varepsilon t}{g}V_s^2 \\ &= n\frac{\varepsilon t}{g}\left(V_b^2 + \frac{V_a^2}{2} - V_s^2 + 2V_b V_a \sin \omega t - \frac{V_a^2}{2}\cos 2\omega t\right) \end{aligned}$$

If $V_a << V_b$, the equation above can be approximated as

$$F_e \cong F_b + F_0 \sin \omega t \tag{9.5}$$

where

$$F_b = n\frac{\varepsilon t}{g}\left(V_b^2 + \frac{V_a^2}{2} - V_s^2\right) \cong n\frac{\varepsilon t}{g}(V_b^2 - V_s^2) \tag{9.6}$$

$$F_0 = 2n\frac{\varepsilon t}{g}V_b V_a \tag{9.7}$$

denote the amplitudes of the constant and sinusoidal forces, respectively. We may use the push-pull driving to generate higher force. For the push-pull operation, a voltage, $V = V_b - V_a \sin \omega t$, is applied to the left set of combs instead of the sensing voltage in Fig. 9.2. Application of equal and opposite sinusoidal voltages with the same bias voltage yields

$$\begin{aligned} F_e &= n\frac{\varepsilon t}{g}(V_b + V_a \sin \omega t)^2 - n\frac{\varepsilon t}{g}(V_b - V_a \sin \omega t)^2 \\ &= F_0 \sin \omega t \end{aligned} \tag{9.8}$$

where

$$F_0 = 4n\frac{\varepsilon t}{g}V_b V_a \tag{9.9}$$

is the amplitude of the time-varying force. It is noted that the push-pull driving gives the pure sinusoidal force. The voltage $V = -V_b + V_a \sin \omega t$ may be applied to the left set of combs for the push-pull driving to generate the same net force as (9.8).

When we substitute the preceding time-varying force into (9.3), we have the linear equation of motion for the displacement of the comb drive actuator as follows:

$$m\ddot{x} + c\dot{x} + kx = F_b + F_0 \sin \omega t \tag{9.10}$$

whose solution was obtained in Section 5.6. The solution of equation (9.10) is given by (5.76) as follows:

$$\begin{aligned} x = {} & \frac{F_b}{k} + e^{-\zeta\omega_n t}(A\cos\omega_d t + B\sin\omega_d t) \\ & + \frac{F_0}{m}\frac{1}{\sqrt{(\omega_n^2 - \omega^2)^2 + (2\zeta\omega_n\omega)^2}}\sin(\omega t - \phi) \end{aligned} \tag{9.11}$$

where the constants A and B are obtained from initial conditions. If $F_0 = 0$, equation (9.11) becomes $x = F_b/k$. If we are interested in the time-varying steady-state response, the foregoing solution reduces to

$$x = x_0 \sin(\omega t - \phi) \tag{9.12}$$

where

$$\begin{aligned} x_0 &= \frac{F_0}{k}\frac{1}{\sqrt{(1-\Omega^2)^2 + (\Omega/Q)^2}} \\ \Omega &= \frac{\omega}{\omega_n} = \frac{f}{f_n} \\ \omega_n &= \sqrt{\frac{k}{m}} \\ \phi &= \tan^{-1}\frac{\Omega}{Q(1-\Omega^2)} \\ Q &= \frac{\sqrt{mk}}{c} \end{aligned} \tag{9.13}$$

represent the vibration amplitude, the frequency ratio, the natural angular frequency, the phase, and the quality factor, respectively. If the steady-state response with bias is interested, the solution is obtained as $x = F_b/k + x_0 \sin(\omega t - \phi)$, which is written as

$$x = \frac{1}{k}\left[F_b + F_0\frac{1}{\sqrt{(1-\Omega^2)^2 + (\Omega/Q)^2}}\sin(\omega t - \phi)\right] \tag{9.14}$$

We can use the preceding solution for the displacement to obtain the output current, which is proportional to the displacement of the mass. Since the capacitance of the left set of combs in Fig. 9.2a is given by

$$C = 2n\frac{\varepsilon t(l-x)}{g} \tag{9.15}$$

where l is the initial overlap of the comb fingers, the electric charge is accumulated and is given by

$$Q = CV_s \tag{9.16}$$

Recalling that the electric current is defined by the time rate of charge, we obtain the current in the form

$$i_s = \frac{dQ}{dt} = \frac{d(CV_s)}{dt} = \frac{V_s dC}{dx} = V_s\frac{dC}{dx}\frac{dx}{dt} = V_s\frac{dC}{dx}\dot{x} \tag{9.17}$$

where $\dot{x}$ denotes the velocity of the plate. Substituting (9.14) into equation (9.17) leads to the steady-state sensing current,

$$i_s = V_s\frac{dC}{dx}\dot{x} = V_s\frac{dC}{dx}x_0\omega\cos(\omega t-\phi) \tag{9.18}$$

From (9.15) we obtain the sensitivity of the capacitance to the displacement as

$$\frac{dC}{dx} = \frac{d}{dx}\left[2n\frac{\varepsilon t(l-x)}{g}\right] = -2n\frac{\varepsilon t}{g} \tag{9.19}$$

Combining the first equation of (9.13) and (9.18) gives us

$$i = \frac{dC}{dx}V_s\frac{F_0}{k}\omega_n\frac{\Omega}{\sqrt{(1-\Omega^2)^2+(\Omega/Q)^2}}\cos(\omega t-\phi) \tag{9.20}$$

Substituting (9.19) and (9.7) into (9.20) leads to

$$\begin{aligned} i &= -\left(2n\frac{\varepsilon t}{g}\right)^2\frac{V_s V_b V_a}{k}\omega_n\frac{\Omega}{\sqrt{(1-\Omega^2)^2+(\Omega/Q)^2}}\cos(\omega t-\phi) \\ &= -\left(\frac{dC}{dx}\right)^2\frac{V_s V_b V_a}{k}\omega_n\frac{\Omega}{\sqrt{(1-\Omega^2)^2+(\Omega/Q)^2}}\cos(\omega t-\phi) \\ &= -i_0\cos(\omega t-\phi) \end{aligned} \tag{9.21}$$

where

$$i_0 = \left(2n\frac{\varepsilon t}{g}\right)^2 \frac{V_s V_b V_a}{k}\omega_n \frac{\Omega}{\sqrt{(1-\Omega^2)^2 + (\Omega/Q)^2}} \tag{9.22}$$

In the preceding time-varying current, the negative sign denotes that if the movable plate moves in the x direction in Fig. 9.2, the current flows out from the comb. At resonance (i.e., $\Omega = 1$ for $Q >> 1$), the preceding current amplitude i_0 is reduced to

$$i_0|_{\Omega=1} = \left(2n\frac{\varepsilon t}{g}\right)^2 \frac{V_s V_b V_a}{k}\omega_n Q \tag{9.23}$$

where the quantity in parentheses represents the sensitivity of the capacitance given by (9.19). Referring to the second, third, and fourth equations of (9.13), we have $\omega_n Q/k = 1/c$, where c denote the damping coefficient. Therefore, the preceding output current at resonance can be written

$$i_0|_{\Omega=1} = \left(2n\frac{\varepsilon t}{g}\right)^2 \frac{V_s V_b V_a}{c} \tag{9.24}$$

It should be noted from equation (9.24) that the output current i_0 is proportional to the square of the capacitance sensitivity and the product of the three voltages involved and inversely proportional to the damping coefficient. If the damping coefficient is reduced (e.g., the gas pressure decreases), the sensing current increases.

Electrostatic Stiffness and Frequency Tuning

Lee et al. (1998) proposed various structures and methods to tune the effective stiffness and natural frequency of structures by changing the electrostatic force, which is linearly proportional to the displacement of the movable comb structure. One of them is illustrated in Fig. 9.3a, where two sets of structures for tailoring the electrostatic force are used, which was dealt with in Section 7.5. The control voltage V_c is applied to the left and right sets of the comb to adjust the electrostatic force, (7.117), and the electrostatic stiffness, while the comb drive actuator, discussed in the preceding paragraphs, vibrates the movable structure in the x direction. Referring to the electrostatic force of the tuning comb, (7.117), we write the electrostatic forces, F_{etr} and F_{etl}, of the left and right sets of tuning combs as follows:

$$F_{etr} = n_t \frac{\varepsilon t}{g_0}\left[\left(\frac{g_0}{g_1} - 1\right)\frac{-x}{l} + 1\right]V_c^2 \tag{9.25}$$

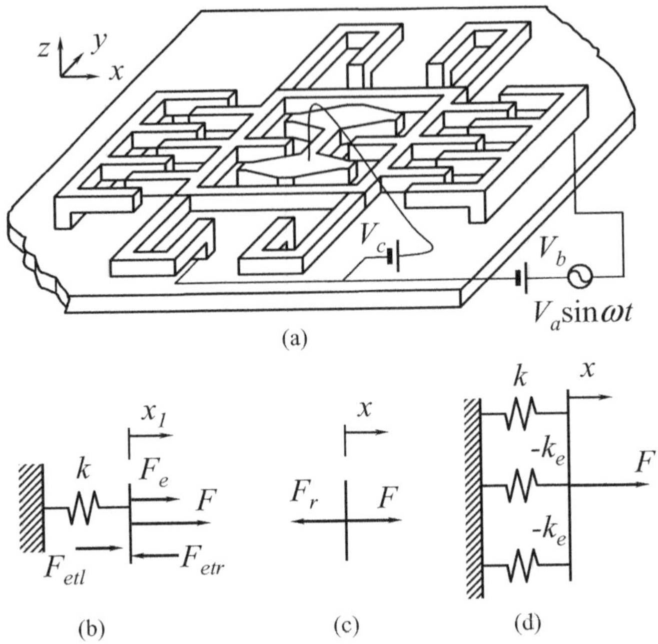

Figure 9.3 Electrostatic frequency tuning.

$$F_{etl} = n_t \frac{\varepsilon t}{g_0}\left[\left(\frac{g_0}{g_1} - 1\right)\frac{x}{l} + 1\right]V_c^2 \tag{9.26}$$

where n_t, g_0, g_1, l, and x denote the number of pairs of comb fingers in each set of tuning combs, the interfinger gaps at the start and end points of the width-varying fingers, the length of the width-varying portion, and the displacement of the movable structure, respectively. The electrostatic force due to the driving voltage V is given by (9.1), which is written

$$F_e = n\frac{\varepsilon t}{g}V^2 \tag{9.27}$$

Let us consider the restoring force due to the flexure and the electrostatic forces given by the preceding equations. If V and V_c are constants, the movable structure is in equilibrium at x_0. With reference to Fig. 9.3b, we can write the force equilibrium equation as follows:

$$F + F_e - F_{etr} + F_{etl} - k(x_0 + x_1) = 0$$

where F denotes a force applied to cause additional displacement x_1 from the equilibrium position x_0. Setting $x = x_0 + x_1$ and referring Fig. 9.3c, we obtain the restoring force as

$$F_r = F = kx - F_e + F_{etr} - F_{etl}$$

Substituting (9.25)–(9.27) into the foregoing equation gives

$$F_r = kx - n\frac{\varepsilon t}{g}V^2 - 2n_t\frac{\varepsilon t}{g_0}\left(\frac{g_0}{g_1} - 1\right)\frac{x}{l}V_c^2 \tag{9.28}$$

The effective stiffness of the frequency tuning structure in Fig. 9.3a is obtained by taking the first derivative with respect to x:

$$\begin{aligned} k_{\text{eff}} &= \frac{\partial F_r}{\partial x} = k - 2n_t\frac{\varepsilon t}{g_0}\left(\frac{g_0}{g_1} - 1\right)\frac{1}{l}V_c^2 \\ &= k - 2k_e \end{aligned} \tag{9.29}$$

where

$$k_e = n_t\frac{\varepsilon t}{g_0}\left(\frac{g_0}{g_1} - 1\right)\frac{1}{l}V_c^2 \tag{9.30}$$

represents the electrostatic stiffness of a set of tuning combs and is independent of the displacement x. The driving comb does not affect the effective stiffness above. It is worth noting that the effective stiffness due to the control voltage decreases as the control voltage increases. Since the preceding electrostatic stiffness can be considered a stiffness, the effective stiffness k_{eff} is obtained by formulas (4.1) and (4.2), for the parallel and series connections of springs. For example, the spring connection shown in Fig. 9.3d represents the springs due to the flexures and the two sets of tuning comb. Formula (4.1) for the parallel springs is written to obtain the effective stiffness,

$$k_{\text{eff}} = k - k_e - k_e = k - 2k_e$$

which is the same as (9.29). If the effective stiffness above is equal to zero or negative, the structure becomes unstable. Thus, the maximum control voltage is obtained by setting $k_{\text{eff}} = 0$:

$$V_{cm} = \sqrt{\frac{kg_0 l}{2n_t\varepsilon t(g_0/g_1 - 1)}} \tag{9.31}$$

The maximum control voltage V_{cm} is proportional to the square root of the stiffness k of the structure. For the frequency tuning structure shown in Fig. 9.3a, the maximum control voltage V_{cm} can be considered as a pull-in voltage at which the movable structure becomes unstable.

In general, if the restoring force is nonlinear, the nonlinear effective stiffness is then expressed as

$$k_{\text{eff}} = \frac{\partial F_r}{\partial x} = \frac{\partial F_s}{\partial x} - \frac{\partial F_e}{\partial x} \tag{9.32}$$

which represents the slope difference of the nonlinear forces at the equilibrium position. It should be noted that as the voltage for the attractive electrostatic increases, the effective stiffness decreases. However, if the electrostatic force acting on a structure is repulsive, the effective stiffness of the structure increases as the applied voltage increases (Lee and Cho, 2001).

Let us obtain the dynamic response of the electrostatic stiffness-tuning structure. Since the effective stiffness (9.29) is linear and the movable mass is actuated by the driving voltage, the equation of motion of the frequency tuning structure in Fig. 9.3a is obtained by substituting k_{eff} for k in (9.10):

$$m\ddot{x} + c\dot{x} + k_{\text{eff}}x = F_b + F_0 \sin \omega t \tag{9.33}$$

where F_b and F_0 are given by (9.6) and (9.7), rewritten as

$$F_b = n\frac{\varepsilon t}{g}V_b^2 \qquad F_0 = 2n\frac{\varepsilon t}{g}V_b V_a$$

Therefore, the nondecaying solution of equation (9.33) is obtained by substituting k_{eff} for k in (9.14) and related equations:

$$x = \frac{1}{k_{\text{eff}}}\left[F_b + F_0 \frac{1}{\sqrt{(1-\Omega^2)^2 + (\Omega/Q)^2}} \sin(\omega t - \phi)\right] \tag{9.34}$$

where Ω, Q, and ϕ are given by (9.13) as follows:

$$\Omega = \frac{\omega}{\omega_n} = \frac{f}{f_n} \qquad \omega_n = \sqrt{\frac{k_{\text{eff}}}{m}} \qquad \phi = \tan^{-1}\frac{\Omega}{Q(1-\Omega^2)} \qquad Q = \frac{\sqrt{mk_{\text{eff}}}}{c} \tag{9.35}$$

It should be noted that since the effective stiffness k_{eff} is adjustable, all the quantities in (9.34) and (9.35) can be changed. Furthermore, the time-varying force is a function of the dc bias and ac voltages. Therefore, the preceding steady-state response of the structure in Fig. 9.3a is expressed as $x = f(V_b, V_a, V_c)$.

Electrothermal Tuning of the Stiffness and Natural Frequency

Joule heating can be used to adjust the stiffness and natural frequency of the resonator shown in Fig. 9.4. The folded beams of the comb drive actuator in

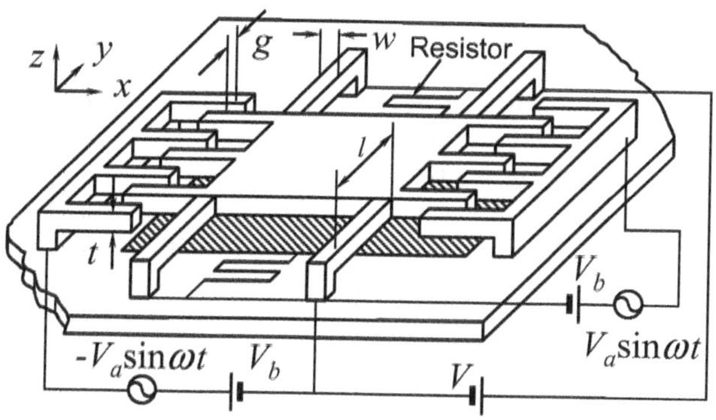

Figure 9.4 Electrothermal frequency tuning.

Fig. 9.2 are replaced with straight beams in Fig. 9.4, which are Joule-heated by applying a heating voltage V. The actuation voltages $V = V_b + V_a \sin \omega t$ and $V = V_b - V_a \sin \omega t$ for push-pull driving are applied to generate the vibration in the x direction as discussed earlier. It should be noted that a lower electrode whose voltage is $V/2$ is placed under the movable structure to avoid the collapse of the movable structure (the collapse phenomenon, known as *pull-in*, is covered in the following section). In Fig. 9.4, two serpentine resistors with large resistance are formed by using the same material as the lower electrode (e.g., polysilicon layer). The stiffness of the straight beam is obtained from (8.169) as follows:

$$k_{\text{eff}} = k_1 + k_3 x^2 \tag{9.36}$$

where k_1 and k_3 denote the linear and cubic stiffnesses given by (8.167) and (8.168), which are rewritten here:

$$k_1 = 48\frac{\tilde{E}I}{l^3} + \frac{24}{5}\frac{N_0}{l} \tag{9.37}$$

$$k_3 = \frac{72}{25}\frac{\tilde{E}A}{l^3} \tag{9.38}$$

$\tilde{E}$, I, l, N_0, and A represent the effective Young's modulus, the moment of inertia defined by $I = tw^3/12$, the beam length, the axial force, and the cross-sectional area given by $A = wt$, respectively. The axial force N_0 is expressed as

$$N_0 = (\sigma_0 + \sigma)A \tag{9.39}$$

where σ_0 and σ denote the residual stress and thermal stress, respectively. When the straight beam is Joule-heated, we can use (8.164) and (8.166) to evaluate the thermal stress σ. As a result, we can adjust the linear stiffness

k_1, (9.37), and the static and dynamic responses by controlling the heating voltage V.

Let us obtain the equation of motion for the static and dynamic response of the thermally controlled resonator in Fig. 9.4. The electrostatic force F_e acting on the movable plate is given by

$$F_e = n\frac{\varepsilon t}{g}\left(V_b - \frac{V}{2} + V_a \sin\omega t\right)^2 - n\frac{\varepsilon t}{g}\left(V_b - \frac{V}{2} - V_a \sin\omega t\right)^2 \\ \cong F_0 \sin\omega t \tag{9.40}$$

where

$$F_0 = 4n\frac{\varepsilon t}{g}\left(V_b - \frac{V}{2}\right)V_a \tag{9.41}$$

denotes the amplitude of the sinusoidal force, which may be written $F_0 = 4n\varepsilon tV_bV_a/g$ for $V << V_b$. Substituting k_{eff} for k in (9.3), we find the equation of motion:

$$m\ddot{x} + c\dot{x} + k_1 x + k_3 x^3 = F_b + F_0 \sin\omega t$$

where F_b and F_0 denote the constant force and the amplitude of the sinusoidal force, respectively. For the push-pull driving shown in Fig. 9.4, the equation above is reduced to

$$m\ddot{x} + c\dot{x} + k_1 x + k_3 x^3 = F_0 \sin\omega t \tag{9.42}$$

where F_0 is given by (9.41). This is Duffing's equation, whose solution was obtained in Section 5.11. If $\varsigma << 1$ and $3|k_r|/4\varsigma^2 << 1$ (where ζ and k_r is the damping factor, defined as $\zeta = c/2\sqrt{mk}$, and the nonlinearity factor, given by $k_r = k_3F_0^2/k_1^3$, both of which were dealt with in Chapter 5), the solution of (9.42) can be expressed as the linear response shown in (5.224) and (5.225); that is, if $\varsigma << 1$ and $3|k_r|/4\varsigma^2 << 1$ are satisfied, the nonlinear effect due to the cubic stiffness k_3 can be neglected. In this case the steady-state response x is obtained by substituting k_1 for k_{eff} in (9.34):

$$x = \frac{1}{k_1}\left[F_b + F_0\frac{1}{\sqrt{(1-\Omega^2)^2 + (\Omega/Q)^2}}\sin(\omega t - \phi)\right] \tag{9.43}$$

where Ω and Q represent the frequency ratio and the quality factor, both of which are defined in (9.35).

If the cubic stiffness k_3 is not negligible but plays a role, the solution of Duffing's equation derived in Section 5.11 may be used. The forcing term in the preceding equation of motion includes the sine function, while that in (5.195) includes the cosine function. Because the sine and cosine functions are periodic functions and we are interested in the steady-state solution, $\cos \omega t_1$ in (5.195) can be replaced by $\sin \omega t$ to obtain the solution of the preceding equation, and the steady-state solution is given by (5.196), rewritten here for convenience:

$$x = x_0 \sin(\omega t - \phi) \tag{9.44}$$

where x_0 and ϕ denote the vibration amplitude given by (5.212) and the phase that can be obtained from equation (5.204). The vibration amplitude x_0 is given by (5.212) as follows:

$$x_0 = X_0 \frac{F_0}{k_1} \tag{9.45}$$

where X_0 denotes the dimensionless amplitude defined in (5.211). Note that the response x_0 is nonlinear and depends on the force amplitude F_0, the mass m, the damping coefficient c, the linear stiffness k_1, and the cubic stiffness k_3. The dimensionless amplitude X_0 and the frequency ratio Ω at the maximum amplitude may be considered as characteristics of Duffing's equation and given by (5.216) and (5.218) respectively. If the system represented by Duffing's equation is slightly damped (i.e., $\zeta << 1$), X_0 and Ω at the maximum amplitude are given by (5.222) and (5.223), which are rewritten here for convenience:

$$X_{0m} = \sqrt{\frac{2}{3k_r}\left(-1 + \sqrt{1 + \frac{3k_r}{4\varsigma^2}}\right)}$$

$$\left(\frac{f}{f_n}\right)_m = \sqrt{1 - 2\varsigma^2 - \frac{1}{2}\left(1 - \sqrt{1 + \frac{3k_r}{4\varsigma^2}}\right)}$$

It is noted that since the stress in the beam is controlled by adjusting the voltage V in Fig. 9.4, the preceding amplitude and frequency are functions of the control voltage V. When the linear stiffness becomes negative, the movable mass vibrates about a new stable position. The vibration associated with this situation was discussed in Section 5.11, (5.226)–(5.232).

9.3 PARALLEL-PLATE ACTUATOR

As an electrostatic gap-closing actuator, the parallel-plate actuator in Fig. 9.5 consists of a movable plate, flexures to suspend the movable plate over a substrate, and a stationary plate placed on the substrate. When a voltage V

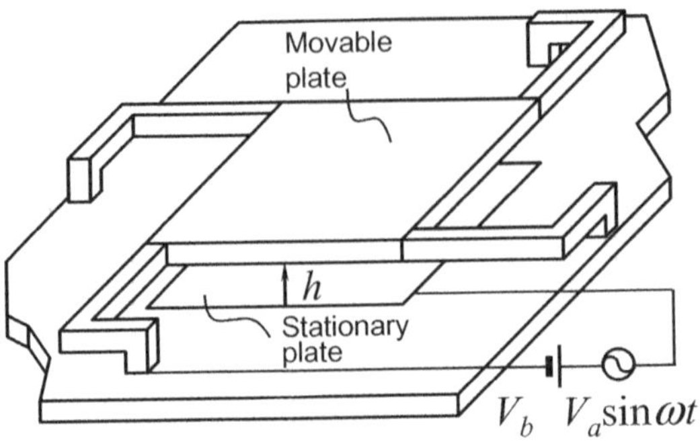

Figure 9.5 Parallel plate.

is applied between the movable and stationary plates, the movable plate subjected to the electrostatic attractive force moves down and then the gap h between the plates is reduced. In parallel-plate actuators, referred to in many publications as *electrostatic parallel plates*, the movable plate is initially set parallel to the stationary plate and moves into the stationary plate as the applied voltage increases (i.e., the movable plate can move without or with angular motion, depending on boundary conditions or the rigidities of the movable plates). For example, the flexures in Fig. 9.5 are much more compliant than the movable plate, and the structure has a 90° rotational symmetry about its center. As a result, the movable plate is considered to move downward to the stationary plate without angular motion upon applying the voltage V. However, if the movable structure of an electrostatic gap-closing actuator is nonsymmetric or the movable plate is flexible, the movable structure moves into the stationary plate and is angularly displaced. Since the governing equation of the electrostatic gap-closing actuators including angular motion is more complicated than that without angular motion, we first limit ourselves to an electrostatic gap-closing actuator without angular motion. Afterward, we develop theory for a gap-closing actuator with angular motion. For convenience, we classify nonlinear electrostatic gap-closing actuators into three groups: a parallel-plate actuator whose movable plate under an electrostatic attractive force moves into a stationary actuator without angular motion, as shown in Fig. 9.5; a torsional actuator whose movable plate moves torsionally; and a continuous structure actuator whose structure is elastically deformed when electrostatic force is applied to the structure. In later sections of this chapter we deal with the fixed–fixed and cantilever beam actuators as continuous structure actuators.

Static Response

Even though a typical parallel-plate actuator with a rectangular plate and L-shaped beams is shown in Fig. 9.5, the shapes of the movable plate and

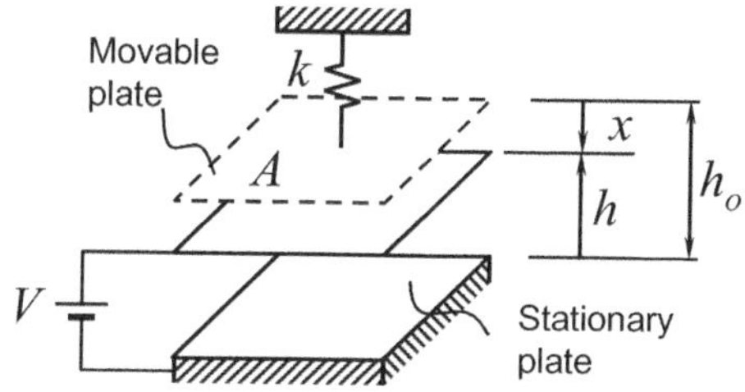

Figure 9.6 Static model of a parallel plate.

flexures may vary depending on purpose (e.g., the movable plate is hexagonal and the flexures are folded). Let us consider the model shown in Fig. 9.6 in order to analyze the static behavior of the parallel-plate actuator shown in Fig. 9.5. The rigid movable plate of area A is suspended by a linear spring of stiffness k and the stationary plate is spaced by an initial gap h_0 from the movable plate. As the voltage V between the plates increases from zero, the movable plate is displaced downward by x and then the interplate gap is defined to be $h = h_0 - x$. The electrostatic force acting on the movable plate is applied in the x direction in Fig. 9.6 and given by (7.88). The downward electrostatic force is rewritten as

$$F_e = \frac{1}{2}\frac{\varepsilon A}{h^2}V^2 \tag{9.46}$$

and the spring force acts as a restoring force, given by

$$F_s = kx = k(h_0 - h) \tag{9.47}$$

Before deriving an analytic solution for h (or x) from the force equilibrium, we first graphically examine the interplate gap h at the voltage V. The electrostatic and spring forces acting on the movable plate are shown in Fig. 9.7 when the displacement x is changed from 0 to h_0. Since the spring force (upward force), F_s, balances the electrostatic force (downward force), F_e, the intersections of the curves in Fig. 9.7 represent the equilibrium positions. When the voltage V is lower than a critical voltage known widely as the pull-in voltage (discussed subsequently), the electrostatic force (a solid curve) makes two intersections with the spring force (a straight line), labeled S and U; position S is the stable position since the effective stiffness (defined as the slope difference of the spring and electrostatic forces at the position) is positive, while position U is the unstable position because of the negative effective stiffness. If the movable plate under the voltage V is placed at position S, the movable plate stays at the position. However, if the movable plate is positioned at $x = x_U + \varepsilon_1$, where ε_1 denotes an infinitely small number, the movable plate will collapse into the stationary plate (since the force equilibrium condi-

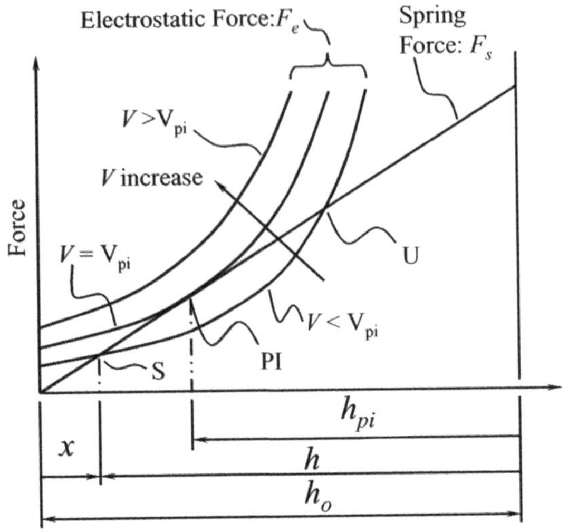

Figure 9.7 Electrostatic and spring forces.

tion is not satisfied and the effective stiffness at the point is negative). At the critical voltage V_{pi}, the electrostatic force curve is tangent to the spring force at $x = x_{pi}$, at which point the movable plate is in critical condition; in other words, if any disturbance occurs, the movable plate collapses into the stationary plate. For $V > V_{pi}$, the movable plate always collapses since there is no intersection or tangent (i.e., the electrostatic force is always greater than the spring force).

Let us obtain closed-form solutions for the stable position S and the critical position in Fig. 9.7. The force equilibrium condition, $F_e = F_s$, at the equilibrium position gives the force equilibrium equation as follows:

$$k(h_0 - h) = \frac{1}{2}\frac{\varepsilon A}{h^2}V^2 \tag{9.48}$$

which is a nonlinear equation representing the static behavior of the movable plate at the given voltage. It is more convenient to express equation (9.48) in dimensionless form:

$$(1 - H) = \frac{G}{H^2}$$

or

$$H^3 - H^2 + G = 0 \tag{9.49}$$

where

$$H = \frac{h}{h_0} \tag{9.50}$$

$$G = \frac{\varepsilon A}{2kh_0^3} V^2 \tag{9.51}$$

denote the dimensionless gap and the dimensionless force, respectively.

It is worth noting that equation (9.49) is a cubic equation which governs the behavior of the parallel-plate actuator. The exact solution of the general cubic equation $x^3 + ax^2 + bx + c = 0$ was described in Section 5.1. To obtain the roots of the cubic equation, we first examine the discriminant, (5.6), which is written here as $\Delta = 4a^3c - a^2b^2 - 18abc + 4b^3 + 27c^2$. Substituting $a = -1$, $b = 0$, and $c = G$ into the preceding discriminant equation gives $\Delta = -4G + 27G^2 = G(-4 + 27G)$. Setting $\Delta = 0$, which corresponds to the pull-in, labeled PI in Fig. 9.7, we find the dimensionless pull-in force G_{pi} as follows:

$$G_{pi} = \frac{4}{27} \tag{9.52}$$

Substituting G_{pi} into (9.51) yields the pull-in voltage:

$$V_{pi} = \sqrt{\frac{8}{27} \frac{kh_0^3}{\varepsilon A}} \tag{9.53}$$

which corresponds to the dimensionless pull-in force G_{pi}.

Substituting the coefficients a, b, and c into (5.15) and (5.16), we find three distinct roots:

$$\begin{aligned} H_1 &= \frac{1}{3}\left(1 + 2\cos\frac{\phi_1}{3}\right) \\ H_2 &= \frac{1}{3}\left[1 + 2\cos\left(\frac{\phi_1}{3} + \frac{2\pi}{3}\right)\right] \\ H_3 &= \frac{1}{3}\left[1 + 2\cos\left(\frac{\phi_1}{3} + \frac{4\pi}{3}\right)\right] \end{aligned} \tag{9.54}$$

where $\phi_1 = \cos^{-1}[1 - (27/2)G]$. In Fig. 9.8 we compare the three distinct roots above to check their suitability. The roots H_1 and H_3 corresponds to the stable and unstable solutions, labeled S and U in Fig. 9.7, respectively. H_2 is a solution without physical meaning. We then take H_1 as the solution of the equilibrium equation (9.49), which may be written as

$$H = \frac{1}{3}\left\{1 + 2\cos\left[\frac{1}{3}\cos^{-1}\left(1 - \frac{27}{2}G\right)\right]\right\} \tag{9.55}$$

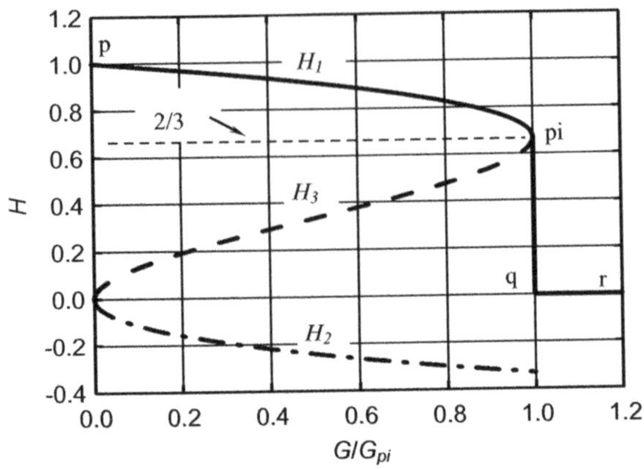

Figure 9.8 Dimensionless gap.

From equation (9.55) and Fig. 9.8, we note that the dimensionless gap H decreases from 1 to 2/3 as the normalized force G/G_{pi} increases from 0 to 1. When the normalized force becomes greater than 1, the gap H goes to zero (i.e., the movable plate collapses into the stationary plate). If we define the dimensionless displacement measured from h_0 as $X = x/h_0$, X is expressed as

$$X = \frac{1}{h_0}(h_0 - h) = 1 - H = \frac{2}{3}\left\{1 - \cos\left[\frac{1}{3}\cos^{-1}\left(1 - \frac{27}{2}G\right)\right]\right\} \tag{9.56}$$

For design of MEMS employing a parallel-plate actuator, we may need to use expressions for the effective stiffness, resonant frequency, and capacitance at the voltage V. Using the preceding exact solution, we can also derive the sensitivities. Since the restoring force F_r at h (or H) is written as

$$F_r = F_s - F_e = kx - \frac{1}{2}\frac{\varepsilon AV^2}{(h_0 - x)^2}$$

the effective stiffness is defined to be the derivative of the restoring force with respect to the displacement x:

$$k_{\text{eff}} = \frac{\partial F_r}{\partial x} = \frac{\partial}{\partial x}\left[kx - \frac{1}{2}\frac{\varepsilon AV^2}{(h_0 - x)^2}\right] = k - \frac{\varepsilon AV^2}{h^3} \tag{9.57}$$

Combining (9.48) and equation (9.57), we obtain the effective stiffness in a dimensionless form:

$$\frac{k_{\text{eff}}}{k} = 3 - \frac{2}{H} \tag{9.58}$$

Therefore, the natural angular frequency ω_n is obtained:

$$\frac{\omega_n}{\omega_{n0}} = \frac{f_n}{f_{n0}} = \sqrt{\frac{k_{\text{eff}}/m}{k/m}} = \sqrt{\frac{k_{\text{eff}}}{k}} = \sqrt{3 - \frac{2}{H}} \tag{9.59}$$

where $\omega_{n0} = \sqrt{k/m}$ denotes the natural angular frequency at $G = 0$ (i.e., $V = 0$) and f_n and f_{n0} represent the natural frequencies corresponding to ω_n and ω_{n0}, respectively. Similarly, we obtain a dimensionless capacitance from the definition of the capacitance, $C = \varepsilon A/h$:

$$\frac{C}{C_0} = \frac{1}{H} \tag{9.60}$$

where C_0 is the initial capacitance, defined as $\varepsilon A/h_0$. It is noted that the dimensionless effective stiffness k_{eff}/k, natural angular frequency ω_n/ω_{n0}, and capacitance C/C_0 are expressed in terms of the dimensionless gap H. Since the gap H of (9.55) is a function of the dimensionless force G, the preceding effective stiffness, natural frequency, and capacitance are also functions of the dimensionless force G (i.e., the voltage V). That is, if the applied voltage V, stiffness k, and geometry (A, h_0) are given, the corresponding gap h, stiffness k_{eff}, natural frequency f_n, and capacitance C are obtained. This procedure is illustrated in Fig. 9.9. The dimensionless stiffness, natural frequency, and capacitance are plotted against the normalized force in Fig. 9.10. When the normalized force G/G_{pi} goes to 1, C/C_0 approaches 3/2 and k_{eff}/k and ω_n/ω_{n0} go to zero; at the pull-in, the maximum capacitance is 3/2 times the initial capacitance C_0, and the effective stiffness and natural frequency become zero.

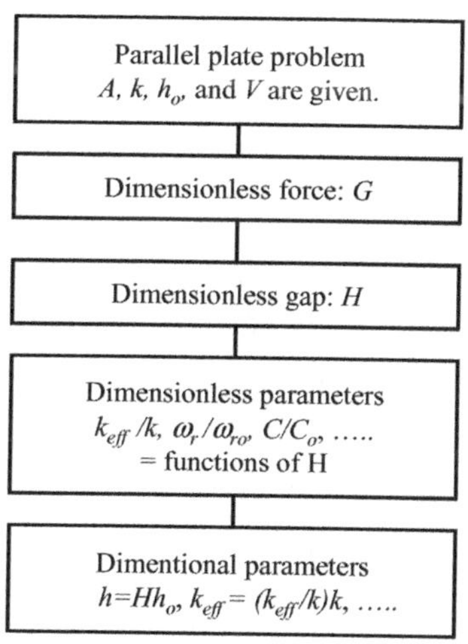

Figure 9.9 Procedure to calculate parameters.

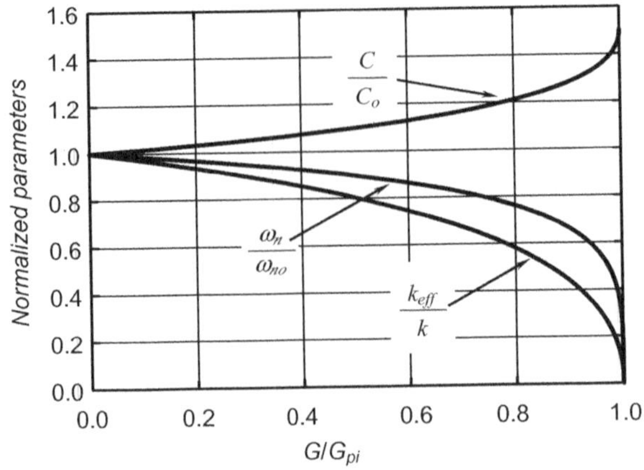

Figure 9.10 Normalized capacitance, natural frequency, and effective stiffness.

The sensitivities of the preceding parameters to the applied voltage are useful for design or optimization of the parallel-plate actuator. We shall first obtain the sensitivities in dimensionless forms and convert them into dimensional forms. The sensitivity of H to G is expressed from the derivative of (9.49) with respect to G:

$$\frac{\partial H}{\partial G} = \frac{-1}{3H^2 - 2H} = \frac{-1}{H^2(3-2/H)} \tag{9.61}$$

We can also obtain a different form of $\partial H/\partial G$ by taking the first derivative (9.55) with respect to G. The different form of $\partial H/\partial G$ is much more complicated than the preceding equation, so we shall use the preceding equation for $\partial H/\partial G$ for further derivation.

From (9.58)–(9.60), the sensitivities of the effective stiffness, natural angular frequency, and the capacitance to the dimensionless force G are also derived as functions of the gap H, respectively:

$$\frac{\partial (k_{\text{eff}}/k)}{\partial G} = \frac{2}{H^2}\frac{\partial H}{\partial G} = \frac{-2}{H^4(3-2/H)} \tag{9.62}$$

$$\frac{\partial (\omega_r/\omega_{r0})}{\partial G} = \frac{\partial (f_r/f_{r0})}{\partial G} = \frac{-1}{H^4(3-2/H)^{3/2}} \tag{9.63}$$

$$\frac{\partial (C/C_0)}{\partial G} = \frac{1}{H^4(3-2/H)} \tag{9.64}$$

These dimensionless sensitivities are illustrated in Fig. 9.11. At the pull-in ($G/G_{pi} = 1$), the capacitance sensitivity goes to positive infinity, but the

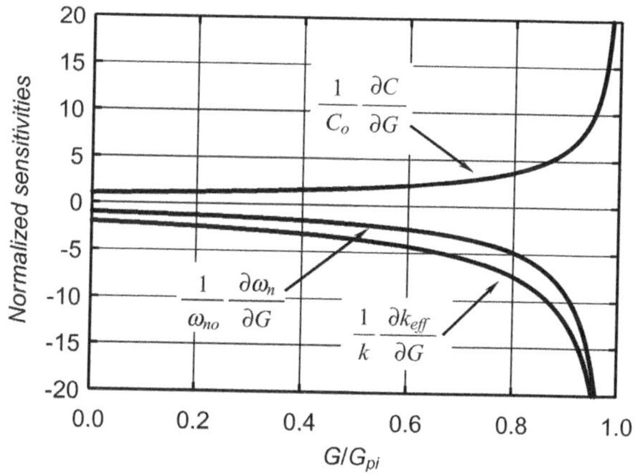

Figure 9.11 Sensitivities of the effective stiffness, resonant frequency, and capacitance to the dimensionless force.

sensitivities of the effective stiffness and natural frequency approach negative infinity.

Let us convert the preceding dimensionless sensitivities into dimensional sensitivities. Dimensional sensitivities are obtained by using the chain rule. When the applied voltage varies, the gap sensitivity to voltage can be expressed as

$$\frac{\partial h}{\partial V} = \frac{\partial h}{\partial H}\frac{\partial H}{\partial G}\frac{\partial G}{\partial V} \tag{9.65}$$

$\partial h/\partial H = h_0$ is obtained from (9.50) and $\partial G/\partial V$ is derived from (9.51):

$$\frac{\partial G}{\partial V} = \frac{\varepsilon A V}{k h_o^3} = 2\frac{G}{V} \tag{9.66}$$

Substituting $\partial h/\partial H = h_0$, (9.61) and (9.66) into (9.65) and recalling (9.49), $G/H^2 = (1 - H)$, the gap sensitivity to the voltage is obtained as

$$\frac{\partial h}{\partial V} = h_0 \frac{-1}{H^2(3-2/H)} 2\frac{G}{V} = \frac{-2}{H^2(3-2/H)}\frac{h_0 G}{V} = \frac{-2(1-H)}{3-2/H}\frac{h_0}{V} \tag{9.67}$$

Similarly, the sensitivities of the effective stiffness, angular frequency, and capacitance to voltage are also derived as follows:

$$\frac{\partial k_{\text{eff}}}{\partial V} = \frac{-4(1-H)}{H^2(3-2/H)}\frac{k}{V} \tag{9.68}$$

$$\frac{\partial \omega_n}{\partial V} = \frac{-2(1-H)}{H^2(3-2/H)^{3/2}} \frac{\omega_{n0}}{V} \tag{9.69}$$

$$\frac{\partial C}{\partial V} = \frac{2}{H^2(3-2/H)} \frac{C_0}{V} \tag{9.70}$$

Example 9.1 A parallel-plate actuator consists of a movable plate of area $A = 3.6 \times 10^{-9}\,\text{m}^2$ and mass $m = 1.7 \times 10^{-11}\,\text{kg}$. The initial gap h_0 between the movable and stationary plates is given as $2\,\mu\text{m}$ and the stiffness k due to the flexures is $1\,\text{N/m}$. It is desired to obtain the pull-in voltage V_{pi} and the gap, effective stiffness, natural frequency, and capacitance at a voltage of $V_{pi}/2$, which is applied between the movable and stationary plates. We also wish to find the sensitivity of natural frequency to voltage at $V = V_{pi}/2$.

Substituting $k = 1\,\text{N/m}$, $A = 3.6 \times 10^{-9}\,\text{m}^2$, $h_0 = 2\,\mu\text{m}$, and $\varepsilon = 8.854 \times 10^{-12}$ F/m into

$$V_{pi} = \sqrt{\frac{8}{27}\frac{kh_0^3}{\varepsilon A}} \tag{a}$$

gives $V_{pi} = 8.62\,\text{V}$. The applied voltage V is then obtained as $V = V_{pi}/2 = 4.31\,\text{V}$. The corresponding dimensionless force is given by

$$G = G_{pi}\left(\frac{V}{V_{pi}}\right)^2 \tag{b}$$

The dimensional force is then obtained as $G = (4/27)(1/2)^2 = 1/27$. Upon substituting $G = 1/27$ into

$$H = \frac{1}{3}\left\{1 + 2\cos\left[\frac{1}{3}\cos^{-1}\left(1 - \frac{27}{2}G\right)\right]\right\} \tag{c}$$

we have

$$H = \frac{1}{3}\left\{1 + 2\cos\left[\frac{1}{3}\cos^{-1}\left(1 - \frac{27}{2}\frac{1}{27}\right)\right]\right\} = \frac{1}{3}\left(1 + 2\cos\frac{\pi}{9}\right) \cong 0.9598$$

and the corresponding dimensional gap, $h = h_0 H = 2 \times 10^{-6} \times 0.9598 = 1.9196 \times 10^{-6}\,\text{m}$. The effective stiffness, natural frequency, and capacitance are obtained as

$$k_{\text{eff}} = k\left(3 - \frac{2}{H}\right) = 1 \times \left(3 - \frac{2}{0.9598}\right) = 0.9162\ \text{N/m}$$

$$f_n = f_{n0}\sqrt{3-\frac{2}{H}} = 3.8601\times10^4\sqrt{3-\frac{2}{0.9598}} = 3.6948\times10^4 \text{ Hz}$$

where

$$f_{n0} = \frac{1}{2\pi}\sqrt{\frac{k}{m}} = \frac{1}{2\pi}\sqrt{\frac{1}{1.7\times10^{-11}}} = 3.8601\times10^4 \text{ Hz}$$

$$C = C_0\frac{1}{H} = \frac{\varepsilon A}{h_0}\frac{1}{H} = \frac{8.854\times10^{-12}\times3.6\times10^{-9}}{2\times10^{-6}}\frac{1}{0.9598} = 1.6605\times10^{-14} \text{ F}$$

The sensitivity of natural frequency to voltage at $V = V_{pi}/2$ is given by

$$\frac{\partial f_n}{\partial V} = \frac{-2(1-H)}{H^2(3-2/H)^{3/2}}\frac{f_{n0}}{V}$$
$$= \frac{-2(1-0.9598)}{0.9598^2(3-2/0.9598)^{3/2}}\frac{3.8601\times10^4}{4.31} = -853.12 \text{ Hz/V}$$

Dynamic Response

In preceding paragraphs we have derived the static response of the parallel-plate actuator to the voltage applied between the movable and stationary plates. In many cases, a time-varying voltage is applied to the plates and then the movable plate vibrates up and down. To analyze the mechanical vibration due to the time-varying voltage, let us consider a dynamic model of the parallel-plate actuator in Fig. 9.12. The movable plate of mass m and area A is suspended by a spring of stiffness k and a damper of damping coefficient c and spaced by an initial gap h_0 from the stationary plate. When a time-varying input voltage V_i with a dc bias voltage V_b is applied between the plates, the plate moves downward and vibrates. If we assume that the dc bias voltage V_b is first applied and then the time-varying voltage V_i is added, the overall dynamic response of the movable plate can be considered as a sum of the static

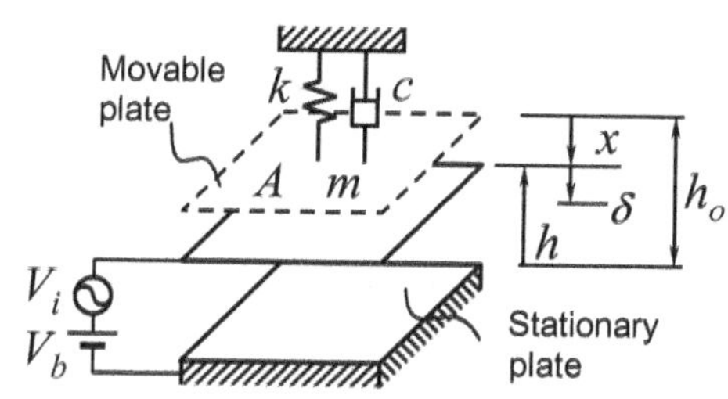

Figure 9.12 Dynamic model of a parallel plate.

displacement x and the dynamic response. Since the electrostatic force is nonlinear, the response δ is a nonlinear function that may be expressed as follows: $\delta = f(h, h_0, A, m, c, k, V_b, V_i)$.

Let us obtain a closed-form expression for the dynamic response δ. Applying Newton's second law to the movable plate in Fig. 9.12, we find the equation of motion as

$$m\ddot{\delta} + c\dot{\delta} + k(h_0 - h + \delta) = F_e \tag{9.71}$$

where F_e is the electrostatic force due to the applied voltage and can be expressed as

$$F_e = \frac{\varepsilon A V^2}{2(h-\delta)^2} = \frac{\varepsilon A (V_b + V_i)^2}{2(h-\delta)^2} \tag{9.72}$$

Assuming that $V_i/V_b << 1$ and $\delta/h << 1$, we expand the preceding force in a Taylor series,

$$F_e = \frac{\varepsilon A V_b^2}{2h^2}\frac{(1+V_i/V_b)^2}{(1-\delta/h)^2} \cong \frac{\varepsilon A V_b^2}{2h^2}\left(1 + 2\frac{V_i}{V_b} + 2\frac{\delta}{h} + \text{high-order terms}\right)$$

Substituting the force above into (9.71) and neglecting the high-order terms gives

$$m\ddot{\delta} + c\dot{\delta} + k(h_0 - h) + \left(k - \frac{\varepsilon A V_b^2}{h^3}\right)\delta = \frac{\varepsilon A V_b^2}{2h^2} + \frac{\varepsilon A V_b V_i}{h^2}$$

Subtracting (9.48) with $V = V_b$ from the preceding equation and using the effective stiffness (9.57), $k_{\text{eff}} = k - \varepsilon A V_b^2/h^3$, we find that

$$m\ddot{\delta} + c\dot{\delta} + k_{\text{eff}}\delta = \frac{\varepsilon A V_b V_i}{h^2} \tag{9.73}$$

Equation (9.73) is highly nonlinear differential equation since the effective stiffness k_{eff} and the gap h are nonlinear functions of the dc bias voltage V_b. However, if the bias voltage V_b remains constant, the gap h also remains constant and the stiffness k_{eff} can be considered unchanged during vibration. The gap h and effective stiffness k_{eff} are obtained from (9.55) and (9.58), respectively. As a result, the preceding equation of motion becomes linear and we can solve it analytically if the input voltage V_i is given. The solution of the linear second-order equation of motion was dealt with in Chapter 5. For example, if the input voltage V_i is a sinusoidal voltage (i.e., $V_i = V_a \sin \omega t$), the preceding equation may be written

$$m\ddot{\delta}+c\dot{\delta}+k_{\text{eff}}\delta = F_0 \sin \omega t \tag{9.74}$$

where

$$F_0 = \frac{\varepsilon A V_b V_a}{h^2} \tag{9.75}$$

denotes the amplitude of the actuating force. If we are interested in the steady-state response δ_s, the response is given by (5.89):

$$\delta_s = \delta_1 \sin(\omega t - \phi) \tag{9.76}$$

where

$$\frac{\delta_1}{\delta_{\text{st}}} = \frac{1}{\sqrt{(1-\Omega_1^2)^2+(\Omega_1/Q)^2}} \tag{9.77}$$

denotes the magnification factor and

$$\begin{aligned}
\delta_{\text{st}} &= \frac{F_0}{m\omega_{n1}^2} = \frac{F_0}{k_{\text{eff}}} \\
\omega_{n1} &= \sqrt{\frac{k_{\text{eff}}}{m}} \\
\Omega_1 &= \frac{\omega}{\omega_{n1}} \\
Q &= \frac{\sqrt{mk_{\text{eff}}}}{c} \\
\phi &= \tan^{-1}\frac{\Omega_1}{Q(1-\Omega_1^2)}
\end{aligned} \tag{9.78}$$

represent the static displacement, the angular frequency, the frequency ratio, the quality factor, and the delay, respectively. Substituting the effective stiffness, (9.58), and the force amplitude, (9.75), into the first equation of (9.78) yields the static displacement δ_{st} as follows:

$$\frac{\delta_{\text{st}}}{h_0} = \frac{2}{H^2(3-2/H)}\frac{V_a}{V_b}G \tag{9.79}$$

Since the effective stiffness k_{eff} is related to the dc bias voltage V_b, the frequency Ω_1 and the other parameters shown in the preceding equations are functions of V_b. It is convenient to express the preceding equations as functions of a new frequency ratio, which is independent of V_b. Noting that $\omega_{n1}/\omega_{n0} = \sqrt{k_{\text{eff}}/k}$, where $\omega_{n0} = \sqrt{k/m}$ denotes the natural frequency at $V_b = 0$, we obtain a new expression of the frequency Ω_1:

$$\Omega_1 = \frac{\omega}{\omega_{n1}} = \frac{\Omega}{\sqrt{3-2/H}} \tag{9.80}$$

where

$$\Omega = \frac{\omega}{\omega_{n0}} \tag{9.81}$$

denotes the ratio of the angular frequency of ac signal to the initial natural angular frequency ω_{n0}, defined by

$$\omega_{n0} = \sqrt{\frac{k}{m}}$$

Substituting (9.79) and (9.80) into (9.76)–(9.78), we find the steady-state response

$$\frac{\delta_s}{h_0} = \frac{\delta_1}{h_0}\sin(\omega t - \phi) \tag{9.82}$$

where

$$\frac{\delta_1}{h_0} = \frac{2}{H^2}\frac{1}{\sqrt{(3-(2/H)-\Omega^2)^2 + (\Omega/Q)^2(3-2/H)}}\frac{V_a}{V_b}G \tag{9.83}$$

$$\phi = \tan^{-1}\frac{\Omega\sqrt{3-2/H}}{Q(3-(2/H)-\Omega^2)} \tag{9.84}$$

represent the dimensionless amplitude of the dynamic response and the phase, respectively. Recalling $G/H^2 = 1 - H$ from (9.49), we further simplify (9.83) as

$$\frac{\delta_1}{h_0} = 2\frac{1-H}{\sqrt{(3-2/H-\Omega^2)^2 + (\Omega/Q)^2(3-2/H)}}\frac{V_a}{V_b} \tag{9.85}$$

If a parallel-plate actuator is operated in a vacuum environment, the quality factor is not related to the interplate gap. However, if a fluid (gas or liquid) is squeezed between the movable and stationary plates, the damping coefficient c may be expressed in the general form

$$c = \frac{\alpha_d}{h^3} \tag{9.86}$$

where α_d denotes a constant which can be determined by the theory developed in Section 6.8. Substituting equation (9.58) into the fourth equation of (9.78), we obtain the quality factor as a function of H:

$$\frac{Q}{Q_0} = H^3\sqrt{\frac{k_{\text{eff}}}{k}} = H^3\sqrt{3 - \frac{2}{H}} \tag{9.87}$$

where $Q_0 = \sqrt{mk}/c_0$ is the initial quality factor at $V_b = 0$ and $c_0 = \alpha_d/h_0^3$ is the initial damping coefficient. It is worth noting that the nonlinear dynamic response, $\delta = f(h_0, A, m, c, k, V_b, V_a, \omega)$, is expressed in (9.82) as a function of three independent parameters: the dimensionless force G, the voltage ratio V_a/V_b, and the frequency ratio Ω, since H and Q are expressed as functions of G, as shown in (9.55) and (9.87). If a capacitance detection circuit is connected between the movable and stationary plates, the capacitance C for the steady-state response is obtained as

$$C = \frac{\varepsilon A}{h - \delta_s} \tag{9.88}$$

where h and δ_s denote the static interplate gap and the steady-state response (9.82), respectively. For $\delta_s/h << 1$, the preceding capacitance can be expressed in a Taylor's series,

$$C = \frac{\varepsilon A}{h}\frac{1}{1 - \delta_s/h} \cong \frac{C_0}{H}\left(1 + \frac{1}{H}\frac{\delta_s}{h_0}\right) \tag{9.89}$$

where C_0 is the initial capacitance, defined as $\varepsilon A/h_0$. Equation (9.89) may be expressed in a different form,

$$\frac{\Delta C}{C_0} = \frac{C - C|_{\delta=0}}{C_0} = \frac{1}{H^2}\frac{\delta_s}{h_0}$$

Substituting (9.82) into the equation above, we find the steady-state response of the capacitance to the time-varying voltage,

$$\frac{\Delta C}{C_0} = \frac{\Delta C_1}{C_0}\sin(\omega t - \phi) \tag{9.90}$$

where

$$\frac{\Delta C_1}{C_0} = \frac{1}{H^2}\frac{\delta_1}{h_0} = 2\frac{1 - H}{H^2\sqrt{(3 - 2/H - \Omega^2)^2 + (\Omega/Q)^2(3 - 2/H)}}\frac{V_a}{V_b} \tag{9.91}$$

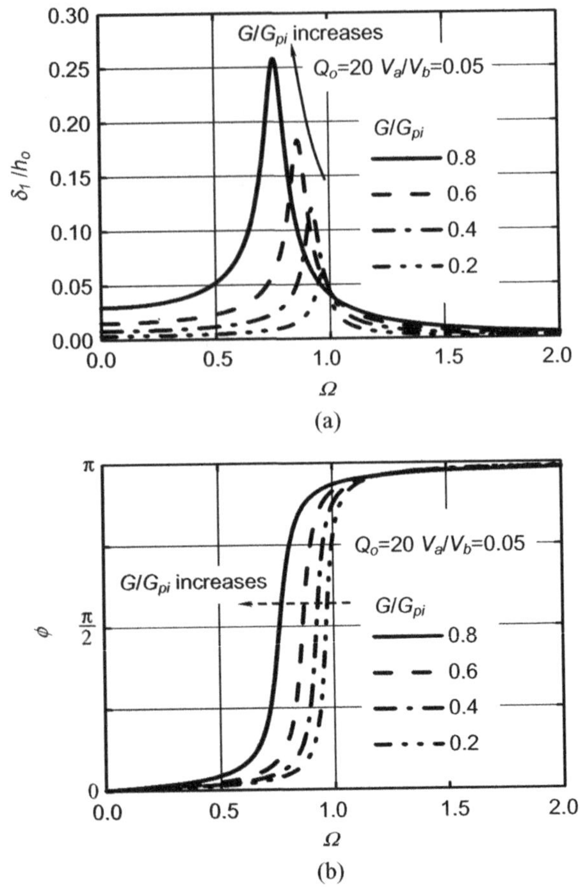

Figure 9.13 Dynamic response of parallel-plate actuator.

represents the capacitance amplitude, which can be measured by a measuring instrument. The preceding capacitance, corresponding to the steady-state mechanical vibration, does not include the effect of parasitic capacitance.

For $V_a/V_b = 0.05$ and $Q_0 = 20$ (as discussed in Chapter 6, the quality factor can be adjusted by perforation or reduced gas pressure), the vibration amplitude, (9.85), and the phase, (9.84), are shown in Fig. 9.13a and b. As G/G_{pi} increases from 0.2 to 0.8 (i.e., the bias voltage V_b increases), the dimensionless amplitude increases, the natural frequency decreases, and the phase curve shifts leftward. For the same range of parameters, the steady-state response of the capacitance is illustrated in Fig. 9.14, in which the capacitance change also increases with increasing G/G_{pi}. For validation of the analytic dynamic response, (9.82), of the parallel-plate actuator to a sinusoidal voltage, the analytic response is compared to a response obtained from a numerical

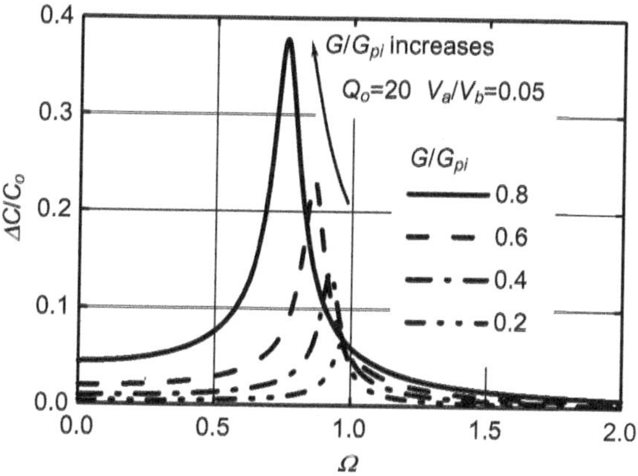

Figure 9.14 Capacitance response.

analysis of the nonlinear differential equation (9.71). For the numerical analysis, introducing a relation $h_q = h - \delta$ from geometry in Fig. 9.12 and substituting the relation into (9.71), we have a more convenient form of the original differential equation:

$$m\ddot{h}_q + c\dot{h}_q + kh_q = kh_0 - \frac{\varepsilon AV^2}{2h_q^2} \tag{9.92}$$

If the parallel-plate actuator is surrounded by a fluid, the damping coefficient (9.86) is used, which may be written

$$c = \frac{\alpha_d}{h_q^3} \tag{9.93}$$

With the initial quality factor Q_0 at $h_q = h_0$, the damping coefficient c may be obtained from the relations $Q_0 = \sqrt{mk}/c_0$ and $c = \alpha_d/h_q^3 = (\alpha_d/h_0^3)(h_0/h_q)^3 = c_0 h_0^3/h_q^3$, as follows:

$$c = \frac{\sqrt{mk}}{Q_0}\frac{h_0^3}{h_q^3} \tag{9.94}$$

The numerical response of the preceding differential equation is obtained by using a stable numerical scheme for numerical analysis of nonlinear second-order differential equations (e.g., Park, 1975). To validate the analytic dynamic response (9.82), the dynamic responses from theory and numerical analysis are compared in Fig. 9.15, where the following data are used: mass $m = 1 \times 10^{-11}$ kg, stiffness $k = 1$ N/m, effective plate area $A = 2.15 \times 10^{-9}$ m, initial

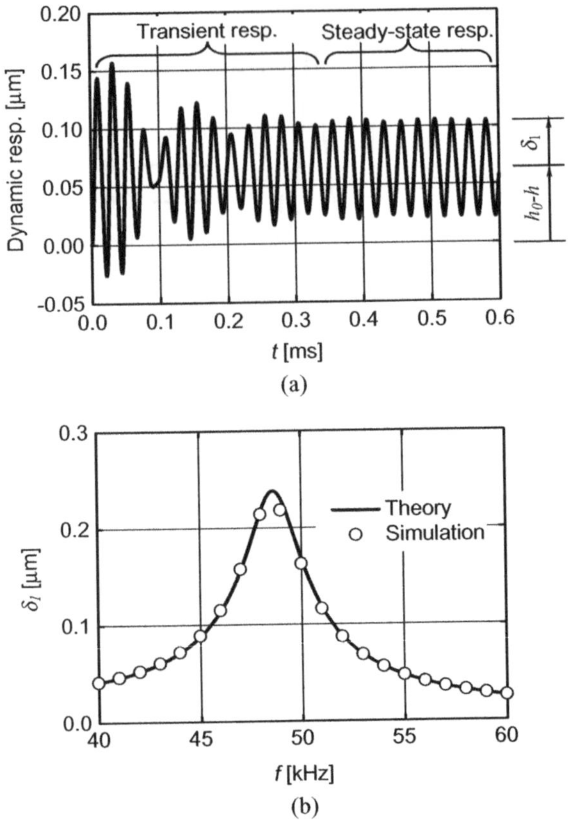

Figure 9.15 Comparison of theoretical and numerical dynamic responses.

quality factor $Q_0 = 20$, initial gap $h_0 = 2 \times 10^{-6}$ m, dc bias voltage $V_b = 5$ V, and ac drive voltage $V_a = 0.5$ V. For the numerical analysis, the initial conditions $(h_q = h_0, \ddot{h}_q = 0, \dot{h}_q = 0)$ at $t = 0$ are imposed on the problem. After the transient response during a short period in Fig. 9.15a, the dynamic response from the numerical analysis becomes the steady-state response, in which $h_0 - h$ and δ_1 correspond to the static displacement from h_0 and the analytic response (9.82), respectively. The analytic vibration amplitude δ_1 (9.85) is compared to the amplitude from the numerical analysis in Fig. 9.15b. The analytic amplitude (9.85) is in good agreement with that obtained from the numerical analysis, and then we can use the analytic dynamic responses (9.82)–(9.91) for analysis of the parallel-plate actuator.

So far we have derived the dynamic responses of the parallel-plate actuator from the nonlinear equation, (9.71). In many cases, the maximum amplitude (i.e., the amplitude at resonance) is important for design purposes. Let us obtain the natural frequency and the amplitude at resonance. We shall use the fact that the derivative of (9.85) with respect to Ω is zero at resonance. Setting $\partial\left[(\delta_1/h_0)^2\right]/\partial\Omega = 0$, the frequency ratio Ω_m at the maximum amplitude is obtained as

$$\Omega_m = \sqrt{\left(3-\frac{2}{H}\right)\left(1-\frac{1}{2Q^2}\right)} \cong \sqrt{3-\frac{2}{H}} \tag{9.95}$$

Thus, substituting equation (9.95) into (9.85), we have the maximum amplitude at Ω_m, given by

$$\left.\frac{\delta_1}{h_0}\right|_m = 2\frac{Q}{3-2/H}\frac{1-H}{\sqrt{1-1/4Q^2}}\frac{V_a}{V_b} \cong 2Q\frac{1-H}{3-2/H}\frac{V_a}{V_b} \tag{9.96}$$

The approximate expressions on the right-hand side of equations (9.95) and (9.96) are valid for $Q >> 1$ and can be interpreted as the frequency ratio and the amplitude at resonance, respectively. Noted that (9.95) for $Q >> 1$ is equal to (9.59), which is obtained directly from the effective stiffness k_{eff}. Therefore, the natural frequency f_n for a large quality factor ($Q >> 1$) is expressed as follows:

$$f_n = f_{n0}\sqrt{3-\frac{2}{H}} \tag{9.97}$$

where f_{n0} is the initial natural frequency defined as $f_{n0} = \sqrt{k/m}/2\pi$. Similarly, upon substituting (9.95) into (9.84), we find the phase ϕ_m at resonance:

$$\phi|_m = \tan^{-1}\left(2Q\sqrt{1-\frac{1}{2Q^2}}\right) \cong \tan^{-1} 2Q \tag{9.98}$$

The right-hand side of equation (9.98) is an approximate expression which is valid for $Q >> 1$. Equation (9.98) states that if the quality factor Q becomes large, the phase delay ϕ_m approaches $\pi/2$.

Dynamic Pull-in

In the preceding paragraphs it is assumed that the dc bias voltage increases slowly to a desired value to avoid the dynamic effect due to the voltage; that is, the period to increase from zero to the desired voltage is much greater than the time constant, τ, which can be defined as $\tau = 1/\zeta\omega_n = 2Q/\omega_n$ for $Q >> 1$, which is discussed following (5.76). However, if the bias voltage is applied within a period smaller than the time constant τ, the dynamic effect due to the dc voltage is not negligible and plays an important role. In this case, the dc bias voltage can be considered as a step input of voltage. We shall

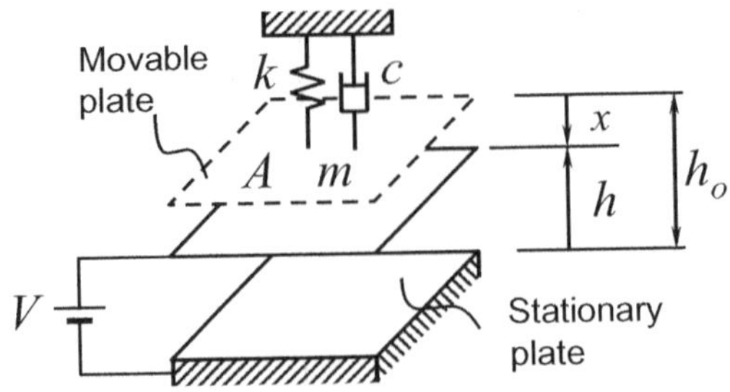

Figure 9.16 Model for dynamic pull-in.

investigate the dynamic effect of the step voltage. Consider the model of a parallel-plate actuator in Fig. 9.16, which is the same as the previous model in Fig. 9.12 except for the voltage source and displacement; the dc bias voltage is connected between the movable and stationary plates at $t = 0$ and the displacement x is measured from the initial gap h_0. When step voltage is applied at $t = 0$, the equation of motion of the movable plate is written as

$$m\ddot{x} + c\dot{x} + kx = \frac{\varepsilon A V^2}{2(h_0 - x)^2} \tag{9.99}$$

Since equation (9.99) is a nonlinear differential equation, we cannot express the displacement x in a closed form. However, if it is assumed that the damping force is neglected (i.e., $c = 0$), we can obtain a relationship between the velocity $\dot{x}$ and the displacement x. The movable plate is at the initial position when the voltage V is applied at $t = 0$ (i.e., the initial conditions $x = 0$ and $\dot{x} = 0$ at $t = 0$). Setting $c = 0$ and noting that

$$\ddot{x} = \frac{d^2x}{dt^2} = \frac{dv}{dt} = \frac{dv}{dx}\frac{dx}{dt} = v\frac{dv}{dx}$$

we have

$$mv\frac{dv}{dx} + kx = \frac{\varepsilon A V^2}{2(h_0 - x)^2} \tag{9.100}$$

Multiplying by dx on both sides and integrating as follows:

$$\int_0^v mv\,dv + \int_0^x kx\,dx = \int_0^x \frac{\varepsilon A V^2}{2(h_0 - x)^2}dx$$

gives a relationship between the velocity and the displacement,

$$\frac{1}{2}mv^2 + \frac{1}{2}kx^2 = \frac{\varepsilon A V^2}{2}\left(\frac{1}{h_0 - x} - \frac{1}{h_0}\right) \tag{9.101}$$

which yields the velocity in terms of the displacement x:

$$v=\sqrt{\frac{2}{m}\left[\frac{\varepsilon AV^2}{2}\left(\frac{1}{h_0-x}-\frac{1}{h_0}\right)-\frac{1}{2}kx^2\right]} \tag{9.102}$$

It is noted that (9.101) can be interpreted as an energy balance equation (i.e., the energy in the system is conserved. We define a potential energy of the parallel-plate actuator shown in Fig. 9.16 as follows:

$$U_p=U_s-U_e=\frac{1}{2}kx^2-\frac{1}{2}\frac{\varepsilon AV^2}{h_0-x} \tag{9.103}$$

where U_s and U_e represent the elastic energy and the electric energy defined by (7.68), respectively. The equation of conservation of energy is written as

$$(U_k+U_p)_1=(U_k+U_p)_2 \tag{9.104}$$

where U_k represents the kinetic energy of the movable plate and the subscripts 1 and 2 denote positions 1 and 2, respectively. If position 1 is $x = 0$ and position 2 is a position x, equation (9.104) may be written as

$$-\frac{1}{2}\frac{\varepsilon AV^2}{h_0}=\frac{1}{2}mv^2+\frac{1}{2}kx^2-\frac{1}{2}\frac{\varepsilon AV^2}{h_0-x}$$

which is the same as (9.101). From the preceding equation, the velocity equation (9.102) is also obtained. For convenience, the potential energy (9.103) is expressed in a dimensionless form,

$$U_{pn}=\frac{U_p}{U_{\text{ref}}}=\frac{1}{2}X^2-\frac{G}{H} \tag{9.105}$$

where

$$U_{\text{ref}}=kh_0^2 \qquad X=\frac{x}{h_0} \qquad H=\frac{h_0-x}{h_0}=\frac{h}{h_0} \tag{9.106}$$

and G is the dimensionless force defined by $G=\varepsilon AV^2/2kh_0^3$. From the preceding energy equation, $X^2/2$ and $-G/H$ can be interpreted, respectively, as the dimensionless potential energies due to the spring and electric field formed between the plates. The dimensionless potential energy U_{pn} is plotted against the dimensionless displacement X in Fig. 9.17. The right side of the potential curve is lowered as the normalized force G/G_{pi}, where G_{pi} is the dimensionless pull-in force, increases from zero. The stable positions are indicated by the

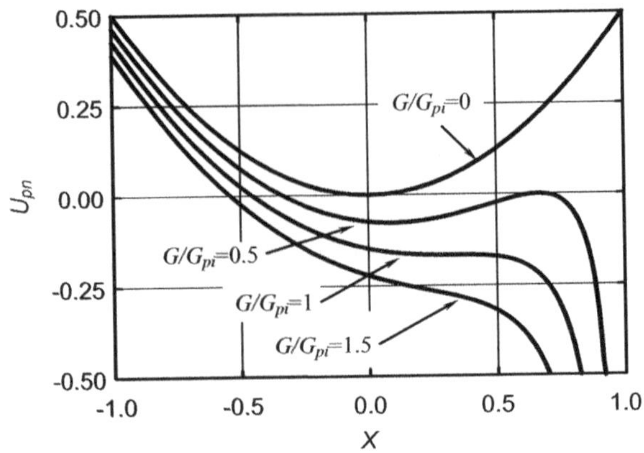

Figure 9.17 Potential energy of an electrostatic parallel plate.

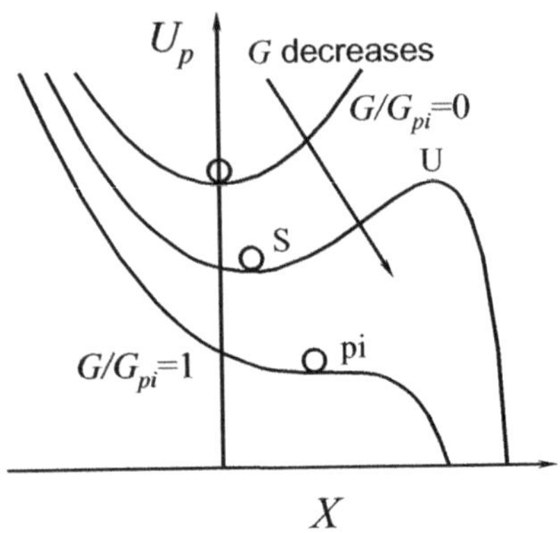

Figure 9.18 Stable positions.

first two circles in Fig. 9.18. If the movable plate is placed at the stable position S, the plate will not move from the stable position. When the plate is placed by a distance from the stable position S, the plate tends to move to the stable position S. However, if the movable plate is located at U, a disturbance may cause a collapse of the movable plate to the stationary plate. The position U is then called an unstable position. When the voltage V reaches the pull-in voltage (i.e., $G/G_{pi} = 1$), the movable plate moves to the pull-in position labeled pi in Fig. 9.18. For $G/G_{pi} > 1$, the movable plate cannot have a stable position but always collapse into the stationary plate.

When the step voltage V in Fig. 9.16 is applied to the movable plate whose potential energy is represented by the curves in Figs. 9.17 and 9.18, the potential curve is changed from the initial curve to a curve corresponding to the voltage V. If the voltage V (or dimensionless force G) reaches a critical value,

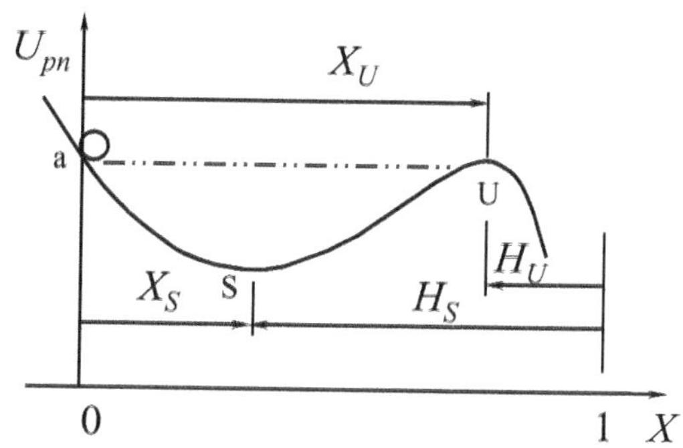

Figure 9.19 Dynamic pull-in.

the movable plate collapses to the stationary plate. This phenomenon is called *dynamic pull-in*, while the pull-in derived from (9.48) is sometimes called *static pull-in*. Let us obtain the dynamic pull-in voltage and the corresponding gap under the assumption that the damping force is neglected. Right after the step voltage V is applied, the movable plate labeled "a" in Fig. 9.19 is still placed at $X = 0$ and will move toward the stationary plate (i.e., $X = 1$ in the figure). The potential energy curve has a valley (well) and a peak as shown in Fig. 9.19. The movable plate has its maximum velocity at the lowest potential energy and has zero velocity at the same potential energy as that at $X = 0$. If the energy at the peak labeled U is greater than that at $X = 0$, the movable plate is bound in the valley formed between the initial position and the peak and vibrates in the valley. However, the energy peak is lower than the energy at $X = 0$; the movable plate has an energy enough to overcome the peak and then collapses into the stationary plate (i.e., $X = 1$). Therefore, we can consider a critical condition at which the initial potential energy is equal to that at the peak. The equation of conservation of energy, (9.104), for the initial and critical positions is written as $(0 + U_{pn})_a = (0 + U_{pn})_{\mathrm{U}}$, which may be expressed as

$$-G_{\mathrm{U}} = \frac{1}{2}X_{\mathrm{U}}^2 - \frac{G_{\mathrm{U}}}{H_{\mathrm{U}}} \tag{9.107}$$

where $H_{\mathrm{U}} = 1 - X_{\mathrm{U}}$ and the subscript U denotes the unstable position. Equation (9.107) may be written as

$$\frac{1}{H_{\mathrm{U}}}G_{\mathrm{U}} = \frac{1}{2}X_{\mathrm{U}} \tag{9.108}$$

We have two unknowns in equation (9.108), G_{U} and H_{U}, and need one more equation. The additional equation is obtained from the fact that the movable plate is in static equilibrium at $X = X_{\mathrm{U}}$ since the velocity and acceleration are zero at the peak. The equilibrium equation (9.49) may be written

$$X_U = \frac{G_U}{H_U^2} \tag{9.109}$$

Combining equations (9.108) and (9.109), we find the dynamic pull-in gap H_U and the dynamic pull-in displacement X_U as follows:

$$H_{pi,d} = H_U = \frac{1}{2} \qquad X_{pi,d} = X_U = 1 - H_U = \frac{1}{2} \tag{9.110}$$

In order to stress that the positions U in Fig. 9.19 are the unstable position for the dynamic pull-in, the subscript d in equations (9.110) was placed after pi. Substituting the preceding equations into (9.109), we find the dimensionless dynamic pull-in force G_U,

$$G_{pi,d} = G_U = X_U H_U^2 = \frac{1}{8} \tag{9.111}$$

Note that the dynamic pull-in force $G_{pi,d} = 1/8 = 0.125$ is smaller than the static pull-in force $G_{pi} = 4/27 = 0.148$ given by (9.52) and that the dynamic pull-in gap $H_U = 1/2$ is smaller than the static pull-in gap $H_{pi} = 2/3$; that is, if step voltage is applied to the plates, the pull-in voltage is smaller than the static pull-in voltage, and the movable plate can be displaced by the distance $X_{pi,d} = 1/2$. The corresponding dimensional values are written

$$h_{pi,d} = h_0 H_{pi,d} = \frac{h_0}{2} \qquad x_{pi,d} = h_0 X_{pi,d} = \frac{h_0}{2} \tag{9.112}$$

$$V_{pi,d} = \sqrt{2G_{pi,d}\frac{kh_0^3}{\varepsilon A}} = \sqrt{\frac{kh_0^3}{4\varepsilon A}} \tag{9.113}$$

where equation (9.113) is obtained from (9.51) and (9.111). The ratio of the dynamic and static pull-in voltages is given by

$$\frac{V_{pi,d}}{V_{pi}} = \frac{\sqrt{kh_0^3/4\varepsilon A}}{\sqrt{8kh_0^3/27\varepsilon A}} = \sqrt{\frac{27}{32}} \cong 0.9186 \tag{9.114}$$

Therefore, if one designs a device using a parallel-plate actuator whose dc operation voltage is smaller than the dynamic pull-in voltage $V_{pi,d}$, the static and dynamic pull-in phenomenon never occurs.

The dynamic pull-in voltage and gap above are derived under the assumption that the damping force is negligible (i.e., damping coefficient $c = 0$); that is, the total energy, after the voltage is applied, is conserved. However, when the damping force is involved in the problem, the nonlinear differential equation (9.99) should be solved to obtain the voltage. It is difficult to solve equation (9.99) for x in a closed form due to the nonlinearity. Instead of solving (9.99), we can consider a particular case in which the damping coefficient c is large enough to dissipate all the kinetic energy (i.e., the movable plate moves quasistatically from $X = 0$ to the stable position labeled S in Fig. 9.19. Thus, in this case, the pull-in voltage is thought of as the static pull-in voltage, given by (9.53), and we can define the ranges of the pull-in voltage and gap, V_p and $H_p = h_p/h_0$, for an arbitrary damping coefficient as

$$\sqrt{\frac{kh_0^3}{4\varepsilon A}} \le V_p \le \sqrt{\frac{8}{27}\frac{kh_0^3}{\varepsilon A}} \tag{9.115}$$

$$\frac{1}{2} \le H_p \le \frac{2}{3} \tag{9.116}$$

9.4 TORSIONAL ACTUATOR

In Section 9.3 we have derived the static and dynamic responses of the parallel-plate actuator that is the simplest gap-closing actuator. If electrostatic gap-closing actuators have more than one degree of freedom, the actuators can move in more than one direction. A two-degree-of-freedom (DOF) torsional actuator is illustrated in Fig. 9.20, where the torsional actuator consists of a movable plate, a stationary lower plate fixed on a substrate, and two spring beams anchored on the substrate. The dimensions for the actuator are illustrated in the figure. When a voltage V, applied between the movable plate and stationary plates, increases from zero, the movable plate both moves downward and tilts due to the electrostatic force. For convenience, we first assume that the movable plate is angularly displaced without the downward displacement, and then discuss the effect of the downward displacement of the movable

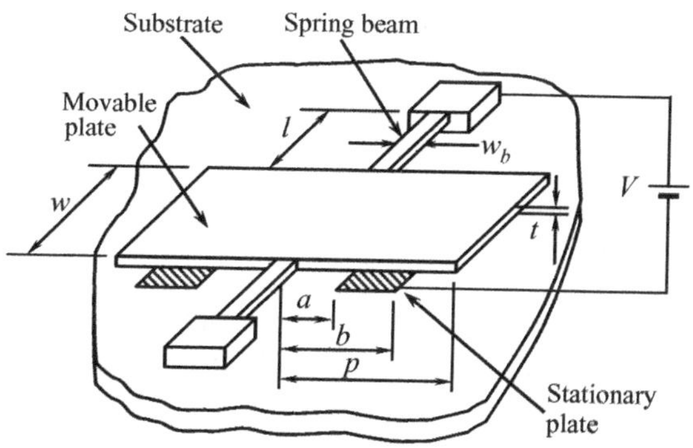

Figure 9.20 Electrostatic torsional actuator.

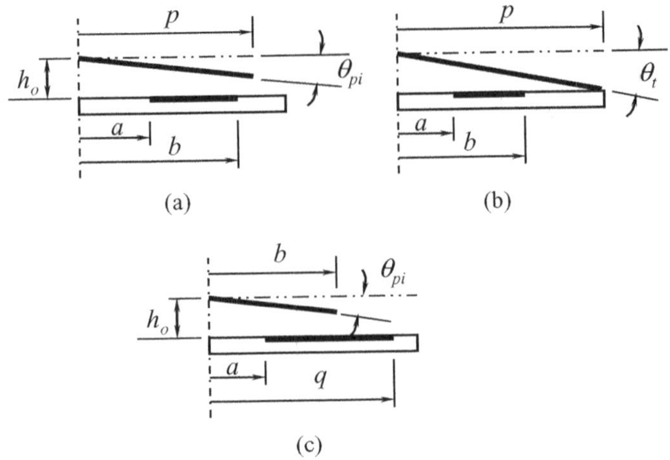

Figure 9.21 Possible final modes of a torsional actuator.

plate. In Fig. 9.21a and b, the starting and end points of the stationary plate are represented, respectively, by a and b, and the movable plate length is denoted by p. In Fig. 9.21c the length of the movable plate is given by b, while the starting and end points are represented by a and q, respectively. These configurations of the actuators can be categorized into two types of actuators: one with a stationary plate covered *fully* by a movable plate, and the other with a stationary plate covered *partially* by a movable plate. Before developing the theory of the torsional actuator, the final modes of the movable plate are examined when the applied voltage reaches the pull-in voltage. As shown in Fig. 9.21, two final modes of the movable plate are possible. When the applied voltage reaches the pull-in voltage, the movable plate may collapse angularly as shown in Fig. 9.21a and c. In the other mode (Fig. 9.21b), the edge of the movable plate can touch the substrate before the voltage reaches the pull-in voltage. When (9.117) is satisfied, the movable plate experiences the pull-in phenomenon before touching the substrate, so the movable plate can collapse as shown in Fig. 9.21a:

$$\frac{h_0}{p} > \theta_{pi} \tag{9.117}$$

where p, h_0, and θ_{pi} denote the plate length of the movable plate measured from the center of the spring beam, the initial gap of the movable plate, and the angle at a pull-in voltage, respectively. As the other criterion, (9.118) shows that the edge of the movable plate touches the substrate as shown Fig. 9.21b before the voltage reaches the pull-in voltage:

$$\frac{h_0}{p} \leq \theta_{pi} \tag{9.118}$$

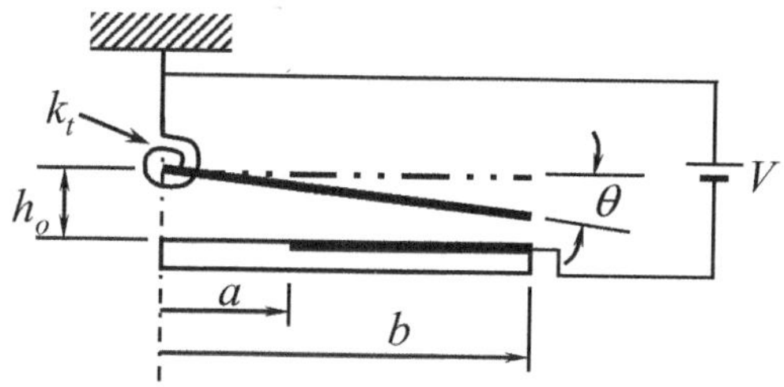

Figure 9.22 Model of a torsional actuator.

Figure 9.21b shows that the movable plate touches the substrate at $\theta_t = h_0/p$ less than the pull-in angle θ_{pi}. Therefore, the plate length, p, of Fig. 9.21 is used only to check the criteria, (9.117) and (9.118), and thus the parameter p will be ruled out in the development of the theory.

A model of the one-DOF torsional actuator, which consists of a rigid plate and a torsional spring, is set up as shown in Fig. 9.22. It is noted in Fig. 9.22 that the length of the movable plate, p, in Fig. 9.21 does not appear. When a voltage V is applied across the movable and stationary plates, the movable plate is displaced angularly by θ. From the fact that the electrostatic moment acting on the movable plate is equal to the restoring moment due to the torsional spring, the moment equilibrium equation is given by

$$k_t\theta = \frac{\varepsilon w V^2}{2\theta^2}\left(\frac{h_0}{h_0 - b\theta} - \frac{h_0}{h_0 - a\theta} + \ln\frac{h_0 - b\theta}{h_0 - a\theta}\right) \tag{9.119}$$

where the left- and right-hand sides denote the moment caused by the linear torsional spring of stiffness k_t and the electrostatic moment given by (7.102), respectively.

For convenience, equation (9.119) is expressed in dimensionless form as

$$\Theta = N\frac{2}{1-\alpha^2}\frac{1}{\Theta^2}\left(\frac{1}{1-\Theta} - \frac{1}{1-\alpha\Theta} + \ln\frac{1-\Theta}{1-\alpha\Theta}\right) \tag{9.120}$$

where

$$\Theta = \frac{b\theta}{h_0} \tag{9.121}$$

$$\alpha = \frac{a}{b} \tag{9.122}$$

$$N = \frac{\varepsilon b(b^2 - a^2)w}{4k_t h_0^3}V^2 \tag{9.123}$$

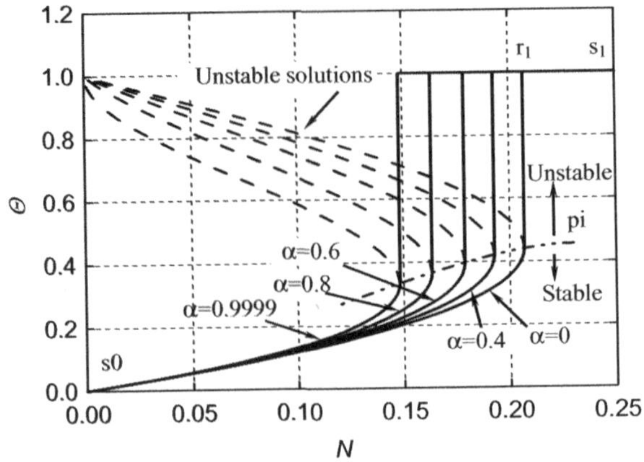

Figure 9.23 Dependence of the angular displacement on the moment.

The dimensionless parameters Θ, α, and N represent the angular displacement, the ratio of the distances of the starting and end points of the stationary plate, and the electrostatic moment, respectively. The dimensionless moment N in (9.123) is taken to make the relation $\Theta = N$ as Θ approaches 0. The torsional stiffness k_t of the two rectangular beams shown in Fig. 9.20 is obtained from (3.32) as

$$k_t = 2\frac{G_1 J}{l} \tag{9.124}$$

where G_1, J, and l denote the shear modulus defined by (3.15), the polar moment of inertia of the spring beam, and the length of the spring beam, respectively. For the rectangular beam, the polar moment of inertia is given by (3.33) and (3.34) depending on the ratio of the width to the thickness of the spring beam, w_b/t. Since the voltage V, the torsional stiffness k_t, and other dimensions are given, the unknown Θ is obtained by solving the nonlinear equation (9.120). We first examine (9.120) graphically and then solve it. To solve the nonlinear equation (9.120) for Θ, we can use a numerical method such as the bisection or Newton–Raphson method (for these methods, refer to other publications, e.g., Atkinson, 1978). In Fig. 9.23, the solution Θ is plotted against the dimensionless moment N. The plot of Fig. 9.23 is also obtained by calculating $N = f(\alpha, \Theta)$ as Θ varies from 0 to 1. As shown in Fig. 9.23, when the dimensionless moment N increases from zero (i.e., the applied voltage V increases), the dimensionless angular displacement Θ starts at zero and increases with N. However, the movable plate undergoes the pull-in phenomenon at pull-in voltages which depend on the ratio α. For $\alpha = 0$, as the dimensionless moment N increases from zero, the angular displacement Θ starts from s_0 (zero), follows the lower curve (stable solution), reaches the pull-in (labeled *pi*), jumps to r_1, and follows the line r_1–s_1, which represents mechanical contact of the movable plate with the stationary plate. For other α values, the same tendency is observed, and the pull-in moment and angular displacement decrease as the ratio increases from 0 to 1.

Approximate Moment Equilibrium Equation

To obtain an approximate solution of the moment equilibrium equation, (9.120), we expand the right-hand side of the equation. Noting the series expansions,

$$\frac{1}{1-x} = 1 + x + x^2 + x^3 + x^4 + \cdots$$

$$-\ln(1-x) = x + \frac{x^2}{2} + \frac{x^3}{3} + \frac{x^4}{4} + \cdots$$

we can express the moment equilibrium equation, (9.120), in series form as follows:

$$\begin{aligned}
\Theta &= N\frac{2}{1-\alpha^2}\frac{1}{\Theta^2}\left[\frac{1}{1-\Theta} - \frac{1}{1-\alpha\Theta} + \ln(1-\Theta) - \ln(1-\alpha\Theta)\right] \\
&= N\left[1 + \frac{2}{1-\alpha^2}\frac{2}{3}(1-\alpha^3)\Theta + \frac{2}{1-\alpha^2}\frac{3}{4}(1-\alpha^4)\Theta^2 + \cdots\right] \\
&= N\sum_{j=0}^{\infty}\beta_j\Theta^j
\end{aligned} \tag{9.125}$$

where β_j denotes the jth coefficient of the series and is given by

$$\beta_j = \frac{2}{1-\alpha^2}\frac{1+j}{2+j}(1-\alpha^{2+j}) \tag{9.126}$$

One may use a partial sum of the preceding series to give an approximation that is acceptable to estimate the right-hand side of (9.120). For example, if one chooses a quadratic equation as the partial sum of the preceding series, the moment equilibrium equation may be given by

$$\Theta = N\left[1 + \frac{2}{1-\alpha^2}\frac{2}{3}(1-\alpha^3)\Theta + \frac{2}{1-\alpha^2}\frac{3}{4}(1-\alpha^4)\Theta^2\right]$$

Solving the preceding quadratic equation, we can obtain the approximate solution Θ. However, the solution of the quadratic equation cannot give a good approximation to the solution shown in Fig. 9.23. The cubic equation also does not yield an acceptable solution. The reason is that the coefficients β_j increase as j increases (e.g., for $\alpha = 0$, $\beta_1 = 4/3$, $\beta_2 = 3/4$, $\beta_3 = 6/5$, ...). Thus, we must find an expression that gives a good approximation to the original moment equilibrium equation even though a relatively simple expression is used for the approximate equation. To find a better partial sum, let us examine an inverse series of series (9.125), which may be expressed in the following form:

$$\Theta = N\sum_{j=0}^{\infty}\beta_j\Theta^j = N\frac{1}{\sum_{j=0}^{\infty}\gamma_j\Theta^j} \tag{9.127}$$

where γ_j represents the coefficients of the inverse series. Multiplying both sides of equation (9.127) by the denominator on the right-hand side gives

$$\begin{aligned}
&\left(\sum_{j=0}^{\infty}\beta_j\Theta^j\right)\left(\sum_{j=0}^{\infty}\gamma_j\Theta^j\right)\\
&=(\beta_0\Theta^0+\beta_1\Theta^1+\beta_2\Theta^2+\cdots)(\gamma_0\Theta^0+\gamma_1\Theta^1+\gamma_2\Theta^2+\cdots)\\
&=\beta_0\gamma_0+(\beta_0\gamma_1+\beta_1\gamma_0)\Theta+(\beta_0\gamma_2+\beta_1\gamma_1+\beta_2\gamma_0)\Theta^2\cdots=1
\end{aligned}$$

Comparing the coefficients yields the coefficients γ_j in terms of the coefficient β_j of the original series:

$$\begin{aligned}
\gamma_0 &= \frac{1}{\beta_0} = 1\\
\gamma_1 &= -\beta_1\\
\gamma_2 &= -\gamma_1\beta_1 - \beta_2\\
\gamma_3 &= -\gamma_2\beta_1 - \gamma_1\beta_2 - \beta_3\\
\gamma_4 &= -\gamma_3\beta_1 - \gamma_2\beta_2 - \gamma_1\beta_3 - \beta_4\\
&\vdots
\end{aligned} \tag{9.128}$$

For $\alpha = 0$, the coefficients are obtained as $\gamma_0 = 1$, $\gamma_1 = -4/3$, $\gamma_2 = 5/18$, $\gamma_3 = 4/135$ and $\gamma_4 = 17/1620$. For $\alpha = 1/2$, the coefficients are $\gamma_0 = 1$, $\gamma_1 = -14/9$, $\gamma_2 = 353/648$, $\gamma_3 = 38/14{,}580$ and $\gamma_4 = 11{,}643/6{,}298{,}560$. As α approaches unity, γ_0, γ_1, and γ_2 go to 1, −2, 1, respectively, and the others become zero. It should be noted that the absolute value of γ_j decreases rapidly with increasing j (in particular, $j \geq 3$), while β_j increases as j increases. This trend in the coefficients γ_j suggests that the third partial sum of the series based on the coefficients γ_j gives a quadratic polynomial that sufficiently represents the moment due to electrostatic force at α and Θ. To confirm the use of the partial sum for the electrostatic moment, we define the nth partial sums, $f_{\beta,n}$ and $f_{\gamma n}$, normalized by the exact electrostatic moment of (9.120),

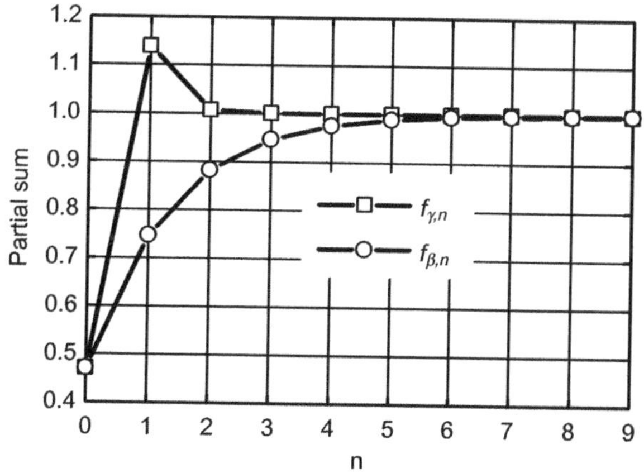

Figure 9.24 Partial sum of the series for the electrostatic moment at $\alpha = 0$ and $\Theta = 0.44$.

$$f_{\beta,n} = \frac{\sum_{j=0}^{n} \beta_j \Theta^j}{\frac{2}{1-\alpha^2}\frac{1}{\Theta^2}\left(\frac{1}{1-\Theta} - \frac{1}{1-\alpha\Theta} + \ln\frac{1-\Theta}{1-\alpha\Theta}\right)} \tag{9.129}$$

$$f_{\gamma,n} = \frac{1}{\sum_{j=0}^{n} \gamma_j \Theta^j} \frac{1}{\frac{2}{1-\alpha^2}\frac{1}{\Theta^2}\left(\frac{1}{1-\Theta} - \frac{1}{1-\alpha\Theta} + \ln\frac{1-\Theta}{1-\alpha\Theta}\right)} \tag{9.130}$$

The above nth partial sums at $\alpha = 0$ and $\Theta = 0.44$ are compared in Fig. 9.24. The partial sum $f_{\beta,n}$ derived from the original series slowly converges to unity as n increases, but the partial sum $f_{\gamma,n}$ based on γ_j rapidly approaches unity. A fifth-order polynomial (i.e., $f_{\beta,5}$) based on β_j gives an error of less than 1%, while a quadratic equation ($f_{\gamma,2}$) based on γ_j is in good agreement within an error of 0.65%. Therefore, the inverse of the quadratic equation represents the electrostatic moment of (9.120) and (9.125). Without loss of generality, the governing equation, (9.125), can be rewritten in a simple form:

$$N = \Theta \sum_{j=0}^{\infty} \gamma_j \Theta^j = \gamma_0 \Theta + \gamma_1 \Theta^2 + \gamma_2 \Theta^3 + \gamma_3 \Theta^4 + \cdots \tag{9.131}$$

Since the absolute value of γ_j for $j > 2$ is much less than that of γ_0, γ_1, and γ_2, there are few terms higher than third order in equation (9.131), and we take the following cubic equation to obtain the approximate analytic solution:

$$r\Theta^3 - s\Theta^2 + \Theta - N = 0 \tag{9.132}$$

where

$$r = \gamma_2 = -\gamma_1\beta_1 - \beta_2 = \beta_1^2 - \beta_2 = \frac{32(1+\alpha+\alpha^2)^2 - 27(1+\alpha)(1+\alpha+\alpha^2+\alpha^3)}{18(1+\alpha)^2} \tag{9.133}$$

$$s = -\gamma_1 = \beta_1 = \frac{4(1+\alpha+\alpha^2)}{3(1+\alpha)} \tag{9.134}$$

are new coefficients for the cubic equation and have positive values. For $\Theta << 1$, equation (9.132) is further simplified as

$$\Theta - N = 0 \quad \text{or} \quad \Theta = N \tag{9.135}$$

which states that if the dimensionless moment N is much less than the pull-in moment (which will be obtained subsequently), the dimensionless angular displacement Θ is equal to the dimensionless moment N. That is, for $V << V_{pi}$ (V_{pi} is obtained later), referring to (9.121) and (9.123), equation (9.135) is expressed as

$$\theta = \frac{\varepsilon(b^2 - a^2)w}{4k_t h_0^2} V^2 \tag{9.136}$$

Therefore, the dimensional angular displacement θ is proportional to the square of the applied voltage V and inversely proportional to the torsional stiffness k_t.

Analytic Solution of the Moment Equilibrium Equation

We can find the solution of the preceding cubic moment equilibrium equation by applying the solution procedure described in Section 5.1. Setting $L = r$, $A_1 = -s$, $B_1 = 1$, and $C_1 = -N$ and assuming that the dimensionless moment is not greater than the pull-in moment (i.e., the discriminant Δ is less than or equal to zero), we obtain the three solutions from (5.17) and (5.18), as follows:

$$\begin{aligned}
\Theta_1 &= \frac{1}{3r}\left(s + 2\sqrt{s^2 - 3r}\cos\frac{\phi_2}{3}\right) \\
\Theta_2 &= \frac{1}{3r}\left[s + 2\sqrt{s^2 - 3r}\cos\left(\frac{\phi_2}{3} + \frac{2\pi}{3}\right)\right] \\
\Theta_3 &= \frac{1}{3r}\left[s + 2\sqrt{s^2 - 3r}\cos\left(\frac{\phi_2}{3} + \frac{4\pi}{3}\right)\right]
\end{aligned} \tag{9.137}$$

where

$$\phi_2 = \cos^{-1}\frac{s^3-(9/2)sr+(27/2)r^2N}{\left(s^2-3r\right)^{3/2}} \tag{9.138}$$

Examining the three solutions, we find that Θ_2 is the stable solution, Θ_3 the unstable solution, and Θ_1 has no physical meaning. Therefore, we take Θ_2 as the solution Θ:

$$\Theta = \frac{1}{3r}\left[s+2\sqrt{s^2-3r}\cos\left(\frac{\phi_2}{3}+\frac{2\pi}{3}\right)\right] \tag{9.139}$$

Using the relation $\cos^{-1}(J) = \cos^{-1}(-J) + \pi$ and the identity $\cos(J + \pi) = -\cos(J)$, the cosine function of equation (9.139) may be written

$$\cos\left(\frac{\cos^{-1}T}{3}+\frac{2\pi}{3}\right) = \cos\left[\frac{\cos^{-1}(-T)+\pi}{3}+\frac{2\pi}{3}\right]$$
$$= \cos\left(\frac{\cos^{-1}(-T)}{3}+\pi\right) = -\cos\frac{\cos^{-1}(-T)}{3}$$

where T denotes the argument of the arcsine function in (9.138). We can write the preceding stable solution Θ in a convenient form as follows:

$$\Theta = \frac{1}{3r}\left(s-2\sqrt{s^2-3r}\cos\frac{\phi}{3}\right) \tag{9.140}$$

where

$$\phi = \cos^{-1}\frac{(9/2)sr-s^3-(27/2)r^2N}{\left(s^2-3r\right)^{3/2}} \tag{9.141}$$

Similarly, if we use $\cos^{-1}(J) = \cos^{-1}(-J) + 3\pi$, the cosine function of the third equation of (9.137) may be written

$$\cos\left(\frac{\cos^{-1}T}{3}+\frac{4\pi}{3}\right) = \cos\left[\frac{\cos^{-1}(-T)+3\pi}{3}+\frac{4\pi}{3}\right]$$
$$= \cos\left[\frac{\cos^{-1}(-T)}{3}+\frac{4\pi}{3}+\pi\right] = -\cos\left[\frac{\cos^{-1}(-T)}{3}+\frac{4\pi}{3}\right]$$

and then we obtain the unstable solution Θ_U in terms of ϕ:

$$\Theta_U = \Theta_3 = \frac{1}{3r}\left[s-2\sqrt{s^2-3r}\cos\left(\frac{\phi}{3}+\frac{4\pi}{3}\right)\right] \tag{9.142}$$

where ϕ is given by (9.141).

When the dimensionless moment N increases from zero (i.e., $V = 0$) to its pull-in value (N_{pi}), which corresponds to the condition that the discriminant of the cubic equation is zero, $\Delta = 0$, ϕ varies from an initial value to π. Substituting $\phi = \pi$ into (9.141), we have the dimensionless pull-in moment, N_{pi}, as

$$N_{pi} = \frac{1}{27r^2}\left[9rs - 2s^3 + 2(s^2 - 3r)^{3/2}\right] \tag{9.143}$$

Upon substitution of $\phi = \pi$ into (9.140), we find the dimensionless pull-in angular displacement,

$$\Theta_{pi} = \frac{1}{3r}\left(s - \sqrt{s^2 - 3r}\right) \tag{9.144}$$

The preceding pull-in angular displacement and moment are also obtained from the infinite sensitivity condition for the pull-in ($\partial\Theta/\partial N = \infty$ or $\partial N/\partial\Theta = 0$); that is, differentiating (9.132) with respect to Θ and setting $\partial N/\partial\Theta = 0$ gives the dimensionless pull-in angular displacement, (9.144), and substituting Θ_{pi} into (9.132) yields the dimensionless pull-in moment, (9.143). The dimensionless angular displacement Θ and moment N are converted into the angular displacement θ in radians and the voltage V by using

$$\theta = \Theta\frac{h}{b} \tag{9.145}$$

$$V = \sqrt{N\frac{4k_t h_0^3}{\varepsilon b(b^2 - a^2)w}} \tag{9.146}$$

Thus, the pull-in voltage corresponding to N_{pi}, (9.143), is obtained as

$$V_{pi} = \sqrt{N_{pi}\frac{4k_t h_0^3}{\varepsilon b(b^2 - a^2)w}}$$

Since the capacitance C formed between the movable and stationary plates is given by (7.100), which is rewritten here for convenience:

$$C = \frac{\varepsilon w}{\theta}\ln\frac{h_0 - a\theta}{h_0 - b\theta}$$

the dimensionless capacitance is expressed as

$$\frac{C}{C_o} = \frac{1}{1-\alpha}\frac{1}{\Theta}\ln\frac{1-\alpha\Theta}{1-\Theta} \tag{9.147}$$

where

$$C_0 = \frac{\varepsilon w(b-a)}{h_0} \tag{9.148}$$

is the initial capacitance before a voltage is applied.

We shall validate the analytic solution for angular displacement by comparing the analytic solution with the numerical solution of the original equation, (9.120). Since the maximum error between the analytic and numerical solutions occurs at pull-in, the pull-in moment N_{pi} and angular displacement Θ_{pi} are plotted in Figs. 9.25 and 9.26. In Fig. 9.25, as α varies from 0 to 1, N_{pi} starts from $4\left(-38+17\sqrt{17/2}\right)/225$ and approaches 4/27. For the same range of α, Θ_{pi} decreases from $2\left(4-\sqrt{17/2}\right)/5$ to 1/3, as shown in Fig. 9.26. The values of the analytic pull-in angular displacement [(9.144) and Fig. 9.26] and moment [(9.143) and Fig. 9.25] are in good agreement with the numerically solved pull-in values (Lee, 2008a) within errors of 1.50% and 0.63% over the entire range of α $(0 \le \alpha \le 1)$, respectively. For $\alpha = 0$, (9.143) and (9.144) give the pull-in moment of $N_{pi} = 4\left(-38+17\sqrt{17/2}\right)/225 \cong 0.2056$ and the corresponding pull-in angular displacement of $\Theta_{pi} = 2\left(4-\sqrt{17/2}\right)/5 \cong 0.4338$, respectively. For the same α range, the numerical analysis yields $N_{pi} = 0.2069$ and $\Theta_{pi} = 0.4404$, respectively. In Fig. 9.27, maximum error in the dimensionless moment and angular displacement occur at $\alpha = 0$ and are 0.63% and 1.50% at $\alpha = 0$, respectively. It is noted that the maximum error in the pull-in voltage, corresponding to the maximum error in the dimensionless moment, is 0.32%. Errors for both the dimensionless pull-in moment and angular displacement approaches zero as the ratio α goes to unity. Using the set of data $a = 8\times10^{-6}\,\text{m}, b = 40\times10^{-6}\,\text{m}, w = 40\times10^{-6}\,\text{m}, h_0 = 2\times10^{-6}\,\text{m}$, and $k_t = 3.37\times10^{-8}$ N · m, the analytic response of a typical one-DOF torsional actuator is

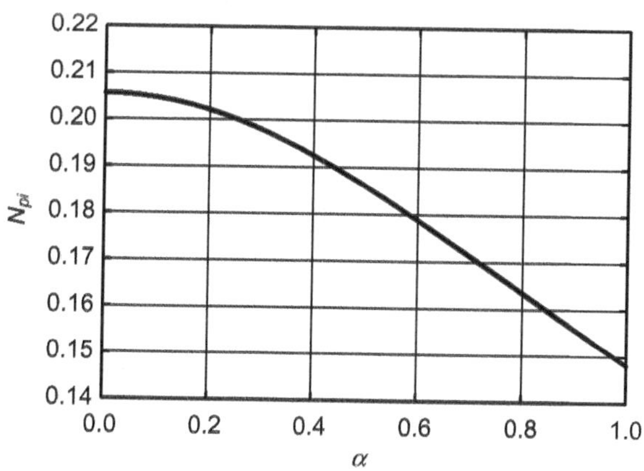

Figure 9.25 Pull-in moment N_{pi} as a function of α.

compared to the numerical response in Fig. 9.28. The analytic angular displacement value is in good agreement with the numerical value within an error of 0.94%, and the maximum error between the analytical and numerical pull-in voltages is estimated as 0.20%. For example, both analytical and numerical responses from Fig. 9.28 provide the same angular displacement of 0.4702° at a voltage of 80V. So far, under the assumption that the movable plate is tilted without downward displacement of the spring beams, we have evaluated the angular displacement of the movable plate. However, if the downward displacement is not negligible, the analytic angular displacement obtained above will be less than that given by (9.140). This situation is described in detail in the following paragraphs.

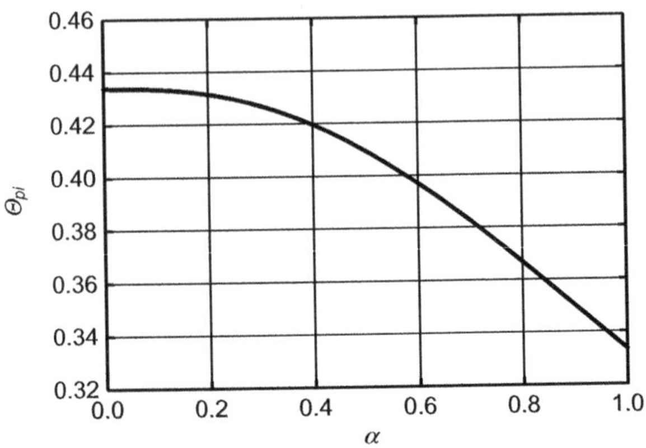

Figure 9.26 Pull-in angular displacement Θ_{pi} as a function of α.

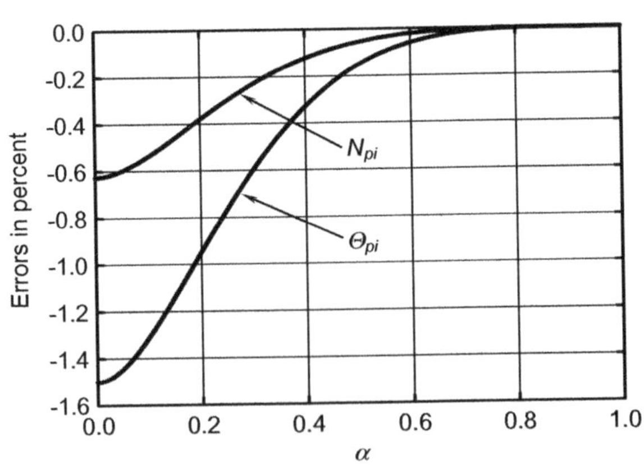

Figure 9.27 Error between theoretical and numerical solutions in the entire range of α.

Two-DOF Torsional Actuator

If the spring beams in the x direction in Fig. 9.29 are very stiff, the electrostatic torsional actuator can be considered as the one-DOF torsional actuator whose analytical solutions have been obtained in the preceding paragraphs. However, the spring beams are compliant in the x direction in Fig. 9.29, and the movable plate under the electrostatic force and moment are both displaced in the x direction and tilted in the θ direction. Therefore, we should consider the torsional actuator in Fig. 9.29 as a two-DOF torsional actuator whose behavior will be expressed in terms of the displacement x and the angular displacement θ. Recalling the electrostatic force, (7.103), and moment, (7.104), we can set up the force and moment equilibrium equations as follows:

$$k(h_0 - h) = F_e = \frac{\varepsilon w V^2}{2\theta}\left(\frac{1}{h - b\theta} - \frac{1}{h - a\theta}\right) \tag{9.149}$$

$$k_t \theta = M_e = \frac{\varepsilon w V^2}{2\theta^2}\left(\frac{h}{h - b\theta} - \frac{h}{h - a\theta} + \ln\frac{h - b\theta}{h - a\theta}\right) \tag{9.150}$$

where k, k_t, h_0, and h represent the linear and torsional stiffnesses due to the two springs, the initial gap at $V = 0$, and the gap at V, respectively. As shown in nonlinear equations (9.149) and (9.150), when a voltage is applied between the movable and stationary plates, the responses h and θ are coupled to each other. Thus, we cannot derive exact solutions of the equations but can obtain numerical solutions by using a numerical scheme. Strategies for obtaining numerical solutions may be found in existing publications. The following

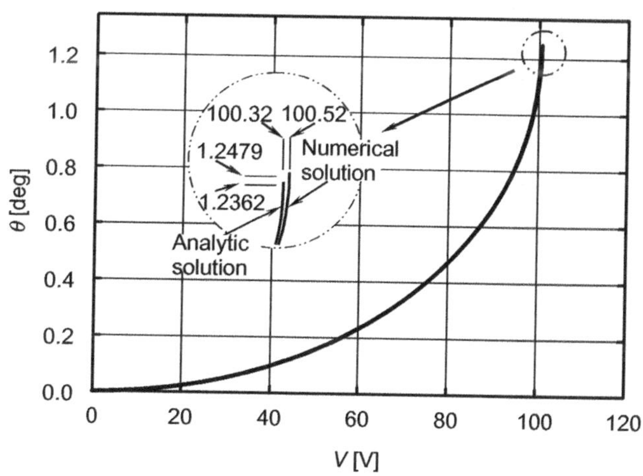

Figure 9.28 Theoretical and numerical responses of a one-DOF torsional actuator.

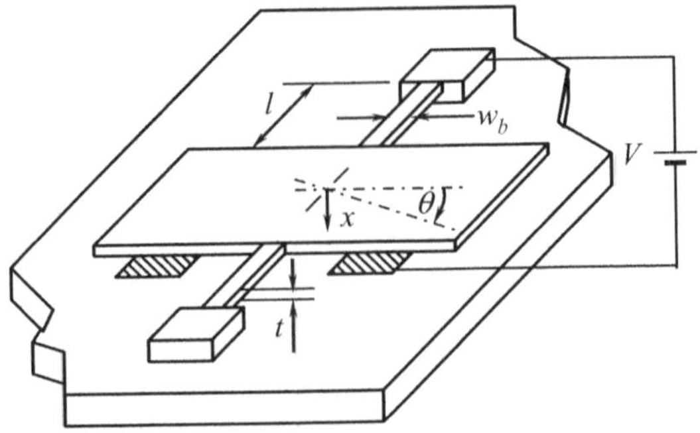

Figure 9.29 Two-degree-of-freedom torsional actuator.

discussion will be useful for analyzing and designing a torsional actuator; for formulation and numerical analysis, the reader is referred to Lee (2007).

To describe the behavior of a two-DOF torsional actuator, we define one more dimensionless parameter:

$$R = \frac{k_t}{kb[(a+b)/2]} = f(E, \nu, \text{geometry}) \tag{9.151}$$

where R is a ratio of the dimensionless force to the dimensionless moment; that is, $R = G/N$, where G is the dimensionless force (9.51) with $A = (b - a)w$ and N is the dimensionless moment defined by (9.123). The ratio R is not a function of the applied voltage V but a function of k, k_t, a, and b. R is then a function of Young's modulus E, Poisson's ratio ν, and the geometry of the actuator because the stiffness k and k_t are related to E, ν, and the spring geometry.

The responses, h and θ, are obtained as functions of the voltage V, the force to moment ratio R, the ratio $\alpha = a/b$, which is defined by (9.122), and geometry. For $\alpha = 0$, the numerical solutions for H_{pi} and Θ_{pi} are shown in Fig. 9.30, in which $H_{pi} = h_{pi}/h_0$ and $\Theta_{pi} = \theta_{pi}b/h_0$. Both H_{pi} and Θ_{pi} rapidly decrease when R varies from 0 to 5. Since the pull-in voltage V_{pi}, gap h_{pi}, and angular displacement θ_{pi} give a design guideline of the two-DOF torsional actuator, we express the pull-in parameters in terms of α, R, and the geometric dimensions, as follows: The pull-in voltage

$$V_{pi} = 0.9097\sqrt{\frac{1}{1 + c_1 + c_2 R}\frac{k_t h_0^3}{\varepsilon b(b^2 - a^2)w}} \tag{9.152}$$

where

$$c_1 = 0.006383\alpha + 0.4757\alpha^2 - 0.08543\alpha^3$$

$$c_2 = 1.3965$$

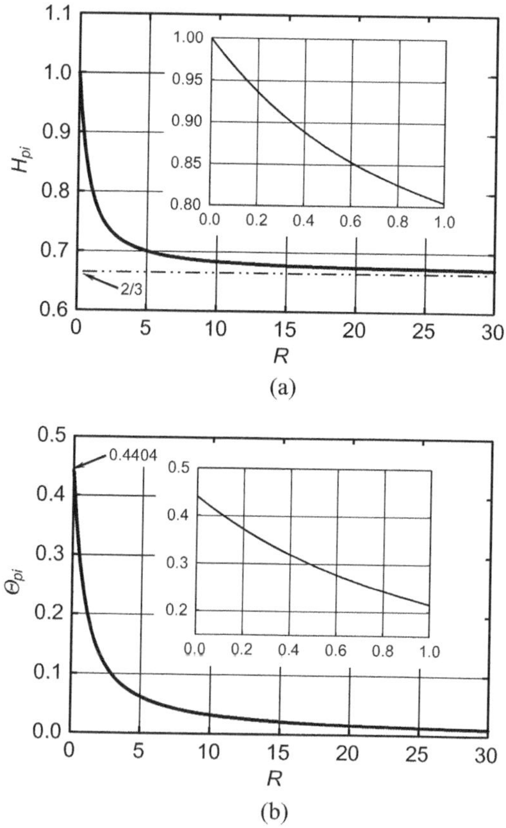

Figure 9.30 Pull-in gap and angular displacement at $\alpha = 0$.

the pull-in gap

$$h_{pi} = h_0 \frac{2(1+R^{c_3})}{2+3R^{c_3}} \tag{9.153}$$

where

$$c_3 = 1.1339 - 0.2404\alpha + 0.2200\alpha^2$$

and the pull-in angular displacement,

$$\theta_{pi} = \frac{h_0}{b} \frac{0.4404}{1+c_4+c_5 R} \tag{9.154}$$

where

$$c_4 = 0.3212\alpha^{2.1095}$$

$$c_5 = 1.2894 + 0.07030\alpha - 0.03997\alpha^2$$

The maximum errors in the range $0 \le \alpha \le 1$ and $0 \le R \le \infty$ are estimated as 3.4% for V_{pi}, 4.2% for h_{pi}, and 11% θ_{pi}, respectively. If one obtains the angular displacement θ at a voltage V under the assumption that a two-DOF torsional actuator is approximated as a one-DOF torsional actuator, the calculation error between one- and two-DOF torsional actuators will increase as the ratio R increases. Referring to the small plots in Figs. 9.30a and b, we can express the pull-in gap H_{pi} and the pull-in angular displacement Θ_{pi} for $\alpha = 0$ and $R \le 1$ in closed forms,

$$H_{pi} = \frac{h}{h_0} = 1 - 0.3362R + 0.1456R^2 \tag{9.155}$$

$$\Theta_{pi} = \frac{\theta}{h_0/b} = 0.4404 - 0.3589R + 0.1388R^2 \tag{9.156}$$

The maximum errors between the data in Fig. 9.30 and the curve-fitting functions above are 0.5434% for H_{pi} and 1.6799% Θ_{pi}, respectively. If we calculate the pull-in angular displacement Θ_{pi} from a one-DOF torsional actuator instead of using a two-DOF torsional actuator with R and α, the following error in the pull-in angular displacement Θ_{pi} may be involved in the angular displacement: 1.8% at $R = 0.02$ and 9.0% at $R = 0.1$. We note from Fig. 9.30a and b that when R increases beyond 30, H_{pi} approaches 2/3 and Θ_{pi} goes to zero (i.e., the static response of a two-DOF torsional actuator approaches that of a parallel-plate actuator, which was dealt with in the preceding section). Therefore, if a two-DOF torsional actuator is operated in the range, $R \le R_1$, where R_1 can be determined depending on allowable error, the actuator can be considered to be a one-DOF actuator whose analytical solution was obtained earlier. For example, if $R_1 = 0.02$ and $\alpha = 0$, Θ_{pi} obtained from a one-DOF torsional actuator is 1.8% greater than that from a two-DOF torsional actuator (Fig. 9.30b). Similarly, if $R = 20$, the gap h is 1.25% greater than the pull-in gap 2/3, which is the pull-in value of the parallel-plate actuator. For $R \ge 20$, a two-DOF torsional actuator can be considered to be a parallel-plate actuator. Therefore, if the two-DOF torsional actuator whose R is in the range

$$R_1 \le R \le R_2 \tag{9.157}$$

where R_1 and R_2 are determined depending on allowable error (e.g., $R_1 = 0.02$ and $R_2 = 20$, as discussed above). Therefore, if $R \le R_1$, we can consider the two-DOF torsional actuator to be a one-DOF torsional actuator and obtain its response from (9.140). If $R > R_2$, the torsional actuator is approximated as the parallel-plate actuator, and the displacement is determined from the analytical solution (9.55).

The capacitance of the two-DOF variable capacitor is given by

$$C_{pi} = \frac{\varepsilon w(b-a)}{h_0} \frac{3D(1+r_1)}{3+2Dr_1} \tag{9.158}$$

where

$$D = 1.3182 + 0.3372\alpha - 0.1052\alpha^2 - 0.05087\alpha^3$$

$$r_1 = c_6 R + c_7 R^2$$

$$c_6 = 1.4887 + 1.7946\alpha - 1.2\alpha^2$$

$$c_7 = 1.3721 + 0.9757\alpha - 1.4\alpha^2$$

represent the normalized pull-in capacitance at $R = 0$, a correction function, and coefficients for r_1, respectively. Capacitance equation (9.158), a curve-fitting expression obtained using numerical analysis, gives a very good approximation of the pull-in capacitance within an error of 0.1% in the range $\alpha = 0$ to 1 and $R = 0$ to ∞.

For convenience, the ratio R, (9.151), may be expressed as follows: For $w_b \geq t$,

$$R = \frac{\gamma_a}{(1+\nu)} \frac{E}{\tilde{E}} \frac{1}{1+\dfrac{6}{5}\dfrac{\sigma_0}{\tilde{E}}\left(\dfrac{l}{t}\right)^2} \frac{l^2}{(a+b)b} \tag{9.159}$$

and for $w_b \leq t$,

$$R = \frac{\gamma_a}{(1+\nu)} \frac{E}{\tilde{E}} \frac{1}{1+\dfrac{6}{5}\dfrac{\sigma_0}{\tilde{E}}\left(\dfrac{l}{t}\right)^2} \frac{l^2}{(a+b)b} \left(\frac{w_b}{t}\right)^2 \tag{9.160}$$

where w_b and t denote the width and thickness of the spring beam as illustrated in Fig. 9.29 and γ_a is given by

$$\gamma_a = \frac{1}{3}(1 - 0.616\beta_a - 0.087\beta_a^2 + 0.125\beta_a^3) \tag{9.161}$$

where β_a is defined as t/w_b for $w_b \geq t$ and w_b/t for $w_b \leq t$, respectively. To obtain R above, the following were substituted in to (9.151): the stiffness (3.75), the torsional stiffness (3.32) with the polar moment of inertia (3.33) and (3.34), the moment of inertia $I = w_b t^3/12$, and the axial force $N_0 = \sigma_0 w_b t$ due to the residual stress σ_0 in the beam. $\tilde{E}$ denotes the effective Young's modulus given by (3.5) and (3.6). It is noted that the R above hold if $\sigma_0 > -(5/6)(t/l)^2\,\tilde{E}$ (i.e., the beam is not buckled).

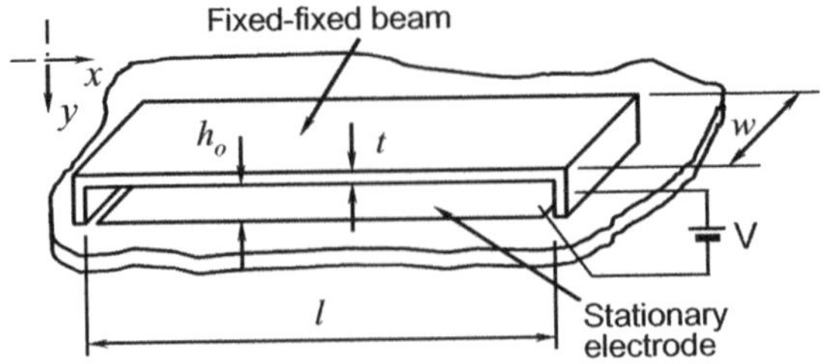

Figure 9.31 Fixed–fixed beam actuator.

9.5 FIXED–FIXED BEAM ACTUATOR

In MEMS, continuous structures subjected to an electrostatic force are used as actuators and sensors. For example, the fixed–fixed beam shown in Fig. 9.31 is used as an electrostatic attractive force actuator which is operated by a voltage V. A cantilever beam under an electrostatic force, which is described in the next section, is also used as an actuator. In Fig. 9.31, the fixed–fixed beam actuator consists of a straight beam of length l, width w, and thickness t, and a stationary electrode placed on a substrate. When a voltage V is applied between the beam and the stationary electrode, the beam is deflected downward and the gap between the beam and the stationary electrode is reduced as shown in Fig. 9.32a. Let x, y, y_m, and h represent the coordinate from the left end of the beam, the beam deflection at x, the midspan deflection, and the gap at x, respectively. We assume that the beam thickness t and the initial gap h_0 are much less than the beam length l. If the beam is subjected to an axial force N_0 (the tensile force is defined to be positive), the force equilibrium of the free body shown in Fig. 9.32b gives the beam equation (3.54), which is rewritten here for convenience:

$$\tilde{E}I\frac{d^4y}{dx^4}-\left[N_0+\frac{1}{2}\frac{\tilde{E}A}{l}\int_0^l\left(\frac{dy}{dx}\right)^2dx\right]\frac{d^2y}{dx^2}=q \tag{9.162}$$

where $\tilde{E}I$, N_0, A, and q denote, respectively, the flexural rigidity of the beam, the axial force, the cross-sectional area of the beam, and the distributed load (force per unit length) caused by the electrostatic force. Let us consider the small element of length dx and width w shown in Fig. 9.32b. For $t << w$, we obtain the downward electrostatic force acting on the element from (7.88) as follows:

$$dF = q\,dx = \frac{\varepsilon_0 dAV^2}{2h^2} = \frac{\varepsilon_0 wV^2}{2h^2}dx$$

which gives the distributed load,

$$q = \frac{\varepsilon_0 wV^2}{2h^2} \tag{9.163}$$

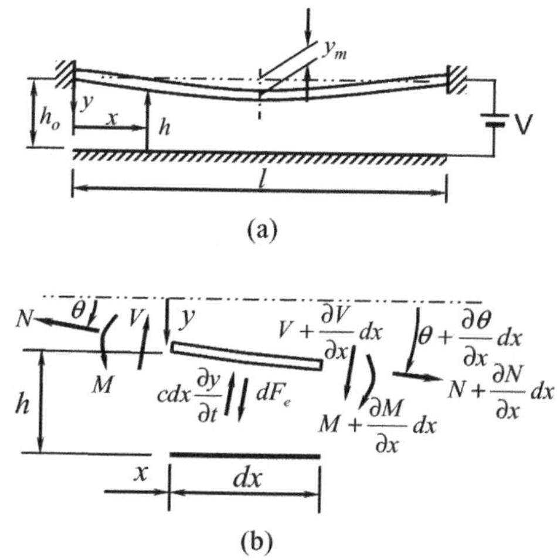

Figure 9.32 Model of a fixed–fixed beam actuator.

Upon substitution of the preceding load into (9.162), we have the following equation, which can be used to understand the behavior of the beam:

$$\tilde{E}I\frac{d^4y}{dx^4}-\left[N_0+\frac{1}{2}\frac{\tilde{E}A}{l}\int_0^l\left(\frac{dy}{dx}\right)^2dx\right]\frac{d^2y}{dx^2}=\frac{\varepsilon_0 wV^2}{2h^2} \tag{9.164}$$

This equation includes two nonlinear terms: the geometric nonlinear term on the left hand side, and the electrostatic force term on the right-hand side.

If a certain condition is satisfied, the geometric nonlinearity becomes negligible. When the fixed–fixed beam satisfies (3.114), which is written

$$\frac{y_m}{t}<<0.8472\sqrt{1+\frac{2}{7}\frac{\sigma_0}{\tilde{E}}\left(\frac{l}{t}\right)^2} \tag{9.165}$$

where σ_0 denotes the stress defined by the axial force divided by the cross-sectional area of the beam, the geometric nonlinearity can be neglected and then differential equation (9.164) can be written

$$\tilde{E}I\frac{d^4y}{dx^4}-N_0\frac{d^2y}{dx^2}=\frac{\varepsilon wV^2}{2(h_0-y)^2} \tag{9.166}$$

where h_0-y denotes the gap h. Since $y_m/h_0=0.3880$ at pull-in, which will be derived in subsequent paragraphs, the preceding condition is expressed as follows:

$$h_0 << 2.1835t\sqrt{1+\frac{2}{7}\frac{\sigma_0}{\tilde{E}}\left(\frac{l}{t}\right)^2} \tag{9.167}$$

When the condition above is satisfied, the geometric nonlinearity from the fixed–fixed beam is negligible and we can solve (9.166) to obtain the deflection y. If condition (9.167) is not satisfied, the geometric nonlinearity cannot be neglected and we should solve the more complicated beam equation (9.164) for y.

Under the assumption that the fixed–fixed beam in Fig. 9.32a satisfies the preceding condition and that the beam thickness t and the initial gap h are much less than the length l, we solve nonlinear differential equation (9.166) with the boundary conditions $y = 0$ and $dy/dx = 0$ at $x = 0$ and $x = l$. Even though the preceding differential equation has the linear terms on the left-hand side, the electrostatic force is still nonlinear. Thus, we cannot obtain an exact solution for y but will obtain an approximate solution based on Galerkin's method, which are described in Section 3.7. Since the electrostatic force is distributed on the lower face of the beam, we shall use the following mode ψ_1 shape, which is given by (3.88):

$$y_1 = y_m\psi_1 \tag{9.168}$$

where

$$\psi_1 = 16\left(\frac{x}{l}\right)^2\left(1-\frac{x}{l}\right)^2 = \eta\psi(X) \tag{9.169}$$

$$\psi(X) = \left(\frac{x}{l}\right)^2\left(1-\frac{x}{l}\right)^2 = X^2(X-1)^2 \tag{9.170}$$

where $\eta = 16$. Substituting (9.168) for y into (9.166), multiplying by ψ_1, and integrating the resulting equation from 0 to l, we have a force equilibrium equation,

$$k_1 y_m = F_e \tag{9.171}$$

where

$$k_1 = \int_0^l \tilde{E}I\frac{d^4\psi_1}{dx^4}\psi_1\,dx - \int_0^l N_0\frac{d^2\psi_1}{dx^2}\psi_1\,dx \tag{9.172}$$

$$F_e = \int_0^l \frac{\varepsilon w V^2}{2(h_0 - y_m\psi_1)^2}\psi_1\,dx \tag{9.173}$$

represent a linear stiffness and an electrostatic force.

For convenience, we introduce the following dimensionless parameters:

$$Y_m = \frac{y_m}{h_0} \qquad X = \frac{x}{l} \tag{9.174}$$

which represent the midspan deflection and the coordinate from along the beam, respectively. Substituting (9.169) into (9.172) and recalling integration by parts, we have the following equation [whose derivation was described in (3.101)–(3.103)]:

$$\begin{aligned} k_1 &= \int_0^l \tilde{E}I \frac{d^4\psi_1}{dx^4}\psi_1\,dx - \int_0^l N_0 \frac{d^2\psi_1}{dx^2}\psi_1\,dx \\ &= \eta^2 \frac{\tilde{E}I}{l^3}\int_0^1 \left(\frac{d^2\psi}{dX^2}\right)^2 dX + \eta^2 \frac{N_0}{l}\int_0^1 \left(\frac{d\psi}{dX}\right)^2 dX \end{aligned} \tag{9.175}$$

Substituting (9.170) and the constant $\eta = 16$ into (9.175), we obtain the stiffness k_1 for the fixed–fixed beam as follows:

$$k_1 = \eta^2 \frac{4\tilde{E}I}{5l^3} + \eta^2 \frac{2N_0}{105l} = \frac{1024}{5}\frac{\tilde{E}I}{l^3} + \frac{512}{105}\frac{N_0}{l} \tag{9.176}$$

where the first and second terms represent the stiffnesses due to the bending moment and axial force acting on the beam, respectively.

The electrostatic force F_e is more complex than the stiffness. To express the integration for the electrostatic force, we postulate that gap-closing actuators (including the fixed–fixed beam) under electrostatic attractive force experience the pull-in phenomenon. We shall first express the electrostatic force F_e in a series and then obtain the inverse series from the series as discussed in the preceding section. We expand the integrand of (9.173) in a series and integrate the series as follows:

$$\begin{aligned} F_e &= \int_0^l \frac{\varepsilon w V^2}{2(h_0 - y_m\psi_1)^2}\psi_1\,dx = \eta\frac{\varepsilon wlV^2}{2h_0^2}\int_0^1 \frac{\psi}{(1-Y_m\eta\psi)^2}dX \\ &= \eta\frac{\varepsilon wlV^2}{2h_0^2}\int_0^1 \psi\left[1+2Y_m\eta\psi+3(Y_m\eta\psi)^2+4(Y_m\eta\psi)^3+5(Y_m\eta\psi)^4+\cdots\right]dX \\ &= \eta\frac{\varepsilon wlV^2}{2h_0^2}(\phi_1+2\eta\phi_2Y_m+3\eta^2\phi_3Y_m^2+4\eta^3\phi_4Y_m^3+5\eta^4\phi_5Y_m^4+\cdots) \\ &= \eta\frac{\varepsilon wlV^2}{2h_0^2}\phi_1(1+\beta_1Y_m+\beta_2Y_m^2+\beta_3Y_m^3+\beta_4Y_m^4+\cdots) \\ &= \eta\frac{\varepsilon wlV^2}{2h_0^2}\phi_1\sum_{j=0}^{\infty}\beta_jY_m^j \end{aligned} \tag{9.177}$$

where $\phi_1, \phi_2, \ldots$ and $\beta_1, \beta_2, \ldots$ denote coefficients for the infinite series, given by

$$\phi_1 = \int_0^1 \psi \, dX, \quad \phi_2 = \int_0^1 \psi^2 \, dX, \quad \phi_3 = \int_0^1 \psi^3 \, dX, \quad \phi_4 = \int_0^1 \psi^4 \, dX,$$
$$\phi_5 = \int_0^1 \psi^5 \, dX, \quad \ldots \tag{9.178}$$

$$\beta_1 = 2\eta \frac{\phi_2}{\phi_1}, \quad \beta_2 = 3\eta^2 \frac{\phi_3}{\phi_1}, \quad \beta_3 = 4\eta^3 \frac{\phi_4}{\phi_1}, \quad \beta_4 = 5\eta^4 \frac{\phi_5}{\phi_1}, \quad \ldots \tag{9.179}$$

Substituting the mode shape ψ (9.170) into equations (9.178) and (9.179), we obtain the coefficients. However, integration may become a time-consuming job, so we need to find a simple formula for the integration. This integration formula may be obtained by applying integration by parts to the preceding integration. The general form of the preceding integration is expressed as

$$\begin{aligned} \int_0^1 X^m (X-1)^n \, dX &= \frac{1}{n+1} X^m (X-1)^{n+1} \bigg|_0^1 - \frac{m}{n+1} \int_0^1 X^{m-1} (X-1)^{n+1} \, dX \\ &= \cdots \\ &= (-1)^n \frac{m!n!}{(m+n+1)!} \end{aligned} \tag{9.180}$$

With formula (9.180), we can easily obtain the coefficients:

$$\begin{aligned} \phi_1 &= \int_0^1 X^2 (X-1)^2 \, dX = (-1)^4 \frac{2! \times 2!}{5!} = \frac{2}{5 \times 4 \times 3} = \frac{1}{30} \\ \phi_2 &= \int_0^1 \left[X^2 (X-1)^2 \right]^2 dX = \frac{4! \times 4!}{9!} = \frac{1}{630} \\ \phi_3 &= 1/12{,}012, \quad \phi_4 = 1/218{,}790, \quad \phi_5 = 1/3{,}879{,}376, \quad \ldots \end{aligned} \tag{9.181}$$

We then obtain the coefficients β_j as follows:

$$\begin{aligned} \beta_1 &= 32/21 \cong 1.5238 \\ \beta_2 &= 1920/1001 \cong 1.9181 \\ \beta_3 &= 16{,}384/7293 \cong 2.2465 \\ \beta_4 &= 819{,}200/323{,}323 \cong 2.5337 \\ &\vdots \end{aligned} \tag{9.182}$$

The coefficients β_j increase as j increases, so we expect that the partial sum of the preceding series will slowly converge.

The foregoing infinite series may be defined as a new function,

$$f_\beta = 1 + \beta_1 Y_m + \beta_2 Y_m^2 + \beta_3 Y_m^3 + \beta_4 Y_m^4 + \cdots = \sum_{j=0}^{\infty} \beta_j Y_m^j \tag{9.183}$$

whose form is the same as that of the infinite series of the electrostatic force, (9.127), of the one-DOF torsional actuator. Therefore, referring (9.127) and (9.128), we write down the inverse of the series above:

$$f_\gamma = f_\beta = \frac{1}{\sum_{j=0}^{\infty} \gamma_j Y_m^j} \tag{9.184}$$

where γ_j denotes the coefficient of the inverse series, which is given by (9.128). Substituting the coefficient β_j, (9.182), into (9.128), we have the coefficients of the inverse series:

$$\begin{aligned}
\gamma_0 &= 1 \\
\gamma_1 &= -32/21 \cong -1.5238 \\
\gamma_2 &= 25{,}472/63{,}063 \cong 0.40391 \\
\gamma_3 &= 1{,}368{,}064/22{,}513{,}491 \cong 0.06077 \\
\gamma_4 &= 28{,}607{,}922{,}176/1{,}284{,}552{,}255{,}987 \cong 0.02227 \\
&\vdots
\end{aligned} \tag{9.185}$$

The coefficients γ_j for the function f_γ decrease rapidly as j goes beyond 2. Figure 9.33 shows the convergence of the functions f_β and f_γ at $Y_m = 0.4$ as the terms for f_β and f_γ are summed from 0 to n. For the nth partial sum, f_β slowly approaches its converged value. However, in the same case, f_γ rapidly goes to its converged value as n increases. Figure 9.33 also shows that for an error of 1%, f_γ needs only the second-order polynomial, while f_β requires the fifth-order polynomial for the same error. Therefore, we take the function f_γ, since the force equilibrium (9.171) can be written in a simple form:

$$f_\gamma = \frac{1}{\sum_{j=0}^{2} \gamma_j Y_m^j} = \frac{1}{rY_m^2 - sY_m + 1} \tag{9.186}$$

where the coefficients r and s are defined as

$$r = \gamma_2 \qquad s = -\gamma_1 \tag{9.187}$$

Note that the partial sum, (9.186), was also obtained for the torsional actuator discussed in the preceding section. The same equation will also appear in the problem regarding an electrostatic cantilever beam actuator that is dealt with

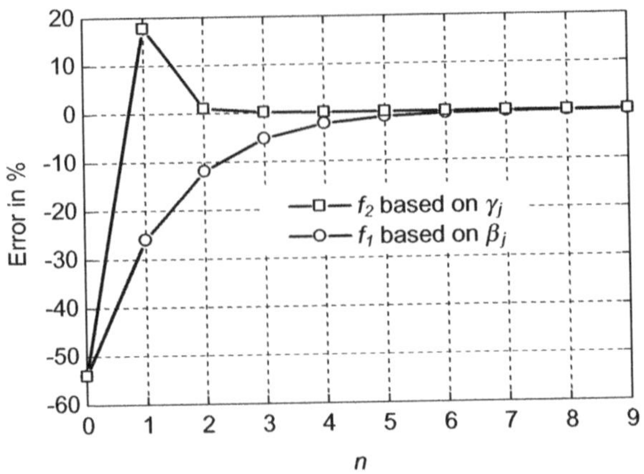

Figure 9.33 Errors in electrostatic forces at $Y_m = 0.4$.

in the following section. Referring to the coefficients γ_j, (9.185), we find the coefficients for the fixed–fixed beam actuator:

$$r = 25{,}472/63{,}063 \cong 0.4039 \quad \text{and} \quad s = 32/21 \cong 1.5238 \tag{9.188}$$

Therefore, referring to (9.177) and (9.184), the force equilibrium equation, (9.171), is approximated as

$$k_1 y_m = \eta\phi_1 \frac{\varepsilon w l V^2}{2h_0^2} \frac{1}{rY_m^2 - sY_m + 1} \tag{9.189}$$

which is a dimensional equation. Introducing the dimensionless deflection, $Y_m = y_m/h_0$ from (9.174), and the parameters

$$G = \eta\phi_1 \frac{\varepsilon w l V^2}{2k_1 h_0^3} = \frac{\varepsilon A_e V^2}{2k_1 h_0^3} \tag{9.190}$$

$$A_e = \eta\phi_1 w l \tag{9.191}$$

which can be considered, respectively, as a dimensionless force and an effective area for the electrostatic force, the preceding force equilibrium equation may be written in the form

$$Y_m = \frac{G}{rY_m^2 - sY_m + 1} \tag{9.192}$$

Substituting $\eta = 16$ and $\phi_1 = 1/30$ into (9.191), the effective area of the fixed–fixed beam is given by

$$A_e = \frac{8}{15} wl$$

Equation (9.192) is a dimensionless force equilibrium equation of the one-DOF model of a fixed–fixed beam actuator; the left- and right-hand sides of (9.192) represent the dimensionless restoring force (spring force) and the dimensionless electrostatic force, respectively. Rearranging the terms in (9.192), we have a cubic equation:

$$rY_m^3 - sY_m^2 + Y_m - G = 0 \tag{9.193}$$

Since the form of equation (9.193) is the same as that of the moment equilibrium equation, (9.132), the solution of equation (9.193) may be written down without further derivation. Substituting Y_m for Θ and G for N into (9.140), we find the stable solution,

$$Y_m = \frac{1}{3r}\left(s - 2\sqrt{s^2 - 3r}\cos\frac{\phi}{3} \right) \tag{9.194}$$

where

$$\phi = \cos^{-1}\frac{(9/2)sr - s^3 - (27/2)r^2 G}{\left(s^2 - 3r\right)^{3/2}} \tag{9.195}$$

The same substitution into (9.143) and (9.144) gives the dimensionless pull-in force and displacement, G_{pi} and Y_{mpi}, as follows:

$$G_{pi} = \frac{1}{27r^2}\left[9rs - 2s^3 + 2\left(s^2 - 3r\right)^{3/2} \right] \tag{9.196}$$

$$Y_{mpi} = \frac{1}{3r}\left(s - \sqrt{s^2 - 3r} \right) \tag{9.197}$$

The midspan deflection y_m is obtained from (9.194) as follows:

$$y_m = h_0 Y_m = h_0 \frac{1}{3r}\left(s - 2\sqrt{s^2 - 3r}\cos\frac{\phi}{3} \right) \tag{9.198}$$

So, from (9.168), we find the approximate beam deflection as

$$y_1 = y_m \psi_1 \tag{9.199}$$

where ψ_1 denotes the mode shape. Referring to ψ_1, (9.169) and (9.170), for the fixed–fixed beam, we have the beam deflection as

$$y_1 = 16y_m X^2(X-1)^2 \tag{9.200}$$

Combining (9.198) and (9.200) gives the beam deflection of the fixed–fixed beam actuator in terms the coordinate X:

$$\frac{y_1}{h_0} = \frac{16}{3}X^2(X-1)^2\frac{1}{r}\left(s - 2\sqrt{s^2-3r}\cos\frac{\phi}{3}\right) \tag{9.201}$$

where $X = x/l$ and ϕ is defined by (9.195). Note that ϕ is a function of r, s, and G, as shown in (9.190), and G is related to the applied voltage and the stiffness k. From the definition of the dimensionless force G, (9.190), the voltage corresponding to G is given by

$$V = \sqrt{\frac{2Gk_1h_0^3}{\varepsilon A_e}} \tag{9.202}$$

We find the pull-in voltage V_{pi} by substituting (9.196) into equation (9.202):

$$V_{pi} = \sqrt{\frac{2G_{pi}k_1h_0^3}{\varepsilon A_e}} = \sqrt{\frac{2}{27r^2}\left(9rs - 2s^3 + 2\left(s^2-3r\right)^{3/2}\right)\frac{k_1h_0^3}{\varepsilon A_e}} \tag{9.203}$$

Substituting $r = 0.4039$ and $s = 1.5238$ into (9.196) and (9.197), we have the dimensionless pull-in deflection and force and the pull-in voltage as

$$Y_{mpi} = 0.3880 \tag{9.204}$$

$$G_{pi} = 0.1822 \tag{9.205}$$

It is noted that the dimensionless pull-in deflection and force above are constants for the fixed–fixed beam actuator shown in Fig. 9.31. From (9.203) the pull-in voltage V_{pi} is expressed as

$$V_{pi} = 0.6036\sqrt{\frac{k_1h_0^3}{\varepsilon A_e}} \tag{9.206}$$

Substituting the stiffness k, (9.176), into equation (9.206) and recalling that $I = wt^3/12$ for a rectangular beam, the pull-in voltage is written as

$$V_{pi} = 3.4147\sqrt{\frac{\tilde{E}t^3h_0^3}{\varepsilon l^4}\left[1 + 0.2857\frac{\sigma_0}{\tilde{E}}\left(\frac{l}{t}\right)^2\right]} \tag{9.207}$$

where $\tilde{E}$, ε, σ_0, t, l and h_0 denote the effective Young's modulus defined by (3.5) and (3.6), the permittivity of the surrounding fluid (if the actuator is in vacuum, the permittivity of vacuum), the axial stress (or residual stress) defined as the axial force divided by the cross-sectional area of the beam, and the thickness and length of the fixed–fixed beam, and the initial gap at $V = 0$, respectively. If the following condition is satisfied:

$$0.2857\frac{|\sigma_0|}{\tilde{E}}\left(\frac{l}{t}\right)^2 << 1 \tag{9.208}$$

the preceding pull-in voltage may be approximated as

$$V_{pi} = 3.4147\sqrt{\frac{\tilde{E}t^3h_0^3}{\varepsilon l^4}}\left[1+0.1429\frac{\sigma_0}{\tilde{E}}\left(\frac{l}{t}\right)^2\right] \tag{9.209}$$

In other words, if condition (9.208) is satisfied, the pull-in voltage is linearly proportional to the stress in the beam. In Fig. 9.34, the analytical response Y_m at G is compared to a numerical result obtained from a numerical analysis of (9.166) with the mode shape (9.169); if interested in the numerical analysis, the reader may refer to existing publications (e.g., Lee, 2008b). The analytical deflection of (9.194) is in good agreement with the deflection obtained numerically from (9.166), within an error of 2.3%. The analytical pull-in force G_{pi} is 0.92% less than that from numerical analysis. Both maximum errors for the pull-in force and deflection occur at the pull-in. It is noted from Fig. 9.34 that the pull-in force and deflection are underestimated by 0.92% and 2.3%, respectively. The discrepancy between the analytical and numerical pull-in parameters comes from 1% overestimation of the electrostatic force (Fig. 9.33) described in (9.185)–(9.189). Then the 1% overestimation of the electrostatic force acting on the beam causes 0.92%

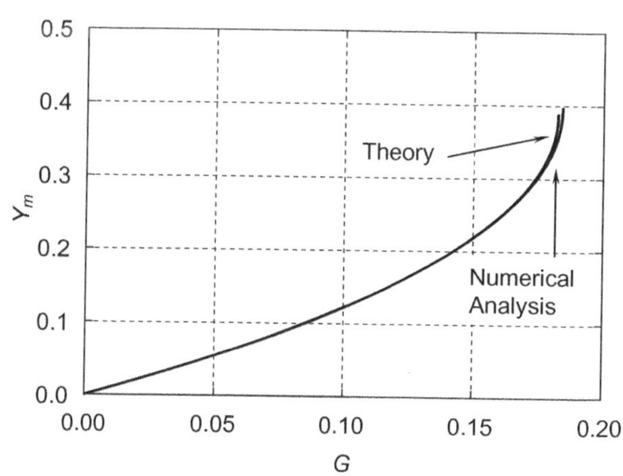

Figure 9.34 Comparison of theoretical and numerical solutions.

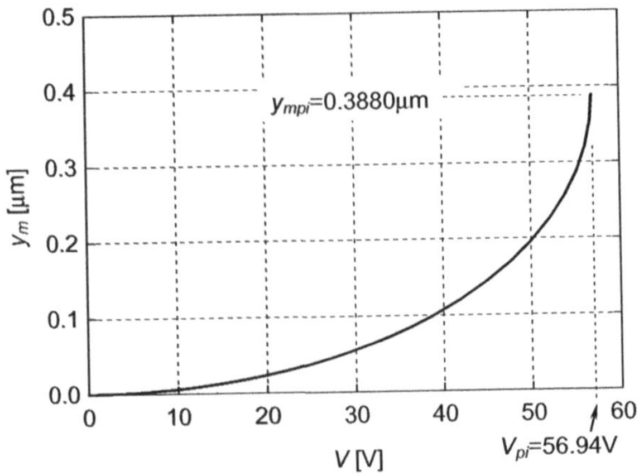

Figure 9.35 Midspan deflection.

and 2.3% underestimations in the dimensionless pull-in force and deflection, respectively.

For further validation, the dimensional midspan deflection y_m is plotted against the voltage V in Fig. 9.35 while the voltage increases from zero to the pull-in voltage (9.207); the data for the analysis are as follows: $l = 250\,\mu$m, $w = 50\,\mu$m, $t = 3\,\mu$m, $h_0 = 1\,\mu$m, Young's modulus $E = 169$ GPa, Poisson's ratio of $\nu = 0.06$, biaxial residual stress $\sigma_{0b} = 100$ Mpa (which is the stress in the structural layer before releasing and can be converted to the uniaxial stress $\sigma_0 = \sigma_{0b}(1 - \nu) = 100(1 - 0.06) = 94$ Mpa); the preceding data for comparison are from a previous study (Osterberg and Senturia, 1997). Since $w \geq 5t$, the effective stiffness (3.6) is given by $\tilde{E} = E/(1-\nu^2) = 169/(1-0.06^2) = 169.61$ GPa. The analytical pull-in voltage (9.207) gives $V_{pi} = 56.94$ V, while simulation based on the finite element method (FEM) yields 57.6 V (Osterberg and Senturia, 1997); the error between the analytical and numerical analyses is 1.2%. Therefore, the theory and the derived closed-form solutions for the deflection, pull-in deflection, and voltage can be used for further research and design of MEMS devices using the fixed–fixed beam actuator.

If a fixed–fixed beam actuator is used as a device to adjust the capacitance of the actuator, we may express the capacitance C as a function of the applied voltage V. The capacitance C in Fig. 9.32a is expressed in form

$$C = \int_0^l \frac{\varepsilon w}{h_0 - y} dx \tag{9.210}$$

Using the same procedure as in the preceding paragraphs, we expand the foregoing capacitance as a series:

$$
\begin{aligned}
C &= \frac{\varepsilon wl}{h_0}\int_0^1 \frac{1}{1-Y_m\eta\psi}\,dX \\
&= C_0\int_0^1\left[1+Y_m\eta\psi+(Y_m\eta\psi)^2+(Y_m\eta\psi)^3+(Y_m\eta\psi)^4+\cdots\right]dX \\
&= C_0\left(1+\eta\phi_1Y_m+\eta^2\phi_2Y_m^2+\eta^3\phi_3Y_m^3+\eta^4\phi_4Y_m^4+\cdots\right) \\
&= C_0\left(1+p_1Y_m+p_2Y_m^2+p_3Y_m^3+p_4Y_m^4+\cdots\right) \\
&= C_0\sum_{j=0}^{\infty}p_jY_m^j
\end{aligned}
\tag{9.211}
$$

where C_0 represents the initial capacitance, defined as $\varepsilon wl/h_0$, and ϕ_j denotes

$$
\begin{aligned}
&\phi_1=\int_0^1\psi\,dX,\quad \phi_2=\int_0^1\psi^2\,dX,\quad \phi_3=\int_0^1\psi^3\,dX,\quad \phi_4=\int_0^1\psi^4\,dX,\\
&\phi_5=\int_0^1\psi^5\,dX,\quad \ldots
\end{aligned}
\tag{9.212}
$$

which are the same as the coefficients given by (9.178), and p_j represents coefficients for the capacitance, which are defined as

$$
p_0=1,\quad p_1=\eta\phi_1,\quad p_2=\eta^2\phi_2,\quad p_3=\eta^3\phi_3,\quad p_4=\eta^4\phi_4,\quad \ldots \tag{9.213}
$$

For the fixed–fixed beam, when we substitute $\eta = 16$ and ϕ_j of (9.181) into equation (9.213), we have the coefficients

$$
\begin{aligned}
p_0 &= 1\\
p_1 &= 16/30 \cong 0.5333\\
p_2 &= (16)^2/630 \cong 0.4063\\
p_3 &= (16)^3/12{,}012 \cong 0.3410\\
p_4 &= (16)^4/218{,}790 \cong 0.2995\\
p_5 &= (16)^5/3{,}879{,}376 \cong 0.2703\\
&\;\vdots
\end{aligned}
\tag{9.214}
$$

Similar to the coefficients β_j for the electrostatic force, given (9.182), the coefficients p_j do not decrease rapidly; that is the partial sum of (9.211) slowly approaches its converged value and we need to sum a few terms to obtain more accurate capacitance. As described in the preceding paragraphs, we can express the preceding series for the capacitance as an inverse series. Let us express the capacitance as follows:

$$
\frac{C}{C_0}=\sum_{j=0}^{\infty}p_jY_m^j=\frac{1}{\sum_{j=0}^{\infty}q_jY_m^j} \tag{9.215}
$$

where the new coefficients q_j are obtained by substituting q_j for γ_j and p_j for β_j into (9.128), as follows:

$$\begin{aligned}
q_0 &= 1/p_0 = 1 \\
q_1 &= -p_1 \\
q_2 &= -q_1 p_1 - p_2 \\
q_3 &= -q_2 p_1 - q_1 p_2 - p_3 \\
q_4 &= -q_3 p_1 - q_2 p_2 - q_1 p_3 - p_4 \\
&\vdots
\end{aligned} \tag{9.216}$$

Using (9.214) and equation (9.216), we have the coefficients q_j for the capacitance of the fixed–fixed beam actuator,

$$\begin{aligned}
q_0 &= 1 \\
q_1 &= -0.5333 \\
q_2 &= -0.1219 \\
q_3 &= -0.0593 \\
q_4 &= -0.0365 \\
q_5 &= -0.0254 \\
&\vdots
\end{aligned} \tag{9.217}$$

Since the absolute values of q_j decrease rapidly with increasing j, we can express the capacitance (9.215) as an approximate formula,

$$\frac{C}{C_0} \cong \frac{1}{\sum_{j=0}^{2} q_j Y_m^j} = \frac{1}{1 - 0.5333\, Y_m - 0.1219\, Y_m^2} \tag{9.218}$$

where Y_m is the dimensionless deflection, which is given by (9.194). When the voltage reaches the pull-in voltage, the preceding capacitance becomes the pull-in capacitance C_{pi}, which is obtained by substituting $Y_{mpi} = 0.3880$, which is given by (9.204), into equation (9.218):

$$\frac{C_{pi}}{C_0} = 1.2908 \tag{9.219}$$

which is less than that of the parallel-plate actuator, $C_{pi}/C_0 = 1.5$, given by (9.60).

9.6 CANTILEVER BEAM ACTUATOR

In the early stage of research on MEMS, some researchers obtained the static deflection and pull-in voltage of the cantilever beam actuator by applying the

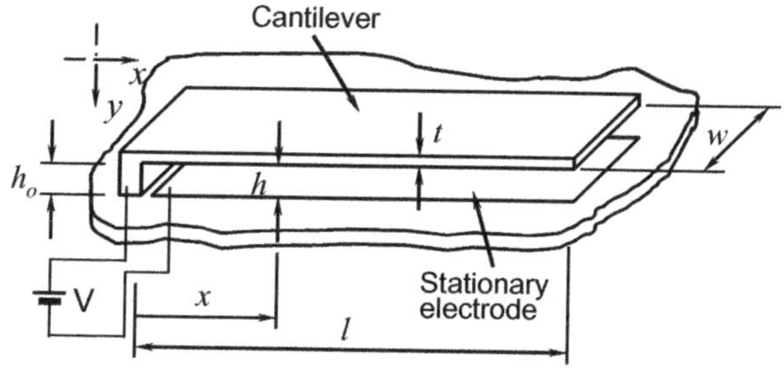

Figure 9.36 Electrostatic cantilever beam actuator.

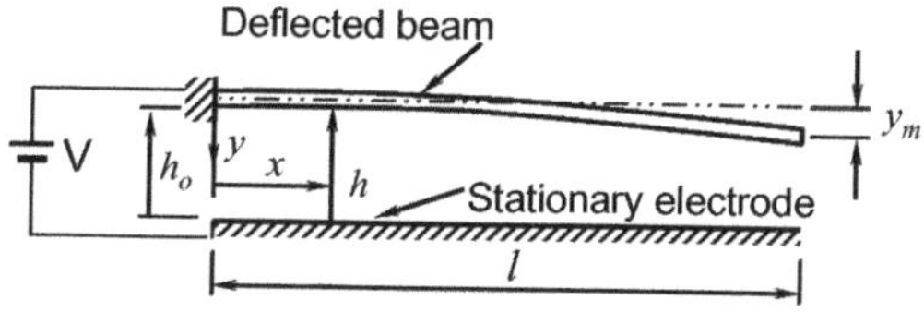

Figure 9.37 Model of a cantilever beam actuator.

principle of superposition to the cantilever beam actuator, shown in Fig. 9.36, which is subjected to a voltage. This approach to the solution for the beam deflection is not correct because the principle of superposition holds only for a linear system; as described in the preceding sections, the gap-closing actuator under a voltage is a nonlinear system. Therefore, the response (e.g., the deflection) of the cantilever beam actuator has been obtained from numerical analyses such as the finite element method (FEM) and the finite difference method (FDM). We obtain a closed-form solution for the beam deflection of the electrostatic cantilever beam actuator by applying the solution procedure, which was described in the preceding section, to the actuator. The model of the cantilever beam actuator is shown in Fig. 9.37. Since there is no constraint in the axial direction, the cantilever beam is free from the axial force. Letting the axial force be zero, the beam equation (9.166) reduces to a simpler equation,

$$\tilde{E}I\frac{d^4y}{dx^4}=\frac{\varepsilon wV^2}{2(h_0-y)^2} \tag{9.220}$$

The following boundary conditions are imposed on the problem:

$$\begin{aligned}&\text{At } x=0\text{: } y=0 \text{ and } dy/dx=0\\&\text{At } x=l\text{: } d^2y/dx^2=0 \text{ and } d^3y/dx^3=0\end{aligned} \tag{9.221}$$

where the last two boundary conditions are based on the fact that the transverse force V and the bending moment M at $x = l$ are zero; the reader should refer to (3.40) and (3.47) for clarity on these boundary conditions.

We use the beam deflection y_1 as a mode shape to obtain the approximate solution; the beam deflection of the cantilever beam is subjected to a uniformly distributed load (Problem 3.2):

$$y_1 = y_m \frac{1}{3}\left(\frac{x}{l}\right)^2 \left[\left(\frac{x}{l}\right)^2 - 4\left(\frac{x}{l}\right) + 6\right] = y_m \psi_1 \tag{9.222}$$

where y_m is the free-end deflection of the cantilever beam and ψ_1 is then the mode shape of the beam, given by

$$\psi_1 = \frac{1}{3}\left(\frac{x}{l}\right)^2 \left[\left(\frac{x}{l}\right)^2 - 4\left(\frac{x}{l}\right) + 6\right] \tag{9.223}$$

It is noted that the foregoing mode shape satisfies the boundary conditions of (9.221). For convenience, the preceding mode shape is written in the form

$$\psi_1 = \eta \psi \tag{9.224}$$

where the constant η is determined as 1/3:

$$\psi = \left(\frac{x}{l}\right)^2 \left[\left(\frac{x}{l}\right)^2 - 4\left(\frac{x}{l}\right) + 6\right] = X^2(X^2 - 4X + 6) \tag{9.225}$$

$$X = \frac{x}{l} \tag{9.226}$$

We also define the dimensionless deflection at the free end as follows:

$$Y_m = \frac{y_m}{h_0} \tag{9.227}$$

Substituting y_1 for y into (9.220), multiplying the resulting equation by ψ_1, and integrating the resulting equation along the beam length gives the following force equilibrium equation that represents the free end deflection y_m at a given voltage V:

$$k_1 y_m = F_e \tag{9.228}$$

where k_1 and F_e denote a stiffness and a force representing the electrostatic force acting on the beam and are given by the following equations:

$$k_1 = \tilde{E}I \int_0^l \frac{d^4\psi_1}{dx^4} \psi_1 \, dx \tag{9.229}$$

$$F_e = \int_0^l \frac{\varepsilon w V^2}{2(h_0 - y_m \psi_1)^2} \psi_1 \, dx \tag{9.230}$$

Referring to the mode shape and dimensionless parameters, (9.223)–(9.227), equations (9.229) and (9.230) are written as

$$k_1 = \eta^2 \frac{\tilde{E}I}{l^3} \int_0^1 \frac{d^4\psi}{dX^4} \psi \, dX \tag{9.231}$$

$$F_e = \eta \frac{\varepsilon w l V^2}{2h_0^2} \int_0^1 \frac{\psi}{(1 - Y_m \eta \psi)^2} dX \tag{9.232}$$

Performing integration by parts and using the four boundary conditions of (9.221) gives a simpler form of the integral of (9.229):

$$\begin{aligned}\int_0^1 \frac{d^4\psi}{dX^4} \psi \, dX &= \left. \frac{d^3\psi}{dX^3} \psi \right|_0^1 - \int_0^1 \frac{d^3\psi}{dX^3} \frac{d\psi}{dX} dX \\ &= \left. \frac{d^3\psi}{dX^3} \psi \right|_0^1 - \left. \frac{d^2\psi}{dX^2} \frac{d\psi}{dX} \right|_0^1 + \int_0^1 \frac{d^2\psi}{dX^2} \frac{d^2\psi}{dX^2} dX = \int_0^1 \left(\frac{d^2\psi}{dX^2} \right)^2 dX\end{aligned}$$

Thus, the stiffness k_1 is expressed as

$$k_1 = \eta^2 \frac{\tilde{E}I}{l^3} \int_0^1 \left(\frac{d^2\psi}{dX^2} \right)^2 dX \tag{9.233}$$

Substituting $\psi = X^4 - 4X^3 + 6X^2$ into equation (9.233), we have the stiffness k_1:

$$k_1 = \eta^2 \frac{144}{5} \frac{\tilde{E}I}{l^3} = \frac{16}{5} \frac{\tilde{E}I}{l^3} \tag{9.234}$$

Since the form of the electrostatic force (9.230) is the same as that of the fixed–fixed beam (9.173), the infinite series of the force is expressed in the same form as (9.177):

$$\begin{aligned}F_e &= \eta \frac{\varepsilon w l V^2}{2h_0^2} \phi_1 (1 + \beta_1 Y_m + \beta_2 Y_m^2 + \beta_3 Y_m^3 + \beta_4 Y_m^4 + \cdots) \\ &= \eta \frac{\varepsilon w l V^2}{2h_0^2} \phi_1 \sum_{j=0}^{\infty} \beta_j Y_m^j\end{aligned} \tag{9.235}$$

where

$$\begin{aligned}&\phi_1 = \int_0^1 \psi \, dX, \quad \phi_2 = \int_0^1 \psi^2 \, dX, \quad \phi_3 = \int_0^1 \psi^3 \, dX, \quad \phi_4 = \int_0^1 \psi^4 \, dX, \\ &\phi_5 = \int_0^1 \psi^5 \, dX, \quad \ldots\end{aligned} \tag{9.236}$$

$$\beta_1 = 2\eta\frac{\phi_2}{\phi_1}, \quad \beta_2 = 3\eta^2\frac{\phi_3}{\phi_1}, \quad \beta_3 = 4\eta^3\frac{\phi_4}{\phi_1}, \quad \beta_4 = 5\eta^4\frac{\phi_5}{\phi_1}, \quad \ldots \tag{9.237}$$

If the preceding electrostatic force in the infinite series does not rapidly approach its convergence, we use the inverse series which is given by (9.128):

$$\begin{aligned}
\gamma_0 &= 1/\beta_0 = 1 \\
\gamma_1 &= -\beta_1 \\
\gamma_2 &= -\gamma_1\beta_1 - \beta_2 \\
\gamma_3 &= -\gamma_2\beta_1 - \gamma_1\beta_2 - \beta_3 \\
\gamma_4 &= -\gamma_3\beta_1 - \gamma_2\beta_2 - \gamma_1\beta_3 - \beta_4 \\
&\vdots
\end{aligned}$$

Therefore, the foregoing electrostatic force is expressed in the inverse series

$$F_e = \eta\frac{\varepsilon wlV^2}{2h_0^2}\phi_1\frac{1}{\sum_{j=0}^{\infty}\gamma_j Y_m^j} \tag{9.238}$$

If the absolute value of γ_j, $j > 2$, is much less than that of γ_1 and γ_2, equation (9.238) is represented by the partial sum

$$F_e = \eta\frac{\varepsilon wlV^2}{2h_0^2}\phi_1\frac{1}{\sum_{j=0}^{2}\gamma_j Y_m^j} = \eta\frac{\varepsilon wlV^2}{2h_0^2}\phi_1\frac{1}{rY_m^2 - sY_m + 1} \tag{9.239}$$

where

$$r = \gamma_2 \qquad s = -\gamma_1 \tag{9.240}$$

Substituting the preceding electrostatic force F_e into (9.228), we have the following force equilibrium equation:

$$k_1 y_m = \eta\phi_1\frac{\varepsilon wlV^2}{2h_0^2}\frac{1}{rY_m^2 - sY_m + 1} \tag{9.241}$$

which is the same as force equilibrium equation (9.189) of a fixed–fixed beam actuator. The mode shapes ψ are different, so the stiffnesses k_1, ϕ_1, r, and s have different values.

Let us obtain the stiffness and coefficients of equation (9.241). Substituting (9.225) and $\eta = 1/3$ into (9.236), we have

$$
\begin{aligned}
\phi_1 &= \int_0^1 X^2(X^2-4X+6)dX = \int_0^1 (X^4-4X^3+6^2)dX \\
&= \frac{1}{5}-1+2=\frac{6}{5} \\
\phi_2 &= \int_0^1 [X^2(X^2-4X+6)]^2 dX = \int_0^1 X^4(X^2-4X+6)^2 dX \\
&= \int_0^1 X^4(X^4-8X^3+28X^2-48X+36)dX \\
&= \int_0^1 (X^8-8X^7+28X^6-48X^5+36X^4)dX \\
&= \frac{1}{9}-8\frac{1}{8}+28\frac{1}{7}-48\frac{1}{6}+36\frac{1}{5}=\frac{104}{45} \\
\phi_3 &= \int_0^1 [X^2(X^2-4X+6)]^3 dX = \int_0^1 X^6(X^2-4X+6)^2(X^2-4X+6)dX \\
&= \int_0^1 X^4(X^4-8X^3+28X^2-48X+36)(X^2-4X+6)dX \\
&\vdots \\
&= \frac{2336}{455}
\end{aligned}
\tag{9.242}
$$

Similarly,

$$
\phi_4 = \frac{1{,}042{,}464}{85{,}085},\ldots
$$

Substituting $\eta = 1/3$ and the preceding coefficients ϕ_j into (9.237) gives β_j:

$$
\begin{aligned}
\beta_1 &= 104/81 \cong 1.2840 \\
\beta_2 &= 1168/819 \cong 1.4261 \\
\beta_3 &= 694{,}976/459{,}459 \cong 1.5126 \\
&\vdots
\end{aligned}
\tag{9.243}
$$

Since the coefficient β_j increases with increasing j, the sixth partial sum of the series based on β_j is needed to obtain a sufficiently accurate electrostatic force. For a simple expression of the electrostatic force, we obtain the coefficients γ_j of the inverse series as follows:

$$
\begin{aligned}
\gamma_0 &= 1 \\
\gamma_1 &\cong -1.2840 \\
\gamma_2 &\cong 0.2224 \\
\gamma_3 &\cong 0.03293 \\
&\vdots
\end{aligned}
\tag{9.244}
$$

The coefficients r and s for the electrostatic force is given by (9.240):

$$r = \gamma_2 \cong 0.2224 \qquad s = -\gamma_1 \cong 1.2840 \tag{9.245}$$

Recalling that $Y_m = y_m/h_0$, (9.227), we express the force equilibrium equation (9.241) in the following dimensional form:

$$Y_m = \frac{G}{rY_m^2 - sY_m + 1} \tag{9.246}$$

where

$$G = \eta\phi_1 \frac{\varepsilon wlV^2}{2k_1h_0^3} = \frac{\varepsilon A_e V^2}{2k_1h_0^3} \tag{9.247}$$

$$A_e = \eta\phi_1 wl \tag{9.248}$$

Dimensionless force equation (9.246) is expressed in the same form as (9.192), which represents the dimensionless force equation of the fixed–fixed beam actuator. Therefore, we know that the solution to the preceding force equilibrium equation is given by (9.194), and the pull-in force and deflection are given by (9.196) and (9.197), respectively.

For the cantilever beam actuator, substituting $\eta = 1/3$ and the first equation of (9.242) into (9.248) gives

$$A_e = \frac{2}{5}wl \tag{9.249}$$

Substituting r and s, (9.245), into (9.196) and (9.197), we have the dimensionless

$$G_{pi} = 0.2104 \tag{9.250}$$

$$Y_{mpi} = 0.4396 \tag{9.251}$$

It is noted that the dimensionless pull-in deflection and force above remain constant for the cantilever beam actuator shown in Fig. 9.36. Since the dimensionless force is a function of the applied voltage V as shown in (9.247), the pull-in voltage V_{pi} is expressed as

$$V_{pi} = \sqrt{\frac{2k_1h_0^3G_{pi}}{\varepsilon A_e}} = 0.6487\sqrt{\frac{k_1h_0^3}{\varepsilon A_e}} \tag{9.252}$$

Substituting (9.234), (9.248) and $I = wt^3/12$ into equation (9.252) yields the pull-in voltage as a function of the effective Young's modulus and the geometry:

$$V_{pi} = 0.5297\sqrt{\frac{\tilde{E}t^3h_0^3}{\varepsilon l^4}} \tag{9.253}$$

In Fig. 9.38, we compare the analytic free-end deflection of the cantilever beam actuator, (9.194), with the deflection obtained from a numerical analysis; the numerical analysis is detailed in Lee (2008b). As the pull-in force G varies from zero and its pull-in value of 0.2104, the deflection Y_m of the free end of the cantilever beam increases from zero to its pull-in value of 0.4396, and the cantilever beam whose mode shape is defined to be the mode shape (9.222) is deflected. The free-end deflection obtained from the theory is in good agreement with that obtained from numerical analysis within an error of 1.7%. When the dimensionless force G is beyond the pull-in force G_{pi}, the cantilever beam collapses to the stationary electrode on the substrate since the electrostatic force is always greater than the restoring force due to the beam stiffness. Figure 9.39a shows the free-end beam deflection y_m plotted against the applied voltage V; for generating the curve, the following data are used: $l = 150\,\mu$m, $w = 50\,\mu$m, $t = 3\,\mu$m, $h_0 = 1\,\mu$m, Young's modulus $E = 169$ GPa, Poisson's ratio $\nu = 0.06$. The analytic solution gives the pull-in voltage of $V_{pi} = 16.93$ V and the pull-in free-end deflection of $0.4396\,\mu$m. The deflection of the cantilever beam at $V = 16$ V is illustrated in Fig. 9.39b, where the free-end deflection is $0.2916\,\mu$m. As described in the present and previous sections, analytical solutions of the fixed–fixed and cantilever beam actuators are obtained from the same theory.

We shall obtain the capacitance of the cantilever beam actuator. Since the capacitance is also given by (9.210), it can be expressed as (9.215)

$$\frac{C}{C_0} = \frac{1}{\sum_{j=0}^{\infty} q_j Y_m^j} \tag{9.254}$$

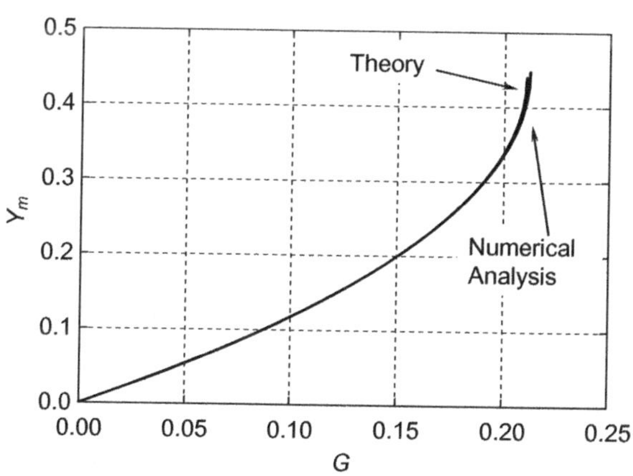

Figure 9.38 Comparison of theoretical and numerical solutions.

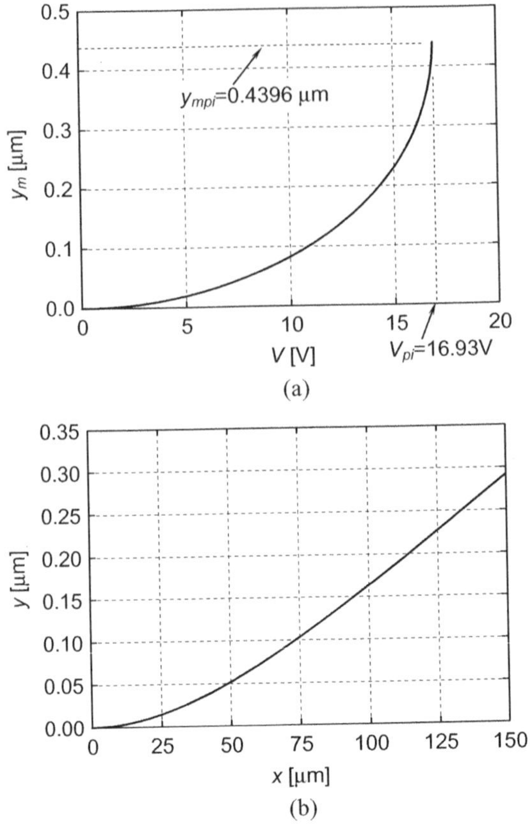

Figure 9.39 Deflection of a cantilever beam.

where q_j represents constants that are a function of ϕ_j. We first evaluate p_j using (9.213) and $\eta = 1/3$:

$$\begin{aligned} p_0 &= 1 \\ p_1 &= \eta\phi_1 = \frac{1}{3}\frac{6}{5} = \frac{2}{5} = 0.4 \\ p_2 &= \eta^2\phi_2 = \frac{1}{3^2}\frac{104}{45} = \frac{104}{405} = 0.2568 \\ p_3 &= 0.1902, \quad p_4 = 0.1513, \quad \ldots \end{aligned} \tag{9.255}$$

The coefficients q_j are obtained from (9.216) as follows:

$$\begin{aligned} q_0 &= 1 \\ q_1 &= -p_1 = -0.4 \\ q_2 &= -q_1 p_1 - p_2 = 0.4^2 - 0.2568 = -0.0968 \\ q_3 &= -0.0487, \quad q_4 = -0.0309, \ldots \end{aligned} \tag{9.256}$$

The capacitance (9.254) may be approximated as

$$\frac{C}{C_0} \cong \frac{1}{1+q_1 Y_m + q_2 Y_m^2} = \frac{1}{1-0.4Y_m - 0.1Y_m^2} \tag{9.257}$$

Substituting $Y_{mp} = 0.4396$ from (9.251) into equation (9.257), we have the pull-in capacitance of the electrostatic cantilever beam actuator:

$$\frac{C_{pi}}{C_0} = 1.2425 \tag{9.258}$$

which is less than that of the parallel-plate actuator, $C_{pi}/C_0 = 1.5$ given by (9.60).

9.7 DYNAMIC RESPONSE OF GAP-CLOSING ACTUATORS

In Section 9.3 we obtained the steady-state response of a parallel-plate actuator, which is considered the simplest gap-closing actuator. We now extend the theory described in Section 9.3 to more complicated gap-closing actuators, which include the torsional actuator, the fixed–fixed actuator, and the cantilever actuator, described in Secstion 9.4 to 9.6. When an input voltage V_i with a dc bias voltage V_b is applied to gap-closing actuators, we can find the following equation of motion by applying Newton's second law to the actuators:

$$m_e\ddot{\delta} + c_e\dot{\delta} + k_{e1}(x_e + \delta) = F_e \tag{9.259}$$

where m_e, c_e, k_{e1}, x_e, δ, and F_e denote generalized parameters that represent the effective mass, the damping coefficient and stiffness, the static displacement due to the bias voltage V_b, the dynamic response due to the time-varying voltage, and the electrostatic force, respectively. Since generalized parameters are used for the equation of motion, the equation may be used for linear and angular motions. That is, for linear motion, m_e, c_e, k_{e1}, x_e, δ, and F_e correspond to the mass, the damping coefficient, the stiffness, the displacement, and force. However, for an angular motion, the same parameters represent the moment of inertia, the damping coefficient for angular motion, the torsional stiffness, the static angular displacement, the time-varying angular displacement, and the moment, respectively. A model for the analysis of the static and dynamic responses is shown in Fig. 9.40, which is similar to that for a parallel-plate actuator Fig. 9.12, but the parameters r and s are added to the model. The electrostatic force F_e may be expressed as

$$F_e = k_{e1}h_0 \frac{G_e}{r(X_e + \delta/h_0)^2 - s(X_e + \delta/h_0) + 1} \tag{9.260}$$

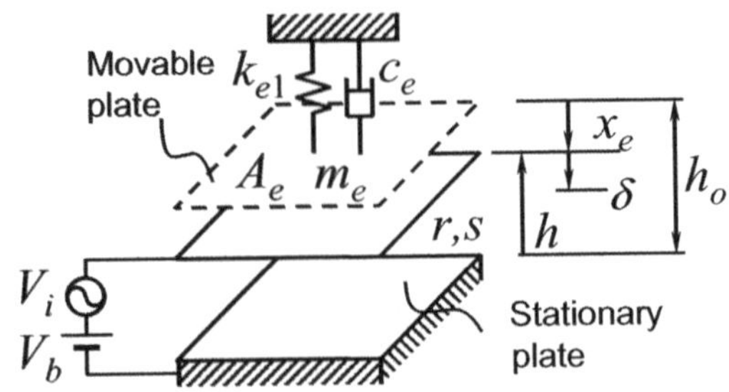

Figure 9.40 Model of a one-DOF gap-closing actuator.

where

$$X_e = \frac{x_e}{h_0}$$
$$G_e = \frac{1}{2}\frac{\varepsilon A_e}{k_{e1}h_0^3}V^2 \qquad (9.261)$$
$$V = V_b + V_i$$

where A_e denotes an effective area (which will be discussed). We first obtain the static response and then determine the dynamic response. Setting $V_i = 0$ and $\delta = 0$, the preceding equation of motion becomes the generalized force equilibrium equation:

$$X_e = \frac{G_e}{rX_e^2 - sX_e + 1} \qquad (9.262)$$

The form of the foregoing equation of motion is the same as that of (9.132); note that (9.132) can be expressed as $\Theta = N/(r\Theta^2 - s\Theta + 1)$. Therefore, one may write the stable solution X_e and the unstable solution $X_{e,\mathrm{U}}$ directly by referring to (9.140) and (9.142):

$$X_e = \frac{1}{3r}\left(s - 2\sqrt{s^2 - 3r}\cos\frac{\phi}{3}\right) \qquad (9.263)$$

$$X_{e,\mathrm{U}} = \frac{1}{3r}\left[s - 2\sqrt{s^2 - 3r}\cos\left(\frac{\phi}{3} + \frac{4\pi}{3}\right)\right] \qquad (9.264)$$

where

$$\phi = \cos^{-1}\frac{(9/2)sr - s^3 - (27/2)r^2G_e}{\left(s^2 - 3r\right)^{3/2}} \qquad (9.265)$$

The pull-in force and displacement are given by (9.143) and (9.144), as follows:

$$G_{e,pi} = \frac{1}{27r^2}\left[9rs - 2s^3 + 2\left(s^2 - 3r\right)^{3/2}\right] \tag{9.266}$$

$$X_{e,pi} = \frac{1}{3r}\left(s - \sqrt{s^2 - 3r}\right) \tag{9.267}$$

Thus, we can convert the preceding pull-in values into dimensional pull-in displacement and voltage:

$$x_{e,pi} = h_0 X_{e,pi} \tag{9.268}$$

$$V_{pi} = \sqrt{G_{e,pi}\frac{2k_{e1}h_0^3}{\varepsilon A_e}} \tag{9.269}$$

Let us return to the equation of motion, (9.259). Since $V = V_b + V_i$, the relation $G_e/G_{e,pi} = (V/V_{pi})^2$ can be expressed as

$$\begin{aligned}\frac{G_e}{G_{e,pi}} &= \left(\frac{V_b + V_i}{V_{pi}}\right)^2 = \left(\frac{V_b}{V_{pi}}\right)^2\left[1 + 2\frac{V_i}{V_b} + \left(\frac{V_i}{V_b}\right)^2\right] \\ &= \frac{G_b}{G_{e,pi}}\left[1 + 2\frac{V_i}{V_b} + \left(\frac{V_i}{V_b}\right)^2\right]\end{aligned} \tag{9.270}$$

where

$$G_b = G_{e,pi}\left(\frac{V_b}{V_{pi}}\right)^2$$

represents the generalized force caused by the bias voltage V_b. Therefore, the preceding relation between G_e and V_i may be written as

$$G_e = G_b\left[1 + 2\frac{V_i}{V_b} + \left(\frac{V_i}{V_b}\right)^2\right] \tag{9.271}$$

Substituting equation (9.271) into (9.260), introducing (9.262), and neglecting the high-order terms, we have the linearized electrostatic force

$$\begin{aligned}\frac{F_e}{k_{e1}h_0} &= \frac{G_b}{(rX_e^2 - sX_e + 1)}\frac{1 + 2V_i/V_b + (V_i/V_b)^2}{1 + \left[(2rX_e - s)\delta/h_0 + r(\delta/h_0)^2\right]/(rX_e^2 - sX_e + 1)} \\ &\cong X_e\left[1 + 2\frac{V_i}{V_b} - \frac{2rX_e - s}{rX_e^2 - sX_e + 1}\delta/h_0\right]\end{aligned} \tag{9.272}$$

Substituting the foregoing force into the equation of motion, (9.259), and rearranging terms, we have an approximate equation of motion for $V_i \ll V_b$,

$$m_e\ddot{\delta}+c_e\dot{\delta}+k_{\text{eff}}\delta=F \tag{9.273}$$

where

$$k_{\text{eff}}=k_{e1}\left(1-X_e\frac{s-2rX_e}{rX_e^2-sX_e+1}\right) \tag{9.274}$$

$$F=2k_{e1}x_e\frac{V_i}{V_b} \tag{9.275}$$

represent the effective stiffness and the applied force. Note from (9.273) that if the bias voltage V_b remains constant and if the displacement δ is much less than h_0, the foregoing equation of motion can be considered linear. However, the response δ is a nonlinear function of V_b since the effective stiffness, k_{eff}, and the static displacement, x_e, are related nonlinearly to the bias voltage V_b. If the damping force relates to x_e nonlinearly, the damping coefficient c_e is also nonlinear.

If the input voltage V_i is a sinusoidal voltage, $V_i=V_a\sin\omega t$, where V_a and ω, respectively, denote the voltage amplitude and the angular frequency, the equation of motion is given by

$$m_e\ddot{\delta}+c_e\dot{\delta}+k_{\text{eff}}\delta=F_0\sin\omega t \tag{9.276}$$

where

$$F_0=2k_{e1}x_e\frac{V_a}{V_b}=2k_{e1}h_0X_e\frac{V_a}{V_b} \tag{9.277}$$

represents the force amplitude. Let δ_s represent the steady-state response of the equation of motion (9.276). For convenience, we divide (9.276) by $k_{e1}h_0$ to give a dimensionless form of the equation:

$$\frac{d^2Z}{dT^2}+\frac{\sqrt{S_{\text{eff}}}}{Q}\frac{dZ}{dT}+S_{\text{eff}}Z=\frac{F_0}{k_{e1}h_0}\sin\Omega T \tag{9.278}$$

where

$$Z=\frac{\delta_s}{h_0}$$
$$T=\omega_{n0}t$$
$$\omega_{n0}=\sqrt{\frac{k_{e1}}{m_e}} \tag{9.279}$$
$$\Omega=\frac{\omega}{\omega_{n0}}=\frac{f}{f_{n0}}$$
$$S_{\text{eff}}=\frac{k_{\text{eff}}}{k_{e1}}$$

represent the displacement, time, the natural frequency at $V_b = 0$, the frequency ratio, and the effective stiffness, respectively. Replacing the right-hand side of the preceding equation of motion by $(F_0/kh_0)e^{j\Omega T}$ and setting $Z = Z_0 e^{j\Omega T}$, we find the steady-state response in complex form,

$$Z_0 = \frac{F_0}{k_{e1}h_0}\frac{1}{S_{\text{eff}} - \Omega^2 + j\left(\sqrt{S_{\text{eff}}}/Q\right)\Omega}$$
$$= \frac{F_0}{k_{e1}h_0}\frac{1}{\sqrt{\left(S_{\text{eff}} - \Omega^2\right)^2 + (\Omega/Q)^2 S_{\text{eff}}}}e^{-j\phi_a}$$

where

$$\phi_a = \tan^{-1}\frac{\Omega\sqrt{S_{\text{eff}}}}{Q\left(S_{\text{eff}} - \Omega^2\right)}$$

The real steady-state response is expressed as the imaginary part of the complex response:

$$\frac{\delta_s}{h_0} = Z = \text{Im}\left(Z_0 e^{j\Omega T}\right)$$

Therefore, the steady-state response δ_s is obtained as

$$\frac{\delta_s}{h_0} = \frac{\delta_1}{h_0}\sin(\Omega T - \phi_a) = \frac{\delta_1}{h_0}\sin(\omega t - \phi_a) \tag{9.280}$$

where

$$\frac{\delta_1}{h_0} = \frac{F_0}{k_{e1}h_0}\frac{1}{\sqrt{\left(S_{\text{eff}} - \Omega^2\right)^2 + (\Omega/Q)^2 S_{\text{eff}}}}$$
$$\phi_a = \tan^{-1}\frac{\Omega\sqrt{S_{\text{eff}}}}{Q\left(S_{\text{eff}} - \Omega^2\right)} \tag{9.281}$$

Referring to (9.277), the preceding vibration amplitude can be written

$$\frac{\delta_1}{h_0} = 2\frac{X_e}{\sqrt{\left(S_{\text{eff}} - \Omega^2\right)^2 + (\Omega/Q)^2 S_{\text{eff}}}}\frac{V_a}{V_b} \tag{9.282}$$

In the preceding equations, X_e is given by (9.263) and S_{eff} is obtained from (9.274) and the last equation of (9.279).

Let us verify the preceding steady-state response. We shall obtain the steady-state response δ_s, (9.85), of the parallel-plate actuator by substituting parameters r and s for the parallel-plate actuator. For the parallel plate we set $m_e = m$, $c_e = c$, and $k_{el} = k$. Dividing the electrostatic force (9.46) by kh_0 and setting $H = 1 - X_e$ and $G_e = G = \varepsilon AV^2/2kh_0^3$, we have the dimensionless electrostatic force:

$$\frac{F_e}{k_{e1}h_0} = \frac{G_e}{H^2} = \frac{G_e}{(1-X_e)^2} = \frac{G_e}{X_e^2 - 2X_e + 1} \tag{9.283}$$

Thus, the force equilibrium equation is given as

$$X_e = \frac{G_e}{X_e^2 - 2X_e + 1}$$

Comparing the equation above with (9.262), we find r and s for the parallel-plate actuator:

$$r = 1 \quad \text{and} \quad s = 2 \tag{9.284}$$

Upon substitution of $X_e = 1 - H$, $r = 1$ and $s = 2$ into (9.274), we have the dimensionless stiffness:

$$\begin{aligned} S_{\text{eff}} &= \frac{k_{\text{eff}}}{k_{e1}} = \frac{k_{\text{eff}}}{k} = 1 - (1-H)\frac{2 - 2(1-H)}{(1-H)^2 - 2(1-H) + 1} \\ &= 3 - \frac{2}{H} \end{aligned} \tag{9.285}$$

which is equal to the effective stiffness, (9.58), of the parallel-plate actuator. Substituting S_{eff} into (9.280) and setting $X_e = 1 - H$ gives the steady-state response of the parallel-plate actuator,

$$\frac{\delta_s}{h_0} = \frac{\delta_1}{h_0}\sin(\omega t - \phi_a) \tag{9.286}$$

where

$$\frac{\delta_1}{h_0} = 2\frac{1-H}{\sqrt{[3-(2/H)-\Omega^2]^2 + (\Omega/Q)^2(3-2/H)}}\frac{V_a}{V_b} \tag{9.287}$$

$$\phi_a = \tan^{-1}\left[\frac{\Omega\sqrt{3-2/H}}{Q(3-(2/H)-\Omega^2)}\right] \tag{9.288}$$

The preceding steady-state response δ_s/h_0 is the same as (9.82). Therefore, we can use (9.280) to evaluate the steady-state response when gap-closing actuators are actuated by an ac voltage with a dc voltage.

As illustrated in the preceding paragraphs, we need the generalized mass, damping coefficient, and stiffness to evaluate the steady-state response δ_s. For the parallel plate, we have used the following: $m_e = m$, $c_e = c$, and $k_{e1} = k$, $r = 1$, and $s = 2$. For the one-DOF torsional actuator, the following are used: $m_e = I$, where I is the moment of inertia about the axis of rotation, $k_{e1} = k_t$, where k_t is the torsional stiffness, $x_e = \theta$, and $X_e = \theta/\theta_0 = b\theta/h_0$, where b and h_0 denote the endpoint and the initial gap, as shown in Fig. 9.20; $G_e = N$ is given by (9.123). Let us obtain the damping coefficients for the torsional actuator. If the damping coefficient is determined by the viscous force due to the fluid squeezed in the torsional actuator and if c_1 represents the damping coefficient per unit length, the damping coefficient per unit length may be expressed at ξ from the center of the spring beam in Fig. 9.20:

$$c_1 = \frac{\beta}{h^3} = \frac{\beta}{(h_0 - \theta\xi)^3}$$

and the damping force acting on the length $d\xi$ is given by

$$dF_d = c_1\, d\xi v = c_1\, dx(\dot{\theta}\xi) = \beta\dot{\theta}\frac{\xi}{(h_0 - \theta\xi)^3} d\xi$$

Since the moment due to the damping of the plate shown in Fig. 9.20 (if β is constant) is written as

$$M_d = \int \xi dF_d = \beta\dot{\theta}\int_{-p}^{p} \frac{\xi^2}{(h_0 - \theta\xi)^3} d\xi$$

we obtain the damping coefficient for the damping moment in the form

$$\begin{aligned} c_e &= \frac{M_d}{\dot{\theta}} = \beta\int_{-p}^{p} \frac{\xi^2}{(h_0 - \theta\xi)^3} d\xi \\ &= \frac{\beta}{\theta^2}\int_{-p}^{p}\left[\frac{h_0}{(h_0 - \theta\xi)^3} - \frac{2h_0}{(h_0 - \theta\xi)^2} + \frac{1}{h_0 - \theta\xi}\right] d\xi \\ &= \frac{\beta b^3}{h_0^3}\frac{1}{\Theta^3}\left[\frac{2\zeta\Theta}{(1-\zeta^2\Theta^2)^2} - \frac{4\zeta\Theta}{1-\zeta^2\Theta^2} - \ln\frac{1-\zeta\Theta}{1+\zeta\Theta}\right] \end{aligned}$$

where $\Theta = b\theta/h_0$ and $\xi = p/b$. The preceding damping coefficient for the moment may be expressed as

$$\frac{c_e}{c_0} = \frac{3}{2\zeta^3\Theta^3}\left[\frac{2\zeta\Theta}{(1-\zeta^2\Theta^2)^2} - \frac{4\zeta\Theta}{1-\zeta^2\Theta^2} - \ln\frac{1-\zeta\Theta}{1+\zeta\Theta}\right] \tag{9.289}$$

where

$$c_0 = 2\frac{\beta b^3}{3h_0^3} \tag{9.290}$$

denotes the damping coefficient as Θ approaches zero.

When we deal with the fixed–fixed and cantilever beam actuators, we set $x_e = y_m$ and $G_e = G$. For the fixed–fixed and cantilever beam actuators, (5.145), (5.146), and (5.147) are used to obtain m_e, c_e, and k_{e1}. Equations (b) to (d) in Example 5.3 of Section 5.9 give m_e, c_e, and k_e for the fixed–fixed beam actuator, respectively, as follows:

$$m_e = \frac{128}{315}m \qquad c_e = \frac{128}{315}c_l \tag{9.291}$$

$$k_{e1} = k_1 = \frac{1024}{5}\frac{\tilde{E}I}{l^3} + \frac{512}{105}\frac{N_0}{l} \tag{9.292}$$

where $m = \rho A l$ denotes the entire mass of the beam and c_l denotes the damping coefficient of the entire beam (i.e., c_l is the damping coefficient of the beam when the beam moves downward without deformation). For the fixed–fixed beam actuator whose width is much less than the length of the beam, referring to (f) of Example. 6.3 in Section 6.7, the preceding equation is written

$$c_e = \frac{128}{315}\mu\left(\frac{w}{h_0}\right)^3 l \tag{9.293}$$

where μ, w, l, and h_0 denote the viscosity, the width and length of the fixed–fixed beam, and the initial gap, respectively. For the cantilever beam actuator, (5.152) gives the following m_e and c_e:

$$m_e = \frac{104}{405}m \qquad c_e = \frac{104}{405}c_l \tag{9.294}$$

$$k_{e1} = k_1 = \frac{16}{5}\frac{\tilde{E}I}{l^3} \tag{9.295}$$

For a narrow beam ($w << l$), equation (9.295) may be expressed as (6.68), which is rewritten here for convenience:

$$c_e = \frac{104}{405}\mu\left(\frac{w}{h_0}\right)^3 l \tag{9.296}$$

9.8 APPROXIMATION OF GAP-CLOSING ACTUATORS

The electrostatic gap-closing actuators have been described in Section 9.3 to 9.6. If certain conditions are satisfied, the actuators may approximate the

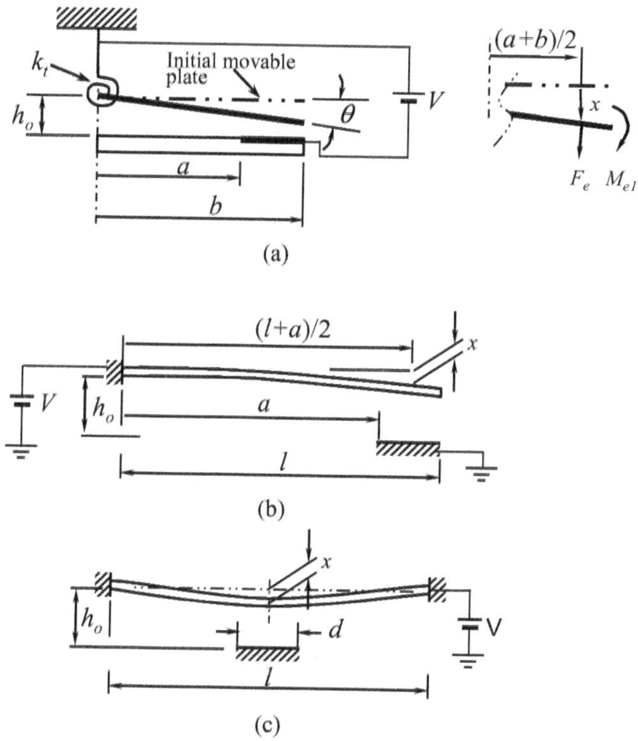

Figure 9.41 Modeling of gap-closing actuators to a parallel-plate actuator.

parallel-plate actuator. We first examine the one-DOF torsional actuator to find the condition. Let us consider the torsional actuator in Fig. 9.41a, where a approaches b. In this case we can consider the electrostatic force F_e acting at $(a + b)/2$ and the electrostatic moment M_{ec} acting on the plate. The electrostatic force and moment were obtained in (7.101) and (7.105), respectively. As discussed below (7.107), the ratio $a/b > 0.5$, the contribution of M_{ec} due to the angular displacement is negligible and the electrostatic moment caused by the voltage V is expressed as

$$M_e \cong F_e \frac{a+b}{2} \tag{9.297}$$

where F_e is the electrostatic force defined by

$$F_e = \frac{\varepsilon A}{2(h_0 - x)^2} V^2 \tag{9.298}$$

where $A = w(b - a)/2$ denote, the overlapping area. We write the moment equilibrium equation using the preceding relation as follows:

$$k_t\theta = F_e \frac{a+b}{2} \tag{9.299}$$

From geometry in Fig. 9.41a, we have a geometric relation, $x = \theta(a + b)/2$, and combining the equations (9.298) and (9.299) gives

$$x = \frac{F_e}{k_t}\left(\frac{a+b}{2}\right)^2 \tag{9.300}$$

Recalling the definition of stiffness, $k = F/x$, the stiffness k may be obtained as

$$k = \frac{F_e}{x} = \frac{4k_t}{(a+b)^2} \tag{9.301}$$

Referring to (9.298), we rewrite (9.300) as follows:

$$kx = \frac{\varepsilon A}{2(h_0 - x)^2} V^2 \tag{9.302}$$

Note that equation (9.302) is the force equilibrium equation of the parallel-plate actuator, (9.48). Therefore, we can obtain the response x from (9.56). The angular displacement θ is obtained by the geometric relation

$$\theta = \frac{2x}{a+b} \tag{9.303}$$

The same argument holds for other gap-closing actuators, such as the cantilever and fixed–fixed beam actuators in Fig. 9.41b and c. For $(l - a) << l$ in Fig. 9.41b, the beam over the lower electrode is not very deformed and then the beam portion can be considered as a straight beam (the right portion of the deformed cantilever is shown in Fig. 9.39b). Therefore, the electrostatic force acting at $(l + a)/2$ can be written as

$$F_e = \frac{\varepsilon A}{2(h_0 - x)^2} V^2$$

where x and A denote the displacement at $(l + a)/2$ and the overlapping given by $w(l - a)$. Since the stiffness at $(l + a)/2$ is given by (3.58),

$$k = \frac{3\tilde{E}I}{[(l+a)/2]^3} \tag{9.304}$$

The force equilibrium equation (9.302) holds and the solution x is given by (9.56).

For the fixed–fixed beam actuator ($d << l$) in Fig. 9.41c, the portion of the beam near the midspan is not much deformed and remains almost straight. Defining the area $A = wd$ and recalling the stiffness, (3.75),

$$k = 2k_1 = 2\left[\frac{12\tilde{E}I}{(l/2)^3} + \frac{6}{5}\frac{N_0}{l/2}\right] = \frac{192\tilde{E}I}{l^3} + \frac{24}{5}\frac{N_0}{l} \tag{9.305}$$

where N_0 denotes the axial force, we can set up the force equilibrium equation (9.302), which in turn leads to a solution of the parallel-plate actuator.

Parallel-Plate Actuators Subjected to Electrostatic Forces in Two Opposite Directions

In some MEMS applications, we may apply two different voltages, V_1 and V_2, to a movable mass as shown in Figs. 9.42 and 9.43. In Fig. 9.42, the movable plate is spaced by h_1 from the lower stationary plate and by h_2 from the upper stationary plate. The movable plate is actuated vertically by the voltages V_1 and V_2 Similarly, in Fig. 9.43, the horizontally movable structure is actuated by the voltage $V_1$and V_2 To analyze both cases, the model shown in Fig. 9.44a is set up, in which the gaps are changed from their initial gap h_0 Referring to the forces acting on the movable mass in Fig. 9.44b, the equation of motion can be written

$$m\frac{d^2\delta}{dt^2} + c\frac{d\delta}{dt} + k(x+\delta) = \frac{\varepsilon A V_1^2}{2(h_{01} - x - \delta)^2} - \frac{\varepsilon A V_2^2}{2(h_{02} + x + \delta)^2} \tag{9.306}$$

where x and δ denote the static displacement due to the dc bias voltages and the dynamic response caused by time-varying voltages. If $V_1 = V_2 = V_b$ and

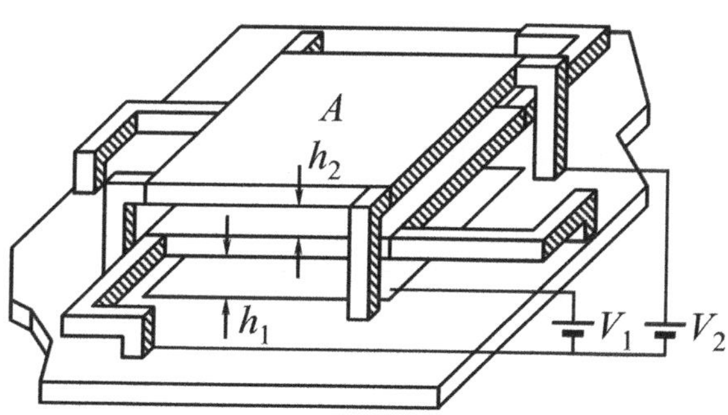

Figure 9.42 Parallel plates subjected to vertical electrostatic forces in two opposite directions.

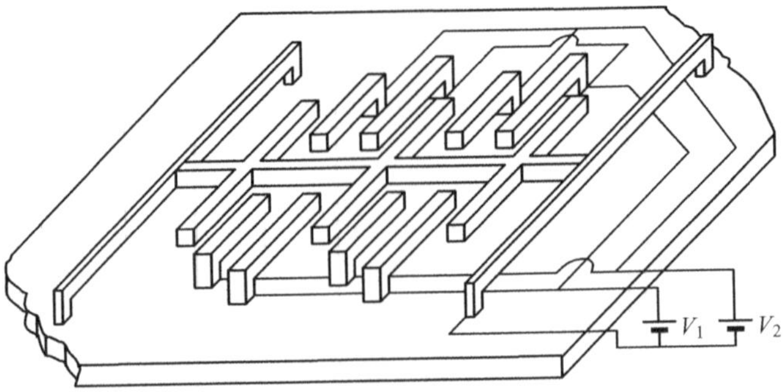

Figure 9.43 Parallel plates subjected to horizontal electrostatic forces in two opposite directions

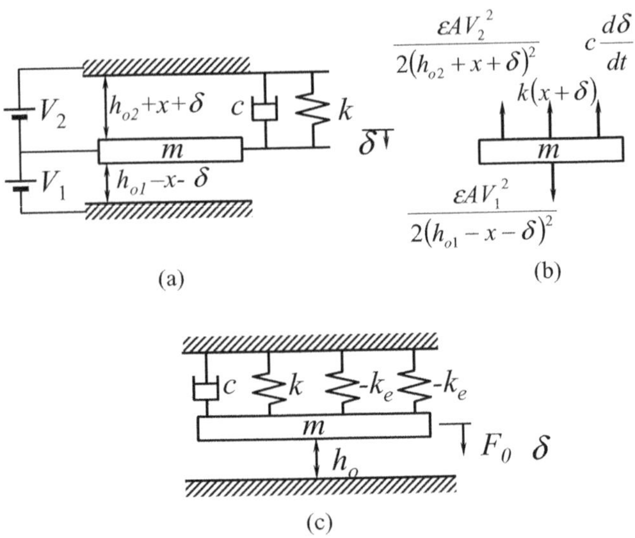

Figure 9.44 Model of a parallel plate subjected to electrostatic forces in two opposite directions.

$h_{01} = h_{02} = h_0$, the preceding equation is reduced to the force equilibrium equation as follows:

$$kx = \frac{\varepsilon A V_b^2}{2(h_0 - x)^2} - \frac{\varepsilon A V_b^2}{2(h_0 + x)^2}$$

Since $x = 0$ satisfies the equation above, the stable position remains at the initial position; in other words, $x = 0$ due to the symmetry. When we set up the voltages

$$V_1 = V_b + V_a \sin \omega t$$
$$V_2 = V_b$$

where $V_a/V_b << 1$, the electrostatic force of (9.306) may be written

$$\begin{aligned}
F_e &= \frac{\varepsilon A(V_b + V_a \sin \omega t)^2}{2(h_0 - \delta)^2} - \frac{\varepsilon A V_b^2}{2(h_0 + \delta)^2} \\
&= \frac{\varepsilon A V_b^2}{2h_0^{\,2}} \left(\frac{1}{(1-\delta/h_0)^2} \left(1 + \frac{V_a}{V_b} \sin \omega t\right)^2 - \frac{1}{(1+\delta/h_0)^2} \right) \\
&\cong \frac{\varepsilon A V_b^2}{2h_0^{\,2}} \left\{ \left[1 + 2\frac{\delta}{h_0} + 3\left(\frac{\delta}{h_0}\right)^2 \cdots \right] \left(1 + 2\frac{V_a}{V_b} \sin \omega t\right) - \left[1 - 2\frac{\delta}{h_0} + 3\left(\frac{\delta}{h_0}\right)^2 - + \cdots \right] \right\} \\
&\cong \frac{\varepsilon A V_b^2}{2h_0^2} \left(4\frac{\delta}{h_0} + 2\frac{V_a}{V_b} \sin \omega t \right) = 2\frac{\varepsilon A V_b^2}{h_0^3} \delta + \frac{\varepsilon A V_b V_a}{h_0^2} \sin \omega t \\
&= 2k_e \delta + F_0 \sin \omega t
\end{aligned}$$

where

$$k_e = \frac{\varepsilon A V_b^2}{h_0^3} \tag{9.307}$$

$$F_0 = \frac{\varepsilon A V_b V_a}{h_0^2} \tag{9.308}$$

represent the electrostatic stiffness and the force amplitude, respectively. Substituting the preceding electrostatic stiffness and force and $x = 0$ into (9.306) leads to

$$m\frac{d^2\delta}{dt^2} + c\frac{d\delta}{dt} + k_{\text{eff}}\delta = F_0 \sin \omega t \tag{9.309}$$

where

$$k_{\text{eff}} = k - 2k_e = k - 2\frac{\varepsilon A V_b^2}{h_0^3} \tag{9.310}$$

represents the effective stiffness. The solution δ of differential equation (9.309) was obtained in Chapter 5. The model corresponding to the preceding equation of motion is illustrated in Fig. 9.44c, where the two electrostatic stiffnesses are added to the model. It is noted that the electrostatic stiffness due to the electrostatic attractive force is negative. If the system is slightly damped (i.e., damping factor $\zeta << 1$ or quality factor $Q >> 1$), the natural angular frequency is given by

$$\omega_n = \sqrt{\frac{k_{\text{eff}}}{m}} \tag{9.311}$$

which is adjustable by changing the bias voltage V_b as shown in (9.310). Furthermore, the quality factor Q is adjustable since Q is expressed as

$$Q = \frac{\sqrt{mk_{\text{eff}}}}{c} \tag{9.312}$$

Setting $k_{\text{eff}} = 0$ in (9.310), we find the pull-in voltage as follows:

$$V_{b,pi} = \sqrt{\frac{kh_0^3}{2\varepsilon A}} \tag{9.313}$$

For $h_0 = h_{01} = h_{02}$, $V_b = V_{b1} = V_{b2}$, and $V_b < V_{b,pi}$, the stable position of the movable mass is $x = 0$ and the movable plate vibrates about $x = 0$ when a time-varying voltage is added. If the bias voltage V_b reaches the pull-in voltage $V_{b,pi}$, the movable plate collapses to one of the stationary plates.

Repeated Structures Subjected to Electrostatic Force

For some applications (e.g., filters), one may make repeated structures whose possible configurations are illustrated in Fig. 9.45a–c, which correspond to the three cases of the resonators shown in Fig. 5.35a–c. In Fig. 9.45, the movable plates of mass m and area A are labeled 1, 2, and 3 and are subjected to the electrostatic force generated by the applied voltage. Their pitch, gap, and width are represented by p, g, and w, respectively. For actuation of one or more movable plate, time-varying voltage may be applied to specific movable plates (the actuation voltage is not shown in Fig. 9.45). Using the formulas developed in Section 5.10, we can evaluate the natural frequencies of the repeated structures. For the structure in Fig. 9.45a in which the left and right plates are immovable, the electrostatic stiffness k_e due to the applied voltage V can be thought of as the coupling stiffness, as shown in Fig. 9.45d. Thus, we use (5.188) to evaluate the natural angular frequencies of the structure. Substituting g for h_0 and V for V_b into (9.307) gives the electrostatic stiffness as

$$k_e = \frac{\varepsilon A V^2}{g^3} \tag{9.314}$$

where A is the area of the movable plate. We obtain the natural angular frequencies of the structure by substituting $-k_e$ for k_c into (5.188),

$$\omega_i = \sqrt{\frac{1}{m}\left\{k - 4k_e\left[\sin\frac{\pi i}{2(n+1)}\right]^2\right\}} \tag{9.315}$$

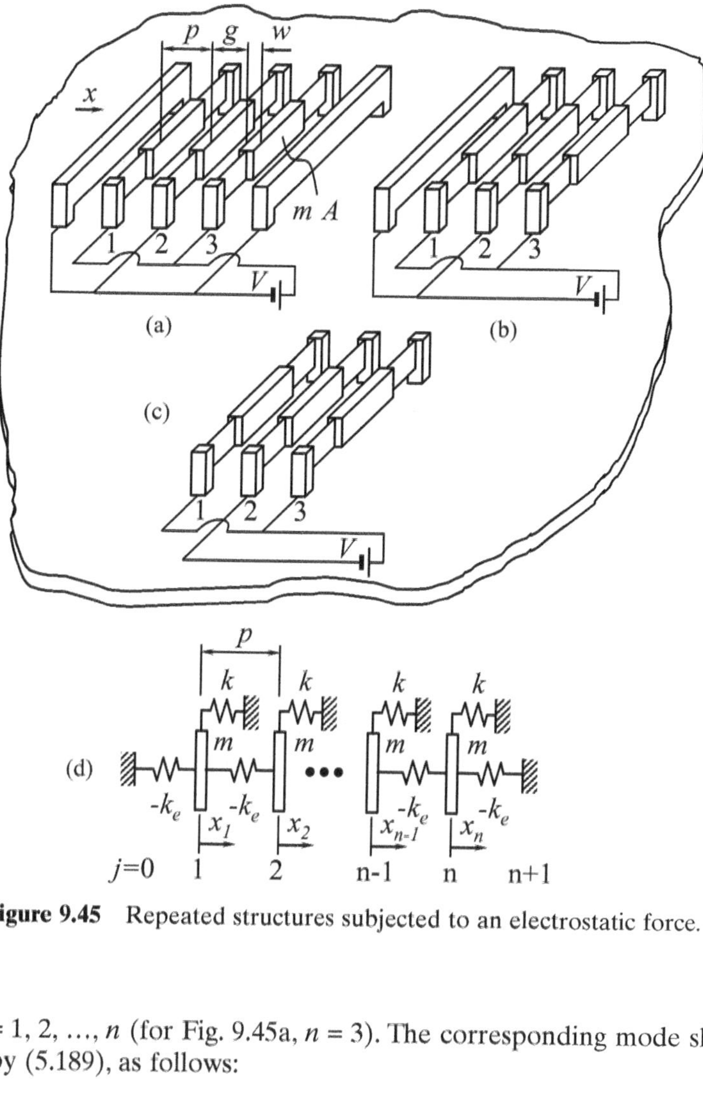

Figure 9.45 Repeated structures subjected to an electrostatic force.

where $i = 1, 2, \ldots, n$ (for Fig. 9.45a, $n = 3$). The corresponding mode shape ψ_i is given by (5.189), as follows:

$$\psi_i = A_1 \sin\left(\frac{i\pi}{n+1} j\right) \tag{9.316}$$

where A_1 is a constant and $j = 1, 2, \ldots, n$ denotes the number to designate the jth movable plate. The preceding mode shape may be written in the form

$$\{\psi\}_i = A_1\left[\sin\left(\frac{i\pi}{n+1}\right), \sin\left(\frac{i\pi}{n+1}2\right), \ldots, \sin\left(\frac{i\pi}{n+1}n\right)\right]^{\mathrm{T}} \tag{9.317}$$

Setting $\omega_i = 0$ in (9.315), we find the pull-in electrostatic stiffness, $k_{e,pi}$:

$$k_{e,pi} = \frac{k}{4\{\sin[\pi i/2(n+1)]\}^2} \tag{9.318}$$

From (9.314) and (9.318) we obtain the ith pull-in voltage as

$$V_{pi,i} = \sqrt{\frac{kg^3}{\varepsilon A} \frac{1}{2\sin[\pi i/2(n+1)]}} \tag{9.319}$$

From equation (9.139), $V_{pi,n}$ is the lowest pull-in voltage, and thus we take the nth pull-in voltage as the pull-in voltage of the repeated structure in Fig. 9.45a:

$$V_{pi} = V_{pi,n} = \sqrt{\frac{kg^3}{\varepsilon A} \frac{1}{2\sin[\pi n/2(n+1)]}} \tag{9.320}$$

Setting $n = 1$ (i.e., one movable mass), the preceding voltage reduces to

$$V_{pi} = \sqrt{\frac{kg^3}{\varepsilon A} \frac{1}{2\sin(\pi/4)}} = \sqrt{\frac{kg^3}{2\varepsilon A}}$$

which is the same as (9.313), the pull-in voltage of the parallel-plate actuator of $h_{01} = h_{02}$ and $V_1 = V_2$ in Fig. 9.44a.

Similarly, for the repeated structures in Fig. 9.45b where the left plate is immovable, (5.191) and (5.192) lead to the ith natural angular frequency and mode shape:

$$\omega_i = \sqrt{\frac{1}{m}\left\{k - 4k_e\left[\sin\frac{\pi(2i-1)}{2(2n+1)}\right]^2\right\}} \tag{9.321}$$

$$\psi_i = A_1 \sin\left[\frac{\pi(2i-1)}{2n+1} j\right] \tag{9.322}$$

Applying the pull-in condition $\omega_i = 0$, angular frequency equation (9.321) gives the pull-in electrostatic stiffness:

$$k_{e,pi} = \frac{k}{4\left[\sin\frac{\pi(2i-1)}{2(2n+1)}\right]^2} \tag{9.323}$$

and the corresponding pull-in voltage for the structure in Fig. 9.45b is then obtained from (9.314) as follows:

$$V_{pi,i} = \sqrt{\frac{kg^3}{\varepsilon A} \frac{1}{2\sin[\pi(2i-1)/2(2n+1)]}} \tag{9.324}$$

The pull-in voltage is obtained as the lowest pull-in voltage,

$$V_{pi} = V_{pi,n} = \sqrt{\frac{kg^3}{\varepsilon A} \frac{1}{2\sin[\pi(2n-1)/2(2n+1)]}} \tag{9.325}$$

For the repeated structure in Fig. 9.45c where the right and left plates are movable, (5.193) and (5.194) give the ith natural angular frequency and mode shape:

$$\omega_i = \sqrt{\frac{1}{m}\left\{k - 4k_e\left[\sin\frac{\pi(i-1)}{2n}\right]^2\right\}} \tag{9.326}$$

$$\psi_i = A_1 \sin\left[\frac{\pi(i-1)}{2n}(2j-1)\right] \tag{9.327}$$

The pull-in elastic stiffness and the ith pull-in voltage are given by

$$k_e = \frac{k}{4\left[\sin\frac{\pi(i-1)}{2n}\right]^2} \tag{9.328}$$

$$V_{pi,i} = \sqrt{\frac{kg^3}{\varepsilon A} \frac{1}{2\sin[\pi(i-1)/2n]}} \tag{9.329}$$

Thus the pull-in voltage for the repeated structure in Fig. 9.45c is defined as

$$V_{pi} = V_{pi,n} = \sqrt{\frac{kg^3}{\varepsilon A} \frac{1}{2\sin[\pi(n-1)/2n]}} \tag{9.330}$$

9.9 ELECTROMAGNETIC ACTUATORS

Actuators employing electromagnetic force are used for some applications: such as the electromagnetic torsional actuator in Fig. 9.46 and electromagnetic relay in Fig. 9.47.

The electromagnetic torsional actuator of Fig. 9.46 consists of a movable plate with planar coils and torsional springs to support the movable plate. When the torsional actuator is exposed to a magnetic flux density $\mathbf{B}$, the electromagnetic moment (7.154) is caused by the current i flowing the coils and is expressed as

$$M_m = 2nr_1 l_1 iB\left[1 + (\alpha+\beta)\frac{n-1}{2} + \alpha\beta\frac{(n-1)(2n-1)}{6}\right] \tag{9.331}$$

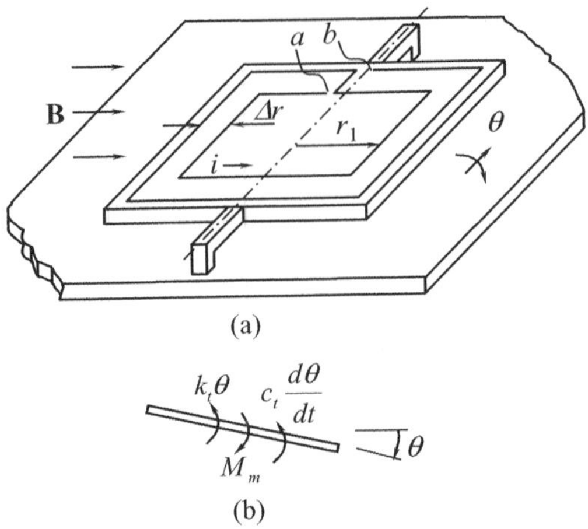

Figure 9.46 Torsional actuator using a multiturn planar coil.

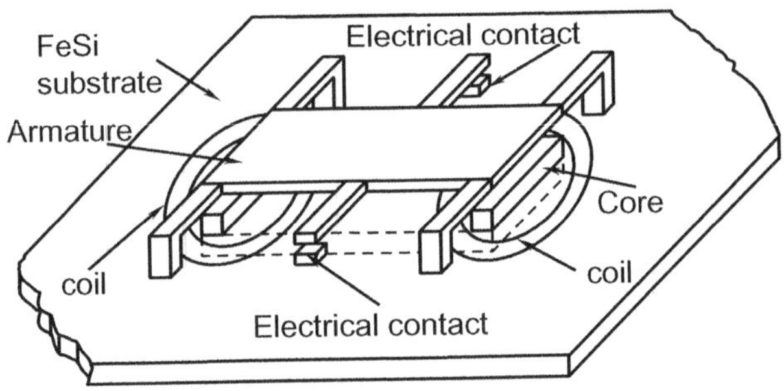

Figure 9.47 Electromagnetic relay.

where

$$\alpha = \frac{\Delta r}{r_1} \qquad \Delta r = \frac{r_n - r_1}{n-1} \qquad \beta = \frac{\Delta l}{l_1} \qquad \Delta l = \frac{l_n - l_1}{n-1} \tag{9.332}$$

and r_1, r_n, l_1, l_n, and n denote the distances of the first and last coils from the axis of rotation, the segment lengths of the first and last segments of the coil (see Fig. 7.39), and the number of coil turns, respectively. Referring to Fig.9.46b, we write the equation of motion of the torsional actuator as follows:

$$I\frac{d^2\theta}{dt^2} + c\frac{d\theta}{dt} + k_t\theta = m_c i \tag{9.333}$$

where

$$m_c = 2nr_1 l_1 B\left[1+(\alpha+\beta)\frac{n-1}{2}+\alpha\beta\frac{(n-1)(2n-1)}{6}\right] \quad (9.334)$$

For time-varying current or time-varying magnetic flux density, the dynamic response θ of the preceding linear differential equation is easily obtained (the reader may refer to Section 5.6 to obtain the response). If the frequency of the applied current is much less than the natural frequency of the actuator, the inertia and damping force are negligible and equation (9.334) reduces to

$$\theta = \frac{m_c}{k_t} i \quad (9.335)$$

Electromagnetic Relay

Electromagnetic relay is used to actuate a movable mass or to switch electric current as shown in Fig. 9.47, where an armature (movable plate) is supported by spring beams and moves downward as electric current flows across the coils. Let us set up a model and obtain an analytic response of the relay. Referring to Figs. 7.35 and 7.36, we can set up the models illustrated in Fig. 9.48, in which parameters and dimensions are specified. In the models, the armature is supported by a spring of stiffness k and can move leftward as the electric current i is applied. The electromagnetic force F_m is attractive and given by (7.148), which is rewritten here for convenience:

$$F_m = \frac{(Ni)^2}{2(R_c + r_d h)^2} r_d \quad (9.336)$$

where h represents the gap between the core and the armature and

$$R_c = \sum_{j=1}^{m} \frac{l_j}{\mu_j A_j} \quad (9.337)$$

$$r_d = \sum_{j=1}^{n} \frac{1}{\mu_0 A_{hj}} \quad (9.338)$$

denote the reluctance of the core and armature and the reluctance change rate, respectively. In equations (9.336) and (9.337), μ_o, μ_j, A_j, l_j, and A_{hj} denote the permeability of the surrounding fluid (gas or liquid), the permeability of the jth element, the cross-sectional area, the length of the jth element of the stationary core and the movable bar in Fig. 9.48a and c and the area of the jth air gap, respectively. For the model of Fig.9.48a and c, the reluctance R_c is given by

$$R_c = \frac{l_1}{\mu A_1} + \frac{l_2}{\mu A_2} + \frac{l_3}{\mu A_3} + \frac{l_4}{\mu A_4} \quad (9.339)$$

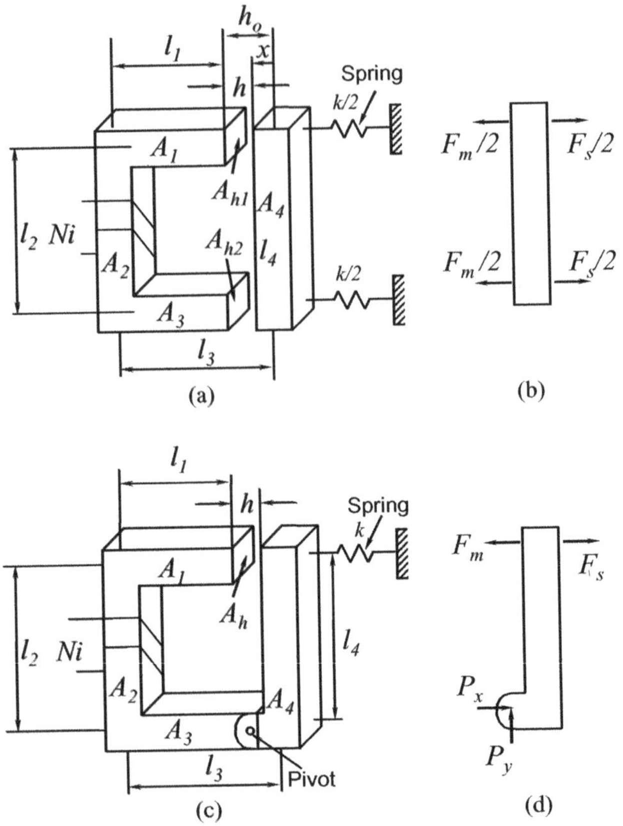

Figure 9.48 Model of an electromagnetic relay.

For the relay with the two gaps in Fig. 9.48a, the reluctance change rate r_d is written as

$$r_d = \frac{1}{\mu_0 A_{h1}} + \frac{1}{\mu_0 A_{h2}} \tag{9.340}$$

and r_d for the model with one gap in Fig. 9.48c is given by

$$r_d = \frac{1}{\mu_0 A_h} \tag{9.341}$$

Referring to the free-body diagrams of Figs. 9.48b and d, the force equilibrium equation is written as $F_s = F_m$:

$$k(h_0 - h) = \frac{(Ni)^2}{2(R_c + r_d h)^2} r_d \tag{9.342}$$

where h_0 is the initial gap. For convenience, the preceding equation may be nondimensionalized as

$$(1-H)(\gamma+H)^2 = G \tag{9.343}$$

where

$$H = \frac{h}{h_0} \tag{9.344}$$

$$G = \frac{(Ni)^2}{2kr_d h_0^3} \tag{9.345}$$

$$\gamma = \frac{R_c}{r_d h_0} \tag{9.346}$$

are dimensionless parameters that represent the gap, the electromagnetic force, and the ratio of the reluctances of the core and gap, respectively. Expanding the preceding force equilibrium gives

$$H^3 - (1-2\gamma)H^2 - (2\gamma-\gamma^2)H - \gamma^2 + G = 0 \tag{9.347}$$

which may be written in the standard form of a cubic equation as follows:

$$H^3 + aH^2 + bH + c = 0 \tag{9.348}$$

where

$$\begin{aligned} a &= -(1-2\gamma) \\ b &= -(2\gamma-\gamma^2) \\ c &= -\gamma^2 + G \end{aligned} \tag{9.349}$$

Since the electromagnetic force is an attractive force that increases as the gap decreases, we assume that the armature experiences the pull-in (i.e., the discriminant $\Delta < 0$). The exact solution of the preceding cubic equation is given by (5.15) as follows:

$$\begin{aligned} H_1 &= \frac{1}{3}\left(1-2\gamma+2(1+\gamma)\cos\frac{\phi}{3}\right) \\ H_2 &= \frac{1}{3}\left[1-2\gamma+2(1+\gamma)\cos\left(\frac{\phi}{3}+\frac{2\pi}{3}\right)\right] \\ H_3 &= \frac{1}{3}\left[1-2\gamma+2(1+\gamma)\cos\left(\frac{\phi}{3}+\frac{4\pi}{3}\right)\right] \end{aligned} \tag{9.350}$$

where

$$\begin{aligned}\phi &= \cos^{-1}\frac{(1+3\gamma+3\gamma^2+\gamma^3)-(27/2)G}{(1+2\gamma+\gamma^2)^{3/2}} \\ &= \cos^{-1}\frac{(1+\gamma)^3-(27/2)G}{(1+\gamma)^3}\end{aligned} \tag{9.351}$$

For convenience, we define $1 + \gamma$ as a new parameter ζ:

$$\zeta = 1+\gamma = 1+\frac{R_c}{r_d h_0} = \frac{R_c + r_d h_0}{r_d h_0} \tag{9.352}$$

which represents the ratio of the whole reluctance to the reluctance of the gaps. Using the preceding definition, we have much simplified solutions:

$$\begin{aligned} H_1 &= \frac{\zeta}{3}\left(1+2\cos\frac{\phi}{3}\right)-\gamma \\ H_2 &= \frac{\zeta}{3}\left[1+2\cos\left(\frac{\phi}{3}+\frac{2\pi}{3}\right)\right]-\gamma \\ H_3 &= \frac{\zeta}{3}\left[1+2\cos\left(\frac{\phi}{3}+\frac{4\pi}{3}\right)\right]-\gamma \end{aligned} \tag{9.353}$$

where

$$\phi = \cos^{-1}\left(1-\frac{27G}{2\zeta^3}\right) \tag{9.354}$$

Plotting the preceding solutions in (9.353) against G (the plot is similar to that in Fig. 9.8), we know that H_1, H_3, and H_2 are the stable and unstable solutions and a solution without any physical meaning, respectively. Therefore, we take H_1 as the solution of the force equilibrium equation, (9.342), as follows:

$$H = \frac{\zeta}{3}\left\{1+2\cos\left[\frac{1}{3}\cos^{-1}\left(1-\frac{27G}{2\zeta^3}\right)\right]\right\}-\gamma \tag{9.355}$$

It is noted that as the ratio γ goes to zero (i.e., the reluctance of the core and armature is neglected), equation (9.355) approaches

$$H = \frac{1}{3}\left\{1+2\cos\left[\frac{1}{3}\cos^{-1}\left(1-\frac{27G}{2}\right)\right]\right\} \tag{9.356}$$

which is the same form as the stable solution, (9.55), of the force equilibrium of a parallel-plate actuator.

Let us obtain the pull-in current and the pull-in gap. From (9.355), the smallest gap is obtained if the following condition is satisfied:

$$1-\frac{27G_{pi}}{2\zeta^3}=-1 \tag{9.357}$$

Equation (9.357) gives the pull-in force G_{pi} as

$$G_{pi}=\frac{4}{27}\zeta^3 \tag{9.358}$$

Combining equation (9.358) and (9.345) gives the pull-in magnetomotive force $(Ni)_{pi}$ as

$$(Ni)_{pi}=\sqrt{2kr_d h_0^3 G_{pi}}=\sqrt{\frac{8}{27}kr_d h_0^3\zeta^3} \tag{9.359}$$

The pull-in current is given by $i_{pi}=(Ni)_{pi}/N$. Substituting (9.358) into (9.355) yields the pull-in gap H_{pi} as

$$H_{pi}=\frac{2}{3}\zeta-\gamma \tag{9.360}$$

It is noted from (9.358) and (9.360) that the pull-in gap and force are functions only of the reluctance ratio γ. As the reluctance of the core and armature is negligible (i.e. $\gamma\to 0$ and $\zeta\to 1$), the dimensionless pull-in force, G_{pi}, and gap, H_{pi}, become 4/27 and 2/3, respectively. The dimensionless gap H, (9.355), and the pull-in gap H_{pi}, (9.360), are plotted against the dimensionless force G in Fig. 9.49 when the reluctance ratio γ and the dimensionless force G vary. H decreases with increasing G and increases with decreasing γ. The pull-in gap H_{pi} becomes zero at $\gamma=2$, and for $\gamma>2$ the pull-in gap H_{pi} goes to negative, results that are unacceptable since the minimum gap is zero (i.e., mechanical contact). Therefore, a dimensionless contact force G_c is introduced to include the mechanical contact of the movable bar to the stationary core. When H in (9.347) approaches zero, the dimensionless contact force G_c is obtained as follows:

$$G_c=\gamma^2 \tag{9.361}$$

Contact force G_c, corresponding to a mechanical contact, is valid only for $\gamma\geq 2$, as shown in Fig. 9.49. If γ is less than 2, the pull-in phenomenon occurs at the pull-in force (9.358) and the pull-in gap (9.360). The same contact force in a different form can be obtained from (9.355) with $H=0$ as follows:

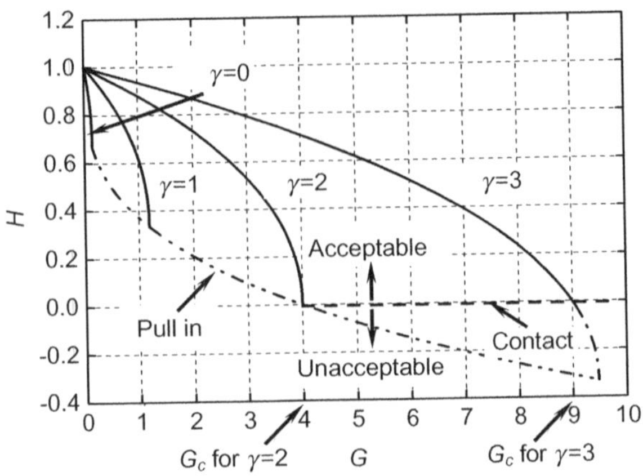

Figure 9.49 Gap H with respect to force G.

$$G_c = \frac{2}{27}\zeta^3\left(1-\cos\left\{3\cos^{-1}\left[\frac{1}{2}\left(\frac{3\gamma}{\varsigma}-1\right)\right]\right\}\right) \tag{9.362}$$

Equations (9.361) and (9.362) for G_c give the same numerical value, as γ is greater than or equal to 2. Therefore, (9.361) is taken to obtain the contact force G_c because of its simplicity. From (9.345) and (9.361), we find the contact electromotive force at which the armature touches the core:

$$(Ni)_c = \gamma\sqrt{2kr_a h_0^3} \tag{9.363}$$

The contact current is given by $i_c = (Ni)_c/N$.

PROBLEMS

9.1 The comb drive actuator in Fig. 9.2a is actuated by a pure sinusoidal voltage of $V = V_a \sin \omega t$ and the sensing voltage remains zero. The mass, damping coefficient, stiffness of the flexures, the interfinger gap, the thickness of comb fingers, and the number of pairs of comb fingers are represented by $m, c, k, g, t,$ and n, respectively. Find the steady-state response of the movable plate.

9.2 Consider the parallel-plate actuator shown in Fig. 9.5: area $A = 5 \times 10^{-9}\,\text{m}^2$ and mass $m = 2 \times 10^{-11}$ kg, and initial gap $h_0 = 2\,\mu\text{m}$. From an experiment, the pull-in voltage V_{pi} is measured as 4 V. The actuator in a vacuum chamber is actuated by $V = V_b + V_a \cos \omega t$, where $V_b = 2\text{V}$ and V_a is much less than V_b. If the mass of the flexures is neglected, determine the stiffness of the flexures and the effective stiffness and natural frequency of the actuator when it is actuated.

9.3 A step voltage V is applied between the movable and stationary plates of a parallel-plate actuator in a vacuum chamber. If the voltage V is less than the dynamic pull-in voltage $V_{pi,d}$ and if the damping force is negligible, find the maximum velocity of the movable plate.

9.4 A one-DOF torsional actuator is fabricated and its design parameters are: $a = 20 \times 10^{-6}\,\text{m}$, $b = 40 \times 10^{-6}\,\text{m}$, $w = 40 \times 10^{-6}\,\text{m}$, $h_0 = 2 \times 10^{-6}\,\text{m}$, and $k_t = 4 \times 10^{-8}\,\text{N}\cdot\text{m}$. Find the pull-in voltage V_{pi}. If $V = V_{pi}/2$ is applied to the actuator, determine the angular displacement and the capacitance.

9.5 A fixed–fixed beam actuator is designed whose design parameters are the following: length $l = 300\,\mu\text{m}$, width $w = 50\,\mu\text{m}$, thickness $t = 2\,\mu\text{m}$, initial gap $h_0 = 2\,\mu\text{m}$, Young's modulus $E = 160\,\text{GPa}$, Poisson's ratio $\nu = 0.28$, and residual stress $\sigma_0 = 100\,\text{MPa}$. Determine the pull-in voltage and deflection at the midspan. Find the midspan deflection and the capacitance when a half pull-in voltage is applied between the beam and the lower electrode.

9.6 The density ρ of the beam in Problem.9.5 is $2330\,\text{kg/m}^3$ and other dimensions and material properties are the same as those in Problem 9.5. At the half pull-in voltage, find the natural frequency of the beam.

9.7 A cantilever beam actuator is fabricated whose date are: $l = 260\,\mu\text{m}$, $w = 50\,\mu\text{m}$, $t = 2\,\mu\text{m}$, $h_0 = 2\,\mu\text{m}$, $a = 160\,\mu\text{m}$, $d = 40\,\mu\text{m}$, Young's modulus $E = 150\text{GPa}$, and Poisson's ratio $\nu = 0.28$ (Fig. P9.7). Find the pull-in voltage of the actuator and the displacement of the free end of the beam at the pull-in.

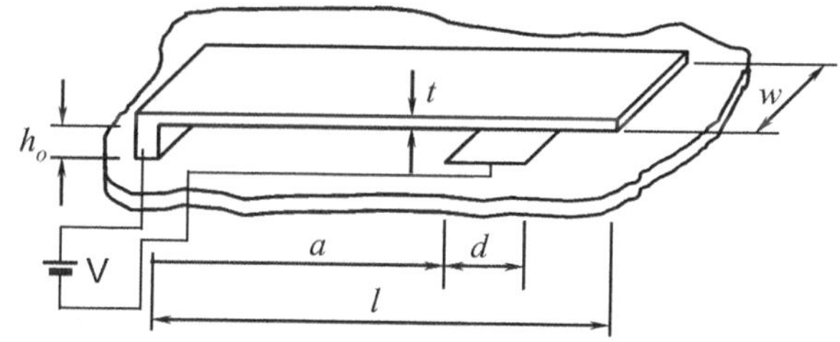

Figure P9.7

CHAPTER 10

SENSORS

In Chapters 1 to 9, we developed theories to explain the mechanical and electrical behavior of MEMS, and described the static and dynamic responses of actuators by combining the theories. In MEMS, sensors are used to measure physical quantities that are usually converted to electrical signals. Combining the theories described in the foregoing chapters, we can develop governing equations for sensors and solve the equations for sensor responses. Since most MEMS sensors measure a force (equivalently, stress or strain) in response to a physical quantity that we wish to measure, we first describe force sensors and then deal with accelerometers and gyroscopes, which are sensors based on the measurement of inertial force. Even though we describe force sensors, accelerometers, and gyroscopes, theories regarding sensors are not limited to these sensors but can be extended to include sensors whose behavior can be described by the theories developed in the previous and present chapters.

10.1 FORCE AND PRESSURE SENSORS

Let us consider the elastic body in Fig. 10.1, where the elastic body is placed on a substrate and subjected to a force F. To maintain equilibrium, stresses inside the body are developed and the body is then deformed. If the body is linear, the stress (or strain) developed can be expressed as a function of the

Principles of Microelectromechanical Systems, By Ki Bang Lee

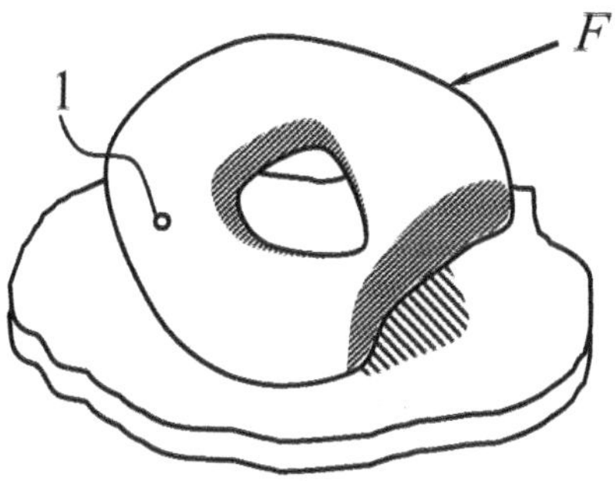

Figure 10.1 Deformed structure subjected to a force.

applied force F, the geometry, and material properties such as Young's modulus and Poisson's ratio. For example, if we measure the strain at point 1 in Fig. 10.1, we can obtain the relation

$$\varepsilon = f_1(F, \text{geometry, mechanical property})$$

where f_1 denotes a function of the quantities in parentheses. Since the elastic body is linear, the strain is proportional to the force and we can express the equation above as

$$\varepsilon = f_2(\text{geometry, mechanical property})F \tag{10.1}$$

If we mount a strain sensor at point 1, we may take the corresponding output voltage V,

$$V = A_1\varepsilon$$

where A_1 denotes a gain, and the voltage may be expressed as

$$V = A_1 f_2 F = f_3 F \tag{10.2}$$

where f_3 is defined as $A_1 f_2$, which represents a proportional constant. According to this argument, we expect that the cantilever beam in Fig. 10.2 will give the following relation:

$$\varepsilon = f_4(E, v, l, w, t, l_s)F \tag{10.3}$$

where E and v denote Young's modulus and Poisson's ratio, respectively, and other parameters are dimensions indicated in Fig. 10.2a. The function f_4 can be determined analytically or experimentally.

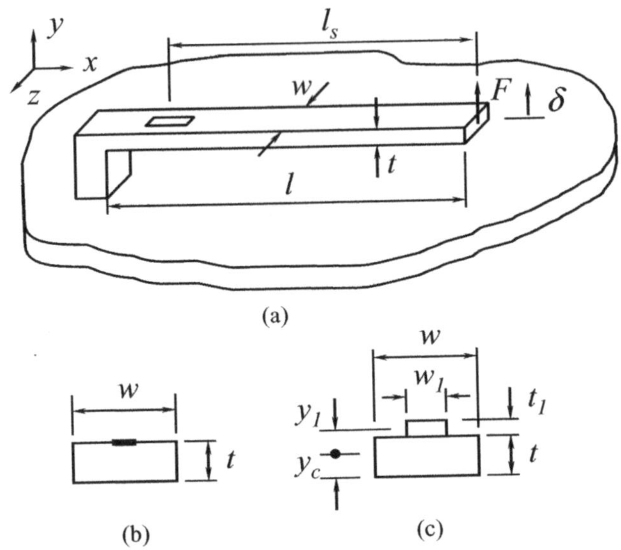

Figure 10.2 Measurement of force.

Stress and Strain in a Cantilever Beam Subjected to Concentrated Force

Let us use the theory described in Chapter 3 to obtain the proportional constant f_4 in equation (10.3). Referring to the dimensions shown in Fig. 10.2a, where we place a sensor at l_s from the free end of the cantilever beam subjected to a force F, we have the moment at l_s as follows:

$$M = Fl_s \tag{10.4}$$

The strain at the sensor site is given by (3.41), which is rewritten here:

$$\varepsilon_x = -\frac{M}{\tilde{E}I} y \tag{10.5}$$

where $\tilde{E}I$ denotes the flexural rigidity of the cantilever beam and y represents the distance from the neutral surface to the sensor site. The stress corresponding to the strain is given by

$$\sigma_x = \tilde{E}\varepsilon_x = -\frac{M}{I} y \tag{10.6}$$

which is not a function of the effective Young's modulus. We assume that the thickness of the sensor (e.g., implanted piezoresistor) is negligible compared to the beam thickness t shown in Fig. 10.2b. Substituting the moment $M = Fl_s$ and $y = t/2$ in equation (10.6) gives the strain at the sensor site in terms of F:

$$\varepsilon_x = -\frac{l_s t}{2\tilde{E}I}F \tag{10.7}$$

where the negative sign indicates that the positive force generates the compressive stress that corresponds to the negative strain. Referring to $I = wt^3/12$, we find the relation between the force and strain as follows:

$$\varepsilon_x = -6\frac{l_s}{\tilde{E}wt^2}F \tag{10.8}$$

Comparing this equation with (10.3), we find the proportional constant f_4 as

$$f_4 = -6\frac{l_s}{\tilde{E}wt^2} = f(E, v, l_s, w, t)$$

Since the stiffness of the cantilever beam is given by (3.58), $k = F/\delta = 3\tilde{E}I/l^3$, the deflection δ is expressed as

$$\delta = \frac{F}{k} = \frac{l^3}{3\tilde{E}I}F \tag{10.9}$$

Combining (10.7) and (10.9) gives the relation between the beam deflection δ and the strain ε_x at l_s as follows:

$$\varepsilon_x = -\frac{3l_s t}{2l^3}\delta \tag{10.10}$$

It should be noted that this equation is a function of δ, l, l_s, and t and that the flexural rigidity $\tilde{E}I$ and beam width w do not appear in this equation. It is also noted that the maximum strain occurs at the fixed end of the cantilever beam. The strain ε_z in the z direction at the sensor site is expressed as follows:

$$\text{For a wide beam, } w \geq 5t\text{: } \varepsilon_z = 0 \quad \text{and} \quad \sigma_z = v\sigma_x \tag{10.11}$$

$$\text{For a narrow beam, } w < 5t\text{: } \varepsilon_z = -v\varepsilon_x \quad \text{and} \quad \sigma_z = 0 \tag{10.12}$$

As mentioned already, the strains, (10.7) and (10.10), are measured by a strain gage and converted into a different type of signal (e.g., voltage) for easy manipulation; we can use the piezoresistive or piezoelectric technology (described in Sections 7.8 and 7.9) for the conversion.

The preceding analysis to evaluate the strains holds for the beam in which the thickness of the sensing element is negligible compared to the beam thickness t, as shown in Fig. 10.2b. When the thickness of the sensing element is comparable to that of the beam as shown in Fig. 10.2c, we should extend the preceding analysis to include the case. Since the beam at the sensor site may be considered as a composite beam, the strain ε_x can be evaluated by (8.9), which is written here for convenience:

$$\varepsilon_x = -\frac{M}{\left(\tilde{E}I\right)_{\text{eff}}} y_1 \tag{10.13}$$

where y_1 denotes the distance from the neutral surface to the center of the sensor layer in Fig. 10.2c. In equation (10.13), $\left(\tilde{E}I\right)_{\text{eff}}$ represents the effective flexural rigidity at the sensor site and is given by (n) of Example 8.1, which is rewritten here:

$$\left(\tilde{E}I\right)_{\text{eff}} = \frac{\tilde{E}I}{1+pqr}\left[(1+pqr)(1+p^3qr)+3pqr(1+p)^2\right] \tag{10.14}$$

where

$$p = \frac{t_1}{t} \qquad q = \frac{w_1}{w}, \qquad r = \frac{\tilde{E}_1}{\tilde{E}} \tag{10.15}$$

and $\tilde{E}I$ denotes the flexural rigidity of the lower layer (beam). In Fig. 10.2c, the distance y_c from the bottom of the lower layer (beam) to the neutral surface is given by (d) of Example 8.1:

$$y_c = \frac{t}{2}\frac{1+pqr(2+p)}{1+pqr} \tag{10.16}$$

where p, q, and r are the ratios defined by (10.15). Referring to Fig. 10.2c, we can define the distance y_1 as follows:

$$\begin{aligned} y_1 &= t_1 + t - y_c - \frac{t_1}{2} = \frac{t}{2}\left(2+p-2\frac{y_c}{t}\right) \\ &= \frac{t}{2}\left[2+p-\frac{1+pqr(2+p)}{1+pqr}\right] = \frac{t}{2}\frac{1+p}{1+pqr} \end{aligned} \tag{10.17}$$

Substituting (10.14) and (10.17) into (10.13), we find the strain ε_x at the center of the sensor layer:

$$\varepsilon_x = -\frac{1}{2}\frac{1+p}{(1+pqr)(1+p^3qr)+3pqr(1+p)^2}\frac{t}{\tilde{E}I}M \tag{10.18}$$

where p, q, r, t, $\tilde{E}I$, and M denote the ratios defined in (10.15), the thickness and flexural rigidity of the lower layer (beam), and the moment at the sensor site, respectively. Substituting the moment at the sensor site, $M = Fl_s$, into equation (10.18) gives the strain as a function of the applied force F, the flexural rigidity, and geometry:

$$\varepsilon_x = -\frac{1}{2}\frac{1+p}{(1+pqr)(1+p^3qr)+3pqr(1+p)^2}\frac{tl_s}{\tilde{E}I}F \tag{10.19}$$

As the thickness of the sensor layer is negligible (i.e., $p \to 0$), equation (10.19) is reduced to (10.7). Under the assumption that the thickness of the sensor layer is much less than that of the beam, we can estimate the strain in the z direction by using (10.11) and (10.12).

Stress and Strain in a Plate and Strip Under Constant Pressure

Plates are used as sensing elements in MEMS; for example, the rectangular plate of length a, width b, and thickness t in Fig. 10.3a may be used as a pressure sensor. Before analyzing the static response of the plate to the pressure p, we first examine the static response of the strip shown in Fig. 10.3b since analysis of the strip is much easier than that of the plate. As b/a goes to infinity, the deflection in Fig. 10.3a approaches that of the strip shown in Fig. 10.3b, where $b \geq 5t$. The uniformly distributed load (force per unit length) acting on a strip of width b is obtained as $q = pb$. To obtain the deflection w in the z direction in Fig. 10.3c, we then use the beam equation (3.105):

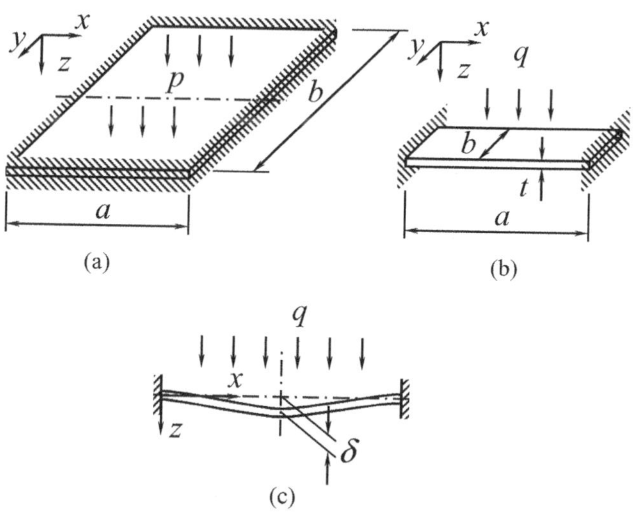

Figure 10.3 Deformations of a plate and strip under pressure.

$$\tilde{E}I\frac{d^4w}{dx^4}=q \tag{10.20}$$

where $\tilde{E}I$ and x represent the flexural rigidity of the strip and the coordinate from the left end of the strip. Integrating equation (10.20) and applying the boundary conditions ($w = 0$ and $dw/dx = 0$ at $x = 0$ and $x = l$), we have strip deflection w at x:

$$w=\frac{q}{24\tilde{E}I}\left(x^4-2ax^3+a^2x^2\right) \tag{10.21}$$

The maximum deflection δ at the midspan is then given by

$$\delta=w|_{x=a/2}=\frac{qa^4}{384\tilde{E}I} \tag{10.22}$$

If the strip is a composite beam, the flexural rigidity $\tilde{E}I$ is replaced by the effective rigidity $\left(\tilde{E}I\right)_{\text{eff}}$, which is defined in (8.12) and in (n) of Example 8.1. Substituting $q = pb$ and $I = wt^3/12$ into equation (10.22), a dimensionless equation to give the deflection δ at the midspan is obtained as

$$\frac{\delta}{t}=\frac{1}{32}\left(\frac{a}{t}\right)^4\frac{p}{\tilde{E}} \tag{10.23}$$

where $\tilde{E}$ denotes the effective Young's modulus defined in (3.5) and (3.6).

The strain ε_x in the x direction is obtained from (3.35) with the curvature $1/\rho = d^2w/dx^2$, as follows:

$$\varepsilon_x=-\frac{z}{\rho}=-\frac{d^2w}{dx^2}z \tag{10.24}$$

Substituting (10.21) into equation (10.24), we find the strain ε_x in the strip:

$$\varepsilon_x=-\frac{q}{12\tilde{E}I}\left(6x^2-6ax+a^2\right)z \tag{10.25}$$

According to this equation, the maximum strain occurs at the fixed ends on the upper or lower face of the strip. Setting $x = 0$ and $z = -t/2$, we obtain the maximum strain on the upper face of the strip at the fixed ends:

$$\varepsilon_x=\frac{qta^2}{24\tilde{E}I} \tag{10.26}$$

Since $q = pb$ and $I = wt^3/12$ for the strip, equation (10.26) may be expressed in a different form,

$$\varepsilon_x = \frac{1}{2}\left(\frac{a}{t}\right)^2 \frac{p}{\tilde{E}} \tag{10.27}$$

and the corresponding stress is given by

$$\sigma_x = \tilde{E}\varepsilon_x = \frac{1}{2}\left(\frac{a}{t}\right)^2 p \tag{10.28}$$

The strain ε_y in the y direction at the fixed ends is expressed as follows:

For a wide strip, $b \geq 5t$: $\varepsilon_y = 0$ and $\sigma_y = \nu\sigma_x$
For a narrow strip, $b < t$: $\varepsilon_y = -\nu\varepsilon_x$ and $\sigma_y = 0$

We now return the rectangular plate problem subjected to pressure in Fig. 10.3a. Considering the force and moment equilibrium of an infinite small plate element leads us to the plate equation,

$$\nabla^4 w = \frac{p}{D} \tag{10.29}$$

where ∇^4 denote the biharmonic operator and D represents the bending rigidity of the plate, which is defined by

$$D = \frac{Et^3}{12(1-\nu^2)} \tag{10.30}$$

Plate equation (10.29) was obtained by Lagrange in 1811 (Timoshenko and Woinowsky-Krieger, 1959). For the rectangular coordinates, the biharmonic operator is given by

$$\begin{aligned}\nabla^4 = \nabla^2\nabla^2 &= \left(\frac{\partial^2}{\partial x^2} + \frac{\partial^2}{\partial y^2}\right)\left(\frac{\partial^2}{\partial x^2} + \frac{\partial^2}{\partial y^2}\right) \\ &= \frac{\partial^4}{\partial x^4} + 2\frac{\partial^4}{\partial x^2 \partial y^2} + \frac{\partial^4}{\partial y^4}\end{aligned} \tag{10.31}$$

The linear plate equation can be solved for the deflection w, but the solution is usually expressed in a complicated form (e.g., an infinite series of functions). Readers interested in the derivation of the plate equation or solutions of the equation may refer to existing publications [e.g., Fung (1965) for the derivation of the plate equation, and Timoshenko and Woinowsky-Krieger (1959) for the derivation and solutions of the plate equation]. For easy design of sensors based on rectangular plates, we need closed-form expressions for the maximum strain and maximum deflection of the plate. Referring to the analysis of the plate equation as well as the analytical and numerical solutions, we can construct closed-form expressions for the maximum strain and maximum deflection of the plate by using a curve-fitting technique. The maximum strain ε_x occurs at the fixed ends along the centerline (dashed-dotted line in Fig. 10.3a for $b \geq a$) on the surface of plate and is given by

$$\varepsilon_x = \alpha_1 \left(\frac{a}{t}\right)^2 \frac{p}{\tilde{E}} \tag{10.32}$$

where

$$\alpha_1 = 0.1063r^3 - 0.6830r^2 + 1.4950r - 0.6105 \tag{10.33}$$

$$r = \frac{b}{a}$$

and $a, b, t,$ and $\tilde{E}$ denote the lengths of the short and long sides of the plate, the thickness, and the effective Young's modulus, which is given by $\tilde{E} = E/(1-\nu^2)$, respectively. The preceding strain equation holds for $1 \leq r \leq 2$. For $r > 2$, the following holds:

$$\alpha_1 \cong \frac{1}{2} \tag{10.34}$$

which is the coefficient for the strain given by (10.27). Note that the form of (10.32) is the same as that of the maximum strain of the strip, which is given in (10.27). At the site of the maximum strain (i.e., at the fixed ends), the strain ε_y vanishes; that is,

$$\varepsilon_y = 0 \tag{10.35}$$

The stresses corresponding to the strains above, ε_x and ε_y, are obtained as

$$\sigma_x = \tilde{E}\varepsilon_x = \alpha_1 \left(\frac{a}{t}\right)^2 p \tag{10.36}$$

$$\sigma_y = \nu\sigma_x$$

where α_1 is given by (10.33) and (10.34). The maximum deflection of the center of plate for $b \geq a$ is obtained as

$$\frac{\delta}{t} = w\big|_{x=a/2} = \alpha_2 \left(\frac{a}{t}\right)^4 \frac{p}{\tilde{E}} \tag{10.37}$$

where

$$\alpha_2 = 0.07595r^4 - 0.4272r^3 + 0.8648r^2 - 0.7282r + 0.2298 \tag{10.38}$$

The preceding maximum deflection equation holds for $1 \le r \le 2$. When r goes beyond 2, the coefficient α_2 approaches 1/32, which is the coefficient for the strip deflection, given by (10.23). For the square plate, substituting $r = 1$ into (10.32) and (10.37) yields the maximum strain and deflection:

$$\varepsilon_x = 0.3078 \left(\frac{a}{t}\right)^2 \frac{p}{\tilde{E}} \tag{10.39}$$

$$\frac{\delta}{t} = 0.01520 \left(\frac{a}{t}\right)^4 \frac{p}{\tilde{E}} \tag{10.40}$$

These formulas for a rectangular plate may be used to design pressure sensors, which are described next.

Pressure Sensor

A typical bulk-micromachined sensor is shown in Fig. 10.4a, where a thin square plate with four identical piezoresistors labeled R_1 to R_4 is used for a pressure-sensing element that measures the pressure difference $p = p_0 - p_i$. The resistors are placed near the edges of the square plate to measure the maximum strain ε_x. To measure the change in strain, we can form the Wheatstone bridge shown in Fig. 10.4b. The output voltage from the Wheatstone bridge is given by (7.29), which is rewritten here for convenience:

$$V_0 = V_1 - V_2 = \left(\frac{R_4}{R_1 + R_4} - \frac{R_3}{R_2 + R_3}\right)V = \frac{R_2R_4 - R_1R_3}{(R_2 + R_3)(R_1 + R_4)}V$$

Assuming that the resistance is measured across the length of all the piezoresistors, we may set $R = R_1 = R_2 = R_3 = R_4$ for $p = 0$. When the pressure varies from zero, the strain ε_x is generated at the positions of the resistors; due to symmetrical structure, the strain along the longitudinal direction of the resistors R_1 and R_3 is the same as that in the transverse direction of the resistors R_2 and R_4. We use the resistance–strain relations, (7.173) and (7.178), to obtain the resistances at pressure p, as follows: For R_1 and R_3,

$$\begin{aligned} R_l + \Delta R_l &= R[1 + (\Delta R/R)_l] = R(1 + \pi_{el}\tilde{E}\varepsilon_x) \\ &= R(1 + \pi_{el}\sigma_x) \end{aligned} \tag{10.41}$$

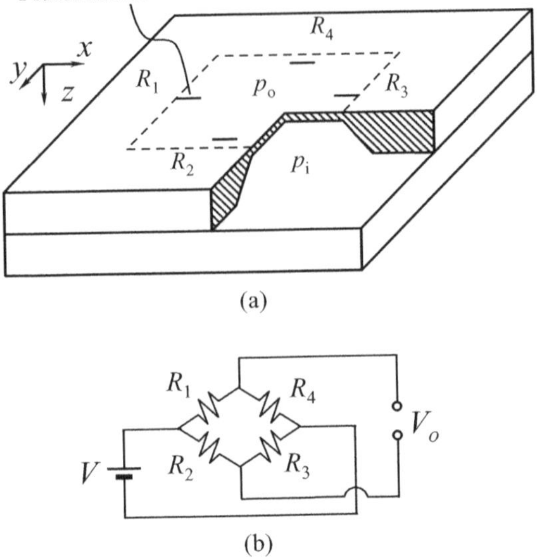

Figure 10.4 Pressure sensor.

and for R_2 and R_4,

$$\begin{aligned} R_t + \Delta R_t &= R[1 + (\Delta R/R)_t] = R(1 + \pi_{et}\tilde{E}\varepsilon_{y,R_2}) \\ &= R(1 + \pi_{et}\tilde{E}\varepsilon_x) = R(1 + \pi_{et}\sigma_x) \end{aligned} \tag{10.42}$$

Substituting equations (10.41) and (10.42) into the foregoing equation for the Wheatstone bridge leads to the voltage–stress relation

$$\begin{aligned} \frac{V_o}{V} &= \frac{R^2(1+\pi_{el}\sigma_x)^2 - R^2(1+\pi_{et}\sigma_x)^2}{[R(1+\pi_{el}\sigma_x) + R(1+\pi_{et}\sigma_x)]^2} \\ &\cong \frac{\pi_{el} - \pi_{et}}{2}\sigma_x \end{aligned} \tag{10.43}$$

which is obtained under the assumption that the resistance change is much less than the initial resistance. Upon substitution of $\sigma_x = \tilde{E}\varepsilon_x$, where ε_x is given by (10.39), into the preceding voltage–stress relation, we find a voltage–pressure relation,

$$\frac{V_0}{V} = 0.1539(\pi_{el} - \pi_{et})\left(\frac{a}{t}\right)^2 p \tag{10.44}$$

which is a function of the applied pressure p, the ratio of side length to thickness, and the effective piezoresistance coefficients, π_{el} and π_{et}. If we use the (100) silicon wafer and if the x and y axes are aligned with [110], piezoresistance coefficients are given by $\pi_{el} = 57.20 \times 10^{-11}\,\text{Pa}^{-1}$ and $\pi_{et} = -50.52 \times 10^{-11}\,\text{Pa}^{-1}$ (which are described in Example 7.1). Substituting these coefficients into equation (10.44) gives the voltage–pressure relation for the piezoresistive pressure sensor made of the (100) silicon wafer as shown in Fig. 10.4a:

$$\frac{V_0}{V} = 16.58\left(\frac{a}{t}\right)^2 p \tag{10.45}$$

If a different material is used to form the plate as the pressure-sensing element, the voltage–pressure relation may be obtained by substituting the piezoresistance coefficients π_{el} and π_{et} of the material into (10.44).

Resonant Structures as Force Sensors

In preceding chapters we have shown that the natural frequency (resonant frequency) of structures in tension or in compression is changed from that of stress-free structures. As an example, we consider a comb drive actuator formed on a surface of a relatively large substrate of uniform cross-sectional area, as shown in Fig. 10.5a. When the substrate is subjected to a tensile force F, the strain of the substrate in the y direction is obtained as

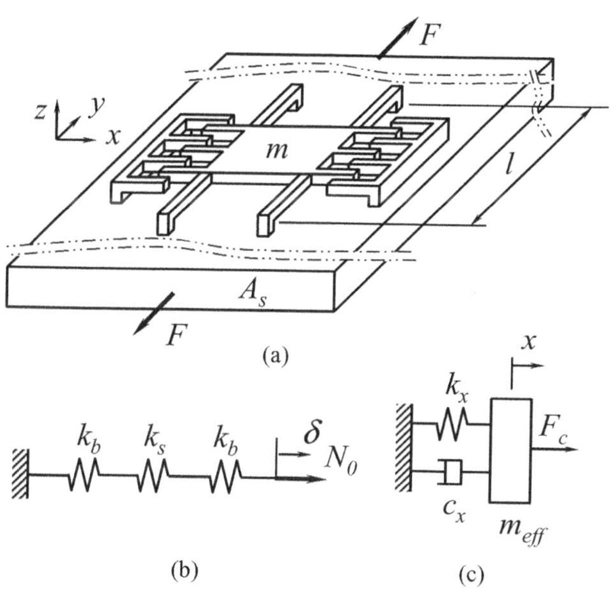

Figure 10.5 Natural frequency change due to force.

$$\varepsilon_y = \frac{F}{\tilde{E}_s A_s} \tag{10.46}$$

Since the comb drive actuator is much more compliant than the substrate, it is assumed that the distance between the anchors of the straight beam is increased in the y direction by $\delta = l\varepsilon_y$, where l denotes the length of the structure as shown in Fig. 10.5a. We can model the right-half structure in Fig. 10.5a as a series of springs in Fig. 10.5b, where k_s denotes the stiffness of the half movable structure, and k_b represents the stiffness of a beam in the longitudinal direction (i.e., in the y direction). The stiffness of the beam is obtained as $k_b = \tilde{E}A_b/l_b$, where $\tilde{E}$, A_b, and l_b denote the effective Young's modulus, cross-sectional area, and length of one straight beam, respectively. The stiffness in the y direction in Fig. 10.5b is then obtained from formula (4.2):

$$\frac{1}{k_y} = \frac{1}{k_b} + \frac{1}{k_s} + \frac{1}{k_b}$$

or

$$k_y = \frac{k_b k_s}{k_b + 2k_s} \tag{10.47}$$

Therefore, the axial force N_0 acting on the end of the straight beam is expressed as

$$N_0 = k_y \delta = \frac{k_b k_s}{k_b + 2k_s} l\varepsilon_y \tag{10.48}$$

The stiffness of the straight beam subjected to the axial force N_0 is given by (3.75) and the stiffness due to the four beams is then expressed as

$$k_x = 4k_1 = \frac{48\tilde{E}_b I}{l_b^3}\left(1 + \frac{1}{10}\frac{l_b^2 N_0}{\tilde{E}_b I}\right) \tag{10.49}$$

where I denotes the moment of inertia of the cross section of the beam. If m_b represents the mass of a straight beam, the effective mass, m_e, of the beam is obtained given by (5.162), $m_e = 13/35\, m_b$. We find the effective mass m_{eff} as follows:

$$m_{\text{eff}} = m + 4m_e = m + 4\frac{13}{35}m_b = m + \frac{52}{35}m_b \tag{10.50}$$

The natural frequency of the comb drive actuator shown in Fig. 10.5a is given by

$$f_n = \frac{1}{2\pi}\sqrt{\frac{k_x}{m_{\text{eff}}}}$$

Substituting (10.49) and (10.50) into the equation above leads us to

$$\frac{f_n}{f_{n0}} = \sqrt{1 + \frac{1}{10}\frac{l_b^2 N_0}{\tilde{E}_b I}} \tag{10.51}$$

where

$$f_{n0} = \frac{1}{2\pi}\sqrt{\frac{48\tilde{E}_b I}{m_{\text{eff}} l_b^3}} \tag{10.52}$$

represents the natural frequency at $N_0 = 0$ and l_b denotes the length of the straight beam. In general, the frequency change is much less than the initial natural frequency f_0, so the preceding frequency may be approximated as

$$\frac{f_n}{f_{n0}} = 1 + \frac{1}{20}\frac{l_b^2 N_0}{\tilde{E}_b I}$$

which is expressed in a different form,

$$\frac{\Delta f_n}{f_n} = \frac{f_n - f_{n0}}{f_n} = \frac{1}{20}\frac{l_b^2}{\tilde{E}_b I} N_0 \tag{10.53}$$

where $\Delta f_n = f_n - f_{n0}$ stands for the frequency change. Combining (10.48) and equation (10.53), we have

$$\frac{\Delta f_n}{f_n} = \frac{1}{20}\frac{l_b^2}{\tilde{E}_b I}\frac{k_b k_s}{k_b + 2k_s} l\varepsilon_y \tag{10.54}$$

Substituting (10.46), $k_b = \tilde{E}_b A_b / l_b$, and $I = tw^3/12$, where t and w denote the thickness and width of the straight beam, respectively, we obtain the frequency change due to the force F:

$$\frac{\Delta f_n}{f_n} = \frac{3}{5}\frac{l_b l}{w^2}\frac{1}{2 + k_b/k_s}\frac{F}{\tilde{E}_s A_s} \tag{10.55}$$

where l_b, l, w, $\tilde{E}_s$, and A_s denote the length of one straight beam, the distance between the anchored points of the beams as shown in Fig. 10.5a, the effective Young's modulus of the substrate, and the cross-sectional area of the substrate, respectively. If the movable structure is considered as a rigid body (i.e., $k_b/k_s \to 0$), equation (10.55) reduces to

$$\frac{\Delta f_n}{f_n} = \frac{3}{10} \frac{l_b l}{w^2} \frac{F}{\tilde{E}_s A_s} \tag{10.56}$$

The dynamic response can be obtained by using the model in Fig. 10.5c, where F_c denotes the actuating force in the x direction; the response was described in Section 9.2. Equations (10.55) and (10.56) hold for a case that satisfies the following condition: The vibrating system can be considered a linear system (i.e., the cubic stiffness is negligible for small vibration amplitude).

Fixed–Fixed Beams Subjected to Axial Force

A fixed–fixed beam of length l, width w, and thickness t on a substrate of cross-sectional area A_s may be used as a resonant force sensor (Fig. 10.6). If the cross-sectional area of the substrate is much greater than that of the beam (i.e., the stiffness of the substrate is much greater than that of the beam), the strain in the substrate is given by $\varepsilon = F/\tilde{E}_s A_s$, which is considered as the strain of the beam. The axial force in the longitudinal direction of the beam is then expressed as

$$N_0 = \sigma_b wt = \tilde{E}_b \varepsilon wt = \tilde{E}_b \frac{F}{\tilde{E}_s A_s} wt = \frac{\tilde{E}_b}{\tilde{E}_s} \frac{wt}{A_s} F \tag{10.57}$$

Let x_e and h_0 represent the midspan deflection of the beam and the initial gap, respectively. We assumed that a voltage is applied to measure the capacitance between the beam and the lower electrode. We may apply to the beam a dc bias voltage or dc bias voltage with ac voltage. The effective stiffness of the beam is given by (9.274), rewritten here for convenience:

$$k_{\text{eff}} = k_{e1}\left(1 - X_e \frac{s - 2rX_e}{rX_e^2 - sX_e + 1}\right) \tag{10.58}$$

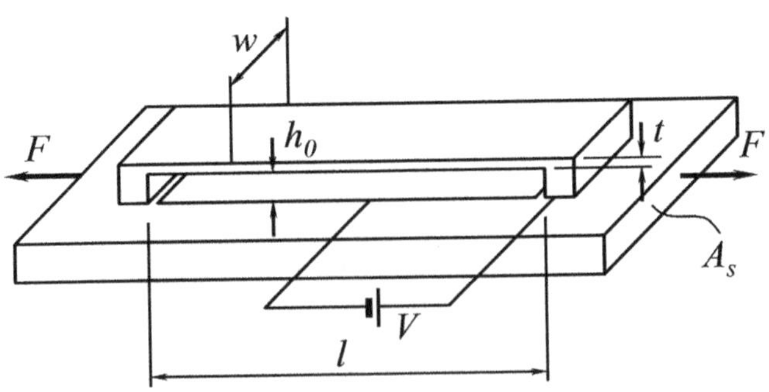

Figure 10.6 Electrostatic fixed–fixed beam actuator subjected to an axial force.

where k_{e1} represents the effective stiffness of the fixed–fixed beam, which is given by (9.292):

$$k_{e1} = \frac{1024}{5}\frac{\tilde{E}I}{l^3} + \frac{512}{105}\frac{N_0}{l} \tag{10.59}$$

and I denotes the moment of inertia of the cross section of the beam, which is given by $I = wt^3/12$. The dimensionless midspan deflection X_e is defined as $X_e = x_e/h_0$, which is evaluated by (9.263):

$$X_e = \frac{1}{3r}\left(s - 2\sqrt{s^2 - 3r}\cos\frac{\phi}{3}\right) \tag{10.60}$$

For the fixed–fixed beam, (9.188) gives the constants r and s:

$$r = 25{,}472/63{,}063 \cong 0.4039 \quad \text{and} \quad s = 32/21 \cong 1.5238 \tag{10.61}$$

and ϕ is given by (9.265) and (9.261):

$$\phi = \cos^{-1}\frac{(9/2)sr - s^3 - (27/2)r^2G_e}{\left(s^2 - 3r\right)^{3/2}} \tag{10.62}$$

$$G_e = \frac{1}{2}\frac{\varepsilon A_e}{k_{e1}h_0^3}V^2 = \frac{1}{2}\frac{\varepsilon}{k_{e1}h_0^3}(\eta\phi_1 wt)V^2 = \frac{4}{15}\frac{\varepsilon wt}{k_{e1}h_0^3}V^2 \tag{10.63}$$

where $\eta = 16$ and $\phi_1 = 1/30$ are used, which were obtained in Section 9.5. The preceding stiffness holds for $V \le V_{pi}$, where V_{pi} denotes the pull-in voltage, which is given by (9.207):

$$V_{pi} = 3.4147\sqrt{\frac{\tilde{E}_b t^3 h_0^3}{\varepsilon l^4}\left[1 + 0.2857\frac{N_0}{\tilde{E}_b wt}\left(\frac{l}{t}\right)^2\right]} \tag{10.64}$$

where $\tilde{E}_b$, ε, N_0, t, l and h_0 denote the effective Young's modulus of the beam, the permittivity of the surrounding fluid (if the actuator is in vacuum, the permittivity of vacuum), the axial force in the beam, the thickness and length of the beam, and the initial gap at $V = 0$, respectively. The effective mass m_{eff} is obtained from (9.291) as follows:

$$m_{\text{eff}} = \frac{128}{315}m \tag{10.65}$$

where $m = \rho wtl$ is the mass of the entire beam. If the quality factor Q is much greater than 1, the damped natural frequency (resonant frequency) is expressed as

$$f_n \cong \frac{1}{2\pi}\sqrt{\frac{k_{\text{eff}}}{m_{\text{eff}}}} \tag{10.66}$$

Substituting (10.58) and (10.65) into equation (10.66), we find the natural frequency as follows:

$$f_n = \frac{1}{2\pi}\sqrt{\frac{504\frac{\tilde{E}_b I}{l^3}\left(1+\frac{1}{42}\frac{N_0 l^2}{\tilde{E}_b I}\right)S_{\text{eff}}}{m}} \tag{10.67}$$

where

$$S_{\text{eff}} = 1 - X_e\frac{s-2rX_e}{rX_e^2 - sX_e + 1} \tag{10.68}$$

denotes the dimensionless stiffness at the voltage V. Substitution of $m = \rho wtl$ and $I = wt^3/12$ into equation (10.67) gives

$$f_n = \frac{1}{2\pi}\frac{t}{l^2}\sqrt{42\frac{\tilde{E}_b}{\rho}\left(1+\frac{1}{42}\frac{N_0 l^2}{\tilde{E}_b I}\right)S_{\text{eff}}}$$

For $N_0 l^2/42\tilde{E}_b I \ll 1$, the preceding equation may be written as

$$f_n = \frac{1}{2\pi}\frac{t}{l^2}\sqrt{42\frac{\tilde{E}_b}{\rho}S_{\text{eff}}}\left(1+\frac{1}{82}\frac{N_0 l^2}{\tilde{E}_b I}\right) \tag{10.69}$$

Combining (10.57) and equation (10.69) and recalling that $I = wt^3/12$, we find the natural frequency in terms of F, geometry, and material properties as follows:

$$\begin{aligned} f_n &= \frac{1}{2\pi}\frac{t}{l^2}\sqrt{42\frac{\tilde{E}_b}{\rho}S_{\text{eff}}}\left[1+\frac{6}{41}\left(\frac{l}{t}\right)^2\frac{F}{\tilde{E}_s A_s}\right] \\ &= f_{ni} + \gamma F \end{aligned} \tag{10.70}$$

where

$$\begin{aligned} f_{ni} &= \frac{1}{2\pi}\frac{t}{l^2}\sqrt{42\frac{\tilde{E}_b}{\rho}S_{\text{eff}}} \\ &\cong 1.0314\frac{t}{l^2}\sqrt{\frac{\tilde{E}_b}{\rho}S_{\text{eff}}} \end{aligned} \tag{10.71}$$

$$\gamma = \frac{3}{41\pi}\frac{1}{t}\sqrt{42\frac{\tilde{E}_b}{\rho}S_{\text{eff}}}\frac{1}{\tilde{E}_s A_s}$$
$$\cong 0.1509\frac{1}{t}\frac{1}{\tilde{E}_s A_s}\sqrt{\frac{\tilde{E}_b}{\rho}S_{\text{eff}}} \tag{10.72}$$

denotes the initial natural frequency at $F = 0$ and the sensitivity of the natural frequency to the applied force, respectively. In the preceding equations, $\tilde{E}_s$, $\tilde{E}_b$, ρ, t, l, S_{eff}, and A represent the effective Young's moduli of the substrate and the beam, the density, the thickness and length of the beam, the effective dimensionless stiffness defined by (10.68), and the cross-sectional area of the substrate, respectively. For voltage $V = 0$ (i.e., $X_e = 0$, $S_{\text{eff}} = 1$), the initial natural frequency and sensitivity of the beam reduce to

$$f_{ni}\big|_{V=0} = 1.0314\frac{t}{l^2}\sqrt{\frac{\tilde{E}_b}{\rho}} \tag{10.73}$$

$$\gamma\big|_{V=0} = 0.1509\frac{1}{t}\frac{1}{\tilde{E}_s A_s}\sqrt{\frac{\tilde{E}_b}{\rho}} \tag{10.74}$$

which is equal to the natural frequency of the fixed–fixed beam, equation (g) of Example 5.3, since $f_n = (3.5730/l^2)\sqrt{\tilde{E}_b I/m_1} = 1.0314t\sqrt{\tilde{E}_b/\rho}/l^2$ (note that E is replaced by $\tilde{E}$). It is worth noting that the initial natural frequency f_{ni} and the sensitivity γ are functions of the applied voltage because the dimensionless stiffness S_{eff} relates to the applied voltage V, as shown in (10.60)–(10.63) and (10.68). Therefore, the natural frequency can be adjusted by controlling both the voltage V and the force F, shown in Fig. 10.6.

10.2 ACCELEROMETERS

The force sensors described in Section 10.1 can be used for inertial sensors such as accelerometers and gyroscopes, since the inertial sensors measure the inertial forces due to acceleration. Let us begin by considering a simple accelerometer, shown in Fig. 10.7. A mass m is suspended by a string from the ceiling of a bus that is at rest. The position of the bus is measured from an inertial coordinate x as shown in Fig. 10.7a. If the bus is accelerated for a long time, during which the transient motion of the mass disappears, the mass in Fig. 10.7b will make an angle θ with the vertical line. Let $\ddot{x}$ represent the acceleration of the bus. Applying Newton's second law to the mass shown in Fig. 10.7c, where T and mg denote the tension in the string and the gravitational force (note that the gravitational force plays a role in the macroworld), we have the equations of motion in the horizontal and vertical directions:

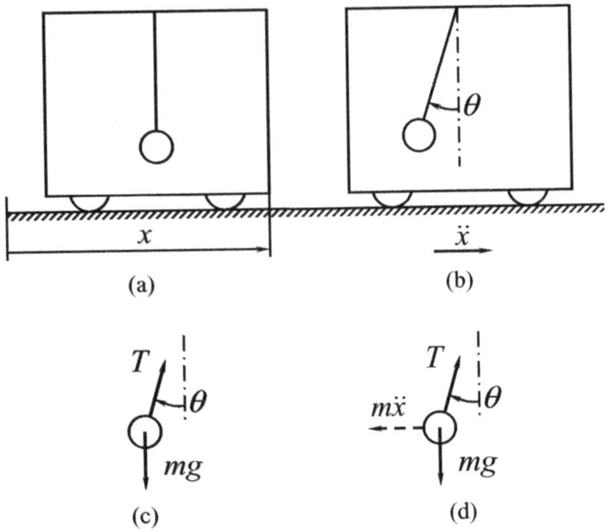

Figure 10.7 Principle of an accelerometer.

$$m\ddot{x} = T\sin\theta \qquad 0 = T\cos\theta - mg$$

From the equation above, we obtain the tension, $T = mg/\cos\theta$, and the acceleration of the bus,

$$\ddot{x} = g\tan\theta \tag{10.75}$$

As a result, we can determine the acceleration if we measure the angle θ. The corresponding velocity and displacement (position) may be obtained from integration of the acceleration. The foregoing procedure to obtain the equation of the mass was simple. However, for complicated systems (e.g., multidegree-of-freedom system), the procedure for setting up equations of motion may be difficult. To simplify the procedure, we may use D'Alembert's principle, which is described below. Rearranging the foregoing equations of motion for the mass in the bus, we have

$$T\sin\theta - m\ddot{x} = 0 \qquad T\cos\theta - mg = 0$$

The equations above are the equations of motion which describe the motion of the mass. However, if we consider $-m\ddot{x}$ as a static force, the preceding equations become a set of equations to determine static force. This situation is shown in Fig. 10.7d, where the arrow for $m\ddot{x}$ is leftward (opposite to the x direction) and indicated by a dashed arrow since the force $m\ddot{x}$ is not a real force but a fictitious force, known as the *inertial force*. This principle to reduce a dynamic problem to the corresponding static problem is known as *D'Alembert's principle*, which is used to simplify dynamic problems.

Accelerometers

A sectional view of a typical large-scale accelerometer is shown in Fig. 10.8, where a proof mass with a force sensor is placed in a housing mounted on a structure (e.g., part of an automobile). When the structure is accelerated in the z_s direction, the accelerometer measures the acceleration $\ddot{z}_s$. A model of the accelerometer of Fig. 10.8 is shown in Fig. 10.9a, where a mass m is suspended by a spring of stiffness k and a damper of damping coefficient c, both of which are attached to a support (housing). The displacements of the mass and support measured from an inertial frame are represented by z_m and z_s, respectively (Fig. 10.9a). Thus, a relative displacement, which represents the output signal of the accelerometer, may be defined as

$$z_a = z_m - z_s \tag{10.76}$$

Let us obtain a relation between the output above and the acceleration $\ddot{z}_s$. Applying Newton's second law to the mass in Fig. 10.9b, we have

$$m\ddot{z}_m = -k(z_m - z_s) - c(\dot{z}_m - \dot{z}_s) \tag{10.77}$$

Combining equations (10.76) and (10.77) gives an equation of motion in terms of the input acceleration and the relative displacement:

$$m\ddot{z}_a + c\dot{z}_a + kz_a = -m\ddot{z}_s \tag{10.78}$$

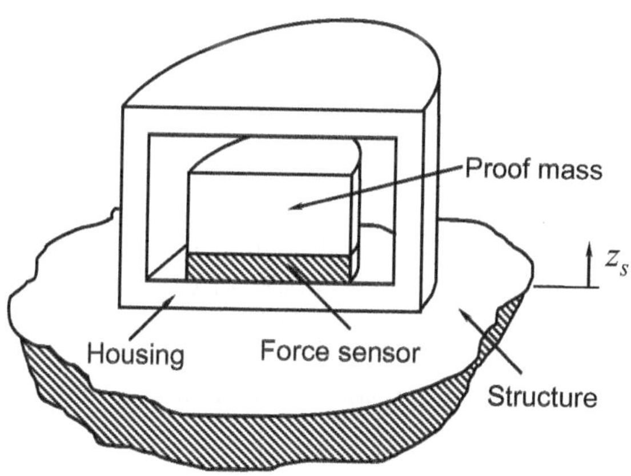

Figure 10.8 Typical accelerometer.

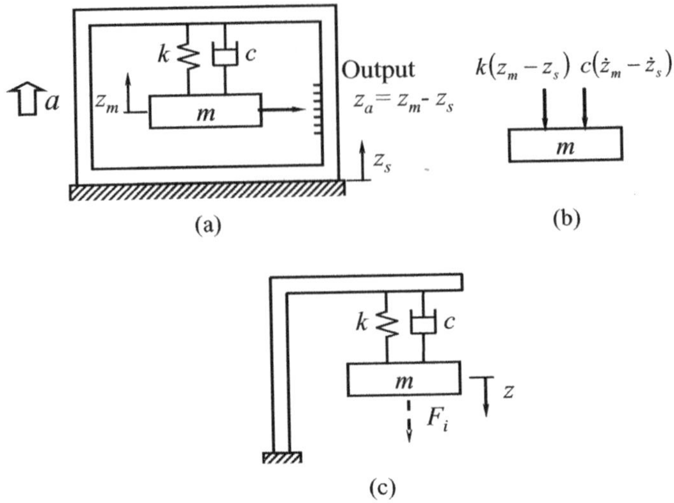

Figure 10.9 Model of accelerometers.

where $-m\ddot{z}_s$ is the inertial force that is applied in the negative z_a direction. For convenience, setting $z = -z_a$, we have a slightly different form of equation (10.78):

$$m\ddot{z} + c\dot{z} + kz = F_i \tag{10.79}$$

where z denotes the relative motion and $F_i = m\ddot{z}_s$ represents the inertial force due to the acceleration. We can interpret this equation as follows: When a mechanical structure in Fig. 10.9a is exposed to an acceleration a in the z_s direction, the inertial force F_i acts as an actuating force to generate a vibration as shown in Fig. 10.9c. If a constant acceleration a is applied for a long period (i.e., much greater than the time constant), the relative acceleration and velocity vanish and equation (10.79) reduces to

$$z = \frac{F_i}{k} = \frac{ma}{k} = \frac{a}{\omega_n^2} \tag{10.80}$$

where $\omega_n = \sqrt{k/m}$ denotes the natural angular frequency. In other words, for a constant acceleration, the dynamic problem can be thought of as a static problem in which the spring force balances the inertial force due to the acceleration.

We shall obtain the dynamic response of the accelerometer. To do this, we assume a periodic acceleration,

$$a = \ddot{z}_b = a_0 \sin \omega t \tag{10.81}$$

which gives the inertial force,

$$F_i = ma = ma_0 \sin \omega t = F_0 \sin \omega t \tag{10.82}$$

where $F_0 = ma_0$ denotes the amplitude of the inertial force. The dynamic response of the second-order differential equation, (10.79), was obtained in Section 5.6. Referring to (5.87)–(5.89), we write the dynamic response z as follows:

$$\begin{aligned} z &= \frac{ma_0}{k} \frac{1}{\sqrt{(1-\Omega^2)^2 + (2\zeta\Omega)^2}} \sin(\omega t - \phi) \\ &= \frac{a_0}{\omega_n^2} \frac{1}{\sqrt{(1-\Omega^2)^2 + (2\zeta\Omega)^2}} \sin(\omega t - \phi) \\ &= z_0 \sin(\omega t - \phi) \end{aligned} \tag{10.83}$$

where

$$\begin{aligned} z_0 &= \frac{a_0}{\omega_n^2} \frac{1}{\sqrt{(1-\Omega^2)^2 + (2\zeta\Omega)^2}} \\ \Omega &= \frac{\omega}{\omega_n} = \frac{f}{f_n} \\ \omega_n &= \sqrt{\frac{k}{m}} \\ \phi &= \tan^{-1} \frac{2\zeta\Omega}{1-\Omega^2} \\ \zeta &= \frac{c}{2\sqrt{mk}} \end{aligned} \tag{10.84}$$

represent the amplitude, the frequency ratio, the natural angular frequency, the phase, and the damping factor, respectively. The dimensionless response, $z_0\omega_n^2/a_0$, is plotted against the frequency ratio, $\Omega = \omega/\omega_n$, in Fig. 10.10 while the damping factor ζ varies from 0.5 to 0.8. As the damping factor decreases (in other words, the quality factor increases), the amplitude of the response increases and the damped natural frequency is close to $\Omega = 1$. To minimize distortion of the acceleration response and to maximize the flat range of the response, a damping ratio ζ near 0.7 is preferred. However, some accelerometers (e.g., a piezoelectric accelerometer), whose natural frequency is very high, have a very low damping ratio (i.e., a high quality factor). If the damping ratio is taken as 0.7, the error at $\omega/\omega_n = 0.4$ is 0.95%. For a frequency much less than the natural frequency (i.e.$\Omega << 1$), the response of (10.83) is reduced to $z = a_0/\omega_n^2 \sin \omega t = (ma_0/k) \sin \omega t$, and then the response can be considered as

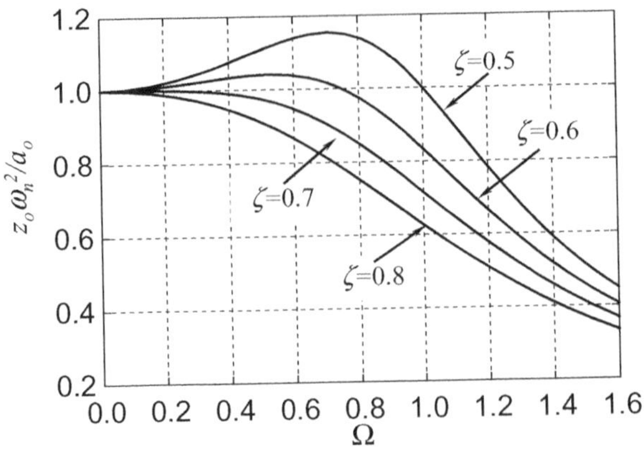

Figure 10.10 Frequency response.

a quasistatic response in which the terms related to the acceleration and velocity of the relative motion effect are neglected; we discussed this phenomenon earlier in association with (10.80). Therefore, the low-frequency response of the mechanical structure is governed by the stiffness of the structure when the structure is exposed to acceleration of a frequency much lower than the natural frequency of the structure. This phenomenon will be used to obtain the response of a variety of structures to acceleration whose frequency is much lower than the natural frequency of the structures.

Let us apply D'Alembert's principle to obtain the response of the microstructures, which may be used as micromachined accelerometers. In Fig. 10.11a, a plate is suspended by a beam whose left end is attached to a substrate. The substrate moves at an acceleration whose frequency is much less than the lowest natural frequency of the structure. If the mass of the beam is much less than that of the plate, the lowest natural frequency of the structure is expressed in closed form, which is given by (5.114). In this case, the inertia force is applied in the opposite direction of the acceleration a, as shown in Fig. 10.11b. From the moment equilibrium of the free-body diagram in Fig. 10.11b (note that the dynamic problem is converted to a static problem),

$$\sum M = -M - ma(l_s + l_c) - m_b \frac{l_s}{l} a \frac{l_s}{2} = 0$$

where m and m_b denote the masses of the plate and beam, respectively. The equation above gives the moment at the sensor site:

$$M = -\left[m(l_s + l_c) + m_b \frac{l_s^2}{2l} \right] a \tag{10.85}$$

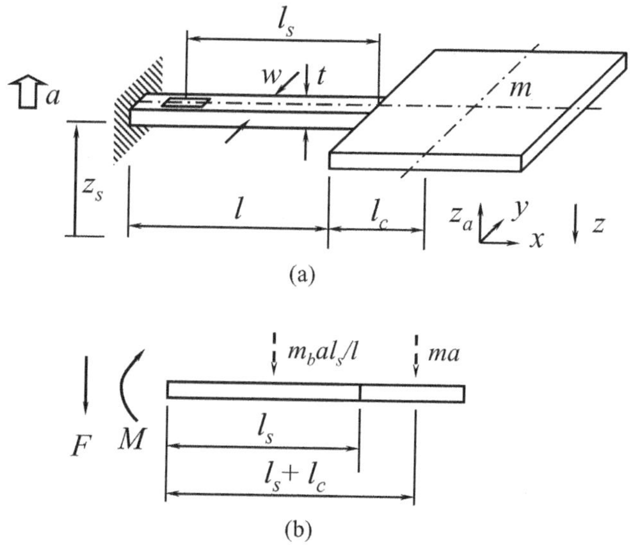

Figure 10.11 Accelerometer using a proof mass supported by a beam.

The strain is obtained from (3.41) by substituting z_1 for y,

$$\varepsilon_x = -\frac{M}{\tilde{E}I} z_1 \tag{10.86}$$

where $\tilde{E}I$ and z_1 denote the flexural rigidity of the beam and the distance from the neural surface of the beam to a point of interest, respectively. Substituting (10.85) into equation (10.86), we find the strain ε_x in terms of the applied acceleration, the masses of the beam and plate, the flexural rigidity, and the geometry, as follows:

$$\varepsilon_x = \left[m(l_s + l_c) + m_b \frac{l_s^2}{2l} \right] \frac{z_1}{\tilde{E}I} a \tag{10.87}$$

If the cross section of the beam is rectangular and if the sensor thickness is negligible compared to the beam thickness, the strain ε_x at the sensor site is obtained by setting $z_1 = t/2$:

$$\varepsilon_x = \frac{1}{2} \left[m(l_s + l_c) + m_b \frac{l_s^2}{2l} \right] \frac{t}{\tilde{E}I} a \tag{10.88}$$

Note that the maximum strain is obtained at $l_s = l$, (i.e., the left end of the beam). Therefore, we should place the sensor as close as to the left end of the beam to obtain maximum strain from the sensor. The stress corresponding to the strain above is then given by

$$\sigma_x = \tilde{E}\varepsilon_x = \frac{1}{2}\left[m(l_s + l_c) + m_b\frac{l_s^2}{2l}\right]\frac{t}{I}a \tag{10.89}$$

which is not a function of the effective Young's modulus. The strain ε_y is estimated as follows:

$$\text{For wide beams, } w \geq 5t\text{: } \varepsilon_y = 0 \quad \text{and} \quad \sigma_y = \nu\sigma_x \tag{10.90}$$

$$\text{For narrow beams, } w < 5t\text{: } \varepsilon_y \cong -\nu\varepsilon_x \quad \text{and} \quad \sigma_y \cong 0 \tag{10.91}$$

The strains above can be measured by sensors such as piezoresistor and piezoelectric material (covered in Sections 7.8 and 7.9) and converted into a voltage signal. For example, when we use a wide beam, (7.173) or (7.178) may be used to measure the strain ε_x. If the sensor thickness is not negligible, (10.18) may be used to calculate the strain ε_x at the center of the sensor layer.

If we can remove the plate of mass m from the preceding case in Fig. 10.11, we have a cantilever beam which is accelerated at acceleration a, as shown in Fig. 10.12. The strain ε_x along the top of the cantilever beam is obtained easily by setting $m = 0$ in (10.88):

$$\varepsilon_x = \frac{1}{4}m_b\frac{l_s^2}{l}\frac{t}{\tilde{E}I}a \tag{10.92}$$

Recalling that $I = wt^3/12$ and $m_b = \rho wtl$, where ρ denotes the density of the beam, we have equation (10.92) in a simple form:

$$\varepsilon_x = 3\frac{\rho}{\tilde{E}}\frac{l_s^2}{t}a \tag{10.93}$$

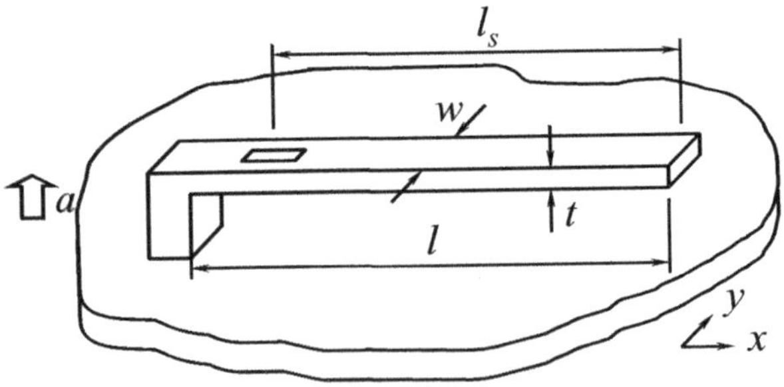

Figure 10.12 Accelerometer using a cantilever beam.

which gives the corresponding stress:

$$\sigma_x = \tilde{E}\varepsilon_x = 3\rho\frac{l_s^2}{t}a \tag{10.94}$$

The strain and stress on the top of the cantilever in the y direction are given by (10.90) and (10.91), depending on the ratio w/t. Equations (10.92) and (10.93) hold for the cantilever beam exposed to an acceleration whose frequency f is much less than the first natural frequency, f_1, which is given by (5.138) with $\beta_1 l = 1.8751$, $I = wt^3/12$, and $m_1 = \rho wt$:

$$f_1 = \frac{(\beta_1 l)^2}{2\pi l^2}\sqrt{\frac{\tilde{E}I}{m_1}} = 0.1615\frac{t}{l^2}\sqrt{\frac{\tilde{E}}{\rho}} \tag{10.95}$$

Symmetric Accelerometers

An accelerometer may be formed symmetrically about a point (e.g., the center of mass). An example is shown in Fig. 10.13a, in which the movable structure and flexures are symmetric about the x and y axes. As a result, the mass moves up and down when an acceleration a is applied parallel to the z_a axis. Let us obtain the response of the structure to the acceleration. Because of the symmetry of the structure, we may consider a quarter of the structure as illustrated in Fig. 10.13b, where the inertial force $ma/4$ is applied to the mass in the negative z_a direction. We assume that the sensor is placed on the top of a beam of length l, and the distance from the mass to the sensor is given by l_s. We also

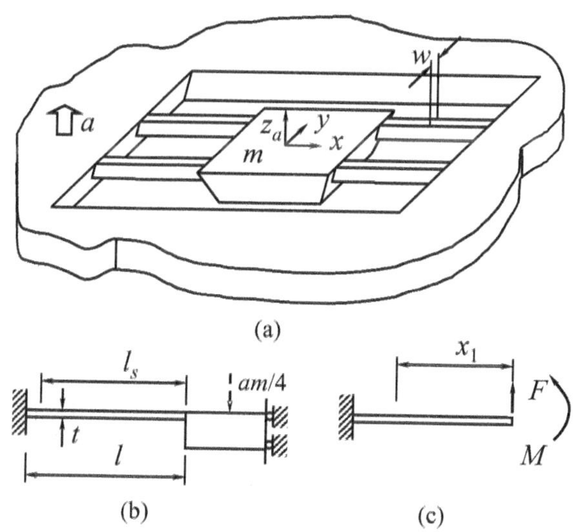

Figure 10.13 Symmetric accelerometer.

assume that the residual stress of bulk-micromachined accelerometer is zero. From Fig. 10.13c, we have the moment in the beam at x_1:

$$M_{x1} = M + Fx_1 \tag{10.96}$$

Thus, we find the elastic energy stored in the beam in terms of the moment, M, and the force, F, acting at the end of the beam:

$$U = \int_0^l \frac{(M + Fx_1)^2}{2\tilde{E}I} dx_1$$

Applying Castigliano's theorem to the equation above leads to the angular displacement,

$$\theta = \frac{\partial U}{\partial M} = \int_0^l \frac{M + Fx_1}{\tilde{E}I} dx_1 = \frac{1}{\tilde{E}I}\left(Ml + \frac{1}{2}Fl^2\right)$$

Since the boundary condition $\theta = 0$ is imposed due to the symmetry shown in Figs. 10.13a and b, the preceding equation gives $M = -Fl/2$. Substituting the moment into the equation below, we have the displacement in the z_a direction:

$$z_a = \frac{\partial U}{\partial F} = \int_0^l \frac{Mx_1 + Fx_1^2}{\tilde{E}I} dx_1 = \frac{1}{\tilde{E}I}\left(\frac{1}{2}Ml^2 + \frac{1}{3}Fl^3\right) = \frac{Fl^3}{12\tilde{E}I}$$

Substituting $F = F_i = -ma/4$, where F_i is the inertial force shown in Fig. 10.13b, we have the displacement of the mass in the direction opposite to the applied acceleration a:

$$z = -z_a = \frac{ml^3}{48\tilde{E}I} a \tag{10.97}$$

Substituting $M = -Fl/2$ into (10.96) yields the moment at the sensor site:

$$M_{x1} = -F\left(\frac{l}{2} - x_1\right) \tag{10.98}$$

Upon substitution of $F = -ma/4$ and $x_l = l_s$ in equation (10.98), we find the moment at the sensor site:

$$M_{x1} = -\frac{ma}{4}\left(l_s - \frac{l}{2}\right) \tag{10.99}$$

Substituting equation (10.99) into the strain–curvature relation, (8.9) leads us to

$$\varepsilon_x = -\frac{M_{x1}}{\tilde{E}I} z_t = \frac{m}{4\tilde{E}I}\left(l_s - \frac{l}{2}\right) z_t a \tag{10.100}$$

where $\tilde{E}I$ and z_t denote the effective flexural rigidity and the distance from the neutral surface to the top of the beam, respectively. For a rectangular beam of one material (thickness t and width w), $\tilde{E}I$ becomes $\tilde{E}wt^3/12$ and z_t is given by $t/2$. Substituting these values into equation (10.100), we obtain

$$\varepsilon_x = \frac{3}{2}\frac{m}{\tilde{E}wt^2}\left(l_s - \frac{l}{2}\right) a \tag{10.101}$$

Strain ε_y in the y direction is given by (10.90) and (10.91), depending on the ratio w/t. Combining (10.97) and (10.100) gives the relative displacement of the mass as a function of the strain, as follows:

$$z = \frac{l^3}{12(l_s - l/2)z_t}\varepsilon_x \tag{10.102}$$

Setting $z_t = t/2$ for the rectangular beam, we have the displacement due to the acceleration:

$$z = \frac{l^3}{6(l_s - l/2)t}\varepsilon_x \tag{10.103}$$

Using equation (10.103), we can obtain the displacement, converted from the measured strain ε_x. It is noted from (10.101) that the maximum strain (or stress) occurs at two points: At the left end of the beam ($l_s = l$),

$$\varepsilon_x = \frac{3}{4}\frac{ml}{\tilde{E}wt^2} a \tag{10.104}$$

and at the right end of the beam ($l_s = 0$),

$$\varepsilon_x = -\frac{3}{4}\frac{ml}{\tilde{E}wt^2} a \tag{10.105}$$

Therefore, in this case, we can place the sensor at the left and right ends of the beam in Fig. 10.13b. It should be noted that equations (10.100)–(10.105) hold only for

$$\frac{18}{25}\left(\frac{z}{t}\right)^2 << 1$$

which is given by equation (s) of Example 3.3. If the displacement z due to the inertial force is comparable to or greater than the thickness of the beam, t, the geometric nonlinearity discussed in Section 3.7 will play a role and the displacement and strain will deviate from those obtained from the preceding equations. We may wish to know an acceleration under which the accelerometer shown in Fig. 10.13a behaves linearly. Substituting (10.97) into the preceding condition for linear operation, we have the acceleration range:

$$|a| << \frac{80}{\sqrt{3}} \frac{\tilde{E}It}{ml^3} \tag{10.106}$$

If the beam is made of a single material and its cross section is rectangular, we can substitute $wt^3/12$ for I into equation (10.106) to obtain

$$|a| << \frac{20}{3\sqrt{3}} \frac{\tilde{E}wt^4}{ml^3} \tag{10.107}$$

Therefore, equations (10.100)–(10.105) hold if (10.106) is satisfied and if the frequency of the acceleration is much less than the natural frequency $f = \sqrt{k/m}/2\pi$, where k denotes the stiffness due to the four straight beams, which is given by $k = 4 \times 12\tilde{E}I/l^3 = 48\tilde{E}I/l^3$.

10.3 ELECTROSTATIC ACCELEROMETERS

In Section 10.2 we stated that deformable structures can be used as accelerometers which measure the strain (or stress) corresponding to the inertial force due to the applied acceleration. Because the electrostatic actuators covered in Chapter 9 are sensitive to acceleration, the performance of MEMS components such as RF tunable switches can be affected by acceleration. If electrostatic actuators are intended to measure the applied acceleration, the actuators are called *electrostatic accelerometers*. To suppress or utilize the effect of acceleration on electrostatic devices, we obtain a relation between the behavior of the structures and the applied acceleration.

Parallel-Plate Accelerometers

Typical electrostatic structures are shown in Figs. 10.14a and 10.15, in which the central masses are suspended by flexures that are anchored on substrates. When the structures are exposed to acceleration a, as shown in Figs. 10.14a and 10.15, whose frequency is much less than the natural frequency of the structure, the masses will be displaced by z for the spring force of the flexures to balance the inertial force. In Fig. 10.14a, a plate of mass m and area A is suspended by L-shaped flexures to reduce the nonlinear factor. The plate may be perforated to control the damping factor (or quality factor) or to etch away

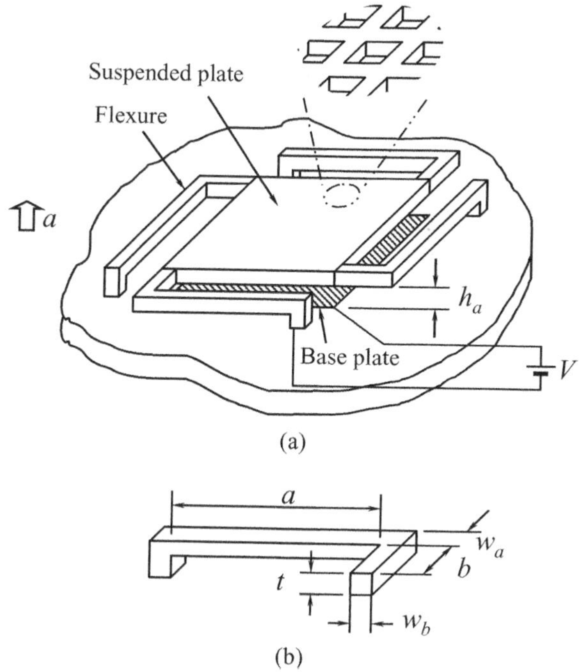

Figure 10.14 Parallel-plate accelerometer.

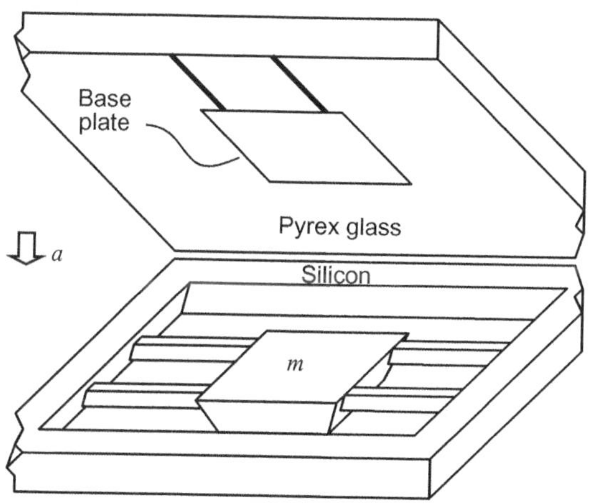

Figure 10.15 Bulk-micromachined parallel-plate accelerometer.

a sacrificial layer (if the plate is made by surface micromachining). The stiffness of the L-shaped beam was obtained in Section 4.2. For the dimensions in Fig. 10.14b, setting $D_\phi = D_\eta = 0$ in (4.50) with (4.59), we find the stiffness of one L-shaped beam as follows:

$$k = \frac{F_z}{\delta_z} = \frac{6EI_{xa}}{a^3} K_{2n,zz} \tag{10.108}$$

where

$$K_{2n,zz} = \frac{1}{2(1+\beta^3 p_1 + 3\beta s_b) - \frac{3}{2}\frac{(1+2\beta s_b)^2}{1+\beta s_b} - \frac{3}{2}\beta^4 p_1^2 \frac{1}{s_a + \beta p_1}} \tag{10.109}$$

$$\begin{aligned} \beta &= \frac{b}{a} & p_1 &= \frac{EI_{xa}}{EI_{xb}} \\ s_a &= \frac{EI_{xa}}{GJ_a} & s_b &= \frac{EI_{xa}}{GJ_b} \\ I_{xa} &= \frac{w_a t^3}{12} & I_{xb} &= \frac{w_b t^3}{12} \\ G &= \frac{E}{2(1+\nu)} \end{aligned} \tag{10.110}$$

and J_a and J_b denote the polar moment of inertial of sections a and b, which are given by (3.33) and (3.34). J_a and J_b can be obtained by substituting w_a or w_b for w into the following: For $w \geq t$,

$$J = \frac{wt^3}{3}\left[1 - 0.616\left(\frac{t}{w}\right) - 0.087\left(\frac{t}{w}\right)^2 + 0.125\left(\frac{t}{w}\right)^3\right]$$

and for $w \leq t$,

$$J = \frac{tw^3}{3}\left[1 - 0.616\left(\frac{w}{t}\right) - 0.087\left(\frac{w}{t}\right)^2 + 0.125\left(\frac{w}{t}\right)^3\right]$$

When β goes to zero (i.e., a is much greater than b), the preceding stiffness is reduced to $k = 12EI_{xa}/a^3$.

If a voltage is applied between the suspended plate and the base plate in Figs. 10.14a and 10.15, we can set up the model (a parallel-plate actuator) shown in Fig. 10.16a to analyze the behavior of the suspended plate. In the model, a suspended plate of mass m and area A is supported by a spring of stiffness k and damper of damping coefficient c, while a voltage less than a pull-in voltage (which will be obtained later) is applied between the suspended and base plates. As discussed in Section 10.2, if the frequency of the applied acceleration $a = \ddot{z}_s$ is much less than the natural frequency at V, the spring force balances the inertial force due to the acceleration, and the gap between the plates is changed from the initial gap h_0 to h_a, as shown in Fig. 10.16a. Referring to the free-body diagram of Fig. 10.16b (note that the problem is

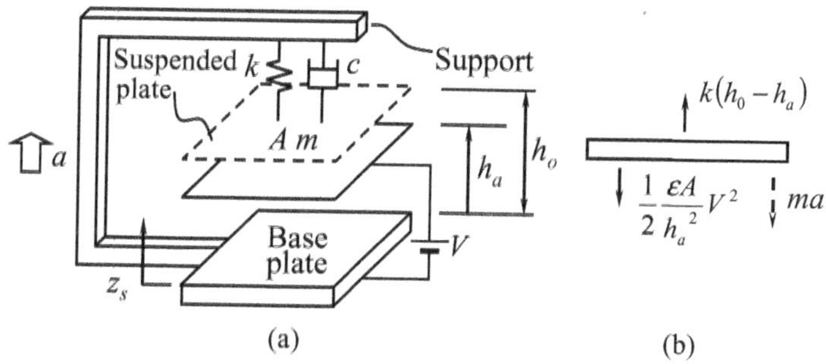

Figure 10.16 Model of a parallel-plate accelerometer.

converted to a static problem for the reason discussed in Section 10.2), we obtain the force equilibrium equation:

$$k(h_0 - h_a) = \frac{1}{2}\frac{\varepsilon A}{h_a^2}V^2 + ma \tag{10.111}$$

For convenience, equation (10.111) may be expressed in a dimensionless form,

$$(1 - H_a) = \frac{G}{H_a^2} + I \tag{10.112}$$

where

$$\begin{aligned} H_a &= \frac{h_a}{h_0} \\ G &= \frac{1}{2}\frac{\varepsilon A}{k h_0^3}V^2 \\ I &= \frac{ma}{k h_0} \end{aligned} \tag{10.113}$$

are dimensionless parameters that represent the interplate gap, the electrostatic force, and the inertial force due to acceleration, respectively. Nonlinear equation (10.112) may be expressed as a cubic equation,

$$H_a^3 - (1 - I)H_a^2 + G = 0 \tag{10.114}$$

The roots of cubic equation (10.114) are described in Section 5.1. Substituting the coefficients $a = -(1 - I)$, $b = 0$, and $c = G$ into (5.15), we have the three distinct roots of the equation (10.114):

$$
\begin{aligned}
H_{a1} &= \frac{1-I}{3}\left(1+2\cos\frac{\phi}{3}\right)\\
H_{a2} &= \frac{1-I}{3}\left\{1+2\cos\left[\frac{1}{3}(\phi+2\pi)\right]\right\}\\
H_{a3} &= \frac{1-I}{3}\left\{1+2\cos\left[\frac{1}{3}(\phi+4\pi)\right]\right\}
\end{aligned}
\tag{10.115}
$$

where

$$
\phi = \cos^{-1}\left[1-\frac{2}{(1-I)^3}\frac{G}{G_{pi}}\right]
$$

$$
G_{pi} = \frac{4}{27}
$$

denote the auxiliary parameter and the dimensionless pull-in force given by (9.52).

Examining the foregoing roots, we find that H_{a1} and H_{a3} are the stable and unstable solutions and H_{a2} is a solution without physical meaning. We thus take H_{a1} as the solution of the dimensionless force equilibrium equation, (10.114):

$$
H_a = \frac{1-I}{3}\left(1+2\cos\left\{\frac{1}{3}\cos^{-1}\left[1-\frac{2}{(1-I)^3}\frac{G}{G_{pi}}\right]\right\}\right)
\tag{10.116}
$$

Noted from equation (10.116) that the two independent parameters I and G affect the dimensionless gap H_a. In Fig. 10.17 the gap H_a is plotted against the inertial force I for the normalized electrostatic force $G/G_{pi} = 0$, 0.2, 0.4, 0.6, and 0.8. For $G/G_{pi} = 0$, the dimensionless gap H_a decreases linearly from unity to zero because there is no electrostatic force acting on the suspended plate. Physically, this means that the mass moves by maximum displacement (i.e., the initial gap h_0) and touches the base electrode of Fig. 10.16a as the maximum acceleration of kh_0/m, corresponding to $I = 1$, is applied. For $G/G_{pi} \neq 0$, the nonlinear electrostatic force is exerted on the suspended plate and then the parallel-plate actuator can experience pull-in. When the normalized electrostatic force G/G_{pi} increases from zero to unity, the starting gap is lowered from unity and the pull-in gap varies from zero (i.e., the suspended plate touches the base plate) to 2/3.

As the inertial or electrostatic force increases from zero to its pull-in value, the argument of the inverse cosine in (10.116) varies from unity to negative unity. Therefore, the suspended plate, exposed to the acceleration a, is in a stable or critical condition if the following condition is satisfied:

$$
1-\frac{2}{(1-I)^3}\frac{G}{G_{pi}} \geq -1
\tag{10.117}
$$

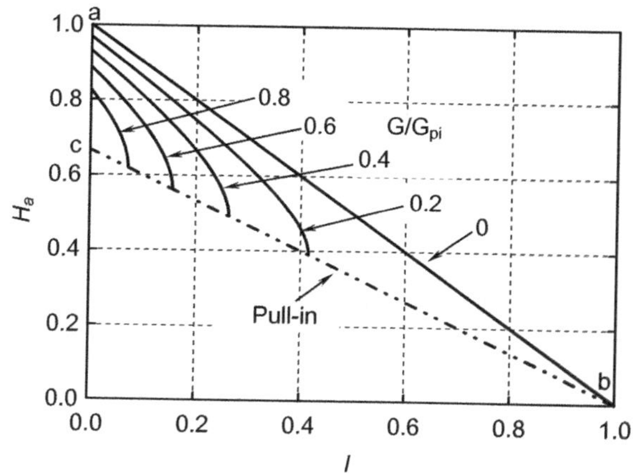

Figure 10.17 Dimensionless gap H_a with respect to the dimensionless inertial force I and the normalized force G/G_{pi}.

Since equation (10.117) includes two independent parameters, I and G, the triangle in Fig. 10.17 defines a stable region in which the suspended plate is stable. It is noted that the lower straight line in the figure can be defined as the pull-in gap, at which the suspended plate experiences pull-in. The pull-in gap H_{pi}, inertial force I_{pi}, and electrostatic force G_{pi} play important roles because they provide a guideline for understanding the nonlinear behavior of a parallel-plate actuator under electrostatic force due to a voltage. When the equality sign of equation (10.117) holds, the pull-in condition is obtained as

$$1-\frac{2}{(1-I)^3}\frac{G}{G_{pi}}=-1 \tag{10.118}$$

This equation includes two independent parameters (I and G), so two conditions for the pull-in are possible: (1) G reaches $G_{pi,a}$ while the acceleration remains unchanged; (2) I reaches I_{pi} while G retains a constant value. Rearranging the terms in equation (10.118), we obtain the dimensionless pull-in electrostatic force $G_{pi,a}$ as a function of the inertial force I applied:

$$G_{pi,a}=G_{pi}(1-I)^3 \tag{10.119}$$

If the dimensionless electrostatic force G is taken as an independent parameter from (10.118), we have the dimensionless pull-in inertial force I_{pi}:

$$I_{pi}=1-\left(\frac{G}{G_{pi}}\right)^{1/3} \tag{10.120}$$

which is a function of the dimensionless electrostatic force G (i.e., the voltage). If the pull-in condition (10.118) is satisfied, (10.116) is reduced to the following simple equation, which is a linear function of the dimensionless inertial force:

$$H_{pi,a} = H_{pi}(1-I) \tag{10.121}$$

where H_{pi} is 2/3 that of the pull-in gap at $I = 0$. In Fig. 10.17, the lower straight line corresponds to the pull-in gap of equation (10.121). Substituting (10.120) into (10.116), the pull-in gap is obtained in terms of the dimensionless electrostatic force G:

$$H_{pi,a} = H_{pi}\left(\frac{G}{G_{pi}}\right)^{1/3} \tag{10.122}$$

Therefore, if the voltage V is unchanged and the acceleration a varies, (10.120) and (10.122) can be used to obtain the pull-in acceleration a_{pi} and the corresponding pull-in gap $h_{pi,a}$, respectively. If the acceleration a remains a constant and if the voltage can be adjusted, (10.119) and (10.121) are suitable to obtain pull-in voltage $V_{pi,a}$ and the corresponding pull-in gap $h_{pi,a}$, respectively.

The gap given by (10.116) is plotted in Fig. 10.18 against the normalized electrostatic force G/G_{pi} as the different inertial force I is applied to the parallel-plate actuator. As the inertial force I varies from –0.2 to 0.4, the gap H_a decreases as the electrostatic force G/G_{pi} increases. It is noted from Fig. 10.18 that the negative dimensionless inertial force increases the gap H_a. In other words, if the parallel-plate actuator is exposed to a negative acceleration (i.e., accelerated downward in Fig. 10.16), the voltage and displacement ranges can be extended. In Fig. 10.19, the pull-in inertial force, (10.120), is shown with

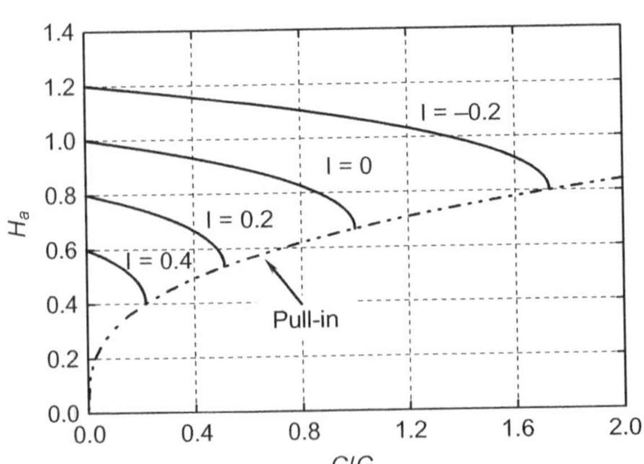

Figure 10.18 Dimensionless gap H_a with respect to the normalized electrostatic force G/G_{pi} and the dimensionless inertial force I.

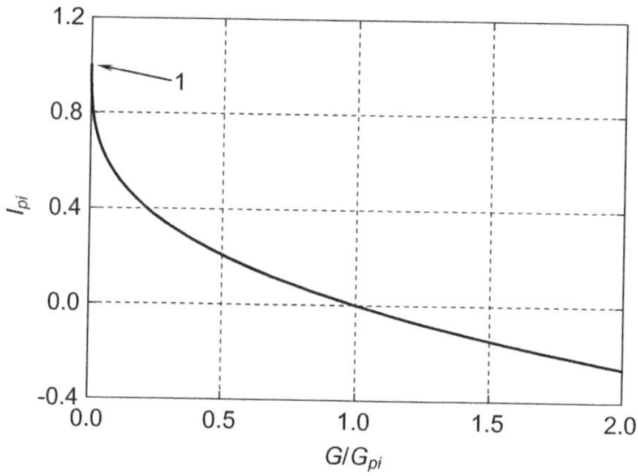

Figure 10.19 Dimensionless pull-in inertial force I_{pi} with respect to normalized force G/G_{pi}.

respect to the electrostatic force G/G_{pi}. The dimensionless pull-in inertial force I_{pi} decreases rapidly in the range $G/G_{pi} = 0$ to 0.2 and afterward decreases slowly with increasing dimensionless electrostatic force. As already mentioned, for negative acceleration, the pull-in voltage $V_{pi,a}$ can be extended beyond the pull-in voltage given by

$$V_{pi} = \sqrt{\frac{8kh_0^3}{27\varepsilon A}} \tag{10.123}$$

which corresponds to the pull-in force $G_{pi} = 4/27$ at zero acceleration. Combining (10.113) and (10.119) gives

$$V_{pi,a} = \sqrt{2G_{pi}\frac{kh_0^3}{\varepsilon A}(1-I)^3} = V_{pi}(1-I)^{3/2} \tag{10.124}$$

The response (10.116) of the suspended mass to the low-frequency acceleration is validated by comparing the analytic solution with a numerical solution of the equation of motion that is obtained by adding the inertial force, $-ma$, to the right-hand side of (9.92). For the numerical simulation, the following data are used: $m = 1 \times 10^{-9}$ kg, $c = 8.85 \times 10^{-5}$ N·s/m, $k = 4.00$ N/m, $A = 2.146 \times 10^{-7}$ m^2, and $h_0 = 2 \times 10^{-6}$ m. Using the data, we have, respectively, the natural frequency $f_n = 10$ kHz at $V = 0$ and $a = 0$, the pull-in voltage $V_{pi} = 2.23$ V at $a = 0$, and the pull-in acceleration $a_{pi} = 3318.5$ m/s^2 at $V = 1$ V. The analytical solution (10.116) and the numerical solution are compared in Figs. 10.20 to 10.22, respectively. Figure 10.20 shows the response of the suspended mass first to the step voltage and then to the step acceleration. For a dc voltage of 1 V, the simulated gap of the suspended plate is changed from the initial gap

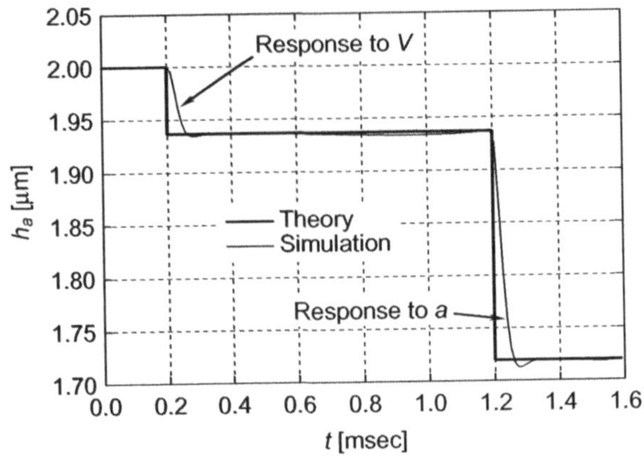

Figure 10.20 Comparison of analytical and numerical responses.

of 2 μm to a stable gap of 1.9367 μm, corresponding to the voltage applied. When the parallel-plate actuator is then exposed to an acceleration of 800 m/s^2 (81.6g), which corresponds to 24.1% of the pull-in acceleration at $V = 1$V, the suspended plate is further displaced from the stable position to another stable position of 1.7197 μm after a short transient response, respectively. The analytic solution of (10.116) gives the same gaps of 1.9367 μm for the step voltage and 1.7197 μm for the step acceleration. The transient period, representing the time for the transient response, is estimated at less than 0.14 ms from Fig. 10.20.

In Figs. 10.21 and 10.22, a dc bias voltage of 1 V is applied at $t = 2$ ms and the input low-frequency acceleration is applied at $t = 3$ ms. In Fig. 10.21, a sinusoidal acceleration of amplitude 800 m/s^2 (81.6g) and an acceleration frequency of 100 Hz are applied at $t = 3$ ms. The portion indicated by the circle is enlarged in the lower figure. For the sinusoidal acceleration, the analytical solution of (10.116) is in excellent agreement with the numerical solution within a maximum error of 0.17%. This discrepancy between the analytical and numerical solutions comes from the phase delay of 0.0153 rad (corresponding to the time delay of 0.0243 ms), due to the inertial and damping forces which were neglected to obtain the analytical solution. The analytical and numerical solutions provide the same maximum value of 2.1485 μm as well as the same minimum value of 1.7197 μm, even though the numerical response to the applied acceleration has a phase delay. As an extreme case, the parallel-plate actuator is exposed to an acceleration of amplitude a_0 of 3280 m/s^2 (334.7g), corresponding to the ratio $a_0/a_{pi} = 3280/3318.5 = 0.9884$ at 1 V, and the analytical and numerical solutions are compared in Fig. 10.22. The same tendency is observed in the figure. For the maximum responses, both solutions give the same numerical value of 2.7895 μm, but for the minimum values, the

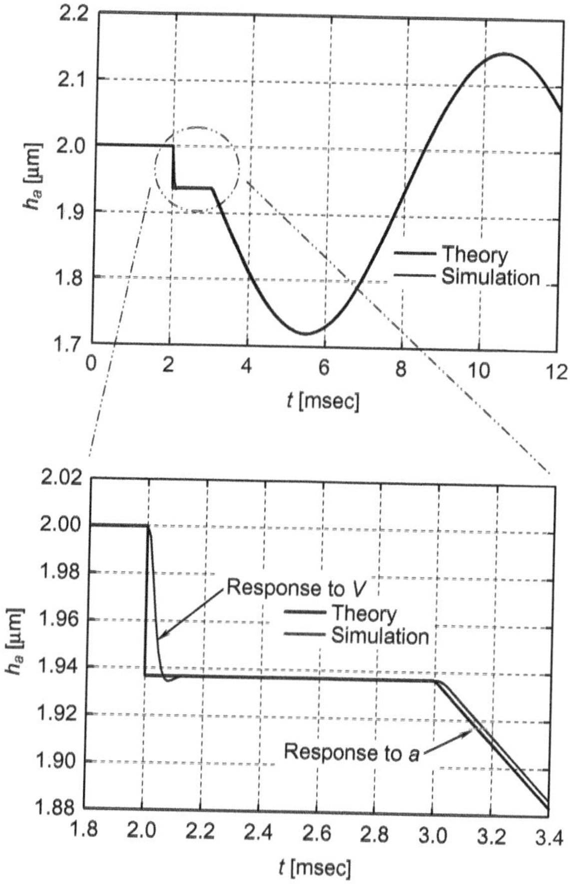

Figure 10.21 Analytical and numerical responses to sinusoidal acceleration: $a = a_0\sin[\omega(t - t_0)]$, where $a_0 = 800\,\text{m/s}^2$, $\omega = 2\pi(100)$ rad/s, and $t_0 = 3\,\text{ms}$.

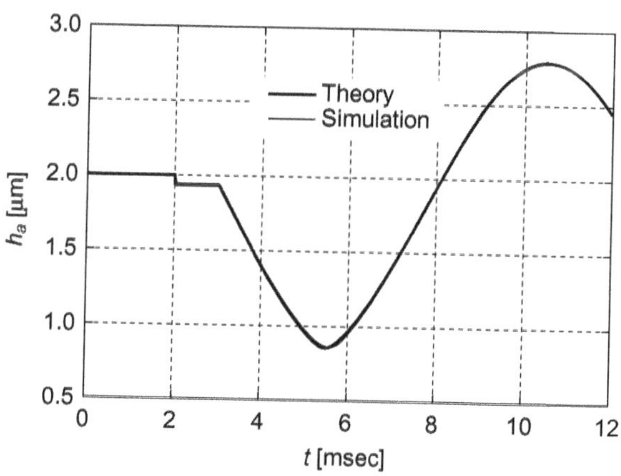

Figure 10.22 Analytical and numerical responses to a much greater sinusoidal acceleration: $a = a_0\sin[\omega(t - t_0)]$, where $a_0 = 3280\,\text{m/s}^2$ (334.7 g), $\omega = 2\pi(100)$ rad/s, and $t_0 = 3\,\text{ms}$.

analytical and numerical responses yield 0.8555 m and 0.8590 μm, respectively. The discrepancy between the analytical and numerical responses at the minimum gap is due to the fact that the effective stiffness of the parallel-plate actuator is greatly reduced in the vicinity of the pull-in gap [0.7803 μm at $V = 1$V from (10.121)], and the inertial and damping forces play a role in the response. As such, the derived analytical solution, (10.116), and the pull-in acceleration and voltage, (10.119)–(10.124), can be used to predict the response of the electrostatic parallel plates when they are exposed to a low-frequency acceleration as well as a constant acceleration.

Force-Balanced Electrostatic Accelerometers

In previous paragraphs, we dealt with parallel-plate accelerometers that use one base plate (electrode). If we use more than one electrode, the suspended mass can be balanced to extend the linear range of operation. As an example, let us consider the mechanical structure in Fig. 10.23 as an accelerometer. This structure was considered as a gap-closing actuator in Section 9.8. The structure is exposed to an acceleration a as shown in Fig. 10.23a, while a dc voltage V_b is applied to the terminals, 1 and 2. The equation of motion of the accelerometer is obtained by adding the inertial force term, $F_i = ma$, to (9.309) and setting $F_0 = 0$, where F_0 is the amplitude for sinusoidal actuation:

$$m\frac{d^2\delta}{dt^2} + c\frac{d\delta}{dt} + k_{\text{eff}}\delta = ma \tag{10.125}$$

where

$$k_{\text{eff}} = k - 2k_e = k - 2\frac{\varepsilon A V_b^2}{h_0^3} \tag{10.126}$$

and A and h_0 denote the total area of electrodes connected to terminal 1 or 2 and the gap between the suspended and base electrodes, respectively. Since the mass is balanced by the identical electrostatic force at $a = 0$, the stable position is $\delta = 0$. The model corresponding to equation of motion (10.125) is shown in Fig. 10.23b. If the angular frequency of the acceleration is much less than the natural angular frequency, $\omega_n = \sqrt{k_{\text{eff}}/m}$, the first and second terms are negligible (as discussed in Section 10.2), and the response δ is written

$$\delta = \frac{ma}{k_{\text{eff}}} = \frac{a}{\omega_n^2} \tag{10.127}$$

which holds for $V_b < V_{b,pi}$, where $V_{b,pi}$ is the pull-in voltage defined by (9.313), rewritten here for convenience:

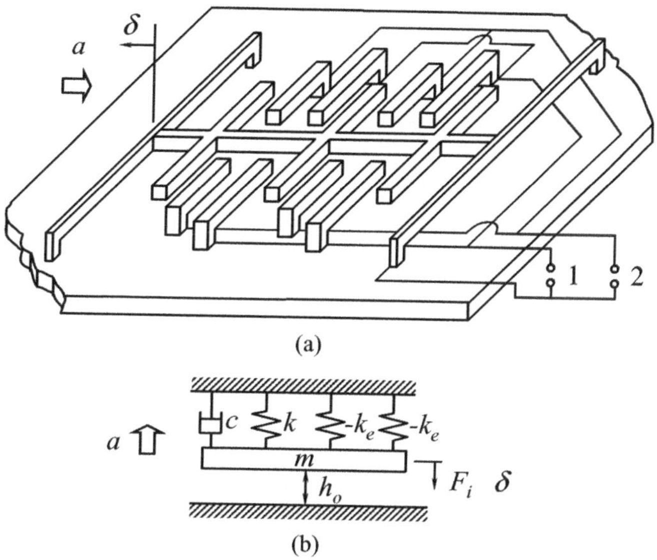

Figure 10.23 Force-balanced electrostatic accelerometer.

$$V_{b,pi} = \sqrt{\frac{kh_0^3}{2\varepsilon A}} \tag{10.128}$$

The response can be measured by a displacement sensor (e.g., electronic devices measuring the capacitance between the suspended and based electrodes). As shown in response equation (10.128), the response of the accelerometer is the displacement δ of the suspended mass. In this type of accelerometer, known as an *open-loop accelerometer*, displacement (or acceleration) is not used to control the position of the suspended mass. If we adjust a force F_c (e.g., the electrostatic force) so that the suspended mass remains at the initial position $\delta = 0$, the preceding equation of motion may be written

$$m\frac{d^2\delta}{dt^2} + c\frac{d\delta}{dt} + k_{\text{eff}}\delta = ma - F_c$$

Since the left-hand terms above vanish, the applied acceleration a is obtained as

$$a = \frac{F_c}{m} \tag{10.129}$$

For example, the voltages $V_1 = V_b + V_i$ and $V_2 = V_b - V_i$ are applied to terminals 1 and 2, respectively, and the control force is given by

$$F_c = \frac{\varepsilon A(V_b + V_i)^2}{2h_0^2} - \frac{\varepsilon A(V_b + V_i)^2}{2h_0^2}$$
$$= \frac{\varepsilon A V_b^2}{2h_0^2}\left[\left(1+\frac{V_i}{V_b}\right)^2 - \left(1-\frac{V_i}{V_b}\right)^2\right]$$
$$= 2\frac{\varepsilon A V_b}{h_0^2} V_i \qquad (10.130)$$

Substituting equation (10.130) into (10.129), we find the acceleration in terms of the applied voltage V_i:

$$a = 2\frac{\varepsilon A V_b}{m h_0^2} V_i \qquad (10.131)$$

It is noted that the acceleration a is a linear function of V_i. In this type of accelerometer, known as a force-balanced or closed-loop accelerometer, displacement or acceleration is used to control the position of the suspended mass. To adjust the control force, we may use analog or digital controllers; readers interested in controllers can find more information in the literature.

10.4 VIBRATORY GYROSCOPES

Unlike a conventional gyroscope, micromachined gyroscopes usually employ vibrating structures which are used to measure the inertial force or moment due to angular motion. These gyroscopes are known as *vibratory* or *vibrating structure gyroscopes*. Before obtaining an analytic response of micromachined vibratory gyroscopes, we must understand the principle of vibratory gyroscopes. In Fig. 10.24a we consider a mass m that is suspended by a beam (anchored on a substrate) and vibrate in the x direction. If the stiffness of the beam is given by k_x in the x direction and k_z in the z direction, the natural frequencies of the structure are expressed in the x and z directions, respectively, as follows:

$$f_{nx} = \frac{1}{2\pi}\sqrt{\frac{k_x}{m}} \qquad f_{nz} = \frac{1}{2\pi}\sqrt{\frac{k_z}{m}}$$

The frequency response curves are shown in Fig. 10.24b for $k_x = k_z$ and in Fig. 10.24c for $k_z > k_x$. If the velocity of the mass is given by $\upsilon = \upsilon_0 \cos \omega t$, and if the substrate rotates at an input angular rate (angular velocity) ω_i in rad/s as shown in Fig. 10.24a, the Coriolis force (inertial force) is given by (5.41), rewritten here:

$$F_i = 2m\omega_i \upsilon = 2m\omega_i \upsilon_0 \cos \omega t = F_{i0} \cos \omega t \qquad (10.132)$$

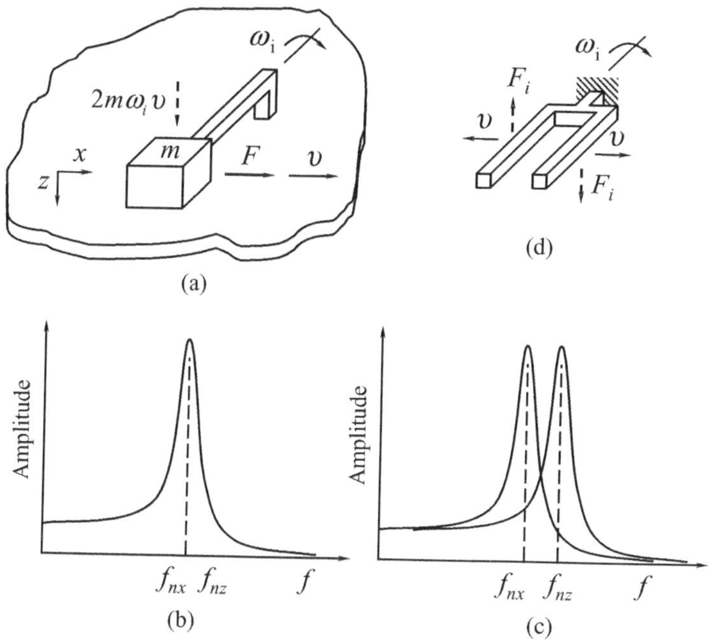

Figure 10.24 Principle of a vibratory gyroscope.

which is shown in Fig. 10.24a and acts as an actuation force in the z direction. As a result, if $f = f_{nz} = f_{nx}$ as shown in Fig. 10.24a (where f is the frequency of the actuation force F), the displacement due to the inertial force will be maximized [i.e., the amplitude becomes $F_{i0}Q_z/k_z$ which is given by (5.100)]. However, if $f_{nx} \neq f_{nz}$, the displacement of the mass is less than $F_{i0}Q_z/k_z$. Therefore, for vibratory gyroscopes, the tuning of the natural frequencies of structures is important to maximize the gyroscope output. For large-scale structures, we may adjust the stiffness in the x and z directions by partial removal of the beam, a process referred to as *frequency tuning*. Instead of frequency tuning by partial removal of the beam, electrostatic frequency tuning is used to adjust the natural frequency of MEMS structures. To increase the quality factor, tuning fork structures in Fig. 10.24d are used to implement vibratory gyroscopes.

As a micromachined gyroscope, the structure shown in Fig. 10.25 is used, in which the movable plate of mass m moves at a velocity v in the x direction since a sinusoidal force F (e.g., the electrostatic force of the comb drive) is applied in the x direction. When an input angular rate ω_i is applied as shown Fig. 10.25 (i.e., the substrate rotates at the angular velocity ω_i), the inertial force F_i is generated in the z direction. To simplify the illustration, the drive causing vibration in the x direction is not shown in the figure. If we use the

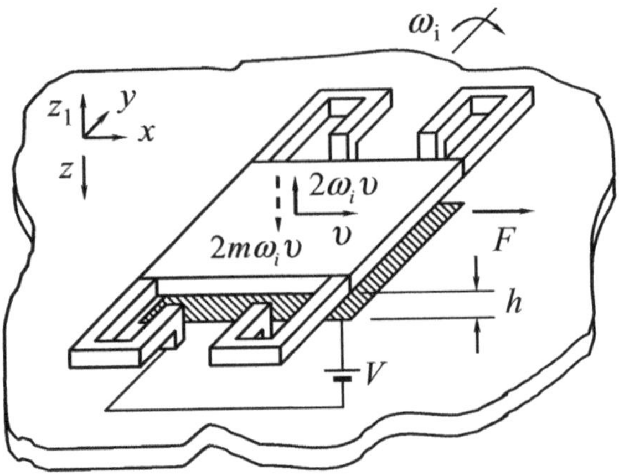

Figure 10.25 Structure of a gyroscope.

comb drive actuator, the electrostatic force is given by (9.5), $F_e = F_b + F_0 \sin \omega t$, where F_b, F_0 and ω denote the force amplitudes given by (9.6) and (9.7), and the angular frequency of the actuation force, respectively. We assume that for $\omega_i = 0$, the movable plate is spaced by an initial gap h_0 from the substrates and is vibrating in the x direction. We set up the gyroscope model shown in Fig. 10.26a. Applying Newton's second law to the mass in the x and z directions, we find the equations of motion as follows:

$$m\ddot{x} + c_x\dot{x} + k_x(x_0 + x) = F_b + F_0 \sin \omega t \tag{10.133}$$

$$m\ddot{\delta} + c_z\dot{\delta} + k_z(z + \delta) = \frac{\varepsilon A V^2}{2(h_0 - z - \delta)^2} + 2m\omega_i\dot{x} \tag{10.134}$$

where c_x, c_z, x_0, x, z, and δ represent the damping coefficients in the x and z directions, the constant displacement due to F_b, the displacement due to the time-varying force, the displacement of the plate due to the electrostatic force caused by the voltage V, and the displacement (response) caused by the Coriolis force, respectively. Since $k_x x_0 = F_b$, the equation of motion in the x direction may be written as

$$m\ddot{x} + c_x\dot{x} + k_x x = F_0 \sin \omega t \tag{10.135}$$

whose response is given by (9.12), rewritten here for convenience:

$$x = x_0 \sin(\omega t - \phi_x) \tag{10.136}$$

where

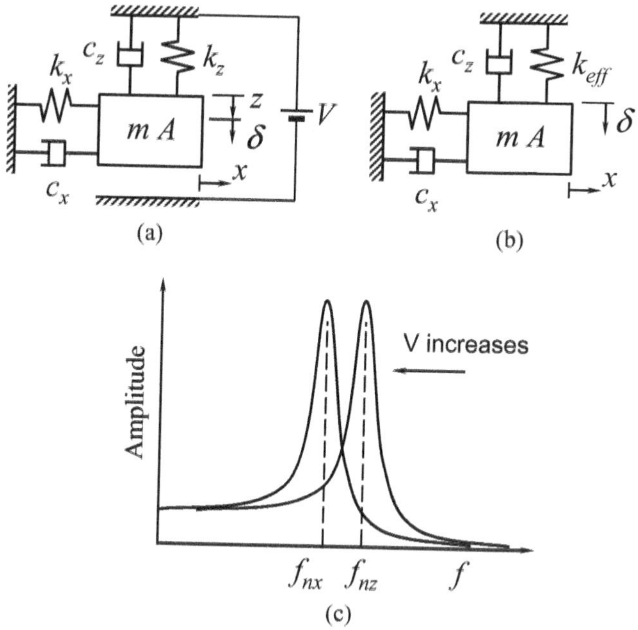

Figure 10.26 Model of a gyroscope.

$$
\begin{aligned}
x_0 &= \frac{F_0}{k_x}\frac{1}{\sqrt{(1-\Omega_x{}^2)^2+(\Omega_x/Q_x)^2}} \\
\Omega_x &= \frac{\omega}{\omega_{nx}} = \frac{f}{f_{nx}} \\
\omega_{nx} &= \sqrt{\frac{k_x}{m}} \\
\phi_x &= \tan^{-1}\frac{\Omega_x}{Q_x(1-\Omega_x^2)} \\
Q_x &= \frac{\sqrt{mk_x}}{c_x}
\end{aligned}
\qquad (10.137)
$$

represent the vibration amplitude, the frequency ratio, the natural frequency, the phase, and the quality factor, respectively.

For $\omega_i = 0$, (10.134) reduces to a force equilibrium equation,

$$k_z z = \frac{\varepsilon A V^2}{2h^2} \qquad (10.138)$$

where $h = h_0 - z$. The exact solution of equation (10.138) is given by (9.55):

$$H = \frac{1}{3}\left\{1 + 2\cos\left[\frac{1}{3}\cos^{-1}\left(1 - \frac{27}{2}G\right)\right]\right\} \tag{10.139}$$

where

$$H = \frac{h}{h_0} \tag{10.140}$$

$$G = \frac{1}{2}\frac{\varepsilon A}{k_z h_0^3}V^2 \tag{10.141}$$

denote the dimensionless gap and the dimensionless force, respectively. For $V < V_{pi}$, where V_{pi} is the pull-in voltage given by (9.53), we can rewrite the electrostatic force of (10.134) as follows:

$$\begin{aligned}\frac{\varepsilon A V^2}{2(h_0 - z - \delta)^2} &= \frac{\varepsilon A V^2}{2h^2}\frac{1}{(1-\delta/h)^2} \\ &= \frac{\varepsilon A V^2}{2h^2}\left[1 + 2\frac{\delta}{h} + 3\left(\frac{\delta}{h}\right)^2 + 3\left(\frac{\delta}{h}\right)^2 + \cdots\right]\end{aligned} \tag{10.142}$$

Neglecting the higher terms of equation (10.142) and recalling the force equilibrium equation, (10.138), equation (10.134) is reduced to the equation of motion to describe the displacement of the plate:

$$m\ddot{\delta} + c_z\dot{\delta} + k_{\text{eff}}\delta = 2m\omega_i\dot{x} \tag{10.143}$$

where

$$k_{\text{eff}} = k_z - \frac{\varepsilon A V^2}{h^3} = k_z\left(3 - \frac{2}{H}\right) \tag{10.144}$$

denotes the effective stiffness at V and the last expression is given by (9.58). The model corresponding to equation (10.144) is shown in Fig. 10.26b. If we divide equation of motion (10.143) by m, we have it in a slightly different form:

$$\ddot{\delta} + \frac{\omega_{nz}}{Q_z}\dot{\delta} + \omega_{nz}^2\delta = 2\omega_i\dot{x} \tag{10.145}$$

where

$$\omega_{nz} = \sqrt{\frac{k_{\text{eff}}}{m}} \tag{10.146}$$

$$Q_z = \frac{\sqrt{mk_{\text{eff}}}}{c_z} \tag{10.147}$$

denote the natural frequency and the quality factor in the z direction at V, respectively. Since the actuation force in the x direction is given by $F = F_0 \sin \omega t$ in (10.135), the force and displacements of the plate can be written in complex form:

$$\begin{aligned} F &= F_0 e^{j\omega t} \\ x &= x_0 e^{j(\omega t - \phi_x)} \\ \delta &= \delta_c e^{j\omega t} \end{aligned} \tag{10.148}$$

where x_0 and ϕ_x denote the amplitude and phase of x, which are given by the first and fourth equations of (10.137), respectively. The response δ can be obtained by $\delta = \text{Im}(\delta_c e^{j\omega t})$, where Im denotes the imaginary part of complex number. When the second and third equations of (10.148) are substituted into (10.145), we have

$$\left(\omega_{nz}^2 - \omega^2 + j\omega \frac{\omega_{nz}}{Q_z}\right)\delta_c = 2j\omega_i x_0 \omega$$

which may be written in the form

$$\begin{aligned} \delta_c &= x_0 \frac{2j\omega_i \omega}{\omega_{nz}^2 - \omega^2 + j\omega(\omega_{nz}/Q_z)} e^{-j\phi_x} \\ &= x_0 \frac{2\omega_i \omega}{\sqrt{(\omega_{nz}^2 - \omega^2)^2 + (\omega_{nz}\omega/Q_z)^2}} e^{j(\pi/2 - \phi_x - \phi_z)} \end{aligned}$$

where the last expression is obtained by using $j = e^{j(\pi/2)}$ and the phase ϕ_z is given by

$$\phi_z = \tan^{-1} \frac{\omega_{nz}\omega}{Q_z(\omega_{nz}^2 - \omega^2)}$$

The displacement δ is thus expressed as

$$\begin{aligned} \delta &= \text{Im}\left[\delta_c e^{j\omega t}\right] \\ &= x_0 \frac{2\omega_i \omega}{\sqrt{(\omega_{nz}^2 - \omega^2)^2 + (\omega_{nz}\omega/Q_z)^2}} \sin\left(\omega t - \phi_x - \phi_z + \frac{\pi}{2}\right) \\ &= 2x_0 \frac{\omega_i \omega}{\omega_{nz}^2} \frac{1}{\sqrt{(1 - \Omega_z^2)^2 + (\Omega_z/Q_z)^2}} \sin\left(\omega t - \phi_x - \phi_z + \frac{\pi}{2}\right) \end{aligned} \tag{10.149}$$

Upon substitution of the first equation of (10.137) into equation (10.149), we find the response of the vibratory gyroscope to the angular velocity ω_i:

$$\delta = \delta_0 \sin(\omega t - \phi) \tag{10.150}$$

where

$$\delta_0 = 2\frac{F_0}{k_x}\frac{\omega_i \omega}{\omega_{nz}^2}\frac{1}{\sqrt{(1-\Omega_x^2)^2 + (\Omega_x/Q_x)^2}}\frac{1}{\sqrt{(1-\Omega_z^2)^2 + (\Omega_z/Q_z)^2}} \tag{10.151}$$

$$\begin{aligned}
\Omega_x &= \frac{\omega}{\omega_{nx}} = \frac{f}{f_{nx}} \qquad \Omega_z = \frac{\omega}{\omega_{nz}} = \frac{f}{f_{nz}} \\
\omega_{nx} &= \sqrt{\frac{k_x}{m}} \qquad \omega_{nz} = \sqrt{\frac{k_{eff}}{m}} \\
Q_x &= \frac{\sqrt{mk_x}}{c_x} \qquad Q_z = \frac{\sqrt{mk_{\text{eff}}}}{c_z} \\
\phi &= \phi_x + \phi_z - \frac{\pi}{2} \\
\phi_x &= \tan^{-1}\frac{\Omega_x}{Q_x(1-\Omega_x^2)} \qquad \phi_z = \tan^{-1}\frac{\Omega_z}{Q_z(1-\Omega_z^2)}
\end{aligned} \tag{10.152}$$

To maximize the response above, we can operate the gyroscope in which the quality factors in the x and y directions are much greater than 1. Furthermore, ω_z is tuned to ω_x while the actuation frequency ω is equal to ω_x [i.e., $\omega = \omega_x = \omega_z$ (or $\Omega_x = \Omega_z = 1$)]. Under this condition, the preceding response of the tuned gyroscope is reduced to

$$\delta_t = \delta_{0t} \sin\left(\omega t - \frac{\pi}{2}\right) = -\delta_{0t} \cos \omega t \tag{10.153}$$

where

$$\delta_{0t} = 2\left(\frac{F_0}{k_x}Q_x\right)\frac{\omega_i}{\omega_{nz}}Q_z = 2x_0\frac{\omega_i}{\omega_{nx}}Q_z \tag{10.154}$$

denotes the amplitude of the output displacement. We know from equation (10.154) that to increase the output of vibratory gyroscopes, both the displacement x_0 and the quality factor Q_z should be increased and the natural frequency of the gyroscope should be decreased. To do this, micromachined vibratory gyroscopes are usually operated in a reduced-pressure environment (e.g., a vacuum package). The input angular rate ω_i may be a constant or low-frequency angular velocity; for example, the frequency, ω_i, of the input angular rate of handheld camcorders (video camera recorders) is less than 10 Hz. If a time-varying input angular rate is applied to the gyroscope, the actuation frequency will be an am (amplitude-modulated) signal; that is, the actuation frequency, f_x, can be considered as a carrier frequency, while the frequency of the input angular rate serves as an input signal. When the am signal passes through a low-pass filter, we obtain the envelope corresponding to the input angular rate.

Let us find the voltage for tuning the natural frequency of the gyroscope. As we discussed in Chapter 9, the effective stiffness in the z direction is tuned electrostatically and thus the corresponding natural frequency is also adjusted. The condition $f_x = f_z$ gives $k_x = k_{\text{eff}}$, which may be expressed as

$$k_x = k_{\text{eff}} = k_z\left(3 - \frac{2}{H_t}\right) \tag{10.155}$$

where H_t represents the dimensionless tuned gap as the gyroscope is tuned. From equation (10.155), we obtain the tuned gap as

$$H_t = \frac{2}{3 - k_x/k_z} \tag{10.156}$$

Since the force equilibrium equation, (10.138), is written as $H^3 - H^2 + G = 0$, the force G_t is written in terms of H_t as follows:

$$G_t = H_t^2(1 - H_t) \tag{10.157}$$

Referring to the definition of G, (10.141), the tuning voltage V_t is obtained as

$$V_t = \sqrt{\frac{2k_z h_0^3}{\varepsilon A} G_t} = H_t\sqrt{2(1 - H_t)\frac{k_z h_0^3}{\varepsilon A}} \tag{10.158}$$

which may be used if k_x/k_z or H_t is given. We can consider a different case; the tuning voltage is given as

$$V_t = \zeta V_{pi} \tag{10.159}$$

where $0 \le \zeta < 1$. Since $G_t/G_{pi} = (V_t/V_{pi})^2$, the voltage relation (10.159) is written as

$$G_t = \zeta^2 G_{pi} \tag{10.160}$$

Substituting equation (10.160) into (10.139) and recalling that $G_{pi} = 4/27$ leads us to the tuning gap H_t:

$$H_t = \frac{1}{3}\left\{1 + 2\cos\left[\frac{1}{3}\cos^{-1}(1-2\zeta^2)\right]\right\} \qquad (10.161)$$

For example, $\zeta = 1/2$ gives $H_t = [1 + 2\cos(\pi/9)]/3 \cong 0.9598$. The stiffness k_z required is then obtained from (10.155) as follows:

$$k_z = \frac{k_x}{3 - 2/H_t} \qquad (10.162)$$

where H_t is the tuning gap given by (10.161). If $k_z > k_x$, the natural frequency in the z direction can be tuned to that in the x direction by adjusting the voltage V.

Tuning Fork Vibratory Gyroscopes

The gyroscope described above measures the input angular rate. Because the mechanical structure of the gyroscope is also sensitive to acceleration (the acceleration effect was discussed in the sections related to accelerometers), the output from a gyroscope with a single mass includes the acceleration effect. To eliminate the acceleration effect and increase the quality factor, we can form a different type of gyroscope. For example, in the dual mass gyroscope illustrated in Fig. 10.27a, two identical mass resonators are coupled via coupling springs. Since the mechanical structure is symmetric about the dashed-dotted line, the masses vibrate out of phase. When the gyroscope is exposed to an input angular rate ω_i, the capacitance changes in the right and left gyroscopes are given, respectively, by

$$\Delta C_r = \Delta C_c + \Delta C_a$$

$$\Delta C_l = -\Delta C_c + \Delta C_a$$

where ΔC and ΔC_a denote the capacitance changes due to the Coriolis force and the acceleration, respectively. If we subtract the capacitance of the left gyroscope from that of the right gyroscope (this operation is performed by an electronic circuit), we obtain the output from the gyroscope as follows:

$$\Delta C = \Delta C_r - \Delta C_l = 2\Delta C_c \qquad (10.163)$$

The output due to the Coriolis force is doubled, but the capacitance change due to the acceleration is eliminated during the subtraction. Since the mechanical structure is symmetric about the dashed-dotted line in Fig. 10.27a, we can consider the half structure shown in Fig. 10.27b. The model in Fig. 10.27b is identical to that shown in Fig. 10.26a, so we can use the closed-form solutions for the response of a single gyroscope.

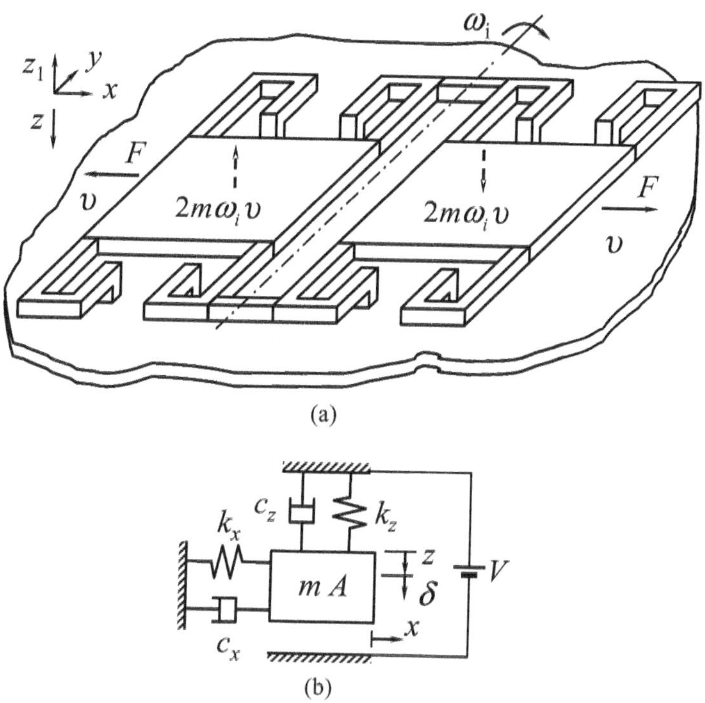

Figure 10.27 Dual mass gyroscope.

10.5 OTHER ISSUES

In the preceding sections of this chapter, we have described sensors. In most microsensors, the measured physical quantities are converted into electrical signals for further processing. When we use an electric circuit to measure physical quantities and to manipulate the corresponding electric signals, noises (unwanted signal) are included. Electrical noises are due to the Brownian motion of electrons in resistors. Similarly, when mechanical structures are used to measure physical quantities, mechanical noises are included. For example, if we design an accelerometer of mass m, damping coefficient c, and stiffness k, the acceleration noise is expressed as

$$\frac{\langle a_n \rangle}{\sqrt{\Delta f}} \approx \sqrt{\frac{4k_B T\omega_n}{mQ}} \tag{10.164}$$

where Δf, k_B, T, ω_n, and Q denote the bandwidth, Boltzmann constant given by 1.3807×10^{-23} J/K, the temperature in kelvin, the angular natural frequency of the structure, and the quality factor defined by $\sqrt{mk}/c$, respectively. If $m = 5 \times 10^{-7}$ kg, $Q = 3$, $k = 1$ N/m, $\omega_n = 1000$ rad/s, and $T = 300$ K, we have $\langle a_n \rangle/\sqrt{\Delta f} = 3.3 \times 10^{-6}\ \text{m}/(\text{s}^2 \cdot \sqrt{\text{Hz}}) = 3.4\ \mu\text{g}/\sqrt{\text{Hz}}$. The acceleration noise $\langle a_n \rangle$ is obtained as $\langle a_n \rangle = \left(\langle a_n \rangle / \sqrt{\Delta f}\right)\sqrt{\Delta f}$, where Δf denotes the bandwidth of the accelerometer. After the acceleration is measured by the accelerometer, the electrical signal corresponding to the acceleration is manipulated by an electric circuit. If the electrical circuit is modeled as the series LCR circuit, the voltage noise is expressed as

$$\frac{\langle v_n \rangle}{\sqrt{\Delta f}} = \sqrt{4k_B TR} \tag{10.165}$$

The minimum detectable limit of the accelerometer is given by the preceding noises. For further information on noise, the reader is referred to existing publications [e.g., Nyquist (1928) for electrical noise and Gabrielson (1993) for mechanical–thermal noise]. All electrostatic actuators and sensors are described in this book under the assumption that electric breakdown does not occur. However, if the electric field strength between conductors reaches beyond a critical level that depends on the type of gas, gas pressure, and the distance between conductors, electric breakdown occurs. This phenomenon was observed by Friedrich Paschen, so the curve showing the breakdown voltage is known as *Paschen curve* (see, e.g., Schaffert, 1975).

PROBLEMS

10.1 A beam at a sensor site (Fig. P10.1) is subjected to a bending moment M. Find the strain at the sensor site as a function of material properties and geometry.

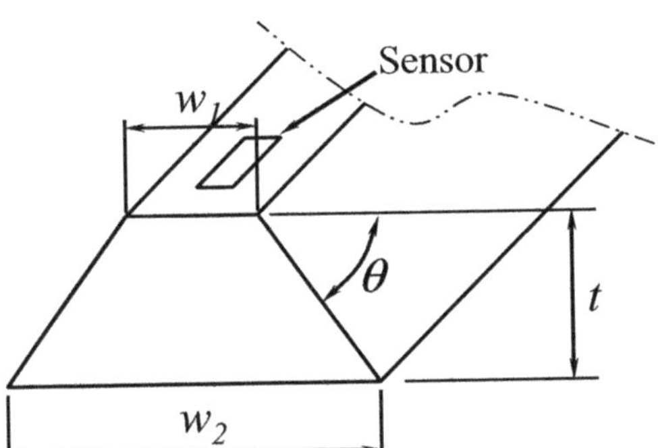

Figure P10.1

10.2 A mechanical structure (Fig. P10.2) is exposed to an acceleration a whose frequency is much less than the lowest natural frequency of the structure. The resistance of identical resistors on the beams is used to measure the acceleration. Find an expression that relates the resistance to the acceleration applied.

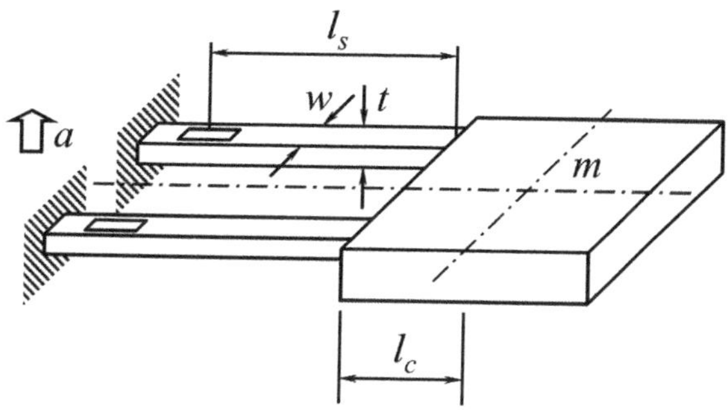

Figure P10.2

10.3 The pull-in voltage of the bulk-micromachined parallel-plate accelerometer shown in Fig. 10.15 is measured as V_{pi} at zero acceleration. The accelerometer mounted on a rocket is accelerated at $a = -kh_0/m$, where k, h_0, and m, respectively, denote the stiffness of flexures, the initial gap between the suspended and base plates, and the proof mass. Determine the pull-in voltage of the accelerometer under the acceleration.

10.4 Consider the single-mass vibratory gyroscope shown in Fig. 10.25. If the stiffness k_z is 10% higher than that of k_x, find the tuning voltage that should be applied between plates to tune the effective stiffness in the z direction to k_x.

10.5 A cantilever beam is actuated by a force. When the beam is exposed to a particular environment, a material is deposited uniformly along the beam. The natural frequency of the beam after deposition is increased by 1% of the initial natural frequency. Assuming that the flexural rigidity of the beam remains unchanged, determine the mass of the deposited material as a fraction of the initial mass of the cantilever beam.

APPENDIX

Consider the set of equations

$$\begin{aligned} a_{11}x_1 + a_{12}x_2 + a_{13}x_3 &= y_1 \\ a_{21}x_1 + a_{22}x_2 + a_{23}x_3 &= y_2 \\ a_{31}x_1 + a_{32}x_2 + a_{33}x_3 &= y_3 \end{aligned} \tag{A.1}$$

which can be expressed as

$$\begin{bmatrix} a_{11} & a_{12} & a_{13} \\ a_{21} & a_{22} & a_{23} \\ a_{31} & a_{32} & a_{33} \end{bmatrix} \begin{Bmatrix} x_1 \\ x_2 \\ x_3 \end{Bmatrix} = \begin{Bmatrix} y_1 \\ y_2 \\ y_3 \end{Bmatrix} \tag{A.2}$$

or

$$\mathbf{AX} = \mathbf{Y} \tag{A.3}$$

where

$$\mathbf{A} = \begin{bmatrix} a_{11} & a_{12} & a_{13} \\ a_{21} & a_{22} & a_{23} \\ a_{31} & a_{32} & a_{33} \end{bmatrix} \tag{A.4}$$

Principles of Microelectromechanical Systems, By Ki Bang Lee

$$\mathbf{X} = \begin{Bmatrix} x_1 \\ x_2 \\ x_3 \end{Bmatrix} \tag{A.5}$$

$$\mathbf{Y} = \begin{Bmatrix} y_1 \\ y_2 \\ y_3 \end{Bmatrix} \tag{A.6}$$

The solution is represented by

$$\mathbf{X} = \mathbf{A}^{-1}\mathbf{Y} = \mathbf{BY} \tag{A.7}$$

where

$$\mathbf{B} = \mathbf{A}^{-1} = \begin{bmatrix} b_{11} & b_{12} & b_{13} \\ b_{21} & b_{22} & b_{23} \\ b_{31} & b_{32} & b_{33} \end{bmatrix} \tag{A.8}$$

denotes the inverse matrix of **A** to be determined. We shall use Cramer's rule to find the inverse matrix. Applying Cramer's rule (Kreyszig, 1979) to (A.2), we have x_1:

$$x_1 = \frac{\begin{vmatrix} y_1 & a_{12} & a_{13} \\ y_2 & a_{22} & a_{23} \\ y_3 & a_{32} & a_{33} \end{vmatrix}}{\begin{vmatrix} a_{11} & a_{12} & a_{13} \\ a_{21} & a_{22} & a_{23} \\ a_{31} & a_{32} & a_{33} \end{vmatrix}} = \frac{y_1 \begin{vmatrix} a_{22} & a_{23} \\ a_{32} & a_{33} \end{vmatrix} - y_2 \begin{vmatrix} a_{12} & a_{13} \\ a_{32} & a_{33} \end{vmatrix} + y_3 \begin{vmatrix} a_{12} & a_{13} \\ a_{22} & a_{23} \end{vmatrix}}{a_{11} \begin{vmatrix} a_{22} & a_{23} \\ a_{32} & a_{33} \end{vmatrix} - a_{21} \begin{vmatrix} a_{12} & a_{13} \\ a_{32} & a_{33} \end{vmatrix} + a_{31} \begin{vmatrix} a_{12} & a_{13} \\ a_{22} & a_{23} \end{vmatrix}}$$
$$= b_{11}y_1 + b_{12}y_2 + b_{13}y_3 \tag{A.9}$$

If we wish to find b_{ij} in terms of a_{ij}, b_{ij} may be expressed as

$$\frac{1}{b_{11}} = a_{11} - a_{21}\frac{\begin{vmatrix} a_{12} & a_{13} \\ a_{32} & a_{33} \end{vmatrix}}{\begin{vmatrix} a_{22} & a_{23} \\ a_{32} & a_{33} \end{vmatrix}} + a_{31}\frac{\begin{vmatrix} a_{12} & a_{13} \\ a_{22} & a_{23} \end{vmatrix}}{\begin{vmatrix} a_{22} & a_{23} \\ a_{32} & a_{33} \end{vmatrix}} \tag{A.10}$$

$$\frac{1}{b_{12}} = -a_{11}\frac{\begin{vmatrix} a_{22} & a_{23} \\ a_{32} & a_{33} \end{vmatrix}}{\begin{vmatrix} a_{12} & a_{13} \\ a_{32} & a_{33} \end{vmatrix}} + a_{21} - a_{31}\frac{\begin{vmatrix} a_{12} & a_{13} \\ a_{22} & a_{23} \end{vmatrix}}{\begin{vmatrix} a_{12} & a_{13} \\ a_{32} & a_{33} \end{vmatrix}} \tag{A.11}$$

$$\frac{1}{b_{13}} = a_{11}\frac{\begin{vmatrix} a_{22} & a_{23} \\ a_{32} & a_{33} \end{vmatrix}}{\begin{vmatrix} a_{12} & a_{13} \\ a_{22} & a_{23} \end{vmatrix}} - a_{21}\frac{\begin{vmatrix} a_{12} & a_{13} \\ a_{32} & a_{33} \end{vmatrix}}{\begin{vmatrix} a_{12} & a_{13} \\ a_{22} & a_{23} \end{vmatrix}} + a_{31} \tag{A.12}$$

Similarly,

$$x_2 = \frac{\begin{vmatrix} a_{11} & y_1 & a_{13} \\ a_{21} & y_2 & a_{23} \\ a_{31} & y_3 & a_{33} \end{vmatrix}}{\begin{vmatrix} a_{11} & a_{12} & a_{13} \\ a_{21} & a_{22} & a_{23} \\ a_{31} & a_{32} & a_{33} \end{vmatrix}} = \frac{-y_1\begin{vmatrix} a_{21} & a_{23} \\ a_{31} & a_{33} \end{vmatrix} + y_2\begin{vmatrix} a_{11} & a_{13} \\ a_{31} & a_{33} \end{vmatrix} - y_3\begin{vmatrix} a_{11} & a_{13} \\ a_{21} & a_{23} \end{vmatrix}}{-a_{12}\begin{vmatrix} a_{21} & a_{23} \\ a_{31} & a_{33} \end{vmatrix} + a_{22}\begin{vmatrix} a_{11} & a_{13} \\ a_{31} & a_{33} \end{vmatrix} - a_{32}\begin{vmatrix} a_{11} & a_{13} \\ a_{21} & a_{23} \end{vmatrix}}$$

$$= b_{21}y_1 + b_{22}y_2 + b_{23}y_3 \tag{A.13}$$

$$\frac{1}{b_{21}} = a_{12} - a_{22}\frac{\begin{vmatrix} a_{11} & a_{13} \\ a_{31} & a_{33} \end{vmatrix}}{\begin{vmatrix} a_{21} & a_{23} \\ a_{31} & a_{33} \end{vmatrix}} + a_{32}\frac{\begin{vmatrix} a_{11} & a_{13} \\ a_{21} & a_{23} \end{vmatrix}}{\begin{vmatrix} a_{21} & a_{23} \\ a_{31} & a_{33} \end{vmatrix}} \tag{A.14}$$

$$\frac{1}{b_{22}} = -a_{12}\frac{\begin{vmatrix} a_{21} & a_{23} \\ a_{31} & a_{33} \end{vmatrix}}{\begin{vmatrix} a_{11} & a_{13} \\ a_{31} & a_{33} \end{vmatrix}} + a_{22} - a_{32}\frac{\begin{vmatrix} a_{11} & a_{13} \\ a_{21} & a_{23} \end{vmatrix}}{\begin{vmatrix} a_{11} & a_{13} \\ a_{31} & a_{33} \end{vmatrix}} \tag{A.15}$$

$$\frac{1}{b_{23}} = a_{12}\frac{\begin{vmatrix} a_{21} & a_{23} \\ a_{31} & a_{33} \end{vmatrix}}{\begin{vmatrix} a_{11} & a_{13} \\ a_{21} & a_{23} \end{vmatrix}} - a_{22}\frac{\begin{vmatrix} a_{11} & a_{13} \\ a_{31} & a_{33} \end{vmatrix}}{\begin{vmatrix} a_{11} & a_{13} \\ a_{21} & a_{23} \end{vmatrix}} + a_{32} \tag{A.16}$$

$$x_3 = \frac{\begin{vmatrix} a_{11} & a_{12} & y_1 \\ a_{21} & a_{22} & y_2 \\ a_{31} & a_{32} & y_3 \end{vmatrix}}{\begin{vmatrix} a_{11} & a_{12} & a_{13} \\ a_{21} & a_{22} & a_{23} \\ a_{31} & a_{32} & a_{33} \end{vmatrix}} = \frac{y_1\begin{vmatrix} a_{21} & a_{22} \\ a_{31} & a_{32} \end{vmatrix} - y_2\begin{vmatrix} a_{11} & a_{12} \\ a_{31} & a_{32} \end{vmatrix} + y_3\begin{vmatrix} a_{11} & a_{12} \\ a_{21} & a_{22} \end{vmatrix}}{a_{13}\begin{vmatrix} a_{21} & a_{22} \\ a_{31} & a_{32} \end{vmatrix} - a_{23}\begin{vmatrix} a_{11} & a_{12} \\ a_{31} & a_{32} \end{vmatrix} + a_{33}\begin{vmatrix} a_{11} & a_{12} \\ a_{21} & a_{22} \end{vmatrix}}$$

$$= b_{31}y_1 + b_{32}y_2 + b_{33}y_3 \tag{A.17}$$

$$\frac{1}{b_{31}} = a_{13} - a_{23}\frac{\begin{vmatrix} a_{11} & a_{12} \\ a_{31} & a_{32} \end{vmatrix}}{\begin{vmatrix} a_{21} & a_{22} \\ a_{31} & a_{32} \end{vmatrix}} + a_{33}\frac{\begin{vmatrix} a_{11} & a_{12} \\ a_{21} & a_{22} \end{vmatrix}}{\begin{vmatrix} a_{21} & a_{22} \\ a_{31} & a_{32} \end{vmatrix}} \tag{A.18}$$

$$\frac{1}{b_{32}} = -a_{13}\frac{\begin{vmatrix} a_{21} & a_{22} \\ a_{31} & a_{32} \end{vmatrix}}{\begin{vmatrix} a_{11} & a_{12} \\ a_{31} & a_{32} \end{vmatrix}} + a_{23} - a_{33}\frac{\begin{vmatrix} a_{11} & a_{12} \\ a_{21} & a_{22} \end{vmatrix}}{\begin{vmatrix} a_{11} & a_{12} \\ a_{31} & a_{32} \end{vmatrix}} \tag{A.19}$$

$$\frac{1}{b_{33}} = a_{13}\frac{\begin{vmatrix} a_{21} & a_{22} \\ a_{31} & a_{32} \end{vmatrix}}{\begin{vmatrix} a_{11} & a_{12} \\ a_{21} & a_{22} \end{vmatrix}} - a_{23}\frac{\begin{vmatrix} a_{11} & a_{12} \\ a_{31} & a_{32} \end{vmatrix}}{\begin{vmatrix} a_{11} & a_{12} \\ a_{21} & a_{22} \end{vmatrix}} + a_{33} \tag{A.20}$$

If the coefficients a_{ij} are given by constants, the following expression is more convenient than the above; combining (A.9), (A.13), and (A.17) gives

$$\mathbf{X} = \frac{1}{|\mathbf{A}|}\begin{bmatrix} \begin{vmatrix} a_{22} & a_{23} \\ a_{32} & a_{33} \end{vmatrix} & -\begin{vmatrix} a_{12} & a_{13} \\ a_{32} & a_{33} \end{vmatrix} & \begin{vmatrix} a_{12} & a_{13} \\ a_{22} & a_{23} \end{vmatrix} \\ -\begin{vmatrix} a_{21} & a_{23} \\ a_{31} & a_{33} \end{vmatrix} & \begin{vmatrix} a_{11} & a_{13} \\ a_{31} & a_{33} \end{vmatrix} & -\begin{vmatrix} a_{11} & a_{13} \\ a_{21} & a_{23} \end{vmatrix} \\ \begin{vmatrix} a_{21} & a_{22} \\ a_{31} & a_{32} \end{vmatrix} & -\begin{vmatrix} a_{11} & a_{12} \\ a_{31} & a_{32} \end{vmatrix} & \begin{vmatrix} a_{11} & a_{12} \\ a_{21} & a_{22} \end{vmatrix} \end{bmatrix}\mathbf{Y} \tag{A.21}$$

where

$$|\mathbf{A}| = \begin{vmatrix} a_{11} & a_{12} & a_{13} \\ a_{21} & a_{22} & a_{23} \\ a_{31} & a_{32} & a_{33} \end{vmatrix} \tag{A.22}$$

represents the determinant of **A**. Comparing (A.7) with (A.21) gives the inverse matrix of **A**:

$$\mathbf{B} = \mathbf{A}^{-1} = \begin{bmatrix} b_{11} & b_{12} & b_{13} \\ b_{21} & b_{22} & b_{23} \\ b_{31} & b_{32} & b_{33} \end{bmatrix}$$
$$= \frac{1}{|\mathbf{A}|}\begin{bmatrix} \begin{vmatrix} a_{22} & a_{23} \\ a_{32} & a_{33} \end{vmatrix} & -\begin{vmatrix} a_{12} & a_{13} \\ a_{32} & a_{33} \end{vmatrix} & \begin{vmatrix} a_{12} & a_{13} \\ a_{22} & a_{23} \end{vmatrix} \\ -\begin{vmatrix} a_{21} & a_{23} \\ a_{31} & a_{33} \end{vmatrix} & \begin{vmatrix} a_{11} & a_{13} \\ a_{31} & a_{33} \end{vmatrix} & -\begin{vmatrix} a_{11} & a_{13} \\ a_{21} & a_{23} \end{vmatrix} \\ \begin{vmatrix} a_{21} & a_{22} \\ a_{31} & a_{32} \end{vmatrix} & -\begin{vmatrix} a_{11} & a_{12} \\ a_{31} & a_{32} \end{vmatrix} & \begin{vmatrix} a_{11} & a_{12} \\ a_{21} & a_{22} \end{vmatrix} \end{bmatrix} \tag{A.23}$$

Equation (A.23) is convenient to use to evaluate the inverse matrix when the coefficients a_{ij} are given by constants. If the coefficients a_{ij} are functions of more than one variable, equations (A.10)–(A.12), (A.14)–(A.16), and (A.18)–(A.20) give closed-form coefficients b_{ij}, which are expressed in terms of the variables.

REFERENCES

Anderson DA, Tannehill JC, Pletcher RH. *Computational Fluid Mechanics and Heat Transfer*. Washington, DC: Hemisphere, 1984.

Atkinson KE. *An Introduction to Numerical Analysis*. New York: Wiley, 1978.

Bathe KJ. *Finite Element Procedures in Engineering Analysis*. Englewood Cliffs, NJ: Prentice-Hall, 1982.

Blech JJ. On isothermal squeeze films. *J. Lubr. Technol.*, 1983;105:615–620.

Buckingham E. On physically similar systems; illustrations of the use of dimensional equations. *Phys. Rev.*, 1914;4:345–376.

Cho Y-H, Pisano AP, Howe RT. Viscous damping model for laterally oscillating microstructures. *J. Microelectromech. Syst.*, 1994;3(2):81–87.

Crandall SH, Dahl NC, Lardner TJ. *An Introduction to the Mechanics of Solids*, 2nd ed. New York: McGraw-Hill, 1978.

Deal BE, Grove AS. General relationship for the thermal oxidation of silicon. *J. Appl. Phys.*, 1965;36(12):3770–3778.

Deal DE. Thermal oxidation kinetics of silicon in pyrogenic H_2O and 5% HCl/H_2O mixtures, *J. Electrochem. Soc.*, 1978;125:576–579.

Demarest KR. *Engineering Electromagnetics*. Upper Saddle Rivew, NJ: Prentice Hall, 1998.

Fung YC. *Foundations of Solid Mechanics*. Englewood Cliffs, NJ: Prentice-Hall, 1965.

Gabrielson TB. Mechanical-thermal noise in micromachined acoustic and vibration sensors, *IEEE Trans. Electron Dev.*, 1993:40(5):903–909.

Principles of Microelectromechanical Systems, By Ki Bang Lee

Jaeger RG. *Introduction to Microelectronic Fabrication*. Reading, MA: Addison-Wesley, 1988.

Kern W, Poutinen DA. Cleaning solutions based upon hydrogen peroxide for use in silicon semiconductor technology. *RCA Rev.*, 1970;31:187–206.

Kovacs GTA. *Micromachined Transducers*. New York: McGraw-Hill, 1998.

Kreyszig E. *Advanced Engineering Mathematics*. New York: Wiley, 1979.

Lee KB. Closed-form expressions for pull-in parameters of two-degree-of-freedom torsional microactuators. *J. Micromechan. Microeng.*, 2007;17:1853–1862

Lee KB. A theoretical solution for the angular displacement of electrostatic one-degree-of-freedom torsional microactuators. *Smart Mater. Struct.*, 2008a;17: 065014.

Lee KB. The theoretical static response of electrostatic fixed–fixed beam microactuators. *Smart Mater. Struc.*, 2008b;17:065017.

Lee KB, Cho Y-H, Song C. Vibratory structure, method for controlling natural frequency thereof, and sensor and actuator adopting the vibratory structure. U.S. patent 5,780,948, July. 14, 1998.

Lee KB, Cho Y-H. Laterally driven electrostatic repulsive-force microactuators using asymmetric field distribution. *J. Microelectromech. Syst.*, 2001;10(1):128–136.

Meijs NV, Fokkema JT. VLSI circuit reconstruction from mask topology. *Integration*, 1984;2:85–119.

Millman J, Grabel A. *Microelectronics*. New York: McGraw-Hill, 1987.

Muller RS, Kamins TI. *Device Electronics for Integrated Circuits*, 2nd ed. New York: Wiley, 1986.

Nathanson HC, Newell WE, Wickstrom RA, Davis JR Jr. The resonant gate transistor. *IEEE Trans. Electron Devices*, 1967:ED-14(3):117–133.

Nyquist N. Thermal agitation of electric charge in conductors, *Phys. Rev.* 1928;32:110–113.

Osterberg PM, Senturia SD. M-test: a test chip for MEMS material property measurement using electrically actuated test structures. *J. Microelectromech. Syst.*, 1997; 6:107–118.

Palmer HB. Capacitance of a parallel-plate capacitor by the Schwartz–Christoffel transformation. *Trans. AIEE*, 1927;56:363.

Park KC. An improved stiffly stable method for direct integration of nonlinear structural dynamics equations. *J. Appl. Mech.*, 1975;42:464–470.

Petersen KE. Silicon as a mechanical material. *Proc. IEEE*, 1982;70(5):420–457.

Schaffert RM, *Electrophotography*, New York: Wiley & Sons, 1975.

Smith CS. Piezoresistance effect in germanium and silicon. *Phys. Rev.*, 1954;94:42–49.

Tang WC, Nguyen T-CH, Howe RT. Laterally driven polysilicon resonant microstructures. *Sensors Actuators*, 1989;20:25–32.

Timoshenko SP, Goodier JN. *Theory of Elasticity*, 3rd ed. New York: McGraw-Hill; 1970.

Timoshenko SP, Woinowsky-Krieger S. *Theory of Plates and Shells*, 2nd ed. New York: McGraw-Hill, 1959.

Tufte ON, Long D. Recent developments in semiconductor piezoresistive devices, *Solid-State Electro.* 1963;6:323–338.

Uspensky JV. *Theory of Equations*. New York: McGraw-Hill, 1948.

Veijola T, Kuisma H, Lahdenpera J, Ryhanen T. Equivalent-circuit model of the squeeze gas film in a silicon accelerometer. *Sensors Actuators A*, 1995;48(3):239–248.

White FM. *Viscous Fluid Flow*. New York: McGraw-Hill, 1974.

Williams KR, Gupta K, Wasil M. Etch rates for micromachining processing: II. *J. Microelectromech. Syst.*, 2003;12(6):761–778.

INDEX

Printed and bound by CPI Group (UK) Ltd, Croydon, CR0 4YY
16/09/2026
14963412-0008